THE FAMILIES
AND GENERA
OF VASCULAR PLANTS

Edited by K. Kubitzki

For further volumes see list at the end of the book and:
http://www.springer.com/series/1306

The Families and Genera of Vascular Plants

Edited by K. Kubitzki

XIII *Flowering Plants · Monocots*

Poaceae

Elizabeth A. Kellogg

With 96 Figures

Springer

Elizabeth A. Kellogg
Donald Danforth Plant Science Center
St. Louis
Missouri
USA

Series Editor

Prof. Dr. Klaus Kubitzki
Universität Hamburg
Biozentrum Klein-Flottbek und Botanischer Garten
22609 Hamburg
Germany

ISBN 978-3-319-35918-2   ISBN 978-3-319-15332-2 (eBook)
DOI 10.1007/978-3-319-15332-2
Springer Cham Heidelberg New York Dordrecht London

Softcover reprint of the hardcover 1st edition 2015

Printed on acid-free paper

Springer is part of Springer Science+Business Media (www.springer.com)

To P.F.S.

Preface

The goal of this book is to present a state-of-the-art view of grass taxonomy and to summarize our current understanding of morphological variation in the grasses. Today, both aspects of agrostology have been engulfed in a flood of molecular sequence data and an equally large influx of developmental genetic data. I have tried throughout to incorporate these companion data as they affect our interpretation of morphological characters and our understanding of phylogeny.

The first sections of the book describe grass morphology, character by character. I also refer to many studies of developmental genetics that illuminate the genetic basis of traditional taxonomic characters. Often the data are incomplete, focused on only a handful of major cereal crops. Where possible, I include information on critical genes underlying each set of characters. As is conventional in the literature, names of genes are written in lowercase italics and proteins in uppercase Roman letters. Today, it is much less common to undertake broad surveys of particular characters across a large group of organisms than it was in the early 20th century. I hope that, by highlighting gaps in our knowledge, such survey work can be encouraged in the future.

In terms of taxonomy, this book represents an effort to update the major treatments of grass genera provided by Clayton and Renvoize (Genera Graminum, 1986) and Watson and Dallwitz (Grass Genera of the World, 1992 onward). In the decades since those publications, the major clades of grasses have been identified by the Grass Phylogeny Working Group (2001), and expanded by Sánchez-Ken et al. (2007) and the Grass Phylogeny Working Group II (2012), leading to the recognition of 12 monophyletic subfamilies. Remarkably for such a large family, all but a handful of species (fewer than ten) are confidently placed in a subfamily. Within the subfamilies, the major monophyletic tribes have been recognized, and the limits of these are largely stable. Within the tribes, broad agreement on subtribal limits is emerging, although a number of genera remain unplaced at this level.

As outlined in more detail in the section Subdivision of the Family, the major innovation of this book is its phylogenetic approach. The work of Clayton and Renvoize (1986) and Clayton et al. (2006 onward) arises from the philosophy of the evolutionary or phenetic school of taxonomy, and Watson and Dallwitz (1992 onward) also use an explicitly phenetic approach. More recent checklists are based on current phylogenies (Simon 2007; Simon et al. 2011 onward; Soreng et al. 2012 and onward), updating the classification frequently as indicated by recent molecular studies. The classification used here is similar but not identical to those in the checklists, and includes the rationale for many of the taxonomic decisions.

Remaining phylogenetic and taxonomic problems of the family are at the generic level. Many genera recognized by Clayton and Renvoize (1986) and currently accepted by Clayton et al. (2006 onward) and Watson and Dallwitz (1992 onward) are para- or polyphyletic. This book, like the online checklists, updates those generic limits based on current molecular phylogenetic studies in an effort to recognize only monophyletic genera. Nonetheless, current phylogenies leave many loose ends and not all generic problems can be resolved by current data. Problems outstanding are discussed throughout the text.

In the formal descriptions I have used several conventions to convey phylogenetic information. I have attempted to make descriptions more or less hierarchical, so that character states are only those that apply at a particular level. This means that the descriptions of the family, subfamilies, tribes and subtribes do not encompass all possible character states, but only include those that are likely to be synapomorphic for the clade and/or applicable to the early-diverging taxa. Thus, for example, Chloridoideae are described as having bisexual flowers because dioecy is derived later in the history of the clade. However, the hierarchical aspect of the descriptions breaks down frequently because the ancestral state of a clade is often uncertain; plant habit and ligule structure are two good examples. In these cases, several states are listed.

Taxa that are clearly not monophyletic are indicated in quotes—e.g., "*Chloris*", or "*Panicum*" s.l., the latter being distinct from *Panicum* s.s., which is monophyletic. Putative synapomorphies are indicated in italics. The strength of the evidence for these varies, so that they should be considered as hypotheses to be tested.

I had initially hoped to avoid many of the arcane grass-specific floral terms, in an effort to make the entire book more accessible to non-agrostologists. This effort was not particularly successful, although I had no trouble describing bamboo inflorescences without the terms iterauctant and semelauctant. As laid out in the section on Flower Structure, recent data on the grass floret suggest that it is simply a zygomorphic monocot flower, and not as peculiar as formerly believed. Accordingly, I have used the term "flower" instead of "floret". This is certain to irritate some people, but may make things clearer to others. As described in the section on Inflorescence Structure, the terms spike, raceme, and panicle are inaccurate and so are not used. Instead, inflorescences are described according to the number of orders of branching and whether axillary branches proliferate, as in some Andropogoneae.

As noted in the section Subdivision of the Family, phylogenies show that the gross morphology of grasses is subject to substantial convergence and is not a good guide to evolutionary history. While many of the well-supported monophyletic groups are marked by strong synapomorphies, these are often characters of micromorphology or even genome structure (e.g., chromosome number) and are thus not useful in the herbarium or in the field. Identification keys are therefore cumbersome and many taxa are keyed out more than once. The presentation here thus illustrates the tension between a fully phylogenetic classification and one that is developed for identification purposes.

In summary, I hope that this book provides food for thought, encouragement for debate, and an impetus for additional research.

St. Louis, MO Elizabeth A. Kellogg

Acknowledgements

This project has required input from many people and I am grateful to everyone who helped. The project depended completely on having full access to the incomparable resources of the Missouri Botanical Garden. For this I am grateful to Peter Raven for being instrumental in bringing me to St. Louis in the first place, and being a staunch supporter of me and this work. I also depended on the help of the superb staff, particularly Jim Solomon for access to the collections and Bob Magill for the Tropicos database. The entire project required a thoroughly well-curated grass collection, which is available at the Garden because of the long-time efforts of Gerrit Davidse; I thank Gerrit for his willingness to be interrupted with questions and for sharing his taxonomic expertise. Ihsan Al-Shehbaz let me share his office for several years and gave me a place to leave books and provided great friendship. Much of this book was written while I held the E. Desmond Lee and Family Professorship in Botanical Studies, at the University of Missouri-St. Louis. That position was designed explicitly to link the university with the Botanical Garden, and this book is one result of that formal collaboration.

Many colleagues contributed expertise in their particular taxonomic groups. Neil Snow and Paul Peterson kindly shared a pre-publication manuscript on *Leptochloa* and the Eleusininae, which relieved a number of chloridoid headaches, and Paul also shared a pre-publication manuscript on placement of a substantial number of chloridoid genera. Neil Snow provided valuable updated information on *Disakisperma* and *Leptochloa*. Rob Soreng shared unpublished data on several pooid taxa and kept me up to date on changes in the very useful online Catalog of World Grasses. Jordan Teisher provided valuable comments on Arundinoideae and Micrairoideae, and was able to correct several errors that had propagated in the literature. I thank Francisco Vazquez and Mary Barkworth for sharing their considerable expertise in Stipeae. Jimmy Triplett contributed helpful comments on Arundinarieae, and shared unpublished data on *Pleioblastus*. Finally, my many friends and colleagues at the Instituto de Botánica Darwinion in Buenos Aires shared their incomparable knowledge of the Panicoideae; I wish to thank particularly my long-time collaborator and director of the Institute, Fernando Zuloaga, as well as the late Osvaldo Morrone, and Liliana Giussani.

Special thanks are due to Lynn Clark who spent a huge amount of time working through the treatment of the Bambusoideae, correcting many faux-pas and clarifying points of morphology. Her input is substantial enough for me to acknowledge her specifically at the beginning of that section. Furthermore, she went through the descriptions of grass morphology at the beginning of the book, and made sure that any statements applied also to the bamboos.

Bryan Simon checked all the taxonomic treatments in detail and carefully compared the numbers of species with those in Grassworld and Grassbase. I thank him in particular for his patience with my insistence on a phylogenetic classification. I also am grateful to Maria Vorontsova and David Simpson for their enthusiasm and encouragement for this project. There were many moments during the writing phase when I wondered how many would read this treatment—the firm support of Bryan, Maria, and David kept me pushing ahead.

The classification and description of morphology presented here incorporates much new information generated by my laboratory with steady support from the National Science Foundation. Former and current graduate students who contributed data and expertise include Emilie Bess, Paulo Camara, Ken Hiser, John Hodge, Daniel Layton, Russell Spangler, Jill Preston, Sarah Youngstrom, Cassiano Welker, and Jinshun Zhong. In addition, this project has benefited from the careful work of postdoctoral fellows Sandra Aliscioni, Janet Barber, Hugo Cota, Andrew Doust, Matt Estep, Pu Huang, Elma Kay, Simon Malcomber, Michael McKain, Roberta Mason-Gamer, Sarah Mathews, Renata Reinheimer, Jimmy Triplett, Tony Verboom, and Michael Zanis.

I am indebted to Klaus Kubitzki, the editor of this Springer book series, for the invitation to undertake this project and I appreciate his patience. Its completion took far longer than he had hoped or expected—he will surely be pleasantly surprised that the end of the tunnel has been reached. I also thank Monique T. Delafontaine for her meticulous copy editing, and Andrea Schlitzberger at Springer (Heidelberg) for guiding this through to publication.

Finally, I heartily thank Peter Stevens who has encouraged this project from the start, reading the entire manuscript at least once and some parts several times, commenting on the contents, and providing a steady stream of relevant references. In addition, he has patiently cooked a very large number of nourishing meals for the two of us.

I should conclude by saying that, notwithstanding this wonderful input from colleagues, all taxonomic decisions and any remaining errors in the book are my own.

Elizabeth A. Kellogg

Contents

Poaceae – General Information

Description of the Family, Vegetative Morphology and Anatomy

Poaceae (R. Br.) Barnh. (1895).
Gramineae Juss. (1789).

Rhizomatous perennials, bisexual or monoecious. Culms herbaceous to somewhat lignified, erect. Leaf blades broad, *with pseudopetioles*, ligules membranous or a fringe of hairs. Inflorescences branched or unbranched, the floral units subtended by bracts. Perianth green to brown or absent. Stamens 6; style branches and stigmas 3. Pollen monoporate, with an annulus, with channels in the exine, *lacking scrobiculi. Ovule 1.* Embryo lateral, *differentiated with clear root and shoot meristems enclosed by sheaths (coleorhiza and coleoptile), several embryonic leaves, and a lateral haustorial organ (scutellum).* Fruit indehiscent, *with one seed, the seed coat fused to the inner wall of the pericarp*, the hilum linear. *Mesophyll with fusoid cells and cells with invaginated cell walls, midrib complex.* Epidermis with multicellular microhairs, with alternating long and short cells, the short cells developing silica bodies. Photosynthetic pathway C_3.

A full description of Poaceae including all character variation is lengthy and obscures the ancestral conditions for the family that are the basis for the description above (GPWG 2001). Other familiar characters such as reduction of the style branches and stigmas to two occurred well after the origin of the family, so are listed in clade or subfamily descriptions below. Modification of the inner perianth to form lodicules may be synapomorphic for the family and have been lost in Anomochlooideae, but it is simpler to assume that the origin of lodicules occurred before divergence of Pharoideae and the remainder of the grasses. The grasses almost certainly originated in shady moist environments; occupation of open habitats occurred several times independently well after the origin of the family.

Vegetative Morphology and Anatomy

Roots

As in most seed plants, the radicle of the grass embryo is the first structure to emerge from the caryopsis at seed germination. Additional roots also form at the scutellar node in some taxa; these have been called "transitory node roots" by Hoshikawa (1969), who notes that they are present in many (but not all) Pooideae and in *Ehrharta*, but absent in rice and in all other grasses investigated. Although the radicle and scutellar node roots (together known as seminal roots) are usually described as short lived, they have been found to survive at least 3.5 to 4 months and extend to depths of two feet (Weaver and Zink 1945). In annuals that have been investigated, including wheat, plants can survive and flower with the seminal roots alone (Weaver and Zink 1945). Roots subsequently form from the mesocotyl and the coleoptile nodes (Hoshikawa 1969), with additional roots arising from subsequent nodes of the main stem and its branches. The roots produce an extensive fibrous network (Fig. 1) (Kutschera and Lichtenegger 1982). Roots may also develop from the lowermost nodes of the plant, from decumbent stems or stolons, and from rhizomes. In the latter case, the roots bind the soil and can stabilize sand dunes. Roots emanating from rhizomes also contribute to formation of sod.

The root apical meristem has been studied in detail in maize and rice, and is presumed to be similar in other grasses. The meristem has a closed organization similar to that in the well-studied eudicot *Arabidopsis*, but unlike the eudicots root

E.A. Kellogg, *Flowering Plants. Monocots*, The Families and Genera of Vascular Plants 13,
DOI 10.1007/978-3-319-15332-2_1, © Springer International Publishing Switzerland 2015

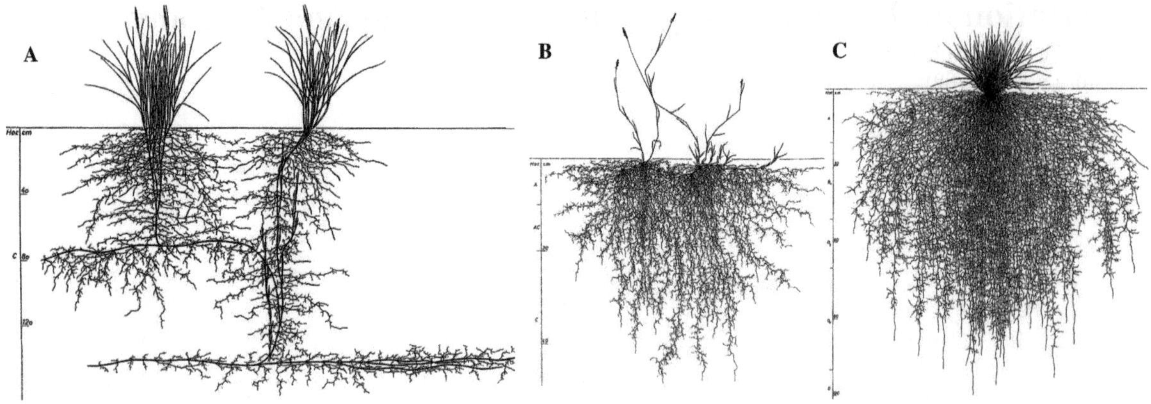

Fig. 1. Grass root systems. **A** Deep rhizomes and roots of *Ammophila arenaria*. **B** Shallow rhizomes and roots of *Poa compressa*. **C** Roots from a caespitose species, *Deschampsia caespitosa*. (From Kutschera and Lichtenegger 1982)

cap initials are wholly separate from the progenitor cells for other tissues (Coudert et al. 2010). The quiescent center, at least in maize and rice, includes hundreds of cells (Hochholdinger et al. 2004). Initiation of lateral roots begins with cell divisions in both the pericycle and endodermis, rather than just in the pericycle as in eudicots.

Cross-sectional anatomy of the mature roots of grasses is similar to that of other monocots (Clark and Fisher 1987). The root has an epidermis, an exodermis, a cortex of variable width, an endodermis, pericycle, and polyarch stele. Both the exodermis and endodermis have suberized cell walls that undergo secondary and even tertiary thickening, most notably on the inner face of the cells. However, a symplastic pathway through these walls is maintained by plasmodesmata, and some apoplastic flow appears to occur through the exodermis (Hose et al. 2001). Most species of grasses have roots with a central pith, but a few (generally annuals) have a central xylem vessel; the pericycle and pith may or may not be sclerified, and the extent of tertiary thickening varies in the endodermal cells (Goller 1977). Silica may be deposited in endodermal cells (Goller 1977; Hose et al. 2001). Goller (1977) notes that variation in root anatomy correlates with subfamilial classification, but his Table II suggests that many characters may be diagnostic for genera or tribes rather than subfamilies. The differences are generally quantitative (see summary in Clark and Fisher 1987). Enlarged storage roots are rare in the grasses (Clark and Fisher 1987).

Grasses are known to develop high root pressure. This is controlled actively by raising the concentration of ions across the suberized cell layers of the root, causing water to flow in to the xylem (Cao et al. 2012; Holloway-Phillips and Brodribb 2011). High root pressure develops at night when leaves are not actively transpiring, and effectively refills xylem vessels that have embolized during the day. Embolism is a particular problem for some grasses, which fail to close their stomata even when the leaf water potential is quite low. For example, *Lolium perenne* is able to continue gas exchange and photosynthesis even when the leaves have lost over 50 % of their hydraulic conductivity (Holloway-Phillips and Brodribb 2011). The ability of xylem to repair embolisms limits the height of plants in general; in bamboos, there is a high correlation ($r^2 = 0.81$) between root pressure and observed height (Cao et al. 2012).

Root hairs form just behind the growing point of the root, in vertical files, and hair cells alternate with non-hair cells (Clowes 2000). In many species, an epidermal cell divides asymmetrically to produce a large daughter cell (an atrichoblast) that will not become a hair and a smaller daughter cell (a trichoblast) that will become one. In the grasses, as well as in their relatives Restionaceae, Juncaceae, and Cyperaceae, the trichoblast is the daughter cell that is closest to the apical meristem, whereas it is the other way around in other monocots (Clowes 2000; Dolan and Costa 2001). In rice, however, the trichoblasts and atrichoblasts do not differ in size immediately after division, but rather undergo differential growth such that the atrichoblasts become larger (Kim and Dolan 2011). In other grasses, the trichoblasts and atrichoblasts are not different in size at all (Rothwell 1966; Row and Reeder 1957).

Particularly in dry environments, many grasses also develop a rhizosheath, a discrete layer of soil particles that is firmly attached to the root and fully separable from the surrounding soil (Price 1911; Thomas 1921; Wullstein et al. 1979; Wullstein and Pratt 1981). Rhizosheaths have been studied in a handful of crop plants (Duell and Peacock 1985; McCully 1995; St. Aubin et al. 1986; Young 1995) and in a small number of xerophytic grasses; there appear to be differences between crops and xerophytes, but the literature is sparse. The most comprehensive description of a rhizosheath is for *Lyginia barbata* (Restionaceae) (Shane et al. 2011), in which the sheath appears to be similar to grasses. In maize, the rhizosheath forms about 1 cm behind the root apex and extends 20–30 cm back from the apex (McCully 1995), whereas in grasses that normally grow in dry sand, the rhizosheath may be much longer, up to several meters (Buckley 1982; Price 1911). In roots with both sorts of rhizosheaths, root hairs are unusually dense, and curl around the sand grains of the sheath. In maize, the root hairs in the region of the rhizosheath are living (McCully 1995), whereas they are persistent and possibly dead in the long rhizosheaths of desert grasses, suggesting that they are no longer taking up water. The rhizosheath of maize appears only early in development, and is lost as the epidermis matures and sloughs off. In contrast, in a mature rhizosheath of a xerophyte, the hypodermis, exodermis, epidermis and root hairs plus sand grains form a layer that is largely impermeable to water. Inside the hypodermis the outer cortex breaks down, creating a long empty tube surrounding the inner cortex and stele (Buckley 1982; Wullstein and Pratt 1981).

Bacteria are found in the rhizosheath (Gochnauer et al. 1989; Wullstein and Pratt 1981) and appear to have several roles. First, they apparently secrete polysaccharides that, along with mucilage secreted from the root tip itself, glue the rhizosheath together (McCully 1995; Price 1911). Second, they fix nitrogen (Bergmann et al. 2009; Wullstein et al. 1979), although it is unclear how much of this is translocated into the plant. Third, they may have antibiotic properties and protect the root from fungi (Shane et al. 2011).

Grasses also develop associations with fungi. Most grasses can develop arbuscular endomycorrhizae, associations that can substantially improve uptake of phosphorus. Development of a functioning symbiosis is under genetic control, involving some genes that are common throughout land plants and others that appear to be grass specific (Yang et al. 2012). Ascomycetes in the family Clavicipitaceae are generally arthropod pathogens, but one clade shifted to form symbiotic associations with grasses (Spatafora et al. 2007). A member of this clade, *Metarhizium robertsii*, will invade the roots of switchgrass (*Panicum virgatum*), and stimulate root hair growth.

The genetic basis of grass root development is only beginning to be explored (Hochholdinger et al. 2004; Hochholdinger and Zimmermann 2008; Smith and De Smet 2012). For example, the gene *Rootless concerning crown and seminal roots* (*Rtcs*) has been cloned from maize, and encodes a transcription factor with a Lateral Organ Boundaries (LOB) domain; *Rtcs* controls formation of all shoot-borne roots, both seminal roots and crown roots (Majer et al. 2012). Homologues have also been cloned from rice, where they appear to have a similar function (Smith and De Smet 2012). Other loci affecting lateral root initiation and elongation have been characterized in both maize and rice (Hochholdinger et al. 2001; Hochholdinger 1998; Kitomi et al. 2011; Liu et al. 2009). Despite progress cloning genes in rice and maize, the size of the plants makes study of root systems difficult. Fortunately, the recent development of *Brachypodium distachyon* as a model system will certainly provide new tools for understanding the controls of root architecture in the grasses (Cochois et al. 2012).

Stems

The shoot apical meristem has a characteristic zonal organization like that of most seed plants; the outer layer (tunica) consists of cells that divide primarily anticlinally, whereas the inner part consists of cells with less consistent patterns of division. Brown et al. (1957) suggested that the "festucoid" grasses (an informal group that at the time included all grasses outside the Panicoideae) have two tunica layers in the meristem, whereas the panicoids have only one. In maize (a panicoid) the outer L1 layer gives rise to the epidermis, whereas other tissues are specified by the inner cells (Jackson 2009).

As the stem matures, cells in the internodes stop dividing and differentiate basipetally. This leaves a small meristematic region, the intercalary meristem, at the base of each internode just above the next lowest node. Although the intercalary meristem is a weak spot on the stem, support is provided by the surrounding leaf sheaths. Protoxylem and protophloem are present in the meristem, so that vascular continuity is maintained (Clark and Fisher 1987 and references therein). Activity of this meristem allows lodged grass stems to right themselves.

Internodes are generally short in early development, and those near ground level often elongate little if at all. The timing of internode elongation varies between species, but is often similar within major taxa. For example, in Pooideae, internode elongation is delayed until just before flowering. In contrast, in Bambuseae and many Panicoideae, internodes elongate apparently independent of flowering. The timing of internode elongation also determines whether the plant is grazing resistant or not (Branson 1953; Holechek et al. 1998); as long as the shoot apical meristem is near ground level it cannot be easily removed by large herbivores.

Internally, internodes may be solid throughout development, or may become hollow. In some bamboos, the internodes at the base of the plant may be solid, whereas those at higher nodes are hollow. The distinction between solid and hollow internodes is not absolute, however, in that some species have aerenchyma in the internode. Variation in the internal anatomy of the internode may be taxonomically diagnostic, although it is highly homoplasious (GPWG 2001), and is probably most useful at the level of genus and species. Most Pooideae tend to have hollow internodes, whereas other subfamilies are more variable (Brown et al. 1959a). Clearly the oft-repeated jingle "sedges have edges, and rushes are round, and grasses are hollow right to the ground" is a serious over-simplification. Pooideae tend to have wider hollows than other taxa, but the ecological and evolutionary significance of hollow internodes is unknown; Brown et al. (1959a) suggest a correlation of hollow internodes with moist habitats. Genetic studies in durum wheat show that having a solid stem is dominant to hollow stem, and that the trait is controlled by a single locus (Houshmand et al. 2007). Breeding for a solid stem confers resistance to the wheat stem sawfly, suggesting that taxa with solid stems have enhanced protection against insects. The wheat mutant *tiller inhibition* (*tin*) also causes the lower three internodes of the culm to become solid, apparently by diverting sucrose away from axillary buds (Kebrom et al. 2012). This result hints that this taxonomically important character may reflect a fundamental difference among species in the way carbohydrates are partitioned.

Stems in most grasses are herbaceous, but become woody in members of Bambuseae and Arundinarieae (the woody bamboos). In these taxa, dense clusters or caps of sclerenchyma cells form both externally and internally around the vascular bundles of the stem (Liese 1998). The bundles themselves are closely spaced, resulting in an extremely hard culm. The height of bamboos is strongly correlated with root pressure, which provides the force needed to refill embolized xylem vessels (Cao et al. 2012). Other large reed-like grasses (e.g., *Phragmites*, *Thysanolaena*) and the handful of shrubby ones (e.g., *Cladoraphis spinosa*) also develop hard woody culms, but whether these are histologically and developmentally similar to bamboo stems is unknown (GPWG 2001).

Most leaves on a grass plant have a single bud in their axils. The buds are under both developmental and environmental control, and their fate also depends on where they form on the plant. Axillary buds from the short basal internodes may grow horizontally to form stolons or rhizomes, or may grow more or less vertically to form axillary branches. When the upright branches occur near ground level they are known as tillers. As the axillary branch develops, it may break through its subtending leaf sheath (an extravaginal branch) or not (an intravaginal branch). In general, rhizomes and stolons are formed from extravaginal branches, whereas tillers may be either extra- or intravaginal.

The extent and nature of basal branching controls the overall architecture of the plant. Grasses that form only tillers develop a clumped architecture (i.e., are caespitose), whereas at the other extreme those that form rhizomes or stolons are spreading and may be sod-forming. Tillers may be geniculate at the base and root from their nodes so that they are scarcely distinct from short rhizomes. Nonetheless, the growth form of any particular species is generally reasonably constant.

The number of tillers is controlled by hormones, particularly by auxin, strigolactones, and brassinosteroids, and by carbohydrate levels. When auxin transport is inhibited, or when the apical meristem of the plant is removed, the number of tillers increases and their angle becomes wider (Li et al. 2007; Xu et al. 2005). Although this has been demonstrated experimentally only in rice, it is likely that the result is general. Proteins in the strigolactone pathway, such as HIGH TILLERING DWARF1 and DWARF10, and in the brassinosteroid pathway, such as DWARF AND LOW-TILLERING, also regulate the number of tillers in rice (Arite et al. 2007; Tong et al. 2012; Zou et al. 2006) and are likely to be involved in other grasses as well. Tiller outgrowth is affected by TEOSINTE BRANCHED1 (TB1) and its orthologues, a cell-cycle regulator that integrates input from hormonal pathways with environmental signals (McSteen 2009; Ramsay et al. 2011; Remigereau et al. 2011). Tiller production is also affected by carbohydrate partitioning; diversion of sucrose from axillary buds to the main stem in wheat, as apparently occurs in *tiller inhibition* (*tin*) mutants, leads to reduced tillering (Kebrom et al. 2012).

Tiller angle is also under genetic control and has been investigated extensively in rice, in which tiller spreading affects yield and pest resistance (Wang and Li 2008a). Plants that spread too much shade their neighbors and thereby reduce grain production per unit area, whereas those that are too upright are susceptible to insect pests and pathogens because of increased contact with other plants and higher humidity within the clump. Several proteins have been identified that control tiller angle, including PROSTRATE GROWTH1 (PROG1) (Jin et al. 2008; Tan et al. 2008), LOOSE PLANT ARCHITECTURE 1 (LPA1) (Wu et al. 2013), LAZY1 (Li et al. 2007), and PIN-FORMED2 (Chen et al. 2012; Xu et al. 2005). The latter two proteins regulate auxin transport, whereas the mechanism of PROG1 influence is unknown. LPA1 affects tiller angle by controlling the growth of cells on the adaxial side of the branch; longer cells in that position lead to a more spreading tiller (Wu et al. 2013). *Tiller Angle Control1* (*TAC1*) is a quantitative trait locus that also affects tiller angle but the underlying gene has not yet been cloned (Yu et al. 2007).

Branches also form on the stem (culm) itself, although this is taxon specific. For example, branched culms are unknown in the Pooideae, whereas they are common in Panicoideae (particularly Andropogoneae) and Olyreae, and almost universal in Bambuseae and Arundinarieae. Genetic studies (Doust et al. 2004; Doust and Kellogg 2006) have shown that the genes that control the formation of culm branches differ from those that control basal branching (tillering). The genetic basis of the trait is thus consistent with the taxonomic observations.

Although the extent of branching – whether as tillers, rhizomes, or culm branches – is taxon specific, it is also controlled by the environment, particularly shade and light, and mediated by auxin, cytokinin and strigolactone, at least in the crop plants studied (Doust 2007a; McSteen 2009; Wang and Li 2008b). Many of the cellular mechanisms controlling branch formation are shared among monocots and eudicots, but others appear to be grass specific (see citations in Doust 2007b).

Not all leaves on all grasses bear axillary buds. For example, the lower culm nodes of some bamboos fail to form buds (Clark and Fisher 1987). In most such cases, whether the bud is specified but simply fails to differentiate, or whether the signal for bud formation is never transmitted or received, is unknown.

A few woody bamboos (e.g., *Chusquea*) have multiple axillary buds, which may be formed by supernumary axillary meristems in the axil of a single leaf, or may represent a highly compressed branch complex. These have never been studied developmentally.

As in most monocots, the branches produced by axillary buds bear an adaxial prophyll. This is generally two-keeled and is particularly prominent in the woody bamboos, where variation in its shape is often taxonomically useful.

The node is complex, both internally and externally. Internally it is marked by a plexus of extensively anastomosing vascular tissue that forms just above the point of insertion of the leaf (Liese 1998; Sharman 1942). Although only a handful of grass species have been investigated, the nodal plexus is a web of transverse vessels that connect the axial vessels (Pizzolato 2000); the overwhelming majority of the latter end at the nodal plexus with only a tiny percentage extending through (Shane et al. 2000; see also André 1998). Even in taxa with hollow internodes, the node is more or less solid. In the bamboos there is a clear woody wall, the diaphragm (Liese 1998).

The vascular structure and function of the node are described in detail for rice and barley, and are presumed to be broadly similar in other grasses (Yamaji and Ma 2014). A vascular bundle that will ultimately extend into a leaf can be traced to two nodes below the leaf (i.e., leaf node minus 2) where is it relatively small in diameter (called a diffuse vascular bundle by Yamaji and Ma 2014). It connects through the next node above (leaf node minus 1; a transit vascular bundle), and at the leaf node appears enlarged, with an increased number of xylem elements and phloem sieve tubes (an enlarged vascular bundle). Thus any given node contains diffuse vascular bundles, transverse vascular bundles and expanded vascular bundles. Xylem transfer cells in the expanded vascular bundles move solutes to the diffuse vascular bundles and thus up to higher nodes in the plant. Such a pathway has been demonstrated for silicon (Si), which accumulates at nodes where it is then distributed among vessels at the nodes (Yamaji et al. 2008, 2012; Yamaji and Ma 2009). A similar pathway exists in the phloem, transporting zinc (Zn) and also the toxic metal cadmium (Cd); both Zn and Cd accumulate in the nodes and are distributed to developing tissues (Satoh-Nagasawa et al. 2012; Yamaguchi et al. 2012). Transporters for copper (Cu) and manganese (Mn) are also located at the nodes but neither mineral accumulates; rather, influx and efflux are tightly controlled to keep levels consistent throughout the plant (Deng et al. 2013; Yamaji et al. 2013).

Externally, the node is often marked by a slightly swollen area, the nodal pulvinus, which is often surrounded by a corresponding area in the sheath. This area is flexible and is involved in reorienting the stem in response to lodging, although in mature stems it may become lignified and lose the capacity to bend (Kaufman et al. 1987). Bending of the pulvinus correlates with changes in levels of auxin and gibberellin, which is consistent with the role of these hormones in cell expansion (Clore 2013; Wolbang et al. 2007), and is regulated in rice by LPA1 (Wu et al. 2013). In general, the pulvinus lacks sclerenchyma and instead is supported by collenchyma (Paiva and Machado 2003). In North American grasses, the sheath pulvinus is almost universally present, whereas the nodal pulvinus occurs primarily in Panicoideae and Chloridoideae, and is generally absent in Pooideae (Brown et al. 1959b; Clore 2013).

In a few grasses the lower nodes of the culm may be enlarged to form storage organs (Burns 1945). Whereas in a few species the thickened organs are leaves (e.g., *Poa bulbosa*), and thus the structure is a true bulb, in other species the storage organ is the stem and so is properly a corm (e.g., *Zuloagaea bulbosa*, *Melica bulbosa*, *Arrhenatherum avenaceum* var. *nodosum*, *Ehrharta capensis* and relatives). Species with corms often occur in areas with low summer rainfall (Burns 1945; Verboom et al. 2003). Most bulb- or corm-bearing species are in subfamily Pooideae, but a few are panicoid (*Zuloagaea bulbosa*; Bess et al. 2006) or ehrhartoid (Verboom et al. 2003).

In terms of life history, the ancestral condition for the grasses is herbaceous, perennial, and rhizomatous (GPWG 2001), but the annual habit has been derived repeatedly. Humans have particularly exploited the annuals (e.g., wheat, maize, rice), in which much of the photosynthate is accumulated in seeds. Multiple genes control the switch between annual and perennial. As shown in sorghum and rice, some of the same loci are involved in both species, suggesting that changes of plant habit can occur relatively easily (Hu et al. 2003).

Leaves

Morphology and development

As in all seed plants, the shoot in grasses is made up of repeating units known as phytomers or phytomeres; each unit consists of a leaf, an internode, and an axillary bud. Whether the internode and bud should be associated with the leaf above or below is a matter of debate, but however defined the phytomer is repeated over and over in the growth of the grass shoot (Clark and Fisher 1987).

Leaves form on the flanks of the shoot apical meristem. The position of a nascent leaf can be identified initially by a change in expression of the meristem identity gene *knotted1*, which is switched off in the cluster of cells that will become the leaf primordium (Jackson et al. 1994). Auxin becomes concentrated in these cells, and their subsequent divisions lead to the formation of a leaf primordium (Reinhardt et al. 2003). The primordium develops both laterally and in the proximo-distal axis to become a broad flat structure that encircles the meristem. In maize, the leaf forms from the outer two layers of the shoot apical meristem, with the L1 layer sometimes contributing to the mesophyll as well as producing the epidermis (Poethig 1984). About 40 cells initially contribute to the circumference of the leaf primordium, and the entire primordium is made up of about 200 cells.

Leaf initiation in the grasses (as in all graminoid Poales) is strictly distichous, except in the tiny moss-like grass *Micraira*. In this species, serial sections through the shoot show a phyllotaxis of 3/8; there is no evidence of twisting of the stem or sheaths (Philipson 1935a). Spiral phyllotaxis has also been reported for the large reed-like species *Arundoclaytonia dissimilis*, but this plant has never been studied in detail. Although Judziewicz and Soderstrom (1989) cite Page (1951) to suggest that leaf initiation is spiral in *Streptochaeta*, Page was in fact referring to the inflorescence. The leaves are distichous.

Cell division and expansion initially occur throughout the young leaf primordium, but actively dividing meristematic cells become increasingly restricted to the base of the young leaf so that the leaf matures from the apex to the base (Sharman 1942). The region of active division, termed the "proliferative zone" by Sylvester and Smith (2009), is later divided into two by the developing ligule and sheath, so that two meristems are formed, one each at the base of the blade and the base of the sheath.

Mature leaves in the grasses generally consist of a distal blade and a proximal sheath. The sides of the leaf blade – on either side of the midrib – are controlled developmentally by the NARROW-SHEATH proteins, which define a lateral compartment of the leaf (Nardmann et al. 2004). The region where the blade and sheath join is known as the collar; in some taxa this corresponds to a region of more flexible tissue and less sclerenchyma (Paiva and Machado 2003). In many grasses there is a wedge-shaped region on either side of the leaf at the collar that can be identified by its color and texture. This region appears to be involved in positioning the blade (Foster and Timmermans 2009), and in rice is controlled by the gene LPA1 (Wu et al. 2013). Curiously for such an obvious morphological feature, the wedge-shaped region has no standard name. It is called a "leaf joint" in some papers on rice (e.g., Wu et al. 2013), whereas is has been called a "dewlap" by some taxonomists (e.g., Martínez-y-Pérez et al. 2008; Pohl 1980) and by agronomists working on sugarcane (Artschwager 1951), although the term is not widely used. Maize geneticists call this region an auricle. Taxonomists, however, reserve the term "auricle" for the tiny prongs or hooks that extend from the wedge-shaped region, and "auricle" is used in this sense here. In this taxonomic sense, maize lacks auricles. Bowden (1970) reports that the dewlaps of *Andropogon gayanus* var. *bisquamulatus* (Hochst.) Hack. secrete sweet nectar, but this observation appears not to have been followed up.

Leaf length and width are variable within and between species, and are often taxonomically informative. Fiorani et al. (2000) found that differences in leaf length in species of *Poa* can be attributed to changes in the rate of growth rather than its duration – i.e., leaves grow faster for about the same length of time, a result extended by Sugiyama (2005) to several species of C_3 grasses. Arredondo and Schnyder (2003) also found a correlation between the size of the meristem at the base of the blade and the rate of leaf elongation in eight species of pooid grasses. If this result is generally true, it may point to phylogenetically correlated differences in regulation of the cell cycle.

Also at the blade-sheath boundary is an adaxial flap of epidermal tissue, the ligule; in most cases this is small, no more than a couple of millimeters long, although it may be several centimeters long in some woody bamboos. During leaf development, the blade-sheath boundary is defined by a region of increased cell division known as the pre-ligular band; this marks the position where the ligule will form (Sharman 1942; Sylvester et al. 1990). Development of the ligule has been studied extensively, especially in maize (reviewed by Foster and Timmermans 2009). Many genes are involved in the proper positioning and development of the ligule, but only a few seem to affect ligules exclusively, whereas others disrupt the entire structure of the leaf.

In a detailed description of ligule development in *Deschampsia*, *Melica* and *Phyllostachys*, Philipson (1935b) noted that the ligule appears to form from separate domains, a central epidermal one, and lateral extensions of the margins of the sheath, thus anticipating by more than 50 years the discoveries of the lateral domains in maize leaves (Scanlon and Freeling 1997). Ligule morphology is remarkably variable and is often taxonomically informative. Membranous ligules predominate among members of the BEP clade, whereas the PACMAD subfamilies often have ligules consisting of a fringe of hairs.

A ligule-like structure, which may be membranous or ciliate, may form on the abaxial side of the collar region, and is variously known as a contraligule, pseudoligule, or external ligule; the development of this structure has never been studied. It occurs in some genera such as *Puelia* (Puelioideae), *Streptogyna*, and in most, if not all, genera of Bambuseae and Arundinarieae, and can be helpful for genus and species identification.

The base of the leaf blade above the ligule may be constricted to form a pseudopetiole. Presence of a pseudopetiole is ancestral and synapomorphic among the grasses (GPWG 2001). All species of Anomochlooideae, Pharoideae, and Puelioideae have pseudopetioles, as do most species of Bambusoideae. The structure also appears in some Panicoideae. Development of pseudopetioles has never been studied and it is unknown how it affects – or is affected by – the meristem at the base of the blade, nor is there information on its contribution to the hydraulic architecture of the leaf.

In taxa such as *Streptogyna* (BEP clade, incertae sedis), most bamboos, some Panicoideae (e.g., *Thysanolaena*, tribe Centotheceae; *Gerritea*, tribe Paspaleae), Micrairoideae (*Micraira*), Arundinoideae (*Molinia*), some species of *Ehrharta*, in *Macrochloa* (Stipeae), and *Aristida* (Aristidoideae), an abscission zone forms at the collar and the leaf blades disarticulate. The anatomy of the abscission zone varies among species (Röser and Heklau 2011). In other species (e.g., in some species of *Rytidosperma*, Danthonioideae), an abscission zone forms at the base of the sheath and the entire leaf is deciduous. In the culm leaves of woody bamboos, even if the blade persists for several seasons, it will ultimately disarticulate from the sheath. Subsequent disarticulation of the sheath from the culm varies among genera or groups of genera.

Leaf angle is under the control of brassinosteroids (Tong et al. 2012), as well as other plant hormones whose effects interact with those of the brassinosteroids (Song et al. 2009). However, other controls of leaf angle operate independently of hormonal pathways, affecting cell division in and around the collar. These include the genes *Leaf inclination2* (Zhao et al. 2010), *Increased leaf angle1* (Ning et al. 2011), and *LPA1* (Wu et al. 2013).

Leaf epidermis

In most grasses, the leaf epidermis consists of a single layer of cells that form in long files parallel to the proximo-distal axis of the leaf (Fig. 2). Because the leaf matures from tip to base, these long files represent a developmental gradient that has helped in the investigation and understanding of differentiation. Many of the late cell divisions are asymmetric, with a single cell giving rise to two differently sized daughter cells (see Fig. 8A in Sylvester et al. 1990). The result of this is a characteristic alternation of long and short cells in the mature epidermis. The alternation is not perfect, and short cells may occur in pairs or files of up to five. Alternation of long and short cells occurs in other monocots as well, but only in the cell files that will produce stomata. Thus the asymmetric cell divisions that occur throughout the epidermis in the grasses represent a change in position of a developmental program (Kellogg 2000). Long-and-short cell alternation is shared with *Joinvillea*, one of the close relatives of the grasses, and provides one piece of evidence for their close relationship (see Affinities); because *Ecdeiocolea* is leafless, the ancestral condition for the grass sister group is unknown (Campbell and Kellogg 1987). The epidermis is often structured differently over the veins (costal region) than it is between them (intercostal regions), a difference that is generally consistent within a species or genus (Fig. 2). The genetic basis of the asymmetric cell divisions that will produce stomata is becoming increasingly well understood (Abrash and Bergmann 2009), but whether this machinery is also activated in other short cells is unknown.

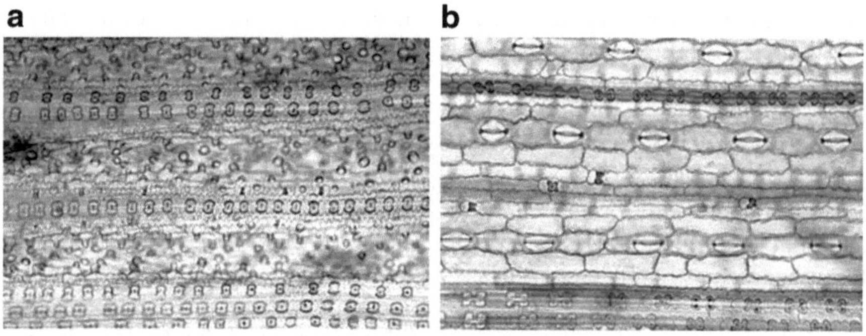

Fig. 2. Leaf epidermis showing files of cells extending along the proximo-distal axis. **A** *Oxychloris scariosa* (Chloridoideae), showing characteristic saddle-shaped silica bodies in short cells over the veins. Intercostal epidermal cells bear papillae. **B** *Monocymbium cerisiiforme* (Panicoideae), showing bilobate silica bodies in short cells over the veins. (From Watson and Dallwitz 1992 onward)

Epidermal long cells vary only slightly in shape, usually being rectangular but sometimes fusiform. Their longitudinal walls, however, may be sinuous and thus interlocking, or straight; the latter condition occurs frequently, but not universally, in Pooideae. The shape of the longitudinal wall is determined in part by localization of microtubules and actin (Frank et al. 2003). In maize, in which the longitudinal walls are normally sinuous, mutations in loci known as *Brick* (because they create brick-like cells when mutated) create long cell shapes reminiscent of those occurring in, for example, species of *Poa* or *Ehrharta*. A survey of grasses staining for tubulin and actin would help determine whether distribution of these cytoskeletal proteins is the primary determinant of taxonomic variation in cell wall morphology.

In the grasses, as in most commelinid monocots (Stevens 2012), the stomata are paracytic – there are two guard cells, with the stoma oriented parallel to the long axis of the leaf, and two subsidiary cells parallel to the guard cells. In stomatal development, a short cell divides longitudinally to give two cells and then these divide, again longitudinally, to give a set of four, more or less parallel, rectangular cells (Abrash and Bergmann 2009). The outer two differentiate into subsidiaries and the inner two into guard cells. The ancestral and most widespread condition is for subsidiaries to be somewhat dome-shaped in surface view, although in some species they are triangular. However, in the pooid clade that includes Poeae, Triticeae, Bromeae and *Brachypodium*, the subsidiaries have parallel walls, a condition that is uniquely derived. The subsidiaries overlap the guard cells in all Pooideae that originated after the divergence of Nardeae.

Epidermal short cells have various fates, depending on the species, position in the plant, and position in the leaf (or leaf-like structure such as the lemma). One common role of short cells is to differentiate as silica-containing cells. The accumulation of silica in the leaf is a character shared by most of the commelinid monocots, and accumulation specifically in the epidermis is a character of the Poales (Stevens 2012). Monosilicic acid, $Si(OH)_4$, is produced by weathering of soils, and is taken up by the plant both actively and passively in a highly regulated process (Yamaji et al. 2008, 2012; Yamaji and Ma 2014). Silica is deposited in an amorphous non-crystalline form as silicon dioxide ($SiO_2.nH_2O$) throughout the plant, particularly in tissues involved in support of the stems and leaves (Isa et al. 2010), but also in the walls of guard cells, subsidiary cells, and epidermal papillae (Ueno and Agaric 2005); the rate and developmental timing of accumulation is specific for different cell types (Sakai and Sanford 1984). In addition, silica commonly accumulates in specialized short epidermal cells (silica cells) in which it is deposited initially in the cell wall and then accumulates centripetally, while the cellular contents break down (reviewed by Prychid et al. 2004). Silica deposition protects the plant from bacterial and fungal pathogens, supports the stems, reduces the uptake of toxic metals, and regulates water loss (Isa et al. 2010; Ma and Yamaji 2006). In addition, silica rapidly wears down the mandibles of insect herbivores and reduces digestibility (Massey and Hartley 2009). Silica accumulation is presumed to be energetically costly, but its effect on plant biomass varies between species; this variation affects susceptibility to herbivores and thus competitive interactions (Garbuzov et al. 2011).

Silica may also be a defense against large mammalian herbivores, although the data on this are less clear. Simpson (1951) famously proposed that the evolution of hypsodont teeth in equids was driven by a shift to diets of grass, and their high concentrations of silica. However, Sanson et al. (2007) have recently questioned whether silica bodies are in fact hard enough to wear down mammalian tooth enamel, as had been suggested (Baker et al. 1959). In addition, Strömberg (2006) showed that hypsodonty appeared well after the spread of grasslands, at least in North America, weakening the hypothesized connection.

The shape of the silica deposits (silica-bodies or phytoliths) is often characteristic of particular taxonomic groups. These shapes were originally described based on their appearance in two dimensions (Metcalfe 1960; Prychid et al. 2004), and early attempts to use silica body characteristics in phylogeny reconstruction found that they were highly homoplasious (Kellogg and Campbell 1987; Kellogg and Watson 1993). The description and classification of silica bodies (phytoliths) is improving due to efforts to describe them in three dimensions rather than two (Piperno 2006; Piperno and Pearsall 1998). For example, bilobate silica bodies occur in many subfamilies of grasses (Fig. 2A, B). However, those of Stipeae are asymmetrical in cross section, while those of Panicoideae are generally symmetrical. Aristidoid bilobates tend to have a long thin shaft between the lobes, whereas in other taxa the shaft is short and in bamboos is lacking altogether. So-called "oryzoid" bilobates are elongated perpendicular to the long axis of the leaf. This form is found in Ehrhartoideae, some bambusoids (Olyreae), and arundinoids (*Eriachne*); careful analysis of these, however, shows that they can in fact be distinguished (Prasad et al. 2011). More detailed description of silica bodies will not wholly solve the problem of homoplasy in the character, however. Individual plants have a range of silica body forms (Piperno and Pearsall 1998) and thus characterization of silica bodies within a species or genus must be somewhat quantitative.

Short cells may also form trichomes. These may be unicellular (prickles, macrohairs) or bicellular (microhairs), and may accumulate silica or not (Prat 1932). Development of prickles and macrohairs begins with enlargement of the cell. As the trichome develops, the outer wall expands, so that the outline of the cell looks more or less like a muffin. The nucleus then moves up in the cell closer to the outer wall. As the outer wall continues to expand, growth becomes asymmetrical, and the cell elongates parallel to the surface of the leaf (Kellogg 1990).

In cells that will become prickles, the tip of the cell develops a sharp point and silica accumulates in the tip (Prat 1932). Silica accumulation appears to occur early before the leaf has fully opened and before silica deposition in other cells (Motomura et al. 2006). Prickles generally point toward the distal end of the structure on which they occur, but sometimes point toward the proximal end (i.e., are retrorse). Presence or absence of prickle hairs on particular structures often is a good field identification character. However, I know of no study in which this character has been evaluated in a phylogenetic context; intuitively, it seems as though it should be highly labile in evolutionary time. The function of prickle hairs is unknown, but could help deter small herbivores such as slugs and nematodes.

Bicellular microhairs are found in all non-pooid grasses (Johnston and Watson 1976), but their development has never been investigated. As their name implies, these trichomes have only two cells, one apical and one basal, but the shape of the cells is often characteristic of particular taxa. In the "panicoid type" microhair, both cells are longer than wide, and internal membranes are not readily visible (Amarasinghe and Watson 1988, 1989). In contrast, in the "chloridoid type" the apical cell is nearly as wide as long. The distinction between the two is not sharp, however, and a graph of the length-width ratio of the apical cells in all microhairs in the family shows that the variation is continuous (GPWG 2001; Kellogg, unpublished observations). Some chloridoid microhairs contain internal membranes in the basal cell and secrete salt (Liphschitz and Waisel 1974; Marcum 1999; Oi et al. 2012), but there is no evidence that the panicoid hairs are secretory (Amarasinghe and Watson 1989). Other chloridoid microhairs, the "*Enneapogon* type" have internal membranes in the apical cell, but appear to be non-secretory (Amarasinghe and Watson 1988, 1989). Lack of microhairs is synapomorphic for all Pooideae after the divergence of *Brachyelytrum*, *Nardus* and *Lygeum*, but ecological consequences of this loss are unknown. Microhairs with more than two cells are reported for *Joinvillea* (Joinvilleaceae), a close outgroup of the grasses, and for *Streptogyna crinita* and several members of Bambusoideae (Soderstrom and Judziewicz 1987).

Macrohairs on the leaf blades are inherited independently of hairs on other plant parts (Moose et al. 2004), supporting their use as a different taxonomic character. Macrohairs are often surrounded by a multicellular and slightly raised set of epidermal cells (Prat 1932). These have been reported to be secretory (Bowden 1971), with the cells containing a variety of sugars and pectic substances. However, the number of species investigated is tiny, and macrohairs would repay closer investigation (Sylvester and Smith 2009).

In areas around some or all of the veins, adaxial epidermal cells may differentiate as bulliform cells. These cells are enlarged in the ab-adaxial direction, extending into the region normally occupied by mesophyll. They may occur as fans of cells or in irregular groups, and may or may not be associated with other colorless mesophyll cells, and thus provide a source of taxonomically informative characters. Bulliforms may occur on either side of the mid-vein only, or on the sides of lateral veins as well. Some species lack bulliform cells entirely.

Bulliform cells can rapidly take up or lose water. By expanding and contracting, bulliform cells are reported to control leaf rolling (Arber 1934; Bidlack and Jansky 2011), but there is surprisingly little evidence to support the hypothesis that changes in bulliform turgor are actually causative (Arber 1934). In the resurrection species *Sporobolus stapfianus* the outer wall of the bulliforms is thick and water is lost to the adjacent mesophyll cells (Dalla Vecchia et al. 1998). While water loss from the bulliforms may lead to leaf rolling in some species, in *S. stapfianus* the bulliforms are involved in maintaining hydration of the mesophyll. In late development of the leaf, bulliform cells may accumulate silica (Motomura et al. 2004).

A number of proteins regulate the number of bulliform cells in each group. Transcription factors that help specify identity of the ad- and abaxial sides of the leaf blade determine whether bulliforms will develop on the adaxial side (as is most common) or abaxial side, and also regulate how many bulliforms form in a cluster (Dai et al. 2007; Zhang et al. 2009; Zou et al. 2011). Various enzymes also affect bulliform cell development (Fujino et al. 2008; Hu et al. 2010; Li et al. 2010; Xiang et al. 2012), in some cases via affecting the differentiation between the ab- and adaxial sides of the leaf (Hibara et al. 2009). Comparative studies have yet to be done on most of these proteins, so it is unclear which components of the bulliform specification network may have been selected to produce the diversity of leaf anatomy observed among the grasses.

Multicellular structures known as "glands" have been observed in some grasses, particularly in Danthonioideae (Linder et al. 1990). Two-celled salt glands have also been described in *Spartina* (now part of *Sporobolus* s.l.; Peterson et al. 2014), in which an epidermal initial cell expands downward into the mesophyll, and later divides asymmetrically to form a small apical cell (Fahn 1979). Salt glands have also been characterized in *Chloris gayana*; the density of glands on leaves increases when the plant is grown in higher concentrations of salt (Oi et al. 2012). The apical cell of the gland has a complex endomembrane system and a high number of mitochondria, suggesting that salt excretion is energetically costly.

The epidermis of many grasses is coated with a layer of wax. This is characteristic of leaves on juvenile plants in maize and the presence of wax has been used as a marker of the transition from juvenile to adult morphology (Moose and Sisco 1996). In some taxa, such as *Sorghum*, the wax forms on the stem and sloughs off as large flakes. In some bambusoids, wax forms only on one portion of the leaf, indicating very precise, cell specific, genetic and developmental control.

Internal anatomy of leaves

The internal anatomy of grass leaves has been studied extensively (Brown 1977; Ellis 1976; Metcalfe 1960; Watson and Dallwitz 1992 onward). Most investigations have focused exclusively on the cross-sectional appearance of the middle portion of the adult leaf blade, so comparative data on sections in other planes, on sheaths and on early development are limited. In most species, the mesophyll cells are not tightly packed and are relatively homogeneous throughout the leaf. However, some Centotheceae have an adaxial palisade layer. In some early-diverging grasses and in virtually all bamboos the mesophyll is interrupted by fusoid cells (Fig. 3A).

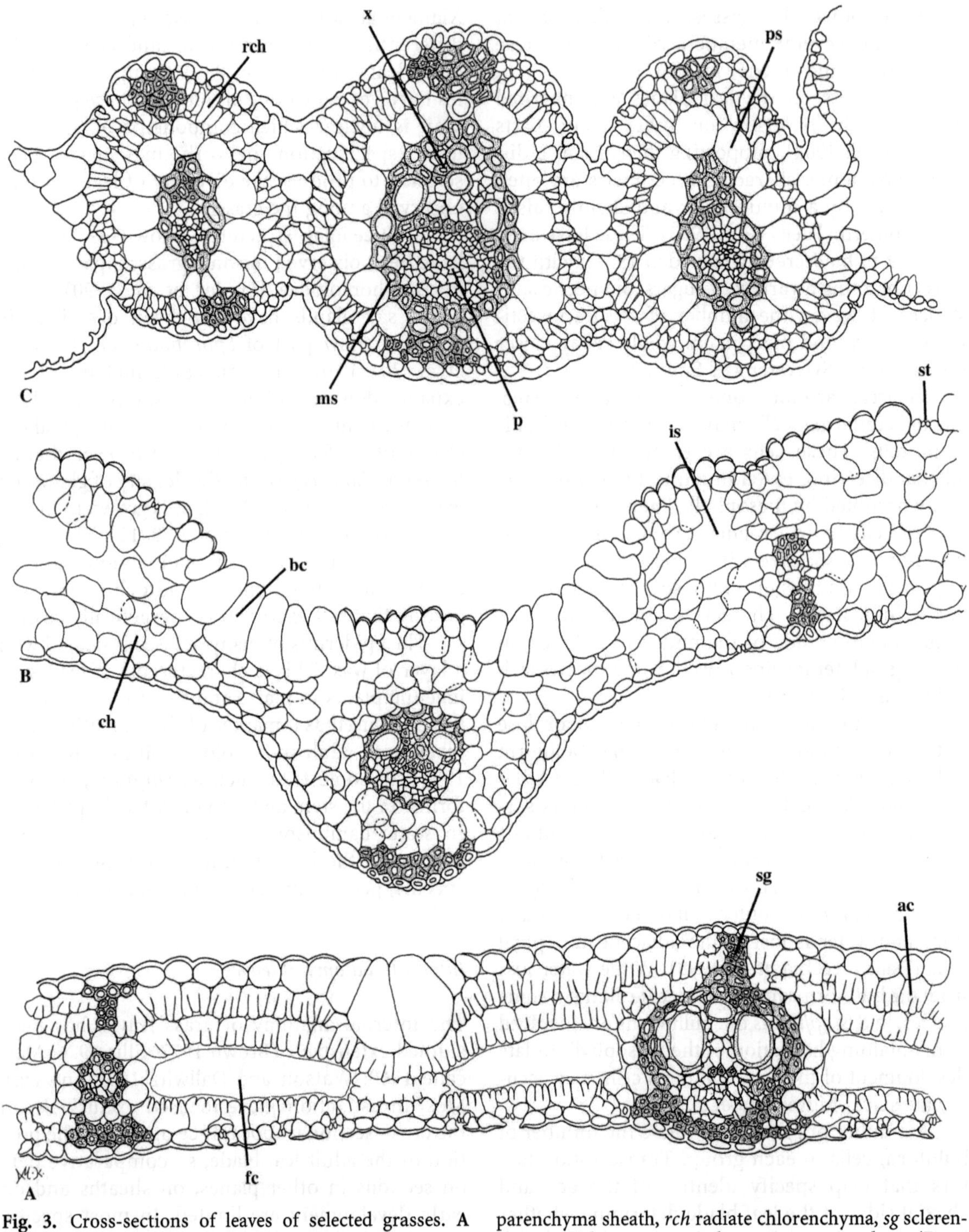

Fig. 3. Cross-sections of leaves of selected grasses. **A** *Dinochloa macclellandii* (Bambusoideae). **B** *Poa* sp. (Pooideae). **C** *Bouteloua* sp. (Chloridoideae). *ac* arm cell, *bc* bulliform cell, *ch* chlorenchyma, *fc* fusoid cell, *is* intercellular space, *ms* mestome sheath, *p* phloem, *ps* outer parenchyma sheath, *rch* radiate chlorenchyma, *sg* sclerenchyma girder, *st* stomatal apparatus, *x* xylem. (From GPWG 2001, p. 405, with permission of the Missouri Botanical Press; drawn by M. Kojima)

In all the early-diverging lineages of grasses, mesophyll cells have obvious invaginations of the cell wall when viewed in cross section; such cells are known in the literature as arm cells (Fig. 3A). Such cell wall invaginations also occur in all bamboos, and in the tribe Oryzeae of subfamily Ehrhartoideae (GPWG 2001). Invaginated cell walls also appear in *Phragmites* (subfamily Arundinoideae), where they constitute a reversal following loss at the base of the PACMAD clade. Longitudinal sections of leaves have revealed additional variation in internal morphology (Sánchez-Ken and Clark, unpubl. data).

Mesophyll cells with invaginated walls are apparently uniquely derived in the grasses. They are lacking in Joinvilleaceae. Because Ecdeiocoleaceae are leafless, there are no data for this family. However, a similar phenotype appears in other more distantly related families. For example, Restionaceae have cells with invaginated cell walls in their stems, where they are known as peg cells (Cutler 1969).

Fusoid cells are large, rectangular to cigar-shaped mesophyll cells, and appear curiously empty in leaf cross sections. They are apparently synapomorphic for the grasses; the plants in which they occur are shade loving and prefer moist habitats. They occur only in the early-diverging lineages, *Streptogyna*, the Bambusoideae, and a handful of Panicoideae in tribe Paspaleae, subtribe Arthropogoninae. Superficially similar cells have been reported in Centotheceae (Panicoideae) as well, but they appear to be laterally expanded bundle sheath cells and are more accurately described as bundle sheath extensions. The function and development of fusoid cells are unknown and it is not even clear whether they are alive or dead, although some data suggest that they may be dead at maturity.

Development of fusoid cells has been described in some detail in *Streptochaeta spicata* (Page 1947). As the primary vascular bundles begin to differentiate, most mesophyll cells are still dividing in all three planes of the leaf (ab-adaxial, lateral, and proximo-distal). However, the mesophyll cells adjacent to the vascular tissue cease to divide in the ab-adaxial and lateral planes and simply enlarge; they continue to divide along the proximo-distal axis. While cell division continues in all mesophyll, epidermis and vascular tissue, the fusoid cells develop large vacuoles. As the leaf blade emerges from the sheath of the leaf below it, the fusoid cells appear to die and their walls apparently collapse. Page (1947) then speculates that some force must be operating on the fusoid cells to preserve their regular shape during development. In long-lived leaves of some bamboos the fusoid cells accumulate silica in their lumens (Motomura et al. 2004). March and Clark (2011) find that shade grown leaves of *Chusquea*, *Phyllostachys*, and *Yushania* all developed fusoid cells, whereas sun-grown leaves did not. They speculate that perhaps fusoid cells are a way to increase light availability inside the leaf.

Grass leaves generally contain sclerenchyma associated at least with the vascular bundles, although there are often clusters of sclerenchyma at the leaf margin as well (Fig. 3A, C). The distribution of sclerenchyma within the leaf is often distinctive and can be helpful in species identification. Sclerenchyma may extend from the vascular bundle to the abaxial epidermis, the adaxial epidermis or both, or may be present as a cap over the bundle (Ellis 1976).

The functional consequences of different patterns of sclerenchyma distribution are largely unknown. Leaves with sclerenchyma girders extending from the bundle sheath to both epidermes are known as "heterobaric", whereas those without such girders are called "homobaric". In heterobaric leaves, such as those in *Hordeum vulgare*, the girders, called bundle sheath extensions in the physiology literature, provide a direct hydraulic connection between the vascular bundle and the epidermis (Buckley et al. 2011) allowing stomatal movements to respond rapidly and easily to water availability. Bundle sheath extensions also divide the leaf into functional compartments that affect the structural and functional aspects of the leaf, although this functional compartmentation has not been investigated in grasses. Investigations in dicotyledonous trees and shrubs suggest that heterobaric leaves have lower leaf mass per unit area, more nitrogen per unit mass, and have higher photosynthetic capacity per unit mass, than homobaric leaves (Liakoura et al. 2009). In homobaric leaves, CO_2 can diffuse laterally for a distance of several millimeters, whereas the bundle sheath extensions of a heterobaric leaf effectively prevent such diffusion (Pieruschka et al. 2008). It seems likely that such relationships will hold in grasses as well.

The phloem of grass leaves contain two distinct sorts of sieve elements (Botha 2013). The canonical sieve element is thin-walled, and is associated with a companion cell. This pair of cells appears to function in a manner similar to that of dicot sieve elements and companion cells, in which the sieve element is responsible for long distance transport, is enucleate a maturity, and is metabolically dependent on the companion cell, to which it is connected by plasmodesmata. The other sort of sieve element is thick-walled, appears not to be involved in transport, and is symplastically isolated from the thin-walled sieve elements; instead it is connected to the vascular parenchyma of the metaxylem. The function of the thick-walled sieve elements is not known, nor is their taxonomic distribution. They occur in grasses and sedges but whether they appear in other monocots is unknown (Botha 2013).

The internal anatomy of the leaf varies considerably depending on the photosynthetic pathway used. Grasses may use either the conventional C_3 pathway, or the high efficiency C_4 pathway. Within the family, the C_4 photosynthetic pathway has arisen multiple times among members of the PACMAD clade (Christin et al. 2007, 2008; Edwards and Smith 2010; GPWG II 2012; Vicentini et al. 2008). C_4 is a mechanism for reducing the amount of oxygen available to ribulose 1,5 bisphosphate carboxylase/oxygenase (Rubisco), thus minimizing photorespiration. By reducing the energy lost to photorespiration, C_4 plants make better use of light energy; Zhu et al. (2008) have calculated that C_4 plants may capture 24 % more energy than C_3 plants. In addition, C_4 plants make more efficient use of water and nitrogen (Sage 1999).

In all C_3 grasses, there are two rings of cells around the vascular bundles, the ancestral condition (Fig. 3A, B). The outer sheath is made up of thin-walled parenchyma cells and is known as the parenchymatous sheath, and the inner sheath has cells with thick walls and is known as the mestome sheath (Carolin and Jacobs 1973); none of the sheath cells have many chloroplasts. The mestome sheath of some species such as rice and the pooid grasses has an osmophilic layer in the cell wall in transmission electron microscopy (TEM), whereas this layer is apparently absent in *Arundo* and *Phragmites* (Arundinoideae) (Carolin and Jacobs 1973). In mestome sheaths in general the outer walls are suberized, minimizing the apoplastic pathway for water.

In C_4 plants, the carbon-fixing enzyme Rubisco is produced only in the bundle sheath cells, rather than in the mesophyll as in C_3 species. Rubisco is then replaced in the mesophyll by PEP carboxylase (Kanai and Edwards 1999; Wang et al. 2011). Atmospheric CO_2 is hydrated to carbonic acid and subsequently fixed as a four-carbon compound oxaloacetate (OAA) in the mesophyll. OAA is then reduced to either malate or aspartate, and is shuttled to the bundle sheath, where it is decarboxylated and the carbon is reduced by the normal Calvin Cycle to make three-carbon sugars. This process leaves the three-carbon phosphoglyceric acid, which is returned to the mesophyll.

Because the C_4 pathway requires constant shuttling of carbon compounds between mesophyll and bundle sheath, every bundle sheath cell is next to and connected directly with a mesophyll cell in C_4 plants. This is accomplished by increasing the density of venation. In all C_4 grasses, the vascular bundles are separated by no more than two mesophyll cells (Fig. 3C) (Hattersley and Watson 1975). In a few C_4 taxa, such as *Arundinella*, the fully developed veins are separated by more than two mesophyll cells, but in the place of minor veins are lines of cells that appear unusual in cross section and were initially given the name "distinctive cells" by Tateoka (1958). These cells function in carbon reduction, produce RuBisCO (Sinha and Kellogg 1996), and appear to represent isolated bundle sheath cells from degenerate minor veins. Photosynthetic pathway correlates perfectly with vein spacing. Leaves of all or nearly all genera of the family have been examined anatomically (Watson and Dallwitz 1992 onward), and a substantial majority of these have also been evaluated using stable isotopes (Brown 1977) (C. Osborne, Univ. of Sheffield, pers. comm.); the ratio of ^{13}C to ^{12}C is a reliable predictor of photosynthetic pathway.

The genetic regulation of vein spacing is not understood, and existing data show that there is no simple switch between C_3 and C_4-like spacing. Vein density in oats (C_3) could be increased by introduction of entire maize (C_4) chromosomes 1, 5, or 9, one at a time (Tolley et al. 2012). While each chromosomal introduction produced a significant reduction in vein spacing, the number of cells between the veins was unchanged suggesting that the cells were merely smaller. Addition of several maize chromosomes also increased the size of the bundle sheath cells.

C_4 species vary in the enzyme used to decarboxylate the four-carbon compound in the bundle sheath. Most grasses use one of several malic enzymes, some of which use NAD as a co-factor and some of which use NADP; these are known as NAD-ME and NADP-ME respectively. In addition, some species use PEP carboxykinase (PCK) as their primary decarboxylating enzyme. Although it is common to classify C_4 species according to their decarboxylating enzyme, many species use more than one. For example, PCK activity has been detected in maize, even though it is generally considered to be an NADP-ME species (Walker et al. 1997).

The structure of the bundle sheath cells generally correlates with the primary decarboxylating enzyme (Hattersley 1987; Prendergast and Hattersley 1987; Prendergast et al. 1987; Sinha and Kellogg 1996). The C_4 species that use predominantly NAD-ME or PCK include all members of Chloridoideae, subtribes Melinidinae and Panicinae (Paniceae, Panicoideae), and the genus *Stipagrostis* (Aristidoideae). These all have the ancestral condition of two bundle sheaths, and the inner bundle sheath retains its ancestral structure as a thick-walled mestome sheath (Fig. 3C) (Brown 1975). The outer parenchyma sheath, however, produces large numbers of chloroplasts and is the site of carbon reduction. The structure of the outer bundle sheath often correlates with decarboxylating enzyme as well (Hattersley and Watson 1992). In NAD-ME species, the outer walls of the parenchyma sheath form an even, more or less circular outline and are suberized, whereas the outer sheath cells in PCK species are often irregular or form a triangular shaped bundle, and are non-suberized (Dengler and Nelson 1999). The organelles in the outer sheath of NAD-ME plants are often clustered along the walls closest to the vascular tissue (centripetal), whereas the organelles in sheaths of PCK species are often clustered along the walls closest to the mesophyll. These correlations are not absolute, however (Prendergast and Hattersley 1987; Prendergast et al. 1987).

The C_4 NADP-ME taxa in Micrairoideae and Aristidoideae are anatomically distinctive. In both subfamilies there is an outer and an inner bundle sheath. However, the two C_4 genera of Micrairoideae, *Eriachne* and *Pheidochloa*, have so many chloroplasts in the outer sheath that no cytoplasm or vacuole is visible in cross section (see illustrations in Sinha and Kellogg 1996). In *Aristida*, both the inner and outer sheaths are carbon reducing and neither has a suberized lamella. Carbon reduction occurs primarily in the inner sheath (Ueno 1992), but carbon may leak out to the outer sheath, where it can also be captured by Rubisco and reduced (Hattersley and Browning 1981).

All other C_4 species occur within subfamily Panicoideae, and use NADP-ME as the decarboxylating enzyme. This group includes the most well-studied of the C_4 plants, maize, and its agronomically important relatives sorghum and sugar cane. These species all have a single bundle sheath, which develops from the same cell lineages as the ancestral mestome sheath, and like the mestome sheath has a suberized cell wall (Dengler et al. 1985). In these NADP-ME species, the chloroplasts in the bundle sheath cells generally lack grana. There is also a reduction in the amount of proteins such as LHCPII that anchor the photosynthetic apparatus in the thylakoids (Sinha and Kellogg 1996).

Radiate mesophyll cells occur in many C_4 taxa, particularly those with the NAD-ME subtype. However, radiate mesophyll is not confined to C_4 species. It is also recorded for members of the tribe Isachneae (Micrairoideae), which are entirely C_3. It is an intriguing possibility that Isachneae may have been derived from C_4 ancestors, or conversely that the C_4 Micrairoideae already had radiate mesophyll that made origin of the C_4 anatomy easier.

Data on the anatomy of leaf sheaths are scarce. The veins in the sheath may be more widely spaced than those in the blade, which affects their photosynthetic function. The mesophyll cells farthest from the veins have C_3 physiology in maize, whereas the ones closest to the veins are physiologically C_4 (Langdale et al. 1988). The anatomical structure at the top of the sheath also controls the angle of the leaf blade. In rice, reduction in the amount of sclerenchyma associated with the vascular tissue results in leaves that are less upright than normal (Ning et al. 2011).

Some grass plants go through a heteroblastic transition between juvenile and adult foliage, known as vegetative phase change. This shift is regulated in the grasses, as in all studied angiosperms, by two opposing microRNAs, miR156 and miR172 (Chuck et al. 2007; Poethig 1984; Yang et al. 2011). High levels of miR156 prevent the transition, whereas miR172 enhances it. Vegetative phase change is particularly well studied in maize, in which juvenile leaves have characteristic patterns of waxes (Moose and Sisco 1994, 1996). Similar developmental shifts have been observed in some woody bamboos, and likely occur in other members of the family, but have never been studied rigorously.

References

Abrash, E.B., Bergmann, D.C. 2009. Asymmetric cell divisions: a view from plant development. Developmental Cell 16: 783–796.

Amarasinghe, V., Watson, L. 1988. Comparative ultrastructure of microhairs in grasses. Bot. J. Linn. Soc. 98: 303–319.

Amarasinghe, V., Watson, L. 1989. Variation in salt secretory activity of microhairs in grasses. Aust. J. Plant Phys. 16: 219–229.

André, J.-P. 1998. A study of the vascular organization of bamboos (Poaceae-Bambuseae) using a microcasting method. IAWA J. 19: 265–278.

Arber, A. 1934. The Gramineae: a study of cereal, bamboo, and grass. New York: Macmillan.

Arite, T., Iwata, H., Ohshima, K., Maekawa, M., Nakajima, M., Kojima, M., Sakakibara, H., Kyozuka, J. 2007. *DWARF10*, an *RMS1/MAX4/DAD1* ortholog, controls lateral bud outgrowth in rice. Plant J. 51: 1019–1029.

Arredondo, J.T., Schnyder, H. 2003. Components of leaf elongation rate and their relationship to specific leaf area in contrasting grasses. New Phytol. 158: 305–314.

Artschwager, E. 1951. Structure and taxonomic value of the dewlap in sugarcane. USDA Technical Bull. 1038: 1–12.

Baker, G., Jones, L.H.P., Wardrop, I.D. 1959. Cause of wear in sheep's teeth. Nature 184: 1583–1584.

Bergmann, D., Zehfus, M., Zierer, L., Smith, B., Gabel, M. 2009. Grass rhizosheaths: associated bacterial communities and potential for nitrogen fixation. Western N. Amer. Nat. 69: 105–114.

Bess, E.C., Doust, A.N., Davidse, G., Kellogg, E.A. 2006. *Zuloagaea*, a new genus of tropical grass within the "bristle clade" (Poaceae: Paniceae). Syst. Bot. 31: 656–670.

Bidlack, J.E., Jansky, S.H. 2011. Stern's Introductory Plant Biology. New York: McGraw-Hill.

Botha, C.E.J. 2013. A tale of two neglected systems - structure and function of the thin- and thick-walled sieve tubes in monocotyledonous leaves. Frontiers Plant Sci. 4: 297.

Bowden, B.N. 1970. The sugars in the extrafloral nectar of *Andropogon gayanus* var. *bisquamulatus*. Phytochemistry 9: 2315–2318.

Bowden, B.N. 1971. Studies on *Andropogon gayanus* Kunth VI: the leaf nectaries of *Andropogon gayanus* var. *bisquamulatus* (Hochst.) Hack. (Gramineae). Bot. J. Linn. Soc. 64: 77–80.

Branson, F.A. 1953. Two factors affecting resistance of grasses to grazing. J. Range Manage. 6: 165–171.

Brown, W.V. 1975. Variations in anatomy, associations, and origins of Kranz tissue. Amer. J. Bot. 62: 395–402.

Brown, W.V. 1977. The Kranz syndrome and its subtypes in grass systematics. Mem. Torrey Bot. Club 23: 1–97.

Brown, W.V., Heimsch, C., Emery, H.P. 1957. The organization of the grass shoot apex and systematics. Amer. J. Bot. 44: 590–595.

Brown, W.V., Harris, W.F., Graham, J.D. 1959a. Grass morphology and systematics. I. The internode. Southwest. Nat. 4: 115–125.

Brown, W.V., Pratt, G.A., Mobley, H.M. 1959b. Grass morphology and systematics. II. The nodal pulvinus. Southwest. Nat. 4: 126–133.

Buckley, R. 1982. Sand rhizosheath of an arid zone grass. Plant Soil 66: 417–421.

Buckley, T.N., Sack, L., Gilbert, M.E. 2011. The role of bundle sheath extension and life form in stomatal responses to leaf water status. Plant Phys. 156: 962–973.

Burns, W. 1945. Corm and bulb formation in plants, with special reference to the Gramineae. Trans. Proc. Bot. Soc. Edinburgh 34: 316–347.

Campbell, C.S., Kellogg, E.A. 1987. Sister group relationships of the Poaceae. In: Soderstrom, T.R., Hilu, K. W., Campbell, C.S., Barkworth, M.E. (eds.) Grass systematics and evolution. Washington, DC: Smithsonian Institution. pp. 217–224.

Cao, K.-F., Yang, S.-J., Zhang, Y.-J., Brodribb, T.J. 2012. The maximum height of grasses is determined by roots. Ecol. Lett. 15: 666–672.

Carolin, R.C., Jacobs, S.W.L. 1973. The structure of the cells of the mesophyll and parenchymatous bundle sheath of the Gramineae. Bot. J. Linn. Soc. 66: 259–275.

Chen, Y., Fan, X., Song, W., Zhang, Y., Xu, G. 2012. Overexpression of OsPIN2 leads to increased tiller numbers, angle and shorter plant height through suppression of OsLAZY1. Plant Biotech. J. 10: 139–149.

Chochois, V., Vogel, J.P., Watt, M. 2012. Application of *Brachypodium* to the genetic improvement of wheat roots. J. Exp. Bot. 63: 3467–3474.

Christin, P.-A., Salamin, N., Savolainen, V., Duvall, M.R., Besnard, G. 2007. C_4 photosynthesis evolved in grasses via parallel adaptive genetic changes. Curr. Biol. 17: 1241–1247.

Christin, P.-A., Besnard, G., Samaritani, E., Duvall, M.R., Hodkinson, T.R., Savolainen, V., Salamin, N. 2008. Oligocene CO_2 decline promoted C_4 photosynthesis in grasses. Curr. Biol. 18: 37–43.

Chuck, G., Cigan, A.M., Saeteurn, K., Hake, S. 2007. The heterochronic maize mutant *Corngrass1* results from

overexpression of a tandem microRNA. Nat. Genet. 39: 544–549.
Clark, L.G., Fisher, J.B. 1987. Vegetative morphology of grasses: shoots and roots. In: Soderstrom, T.R., Hilu, K.W., Campbell, C.S., Barkworth, M.E. (eds.) Grass systematics and evolution. Washington, DC: Smithsonian Institution Press. pp. 37–45
Clore, A.M. 2013. Cereal grass pulvini: agronomically significant models for studying gravitropism signaling and tissue polarity. Amer. J. Bot. 100: 101–110.
Clowes, F.A.L. 2000. Pattern in root meristem development in angiosperms. New Phytol. 146: 83–94.
Coudert, Y., Périn, C., Courtois, B., Khong, N.G., Gantet, P. 2010. Genetic control of root development in rice, the model cereal. Trends Pl. Sci. 15: 219–226.
Cutler, D.F. 1969. IV. Juncales. In: Metcalfe, C.R. (ed.) Anatomy of the Monocotyledons. Oxford: Clarendon Press.
Dai, M., Zhao, Y., Ma, Q., Hu, Y., Hedden, P., Zhang, Q., Zhou, D.X. 2007. The rice *YABBY1* gene is involved in the feedback regulation of gibberellin metabolism. Plant Phys. 144: 121–133.
Dalla Vecchia, F., El Asmar, T., Calamassi, R., Rascio, N., Vazzana, C. 1998. Morphological and ultrastructural aspects of dehydration and rehydration in leaves of *Sporobolus stapfianus*. Pl. Growth Reg. 24: 219–228.
Deng, F., Yamaji, N., Xia, J., Ma, J.F. 2013. A member of the heavy metal P-type ATPase OsHMA5 is involved in xylem loading of copper in rice. Plant Physiol. 163: 1353–1362.
Dengler, N.G., Nelson, T. 1999. Leaf structure and development in C_4 plants. In: Sage, R.F., Monson, R.K. (eds.) C_4 plant biology. San Diego: Academic Press. pp. 133–172
Dengler, N.G., Dengler, R.E., Hattersley, P.W. 1985. Differing ontogenetic origins of PCR ("Kranz") sheaths in leaf blades of C_4 grasses (Poaceae). Amer. J. Bot. 72: 284–302.
Dolan, L., Costa, S. 2001. Evolution and genetics of root hair stripes in the root epidermis. J. Exp. Bot. 52: 413–417.
Doust, A.N. 2007a. Grass architecture: genetic and environmental control of branching. Curr. Opinion Pl. Biol. 10: 21–25.
Doust, A. 2007b. Architectural evolution and its implications for domestication in grasses. Ann. Bot. 100: 941–950.
Doust, A.N., Kellogg, E.A. 2006. Genotype-environment interactions for branching in the weed green millet (*Setaria viridis*) and the crop foxtail millet (*S. italica*) (Poaceae). Mol. Ecol. 15: 1335–1349.
Doust, A.N., Devos, K.M., Gadberry, M., Gale, M.D., Kellogg, E.A. 2004. Genetic control of branching in foxtail millet. Proc. Natl. Acad. Sci. USA 101: 9045–9050.
Duell, R.W., Peacock, G.R. 1985. Rhizosheaths on mesophytic grasses. Crop Sci. 25: 880–883.
Edwards, E.J., Smith, S.A. 2010. Phylogenetic analyses reveal the shady history of C_4 grasses. Proc. Natl. Acad. Sci. USA 107: 2532–2537.
Ellis, R.P. 1976. A procedure for standardizing comparative leaf anatomy in the Poaceae. I. The leaf-blade as viewed in transverse section. Bothalia 12: 65–109.
Fahn, A. 1979. Secretory tissues in plants. London: Academic Press.
Fiorani, F., Beemster, G.T.S., Bultynck, L., Lambers, H. 2000. Can meristematic activity determine variation in leaf size and elongation rate among four *Poa* species? a kinematic study. Plant Phys. 124: 845–856.
Foster, T.M., Timmermans, M.C.P. 2009. Axial patterning in the maize leaf. In: Bennetzen, J.L., Hake, S.C. (eds.) Handbook of maize: Its biology. Heidelberg: Springer.
Frank, M.J., Cartwright, H.N., Smith, L.G. 2003. Three *Brick* genes have distinct functions in a common pathway promoting polarized cell division and cell morphogenesis in the maize leaf epidermis. Development 130: 753–762.
Fujino, K., Matsuda, Y., Ozawa, K., Nishimura, T., Koshiba, T., Fraaije, M.W., Sekiguchi, H. 2008. NARROW LEAF 7 controls leaf shape mediated by auxin in rice. Mol. Genet. Genomics 279: 499–507.
Garbuzov, M., Reidinger, S., Hartley, S.E. 2011. Interactive effects of plant-available soil silicon and herbivory on competition between two grass species. Ann. Bot. 108: 1355–1363.
Gochnauer, M.B., McCully, M.E., Labbé, H. 1989. Different populations of bacteria associated with sheathed and bare regions of roots of field-grown maize. Plant Soil 114: 107–120.
Goller, H. 1977. Beiträge zu Anatomie adulter Gramineenwurzeln im Hinblick auf taxonomische Verwendbarkeit. Beitr. Biol. Pflanzen 53: 217–307.
GPWG 2001. Grass Phylogeny Working Group. Phylogeny and subfamilial classification of the Poaceae. Ann. Missouri Bot. Gard. 88: 373–457.
GPWG II 2012. Grass Phylogeny Working Group II. New grass phylogeny resolves deep evolutionary relationships and discovers C_4 origins. New Phytol. 193: 304–312.
Hattersley, P.W. 1987. Variations in photosynthetic pathway. In: Soderstrom, T.R., Hilu, K.W., Campbell, C. S., Barkworth, M.E. (eds.) Grass systematics and evolution. Washington, DC: Smithsonian Institution Press. pp. 49–64
Hattersley, P.W., Browning, A.J. 1981. Occurrence of the suberized lamella in leaves of grasses of different photosynthetic types. I. In parenchymatous bundle sheaths and PCR ('Kranz') sheaths. Protoplasma 109: 371–401.
Hattersley, P.W., Watson, L. 1975. Anatomical parameters for predicting photosynthetic pathways of grass leaves: The "maximum lateral cell count" and the "maximum cells distant count". Phytomorphology 25: 325–333.
Hattersley, P.W., Watson, L. 1992. Diversification of photosynthesis. In: Chapman, G.P. (ed.) Grass evolution and domestication. Cambridge: Cambridge University Press.
Hibara, K.-i., Obara, M., Hayashida, E., Abe, M., Ishimaru, T., Satoh, H., Itoh, J.-i., Nagato, Y. 2009. The *ADAXIALIZED LEAF1* gene functions in leaf and embryonic pattern formation in rice. Devel. Biol. 334: 345–354.
Hochholdinger, F.F.G. 1998. Early post-embryonic root formation is specifically affected in the maize mutant *lrt1*. Plant J. 16: 247–255.
Hochholdinger, F., Zimmermann, R. 2008. Conserved and diverse mechanisms in root development. Curr. Opinion Pl. Biol. 11: 70–74.

Hochholdinger, F., Park, W.J., Felix, G.H. 2001. Cooperative action of SLR1 and SLR2 is required for lateral root-specific cell elongation in maize. Plant Phys. 125: 1529–1539.

Hochholdinger, F., Park, W.J., Sauer, M., Woll, K. 2004. From weeds to crops: genetic analysis of root development in cereals. Trends Pl. Sci. 9: 42–48.

Holechek, J.L., Pieper, R.D., Herbel, C.H. 1998. Range management: principles and practices. Upper Saddle River, New Jersey: Prentice Hall.

Holloway-Phillips, M.-M., Brodribb, T.J. 2011. Minimum hydraulic safety leads to maximum water-use efficiency in a forage grass. Plant Cell Environ. 34: 302–313.

Hose, E., Clarkson, D.T., Steudle, E., Schreiber, L., Hartung, W. 2001. The exodermis: a variable apoplastic barrier. J. Exp. Bot. 52: 2245–2264.

Hoshikawa, K. 1969. Underground organs of the seedlings and the systematics of Gramineae. Bot. Gaz. 130: 192–203.

Houshmand, S., Knox, R.E., Clarke, F.R., Clarke, J.M. 2007. Microsatellite markers flanking a stem solidness gene on chromosome 3BL in durum wheat. Mol. Breeding 20: 261–270.

Hu, F.Y., Tao, D.Y., Sacks, E., Fu, B.Y., Xu, P., li, J., Yang, Y., McNally, K., Khush, G.S., Paterson, A.H., Li, Z.-K. 2003. Convergent evolution of perenniality in rice and sorghum. Proc. Natl. Acad. Sci. USA 100: 4050–4054.

Hu, J., Zhu, L., Zeng, D., Gao, Z., Guo, I., Fang, Y., Zhang, G., Dong, G., Yan, M., Liu, J., Qian, Q. 2010. Identification and characterization of *NARROW AND ROLLED LEAF 1*, a novel gene regulating leaf morphology and plant architecture in rice. Plant Mol. Biol. 73: 283–292.

Isa, M., Bai, S., Yokoyama, T., Ma, J.F., Ishibashi, Y., Yuasa, T., Iwaya-Inoue, M. 2010. Silicon enhances growth independent of silica deposition in a low-silica rice mutant, *lsi1*. Plant Soil 331: 361–375.

Jackson, D. 2009. Vegetative shoot meristems. In: Bennetzen, J.L., Hake, S.C. (eds.) Handbook of maize: Its biology: Springer Science & Business Media.

Jackson, D., Veit, B., Hake, S. 1994. Expression of the maize *KNOTTED-1* related homeobox genes in the shoot apical meristem predicts patterns of morphogenesis in the vegetative shoot. Development 120: 405–413.

Jin, J., Huang, W., Gao, J.-P., Yang, J., Shi, M., Zhu, M.-Z., Luo, D., Lin, H.-X. 2008. Genetic control of rice plant architecture under domestication. Nat. Genet. 40: 1365–1369.

Johnston, C.R., Watson, L. 1976. Microhairs: a universal characteristic of non-festucoid grass genera? Phytomorphology 26: 297–301.

Judziewicz, E.J., Soderstrom, T.R. 1989. Morphological, anatomical, and taxonomic studies in *Anomochloa* and *Streptochaeta* (Poaceae: Bambusoideae). Smithsonian Contr. Bot. 68: 1–52.

Kanai, R., Edwards, G.E. 1999. The biochemistry of C_4 photosynthesis. In: Sage, R.F., Monson, R.K. (eds.) C_4 plant biology. Amsterdam: Elsevier. pp. 49–87

Kaufman, P.B., Brock, T.G., Song, I., Rho, Y.B., Ghosheh, N.S. 1987. How cereal grass shoots perceive and respond to gravity. Amer. J. Bot. 74: 1446–1457.

Kebrom, T.H., Chandler, P.M., Swain, S.M., King, R.W., Richards, R.A., Spielmeyer, W. 2012. Inhibition of tiller bud outgrowth in the *tin* mutant of wheat is associated with precocious internode development. Plant Phys. 160: 308–318.

Kellogg, E.A. 1990. Ontogenetic studies of florets in *Poa* (Gramineae): allometry and heterochrony. Evolution 44: 1978–1989.

Kellogg, E.A. 2000. The grasses: a case study in macroevolution. Ann. Rev. Ecol. Syst. 31: 217–238.

Kellogg, E.A., Campbell, C.S. 1987. Phylogenetic analyses of the Gramineae. In: Soderstrom, T.R., Hilu, K.W., Campbell, C.S., Barkworth, M.E. (eds.) Grass systematics and evolution. Washington, DC: Smithsonian Institution Press. pp. 310–322

Kellogg, E.A., Watson, L. 1993. Phylogenetic studies of a large data set. I. Bambusoideae, Pooideae, and Andropogoneae (Gramineae). Bot. Rev. 59: 273–343.

Kim, C.M., Dolan, L. 2011. Root hair development involves asymmetric cell division in *Brachypodium distachyon* and symmetric division in *Oryza sativa*. New Phytol. 192: 601–610.

Kitomi, Y., Ito, H., Hobo, T., Aya, K., Kitano, H., Inukai, Y. 2011. The auxin responsive AP2/ERF transcription factor *CROWN ROOTLESS 5* is involved in crown root initiation in rice through the induction of *OsRR1*, a type-A response regulator of cytokinin signaling. Plant J. 67: 472–484.

Kutschera, L., Lichtenegger, E. 1982. Wurzelatlas mitteleuropäischer Grünlandpflanzen. Band I. Monocotyledoneae. Stuttgart: Gustav Fischer Verlag.

Langdale, J.A., Zelitch, I., Miller, E., Nelson, T. 1988. Cell position and light influence C_4 versus C_3 patterns of photosynthetic gene expression in maize. EMBO J. 7: 3643–3651.

Li, P.J., Wang, Y.H., Qian, Q., Fu, Z.M., Wang, M., Zeng, D.L., Li, B.H., Wang, X.J., Li, J.Y. 2007. *LAZY1* controls rice shoot gravitropism through regulating polar auxin transport. Cell Res. 17: 402–410.

Li, L., Shi, Z.Y., Li, L., Shen, G.Z., Wang, X.Q., An, L.S., Zhang, J.L. 2010. Overexpression of *ACL1* (*abaxially curled leaf 1*) increased bulliform cells and induced abaxial curling of leaf blades in rice. Mol. Plant 3: 807–817.

Liakoura, V., Fotelli, M.N., Renneberg, H., Karabourniotis, G. 2009. Should structure-function relations be considered separately for homobaric vs. heterobaric leaves? Amer. J. Bot. 96: 612–619.

Liese, W. 1998. The anatomy of bamboo culms. Technical Report 18. Beijing: International Network for Bamboo and Rattan.

Linder, H.P., Thompson, J.F., Ellis, R.P., Perold, S.M. 1990. The occurrence, anatomy, and systematic implications of the glands in *Pentaschistis* and *Prionanthium* (Poaceae, Arundinoideae, Arundineae). Bot. Gaz. 151: 221–233.

Liphschitz, N., Waisel, Y. 1974. Existence of salt glands in various genera of the Gramineae. New Phytol. 73: 507–513.

Liu, S., Wang, J., Wang, L., Wang, X., Xue, Y., Wu, P., Shou, H. 2009. Adventitious root formation in rice requires OsGNOM1 and is mediated by the OsPINs family. Cell Res. 19: 1110–1119.

Ma, J.F., Yamaji, N. 2006. Silicon uptake and accumulation in higher plants. Trends Pl. Sci. 11: 392–397.

Majer, C., Xu, C., Berendzen, K.W., Hochholdinger, F. 2012. Molecular interactions of ROOTLESS CONCERNING CROWN AND SEMINAL ROOTS, a LOB domain protein regulating shoot-borne root initiation in maize (*Zea mays* L.). Phil. Trans. R. Soc. London B 367: 1542–1551.

March, R.H., Clark, L.G. 2011. Sun-shade variation in bamboo (Poaceae: Bambusoideae) leaves. Telopea 13: 93–104.

Marcum, K.B. 1999. Salinity tolerance mechanisms of grasses in the subfamily Chloridoideae. Crop Sci. 39: 1153–1160.

Martínez-y-Pérez, J.L., Mejía-Saulés, T., Sosa, V. 2008. A taxonomic revision of *Luziola* (Poaceae: Oryzeae). Syst. Bot. 33: 702–718.

Massey, F.P., Hartley, S.E. 2009. Physical defences wear you down: Progressive and irreversible impacts of silica on insect herbivores. J. Animal Ecol. 78: 281–291.

McCully, M. 1995. How do real roots work? Plant Phys. 109: 1–9.

McSteen, P. 2009. Hormonal regulation of branching in grasses. Plant Phys. 149: 46–55.

Metcalfe, C.R. 1960. Anatomy of the monocotyledons. I. Gramineae. Oxford: Clarendon Press.

Moose, S.P., Sisco, P.H. 1994. *Glossy15* controls the epidermal juvenile-to-adult phase transition in maize. Plant Cell 6: 1343–1355.

Moose, S.P., Sisco, P.H. 1996. *Glossy15*, an APETALA2-like gene from maize that regulates leaf epidermal cell identity. Genes Devel. 10: 3018–3027.

Moose, S.P., Lauter, N., Carlson, S.R. 2004. The maize *macrohairless1* locus specifically promotes leaf blade macrohair initiation and responds to factors regulating leaf identity. Genetics 166: 1451–1461.

Motomura, H., Fujii, T., Suzuki, M. 2004. Silica deposition in relation to ageing of leaf tissues in *Sasa veitchii* (Carrière) Rehder (Poaceae: Bambusoideae). Ann. Bot. 93: 235–248.

Motomura, H., Fujii, T., Suzuki, M. 2006. Silica deposition in abaxial epidermis before the opening of leaf blades of *Pleioblastus chino* (Poaceae, Bambusoideae). Ann. Bot. 97: 513–519.

Nardmann, J., Ji, J., Werr, W., Scanlon, M.J. 2004. The maize duplicate genes *narrow sheath1* and *narrow sheath2* encode a conserved homeobox gene function in a lateral domain of shoot apical meristems. Development 131: 2827–2839.

Ning, J., Zhang, B., Wang, N., Zhou, Y., Xiong, L. 2011. *Increased leaf angle1*, a Raf-like MAPKKK that interacts with a nuclear protein family, regulates mechanical tissue formation in the lamina joint of rice. Plant Cell 23: 4334–4347.

Oi, T., Taniguchi, M., Miyake, H. 2012. Morphology and ultrastructure of the salt glands on the leaf surface of Rhodes grass (*Chloris gayana* Kunth). Int. J. Plant Sci. 173: 454–463.

Page, V.M. 1947. Leaf anatomy of *Streptochaeta* and the relation of this genus to the bamboos. Bull. Torrey Bot. Club 74: 232–239.

Page, V.M. 1951. Morphology of the spikelet of *Streptochaeta*. Bull. Torrey Bot. Club 78: 22–37.

Paiva, E.A.S., Machado, S.R. 2003. Collenchyma in *Panicum maximum* (Poaceae): localisation and possible role. Aust. J. Bot. 51: 69–73.

Peterson, P.M., Romaschenko, K., Herrera Arrieta, Y., Saarela, J.M. 2014. A molecular phylogeny and new subgeneric classification of *Sporobolus* (Poaceae: Chloridoideae: Sporobolinae). Taxon 63: 1212–1243.

Philipson, W.R. 1935a. A grass with spiral phyllotaxis: *Micraira subulifolia*. Bull. Misc. Inf. (Kew) 1935: 324–326.

Philipson, W.R. 1935b. The development and morphology of the ligule in grasses. New Phytol. 34: 310–325.

Pieruschka, R., Chavarría-Krauser, A., Cloos, K., Scharr, H., Schurr, U., Jahnke, S. 2008. Photosynthesis can be enhanced by lateral CO_2 diffusion inside leaves over distances of several millimeters. New Phytol. 178: 335–347.

Piperno, D.R. 2006. Phytoliths: a comprehensive guide for archaeologists and paleoecologists. New York: Alta Mira.

Piperno, D.R., Pearsall, D.M. 1998. The silica bodies of tropical American grasses: morphology, taxonomy, and implications for grass systematics and fossil phytolith identification. Smithsonian Contr. Bot. 85: 1–40.

Pizzolato, T.D. 2000. A systematic view of the development of vascular systems in culms and inflorescences of grasses. In: Jacobs, S.W.L., Everett, J. (eds.) Grasses: Systematics and evolution. Melbourne: CSIRO. pp. 8–28

Poethig, S. 1984. Cellular parameters of leaf morphogenesis in maize and tobacco. In: White, R., Dickinson, W.C. (eds.) Contemporary problems in plant anatomy. New York: Academic Press. pp. 235–259

Pohl, R.W. (ed.) 1980. Family #15, Gramineae. Flora Costaricensis. Chicago: Field Museum of Natural History.

Prasad, V., Strömberg, C.A.E., Leaché, A.D., Samant, B., Patnaik, R., Tang, L., Mohabey, D.M., Ge, S., Sahni, A. 2011. Late Cretaceous origin of the rice tribe provides evidence for early diversification in Poaceae. Nature Comm. 2: 480.

Prat, H. 1932. L'épiderme des graminées: étude anatomique et systématique. Ann. Sci. Nat.: Botanique, Séries 10 14: 117–324.

Prendergast, H.D.V., Hattersley, P.W. 1987. Australian C_4 grasses (Poaceae): leaf blade anatomical features in relation to C_4 acid decarboxylation types. Aust. J. Bot. 35: 355–382.

Prendergast, H.D.V., Hattersley, P.W., Stone, N.E. 1987. New structural/biochemical associations in leaf blades of C_4 grasses (Poaceae). Aust. J. Plant Phys. 14: 403–420.

Price, S.R. 1911. The roots of some North African desert-grasses. New Phytol. 10: 328–340.

Prychid, C.J., Rudall, P.J., Gregory, M. 2004. Systematics and biology of silica bodies in monocotyledons. Bot. Rev. 69: 377–440.

Ramsay, L., Comadran, J., Druka, A., Marshall, D.F., Thomas, W.T.B., Macaulay, M., MacKenzie, K., Simpson, C., Fuller, J., Bonar, N., Hayes, P.M., Lundqvist, U., Franckowiak, J.D., Close, T.J., Muehlbauer, G.J., Waugh, R. 2011. *INTERMEDIUM-C*, a modifier of lateral spikelet fertility in barley, is an ortholog of the maize domestication gene *TEOSINTE BRANCHED 1*. Nat. Genet. 43: 169–172.

Reinhardt, D., Pesce, E.R., Stieger, P., Mandel, T., Baltensperger, K., Benett, M., Traas, J., Friml, J., Kuhlemeier, C. 2003. Regulation of phyllotaxis by polar auxin transport. Nature 426: 255–260.

Remigereau, M.-S., Lakis, G., Rekima, S., Leveugle, M., Fontaine, M.C., Langin, T., Sarr, A., Robert, T. 2011. Cereal domestication and evolution of branching: evidence for soft selection in the *Tb1* orthologue of pearl millet (*Pennisetum glaucum* [L.] R. Br.). PLoS ONE 6: e22404.

Röser, M., Heklau, H. 2011. Abscission of leaf laminas - an unnoticed factor in tussock grass formation. Flora 206: 32–37.

Rothwell, N.V. 1966. Evidence for diverse cell types in the apical region of the root epidermis of *Panicum virgatum*. Amer. J. Bot. 53: 7–11.

Row, H.C., Reeder, J.R. 1957. Root-hair development as evidence of relationships among genera of the Gramineae. Amer. J. Bot. 44: 596–601.

Sage, R.F. 1999. Why C_4 photosynthesis? In: Sage, R.F., Monson, R.K. (eds.) C_4 plant biology. San Diego: Academic Press. pp. 3–14.

Sakai, W.S., Sanford, W.G. 1984. A developmental study of silicification in the abaxial epidermal cells of sugarcane leaf blades using scanning electron microscopy and energy dispersive X-ray analysis. Amer. J. Bot. 71: 1315–1322.

Sanson, G.D., Kerr, S.A., Gross, K.A. 2007. Do silica phytoliths really wear mammalian teeth? J. Archaeol. Sci. 34: 526–531.

Satoh-Nagasawa, N., Mori, M., Nakazawa, N., Kawamoto, T., Nagato, Y., Sakurai, K., Takahashi, H., Watanabe, A., Akagi, H. 2012. Mutations in rice (*Oryza sativa*) heavy metal ATPase2 (*OsHMA2*) restrict the translocation of zinc and cadmium. Plant Cell Physiol. 53: 213–224.

Scanlon, M.J., Freeling, M. 1997. Clonal sectors reveal that a specific meristematic domain is not utilized in the maize mutant *narrow sheath*. Devel. Biol. 182: 52–66.

Shane, M.W., McCully, M.E., Canny, M.J. 2000. The vascular system of maize stems revisited: implications for water transport and xylem safety. Ann. Bot. 86: 245–258.

Shane, M.W., McCully, M.E., Canny, M.J., Pate, J.S., Lambers, H. 2011. Development and persistence of sandsheaths of *Lyginia barbata* (Restionaceae): relation to root structural development and longevity. Ann. Bot. 108: 1307–1322.

Sharman, B.C. 1942. Developmental anatomy of the shoot of *Zea mays* L. Ann. Bot. 6: 245–282.

Simpson, G.G. 1951. Horses: The story of the horse family in the modern world and through sixty million years of history. Oxford: Oxford University Press.

Sinha, N.R., Kellogg, E.A. 1996. Parallelism and diversity in multiple origins of C_4 photosynthesis in the grass family. Amer. J. Bot. 83: 1458–1570.

Smith, S., De Smet, I. 2012. Root system architecture: insights from Arabidopsis and cereal crops. Phil. Trans. R. Soc. London B 367: 1441–1452.

Soderstrom, T.R., Judziewicz, E.J. 1987. Systematics of the amphi-Atlantic bambusoid genus *Streptogyna* (Poaceae). Ann. Missouri Bot. Gard. 74: 871–888.

Song, Y., You, J., Xiong, L. 2009. Characterization of *OsIAA1* gene, a member of rice Aux/IAA family involved in auxin and brassinotsteroid hormone responses and plant morphogenesis. Plant Mol. Biol. 70: 297–309.

Spatafora, J.W., Sung, G.-H., Sung, J.-M., Hywel-Jones, N. L., White, Jr., J.F. 2007. Phylogenetic evidence for an animal pathogen origin of ergot and the grass endophytes. Mol. Ecol. 16: 1701–1711.

St. Aubin, G., Canny, M.J., McCully, M.E. 1986. Living vessel elements in the late metaxylem of sheathed maize roots. Ann. Bot. 58: 577–588.

Stevens, P.F. 2012. Angiosperm phylogeny website. Version 12, July 2012 [and more or less continuously updated since]. 2001 onward.

Strömberg, C.A.E. 2006. Evolution of hypsodonty in equids: testing a hypothesis of adaptation. Paleobiology 32: 236–258.

Sugiyama, S. 2005. Developmental basis of interspecific differences in leaf size and specific leaf area among C_3 grass species. Func. Ecol. 19: 916–924.

Sylvester, A.W., Smith, L.G. 2009. Cell biology of maize leaf development. In: Bennetzen, J.L., Hake, S.C. (eds.) Handbook of maize: its biology: Springer Science and Business Media.

Sylvester, A.W., Cande, W.Z., Freeling, M. 1990. Division and differentiation during normal and *liguleless-1* maize leaf development. Development 110: 985–1000.

Tan, L., Li, X., Liu, F., Sun, X., Li, C., Zhu, Z., Fu, Y., Cai, H., Wang, X., Xie, D., Sun, C. 2008. Control of a key transition from prostrate to erect growth in rice domestication. Nat. Genet. 40: 1360–1364.

Tateoka, T. 1958. Notes on some grasses. VIII. On leaf structure of *Arundinella* and *Garnotia*. Bot. Gaz. 120: 101–109.

Thomas, H.H. 1921. Some observations on plants in the Libyan desert. J. Ecol. 9: 75–89.

Tolley, B.J., Sage, T.L., Langdale, J.A., Hibberd, J.M. 2012. Individual maize chromsomes in the C_3 plant oat can increase bundle sheath cell size and vein density. Plant Phys. 159: 1418–1427.

Tong, H., Liu, L., Jin, Y., Du, L., Yin, Y., Qian, Q., Zhu, L., Chu, C. 2012. DWARF AND LOW-TILLERING acts as a direct downstream target of a GSK3-SHAGGY-like kinase to mediate brassinosteroid responses in rice. Plant Cell 24: 2562–2577.

Ueno, O. 1992. Immunogold localization of photosynthetic enzymes in leaves of *Aristida latifolia*, a unique C_4 grass with a double chlorenchymatous bundle sheath. Phys. Plant. 85: 189–196.

Ueno, O., Agaric, S. 2005. Silica deposition in cell walls of the stomatal apparatus of rice leaves. Plant Prod. Sci. 8: 71–73.

Verboom, G.A., Linder, H.P., Stock, W.D. 2003. Phylogenetics of the grass genus *Ehrharta*: evidence for radiation in the summer-arid zone of the South African Cape. Evolution 57: 1008–1021.

Vicentini, A., Barber, J.C., Giussani, L.M., Aliscioni, S.S., Kellogg, E.A. 2008. Multiple coincident origins of C_4 photosynthesis in the Mid- to Late Miocene. Global Change Biol. 14: 2963–2977.

Walker, R.P., Acheson, R.M., Técsi, L.I., Leegood, R.C. 1997. Phosphoenolpyruvate carboxykinase in C_4 plants: its role and regulation. Aust. J. Plant Phys. 24: 459–468.

Wang, Y., Li, J. 2008a. Rice, rising. Nat. Genet. 40: 1273–1275.

Wang, Y., Li, J. 2008b. Molecular basis of plant architecture. Ann. Rev. Plant Biol. 59: 253–279.

Wang, L., Peterson, R.B., Brutnell, T.P. 2011. Regulatory mechanisms underlying C_4 photosynthesis. New Phytol. 190: 9–20.

Watson, L., Dallwitz, M.J. 1992 onward. The grass genera of the world: descriptions, illustrations, identification, and information retrieval; including synonyms, morphology, anatomy, physiology, phytochemistry, cytology, classification, pathogens, world and local distribution, and references, 23rd April 2010 edn. Wallingford: CAB International.

Weaver, J.E., Zink, E. 1945. Extent and longevity of the seminal roots of certain grasses. Plant Phys. 20: 359–379.

Wolbang, C.M., Davies, N.W., Taylor, S.A., Ross, J.J. 2007. Gravistimulation leads to asymmetry of both auxin and gibberellin levels in barley pulvini. Phys. Plant. 131: 140–148.

Wu, X., Tang, D., Li, M., Wang, K., Cheng, Z. 2013. *Loose Plant Architecture1*, an INDETERMINATE DOMAIN protein involved in shoot gravitropism, regulates plant architecture in rice. Plant Phys. 161: 317–329.

Wullstein, L.H., Pratt, S.A. 1981. Scanning electron microscopy of rhizosheaths of *Oryzopsis hymenoides*. Amer. J. Bot. 68: 408–419.

Wullstein, L.H., Bruening, M.L., Bollen, W.B. 1979. Nitrogen fixation associated with sand grain root sheaths (rhizosheaths) of certain xeric grasses. Phys. Plant. 46: 1–4.

Xiang, J.-J., Zhang, G.-H., Qian, Q., Xue, H.-W. 2012. *SEMI-ROLLED LEAF1* encodes a putative glycosylphosphatidylinositol-anchored protein and modulates rice leaf rolling by regulating the formation of bulliform cells. Plant Phys. 159: 1488–1500.

Xu, M., Zhu, L., Shou, H.X., Wu, P. 2005. A PIN1 family gene, *OsPIN1*, involved in auxin-dependent adventitious root emergence and tillering in rice. Plant Cell Phys. 46: 1674–1681.

Yamaguchi, N., Ishikawa, S., Abe, T., Baba, K., Arao, T., Terada, Y. 2012. Role of the node in controlling traffic of cadmium, zinc, and manganese in rice. J. Exp. Bot. 63: 2729–2737.

Yamaji, N., Ma, J.F. 2009. A transporter at the node responsible for intervascular transfer of silicon in rice. Plant Cell 21: 2878–2883.

Yamaji, N., Ma, J.F. 2014. The node, a hub for mineral nutrient distribution in graminaceous plants. Trends in Plant Science 19: 556–563.

Yamaji, N., Mitatni, N., Ma, J.F. 2008. A transporter regulating silicon distribution in rice shoots. Plant Cell 20: 1381–1389.

Yamaji, N., Chiba, Y., Mitani-Ueno, N., Ma, J.F. 2012. Functional characterization of a silicon transporter gene implicated in silicon distribution in barley. Plant Physiol. 160: 1491–1497.

Yamaji, N., Sasaki, A., Xia, J.X., Yokosho, K., Ma, J.F. 2013. A node-based switch for preferential distribution of manganese in rice. Nature Commun. 4: 2442.

Yang, L., Conway, S.R., Poethig, R.S. 2011. Vegetative phase change is mediated by a leaf-derived signal that represses the transcription of miR156. Development 138: 245–249.

Yang, S.-Y., Grønlund, M., Jakobsen, I., Grotemeyer, M.S., Rentsch, D., Miyao, A., Hirochika, H., Kumar, C.S., Sundaresan, V., Salamin, N., Catausan, S., Mattes, N., Heuer, S., Paszkowski, U. 2012. Nonredundant regulation of rice arbuscular mycorrhizal symbiosis by two members of the *PHOSPHATE TRANSPORTER1* gene family. Plant Cell 24: 4236–4251.

Young, I.M. 1995. Variation in moisture contents between bulk soil and the rhizosheath of wheat (*Triticum aestivum* L. cv. Wembley). New Phytol. 130: 135–139.

Yu, B., Lin, Z., Li, H., Li, X., Li, J., Wang, Y., Zhang, X., Ahu, Z., Zhai, W., Wang, X., Xie, D., Sun, C. 2007. *TAC1*, a major quantitative trait locus controlling tiller angle in rice. Plant J. 52: 891–898.

Zhang, G.H., Xu, Q., Zhu, X.D., Qian, Q., Xue, H.W. 2009. SHALLOT-LIKE1 is a KANADI transcription factor that modulates rice leaf rolling by regulating leaf abaxial cell development. Plant Cell 21: 719–735.

Zhao, S.-Q., Hu, J., Guo, L.-B., Qian, Q., Xue, H.-W. 2010. *Rice leaf inclination2*, a VIN3-like protein, regulates leaf angle through modulating cell division of the collar. Cell Res. 20: 935–947.

Zhu, X.G., Long, S.P., Ort, D.R. 2008. What is the maximum efficiency with which photosynthesis can convert solar energy into biomass? Curr. Opin. Biotechnol. 19: 153–159.

Zou, J.H., Zhang, S.Y., Zhang, W.P., Li, G., Chen, Z.X., Zhai, W.X., Zhao, X.F., Pan, X.B., Xie, Q., Zhu, L.H. 2006. The rice *HIGH-TILLERING DWARF1* encoding an ortholog of Arabiodpsis *MAX3* is required for negative regulation of the outgrowth of axillary buds. Plant J. 48: 687–696.

Zou, L., Sun, X., Zhang, Z., Liu, P., Wu, J., Tian, C., Qiu, J., Lu, T. 2011. Leaf rolling controlled by the homeodomain leucine zipper class IV gene *Roc5* in rice. Plant Phys. 156: 1589–1602.

Inflorescence Structure

Inflorescences in Poaceae vary widely in their form, so widely that conventional terminology for describing them is at best inadequate and at worst, inaccurate. The problems with descriptive terminology have two sources. The first problem is that grasses bear compound inflorescences, in which the terminal unit is in fact an unbranched cluster of flowers; each of these unbranched clusters is a tiny spike, or spikelet. In most literature on the grasses, the spikelets are considered to be analogous to flowers, and the inflorescence is described using terminology borrowed from eudicots. Thus, an axis bearing sessile spikelets is called a spike (e.g., *Triticum*, *Lolium*), as though it were an axis bearing sessile flowers. An axis with spikelets on pedicels is called a raceme (e.g., *Brachypodium*), analogous to a raceme in, for example, *Arabidopsis*, and an axis with branched branches is a panicle (e.g., *Oryza*, *Avena*, *Panicum* s.s., and many others), and so forth. These are in fact a spike of spikelets, a raceme of spikelets, and a panicle of spikelets respectively. (Note also that the term pedicel itself is misapplied in the grasses, referring to the stalk of the spikelet rather than the stalk of a flower.) Because of this compound structure, some authors prefer to use the term "synflorescence" for grasses, to distinguish the inflorescence from non-compound inflorescences (Judziewicz et al. 1999; Stapleton 1997). Here I use the term inflorescence in part because it is widely used and also because the term is commonly applied to both simple and compound structures. The existence of compound inflorescences is not in itself a problem. Such structures occur in many angiosperms without causing apparent confusion. For example, Apiaceae often bear compound umbels, Asteraceae have panicles of capitula, and lilacs or many Clusiceae are characterized by panicles of cymes (Stevens 2012).

The second, and more serious, problem is that inflorescences in the grasses are more complex than in most other flowering plants, and that often even the term "panicle of spikelets" fails to describe the morphology precisely (Kellogg 2000; Perreta et al. 2009). Authors rapidly devolve into using intricate terminology, such as "a panicle of racemes", "a complex of partial inflorescences ending in racemes" (e.g., Watson and Dallwitz 1992 onward). These terms respect the complexity of the structures, but make it difficult to compare parts and thus to make statements about inflorescence evolution.

One proposed solution is the typology first suggested by Troll (1964) and developed more extensively by Weberling et al. (1993), Hernández and Rua (1992), Vegetti (1991), and Vegetti and Anton (1996). In this system, any group of flowers that is borne on the main axis (i.e., is the product of the inflorescence meristem) is known as the florescence, whereas axillary branches bearing groups of flowers are known as paraclades and their flowers called co-florescences. Thus, in the grasses, the spikelet at the apex of the main inflorescence axis (if present) is called the florescence, and all other spikelets are co-florescences. The complex systems of florescences and co-florescences are then considered to be polytelic synflorescences. If the arrangement of florescences and co-florescences on all paraclades is the same, the synflorescence is described as "homogenized". Loss of the main florescence is described as "truncation". This typological approach may be hard to connect to

E.A. Kellogg, *Flowering Plants. Monocots*, The Families and Genera of Vascular Plants 13,
DOI 10.1007/978-3-319-15332-2_2, © Springer International Publishing Switzerland 2015

known patterns of development; for example, the term "homogenization" does not refer to any obvious biological process. In addition, there seems to be little biological distinction between the florescence and co-florescences (i.e., a spikelet terminating the inflorescence and a spikelet terminating a lateral branch). However, the effort to apply a typological description to multiple species of grasses has led to many publications in which adult inflorescence morphology is diagrammed and described carefully and precisely (e.g., Gasser and Vegetti 1997; Kern et al. 2008; Perreta et al. 2000; Reinheimer et al. 2009; Vegetti and Pensiero 1999). Whether or not one chooses to apply Weberling's (1989) typology, the illustrations in these papers provide valuable comparative data, and can be a major improvement over attempts to stretch the words "panicle" and "raceme".

Another way to view inflorescence morphology is to focus on meristems and their fate, beginning with the apical meristem of the plant as it shifts from producing leaves to producing bracts and higher-order meristems that will ultimately lead to spikelet and, hence, flower production. In most grasses, the inflorescence is easily identified as the product of the shoot apical meristem after the transition to flowering, and is a discrete structure that is terminal on the culm. The problematical situations that occur in some woody bamboos (Bambuseae and Arundinarieae) and some Andropogoneae are addressed in more detail below. In this work any spikelet-bearing axis that is either terminal on the plant or on an axillary branch with a subtending leaf or prophyll is called an inflorescence; this terminology follows Clayton (1969). Under this definition, an inflorescence (as used here) is equivalent to a synflorescence as defined by Judziewicz et al. (1999). Aggregates of such units are simply described but not given particular names.

Recent work in developmental genetics has uncovered the effects of many proteins that control inflorescence development. As we begin to understand the specific gene-level changes that affect the inflorescence, we can begin to see the basis for many of the morphological characters that have been identified by taxonomists. Descriptions that take into account development and genetic controls allow more accurate assessment of homology and more precise definition of characters and character states.

Transition to Flowering

Early in development, the shoot apical meristem produces only leaves, each of which develops a meristem in its axil. After receiving an endogenous (hormonal) or exogenous (light, heat) signal, the meristem acquires the capacity to produce an inflorescence (Colasanti and Coneva 2009; Wellmer and Riechmann 2010). In all flowering plants, the transition to flowering is controlled by a mobile signal, or florigen, known as Flowering locus T or FT, that is produced in the leaves and transported to the shoot apical meristem (Zeevaart 2008). In the meristem, FT binds to another regulator, FD. Homologues of both FT and FD have been cloned from maize and rice, although there may be more copies of each in the grasses than in other angiosperms; the function of both sets of proteins appears to be conserved (Danilevskaya et al. 2008; Kojima et al. 2002; Komiya et al. 2008; Lazakis et al. 2011; Muszynski et al. 2006; Tamaki et al. 2007; Yan et al. 2006).

While the FT-FD module is shared among flowering plants, the proteins that regulate it are not. The maize protein INDETERMINATE1 (ID1) and its rice homologue RID1 (= OsID1) are only found in grasses (Baumel et al. 2002). These proteins are genetically upstream of the FT homologues, although they appear not to interact directly with FT-like genes (Lazakis et al. 2011; Wu et al. 2008). Downregulation of ID1 or RID1 delays flowering or blocks it entirely. Thus, the internal regulation of flowering in the grasses proceeds via mechanisms different from those in other plants.

In the leaves of many grasses, the FT homologues are also regulated by proteins similar to FRUITULL in Arabidopsis. The FUL-like proteins plus the SEPALLATA (SEP)-like protein PANICLE PHYTOMER2 (PAP2) are responsible for integrating environmental signals such as vernalization and increasing the expression of FT homologues (Preston and Kellogg 2008; Yan et al. 2003, 2006). Once in the inflorescence, the regulatory roles appear reversed, with the FT-FD complex activating *Ful-like* and *pap2* (Kobayashi et al. 2012). This general regulatory pathway has been pieced together from work on several cereal crops; it will be interesting to learn whether it is indeed universal in the family as it now appears.

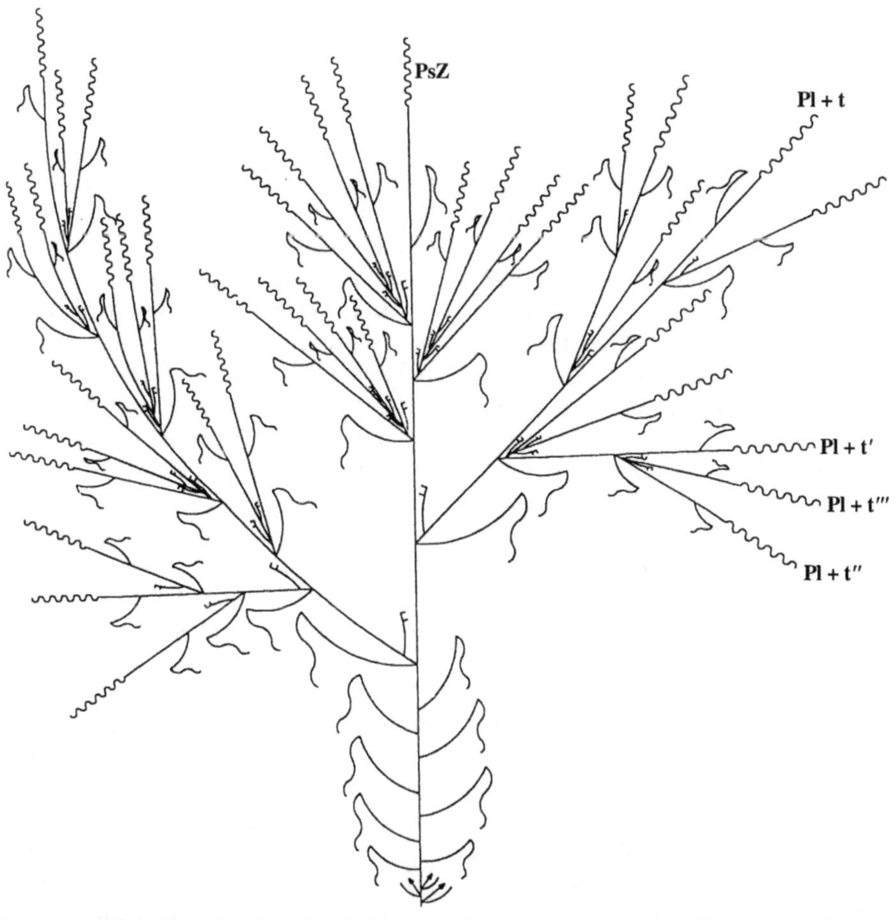

Fig. 4. Proliferating axillary branches in tribe Andropogoneae, subfamily Panicoideae (*Coelorachis aurita* (Steud.) A. Camus). Pl, long paracladium; in this text these are simply called inflorescences. Pl + t, long paracladium with trophotagma, where trophotagma denotes bracts plus prophylls. Pl + t', Pl +t", etc., higher-order long paracladia with trophotagma. PsZ, short paracladia subzone; this corresponds to the terminal inflorescence, which produces spikelet pairs from its apical meristem. (From Vegetti 1999)

In many grasses, it is possible to identify three zones or parts of the plant in which axillary buds have different fates. The lowest zone is described as the "innovation zone" (Vegetti and Weberling 1996), in which buds from the lowest nodes on the plant can produce tillers that reiterate the structure of the primary axis. Above this is an area called the "zone of inhibition" in which axillary buds generally do not elongate at all (Vegetti and Weberling 1996). The upper part of the plant is then the "paracladial zone" (Vegetti 1991; Vegetti and Anton 1996) in which axillary buds can grow out and produce paraclades, inflorescences whose structure is indistinguishable from that of the terminal inflorescence.

The basic three-zone pattern is modified in several subfamilies. In Pooideae the zone of inhibition extends the entire length of the culm below the inflorescence; culm branches rarely form in this subfamily, so that inflorescences are terminal on either the primary axis or on axes developing from tillers. In some genera of Andropogoneae (Panicoideae) (e.g., *Hyparrhenia*, *Andropogon*, *Schizachyrium*), axillary branches in the paracladial zone proliferate. Each axillary branch itself produces an axillary branch, which produces another axillary branch, leading to multiple spikelet-bearing axes at each node. Each of the spikelet-bearing axes produces an adaxial prophyll and a single leaf, which has a reduced blade and a sheath enclosing the part of the axis that bears the spikelets; the internode between the prophyll and the leaf is generally elongated (Fig. 4). At maturity then it appears that the entire upper part of the plant has become a diffuse highly bracteate "inflorescence" in which the terminal units are repeated complex branching structures. Branch proliferation has been studied genetically in maize, and is regulated by the protein GRASSY TILLERS1 (GT1) (Wills et al. 2013); higher expression of *gt1* reduced the number of proliferating branches.

In Bambusoideae, Judziewicz et al. (1999) have defined the synflorescence (equivalent to "inflorescence" in this work) as "the flowering structure that emerges from the apex of a leafy branch as that branch usually grows from a major axis (or the culm apex)." In many woody bamboos, this inflorescence is similar to an inflorescence in any other grass; the inflorescence axis may be unbranched or branched, and the branches may be elongate or not (capitate). Such conventional inflorescences have been described as determinate or semelauctant (McClure 1966; Stapleton 1997); neither term is used here.

A more complex situation occurs in some of the woody bamboos, in which the transition from vegetative to inflorescence growth appears to be more gradual. In these plants, the branch meristem ceases to produce foliage leaves but instead produces a set of bracts, some of which have buds in their axils, and then eventually begins to produce flowers before terminating with a blind rachilla extension. The bracts are called glumes or gemmiparous bracts; it is not clear whether they are homologous with the glumes of more conventional spikelets. The buds in the bract axils can then grow out, themselves producing bracts, some with buds, then flowers, and then a rachilla extension. Depending on the species, this can continue for up to five rounds of branching. Such reiterating structures have been called pseudospikelets, because they have some aspects of spikelet identity and some of branch identity, and have been described as indeterminate or iterauctant (McClure 1966; Stapleton 1997).

Bract Formation and Inflorescence Phyllotaxis

The inflorescence meristem initiates bracts on its flanks; in many species (e.g., *Hordeum vulgare*, *Pennisetum glaucum*, *Trisetum spicatum*) these are clearly visible in early development, but their development is suppressed (Evans 1940; Fraser and Kokko 1993; Latting 1972; Whipple et al. 2010). Even in species in which bracts are not easily visible under the microscope, molecular markers such as TASSELSHEATH1 (Whipple et al. 2010) and TASSELSHEATH4 (Chuck et al. 2010) indicate that cells have been set aside to form the bract. In the axil of each suppressed bract is a primary branch meristem, which quickly enlarges and obscures the bract meristem.

In the transition to flowering, the shoot apical meristem often changes phyllotaxis from the distichous pattern of leaf production to a spiral pattern of inflorescence branch production. This phyllotactic shift appears to be plesiomorphic in the family (Kellogg et al. 2013). The inflorescence meristem of the grass sister group, *Ecdeiocolea monostachya* (Ecdeiocoleaceae) produces bracts and flowers arranged in a spiral (Rudall et al. 2005). In subfamily Anomochlooideae *Streptochaeta* produces primary branches that terminate in flowers and appear to be arranged in a spiral (Sajo et al. 2008); *Anomochloa* is reported to be primarily distichous (Judziewicz and Soderstrom 1989), but data from early inflorescence development are not available. Spiral inflorescence phyllotaxis also appears in *Pharus* (Pharoideae) (Sajo et al. 2007), in *Oryza* and *Zizania* (Ehrhartoideae) (Liu et al. 1998; Moncur 1981; Weir and Dale 1960), and many Panicoideae (Bess et al. 2005; Brown et al. 2006; Doust and Kellogg 2002; Giulini et al. 2004; Kellogg et al. 2004; Moncur 1981; Reinheimer et al. 2005) and Chloridoideae (Liu et al. 2007; Moncur 1981). Distichous inflorescences do occur in Panicoideae (e.g., *Urochloa*; Reinheimer et al. 2005) and some Andropogoneae (Kellogg 2000), but are clearly derived.

In contrast to the spiral phyllotaxis in most subfamilies, the inflorescence branches of all members of subfamily Pooideae except *Brachyelytrum* are two-ranked, although strict distichy (branches separated by an angle of 180°) appears to have been derived after the divergence of Nardeae (Kellogg et al. 2013). The two-ranked inflorescence is quite obvious even at maturity in members of the tribe Triticeae such as *Hordeum vulgare* (barley) and *Triticum aestivum* (wheat), but also can be observed in early development in many Poeae (Evans 1940; Fraser and Kokko 1993; Latting 1972; Moncur 1981). Recently the two-ranked inflorescence has also been demonstrated in Nardeae, Stipeae, Diarrheneae, Meliceae, and Brachpodieae

(Kellogg et al. 2013). The failure of the inflorescence meristem to shift to spiral phyllotaxis in Pooideae is thus a synapomorphy for the large clade that is sister to *Brachyelytrum*, and appears to correlate with a shift from shaded to open habitats.

Mutations in the rice gene *Aberrant Panicle Organization1* (*apo1*), which encodes an F-box protein homologous to UNUSUAL FLORAL ORGANS (UFO) in *Arabidopsis*, display altered phyllotaxis (Ikeda et al. 2005, 2007). While the inflorescence of rice normally produces primary branches in a spiral, *apo1* mutants produce two-ranked primary branches. The change in phyllotaxis correlates with a change in the shape of the meristem. It is possible that changes in *apo1* expression underlie the pooid synapomorphy.

Virtually all primary branch meristems produce secondary meristems in two ranks (Doust and Kellogg 2002). Although the two ranks are often separated by an angle of 180° (i.e., are strictly distichous), in many taxa the secondary branches form on one side of the primary branch. For example, in *Paspalum*, the primary branches each bear two ranks of secondaries, which always form on the abaxial side of the primary. The genetic control of this pattern is unknown. In such unilateral primary branches, the branch axis is often expanded laterally as though it has acquired some aspects of leaf identity. This is particularly noticeable in some taxa such as *Chlorocalymma*, in which the broad rachis of the primary branch droops around and encloses the secondaries.

Numbers of Orders of Branching

Grass inflorescences may be unbranched, or may produce primary branches only, or may have up to six or more orders of branching. Whether branches are produced or not is determined by the fate of the inflorescence meristem, which varies between species and genera (reviewed by Kellogg 2000, 2006; Malcomber et al. 2006; Perreta et al. 2009). Inflorescence meristems may produce spikelet meristems directly on their flanks, and the spikelets may remain sessile (e.g., *Triticum*, *Ctenium*), creating a "spike-like" structure, or the spikelet may develop a pedicel (e.g., *Brachypodium*, some *Perotis*), and the inflorescence thus be "raceme-like". However, the underlying development is similar; in both cases the axis is unbranched.

Many other taxa produce only primary branches. In these genera, each primary branch meristem gives rise to spikelet meristems; again the spikelets may be sessile (e.g., *Heteranthoecia*, *Eleusine*) or pedicellate (e.g., *Paspalidium*, *Gymnopogon*). Finally, the primary branches may themselves branch, producing second-, third-, or higher-order branches.

The main inflorescence meristem may terminate in a spikelet (e.g., *Avena sativa*, *Triticum* spp., *Sorghum bicolor*) or not (e.g., *Setaria italica*, *Zea mays*), and may abort soon after forming primary branches (e.g., *Oryza sativa*, *Urochloa plantaginea*, *Brachiaria decumbens*) (Kellogg et al. 2013; Moncur 1981; Reinheimer et al. 2005). The fate of the inflorescence meristem often varies among closely related species (see, for example, Kellogg et al. 2013; Reinheimer and Vegetti 2008; Reinheimer et al. 2009), suggesting that it may be relatively easy to shift in evolutionary time, a point also made by Butzin (1979). The fate of the inflorescence meristem is independent of that of the primary-, secondary-, and higher-order branches, which may also terminate in spikelets or not.

The genetic control of branching is agronomically important and thus has been studied extensively in the cereals (Bommert et al. 2005; Kellogg 2007; Thompson and Hake 2009). Formation of branch and floral meristems depends on auxin transport and biosynthesis, as it does in eudicots. The auxin efflux carrier PINFORMED1 (PIN1) transports auxin in maize and rice as it does in *Arabidopsis thaliana* (Gallavotti et al. 2008), and is phosphorylated by BARREN INFLORESCENCE2 (BIF2 = PINOID in Arabidopsis), an interaction that is also conserved among flowering plants. However, BIF2 also phosphorylates an unusual basic helix-loop-helix transcription factor, BARREN STALK1 (in maize), which is orthologous to LAX PANICLE in rice, an interaction not reported for dicots (Skirpan et al. 2008). The expression pattern and presumed function of BA1/LAX is similar among all grasses, but somewhat different from the patterns in eudicots (Woods et al. 2011). Because comparable data are not available for other monocots, it is not clear whether the role of BA1/LAX is unique to the grasses or to a larger clade.

Inflorescence maturation, maturation of each individual branch, and maturation of flowers within a spikelet may be acropetal or basipetal, and the direction of maturation in different components of the inflorescence may vary (Reinheimer et al. 2009). For example, the inflorescence of *Hyparrhenia* is made up of two branches, the main axis and a lateral branch; one matures acropetally and the other basipetally (Le Roux and Kellogg 1999). In species such as *Urochloa plantaginea* and *Brachiaria decumbens*, branch initiation appears to be basipetal, and the primary branches may initiate below already formed branches (Reinheimer et al. 2005; Stür 1986), suggesting that meristems form de novo.

Numbers of Branches or Spikelets at Each Order of Branching

The range of values for primary, secondary, and tertiary branch number varies extensively between species, although individual species often have a characteristic and consistent range of values. The high variability of these characters is obvious simply by considering illustrations of closely related species in almost any grass genus. There is often species-specific variation in whether the inflorescence is dense (with many primary branches) or sparse (with few primary branches), whether the primary branches are densely covered with secondaries or not, and whether tertiary and quaternary branches are common or rare. The few studies that record comparative data on numbers of primary branches find a broad range of variation among species (*Spartina*, Kern et al. 2008; Melinidae, Reinheimer and Vegetti 2008; *Setaria*, Doust and Kellogg 2002). In *Spartina* (now part of *Sporobolus* s.l.; Peterson et al. 2014), the number of primary branches does not correlate well with the phylogeny (Baumel et al. 2002; Fortune et al. 2006), indicating that the number of branches is easily modified over evolutionary time.

Perhaps not surprisingly, many genes control branch number. Many of these have pleiotropic effects on other aspects of inflorescence development, and thus must be responding to a variety of selective pressures. In general, branch number controls spikelet number, and hence controls the number of seeds; there is thus a direct effect on fitness. As noted by Ikeda-Kawakatsu et al. (2009), the architecture of the inflorescence is determined by the precise developmental timing of the conversion of a branch meristem to a spikelet meristem. If the conversion happens early in development, then few higher-order meristems are produced, whereas if conversion from branch meristem identity to spikelet meristem identity is delayed, then there is an opportunity for producing more branches.

The number of primary branches varies independently of the number of secondaries, an observation that is supported by extensive genetic data. The number of primary branches is positively correlated with the amount of transcript of the gene *Apo1*, mentioned above for its role in control of phyllotaxis. *Apo1* is the gene underlying the QTL *Primary Branch Number* in rice; higher levels of *Apo1* transcript lead to more primary branches and lower levels lead to fewer (Terao et al. 2009). While APO1 positively regulates the number of primary branches, the proteins DENSE AND ERECT PANICLE1 (DEP1), OsMADS34 (= PANICLE PHYTOMER2, or PAP2), and ERECT PANICLE3 (also called LARGER PANICLE) restrict the number (Gao et al. 2010; Huang et al. 2009; Kobayashi et al. 2010; Li et al. 2011; Piao et al. 2009). The four proteins differ, however, in their effect on secondary branches. APO1 positively regulates the number of secondary branches per primary, DEP1 restricts the number, and OsMADS34 and EP3 have little effect at all on the number of secondary branches per primary. Three other proteins, ERECT PANICLE2 (EP2) and rice CENTRORADIALIS1 and 2 (RCN1 and RCN2) have no effect on the number of primary branches but restrict the number of secondaries (Nakagawa et al. 2002; Zhu et al. 2010). Comparative studies on these genes and proteins are limited, but DEP1 affects the number of primary and secondary branches, and inflorescence elongation in rice, wheat and barley, suggesting a general mechanism (at least in the BEP clade) for branching control (Huang et al. 2009).

The number of primary branches also correlates with the number of vascular bundles in the peduncle, suggesting that the number is limited by carbohydrate supply. EP2 and EP3 control vasculature and sclerenchyma in the inflorescence and thus affect branch number, along with whether the inflorescence is upright or nodding (Piao et al. 2009; Zhu et al. 2010).

Many of the proteins that regulate branch number do so by interacting directly or indirectly with the cytokinin pathway. The locus *Grain number 1* (*Gn1*) in rice encodes CYTOKININ OXIDASE2 (CKX2), which breaks down cytokinin and limits the number of spikelets and panicle branches (Ashikari et al. 2005), as does its ortholog in wheat (Zhang et al. 2012). Conversely the cytokinin biosynthetic protein LONELY GUY (LOG) increases the number of branches (Sakakibara 2006). Also regulating CKX2 (directly or indirectly) is EP3, which encodes an F-box protein (Piao et al. 2009); levels of EP3 expression correlate with levels of CKX2 expression (Li et al. 2011). Spikelet number per panicle is also affected by RICE OUTERMOST CELL-SPECIFIC GENE5 (ROC5) (Zou et al. 2011).

In Andropogoneae, the spikelets are borne in pairs; one member of the pair is pedicellate whereas the other is sessile or nearly so. Studies of developmental morphology, developmental genetics, and gene expression all indicate that the pedicellate spikelet is terminal on a very short branch, and the sessile spikelet terminates the sole lateral branch. The proteins RAMOSA1 (RA1), RAMOSA2 (RA2), RAMOSA3 (RA3), and RAMOSA ENHANCER LOCUS2 (REL2) working together prevent formation of more than one sessile spikelet and are thus responsible for the formation of the spikelet pair (Bommert et al. 2005; Gallavotti et al. 2008, 2010; Satoh-Nagasawa et al. 2006). The genes encoding RA1 and RA3 are only present in the PACMAD clade (and thus may be synapomorphic) and are completely absent from the genomes of rice and *Brachypodium* (Satoh-Nagasawa et al. 2006; Vollbrecht et al. 2005; Reinheimer and Kellogg, unpublished); thus, the RA1-RA2-RA3-REL2 module must also be unique to the PACMAD group. However, RA2 and REL2 are present in all grasses. Therefore, they must bind with different proteins in the BEP clade since RA1 and RA3 are not available.

Branch Angle

The angle of inflorescence branches, whether spreading or upright, varies considerably. The erect panicle phenotype has been of particular interest in rice because more erect panicles correlate with increased yield. In DEP1, EP2 and EP3 mutants of rice, the normally lax inflorescence of rice is borne upright; this appears to be due to increases in the number of vascular bundles, particularly small vascular bundles, and also in sclerenchyma in the peduncle (Huang et al. 2009; Piao et al. 2009; Zhu et al. 2010). Taxonomists have generally not explored this character in a comparative context, but it would be of interest to know whether vascular supply and amount of sclerenchyma generally correlate with the width of the inflorescence.

Branch angle is also controlled by the activity of the pulvinus at the base of each primary inflorescence branch. The pulvinus is a prominent swelling that becomes turgid and forces the branches away from the main axis (Arber 1934; Galinat 1959). RA2 and REL2 contribute to normal cell structure of the pulvinus, whereas BARREN STALK FASTIGIATE and BRANCH ANGLE DEFECTIVE1 control the number of cells (Bai et al. 2012; Gallavotti et al. 2010, 2011). The presence, size and vasculature of the pulvinus have received little or no attention in the comparative literature.

Axis Elongation

Elongation of inflorescence axes occurs after branching is complete, spikelet meristems have been specified, and glumes and lemmas are initiated. Elongation is thus developmentally distinct from branching. Genetic studies in a few grasses also show that it is controlled by different genes (Doust 2007; Doust et al. 2004). In *Triticum aestivum*, an *APETALA2*-like gene known as Q affects elongation of the rachis, and was involved in domestication (Simons et al. 2006); higher levels of Q make the inflorescence more compact (see also below). In *Hordeum vulgare*, another *APETALA2*-like gene, *HvAP2*, which is distinct from Q, also controls rachis elongation and thereby number of spikelets (Houston et al. 2013). *HvAP2* is regulated by microRNA172 (miR172); when *HvAP2* is mutated in a way that reduces the interaction with miR172, the inflorescence axis continues initiating spikelets and fails to elongate. The result is a short inflorescence with an abnormally high number of spikelets. Because of the high HvAP2 levels, spikelet differentiation is delayed, and the window for inflorescence elongation is shortened (Houston et al. 2013).

Elongation of each order of axis is controlled independently of the others. Because of this, inflorescences with the same number of orders of branches and the same number of branches at each order can still look quite different at maturity depending on which axes elongate. For example, the internodes of both the main axis and the primary branch axes may remain close together in digitate inflorescences such as *Digitaria* or *Cynodon*, or the main axis may elongate while the primary branch axes do not as in *Leptochloa* or *Bothriochloa*, or internodes may elongate on both orders of branching as in *Panicum* or *Avena*.

In some taxa such as *Urochloa maxima* (=*Megathyrsus maximus*), internode elongation is not consistent along an axis. In this case, a group of short internodes is followed by one long one, followed by another cluster of short internodes (Reinheimer et al. 2005). At maturity the primary branches of such inflorescences appear whorled or verticillate. If the internode below the secondaries fails elongate, spikelets on secondary branches appear clustered as in *Hordeum*, or the secondaries themselves cluster as in *Agrostis* or other Pooideae. In some Pooideae such as *Nephelochloa*, inflorescence branches are described as whorled; in these species the primary branches are almost certainly distichous as they are in all pooids, although development has never been investigated for this genus. Apparent whorls are created when a large number of higher-order branches, each with contracted basal internodes, originates from the distichous primaries.

Spikelets

Within the spikelet, flowers occur in distichous phyllotaxis on a short axis, the rachilla (literally, a little rachis). At the base of the spikelet below the flowers are two more bracts, known as glumes. (These give the grass family the classical name Glumiflorae, transliterated to gloomy-florae by bored agrostology students.) The glumes are defined as bracts with nothing in their axils.

The grass spikelet is an evolutionary novelty. It shares some characteristics with a branch system and some with a flower, but is clearly neither. The bracteate structure of the spikelet leads to its interpretation as a contracted inflorescence axis. However, this is likely to be an over-simplification; the mutant phenotype of BRANCHED SILKLESS1 (BD1) in maize and FRIZZY PANICLE (FZP) in rice hints that a spikelet is not simply an abnormally short spike (Chuck et al. 2002; Komatsu et al. 2003). If the proteins simply kept the spikelet axis short, one would expect the mutant phenotype to lead to an elongate branch, but with the normal number of flowers. However, mutations in *bd1/fzp* cause the production of additional branches and flowers (see below), with the precise phenotype differing between the tassel and the ear of maize and the branched inflorescence of rice; in addition, axillary meristems are produced in the axils of the glumes in a pattern reminiscent of a pseudospikelet in bamboos. *Bd1/fzp* is expressed in the axils of glumes in all grasses investigated, suggesting that its normal function is to repress branching in that position.

In other respects, the spikelet is similar to a flower. The spikelet is a developmentally integrated unit and is determinate in the sense that a flower is – a particularly canalized developmental program. Although the rachilla ends blindly in the spikelets of many species, the spikelet is not an endlessly reiterating structure. Once the "spikelet developmental program" is switched on, the axis ends. (Note that the word "determinate" is used in two different ways in the grass developmental literature. Among classical morphologists, a determinate axis is one that ends in a flower. By this definition, many if not all grass spikelets would have indeterminate axes because the flowers appear to be lateral. However, among developmental geneticists, a determinate structure is one with a fixed number of parts. By this definition, the spikelet is determinate because the number of flowers is either fixed or varies within a narrow range.)

A spikelet is also similar to a flower in terms of expression of genes that are flower-specific in other families. In most angiosperms, including the grass outgroups, the proteins SQUAMOSA and FRUITFULL (FUL), and certain SQUAMOSA PROMOTER BINDING PROTEIN-LIKE (SPL) proteins are expressed throughout the flower but generally not in subtending bracts; they are thus markers for floral identity (Preston and Hileman 2010; Preston et al. 2012); the SPL proteins are direct regulators of FUL/SQUA (Klein et al. 1996). In grasses, SQUAMOSA (= APETALA1) is absent, whereas its paralogue FUL is duplicated to form FUL1 (= VRN1) and FUL2 (Preston and Kellogg 2006); the best characterized of the SPL proteins is TEOSINTE GLUME ARCHITECTURE. FUL1, FUL2 and TGA are expressed not only in the flowers but also in the glumes (Preston et al. 2009, 2012; Preston and Kellogg 2007; Wang et al. 2005). Homologues of these same proteins are present only in floral organs of the grass outgroup *Joinvillea*. Thus, the expression domains of FUL1, FUL2, and TGA are expanded in the grasses to include the bracts subtending the flowers; this change in expression is another synapomorphy of the family.

The pseudospikelets of bamboos are peculiar in that they share some aspects of spikelets and some aspects of branches. They are usually interpreted as spikelets in which the most proximal glumes have buds in their axils; these buds develop into new branches with new glume-like structures in their axils. The bud-bearing glumes are often followed by one or more conventional glumes without buds before the axis produces a flower. It would be of interest to determine whether bamboos with pseudospikelets have alterations in the structure or expression of *Bd1/Fzp*. A mutation in durum wheat (*Triticum turgidum* ssp. *durum*), *branched head*, also produces a structure similar to a pseudospikelet, with a secondary spikelet forming in the axil of the glume (Shitsukawa et al. 2009); this gene has not been cloned, however.

Spikelet compression varies in Poaceae, and generally correlates with phylogeny. Spikelets may be dorsiventrally compressed, laterally compressed or terete. In laterally compressed spikelets, the glumes and lemmas are folded around their midribs (conduplicate); this compression pattern is nearly universal in Ehrhartoideae, Bambusoideae, Pooideae, Chloridoideae, and Danthonioideae. In dorsiventrally compressed spikelets, in contrast, the glumes and lemmas are flat to slightly curved, and spread out on either side of the midrib; this pattern appears most obviously in Panicoideae, but also in Pharoideae and in the herbaceous bamboos. The third pattern, less common in terms of numbers of species, is terete spikelets, which are found in Aristidoideae and Stipeae (Pooideae).

Spikelet size varies by more than an order of magnitude, from minute (1 mm) to relatively large (1 to 2 cm), to enormous (e.g., up to 10 cm in *Colanthelia*, Bambuseae). Spikelet size is not a direct reflection of flower number, in that some of the largest spikelets (e.g., *Hesperostipa*) have only one flower and some very small spikelets (e.g., some *Eragrostis* species) have several. The controls of spikelet size are largely unknown; it is possible that spikelet size is simply a direct correlate of grain size. If this is true, then ongoing research in rice on grain size will likely uncover some of the regulatory genes.

The number of flowers per spikelet varies among grasses, but is often fixed within a clade. For example, members of Ehrhartoideae have three flowers (only one fertile), Panicoideae have two, and some clades of Pooideae have only one. The number of flowers in turn is controlled by the length of time that the spikelet meristem actively produces lateral organs. In both rice and maize, spikelet meristem activity is regulated by proteins known as INDETERMINATE SPIKELET1 (IDS1) in maize and its paralogue SISTER OF IDS (SID) (Chuck et al. 1998) (or IDS1-like; Malcomber et al. 2006); related proteins in rice are OsIDS and SUPERNUMERARY BRACT (SNB). Mutations in *ids1* lead to production of extra flowers in the spikelet (Chuck et al. 2008), and mutation of both *ids1* and *sid* together causes production of extra bracts. Likewise mutation of both *Osids* and *snb* lead to increased numbers of rudimentary glumes (Lee and An 2012). IDS1, SID1, OsIDS, and SNB are all AP2-like proteins, similar to the wheat domestication gene Q (Simons et al. 2006); like Q they are controlled by a microRNA in the miR172 family (Chuck et al. 2007, 2008). Thus, variation in number of flowers per spikelet observed in grasses could be regulated by the precise timing of expression of miR172.

Flower number is also specified by genes acting later in spikelet development. In rice the number of flowers is controlled by MULTI-FLORET SPIKELET1 (MFS1) and TONGARI-BOUSHI1 (TOB1), whose effect is obvious after specification of glumes and sterile lemmas (Ren et al. 2013). MFS1 is also an AP2-like protein and is apparently regulated by microRNAs.

Glume morphology is often taxonomically informative and forms a standard part of most species descriptions. Glumes may be shorter than the adjoining lemma or longer, sometimes being long enough to enclose the entire spikelet. One or both glumes may be suppressed in development and reduced to a vestigial flap or a ridge. The number of veins varies from zero to many, and these may be prominent or obscure. The texture of the glumes can be similar to that of the lemmas, or they can be harder or softer. In some taxa (e.g., *Mnesithea* or *Hackelochloa*, Andropogoneae), the glumes are not only hard but also elaborately sculptured to form warts, hooks, or other coruscations. The glumes of most grasses lack awns, but in some species awns are present and can be longer than the entire spikelet. The products of the genes Q and *Tenacious glumes* in wheat control the amount of sclerenchyma at the base of the glume (Nalam et al. 2007); in wild wheats the glumes are hard and tightly enclose the flower and caryopsis, whereas in domesticated wheats there is less sclerenchyma and the glumes break easily during threshing.

Maturation of flowers within a spikelet may be acropetal, as in Pooideae, Bambusoideae and Chloridoideae, or basipetal as in Ehrhartoideae and Panicoideae s.s.; this maturation pattern is independently derived but synapomorphic in each of the latter two subfamilies. Expression of the SEPALLATA-like gene, *Leafy hull sterile1* (*Lhs1*/*OsMADS1*) correlates with the direction of flower maturation (Malcomber and Kellogg 2004; Reinheimer et al. 2006), and thus underlies the developmental synapomorphies for Ehrhartoideae and Panicoideae. In the two subfamilies in which the distalmost flower matures first (Ehrhartoideae and Panicoideae s.s.), *Lhs1* expression is confined to the distal flower. In contrast, in spikelets with acropetal maturation, *Lhs1* is expressed in all flowers.

Some authors have interpreted the distalmost flower of the spikelet as being terminal, whereas others infer that all flowers are lateral on the rachis. Evidence is accumulating for the latter interpretation. If development is acropetal, the more distal flowers of the spikelet are often reduced in size, in some cases being little more than a rudiment with a minute lemma, and often the rachilla is prolonged beyond the distalmost flower. In this sort of spikelet, all flowers are clearly lateral. When development is basipetal, the position of the distalmost flower is less clear. However, SEM data show that a vestige of the rachilla apex is often visible in early development (see Bess et al. 2005; Doust and Kellogg 2002).

The rachilla may be variously elongate or contracted between the glumes and lemmas. If it is longer between the upper glume and lowest lemma, it is often described as a stipe (e.g., in *Calamovilfa* (now part of *Sporobolus* s.l., Peterson et al. 2014)). No data are available on the control of rachilla elongation.

Disarticulation

The grass inflorescence is clearly adapted for seed dispersal (see section on Dispersal below), and there is considerable variation in the pattern of disarticulation (Doust et al. 2014); the disarticulation pattern is often consistent within clades and so is commonly used as a taxonomic character. Most commonly, an abscission zone forms at the base of the flower, so that the caryopsis falls with the lemma, palea, and a short rachilla segment. In such species, the glumes generally remain on the parent plant. There is also variation in whether the palea and lemma adhere firmly to the caryopsis or whether the caryopsis is ultimately shed. In the other common pattern (e.g., in virtually all Paniceae and Paspaleae), disarticulation is below the glumes, such that the entire spikelet is shed as a unit; subsequently the rachilla breaks up, releasing the upper flower. In many members of Andropogoneae and Triticeae, and in the genera with narrow "rat-tail" inflorescences (e.g., *Hainardia*), disarticulation occurs at each node of the inflorescence axis so that the dispersal unit is the spikelet plus a rachis internode. In a few taxa (e.g., some species of *Aegilops*, *Eragrostis sessilispica*), the entire inflorescence forms the dispersal unit, with disarticulation at the most proximal node.

Given the variation in the location of abscission zones, it is not surprising that abscission is under the control of several independent genes. Most current data come from cereal crops, in which disarticulation is called "shattering" and is an undesirable trait that was selected against early in domestication (see Domestication). Thus, the Q gene in wheat prevents disarticulation of the rachis, suggesting that its recessive wild-type alleles in wild ancestors may permit disarticulation (Simons et al. 2006). Likewise, shattering loci in rice regulate disarticulation of the rachilla below the two proximal sterile flowers but above the glumes (Konishi et al. 2006; Li et al. 2006; Zhou et al. 2012). Larson and Kellogg (2009) have identified overlapping QTL for shattering among *Leymus* hybrids, *Zizania palustris*, *Oryza sativa*, and *Triticum aestivum*, suggesting common genetic programs, but the corresponding genes have not been cloned. In contrast, many studies find unique genes controlling shattering in each species investigated (Doust et al. 2014).

References

Arber, A. 1934. The Gramineae: a study of cereal, bamboo, and grass. New York: Macmillan.

Ashikari, M., Sakakibara, H., Lin, S.Y., Yamamoto, T., Takashi, T., Nishimura, A., Angeles, E.R., Qian, Q., Kitano, H., Matsuoka, M. 2005. Cytokinin oxidase regulates rice grain production. Science 309: 741–745.

Bai, F., Reinheimer, R., Durantini, D., Kellogg, E.A., Schmidt, R.J. 2012. A TCP transcription factor, *BRANCH ANGLE DEFECTIVE1* (*BAD1*), is required for normal tassel branch angle formation in maize. Proc. Natl. Acad. Sci. USA. 109: 12225–12230.

Baumel, A., Ainouche, M.L., Bayer, R.J., Ainouche, A.K., Misset, M.T. 2002. Molecular phylogeny of hybridizing species from the genus *Spartina* Schreb. (Poaceae). Mol. Phyl. Evol. 22: 303–314.

Bess, E.C., Doust, A.N., Kellogg, E.A. 2005. A naked grass in the "bristle clade": a phylogenetic and developmental study of *Panicum* section Bulbosa (Paniceae: Poaceae). Int. J. Plant Sci. 166: 371–381.

Bommert, P., Satoh-Nagasawa, N., Jackson, D., Hirano, H. Y. 2005. Genetics and evolution of grass inflorescence and flower development. Plant Cell Physiol. 46: 69–78.

Brown, P.J., Klein, P.E., Bortiri, E., Acharya, C.B., Rooney, W.L., Kresovich, S. 2006. Inheritance of inflorescence architecture in sorghum. Theor. Appl. Genet. 113: 931–942.

Butzin, R. 1979. Apikale Reduktionen im Infloreszenzbereich der Gramineae. Willdenowia 9: 161–167.

Chuck, G., Meeley, R., Hake, S. 1998. The control of maize spikelet meristem identity by the *APETALA-2*-like gene *indeterminate spikelet1*. Genes Devel. 12: 1145–1154.

Chuck, G., Muszynski, M., Kellogg, E., Hake, S., Schmidt, R.J. 2002. The control of spikelet meristem identity by the *branched silkless1* gene in maize. Science 298: 1238–1241.

Chuck, G., Meeley, R., Irish, E., Sakai, H., Hake, S. 2007. The maize *tasselseed4* microRNA controls sex determination and meristem cell fate by targeting *Tasselseed6/indeterminate spikelet1*. Nat. Genet. 39: 1517–1521.

Chuck, G., Meeley, R., Hake, S. 2008. Floral meristem initiation and meristem cell fate are regulated by the maize *AP2* genes *ids1* and *sid1*. Development 135: 3013–3019.

Chuck, G., Whipple, C., Jackson, D., Hake, S. 2010. The maize SBP-box transcription factor encoded by *tasselsheath4* regulates bract development and the establishment of meristem boundaries. Development 137: 1243–1250.

Clayton, W.D. 1969. A revision of the genus *Hyparrhenia*. Kew: Her Majesty's Stationery Office.

Colasanti, J., Coneva, V. 2009. Mechanisms of floral induction in grasses: something borrowed, something new. Plant Phys. 149: 56–62.

Danilevskaya, O.N., Meng, X., Hou, Z.L., Ananiev, E.V., Simmons, C.R. 2008. A genomic and expression compendium of the expanded PEBP gene family from maize. Plant Phys. 146: 250–264.

Doust, A.N. 2007. Grass architecture: genetic and environmental control of branching. Curr. Opinion Pl. Biol. 10: 21–25.

Doust, A.N., Kellogg, E.A. 2002. Inflorescence diversification in the panicoid "bristle grass" clade (Paniceae, Poaceae): evidence from molecular phylogenies and developmental morphology. Amer. J. Bot. 89: 1203–1222.

Doust, A.N., Devos, K.M., Gadberry, M., Gale, M.D., Kellogg, E.A. 2004. Genetic control of branching in foxtail millet. Proc. Natl. Acad. Sci. USA 101: 9045–9050.

Doust, A.N., Mauro-Herrera, M., Francis, A.D., Shand, L. C. 2014. Morphological diversity and genetic regulation of inflorescence abscission zones in grasses. Amer. J. Bot. 101: 1759–1769.

Evans, M.W. 1940. Developmental morphology of the growing point of the shoot and the inflorescence in grasses. J. Agric. Res. 61: 481–520.

Fortune, P.M., Schierenbeck, K.A., Ainouche, A.K., Jacquemin, J., Wendel, J.F., Ainouche, M.L. 2006. Evolutionary dynamics of *Waxy* and the origin of hexaploid *Spartina* species (Poaceae). Mol. Phyl. Evol. 43: 1040–1055.

Fraser, J., Kokko, E.G. 1993. Panicle, spikelet, and floret development in orchardgrass (*Dactylis glomerata*). Can. J. Bot. 71: 523–532.

Galinat, W. 1959. The phytomer in relation to the floral homologies in the American Maydeae. Bot. Mus. Leafl. Harvard Univ. 19: 1–32.

Gallavotti, A., Yang, Y., Schmidt, R.J., Jackson, D. 2008. The relationship between auxin transport and maize branching. Plant Phys. 147: 1913–1923.

Gallavotti, A., Long, J.A., Stanfield, S., Yang, X., Jackson, D., Vollbrecht, E., Schmidt, R.J. 2010. The control of axillary meristem fate in the maize *ramosa* pathway. Development 137: 2849–2856.

Gallavotti, A., Malcomber, S., Gaines, C., Stanfield, S., Whipple, C., Kellogg, E., Schmidt, R.J. 2011. BARREN STALK FASTIGIATE1 is an AT-hook protein required for the formation of maize ears. Plant Cell 23: 1756–1771.

Gao, X., Liang, W., Yin, C., Ji, S., Wang, H., Su, X., Guo, C., Kong, H., Xue, H., Zhang, D. 2010. The *SEPALLATA*-like gene *OsMADS34* is required for rice inflorescence and spikelet development. Plant Phys. 153: 728–740.

Gasser, M., Vegetti, A.C. 1997. Inflorescence typology in *Eleusine indica* and *Eleusine tristachya* (Poaceae). Flora 192: 17–20.

Giulini, A., Wang, J., Jackson, D. 2004. Control of phyllotaxy by the cytokinin-inducible response regulator homologue *ABPHYL1*. Nature 430: 1031–1034.

Hernández, J.C., Rua, G. 1992. The synflorescence of Poaceae. Beitr. Biol. Pflanzen 66: 297–311.

Houston, K., McKim, S.M., Comadran, J., Bonar, N., Druka, I., Uzrek, N., Cirillo, E., Guzy-Wrobelska, J., Collins, N.C., Halpin, C., Hansson, M., Dockter, C., Druka, A., Waugh, R. 2013. Variation in the interaction between alleles of *HvAPETALA2* and microRNA172 determines the density of grains on the barley inflorescence. Proc. Natl. Acad. Sci. USA 110: 16675–16680.

Huang, X., Qian, Q., Liu, Z., Sun, H., He, S., Luo, D., Xia, G., Chu, C., Li, J., Fu, X. 2009. Natural variation at the *DEP1* locus enhances grain yield in rice. Nat. Genet. 41: 494–497.

Ikeda, K., Nagasawa, N., Nagato, Y. 2005. *ABERRANT PANICLE ORGANIZATION1* temporally regulates meristem identity in rice. Devel. Biol. 282: 349–360.

Ikeda, K., Ito, M., Nagasawa, N., Kyozuka, J., Nagato, Y. 2007. Rice *ABERRANT PANICLE ORGANIZATION1*, encoding an F-box protein, regulates meristem fate. Plant J. 51: 1030–1040.

Ikeda-Kawakatsu, K., Yasuno, N., Oikawa, T., Iida, S., Nagato, Y., Maekawa, M., Kyozuka, J. 2009. Expression level of *ABERRANT PANICLE ORGANIZATION1* determines rice inflorescence form through control of cell proliferation in the meristem. Plant Phys. 150: 736–747.

Judziewicz, E.J., Soderstrom, T.R. 1989. Morphological, anatomical, and taxonomic studies in *Anomochloa* and *Streptochaeta* (Poaceae: Bambusoideae). Smithsonian Contr. Bot. 68: 1–52.

Judziewicz, E.J., Clark, L.G., Londoño, X., Stern, M.J. 1999. American bamboos. Washington, DC: Smithsonian Institution Press.

Kellogg, E.A. 2000. A model of inflorescence development. In: Wilson, K.L., Morrison, D.A. (eds.) Monocots: systematics and evolution. Melbourne, Australia: CSIRO. pp. 84–88

Kellogg, E.A. 2006. Beyond taxonomy: prospects for understanding morphological diversity in the grasses (Poaceae). Darwiniana 44: 7–17.

Kellogg, E.A. 2007. Floral displays: genetic control of grass inflorescences. Curr. Opinion Pl. Biol. 10: 26–31.

Kellogg, E.A., Hiser, K.M., Doust, A.N. 2004. Taxonomy, phylogeny, and inflorescence development of the genus *Ixophorus* (Panicoideae: Poaceae). Int. J. Plant Sci. 165: 1089–1105.

Kellogg, E.A., Camara, P.E.A.S., Rudall, P.J., Ladd, P., Malcomber, S.T., Whipple, C.J., Doust, A.N. 2013. Early inflorescence development in the grasses (Poaceae). Frontiers Plant Sci. 4. doi: 10.3389/fpls.2013.00250.

Kern, V.G., Guarise, N.J., Vegetti, A.C. 2008. Inflorescence structure in species of *Spartina* Schreb. (Poaceae: Chloridoideae: Cynodonteae) Plant Syst. Evol. 273: 51–61.

Klein, J., Saedler, H., Huijser, P. 1996. A new family of DNA binding proteins includes putative transcriptional regulators of the *Antirrhinum majus* floral meristem identity gene *SQUAMOSA*. Mol. Genet. Genomics 250: 7–16.

Kobayashi, K., Maekawa, M., Miyao, A., Hirochika, H., Kyozuka, J. 2010. *PANICLE PHYTOMER2* (*PAP2*), encoding a SEPALLATA subfamily MADS-box protein, positively controls spikelet meristem identity in rice. Plant Cell Physiol. 51: 47–57.

Kobayashi, K., Yasuno, N., Sato, Y., Yoda, M., Yamazaki, R., Kimizu, M., Yoshida, H., Nagamura, Y., Kyozuka, J. 2012. Inflorescence meristem identity in rice is specified by overlapping functions of three *AP1/FUL*-like MADS box genes and *PAP2*, a *SEPALLATA* MADS box gene. Plant Cell 24: 1848–1859.

Kojima, S., Takahashi, Y., Kobayashi, Y., Monna, L., Sasaki, T., Araki, T., Yano, M. 2002. *Hd3a*, a rice ortholog of the *Arabidopsis* FT gene, promotes transition to flowering downstream of *Hd1* under short-day conditions. Plant Cell Physiol. 43: 1096–1105.

Komatsu, M., Chujo, A., Nagato, Y., Shimamoto, K., Kyozuka, J. 2003. FRIZZY PANICLE is required to prevent the formation of axillary meristems and to establish floral meristem identity in rice spikelets. Development 130: 3841–3850.

Komiya, R., Ikegami, A., Tamaki, S., Yokoi, S., Shimamoto, K. 2008. *Hd3a* and *RFT1* are essential for flowering in rice. Development 135: 767–774.

Konishi, S., Izawa, T., Lin, S.Y., Ebana, K., Fukuta, Y., Sasaki, T., Yano, M. 2006. An SNP caused loss of seed shattering during rice domestication. Science 312: 1392–1396.

Larson, S.R., Kellogg, E.A. 2009. Genetic dissection of seed production traits and identification of a major-effect seed retention QTL in hybrid *Leymus* (Triticeae) wildryes. Crop Sci. 49: 29–40.

Latting, J. 1972. Differentiation in the grass inflorescence. In: Youngner, V.B., McKell, C.M. (eds.) The biology and utilization of grasses. New York: Academic Press. pp. 365–399

Lazakis, C.M., Coneva, V., Colasanti, J. 2011. *ZCN8* encodes a potential orthologue of *Arabidopsis* FT florigen that integrates both endogenous and photoperiod flowering signals in maize. J. Exp. Bot. 62: 4833–3842.

Lee, D.Y., An, G. 2012. Two AP2 family genes, *SUPERNUMERARY BRACT* (*SNB*) and *OsINDETERMINATE SPIKELET1* (*OsIDS1*), synergistically control inflorescence architecture and floral meristem establishment in rice. Plant J. 69: 445–461.

Le Roux, L.G., Kellogg, E.A. 1999. Floral development and the formation of unisexual spikelets in the Andropogoneae (Poaceae). Amer. J. Bot. 86: 354–366.

Li, C., Zhou, A., Sang, T. 2006. Rice domestication by reducing shattering. Science 311: 1936–1939.

Li, M., Tang, D., Wang, K., Wu, X., Lu, L., Yu, H., Gu, M., Yan, C., Cheng, Z. 2011. Mutations in the F-box gene *LARGER PANICLE* improve the panicle architecture and enhance the grain yield in rice. Plant Biotech. J. 9: 1002–1013.

Liu, Q., Oelke, E.A., Porter, R.A., Reuter, R. 1998. Formation of panicles and hermaphroditic florets in wild-rice. Int. J. Plant Sci. 159: 550–558.

Liu, Q., Peterson, P.M., Columbus, J.T., Zhao, N., Hao, G., Zhang, D. 2007. Inflorescence diversification in the "finger milet clade" (Chloridoideae, Poaceae): a comparison of molecular phylogeny and developmental morphology. Amer. J. Bot. 94: 1230–1247.

Malcomber, S.T., Kellogg, E.A. 2004. Heterogeneous expression patterns and separate roles of the SEPALLATA gene *LEAFY HULL STERILE* (*LHS1*) in grasses. Plant Cell 16: 1692–1706.

Malcomber, S.T., Preston, J.C., Reinheimer, R., Kossuth, J., Kellogg, E.A. 2006. Developmental gene evolution and the origin of grass inflorescence diversity. Adv. Bot. Res. 44: 425–481.

McClure, F.A. 1966. The bamboos: a fresh perspective. Cambridge, Massachusetts: Harvard University Press.

Moncur, M.W. 1981. Floral initiation in field crops: an atlas of scanning electron micrographs. Melbourne, Australia: CSIRO.

Muszynski, M.G., Dam, T., Li, B., Shirbroun, D.M., Hou, Z., Bruggemann, E., Archibald, R., Ananiev, E.V., Danilevskaya, O.N. 2006. *delayed flowering1* encodes a basic leucine zipper protein that mediates floral inductive signals at the shoot apex in maize. Plant Phys. 142: 1523–1536.

Nakagawa, M., Shimamoto, K., Kyozuka, J. 2002. Overexpression of *RCN1* and *RCN2*, rice *TERMINAL FLOWER/CENTRORADIALIS* homologs, confers delay of phase transition and altered panicle morphology in rice. Plant J. 29: 743–750.

Nalam, V.J., Vales, M.I., Watson, C.J.W., Johnson, E.B., Riera-Lizarazu, O. 2007. Map-based analysis of genetic loci on chromosome 2D that affect glume tenacity and threshability, components of the free-threshing habit in common wheat (*Triticum aestivum* L.). Theor. Appl. Genet. 116: 135–145.

Perreta, M.G., Tivano, J.C., Vegetti, A.C. 2000. Forma de crecimiento en *Leptochloa chloridiformis* (Poaceae). Darwiniana 38: 219–226.

Perreta, M.G., Ramos, J.C., Vegetti, A.C. 2009. Development and structure of the grass inflorescence. Bot. Rev. 75: 377–396.

Peterson, P.M., Romaschenko, K., Herrera Arrieta, Y., Saarela, J.M. 2014. A molecular phylogeny and new subgeneric classification of *Sporobolus* (Poaceae: Chloridoideae: Sporobolinae). Taxon 63: 1212–1243.

Piao, R., Jiang, W., Ham, T.-H., Choi, M.-S., Qiao, Y., Chu, S.-H., Park, J.-H., Woo, M.-O., Jin, Z., An, G., Lee, J., Koh, H.-J. 2009. Map-based cloning of the *ERECT PANICLE3* gene in rice. Theor. Appl. Genet. 119: 1497–1506.

Preston, J.C., Hileman, L.C. 2010. SQUAMOSA-PROMOTER BINDING PROTEIN 1 initiates flowering in *Antirrhinum majus* through the activation of meristem identity genes. Plant J. 62: 704–712.

Preston, J.C., Kellogg, E.A. 2006. Reconstructing the evolutionary history of paralogous *APETALA1/FRUITFULL*-like genes in grasses (Poaceae). Genetics 174: 421–437.

Preston, J.C., Kellogg, E.A. 2007. Conservation and divergence of *APETALA1/FRUITFULL*-like gene function in grasses: evidence from gene expression analyses. Plant J. 52: 69–81.

Preston, J.C., Kellogg, E.A. 2008. Discrete developmental roles for temperate cereal grass *AP1/FUL*-like genes in flowering competency and the transition to flowering. Plant Phys. 146: 265–276.

Preston, J.C., Christensen, A., Malcomber, S.T., Kellogg, E. A. 2009. MADS-box gene expression and implications for developmental origins of the grass spikelet. Amer. J. Bot. 96: 1419–1429.

Preston, J.C., Wang, H., Doebley, J., Kellogg, E.A. 2012. The role of *teosinte glume architecture* (*tga1*) in coordinated regulation and evolution of grass glumes and inflorescence axes. New Phytol. 193: 204–215.

Reinheimer, R., Vegetti, A.C. 2008. Inflorescence diversity and evolution in the PCK clade (Poaceae: Panicoideae: Paniceae). Plant Syst. Evol. 275: 133–167.

Reinheimer, R., Pozner, R., Vegetti, A.C. 2005. Inflorescence, spikelet, and floral development in *Panicum maximum* and *Urochloa plantaginea* (Poaceae). Amer. J. Bot. 92: 565–575.

Reinheimer, R., Malcomber, S.T., Kellogg, E.A. 2006. Distinct roles of the SEPALLATA gene *LEAFY HULL STERILE1* in *Eleusine indica* and *Megathyrsus maximus* (Poaceae). Evol. Devel. 8: 293–303.

Reinheimer, R., Zuloaga, F.O., Vegetti, A.C., Pozner, R. 2009. Diversification of inflorescence development in the PCK clade (Poaceae: Panicoideae: Paniceae). Amer. J. Bot. 96: 549–564.

Ren, D., Li, Y., Zhao, F., Sang, X., Shi, J., Wang, N., Guo, S., Ling, Y., Zhang, C., Yang, Z., He, G. 2013. *MULTI-FLORET SPIKELET1*, which encodes an AP2/ERF protein, determines spikelet meristem fate and sterile lemma identity in rice. Plant Phys. 162: 872–884.

Rudall, P.J., Stuppy, W., Cunniff, J., Kellogg, E.A., Briggs, B.G. 2005. Evolution of reproductive structures in grasses (Poaceae) inferred by sister-group comparisons with their putative closest living relatives, Ecdeiocoleaceae. Amer. J. Bot. 92: 1432–1443.

Sajo, M.G., Longhi-Wagner, H., Rudall, P.J. 2007. Floral development and embryology in the early-divergent grass *Pharus*. Int. J. Plant Sci. 168: 181–191.

Sajo, M.G., Longhi-Wagner, H.M., Rudall, P.J. 2008. Reproductive morphology of the early-divergent grass *Streptochaeta* and its bearing on the homologies of the grass spikelet. Plant Syst. Evol. 275: 245–255.

Sakakibara, H. 2006. Cytokinins: activity, biosynthesis, and translocation. Ann. Rev. Plant Biol. 57: 431–449.

Satoh-Nagasawa, N., Nagasawa, N., Malcomber, S., Sakai, H., Jackson, D. 2006. A trehalose metabolic enzyme controls inflorescence architecture in maize. Nature 441: 227–230.

Shitsukawa, N., Kinjo, H., Takumi, S., Murai, K. 2009. Heterochronic development of the floret meristem determines grain number per spikelet in diploid, tetraploid and hexaploid wheats. Ann. Bot. 104: 243–251.

Simons, K.J., Fellers, J.P., Trick, H.N., Zhang, Z., Tai, Y.-S., Gill, B.S., Faris, J.D. 2006. Molecular characterization of the major wheat domestication gene Q. Genetics 172: 547–555.

Skirpan, A., Wu, X., McSteen, P. 2008. Genetic and physical interaction suggest that BARREN STALK1 is a target of BARREN INFLORESCENCE2 in maize inflorescence development. Plant J. 55: 787–797.

Stapleton, C.M.A. 1997. The morphology of woody bamboos. In: Chapman, G.P. (ed.) The bamboos. New York: Academic Press. pp. 251-267

Stevens, P.F. 2012. Angiosperm phylogeny website. Version 12, July 2012 [and more or less continuously updated since]. 2001 onward.

Stür, W.W. 1986. Reproductive development of the apex of *Brachiaria decumbens* Stapf. Ann. Bot. 58: 569–575.

Tamaki, S., Matsuo, S., Wong, H.L., Yokoi, S., Shimamoto, K. 2007. *Hd3a* protein is a mobile flowering signal in rice. Science 316: 1033–1103.

Terao, T., Nagata, K., Morino, K., Hirose, T. 2009. A gene controlling the number of primary rachis branches also controls the vascular bundle formation and hence is responsible to increase the harvest index and grain yield of rice. Theor. Appl. Genet. 120: 875–893.

Thompson, B.E., Hake, S. 2009. Translational biology: from Arabidopsis flowers to grass inflorescence architecture. Plant Phys. 149: 38–45.

Troll, W. 1964. Die Infloreszenzen, Typologie und Stellung im Aufbau des Vegetationskörpers. I. Jena: Gustav Fischer.

Vegetti, A.C. 1991. Sobre politelia en las inflorescencias de Poaceae. Kurtziana 21: 275–278.

Vegetti, A.C. 1999. Typology of the synflorescence of Andropogoneae (Poaceae), additional comments. Feddes Repert. 110: 111-126.

Vegetti, A.C., Anton, A.M. 1996. The synflorescence concept in Poaceae. Flora 191: 231–234.

Vegetti, A.C., Pensiero, J.F. 1999. Tipología de la inflorescencia de *Zizaniopsis bonariensis* y *Z. villanensis* (Poaceae, Oryzeae). Darwiniana 37: 345–349.

Vegetti, A.C., Weberling, F. 1996. The structure of the paracladial zone in Poaceae. Taxon 45: 453–460.

Vollbrecht, E., Springer, P.S., Goh, L., Buckler, E.S., Martienssen, R. 2005. Architecture of floral branch systems in maize and related grasses. Nature 436: 1119–1126.

Wang, H., Nussbaum-Wagler, T., Li, B., Zhao, Q., Vigouroux, Y., Faller, M., Bomblies, K., Lukens, L., Doebley, J.F. 2005. The origin of the naked grains of maize. Nature 436: 714–719.

Watson, L., Dallwitz, M.J. 1992 onward. The grass genera of the world: descriptions, illustrations, identification, and information retrieval; including synonyms, morphology, anatomy, physiology, phytochemistry, cytology, classification, pathogens, world and local distribution, and references, 23rd April 2010 edn. Wallingford: CAB International.

Weberling, F. 1989. Morphology of flowers and inflorescences. Cambridge: Cambridge University Press.

Weberling, F., Müller-Doblies, U., Müller-Doblies, D. 1993. Zur deskriptiven und vergleichend morphologischen Terminologie komplexer Infloreszenzen. Beitr. Biol. Pflanzen 67: 453–473.

Weir, C.E., Dale, H.M. 1960. A developmental study of wild rice, *Zizania aquatica* L. Can. J. Bot. 38: 719–739.

Wellmer, F., Riechmann, J.L. 2010. Gene networks controlling the initiation of flower development. Trends Genet. 26: 519–527.

Whipple, C.J., Hall, D.H., DeBlasio, S., Taguchi-Shiobara, F., Schmidt, R.J., Jackson, D.P. 2010. A conserved mechanism of bract suppression in the grass family. Plant Cell 22: 565–578.

Wills, D.M., Whipple, C.J., Takuno, S., Kursel, L.E., Shannon, L.M., Ross-Ibarra, J., Doebley, J.F. 2013. From many, one: genetic control of prolificacy during maize domestication. PLoS Genet. 9: e1003604.

Woods, D.P., Hope, C.L., Malcomber, S.T. 2011. Phylogenomic analyses of the BARREN STALK1/LAX PANICLE1 (BA1/LAX1) genes and evidence for their roles during axillary meristem development. Mol. Biol. Evol. 28: 2147–2159.

Wu, C.Y., You, C.J., Li, C.S., Long, T., Cgeb, G.X., Byrne, M.E., Zhang, Q.F. 2008. RID1, encoding a Cys2/His2-type zinc finger transcription factor, acts as a master switch from vegetative to floral development in rice. Proc. Natl. Acad. Sci. USA 105: 12915–12920.

Yan, L., Loukolanov, A., Tranquilli, G., Helguera, M., Fahima, T., Dubcovsky, J. 2003. Positional cloning of the wheat vernalization gene *VRN1*. Proc. Natl. Acad. Sci. USA 100: 6263–6268.

Yan, L., Fu, D., Li, C., Blechl, A., Tranquilli, G., Bonafede, M., Sanchez, A., Valarik, M., Yasuda, S., Dubcovsky, J. 2006. The wheat and barley vernalization gene *VRN3* is an orthologue of FT. Proc. Natl. Acad. Sci. USA 103: 19581–19586.

Zeevaart, J.A.D. 2008. Leaf-produced floral signals. Curr. Opinion Pl. Biol. 11: 541–547.

Zhang, L., Zhao, Y.-L., Gao, L.-F., Zhao, G.-Y., Zhou, R.-H., Zhang, B.-S., Jia, J.-Z. 2012. *TaCKX6-D1*, the ortholog of rice *OsCKX2*, is associated with grain weight in hexaploid wheat. New Phytol. 195: 574–584.

Zhou, Y., Lu, D., Li, C., Luo, J., Zhu, B.-F., Zhu, J., Shangguan, Y., Wang, Z., Sang, T., Zhou, B., Han, B. 2012. Genetic control of seed shattering in rice by the APETALA2 transcription factor *SHATTERING ABORTION1*. Plant Cell 24: 1034–1048.

Zhu, K., Tang, D., Yan, C., Chi, Z., Yu, H., Chen, J., Liang, J., Gu, M., Cheng, Z. 2010. *ERECT PANICLE2* encodes a novel protein that regulates panicle erectness in *Indica* rice. Genetics 184: 343–350.

Zou, L., Sun, X., Zhang, Z., Liu, P., Wu, J., Tian, C., Qiu, J., Lu, T. 2011. Leaf rolling controlled by the homeodomain leucine zipper class IV gene *Roc5* in rice. Plant Phys. 156: 1589–1602.

Flower Structure

Except in *Anomochloa* and *Streptochaeta*, grass flowers are made up of a gynoecium, androecium, lodicules, a palea and a lemma (Fig. 5). While the structure and homologies of the gynoecium and androecium are similar to those of other angiosperms, the homologies of the surrounding structures are less obvious and the terminology more confusing. Accordingly, flowers are described here beginning with the gynoecium and working outward.

The grass outgroups have a gynoecium with two or three carpels and the corresponding number of locules; the ancestral state is almost certainly three. Although the apex of the carpels forms a transmitting tract, the length of this varies considerably. Therefore, if a style is defined as a structure borne of the fusion of several carpels, the length of the style is variable. Distal to the style are stylar branches; these are generally extremely short. Most of the stylar branch is receptive and is therefore generally called a stigma. *Joinvillea* (Joinvilleaceae) has a three-carpellary, three-locular ovary, with three virtually sessile stigmas (Whipple et al. 2007), as does *Georgeantha* (Edeiocoleaceae) (Rudall et al. 2005); in *Ecdeiocolea* the number of carpels and stigmas is reduced to two (Rudall et al. 2005). In both Joinvilleaceae and Ecdeiocoleaceae, there is little tissue that could be called a style.

In Poaceae, the number of locules in the ovary is reduced to one, and is invariable and synapomorphic for the family. However, the structure of the stylar tissue, the number of stylar branches, and therefore the number of obvious stigmas, is variable. The distinction between styles and style branches is rarely made in the literature on grasses. In general both the style and branches are very short to virtually non-existent and dominated by the feathery stigmatic tissue. It is thus common to count the number of stigmas, rather than the number of stylar branches. In several grass subfamilies, the stylar branches appear to originate from separate positions at the apex of the ovary, as though the style itself has become subsumed into the ovary. In this case the stylar branches are often referred to as "style bases" or simply "styles"; thus, a distinction is rarely made between the obviously compound stylar structure in some subfamilies versus the outgrowth of a (presumably) single carpel in many others.

In subfamily Anomochlooideae, *Streptochaeta* has one style, with three branches each ending in a stigma (Judziewicz and Soderstrom 1989; Preston et al. 2009; Sajo et al. 2008). In contrast, *Anomochloa* has a single structure that presumably corresponds to a single stylar branch and stigma (Sajo et al. 2012). All Pharoideae have one style with three branches and three stigmas, as does *Puelia* (Puelioideae), but *Guaduella* (also Puelioideae) has only two branches and two stigmas. Most descendants of the common ancestor of the BEP and PACMAD clades have two stylar branches and thus two stigmas, but three are present in many bamboos. The GPWG (2001) called all descendants of the common ancestor of Puelioideae and the remaining grasses "the Bistigmatic clade", implying that the reduction to two stylar branches is synapomorphic; this interpretation requires the three branches in *Puelia* and some Bambusoideae to be derived, but it seems equally plausible that stylar branch number was not fixed until after the common ancestor of the BEP and PACMAD clades. In some species,

DOI 10.1007/978-3-319-15332-2_3,

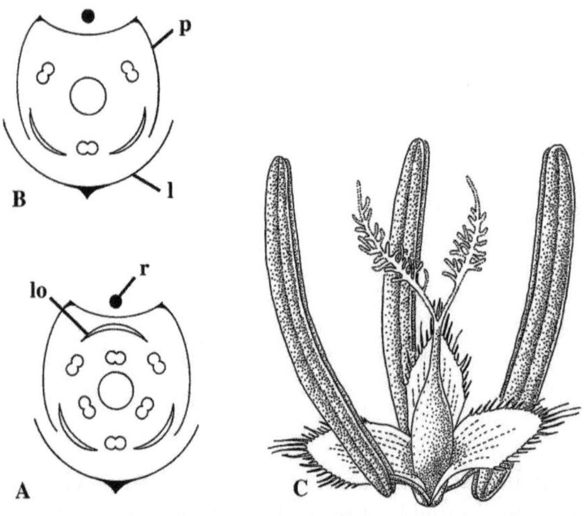

Fig. 5. Flowers. **A** Floral diagram of a grass with three lodicules and six stamens. **B** Floral diagram of a grass with two lodicules and three stamens. **C** Abaxial view of flower of *Yushania* (Arundinarieae) showing pistil with two stylar branches and two stigmas, three stamens, and three vascularized and ciliolate lodicules; lemma and palea have been removed. (From GPWG 2001, p. 402, with permission from Missouri Botanical Garden Press; **A** and **B** drawn by M. Kojima, **C** by D. Friedrick)

the third stylar branch can sometimes be seen in scanning electron micrographs as a tiny apical flap (Le Roux and Kellogg 1999). Further reductions are seen in some species; for example in cultivated maize, the two branches form but are congenitally fused to form the familiar silk. In subfamily Pooideae, the stylar branches are separate in the earliest diverging lineage, *Brachyelytrum*. In the next lineage, Nardeae, there is a single structure; it is unclear whether this represents an unbranched style, or a single stylar branch. After divergence of Nardeae, the style branches originate from a clear style, whereas the stylar branches are once more apparently independent in Diarrheneae plus its large sister clade.

Styles in most grasses are solid, a condition rarely found in monocots (Rudall et al. 2005; Sajo et al. 2012), although open styles are described in *Pharus* (Pharoideae) (Sajo et al. 2007). Pollen tubes are then transmitted through the "stylar core" (Arber 1934). Grass stigmas are plumose and the receptive branches are multicellular, a character shared with the other families of the graminid Poales.

Developmental genetic controls of the grass gynoecium appear to be grass-specific. Whereas eudicots have a single MADS-box transcription factor (AGAMOUS) that specifies both carpel identity and floral meristem determinacy, the grasses have two AG-like genes, originating from a duplication early in grass evolution. This duplication was followed by divergence in gene function, such that one copy of the gene (ZAG1 in maize, OsMADS58 in rice) specifies carpel identity and the other copy (ZMM2, OsMADS3) is preferentially expressed in stamens (Dreni et al. 2011; Kramer et al. 2004; Mena et al. 1996; Yamaguchi et al. 2006); other genes are also involved (e. g., DROOPING LEAF (DL), a YABBY-class transcription factor, Li et al. 2011; Yamaguchi et al. 2004).

Grass stamens are conventional, and have two thecae and four sporangia. The filament in *Streptochaeta* and *Anomochloa* is centrifixed (= pseudobasifixed), such that the filament is completely surrounded by the connective at the point of insertion (Arber 1929; Judziewicz and Soderstrom 1989; Sajo et al. 2009, 2012); such anther attachment is often associated with buzz pollination. In most other grasses, the filament is attached to the abaxial side of the anther and elongates swiftly at anthesis so that the anther is exserted from the flower. Anthers are versatile and tremble in response to any breeze, dispersing copious pollen. Wind pollination appears to be derived in the family, rather than ancestral (Givnish et al. 2010), but this depends on interpretation of Anomochlooideae as insect pollinated; the latter has not been confirmed by observation but is merely inferred from anther morphology.

Anther dehiscence is consistently longitudinal. Dehiscence is latrorse in *Streptochaeta* (Arber 1929; Sajo et al. 2009), whereas in *Anomochloa* it is introrse proximally and latrorse distally (Sajo et al. 2012). In contrast, in grass species with three anthers, the abaxial anther is latrorse or introrse, whereas the lateral anthers are consistently extrorse (Anton and Astegiano 1973).

The earliest grasses and their ancestors had six stamens, a condition that appears today in *Streptochaeta*, Pharoideae, Puelioideae, and in some Bambusoideae and Ehrhartoideae (Fig. 5A). In a few Bambusoideae, and in *Luziola* of Oryzeae, stamen number is higher than six. Reduction of stamen number to four has occurred independently in *Anomochloa* and in some species of *Ehrharta*. Most of the grasses, however, have only three stamens (Fig. 5B); this number might be synapomorphic but independently derived in Pooideae and the PACMAD clade, although some Bambusoideae also have three stamens which may affect character optimization in the BEP clade. Further reduction to two or even one stamen has occurred repeatedly and independently, and often is associated with flowers that are obligately self-pollinated and fail to open at maturity.

Outside the stamens are two or three small structures known as lodicules, which correspond to the inner perianth whorl of other monocots (Fig. 5A–C). As the flower nears maturity, the lodicules increase turgor pressure and force the flower open. The lodicules are generally vascularized, although in all Pooideae except for *Brachyelytrum* the vasculature has been lost (GPWG 2001; Jirásek and Jozífóva 1968).

The lodicules represent modified inner perianth organs. Transcription factors that specify petal and stamen identity in other angiosperms (the so-called B class genes) have been identified and characterized in multiple grasses (Ambrose et al. 2000; Nagasawa et al. 2003; Whipple et al. 2004, 2007). These genes are expressed in both stamens and lodicules, as would be expected if lodicules were highly modified petals. Importantly, Whipple et al. (2007) find that the B-class genes are also expressed in the inner tepals of *Joinvillea* and *Elegia* in Restionaceae, indicating continuity of information between the grasses and their close relatives. In addition, when any of these genes is mutated, the stamens are replaced by carpel-like structures and the lodicules by palea-like structures. In eudicots, mutants in orthologous genes replace stamens by carpels and petals by sepals. Thus, the replacement of lodicules by organs in the next outer whorl would be expected if they were indeed of petaloid origin.

Previous authors have speculated that lodicules might be staminodes, sepals, or nectaries, but the evidence does not support this. If lodicules were staminodial, then in mutants they should be converted to carpels, which does not happen. If they were sepaloid or nectaries similar to those in eudicots, then mutations in petal-identity genes would not affect them; since they are affected, sepaloid or nectary origin is not supported. In summary, the simplest interpretation of the data is that the lodicules are governed by the same genetic controls that regulate the inner tepals of non-grass Poales. The downstream genetic pathways have clearly shifted so that instead of activating tepal-like development and photosynthesis, the regulatory genes activate lodicule development.

In *Streptochaeta*, the three bracts outside the stamens appear to be governed by the same regulatory genes that control lodicules (Preston et al. 2009; Whipple et al. 2007); the three bracts thus share with lodicules and inner tepals similarity in position, gene expression and number. In *Anomochloa*, the stamens are surrounded by a ring of fimbriate structures, which appear to be positionally homologous to lodicules; genetic controls of their development are unknown (Sajo et al. 2012). Three lodicules are found in Pharoideae, Bambusoideae, and some Pooideae (in tribes Phaenospermateae and Stipeae), although one of the three appears in a whorl outside the other two. The significance of this is not known. It has been suggested that two of the lodicules might be modified petals, but that the third might be a sepal; since there are no genetic or developmental data on species with three lodicules, this remains an open question. However, because most monocots have two adaxial sepals rather than one, it seems more likely that the third lodicule is a member of the inner perianth, despite its position on the floral axis.

Outside the lodicules is an adaxial bract, the palea (Fig. 5A, B). Its position outside the lodicules suggests that it could correspond to the fusion of the two outer tepals, a suggestion made by Stebbins (1956a, b). A more common interpretation of the palea, however, is that it is a prophyll. Prophylls in monocots are consistently adaxial, and are generally two keeled; the palea is adaxial and is two-keeled in most (but not all) grasses. However, in most monocots including the grass outgroups, prophylls appear on lateral branches, but not on pedicels of flowers. Conversely, most monocots including the grass outgroups do have a pair of adaxial tepals. In *Ecdeiocolea* these are keeled, so one could speculate that their fusion might lead to a bicarinate structure.

If the palea were derived from fused outer tepals, it would constitute a novel structure in the grasses; the origin of the palea maps to the "spikelet clade", including all grasses except

Anomochlooideae. This hypothesis is consistent with recent studies of gene expression (Preston et al. 2009; Reinheimer and Kellogg 2009) and developmental genetics (Li et al. 2011). Palea identity is controlled by AGL6-like proteins (e.g., MOSAIC FLORAL ORGANS1/OsMADS6 in rice), which suppress DL; when both are mutated, the palea is replaced by two lemma-like organs (Li et al. 2011). AGL6-like expression in the palea appears only in grasses with spikelets but not in *Streptochaeta*, in which the outer bracts are not fused. Palea identity is also controlled by RETARDED PALEA1 (REP1), a CYCLOIDEA-like protein related to proteins that in dicots specify zygomorphy of the perianth (Yuan et al. 2009). In rice, the palea margin is controlled by a set of proteins (e.g., the MADS-box proteins CHIMERIC FLORAL ORGANS/OsMADS32 and MOSAIC FLORAL ORGANS/OsMADS6) different from those controlling the central part of the palea (REP1, MULTI-FLORET SPIKELET1, DEPRESSED PALEA1), which has led some authors to suggest that the marginal region of the palea is sepaloid, whereas the central region is prophyllar (Ren et al. 2013; Sang et al. 2012); there are no data extending this particular model to other grasses. In summary, accumulating evidence is pointing to the somewhat heterodox idea that the palea corresponds to two fused adaxial tepals.

Subtending the entire flower is a bract-like structure, the lemma (Fig. 5A, B). The lemma is similar to a modified leaf. In studies of floral mutants and gene expression, the lemma shares some aspects of floral (i.e., presumed perianth) identity and some of bract identity. It is possible that it represents a novel organ type, resulting from the fusion of two developmental programs.

In summary, the data suggest a much simplified interpretation of the grass floret, which can be interpreted as a highly modified flower, a term that will be used throughout this book. The grass flower is zygomorphic, usually with two or three carpels and a single seed, and 3 or 6 stamens. The inner perianth ("corolla" = lodicules) has two or three members; when only two are present, they are the two abaxial ones. The corolla serves a mechanical rather than attractive function. The outer perianth ("calyx" = palea) is comprised of the two lateral adaxial tepals, which are fused. The abaxial outer tepal is either lost or forms a common primordium with the bract (lemma), which is thus either a floral bract or a bract-like abaxial sepal.

Much of grass diversity is reflected in morphological variation in lemmas, and to a lesser extent, paleas. Lemmas vary as much in size and texture as the glumes (see above). They differ in the number of veins, in their overall shape, in the shape of the apex, and the nature of their attachment to the rachilla (the callus). Indumentum is common as is ornamentation of various sorts. The morphology of the cells in the lemma epidermis is often diagnostic for species (although it is an inconvenient field character).

Awns may form on lemmas or glumes or both, and may be terminal, or originate from the sinus between two apical lobes, or may appear attached to the abaxial side; rarely awns form on paleas. Most awned grasses have a single awn on the lemma, but lemmas with multiple awns have evolved several times. The awns may be straight, or twisted, or geniculate, or both twisted and geniculate. Rather surprisingly, awns are major contributors to photosynthesis in the inflorescence and supply an appreciable amount of photosynthate to the developing caryopsis, at least in wheat and barley where their role has been studied most extensively (Tambussi et al. 2007). Because many wheat cultivars have awned and unawned forms, it is possible to make comparisons in a near isogenic background. Awns are reported to improve grain yield when water is limiting, but appear to have little effect when water is plentiful (Grundbacher 1963; Motzo and Giunta 2002; Tambussi et al. 2007). They also accumulate silica (Grundbacher 1963). Further discussion of the role of awns is under Fruit and seed - Dispersal.

References

Ambrose, B.A., Lerner, D.R., Ciceri, P., Padilla, C.M., Yanofsky, M.F., Schmidt, R.J. 2000. Molecular and genetic analyses of the *silky1* gene reveal conservation in floral organ specification between eudicots and monocots. Mol. Cell 5: 569–579.

Anton, A.M., Astegiano, M.E. 1973. Notas sobre la morfologia floral de Gramineas. Kurtziana 7: 49–53.

Arber, A. 1929. Studies in the Gramineae. VI. 1. *Streptochaeta*. 2. *Anomochloa*. 3. *Ichnanthus*. Ann. Bot. 43: 35–53.

Arber, A. 1934. The Gramineae: a study of cereal, bamboo, and grass. New York: Macmillan.

Dreni, L., Pilatone, A., Yun, D.P., Erreni, S., Pajoro, A., Caporali, E., Zhang, D.B., Kater, M.M. 2011. Functional analysis of all AGAMOUS subfamily members in rice reveals their roles in reproductive organ identity determination and meristem determinacy. Plant Cell 23: 2850–2863.

Givnish, T.J., Ames, M., McNeal, J.R., McKain, M.R., Steele, P.R., dePamphilis, C.W., Graham, S.W., Pires, J.C., Stevenson, D.W., Zomlefer, W.B., Briggs, B.G., Duvall, M.R., Moore, M.J., Heaney, J.M., Soltis, D.E., Soltis, P.S., Thiele, K., Leebens-Mack, J.H. 2010. Assembling the tree of monocotyledons: plastome sequence phylogeny and evolution of Poales. Ann. Missouri Bot. Gard. 97: 584–616.

GPWG 2001. Grass Phylogeny Working Group. Phylogeny and subfamilial classification of the Poaceae. Ann. Missouri Bot. Gard. 88: 373–457.

Grundbacher, F.J. 1963. The physiological function of the cereal awn. Bot. Rev. 29: 366–381.

Jirásek, V., Jozífóva, M. 1968. Morphology of lodicules, their variability and importance in the taxonomy of the Poaceae family. Bol. Soc. Argentina Bot. 12: 324–349.

Judziewicz, E.J., Soderstrom, T.R. 1989. Morphological, anatomical, and taxonomic studies in *Anomochloa* and *Streptochaeta* (Poaceae: Bambusoideae). Smithsonian Contr. Bot. 68: 1–52.

Kramer, E.M., Jaramillo, M.A., Di Stilio, V.S. 2004. Patterns of gene duplication and functional evolution during the diversification of the AGAMOUS subfamily of MADS box genes in angiosperms. Genetics 166: 1011–1023.

Le Roux, L.G., Kellogg, E.A. 1999. Floral development and the formation of unisexual spikelets in the Andropogoneae (Poaceae). Amer. J. Bot. 86: 354–366.

Li, H., Liang, W., Hu, Y., Zhu, L., Yin, C., Xu, J., Dreni, L., Kater, M.M., Zhang, D. 2011. Rice *MADS6* interacts with the floral homeotic genes *SUPERWOMAN1*, *MADS3*, *MADS58*, *MADS13*, and *DROOPING LEAF* in specifying floral organ identities and meristem fate. Plant Cell 23: 2536–2552.

Mena, M., Ambrose, B.A., Meeley, R.B., Briggs, S.P., Yanofsky, M.F., Schmidt, R.J. 1996. Diversification of C-function activity in maize flower development. Science 274: 1537–1540.

Motzo, R., Giunta, F. 2002. Awnedness affects grain yield and kernel weight in near-isogenic lines of durum wheat. Aust. J. Agric. Res. 53: 1285–1293.

Nagasawa, N., Miyoshi, M., Sano, Y., Satoh, H., Hirano, H., Sakai, H., Nagato, Y. 2003. *SUPERWOMAN1* and *DROOPING LEAF* genes control floral organ identity in rice. Development 130: 705–718.

Preston, J.C., Christensen, A., Malcomber, S.T., Kellogg, E. A. 2009. MADS-box gene expression and implications for developmental origins of the grass spikelet. Amer. J. Bot. 96: 1419–1429.

Reinheimer, R., Kellogg, E.A. 2009. Evolution of AGL6-like MADS box genes in grasses (Poaceae): ovule expression is ancient and palea expression is new. Plant Cell 21: 2591–2605.

Ren, D., Li, Y., Zhao, F., Sang, X., Shi, J., Wang, N., Guo, S., Ling, Y., Zhang, C., Yang, Z., He, G. 2013. *MULTI-FLORET SPIKELET1*, which encodes an AP2/ERF protein, determines spikelet meristem fate and sterile lemma identity in rice. Plant Phys. 162: 872–884.

Rudall, P.J., Stuppy, W., Cunniff, J., Kellogg, E.A., Briggs, B.G. 2005. Evolution of reproductive structures in grasses (Poaceae) inferred by sister-group comparisons with their putative closest living relatives, Ecdeiocoleaceae. Amer. J. Bot. 92: 1432–1443.

Sajo, M.G., Longhi-Wagner, H., Rudall, P.J. 2007. Floral development and embryology in the early-divergent grass *Pharus*. Int. J. Plant Sci. 168: 181–191.

Sajo, M.G., Longhi-Wagner, H.M., Rudall, P.J. 2008. Reproductive morphology of the early-divergent grass *Streptochaeta* and its bearing on the homologies of the grass spikelet. Plant Syst. Evol. 275: 245–255.

Sajo, M.D.G., Furness, C.A., Rudall, P.J. 2009. Microsporogenesis is simultaneous in the early-divergent grass *Streptochaeta*, but successive in the closest grass relative, *Ecdeiocolea*. Grana 48: 27–37.

Sajo, M.G., Pabón-Mora, N., Jardim, J., Stevenson, D.W., Rudall, P.J. 2012. Homologies of the flower and inflorescence in the early-divergent grass *Anomochloa* (Poaceae). Amer. J. Bot. 99: 614–628.

Sang, X., Li, Y., Luo, Z., Ren, D., Fang, L., Wang, N., Zhao, F., Ling, Y., Yang, Z., Liu, Y., He, G. 2012. *CHIMERIC FLORAL ORGANS1*, encoding a monocot-specific MADS box protein, regulates floral organ identity in rice. Plant Phys. 160: 788–807.

Stebbins, G.L. 1956a. Cytogenetics and evolution of the grass family. Amer. J. Bot. 43: 890–905.

Stebbins, G.L. 1956b. Taxonomy and the evolution of genera, with special reference to the family Gramineae. Evolution 10: 235–245.

Tambussi, E.A., Bort, J., Guiamet, J.J., Nogués, S., Araus, J. L. 2007. The photosynthetic role of ears in C_3 cereals: metabolism, water use efficiency and contribution to grain yield. Crit. Rev. Plant Sci. 26: 1–16.

Whipple, C.J., Ciceri, P., Padilla, C.M., Ambrose, B.A., Bandong, S.L., Schmidt, R.J. 2004. Conservation of B-class floral homeotic gene function between maize and *Arabidopsis*. Development 131: 6083–6091.

Whipple, C.J., Zanis, M.J., Kellogg, E.A., Schmidt, R.J. 2007. Conservation of B class gene expression in the second whorl of a basal grass and outgroups links the origin of lodicules and petals. Proc. Natl. Acad. Sci. USA 104: 1081–1086.

Yamaguchi, T., Nagasawa, N., Kawasaki, S., Matsuoka, M., Nagato, Y., Hirano, H.Y. 2004. The *YABBY* gene *DROOPING LEAF* regulates carpel specification and midrib development in *Oryza sativa*. Plant Cell 16: 500–509.

Yamaguchi, T., Lee, D.Y., Miyao, A., Hirochika, H., An, G., Hirano, H.-Y. 2006. Functional diversification of the two C-class MADS box genes *OSMADS3* and *OSMADS58* in *Oryza sativa*. Plant Cell 18: 15–28.

Yuan, Z., Gao, S., Xue, D.W., Luo, D., Li, L.T., Ding, S.Y., Yao, X., Wilson, Z.A., Qian, Q., Zhang, D.B. 2009. RETARDED PALEA1 controls palea development and floral zygomorphy in rice. Plant Phys. 149: 235–244.

Embryology

Ovule

The gynoecium initially forms as a bulge at the top of the floral meristem, which enlarges into a dome-like structure. On the abaxial side of the dome (i.e., the side toward the lemma) a circumferential rim forms and grows upward, and soon encircles the entire gynoecium; this leaves the ovule primordium looking like a ball in a cup, but displaced toward the adaxial side (Le Roux and Kellogg 1999; Percival 1921; Zaitchik et al. 2000). In other words, the ovule is lateral from early stages of development. Although not illustrated, development of the ovule in *Sorghum* is described in a similar way (Artschwager and McGuire 1949). In the early diverging grasses *Streptochaeta*, *Anomochloa* and *Pharus*, the ovule is said to initiate basally, but is then displaced to a lateral position by differential growth of the ovary wall (Sajo et al. 2007, 2008, 2012); another interpretation of their photos is that initiation is lateral. Published descriptions of the grass ovule imply that there is great variation in its morphology, variously describing it as anatropous, campylotropous, or amphitropous and everything in between. Anton de Triquell (1987) reviews this literature and notes that in fact grass ovule morphology is quite uniform, and that the "variation" reflects simple misapplication of ovule typology. The funiculus is virtually absent and the outer integument often fairly short until late development, so the ovule type is not always clear; Anton de Triquell (1987) argues that the ovule is best described as hemicampylotropous. At maturity the ovule is described by Sajo et al. (2007, 2008) as hemianatropous, a term also applied to a set of arundinoid taxa by Verboom et al. (1994). The distinction between hemianatropous and hemicampylotropous is slight, particularly in the absence of a clear funiculus. Whatever typology is used to describe the ovule, the micropyle always ends up pointing toward the base of the ovary, and the ovule is attached to the ovary wall for a substantial portion (often more than ½) of its length.

The grass ovule has two integuments. In most grasses, only the inner integument forms the micropyle (Anton de Triquell 1987; Bhanwra et al. 1991; Kam and Maze 1974; Sajo et al. 2008; Verboom et al. 1994) and the ovule is described as endostomic; this developmental pattern has been suggested to be synapomorphic for Poaceae (Linder and Kellogg 1995). In *Pharus* (Sajo et al. 2007) and in *Panicum miliaceum* (Davis 1966), both integuments are reported to form the micropyle, suggesting a reversal in the character; however, the figure in Sajo et al. (2007) shows a normal endostomic ovule. The inner integument may be more than two cells thick near the micropyle (Verboom et al. 1994). There are some reports of taxa in which the inner integument does not fully enclose the nucellus (i.e., a "naked nucellus"), so that a normal narrow micropyle fails to form (Bhanwra et al. 1991; Choda et al. 1982); the effect of this on fertilization and ovule development is unknown.

In most grasses, the outer integument develops more slowly than the inner one (Bhanwra et al. 1991; Davis 1966; Maze and Bohm 1973; Verboom et al. 1994), and often is much shorter than the inner integument at anthesis. However, both integuments develop simultaneously in *Streptochaeta* (Sajo et al. 2008). The outer integument generally degenerates after fertilization (Bhanwra et al. 1991; Sajo et al. 2007, 2008).

E.A. Kellogg, *Flowering Plants. Monocots*, The Families and Genera of Vascular Plants 13,
DOI 10.1007/978-3-319-15332-2_4, © Springer International Publishing Switzerland 2015

Megagametophyte

The megaspore mother cell differentiates from a cell immediately inside the nucellar epidermis and no parietal cell forms; occasionally, however, two megaspore mother cells are observed (Artschwager and McGuire 1949). The lack of a parietal cell means that ovules are tenuinucellate, but in some taxa ovules become pseudo-crassinucellate by periclinal division of the epidermis (Artschwager and McGuire 1949; Aulbach-Smith and Herr, Jr. 1984; Bhanwra 1988; Bhanwra et al. 1991; Verboom et al. 1994). The phylogenetic distribution of pseudocrassinucellate ovules is unclear. Bhanwra (1988) suggests that the condition is common in Panicoideae, but it is also reported in Bambusoideae (Bhanwra et al. 2001) and Pooideae (Maze and Bohm 1973). Meiosis I leads to a pair of cells, one of which is closer to the micropyle and the other to the chalazal end of the ovule; the latter tends to be larger. At Meiosis II, the chalazal cell divides in a plane crosswise to the long axis of the developing megagametophyte, whereas the micropylar cell may divide either crosswise or longitudinally. These divisions result in a linear (Bhanwra et al. 1991; Wu et al. 2011) or T-shaped tetrad of megaspores (Sajo et al. 2007; Stover 1937), respectively. Variation in tetrad pattern is reported within a single species (Aulbach-Smith and Herr, Jr. 1984), suggesting that the pattern is not of phylogenetic significance and may in fact be random. In some cases, the second cell division fails entirely in the micropylar member of the dyad, leading to a triad of cells instead of a tetrad (see references in Anton de Triquell 1987, Bhanwra 1988, Choda et al. 1982); it is not clear whether this is characteristic of particular species, or is simply a common but sporadic occurrence.

The megagametophyte is monosporic and develops from the chalazal megaspore (Anton de Triquell 1987; Evans and Grossniklaus 2009; Stover 1937; Wu et al. 2011). In most species, the megagametophyte is of the Polygonum-type, with three mitoses followed by cellularization leading to seven cells; the central cell contains two nuclei. In most grasses, the antipodal cells continue to divide after cellularization, although the number of extra divisions varies (Anton and Cocucci 1984; Evans and Grossniklaus 2009; Shadowsky 1926; Stover 1937; Wu et al. 2011); Yamaura (1933) reports as many as 300 antipodals in *Sasa paniculata*, although only 16 are shown in the figure. Proliferation of the antipodals is not unique to Poaceae. It also occurs in the Restionaceae (Campbell and Kellogg 1987) but nothing is known about other Poalean families, so it is unclear if it is independently derived in Restionaceae and Poaceae, or if it is synapomorphic for the entire clade. Proliferation of antipodals does not occur in the early diverging grass genera *Streptochaeta* and *Pharus* (Sajo et al. 2007, 2008), suggesting that it could be a synapomorphy for the BEP plus PACMAD clades. However, data on this character are too scattered to draw any conclusions about its consistency within major groups.

The overall shape of the embryo sac varies among species. In the panicoids, the axis from the egg cell through the antipodals is straight and in line with the proximo-distal axis of the ovule, whereas in the other subfamilies the egg-antipodal axis is curved so that the antipodal cells appear more or less lateral at maturity (Anton de Triquell 1987; Bhanwra 1988). As with the preceding characters, the number of taxa sampled is small so it is not clear how widely these generalizations apply.

In subfamily Danthonioideae, the synergids appear to function as haustoria, growing out through the micropyle (Philipson and Connor 1984; Verboom et al. 1994); this character has been interpreted as synapomorphic for the subfamily. Haustorial synergids have been observed in all studied genera of the subfamily (nine of 17 genera, 39 of 281 species), although the character appears to be polymorphic in *Pentameris*.

Apomixis occurs in many grasses and is discussed under Reproductive systems.

Embryo

The first division of the zygote is transverse; the basal cell then divides again transversely, then the apical cell divides vertically or obliquely, and finally the basal cell divides again, giving rise to the suspensor (Artschwager and McGuire 1949; Jones and Rost 1989). Early embryo development in the grasses is usually described to be of the "asterad" type (Bhanwra 1988; Febulaus and Pullaiah 1994; Sajo et al. 2007), meaning that both the basal and the apical cell of the two-celled embryo contribute to formation of the embryo proper. In contrast, both Nardmann and Werr (2009) and Jones and Rost (1989) suggest that only the apical cell contributes to the embryo proper in maize and rice, respectively, and Taylor and Vasil (1995) report no particular pattern of cell division in *Cenchus americanus* (= *Pennisetum glaucum*, pearl millet).

The embryo proper enlarges, first becoming club shaped, and then elongating. As the embryo elongates, it begins to differentiate into a bipolar embryo with shoot and root apical meristems (Takacs et al. 2012). On one side of the embryonic axis, a broad shield-shaped and peltate structure forms, the scutellum, which is generally longer and broader than the embryo itself. As the apex of the scutellum begins to elongate, the shoot apical meristem is seen initially as a densely packed group of cells on one side of the embryo, above which a notch forms. Above the notch, a bulge of cells forms, and these ultimately enlarge to form a ring or collar around the developing meristem (Artschwager and McGuire 1949; Nardmann and Werr 2009; Taylor and Vasil 1995); this ring then elongates into a tube like structure, the coleoptile. The primary root meristem is organized endogenously. Opposite the scutellum and just above the root meristem, an epidermal outgrowth may form; this is the epiblast. In some grasses, the scutellum elongates basipetally alongside, but separate from the coleorhiza; in the pooid grasses, however, this elongation scarcely occurs (Reeder 1957). In the former case, when the embryo is viewed in longitudinal section there is a clear space between the scutellum and the base of the embryo known as the scutellar cleft. The functional significance of the different scutellar structures, if any, is unknown.

The homologies of the scutellum and the coleoptile have been the subject of much discussion, backed, until recently, by remarkably little data. As summarized by Brown (1960), the scutellum has been variously considered a cotyledon, a part of a cotyledon, an outgrowth of the radicle, the embryonic axis, a leaf sheath, or an aborted bud. The coleoptile, while clearly a sheathing structure of some sort, has also been considered a cotyledon or part of it, or two or more fused leaves, or fused cotyledons, or a prophyll, or a bud. Brown (1960), citing developmental evidence, concludes that the scutellum is a modified cotyledon, that the coleoptile is a structure with no homology to organs in other monocots ("caenogenetic"), and that the epiblast is an outgrowth of the coleorhiza. His view of the coleoptile echoes that of Boyd (1931). Weatherwax (1920), in contrast, suggested that the coleoptile is a leaf.

Recent evidence suggests that the coleoptile plus scutellum together constitute the cotyledon (Takacs et al. 2012). The scutellum functions like cotyledons in many plants to take up nutrients from the endosperm during germination. Because the coleoptile develops on the same side of the shoot apical meristem as the scutellum, it appears to be the sheathing portion of the cotyledonary structure, whereas the scutellum may be the distal tip (Takacs et al. 2012). The first leaf forms on the opposite side of the embryo, thus establishing the distichous phyllotaxy that will characterize the plant. This interpretation is supported by global analyses of gene expression, which show high similarity between the coleoptile and scutelllum, with few genes expressed uniquely in the coleoptile (Takacs et al. 2012).

As the first leaf begins to form, the scutellum + coleoptile (i.e., cotyledon) begins to acquire some leaf-like characteristics. A distinct set of genes is expressed in the scutellum + coleoptile structure plus the first leaf (Takacs et al. 2012). The tubular coleoptile apparently has a protective function for the developing seedling, but also provides photosynthate (Jackson 2009). Nardmann et al. (2004) show that in maize the *narrowsheath* genes are expressed in the margins of the coleoptile as it begins to encircle the nascent meristem. These genes are expressed in the margins of all leaf-like organs in the plant.

During embryonic development, a vascular trace differentiates to the scutellum and one to the coleoptile. In some species, these two traces are separated by a clear internode, known as the mesocotyl internode. Brown (1960) felt that the mesocotyl had no homology to organs in other monocots, but that it might be similar to the hypocotyl in dicots; this makes little sense if the scutellum is the cotyledon, since a hypocotyl would have to be below the cotyledon, whereas the mesocotyl is clearly above the scutellum.

After formation of the coleoptile, the apical meristem produces one or more true leaves before seed maturation. The epiblast, which is so prominent particularly in pooid grasses, might correspond to a leaf, but no data exist to

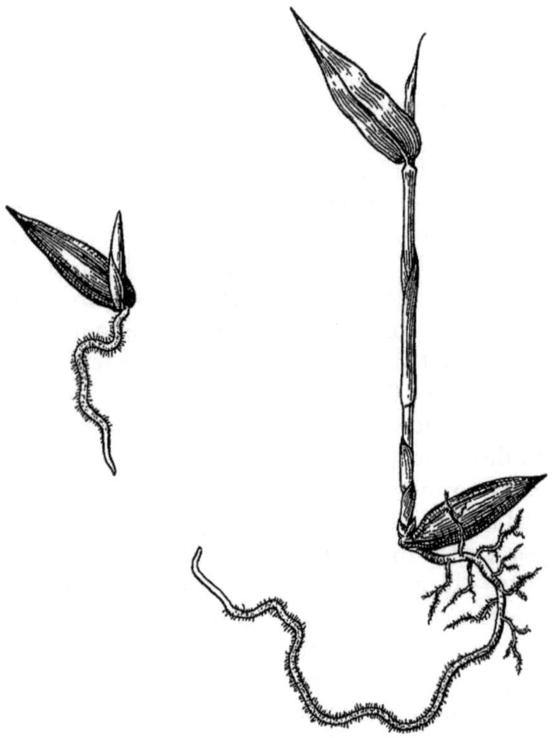

Fig. 6. *Bambusa multiplex*. Seedling in two stages of development: *left* fruit with radicle and plumule just beginning rapid elongation; *right* fruit with attached young plant showing branching root, and culm shoot with first foliage leaf. (From McClure 1966)

test this hypothesis. In *Oryza sativa* and *Poa pratensis*, the first leaf formed after the coleoptile lacks a blade, and only one more embryonic leaf is formed (Sylvester et al. 2001). This presents an interesting comparison with members of the Bambusoideae, in which the first seedling leaf lacks a blade, and the first few internodes of the seedling often fail to elongate (Soderstrom 1981) (Fig. 6). It is possible that the first leaf after the coleoptile is bladeless in all members of the BEP clade; in contrast, in *Zea mays* (PACMAD clade) the blade is present but reduced. *Oryza* and *Poa* develop only one more embryonic leaf, whereas *Zea mays* produces four more leaves in the embryo.

The embryo is lateral within the developing seed, but the mature embryo structure with its highly differentiated tissues and organs is a synapomorphy for the family. In most other monocots, including members of Poales, the mature embryo is a small cluster of cells, lacking any obvious structures. In contrast, in the grass embryo both root and shoot apical meristems are fully organized before seed maturation; vascular tissue is visible. In evolutionary time, a heterochronic shift must have occurred such that embryo development was accelerated relative to seed maturation, or seed maturation was delayed relative to embryo development (Kellogg 2000).

Embryo characters have been used to help determine relationships among grasses and were among the most consistent in an early cladistic study (Kellogg and Campbell 1987). In particular, the scutellar cleft may be present or absent, an epiblast may be present or absent, the mesocotylar internode may be short or long, and the embryonic leaf margins may meet or overlap (Reeder 1957) (Fig. 7). However, the latter character in particular varies within many higher taxa and may be less diagnostic than previously believed. A long mesocotyl internode appears to be the sole morphological synapomorphy for the large PACMAD clade (GPWG 2001).

Endosperm

Endosperm formation has been studied extensively because of its agronomic importance, but the focus has been on only a handful of taxa. Most data come from maize, with additional observations from wheat, barley, rice, sorghum, and more recently *Brachypodium*. (An additional problem is that endosperm development in wheat is generally presented in transverse section, whereas that in maize is presented in longitudinal section; this makes drawings and photomicrographs difficult to compare.) Details of endosperm development differ, but it is unclear whether the differences are specific for entire taxonomic groups or whether they are unique to the few species studied. Given the high fitness consequences of provisioning a seed, selection might have shaped endosperm development in different ways in different lineages. In describing endosperm below, I hope to encourage more comparative studies that might illuminate the consistency of these characters across many taxonomic groups.

In all grasses that have been studied, the central cell of the megagametophyte contains two nuclei, which are unfused until fertilization

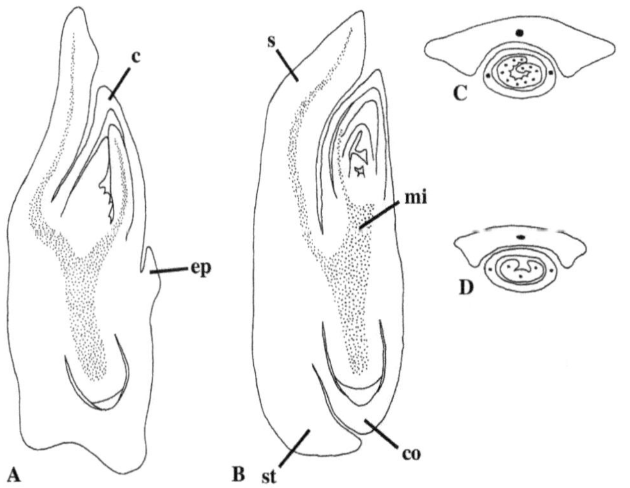

Fig. 7. Embryos. **A** Longitudinal section of a pooid embryo showing presence of an epiblast (*ep*) and coleoptile (*c*). **B** Longitudinal section of a panicoid embryo showing presence of a scutellum (*s*), a scutellar tail (*st*), and elongated mesocotyl internode (*mi*). **C** Cross section of a panicoid embryo apex showing overlapping embryonic leaf margins. **D** Cross section of a pooid embryo apex showing embryonic leaf margins that meet. (From GPWG 2001, p. 402, with permission from Missouri Botanical Garden Press; drawn by M. Kojima)

(Jones and Rost 1989). One of the two sperm fertilizes the central cell to form a triploid nucleus, which then begins to divide within hours, undergoing several rounds of division before the zygote divides at all (Artschwager and McGuire 1949; Brown and Lemmon 2007; Percival 1921). The nuclei divide initially without cytokinesis; the preprophase band does not form, nor do cell walls (Brown and Lemmon 2007; Opanowicz et al. 2011; Sabelli and Larkins 2009a; Leroux et al. 2014). The endosperm nuclei migrate to the chalazal end of the embryo sac, and a large central vacuole develops so that the nuclei are pushed toward the periphery of the endosperm cell (Opanowicz et al. 2011; Weatherwax 1930; Leroux et al. 2014). After nuclear proliferation, microtubules form, radiating from the surface of each nucleus, and cell walls are laid down between the nuclei in the peripheral layer. Anticlinal walls are then formed between the next layer of nuclei to the inside, creating a sort of three-sided box or alveolus surrounding each nucleus. The nucleus then divides and undergoes cytokinesis toward the center of the cell, thus moving the layer of partially enclosed nuclei inward by one. By about 3 to 6 days after pollination in the cereals, the entire central cavity is fully cellularized. Some aspects of developmental timing differ between *Brachypodium* and wheat and may be worth exploring in other pooid grasses (Opanowicz et al. 2011). Among other characteristics, the nucellar epidermis in wheat breaks down while the aleurone differentiates (Percival 1921), whereas the nucellar epidermis is persistent in *Brachypodium*, as it is in rice (Ellis and Chaffey 1987).

At the same time that the endosperm is developing, the nucellus breaks down. Soon after pollination, the cells of the nucellus undergo programmed cell death; the space that the cells had occupied is filled by the developing endosperm (Nguyen et al. 2007).

As cellularization proceeds, the endosperm differentiates into four distinct tissues, the aleurone, the starchy endosperm, transfer cells, and the layer surrounding the embryo, called the embryo-surrounding region or ESR (Sabelli and Larkins 2009a; Leroux et al. 2014). The latter is evanescent, becoming recognizable as soon as cellularization occurs, but then breaking down (Cossegal et al. 2007; Opsahl-Ferstad et al. 1997; Weatherwax 1930). The cells in the ESR appear to be metabolically active and may be involved in transfer of sucrose to the developing embryo, although there are no symplastic connections between the two. In maize the ESR cells express a specific set of genes (the ESR genes) that may be involved in signaling and/or pathogen response (Opsahl-Ferstad et al. 1997). Unfortunately for comparative biologists, the ESR has been described in detail only in maize. Cytologically similar sets of cells have been reported in wheat (Smart and O'Brien 1983), and were apparently observed by Percival (1921), and barley (Brown et al. 1994), but not characterized in detail. However, proteins that are specific to the ESR in maize are apparently not present in rice, which raises questions about the generality of this cell type.

The outer layer of endosperm cells (in the grasses as well as other angiosperms) is known as the aleurone layer. These cells are morphologically different from other endosperm cells and are generally cuboidal. The thickness of the aleurone varies among taxa, although it has not been surveyed in any systematic fashion; maize and wheat have aleurone that is only one cell layer thick whereas there are three layers in barley; the thickness in rice and *Brachypodium* is variable (Ellis and Chaffey 1987; Opanowicz et al. 2011; Sabelli and Larkins 2009a). The aleurone responds to gibberellin secreted by the developing embryo, by synthesizing enzymes that mobilize starch in the endosperm. Humans have taken advantage of this process to create malted barley, the starting point for beer and whiskey.

At germination the aleurone secretes enzymes that break down the macromolecules in the starchy endosperm. As germination proceeds however, the aleurone undergoes programmed cell death (Nguyen et al. 2007).

The vascular strand on the adaxial side of the ovary projects into the nucellus in some taxa and is surrounded by a set of cells known as the nucellar projection. The aleurone surrounding the nucellar projection is sometimes modified cytologically. In Triticeae in particular, the endosperm expands on either side of the vascular strand and nucellar projection so that the latter structures end up at the bottom of a deep groove known as the crease. In contrast to the Triticeae, the vascular strand of the ovary is quite short in *Cenchrus americanus* (=*Pennisetum americanum*, =*P. glaucum*) (Fussell and Dwarte 1980), and scarcely extends past the hilum.

Transfer cells differentiate in the region of the endosperm adjacent to the placenta (Artschwager and McGuire 1949; Charlton et al. 1995; Fussell and Dwarte 1980; Jones and Rost 1989; Rost et al. 1984; Weatherwax 1930). They have highly invaginated cell walls and correspondingly extensive plasma membranes (Schiel et al. 1984); this increased surface area promotes transfer of carbohydrates and amino acids from the maternal tissue (Sabelli and Larkins 2009a). Although Rost et al. (1990) did not find transfer cells in the fruit of *Briza maxima*, it is possible that the stage of development was too late to observe them.

It is no exaggeration to say that cereal endosperm feeds the world, and that the cells of the starchy endosperm are the center of it. Soon after cellularization is complete, the cells of the starchy endosperm undergo remarkable alterations of the normal cell cycle, in which the chromosome complement of each nucleus undergoes endoreduplication, creating cells with as much as 96 times the haploid amount of DNA (Kowles and Phillips 1985; Nguyen et al. 2007; Sabelli and Larkins 2009a, 2009b). The ploidy levels of the starchy endosperm vary between crops, with lower ploidy in wheat than in other species that have been studied (Nguyen et al. 2007). Endoreduplication begins in the center of the endosperm and extends gradually to the periphery, a process that in maize takes about a week. Endoreduplication generally does not extend to the aleurone, which remains diploid, except in barley (Keown et al. 1977). During this process of endoreduplication, the cells begin to accumulate starch and proteins.

Starch granules accumulate in amyloplasts, under the control of starch synthases, starch branching enzymes and starch de-branching enzymes. A single amyloplast may produce one or several starch granules; if several are produced the granules will coalesce at maturity and appear compound, whereas a single large granule will appear simple. Species with simple starch granules are found in Panicoideae s.s. and some of the early-diverging panicoid tribes (Sánchez-Ken and Clark 2010; Tateoka 1962), whereas those of other subfamilies tend to be compound. Starch granules are bimodal in size in Triticeae (Tomlinson et al. 2003), whereas they are uniform and small in *Brachypodium* (Opanowicz et al. 2011).

Starch biosynthesis involves a substantial number of grass-specific proteins, probably contributing to the high starch content in the endosperm in the family. Genes encoding major enzymes of the starch biosynthetic pathway were all duplicated as part of the whole genome duplication that occurred coincident with the origin of the family (Comparot-Moss and Denyer 2009; Li et al. 2012). Following duplication, the genes diverged in function, and exhibited an accelerated rate of evolution with positive selection observed on a number of amino acids in the proteins. For example, the rate-limiting enzyme in starch biosynthesis is ADP-glucose pyrophosphorylase. This enzyme has distinct forms that occur in the cytosol and plastids of plants; whereas the plastid form is common in most starch storage organs of most plants, the cytosolic form is apparently unique to grasses (Comparot-Moss and Denyer 2009) and is expressed predominantly in the endosperm (Tetlow et al. 2004). A protein kinase known as SnRK1b also appears to be restricted to cereals, and is important in the process of sucrose transport and storage (Jain et al. 2008). However, other Poales have not been studied.

In addition to starch, the endosperm includes proteins, first discovered nearly three centuries ago because they did not dissolve in water (Beccari 1745). The endosperm proteins that Beccari discovered are the hydrophobic prolamins that are soluble in alcohol. We now know that there are also hydrophilic globulins that are soluble in water. Understanding the evolution of the proteins is complicated by the nomenclature, in that prolamins tend to be named for the species in which they are found (e.g., zeins in maize, kaffirins in sorghum, hordeins in barley, etc.) rather than for their phylogenetic relationships. Not only are the proteins variable among the major cereals, but there is also variation in where they are stored. In all grasses, endosperm protein accumulates in the rough endoplasmic reticulum; in Triticeae, at least, it is subsequently moved to storage vacuoles (Sabelli and Larkins 2009a).

As the processes of endoreduplication and starch and protein accumulation end, the cells of the starchy endosperm undergo programmed cell death. By the time the seed is mature the only living endosperm cells are in the aleurone (Nguyen et al. 2007). Cell death in maize begins near the center of the endosperm and also at the top of the ovary and extends outward from these two foci. In wheat, however, the first cells to die are scattered throughout the endosperm.

The consistency of the endosperm has been studied for its possible taxonomic importance. Liquid endosperm has been reported in members of subfamily Pooideae by Terrell (1971); according to Davis (1966), "... in *Helictotrichon*, *Koeleria*, *Sphenopholis* and *Trisetum* [the endosperm] remains liquid for years." The occurrence of liquid endosperm is sporadic and does not seem to follow phylogenetic lines. However, this study relied on a subjective assessment of endosperm condition; it would be interesting to assess the protein and lipid composition of the endosperm in a more rigorous way now that appropriate analytical tools are available for doing this. Liquid endosperm is also reported in the bamboo *Melocanna* (Ramanayake and Weerawardene 2003), but its chemical composition was not investigated.

References

Anton, A.M., Cocucci, A.E. 1984. The grass megagametophyte and its possible phylogenetic implications. Plant Syst. Evol. 146: 117–121.

Anton de Triquell, A. 1987. Grass gametophytes: their origin, structure, and relation with the sporophyte. In: Soderstrom, T.R., Hilu, K.W., Campbell, C.S., Barkworth, M.E. (eds.) Grass systematics and evolution. Washington, DC: Smithsonian Institution Press. pp. 11–20

Artschwager, E., McGuire, R.C. 1949. Cytology of reproduction in *Sorghum vulgare*. J. Agric. Res. 78: 659–673.

Aulbach-Smith, C.A., Herr, Jr., J.M. 1984. Development of the ovule and female gametophyte in *Eustachys petraea* and *E. glauca* (Poaceae). Amer. J. Bot. 71: 427–438.

Beccari, J.B. 1745. De Frumento. De Bononiensi Scientiarum et Artium Instituto atque Academia Commentarii II. Part I: 122–127.

Bhanwra, R.K. 1988. Embryology in relation to systematics of Gramineae. Ann. Bot. 62: 215–233.

Bhanwra, R.K., Kaur, N., Kaur, N., Garg, A. 1991. Embryological studies in some grasses and their taxonomic significance. Bot. J. Linn. Soc. 107: 405–419.

Bhanwra, R.K., Sharma, M.L., Vij, S.P. 2001. Comparative embryology of *Bambusa tulda* Roxb. and *Thyrsostachys siamensis* Gamble (Poaceae: Bambusoideae). Bot. J. Linn. Soc. 135: 113–124.

Boyd, L. 1931. Evolution in the monocotyledonous seedling, a new interpretation of the grass embryo. Trans. Bot. Soc. Edinburgh 30: 286–302.

Brown, W.V. 1960. The morphology of the grass embryo. Phytomorphology 10: 215–223.

Brown, R.C., Lemmon, B.E. 2007. The developmental biology of cereal endosperm. In: Olsen, O.-A. (ed.) Endosperm. Berlin: Springer. pp. 1–20

Brown, R.C., Lemmon, B.E., Olsen, O.-A. 1994. Endosperm development in barley: microtubule involvement in the morphogenetic pathway. Plant Cell 6: 1241–1252.

Campbell, C.S., Kellogg, E.A. 1987. Sister group relationships of the Poaceae. In: Soderstrom, T.R., Hilu, K.

W., Campbell, C.S., Barkworth, M.E. (eds.) Grass systematics and evolution. Washington, DC: Smithsonian Institution. pp. 217–224

Charlton, W.L., Keen, C.L., Merriman, C., Lynch, P., Greenland, A.J., Dickinson, H.G. 1995. Endosperm development in *Zea mays*; implication of gametic imprinting and paternal excess in regulation of transfer layer development. Development 121: 3089-3097.

Choda, S.P., Mitter, H., Bhanwra, R.K. 1982. Embryological studies in three species of *Cymbopogon* Spreng. (Poaceae). Proc. Indian Acad. Sci. Plant Sci. 91: 55–60.

Comparot-Moss, S., Denyer, K. 2009. The evolution of the starch biosynthetic pathway in cereals and other grasses. J. Exp. Bot. 60: 2481–2492.

Cossegal, M., Vernoud, V., Depège, N., Rogowsky, P.M. 2007. The embryo surrounding region. In: Olsen, O.-A. (ed.) Endosperm. Berlin: Springer. pp. 57–71

Davis, G.L. 1966. Systematic embryology of the angiosperms. New York: John Wiley & Sons.

Ellis, J.R., Chaffey, N.J. 1987. Structural differentiation of the nucellar epidermis in the caryopsis of rice (*Oryza sativa*). Ann. Bot. 60: 671–675.

Evans, M.M.S., Grossniklaus, U. 2009. The maize megagametophyte. In: Bennetzen, J.L., Hake, S.C. (eds.) Handbook of maize: its biology. Heidelberg: Springer. pp. 79–104

Febulaus, G.N.V., Pullaiah, T. 1994. Embryology of *Aristida* (Poaceae). Taiwania 38: 38–48.

Fussell, L.K., Dwarte, D.M. 1980. Structural changes of the grain associated with black region formation in *Pennisetum americanum*. J. Exp. Bot. 31: 645–654.

GPWG 2001. Grass Phylogeny Working Group. Phylogeny and subfamilial classification of the Poaceae. Ann. Missouri Bot. Gard. 88: 373–457.

Jackson, D. 2009. Vegetative shoot meristems. In: Bennetzen, J.L., Hake, S.C. (eds.) Handbook of maize: Its biology. Heidelberg: Springer.

Jain, M., Li, Q.-B., Chourey, P.S. 2008. Cloning and expression analyses of *sucrose non-fermenting-1-related kinase 1* (*SnRK1b*) gene during development of sorghum and maize endosperm and its implicated role in sugar-to-starch metabolic transition. Phys. Plant. 134: 161–173.

Jones, T.J., Rost, T.L. 1989. Histochemistry and ultrastructure of rice (*Oryza sativa*) zygotic embryogenesis. Amer. J. Bot. 76: 504–520.

Kam, Y.K., Maze, J. 1974. Studies on the relationships and evolution of supraspecific taxa utilizing developmental data. II. Relationships and evolution of *Oryzopsis hymenoides*, *O. virescens*, *O. kingii*, *O. micrantha*, and *O. asperifolia*. Bot. Gaz. 135: 227–247.

Kellogg, E.A. 2000. The grasses: a case study in macroevolution. Ann. Rev. Ecol. Syst. 31: 217–238.

Kellogg, E.A., Campbell, C.S. 1987. Phylogenetic analyses of the Gramineae. In: Soderstrom, T.R., Hilu, K.W., Campbell, C.S., Barkworth, M.E. (eds.) Grass systematics and evolution. Washington, DC: Smithsonian Institution Press. pp. 310–322

Keown, A.C., Taiz, L., Jones, R.L. 1977. The nuclear content of developing barley aleurone cells. Amer. J. Bot. 64: 1248–1253.

Kowles, R.V., Phillips, R.L. 1985. DNA amplification patterns in maize endosperm nuclei during kernel development. Proc. Natl. Acad. Sci. USA 82: 7010–7014.

Le Roux, L.G., Kellogg, E.A. 1999. Floral development and the formation of unisexual spikelets in the Andropogoneae (Poaceae). Amer. J. Bot. 86: 354–366.

Leroux, B.M., Goodyke, A.J., Schumcher, K.I., Abbott, C. P., Clore, A.M., Yadegari, R., Larkins, B.A., Dannenhoffer, J.M. 2014. Maize early endosperm growth and development: from fertilization through cell type differentiation. Amer. J. Bot. 101: 1259–1274.

Li, C., Li, Q.-G., Dunwell, J.M., Zhang, Y.-M. 2012. Divergent evolutionary pattern of starch biosynthetic pathway genes in grass and dicots. Mol. Biol. Evol. 29: 3227–3236.

Linder, H.P., Kellogg, E.A. 1995. Phylogenetic patterns in the commelinid clade. In: Rudall, P.J., Cribb, P.J., Cutler, D.F., Humphries, C.J. (eds.) Monocotyledons: systematics and evolution. Kew: Royal Botanic Gardens. pp. 473–496

Maze, J., Bohm, L.R. 1973. Comparative embryology of *Stipa elmeri* (Gramineae). Can. J. Bot. 51: 235–247.

McClure, F.A. 1966. The bamboos: a fresh perspective. Cambridge, Massachusetts: Harvard University Press.

Nardmann, J., Werr, W. 2009. Patterning of the maize embryo and the perspective of evolutionary developmental biology. In: Bennetzen, J.L., Hake, S.C. (eds.) Handbook of maize: Its biology. Heidelberg: Springer. pp. 105–119

Nardmann, J., Ji, J., Werr, W., Scanlon, M.J. 2004. The maize duplicate genes *narrow sheath1* and *narrow sheath2* encode a conserved homeobox gene function in a lateral domain of shoot apical meristems. Development 131: 2827–2839.

Nguyen, H.N., Sabelli, P.A., Larkins, B.A. 2007. Endoreduplication and programmed cell death in the cereal endosperm. In: Olsen, O.-A. (ed.) Endosperm. Berlin: Springer. pp. 21–43

Opanowicz, M., Hands, P., Betts, D., Parker, M.L., Toole, G.A., Mills, E.N.C., Doonan, J.H., Drea, S. 2011. Endosperm development in *Brachypodium distachyon*. J. Exp. Bot. 62: 735–748.

Opsahl-Ferstad, H.G., Le Deunff, E., Dumas, C., Rogowsky, P.M. 1997. *ZmEsr*, a novel endosperm-specific gene expressed in a restricted region around the maize embryo. Plant J. 12: 235–246.

Percival, J. 1921. The wheat plant: a monograph. New York: E. F. Dutton & Co.

Philipson, M.N., Connor, H.E. 1984. Haustorial synergids in danthonioid grasses. Bot. Gaz. 145: 78–82.

Ramanayake, S.M.S.D., Weerawardene, T.E. 2003. Flowering in a bamboo, *Melocanna baccifera* (Bambusoideae: Poaceae). Bot. J. Linn. Soc. 143: 287–291.

Reeder, J.R. 1957. The embryo in grass systematics. Amer. J. Bot. 44: 756–768.

Rost, T.L., Izaguirre de Artucio, P., Risley, E.B. 1984. Transfer cells in the placental pad and caryopsis coat of *Pappophorum subbulbosum* Arech. (Poaceae). Amer. J. Bot. 71: 948–957.

Rost, T.L., Izaguirre de Artucio, P., Risley, E.B. 1990. Anatomy of the caryopsis of *Briza maxima* (Poaceae). Amer. J. Bot. 77: 69–76.

Sabelli, P.A., Larkins, B.A. 2009a. The development of endosperm in grasses. Plant Phys. 149: 14–26.

Sabelli, P.A., Larkins, B.A. 2009b. The contribution of cell cycle regulation to endosperm development. Sex. Plant Reprod. 22: 207–219.

Sajo, M.G., Longhi-Wagner, H., Rudall, P.J. 2007. Floral development and embryology in the early-divergent grass *Pharus*. Int. J. Plant Sci. 168: 181–191.

Sajo, M.G., Longhi-Wagner, H.M., Rudall, P.J. 2008. Reproductive morphology of the early-divergent grass *Streptochaeta* and its bearing on the homologies of the grass spikelet. Plant Syst. Evol. 275: 245–255.

Sajo, M.G., Pabón-Mora, N., Jardim, J., Stevenson, D.W., Rudall, P.J. 2012. Homologies of the flower and inflorescence in the early-divergent grass *Anomochloa* (Poaceae). Amer. J. Bot. 99: 614–628.

Sánchez-Ken, J.G., Clark, L.G. 2010. Phylogeny and a new tribal classification of the Panicoideae s.l. (Poaceae) based on plastid and nuclear sequence data and structural data. Amer. J. Bot. 97: 1732–1748.

Schiel, J.H.N., Kieft, H., Van Lammeren, A.A.M. 1984. Interactions between embryo and endosperm during early developmental stages of maize caryopses (*Zea mays*). Can. J. Bot. 62: 2842–2853.

Shadowsky, A.E. 1926. Der antipodale Apparat bei Gramineen. Flora 120: 344–370.

Smart, M.G., O'Brien, T.P. 1983. The development of the wheat embryo in relation to the neighbouring tissues. Protoplasma 114: 1–13.

Soderstrom, T.R. 1981. Some evolutionary trends in the Bambusoideae (Poaceae). Ann. Missouri Bot. Gard. 68: 15–47.

Stover, E.L. 1937. The embryo sac of *Eragrostis cilianensis* (All.) Link. Ohio J. Sci. 37: 172–184.

Sylvester, A.W., Parker-Clark, V., Murray, G.A. 2001. Leaf shape and anatomy as indicators of phase change in the grasses: comparison of maize, rice and bluegrass. Amer. J. Bot. 88: 2157–2167.

Takacs, E.M., Li, J., Du, C., Ponnala, L., Janick-Buckner, D., Yu, J., Muehlbauer, G.J., Schnable, P.S., Timmermans, M.C.P., Sun, Q., Nettleton, D., Scanlon, M.J. 2012. Ontogeny of the maize shoot apical meristem. Plant Cell 24: 3219–3234.

Tateoka, T. 1962. Starch grains of endosperm in grass systematics. Bot. Mag. (Tokyo) 75: 336–343.

Taylor, M.G., Vasil, I.K. 1995. The ultrastructure of zygotic embryo development in pearl millet (*Pennisetum glaucum*: Poaceae). Amer. J. Bot. 82: 205–219.

Terrell, E.E. 1971. Survey of occurrences of liquid or soft endosperm in grass genera. Bull. Torrey Bot. Club 98: 264–268.

Tetlow, I.J., Morell, M.K., Emes, M.J. 2004. Recent developments in understanding the regulation of starch metabolism in higher plants. J. Exp. Bot. 55: 2131–2145.

Tomlinson, K., Denyer, K., Callow, J.A. 2003. Starch synthesis in cereal grains. In: Callow, J.A. (ed.) Adv. Bot. Res. London: Academic Press. pp. 1-61

Verboom, G.A., Linder, H.P., Barker, N.P. 1994. Haustorial synergids: An important character in the systematics of danthonioid grasses (Arundinoideae: Poaceae). Amer. J. Bot. 81: 1601–1610.

Weatherwax, P. 1920. Position of the scutellum and homology of coleoptile in maize. Bot. Gaz. 69: 179–182.

Weatherwax, P. 1930. The endosperm of *Zea* and *Coix*. Amer. J. Bot. 17: 371–380.

Wu, C.-C., Diggle, P.K., Friedman, W.E. 2011. Female gametophyte development and double fertilization in Balsas teosinte, *Zea mays* subsp. *parviglumis* (Poaceae). Sex. Plant Reprod. 24: 219–229.

Yamaura, A. 1933. Karyologische und Embryologische Studien über einige Bambus-Arten. Bot. Mag. (Tokyo) 47: 551–555.

Zaitchik, B.F., le Roux, L.G., Kellogg, E.A. 2000. Development of male flowers in *Zizania aquatica* (North American wild-rice; Gramineae). Int. J. Plant Sci. 161: 345–351.

Karyology and Genome Structure

Chromosome Number

The grasses, and in particular the cereals, have long been investigated as models for studies of chromosome number, morphology, dynamics, and evolution (e.g., Avdulov 1931; Stebbins 1956). The number of chromosomes varies from $2n = 4$ in species of *Zingeria* and *Colpodium* (Kim et al. 2009) to $2n =$ ca. 266 in *Poa litorosa* (Hair and Beuzenberg 1961). Based on a parsimony optimization, the chromosome base number of $x = 12$ is synapomorphic for the spikelet clade, i.e., all grasses except Anomochlooideae (GPWG 2001).

The chromosome number for the common ancestor of the family is uncertain. In the outgroups, $n = 18$ for *Joinvillea plicata* (Newell 1969) and $2n =$ ca. 38 (i.e., $n = 19$) for *Ecdeiocolea monostachya* (Hanson et al. 2005). The haploid number in *Anomochloa* is $n = 18$ (Judziewicz and Soderstrom 1989); this number combined with those of the immediate outgroups would suggest that the common ancestor also had $n = 18$. However, *Streptochaeta* is $n = 11$ (Judziewicz and Soderstrom 1989). Using methods of comparative genomics, estimates of the ancestral chromosome base number are $x = 10$ (Salse et al. 2008), or $x = 7$ (Devos 2010); these were postulated to have undergone allotetraploidization followed by rearrangements, leading ultimately to an ancestor with $x = 12$. This estimate converges on that suggested by several authors (Avdulov 1931; GPWG 2001; Raven 1975). Stebbins (1985) suggested a base number of 5, 6 or 7, Mehra et al. (1968) and Sharma (1979) argued for $x = 6$, and Flovik (1938) suggested $x = 5$. If a resolution is possible, it will only be provided by genomic data from Anomochlooideae, Pharoideae, and the grass sister taxa.

Base chromosome number is fairly stable in many clades of grasses and for some is the shared character that provides evidence for monophyly. The core Pooideae (Poeae + Triticeae + Bromeae) are marked by having their genes arranged in seven large chromosomes (GPWG 2001; Kellogg and Campbell 1987). The Panicoideae are divided into three major clades, corresponding to chromosome base numbers of $x = 10$ (Andropogoneae and Paspaleae) and $x = 9$ (Paniceae s. s.) (Giussani et al. 2001). A base number of $x = 11$ is synapomorphic for Olyreae (Bambusoideae) and is independently derived and synapomorphic for Aristidoideae (GPWG 2001). Despite this overall pattern of stability, base chromosome number does vary within some genera; examples include *Phalaris* ($x = 6$ or 7; Baldini 1995), *Brachypodium* ($x = 5$, 7, 8, 9; Catalán et al. 2012), and *Cenchrus* (*Pennisetum*) ($x = 5$, 7, 8, 10, 17; Chemisquy et al. 2010, among many others).

B chromosomes have been reported for over 200 species of grasses (Hunziker and Stebbins 1987). These are supernumerary chromosomes, derived from the normal ("A") chromosome complement; they tend to accumulate in the male gametes by non-disjunction at meiosis (Jones and Rees 1982). B chromosomes have been studied in particular detail in maize (Birchler and Han 2009; Jones et al. 2008) and *Secale* (Jones and Rees 1967, 1982), among others. Many aneuploid chromosome numbers reported in the literature may be caused by mistaking B chromosomes for A chromosomes.

E.A. Kellogg, *Flowering Plants. Monocots*, The Families and Genera of Vascular Plants 13,
DOI 10.1007/978-3-319-15332-2_5, © Springer International Publishing Switzerland 2015

Changes in chromosome number reflect rearrangements of existing chromosomes that reduce or increase the number of centromeres. Centromeres are generally stable within a species, but neocentromeres form relatively easily, and conversely former centromeres can be lost when chromosomes combine. The mechanisms by which this happens are only beginning to be understood (Birchler and Han 2009). At the DNA level, a centromere consists of a short repetitive sequence, interspersed with retrotransposons and occasional transcribed genes (Fan et al. 2011). The repetitive sequence is generally species-specific and may affect the viability of hybrids, but it has not been explored in a systematic context.

One apparently common mechanism of reduction in chromosome number has been documented in *Aegilops tauschii* (Triticeae, Pooideae), in which genome architecture has been compared to that of rice (Luo et al. 2009). *Aegilops* chromosome 1 is collinear with rice chromosomes 5 and 10. The genes found on rice 10 appear in the middle of the *Aegilops* chromosome, whereas the genes of one arm of rice 5 appear to have been attached at one end of rice chromosome 10, with the other arm of rice 5 attached at the other end of rice chromosome 10. It thus appears that an entire chromosome (rice 10) has been inserted into the centromeric region of another (rice 5). The gene content and order of the inserted chromosome is maintained, and the two individual arms of the chromosome into which it is inserted become the distal ends of the new chromosome. The centromere of the inserted chromosome remains functional, whereas the centromere into which it is inserted is inactivated.

Segmental duplications appear to be less common, but have been documented in members of the genus *Zizania* (Oryzeae, Ehrhartoideae), which have $2n = 30$ or 34. A genome map of *Zizania palustris* shows that the base number of $x = 15$ reflects duplication of portions of rice chromosomes 1, 4, and 9 (Kennard et al. 2000). These duplicate regions must have somehow acquired functional centromeres, but the mechanism involved is unknown.

Because of general interest in the ancestry of cultivated maize, there is considerable literature on chromosomal evolution in its relatives in the tribe Andropogoneae. The ancestral base chromosome number (n) for the tribe is most likely 10 (Spangler et al. 1999; Wilson et al. 1999). One popular idea is that the ancestor of the tribe had a $2n$ number of 10, and thus a haploid number of five (Celarier 1956; Garber 1950). This idea is supported by the handful of species in Andropogoneae that exhibit $n = 5$ (*Coix aquatica*, and members of *Sorghum* subg. *Parasorghum* (=*Sarga*)); however, these taxa are clearly derived in the phylogeny (Spangler et al. 1999), pointing to a secondary reduction in chromosome number. Garber (1950) and Price et al. (2005) found that chromosomes of the $x = 5$ species of *Sorghum* were considerably larger than those of the $x = 10$ taxa; the $x = 5$ sorghums also have a higher 2C DNA content than the $x = 10$ species (Price et al. 2005). These observations raise the possibility that $x = 5$ represents an intriguing chromosomal fusion event, rather than preservation of the ancestral base chromosome number. One can imagine that a burst in retrotransposon activity could have led simultaneously to the abrupt increase in genome size seen in this group, as well as the genome rearrangements that led to five large rather than 10 small chromosomes.

Polyploidy

Allopolyploidy is common in Poaceae, leading Hunziker and Stebbins (1987) to claim that "Gramineae is the only large family in which high frequency of polyploids prevails throughout the family." One of the earliest allopolyploids to be studied in detail was wheat (Kihara 1982). Pasta wheat (*Triticum durum*) is an allotetraploid derived from the diploid *Triticum urartu* and a species of goatgrass (probably similar to *Aegilops speltoides*) (see Domestication). This tetraploid then crossed with another diploid species of goatgrass (*Aegilops tauschii*) to produce a sterile hybrid. In experimental hybrids of the two species, non-reduced gametes are formed at high frequencies in both the male and female gametophytes, so that self-pollination can lead to fertile hexaploids (Matsuoka 2011).

Other similar cases have been demonstrated. For example, the polyploid species of *Elymus* are clearly derived from species of *Pseudoroegneria* and *Hordeum*, as shown by cytogenetic studies (Dewey 1984) and molecular sequence data (Kellogg et al. 1996; Mason-Gamer et al. 2002, 2010a,

2010b). The woody bamboos are all polyploid, with most species being tetraploid (Calderón and Soderstrom 1973; Peng et al. 2013; Stebbins 1971). Switchgrass (*Panicum virgatum*) is an allotetraploid, and common millet (*P. miliaceum*) is the result of an independent polyploidization event (Triplett et al. 2012). The genus *Zea* is an ancient allotetraploid (Blanc and Wolfe 2004; Schnable et al. 2009; Swigonova et al. 2004). Sugar cane is a complex polyploid, with an uncertain base chromosome number, although phylogenetic data suggest that the base number might be 10 (Kellogg 2012).

Within every group of Poaceae, it is possible to identify well-supported clades that are clearly similar morphologically but within which hybridization and polyploidy are rampant ("... big messy geographically conditioned splodges of divergence as yet unsanitized by extinction ..."; S. Mayo, pers. comm., referring to *Anthurium*, but could be any of several groups of grasses). For example, in the tribe Paniceae, the *Cenchrus* clade and most likely the *Setaria* clade are easily distinguished, but within them relationships are complex, polyploids are common, and morphology is highly homoplasious. Other such groups include the Triticeae in subfamily Pooideae, and the genus *Oryza* in subfamily Ehrhartoideae. Within the tribe Poeae, it is likely that subtribes Airinae, Poinae, and Loliinae also represent such lineages within which reticulate evolution is common, but between which hybridization is rare.

Polyploidy does not correlate with domestication in angiosperms (Hilu 1993), nor is there an obvious correlation in the grasses. Of the "Big Three" - maize, wheat and rice - the first two are polyploid, but rice is diploid. Common millet and finger millet are tetraploids but pearl millet and foxtail millet are diploid. Oats are hexaploid but barley and rye are diploid. Sugar cane is polyploid but sorghum is diploid. It thus seems unlikely that polyploidy per se confers any particular advantage for domestication.

Genome Size

The sizes of grass genomes vary considerably (Caetano-Anolles 2005; Leitch et al. 2010), with the smallest diploid genomes reported for *Oropetium thomaeum* (haploid size (1C) = 0.25 pg DNA or ca. 205 Mbp) and the largest for *Secale cereale* (8.28 pg DNA or ca. 6790 Mbp); polyploid genomes are of course commensurately larger. The ancestral size of a 1C genome in the family is estimated to be 1.5–2.6 pg DNA (Caetano-Anolles 2005), although this estimate was published before the estimation of genome size of the outgroup *Ecdeiocolea monostachya* at 1 pg of DNA (Hanson et al. 2005); this small value might make the ancestral grass estimate slightly lower. Phylogenetic analyses have identified a major increase in genome size in the Pooideae, with a particularly dramatic increase in the Triticeae (Bennetzen and Kellogg 1997; Kellogg and Bennetzen 2004). The protein-coding genes in the grass genomes are largely in the same order, and variation in genome size thus reflects variation in the space between genes. The "space" is occupied by retrotransposons, which are small pieces of DNA that copy themselves into RNA and then are copied back to DNA and reinserted (Devos 2010; SanMiguel et al. 1996). The retrotransposons preferentially reinsert into themselves so they form a nested arrangement; the ends of retrotransposons are identical (Long Terminal Repeats, or LTRs), and can be identified even after multiple copies of the transposon are inserted between them.

Chromosome numbers do not correlate well with genome size. For example, rice has 12 pairs of chromosomes and the haploid DNA content is about 0.5 picograms (pg) of DNA. In contrast, barley has only seven pairs of chromosomes but a haploid DNA content of about 5.55 pg (Kellogg and Bennetzen 2004). Likewise, *Sorghum biolor* has $2n = 20$ and about 0.83 pg of DNA, but the closely related *Sarga angusta* has $2n = 10$ and more than twice as much DNA per 1C nucleus (1.85 pg). A corollary is that two plants could have the same total amount of DNA in their genomes, but it might be divided into quite different numbers of chromosomes.

Although there has been a suggestion in the literature that genome size may correlate with effective population size (Lynch and Conery 2003), this correlation has not been found in *Oryza*, the only grass genus investigated to date (Ai et al. 2012).

Genome Mapping, Sequencing and Conservation of Gene Order

The rice genome was the first of the grasses to be sequenced in its entirety, and contains ca. 389 mega base pairs (Mb) of DNA (International Rice Genome Sequencing Project 2005). Although compact, it still contains a high proportion (35 %) of transposable elements, or their remains. In addition to that of rice, whole genome sequences are now available for a bambusoid grass, *Phyllostachys heterocycla* (Peng et al. 2013), four members of Pooideae, *Brachypodium distachyon* (The International Brachypodium Initiative 2010), *Hordeum vulgare* (The International Barley Genome Sequencing Consortium 2012), *Triticum urartu* (Ling et al. 2013) and *Aegilops tauschii* (Jia et al. 2013), plus four members of Panicoideae, *Setaria italica* and *S. viridis* (Bennetzen et al. 2012), *Sorghum bicolor* (Paterson et al. 2009), and *Zea mays* (Schnable et al. 2009). Although the size and highly repetitive nature of the wheat genome makes it a particularly difficult challenge for whole genome sequencing, details of its structure and sequence are emerging rapidly (Brenchley et al. 2012). With continuing rapid developments in sequencing technology, it is likely that many additional grass genomes will be available even by the time this book is printed.

One of the most stunning results of early genome mapping efforts was the demonstration that many of the genes in these genomes are in the same order; that is, the genomes are largely collinear (Gale and Devos 1998; Moore et al. 1995). For example, a comparison of genome regions homologous to chromosome 1H (i.e., 1 *Hordeum*) in *Hordeum, Triticum, Brachypodium, Oryza*, and sorghum showed that about 83 % of the genes were in the same order (Mayer et al. 2011).

Despite such gross-level collinearity, however, the distribution of blocks of genes over chromosomes varies and periodic large-scale genome rearrangements can be identified, many of which correlate with phylogeny (Kellogg 2001). For example, the three genomes of hexaploid wheat have a number of structural similarities in common that are not shared with barley, whereas gene order in barley is more similar to *Brachypodium* than it is to rice or sorghum (Mayer et al. 2011). Likewise foxtail millet and sorghum each have several species-specific inversions, whereas the two genomes share a set of panicoid-specific inversions and translocations that distinguish them from rice (Bennetzen et al. 2012).

In addition to translocations and inversions, smaller scale rearrangements cause periodic violations of the overall collinearity of the genomes. Genes are moved from their original locations by the action of transposable elements and double stranded break repair (Wicker et al. 2011). Although in most cases these processes simply generate pseudogenes, functional genes do also appear from time to time, leading to species-specific or clade-specific gene duplicates. This is an obvious source of genetic novelty, which could lead to new phenotypes. Some genes that are known to be important for morphological variation fall in non-collinear regions of the genome, hinting at a link between genomic and morphological change. For example, the barley gene *Vrs1*, which controls development and fertility of the lateral spikelets at each node, is in a different genomic position in rice and barley (Pourkheirandish et al. 2006). Barley chromosome 2 is collinear with rice chromosome 4 for all markers surrounding *Vrs1*, but the *Vrs1* orthologue in rice is on chromosome 7, indicating that the position of the gene has changed in evolutionary time.

As genome sequencing becomes increasingly common, it is likely that structural characteristics of genes and genomes will be found to characterize many clades of grasses. It is also likely that genes will be identified underlying each of the morphological synapomorphies.

GC Content

Some grass genomes have a higher frequency of G-C base pairs than do other plants; in other words, the relative proportions of A-T and G-C base pairs are not equal throughout the genome, as might be expected if genome composition were entirely random. The extent of GC bias is most pronounced at the third positions of codons, and this third-position GC bias varies among species; for example *Phyllostachys edulis* is about 47 % GC on average, whereas *Secale cereale* is about 68 %

(Serres-Giardi et al. 2012). Within a genome, some grasses such as sorghum, rice, and switchgrass show a bimodal distribution of genes, with one set having a GC content around 40 % and a second set with a GC content over 80 % (Serres-Giardi et al. 2012). The extent to which the genes fit this bimodal pattern (as measured by heterogeneity of GC content) correlates with the overall GC content at third positions. Although the pattern of variation in GC content might reflect codon usage in highly expressed genes, in which codons ending in G or C are preferred, the data fit better with an explanation having to do with G-biased gene conversion. In this mechanism, recombination between heterozygous regions of the genome is preferentially resolved in favor of G or C. This counteracts the normal mutation bias toward A-T base pairs caused by deamination of cytosine or methylcytosine. GC content does correlate with recombination frequency, as would be expected if the hypothesis of G-biased gene conversion is correct (Serres-Giardi et al. 2012). Also consistent with the hypothesis, high GC genes have a higher mutation rate and provide more opportunity for methylation (Tatarinova et al. 2010).

References

Ai, B., Wang, Z.-S., Ge, S. 2012. Genome size is not correlated with effective population size in the *Oryza* species. Evolution 66: 3302–3310.

Avdulov, N.P. 1931. Kario-sistematicheskoye issledovaniye semeystva zlakov (Karyosystematic studies in the grass family.). Trudy Prikl. Bot. Prilozheniye (Bull. Appl. Bot. Genet. P1.- Breed., Leningrad) 44: 1–428.

Baldini, R.M. 1995. Revision of the genus *Phalaris* L. (Gramineae). Webbia 49: 265–329.

Bennetzen, J.L., Kellogg, E.A. 1997. Do plants have a one-way ticket to genomic obesity? Plant Cell 9: 1509–1514.

Bennetzen, J.L., Schmutz, J., Wang, H., Percifield, R., Hawkins, J., Pontaroli, A.C., Estep, M., Feng, L., Grimwood, J., Jenkins, J., Barry, K., Lindquist, E., Hellsten, U., Deshpande, S., Wang, X., Wu, X., Mitros, T., Triplett, J., Yang, X., Ye, C.-Y., Mauro-Herrara, M., Wang, L., Li, P., Sharma, M., Sharma, R., Ronald, P.C., Panaud, O., Kellogg, E.A., Brutnell, T., Doust, A., Tuskan, G.A., Rokhsar, D., Devos, K.M. 2012. Grass genome structure, evolution and adaptation uncovered by sequence analysis of *Setaria*. Nature Biotech. 30: 555–561.

Birchler, J.A., Han, F. 2009. Maize centromeres: structure, function, epigenetics. Ann. Rev. Genet. 43: 287–303.

Blanc, G., Wolfe, K.H. 2004. Widespread paleopolyploidy in model plant species inferred from age distributions of duplicate genes. Plant Cell 16: 1667–1678.

Brenchley, R., Spannag, M., Pfeife, M., Barker, G.L.A., D'Amore, R., A. M, A., McKenzie, N., Kramer, M., Kerhornou, A., Bolser, D., Kay, S., Waite, D., Trick, M., Bancroft, I., Gu, Y., Huo, N., Luo, M.-C., Sehgal, S., Gill, B., Kianian, S., Anderson, O., Kersey, P., Dvorak, J., McCombie, W.R., Hall, A., Mayer, K.F. X., Edwards, K.J., Bevan, M.W., Hall, N. 2012. Analysis of the bread wheat genome using whole-genome shotgun sequencing. Nature 491: 705–710.

Caetano-Anolles, G. 2005. Evolution of genome size in the grasses. Crop Sci. 45: 1809–1816.

Calderón, C.E., Soderstrom, T.R. 1973. Morphological and anatomical considerations of the grass subfamily Bambusoideae based on the genus *Maclurolyra*. Smithsonian Contr. Bot. 11: 1–55.

Catalán, P., Müller, J., Hasterok, R., Jenkins, G., Mur, L.A. J., Langdon, T., Betekhtin, A., Siwinska, D., Pimentel, M., López-Alvarez, D. 2012. Evolution and taxonomic split of the model grass *Brachypodium distachyon*. Ann. Bot. 109: 385–405.

Celarier, R.P. 1956. Additional evidence for five as the basic chromosome number of the Andropogoneae. Rhodora 58: 135–143.

Chemisquy, M.A., Giussani, L.M., Scataglini, M., Kellogg, E.A., Morrone, O. 2010. Phylogenetic studies favor the unification of *Pennisetum*, *Cenchrus* and *Odontelytrum*: a nuclear, chloroplast and morphological combined analysis. Ann. Bot. 106: 107–130.

Devos, K.M. 2010. Grass genome organization and evolution. Curr. Opinion Pl. Biol. 13: 139–145.

Dewey, D.R. 1984. The genomic system of classification as a guide to intergeneric hybridization in the perennial Triticeae. In: Gustafson, J.P. (ed.) Gene manipulation in plant improvement. New York: Plenum Press. pp. 209–279

Fan, C., Walling, J.G., Zhang, J., Hirsch, C.D., Jiang, J., Wing, R.A. 2011. Conservation and purifying selection of transcribed genes located in a rice centromere. Plant Cell 23: 2821–2830.

Flovik, K. 1938. Cytological studies of Arctic grasses. Hereditas 24: 265–376.

Gale, M.D., Devos, K.M. 1998. Plant comparative genetics after 10 years. Science 282: 656–659.

Garber, E.D. 1950. Cytotaxonomic studies in the genus *Sorghum*. Univ. California Publ. Bot. 23: 283–361.

Giussani, L.M., Cota-Sánchez, J.H., Zuloaga, F.O., Kellogg, E.A. 2001. A molecular phylogeny of the grass subfamily Panicoideae (Poaceae) shows multiple origins of C_4 photosynthesis. Amer. J. Bot. 88: 1993–2012.

GPWG 2001. Grass Phylogeny Working Group. Phylogeny and subfamilial classification of the Poaceae. Ann. Missouri Bot. Gard. 88: 373–457.

Hair, J.B., Beuzenberg, E.J. 1961. High polyploidy in a New Zealand *Poa*. Nature 189: 160.

Hanson, L., Boyd, A., Johnson, M.A.T., Bennett, M.D. 2005. First nuclear DNA C-values for 18 eudicot families. Ann. Bot. 96: 1315–1320.

Hilu, K.W. 1993. Polyploidy and the evolution of domesticated plants. Amer. J. Bot. 80: 1494–1499.

Hunziker, J.H., Stebbins, G.L. 1987. Chromosomal evolution in the Gramineae. In: Soderstrom, T.R., Hilu, K.

W., Campbell, C.S., Barkworth, M.E. (eds.) Grass systematics and evolution. Washington, DC: Smithsonian Institution. pp. 179–187

International Rice Genome Sequencing Project 2005. The map-based sequence of the rice genome. Nature 436: 793–800.

Jia, J., Zhao, S., Kong, X., Li, Y., Zhao, G., He, W., Appels, R., Pfeifer, M., Tao, Y., Zhang, X., Jing, R., Zhang, C., M, Y., Gao, L., Gao, C., Shannagl, M., Mayer, K.F.X., Li, D., Pan, S., Zheng, F., Hu, Q., Xia, X., Li, J., Liang, Q., Chen, J., Wicker, T., Gou, C., Kuang, H., He, G., Luo, Y., Keller, B., Zia, Q., Lu, P., Wang, J., Zou, H., Zhang, R., Xu, J., Gao, J., Middleton, C., Quan, Z., Liu, G., Want, J., International Wheat Genome Sequencing Consortium, Yang, H., Liu, X., He, Z., Mao, L., Wang, J. 2013. *Aegilops tauschii* draft genome sequence reveals a gene repertoire for wheat adaptation. Nature 496: 91–95.

Jones, R.N., Rees, H. 1967. Genotypic control of chromosome behaviour in rye. XI. The influence of B chromosomes on meiosis. Heredity 22: 333–347.

Jones, R.N., Rees, H. 1982. B chromosomes. London, New York: Academic Press.

Jones, R.N., Viegas, W., Houben, A. 2008. A century of B chromosomes in plants: so what? Ann. Bot. 101: 767–775.

Judziewicz, E.J., Soderstrom, T.R. 1989. Morphological, anatomical, and taxonomic studies in *Anomochloa* and *Streptochaeta* (Poaceae: Bambusoideae). Smithsonian Contr. Bot. 68: 1–52.

Kellogg, E.A. 2001. Evolutionary history of the grasses. Plant Phys. 125: 1198–1205.

Kellogg, E.A. 2012. Phylogenetic relationships in Saccharinae. In: Paterson, A. (ed.) Genetics and genomics of Saccharinae. Berlin: Springer. pp. 3–21.

Kellogg, E.A., Bennetzen, J.L. 2004. The evolution of nuclear genome structure in seed plants. Amer. J. Bot. 91: 1709–1725.

Kellogg, E.A., Campbell, C.S. 1987. Phylogenetic analyses of the Gramineae. In: Soderstrom, T.R., Hilu, K.W., Campbell, C.S., Barkworth, M.E. (eds.) Grass systematics and evolution. Washington, DC: Smithsonian Institution Press. pp. 310–322

Kellogg, E.A., Appels, R., Mason-Gamer, R.J. 1996. When genes tell different stories: the diploid genera of Triticeae (Gramineae). Syst. Bot. 21: 321–347.

Kennard, W.C., Phillips, R.L., Porter, R.A., Grombacher, A.W. 2000. A comparative map of wild rice (*Zizania palustris* L. $2n=2x=30$). Theor. Appl. Genet. 101: 677–684.

Kihara, H. 1982. Wheat studies - retrospect and prospects. Amsterdam: Elsevier.

Kim, E.S., Bolsheva, N.L., Samatadze, T.E., Nosov, N.N., Nosova, I.V., Zelenin, A.V., Punina, E.E., Muravenko, O.V., Rodionov, A.V. 2009. The unique genome of two-chromosome grasses *Zingeria* and *Colpodium*, its origin, and evolution. Russian J. Genet. 45: 1329–1337.

Leitch, I.J., Beaulieu, J.M., Chase, M.W., Leitch, A.R., Fay, M.F. 2010. Genome size dynamics and evolution in monocots. J. Bot. 201: Article ID 862516: 1–18.

Ling, H.-Q., Zhao, S., Liu, D., Wang, J., Sun, H., Zhang, C., Fan, H., Li, D., Dong, L., Tao, Y., Gao, C., Wu, H., Li, Y., Cui, Y., Guo, X., Zheng, S., Wang, B., Yu, K., Liang, Q., Yang, W., Lou, X., Chen, J., Feng, M., Jian, J., Zhang, X., Luo, G., Jiang, Y., Liu, J., Wang, Z., Sha, Y., Zhang, B., Wu, H., Tang, D., Shen, Q., Xue, P., Zou, S., Wang, X., Liu, X., Wang, F., Yang, Y., An, X., Dong, Z., Zhang, K., Zhang, X., Luo, M.-C., Dvorak, J., Tong, Y., Wang, J., Yang, H., Li, Z., Wang, D., Zhang, A., Wang, J. 2013. Draft genome of the wheat A-genome progenitor *Triticum urartu*. Nature 496: 87–90.

Luo, M.C., Deal, K.R., Akhunov, E.D., Akhunova, A.R., Anderson, O.D., Anderson, J.A., Blake, N., Clegg, M. T., Coleman-Derr, D., Conley, E.J., Crossman, C.C., Dubcovsky, J., Gill, B.S., Gu, Y.Q., Hadam, J., Heo, H. Y., Huo, N., Lazo, G., Ma, Y., Matthews, D.E., McGuire, P.E., Morrell, P.L., Qualset, C.O., Renfro, J., Tabanao, D., Talbert, L.E., Tian, C., Toleno, D.M., Warburton, M.L., You, F.M., Zhang, W., Dvorak, J. 2009. Genome comparisons reveal a dominant mechanism of chromosome number reduction in grasses and accelerated genome evolution in Triticeae. Proc. Natl. Acad. Sci. USA 106: 15780–15785.

Lynch, M., Conery, J.S. 2003. The origins of genome complexity. Science 302: 1401–1404.

Mason-Gamer, R.J., Orme, N.L., Anderson, C.M. 2002. Phylogenetic analysis of North American *Elymus* and the monogenomic Triticeae (Poaceae) using three chloroplast DNA data sets. Genome 45: 991–1002.

Mason-Gamer, R.J., Burns, M.M., Naum, M. 2010a. Phylogenetic relationships and reticulation among Asian *Elymus* allotetraploids: analyses of three nuclear gene trees. Mol. Phyl. Evol. 54: 10–22.

Mason-Gamer, R.J., Burns, M.M., Naum, M. 2010b. Reticulate evolutionary history of a complex group of grasses: phylogeny of *Elymus* StStHH allotetraploids based on three nuclear genes. PLoS ONE 5: e10989.

Matsuoka, Y. 2011. Evolution of polyploid *Triticum* wheats under cultivation: the role of domestication, natural hybridization and allopolyploid speciation in their diversification. Plant Cell Physiol. 52: 750–764.

Mayer, K.F.X., Martis, M., Hedley, P.I., Simková, H., Liu, H., Morris, J.A., Steuemagel, B., Taudien, S., Roessner, S., Gundlach, H., Kubaláková, M., Suchánková, P., Murat, F., Felder, M., Nussbaumer, T., Graner, A., Salse, J., Endo, T., Sakai, H., Tanaka, T., Itoh, T., Sato, K., Platzer, M., Matsumoto, T., Scholz, U., Dolezel, J., Waugh, R., Stein, N. 2011. Unlocking the barley genome by chromosomal and comparative genomics. Plant Cell 23: 1249–1263.

Mehra, P.N., Khosla, P.K., Kohli, B.L., Kooner, J.S. 1968. Cytological studies in the North Indian grasses (Part I). Research Bulletin of Punjab University 19: 157–230.

Moore, G., Devos, K.M., Wang, Z., Gale, M.D. 1995. Cereal genome evolution. Grasses, line up and form a circle. Curr. Biol. 5: 737–739.

Newell, T.K. 1969. A study of the genus *Joinvillea* (Flagellariaceae). J. Arnold Arbor. 50: 527–555.

Paterson, A.H., Bowers, J.E., Bruggmann, R., Dubchak, I., Grimwood, J., Gundlach, H., Haberer, G., Hellsten, U., Mitros, T., Poliakov, A., Schmutz, J., Spannagl, M., Tang, H., Wang, X., Wicker, T., Bharti, A.K., Chapman, J., Feltus, F.A., Gowik, U., Grigoriev, I.V., Lyons, E., Maher, C.A., Martis, M., Narechania, A.,

Otillar, R.P., Penning, B.W., Salamov, A.A., Wang, Y., Zhang, L., Carpita, N.C., Freeling, M., Gingle, A.R., Hash, C.T., Keller, B., Klein, P., Kresovich, S., McCann, M.C., Ming, R., Peterson, D.G., Mehboob-ur-Rahman, Ware, D., Westhoff, P., Mayer, K.F.X., Messing, J., Rokhsar, D.S. 2009. The *Sorghum bicolor* genome and the diversification of grasses. Nature 457: 551–556.

Peng, Z., Lu, Y., Li, L., Zhao, Q., Feng, Q., Gao, Z., Lu, H., Hu, T., Yao, N., Liu, K., Li, Y., Fan, D., Guo, Y., Li, W., Lu, Y., Weng, Q., Zhou, C., Zhang, L., Huang, T., Zhao, Y., Zhu, C., Liu, X., Yang, X., Wang, T., Miao, K., Zhuang, C., Cao, X., Tang, W., Liu, G., Liu, Y., Chen, J., Liu, Z., Yuan, L., Liu, Z., Huang, X., Lu, T., Fei, B., Ning, Z., Han, B., Jiang, Z. 2013. The draft genome of the fast-growing non-timber forest species moso bamboo (*Phyllostachys heterocycla*). Nat. Genet. 45: 456–461.

Pourkheirandish, M., Wicker, T., Stein, N., Fujimora, T., Komatsuda, T. 2006. Analysis of the barley chromosome 2 region containing the six-rowed spike gene *vrs1* reveals a breakdown of rice-barley micro collinearity by a transposition. Theor. Appl. Genet. 114: 1357–1365.

Price, H.J., Dillon, S.L., Hodnett, G., Rooney, W.L., Ross, L., Johnston, J.S. 2005. Genome evolution in the genus *Sorghum* (Poaceae). Ann. Bot. 95: 219–227.

Raven, P.H. 1975. The bases of angiosperm phylogeny: cytology. Ann. Missouri Bot. Gard. 62: 724–764.

Salse, J., Bolot, S., Throude, M., Jouffe, V., Plegu, B., Quraishi, U.M., Calcagno, T., Cooke, R., Delseny, M., Feuillet, C. 2008. Identification and characterization of shared duplications between rice and wheat provide new insight into grass genome evolution. Plant Cell 20: 11–24.

SanMiguel, P., Tikhonov, A., Jin, Y.K., Motchoulskaia, N., Zakharov, D., Melake-Berhan, A., Springer, P.S., Edwards, K.J., Lee, M., Avramova, Z., Bennetzen, J. L. 1996. Nested retrotransposons in the intergenic regions of the maize genome. Science 274: 765–768.

Schnable, P.S., Ware, D., Fulton, R.S., Stein, J.C., Wei, F., Pasternak, S., Liang, C., Zhang, J., Fulton, L., Graves, T.A., Minx, P., Felly, A.D., Courtney, L., Kruchowski, S.S., Tomlinson, C., Strong, C., Delehaunty, K., Fronick, C., Courtney, B., Rock, S.M., Belter, E., Du, F., Kim, K., Abbott, R.M., Cotton, M., Levy, A., Marchetto, P., Ochoa, K., Jackson, S.M., Gillam, B., Chen, W., An, L., Higginbotham, J., Cardenas, M., Waligorski, J., Applebaum, E., Phelps, L., Falcone, J., Kanchi, K., Thane, T., Scimone, A., Thane, N., Henke, J., Wang, T., Ruppert, J., Shah, N., Rotter, K., Hodges, J., Ingenthron, E., Cordes, M., Kohlberg, S., Sgro, J., Delgado, B., Mead, K., Chinwalla, A., Leonard, S., Crouse, K., Collura, K., Kudrna, D., Currie, J., He, R., Angelova, A., Rajasekar, S., Mueller, T., Lomeli, R., Scara, G., Ko, A., DeLaney, K., Wissotski, M., Lopez, G., Campos, D., Braidotti, M., Ashley, E., Golser, W., Kim, H., Lee, S., Lin, J., Dujmic, Z., Kim, W., Talag, J., Zuccolo, A., Fan, C., Sebastian, A., Kramer, M., Spiegel, L., Nascimento, L., Zutavern, T., Miller, B., Ambroise, C., Muller, S., Spooner, W., Narechania, A., Ren, L., We, S., Kumari, S., Faga, B., Levy, M.J., McMahan, L., Van Buren, P., Vaughn, M.W., Ying, K., Yeh, C.-T., Emrich, S.J., Jia, Y., Kalyanaraman, A., Hsia, A.-P., Barbazuk, W.B., Baucom, R.S., Brutnell, T.P., Carpita, N.C., Chaparro, C., Chia, J.-M., Deragon, J.-M., Estil, J.C., Fu, Y., Jeddeloh, J.A., Han, Y., Lee, H., Li, P., Lisch, D.R., Liu, S., Liu, Z., Nagel, D.H., McCann, M.C., SanMiguel, P., Myers, A.M., Nettleton, D., Nguyen, J., Penning, B.W., Ponnala, L., Schneider, K.L., Schwartz, D. C., Sharma, A., Soderlunc, C., Springer, N. M., Sun, Q., Wang, H., Waterman, M., Westerman, R., Wolfgruber, T.K., Yang, L., Yu, Y., Zhang, L., Zhou, S., Zhu, Q., Bennetzen, J.L., Dawe, R.K., Jiang, J., Jiang, N., Presting, G.G., Wessler, S.R., Aluru, S., Martienssen, R.A., Clifton, S.W., McCombie, W.R., Wing, R. A., Wilson, R.K. 2009. The B73 maize genome: complexity, diversity, and dynamics. Science 326: 1112–1115.

Serres-Giardi, L., Belkhir, K., David, J., Glémin, S. 2012. Patterns and evolution of nucleotide landscapes in seed plants. Plant Cell 24: 1379–1397.

Sharma, M.L. 1979. Some considerations on the phylogeny and chromosomal evolution in grasses. Cytologia 44: 679–685.

Spangler, R., Zaitchik, B., Russo, E., Kellogg, E. 1999. Andropogoneae evolution and generic limits in *Sorghum* (Poaceae) using *ndhF* sequences. Syst. Bot. 24: 267–281.

Stebbins, G.L. 1956. Cytogenetics and evolution of the grass family. Amer. J. Bot. 43: 890–905.

Stebbins, G.L. 1971. Chromosomal evolution in higher plants. London: Edward Arnold.

Stebbins, G.L. 1985. Polyploidy, hybridization and the invasion of new habitats. Ann. Missouri Bot. Gard. 72: 824–832.

Swigonova, Z., Lai, J., Ma, J., Ramakrishna, W., Llaca, V., Bennetzen, J.L., Messing, J. 2004. Close split of sorghum and maize genome progenitors. Genome Res. 14: 1916–1923.

Tatarinova, T.V., Alexandrov, N.N., Bouck, J.B., Feldman, K.A. 2010. GC_3 biology in corn, rice, sorghum and other grasses. BMC Genomics 11: 308.

The International Barley Genome Sequencing Consortium 2012. A physical, genetic and functional sequence assembly of the barley genome. Nature 491: 711–716.

The International Brachypodium Initiative 2010. Genome sequencing and analysis of the model grass *Brachypodium distachyon*. Nature 463: 763–768.

Triplett, J.K., Wang, Y., Zhong, J., Kellogg, E.A. 2012. Five nuclear loci resolve the polyploid history of switchgrass (*Panicum virgatum* L.) and relatives. PLoS ONE 7: e38702.

Wicker, T., Mayer, K.F.X., Gundlach, H., Martis, M., Steuemagel, B., Scholz, U., Simková, H., Kubaláková, M., Choulet, F., Taudien, S., Platzer, M., Feuillet, C., Fahima, T., Budak, H., Dolezel, J., Keller, B., Stein, N. 2011. Frequent gene movement and pseudogene evolution is common to the large and complex genomes of wheat, barley, and their relatives. Plant Cell 23: 1706–1718.

Wilson, W.A., Harrington, S.E., Woodman, W.L., Lee, M., Sorrells, M.E., McCouch, S.R. 1999. Inferences on the genome structure of progenitor maize through comparative analysis of rice, maize and the domesticated panicoids. Genetics 153: 453–473.

Pollen

Morphology and Anther Structure

Stamen morphology is largely uniform throughout the Poaceae. Anthers are consistently tetrasporangiate, although Bhanwra (1988) reports bisporangiate anthers in *Bromus unioloides*. In all grasses, the anther wall consists of an epidermis, an endothecium, and a tapetum. In all known Poaceae, the tapetum is of the secretory type, as is found in the close relatives of the grasses (Furness and Rudall 1998). Tapetal cells generally have two nuclei, but in some taxa have only one (Anton de Triquell 1987; Nakamura et al. 2010). Grasses in the Bistigmatic clade (i.e., all grasses except for Anomochlooideae and Pharoideae) also have a middle layer of cells between the endothecium and the tapetum; the middle layer breaks down early in anther development. Presence of the middle layer is synapomorphic for the Bistigmatic clade; also synapomorphic is the fate of the inner walls of the endothecial cells, which become fibrous at maturity (Artschwager and McGuire 1949; Bhanwra 1988; Furness and Rudall 1998). In Anomochlooideae and Pharoideae, in contrast, the anther wall lacks the middle layer entirely and thus has only three layers of cells (Sajo et al. 2007, 2009, 2012). The inner walls of the endothecium fail to become fibrous in these two subfamilies (Sajo et al. 2007, 2009, 2012).

Microsporogenesis in most grasses is successive, with cytokinesis taking place after meiosis I; this character state is also the most common in the Poales (Furness et al. 2002) and has recently been documented in *Ecdeiocolea* (Sajo et al. 2009). In *Streptochaeta* and *Anomochloa*, however, simultaneous microsporogenesis, with cytokinesis occurring only after meiosis II is complete, has been noted in addition to the successive type (Sajo et al. 2009, 2012). Mature pollen is three-celled, with one vegetative and two sperm nuclei (Nakamura et al. 2010).

In most grasses, after microsporogenesis the microspores form a single layer around the inside of the anther locule, so that the single pore of each microspore/pollen grain is in direct contact with the tapetum (Christensen and Horner, Jr. 1974; Kirpes et al. 1996). In contrast, in *Pharus lappulaceus*, pollen fills the anther locule and is not restricted to the peripheral region (Kirpes et al. 1996); however, Sajo et al. (2007) report peripheral microsporocytes and centrally located pollen grains. Findings on *Streptochaeta* and *Anomochloa* are contradictory in terms of pollen position. Kirpes et al. (1996) report that *Streptochaeta* pollen is peripheral and *Anomochloa* is not; in contrast, Sajo et al. (2009, 2012) report exactly the opposite. The discrepancy may be due to examining different stages of development. It is thus impossible to determine if peripheral pollen is a synapomorphy and if so, at what level in the phylogeny. Peripheral pollen is derived independently in Cyperaceae (Kirpes et al. 1996).

Grass pollen is remarkably uniform in morphology. Pollen grains are more or less spherical, with a single germination pore, and in this respect are similar to those of other monocots. The pore is surrounded by a raised rim, or annulus; within the pore is a circular bit of exine, the operculum. On the inside (cytoplasmic side) of the annulus, the foot layer is thickened and layered (Christensen and Horner, Jr. 1974; Linder

E.A. Kellogg, *Flowering Plants. Monocots*, The Families and Genera of Vascular Plants 13,
DOI 10.1007/978-3-319-15332-2_6, © Springer International Publishing Switzerland 2015

and Ferguson 1985; Skvarla et al. 2003). The pore forms on the side of the microgametophyte in contact with the tapetum, and this position may be important for proper maturation of the pollen grain (Christensen and Horner, Jr. 1974). In the bambusoid genus *Pariana*, the thickened foot layer is present, but the annulus and operculum are absent (Skvarla et al. 2003). Pollen apertures similar to those in Poaceae are found in other graminid Poales, although Flagellariaceae, Joinvilleaceae and Restionaceae may not form an operculum. The pollen wall of Poaceae is columellate, with a clear foot layer and tectum. The cavities in the pollen wall appear to contain little or no material deposited by the tapetum (Heslop-Harrison and Heslop-Harrison 1987). Unlike most other Poales, the grasses lack channels (scrobiculi) perforating the pollen wall (Linder and Ferguson 1985). The lack of scrobiculi appears also to be shared with Ecdeiocoleaceae and may indicate a sister relationship between the two families. Within the foot layer and tectum, however, narrow channels are visible in transmission electron micrographs (Linder and Ferguson 1985); these may be unique for Poaceae.

The outer surface of the pollen grain (tectum) is covered with tiny pointed bumps, variously called spinules or granules; these may form clusters or islands in pollen of some species, whereas the granules are isolated in others (Andersen and Bertelsen 1972; Christensen and Horner, Jr. 1974; Watson and Bell 1975). Differences in pollen morphology between species, genera, or larger taxonomic groups are subtle. Although various authors have tried to find characters to distinguish species or genera, and to distinguish wild and cultivated plants, these efforts have met with limited success. In broad taxonomic surveys of grass pollen morphology, Page (1975) and Watson and Bell (1975) find few characters to distinguish major groups. In addition, Watson and Bell (1975) find variation within a single grain that mirrors the variation between species. Page (1975) notes that "[t]he overall taxonomic significance of sexine pattern in grass pollen is very limited," and "there is no obvious correlation [of sexine patterns] with habitat, pollination biology or taxonomy."

A recent attempt to quantify differences in pollen grain surface morphology has confirmed previous studies (Mander et al. 2013). A classification algorithm was developed that identified four morphogroups of pollen. While the algorithm could accurately assign pollen to a morphogroup in 77 % of the cases, the groups themselves did not correlate with taxonomy or phylogeny; for example, groups 1 and 2 each contained a representative of Pooideae, Chloridoideae, and Panicoideae. It remains to be seen whether the pollen morphogroups correlate with other aspects of biology such as environment.

Pollination

Pollen transfer

Most grasses are wind-pollinated. The tiny flowers, feathery stigmas, versatile anthers, and lack of nectar are all characteristic of the wind pollination syndrome, and anyone who suffers from hay fever is painfully aware of the large amount of grass pollen that may be airborne at some times of year. In addition to the well-documented floral characteristics, wind-pollinated species often produce a single seed per flower and tend to have unisexual flowers (Givnish et al. 2010). The single seed per flower is characteristic of and synapomorphic for all grasses, and unisexual flowers occur sporadically throughout the family.

The great diversity of grass flowers, spikelets and inflorescences has sparked speculation on whether there has been selection on the grasses for pollination efficiency. Discussion in the literature has focused on the relative importance of simply intercepting pollen on the windward side of the inflorescence versus capturing pollen through vortices created on the leeward side. In a detailed study of 20 species of pooid grasses, Friedman and Harder (2005) measured traits of inflorescence and flower size. They found a negative correlation between numbers of flowers and flower length. The species that they studied could be classified as having large or small flowers, and dense or open inflorescences; all four combinations of these variables were observed suggesting that there was no optimum. Pollen size is positively correlated with the number of pollen grains and total volume of pollen per flower, indicating that there is no trade-off between pollen size and amount. Pollen size also correlates positively with stigma length.

Cresswell et al. (2010) used both modeling and experimental approaches to suggest that most pollen is captured on the windward side of grass inflorescences and thus that the formation of vortices is largely irrelevant to pollination. The one exception that they were able to find in the literature was *Setaria parviflora* (= *S. geniculata*), in which over a third of the pollen was captured on the leeward side of the inflorescence (Niklas 1987). However, *S. parviflora* oscillates in the wind, and thus the leeward side of the inflorescence may end up facing the wind as it bends. In addition, the sterile branches (bristles) characteristic of the genus could presumably have some effect on the boundary layer and may thus cause some of the discrepancy.

For selection to operate on pollination efficiency, pollen must frequently be limiting in natural habitats; however, data on pollen limitation in grasses are sparse. McKone et al. (1998) report that *Andropogon gerardii* and *Sorghastrum nutans* are not pollen limited. Davis et al. (2004) report that pollen is not limiting in *Spartina alterniflora* when it is growing in dense stands but, when density of conspecifics is low, then supplemental pollination raises seed set several fold.

Despite the prevalence of wind pollination in the family, a few taxa may be insect-pollinated. For example, the anthers in *Streptochaeta* and *Anomochloa* are basifixed or centrifixed, rather than versatile, and the stigmas are not plumose (Sajo et al. 2009). Both characteristics suggest insect pollination rather than wind pollination. Insects have been observed on inflorescences of the understory grasses *Olyra* and *Pariana* in Brazil (Soderstrom and Calderón 1971). The most common visitors were gall midges and phorid flies; the insects were observed consuming pollen, but evidence for effective pollen transfer is still lacking. Huang et al. (2002) report visits of honeybees to *Phyllostachys nidularia*, and cite other references of insect visitors to bamboo flowers. Adams et al. (1981) observed halictid bees visiting *Paspalum dilatatum* in south-central Oklahoma, and followed up with exclusion experiments; these showed that more pollen was transferred and more seed was set when flowers were exposed to bees plus wind than to wind alone.

In summary, data on grass pollination are remarkably sparse considering the abundance of grass species. Evidence that the diversity of grass inflorescences has been driven by selection for pollination efficiency is weak. Pollen may not often be limiting. Pollen may be collected primarily on the windward side of the inflorescence, meaning that the formation of eddies by complex architecture may not be relevant. Insect pollination may occur but its importance is unknown.

Pollen germination and pollen tube growth

Lausser and Dresselhaus (2010) divide pollination in grasses into five phases, based on their investigations with *Zea mays*. Phase I is germination, in which pollen grains germinate on the stigmatic papillae. In describing the details of this process, Heslop-Harrison and Heslop-Harrison (1987) provide a thorough description of the ultrastructure and histology of pollen grains and stigmas in the grasses. Grass stigmas are "dry", without obvious liquid produced. The stigmatic cells have a thin cuticle that is interrupted in places; this is covered with a layer of protein and mucilage. When the pollen grain lands on the stigmatic surface, water quickly (20–25 seconds in rye) moves osmotically from the stigmatic cell to the pollen grain, forming a meniscus. Hydration of the pollen grain then proceeds over a period of several minutes. Unlike pollen of many eudicots, grass pollen retains a water content of 15–20 %, so it is partially hydrated when it arrives on a stigma, allowing it to germinate almost instantly (Heslop-Harrison and Heslop-Harrison 1992; Heslop-Harrison et al. 1997; Nepi et al. 2001). A single stigmatic cell can rehydrate four or five pollen grains (Heslop-Harrison and Heslop-Harrison 1987). As germination proceeds, the pollen tube forms and protrudes through what had been the pollen aperture.

The grass stigma appears to exert little selective control over pollen germination (Heslop-Harrison and Heslop-Harrison 1987). Lausser et al. (2010) applied pollen from *Lolium multiflorum*, *Poa nemoralis*, *Oryza sativa*, *Zea mays*, *Tripsacum dactyloides*, *Lilium longiflorum*, and *Arabidopsis thaliana* to stigmas of *Z. mays* and *T. dactyloides*; all pollen except that from *Lilium* germinated. Other more circumstantial evidence for the possibility of indiscriminate germination comes from data on wide crosses. Allopolyploidy is common in the grasses, as noted in the section above on Karyology and genome structure.

In addition, successful pollination has been demonstrated between wheat and maize, wheat and sorghum, and barley and maize (Laurie and Bennett 1988), as well as between maize and oats (Riera-Lizarazu et al. 1996). Given these results, it seems likely that pollen from nearly any grass can germinate on the stigma of nearly any other grass.

Phase II of pollination (sensu Lausser et al. 2010) is stigma invasion, in which the pollen tube grows into the intercellular spaces in the stigmatic papillae (Heslop-Harrison 1982). Penetration is aided by secretion of beta-expansins and xylanases, both of which appear to loosen the cell walls (Cosgrove et al. 1997; Valdivia et al. 2009). Expansins are also known as Group I allergens and are particularly copious; they are responsible for the common human allergic reaction to grass pollen. Cells not involved in pollen tube penetration quickly lose turgor, release water and die, providing a possible mechanism to block further pollination or fungal invasion (Heslop-Harrison and Heslop-Harrison 1987).

In Phase III, transmitting tract growth, the pollen tube grows down the style toward the ovule. In the unusually long styles of maize (known as "silks"), this occurs at the remarkable speed of 0.5 cm per hour (Bedinger and Russell 1994); approximately the same speed is achieved by rye pollen traversing a much shorter pathway (Heslop-Harrison and Heslop-Harrison 1987). Transmitting tract growth is under sporophytic control, and likely involves signaling between the style and the pollen tube, but does not depend on signals from the ovary (Lausser et al. 2010). Experiments in maize and pearl millet indicate that pollen can grow basipetally as efficiently as acropetally, suggesting that no particular directional signal is needed (Heslop-Harrison and Heslop-Harrison 1987).

Phase IV, ovarial cavity growth, appears to be controlled entirely mechanically, unlike the situation in *Arabidopsis* in which both sporophytic and gametopytic signals are involved (Lausser and Dresselhaus 2010). Once the pollen tube(s) enter the ovary, the base of the styles loses turgor, which prevents passage of additional tubes (Heslop-Harrison and Heslop-Harrison 1987). The pollen tube then enters the ovary through the ovary wall, and grows between the inner epidermis of the ovary and the inner integument. (The outer integument is quite short at pollination.) Phase V, gametophytic interactions, appears to be the only stage in maize in which signals are required from the female gametophyte (Lausser and Dresselhaus 2010).

Pollen tube growth in connection with self- or cross-compatibility is discussed under Reproductive systems.

References

Adams, D.E., Perkins, W.E., Estes, J.R. 1981. Pollination systems in *Paspalum dilatatum* Poir. (Poaceae): an example of insect pollination in a temperate grass. Amer. J. Bot. 68: 389–394.

Andersen, S.T., Bertelsen, F. 1972. Scanning electron microscope studies of pollen of cereals and other grasses. Grana 12: 79–86.

Anton de Triquell, A. 1987. Grass gametophytes: their origin, structure, and relation with the sporophyte. In: Soderstrom, T.R., Hilu, K.W., Campbell, C.S., Barkworth, M.E. (eds.) Grass systematics and evolution. Washington, DC: Smithsonian Institution Press. pp. 11–20

Artschwager, E., McGuire, R.C. 1949. Cytology of reproduction in *Sorghum vulgare*. J. Agric. Res. 78: 659–673.

Bedinger, P., Russell, S.D. 1994. Gametogenesis in maize. In: Freeling, M., Walbot, V. (eds.) The maize handbook. New York: Springer. pp. 48–61

Bhanwra, R.K. 1988. Embryology in relation to systematics of Gramineae. Ann. Bot. 62: 215–233.

Christensen, J.E., Horner, Jr., H.T. 1974. Pollen pore development and its spatial orientation during microsporogenesis in the grass *Sorghum bicolor*. Amer. J. Bot. 61: 604–623.

Cosgrove, D.J., Bedinger, P., Durachko, D.M. 1997. Group I allergens of grass pollen as cell wall-loosening agents. Proc. Natl. Acad. Sci. USA 94: 6559–6564.

Cresswell, J.E., Krick, J., Patrick, M.A., Lahoubi, M. 2010. The aerodynamics and efficiency of wind pollination in grasses. Func. Ecol. 24: 706–713.

Davis, H.G., Taylor, C.M., Lambrinos, J.G., Strong, D.R. 2004. Pollen limitation causes an Allee effect in a wind-pollinated invasive grass (*Spartina alterniflora*). Proc. Natl. Acad. Sci. USA 101: 13804–13807.

Friedman, J., Harder, L.D. 2005. Functional associations of floret and inflorescence traits among grass species. Amer. J. Bot. 92: 1862–1870.

Furness, C.A., Rudall, P.J. 1998. The tapetum and systematics in monocotyledons. Bot. Rev. 64: 201–239.

Furness, C.A., Rudall, P.J., Sampson, F.B. 2002. Evolution of microsporogenesis in angiosperms. Int. J. Plant Sci. 163: 235–260.

Givnish, T.J., Ames, M., McNeal, J.R., McKain, M.R., Steele, P.R., dePamphilis, C.W., Graham, S.W., Pires, J.C., Stevenson, D.W., Zomlefer, W.B., Briggs, B.G., Duvall, M.R., Moore, M.J., Heaney, J.M., Soltis, D.E., Soltis, P.S., Thiele, K., Leebens-Mack, J.H. 2010. Assembling the tree of monocotyledons: plastome sequence phylogeny and evolution of Poales. Ann. Missouri Bot. Gard. 97: 584–616.

Heslop-Harrison, J. 1982. Pollen-stigma interaction and cross-incompatibility in the grasses. Science 215: 1358–1364.

Heslop-Harrison, J., Heslop-Harrison, Y. 1987. Pollen-stigma interaction in the grasses. In: Soderstrom, T. R., Hilu, K.W., Campbell, C.S., Barkworth, M.E. (eds.) Grass systematics and evolution. Washington, DC: Smithsonian Institution. pp. 133–142

Heslop-Harrison, J., Heslop-Harrison, Y. 1992. Intracellular motility, the actin cytoskeleton and germinability in the pollen of wheat (*Triticum aestivum* L.). Sex. Plant Reprod. 5: 247–255.

Heslop-Harrison, J., Heslop-Harrison, Y., Heslop-Harrison, J.S. 1997. Motility in ungerminated grass pollen: association of myosin with polysaccharide-containing wall-precursor bodies (P-particles). Sex. Plant Reprod. 10: 65–66.

Huang, S.-Q., Yang, H.-F., Lu, I., Takahashi, Y. 2002. Honeybee-assisted wind pollination in bamboo *Phyllostachys nidularia* (Bambusoideae: Poaceae)? Bot. J. Linn. Soc. 138: 1–7.

Kirpes, C.C., Clark, L.G., Lersten, N.R. 1996. Systematic significance of pollen arrangement in microsporangia of Poaceae and Cyperaceae. Amer. J. Bot. 83: 1609–1622.

Laurie, D.A., Bennett, M.D. 1988. Chromosome behaviour in wheat x maize, wheat x sorghum and barley x maize crosses. In: Brandham, P.E. (ed.) Kew chromosome conference III. London: Her Majesty's Stationery Office. pp. 167–177

Lausser, A., Dresselhaus, T. 2010. Sporophytic control of pollen tube growth and guidance in grasses. Biochem. Soc. Trans. 38: 631–634.

Lausser, A., Kliwer, I., Srilunchang, K., Dresselhaus, T. 2010. Sporophytic control of pollen tube growth and guidance in maize. J. Exp. Bot. 61: 673–682.

Linder, H.P., Ferguson, I.K. 1985. On the pollen morphology and phylogeny of the Restionales and Poales. Grana 24: 65–76.

Mander, L., Li, M., Mio, W., Fowlkes, C.C., Punyasena, S. W. 2013. Classification of grass pollen through the quantitative analysis of surface ornamentation and texture. Proc. Roy. Soc. B 280: 20131905.

McKone, M.J., Lund, C.P., O'Brien, J.M. 1998. Reproductive biology of two dominant prairie grasses (*Andropogon gerardii* and *Sorghastrum nutans*, Poaceae): male-biased sex allocation in wind-pollinated plants? Amer. J. Bot. 85: 776–783.

Nakamura, A.T., Longhi-Wagner, H.M., Scatena, V.L. 2010. Anther and pollen development in some species of Poaceae (Poales). Braz. J. Bot. 70: 351–360.

Nepi, M., Franchi, G.G., Pacini, E. 2001. Pollen hydration status at dispersal: cytophysiological features and strategies. Protoplasma 216: 171–180.

Niklas, K.J. 1987. Pollen capture and wind-induced movement of compact and diffuse grass panicles - implications for pollination efficiency. Amer. J. Bot. 74: 74–89.

Page, J.S. 1975. A scanning electron microscope survey of grass pollen. Kew Bull. 32: 313–319.

Riera-Lizarazu, O., Rines, H.W., Phillips, R.L. 1996. Cytological and molecular characterization of oat x maize partial hybrids. Theor. Appl. Genet. 93: 123–135.

Sajo, M.G., Longhi-Wagner, H., Rudall, P.J. 2007. Floral development and embryology in the early-divergent grass *Pharus*. Int. J. Plant Sci. 168: 181–191.

Sajo, M.D.G., Furness, C.A., Rudall, P.J. 2009. Microsporogenesis is simultaneous in the early-divergent grass *Streptochaeta*, but successive in the closest grass relative, *Ecdeiocolea*. Grana 48: 27–37.

Sajo, M.G., Pabón-Mora, N., Jardim, J., Stevenson, D.W., Rudall, P.J. 2012. Homologies of the flower and inflorescence in the early-divergent grass *Anomochloa* (Poaceae). Amer. J. Bot. 99: 614–628.

Skvarla, J.J., Rowley, J.R., Hollowell, V.C., Chissoe, W.F. 2003. Annulus-pore relationship in Gramineae (Poaceae) pollen: the pore margin of *Pariana*. Amer. J. Bot. 90: 924–930.

Soderstrom, T.R., Calderón, C.E. 1971. Insect pollination in tropical rain forest grasses. Biotropica 3: 1–16.

Valdivia, E.R., Stephenson, A.G., Durachko, D.M., Cosgrove, D. 2009. Class B ß-expansins are needed for pollen separation and stigma penetration. Sex. Plant Reprod. 22: 141–152.

Watson, L., Bell, E.M. 1975. A surface-structural survey of some taxonomically diverse grass pollens. Aust. J. Bot. 23: 981–990.

Fruit and Seed

The fruit of Poaceae is a caryopsis, in which the seed coat is firmly attached to the inner wall of the ovary. This structure is uniquely derived in the family. Within this general category, however, there is variation in the structure of the pericarp and the ultimate fate of the integuments and nucellus; in particular the inner integument of the ovule, which generally constitutes the seed coat, may be intact or crushed at maturity. Only a few taxa have been studied in detail, and these are so widely scattered taxonomically that the data appear somewhat anecdotal; a broader survey would help clarify the generality of the observations.

In wheat (subfamily Pooideae), the outer and inner epidermes of the pericarp consist of cells with the long axes parallel to the apical-basal axis of the grain (Percival 1921). Between them are one or two layers of parenchyma cells with thick walls plus the "cross layer", which is made up of cells with their long axes more or less perpendicular to the epidermal cells. The inner epidermis is discontinuous, with the long cells widely separated from each other; at maturity these are often crushed. The seed coat is the product of the inner integument, which is two layered at maturity although these layers are also crushed in late development, along with the nucellus. In *Sorghum halepense*, *Setaria italica*, and *Cenchrus americanus* (= *Pennisetum glaucum*) (subfamily Panicoideae) the mature pericarp consists of the outer epidermis plus crushed and broken fragments of the inner epidermis; layers in between break down and are crushed in the mature fruit (Fussell and Dwarte 1980; Harrington and Crocker 1923; Narayanaswami 1956). The inner integument is intact but the outer integument and nucellus are broken down. The pericarp in *Pappophorum subbulbosum* (subfamily Chloridoideae) is similar, but the inner integument, outer integument and nucellus are all collapsed (Rost et al. 1984).

In a number of bamboos and chloridoid grasses, the seed coat is free from the pericarp, or the two are only weakly adherent. This is a derived condition within the family. The development and cellular structure of such fruits have not been studied, so it is not clear what aspects of development are modified, or even if the derivation of the free seed occurs by the same mechanism in all cases.

Several bamboo genera have fleshy fruits. In the two that have been studied closely, *Dinochloa* and *Ochlandra*, embryo and endosperm development are similar to that of other grasses, but the pericarp becomes thick and fleshy (Rudall and Dransfield 1989). The structure of the scutellum is normal, with the apical cells elongated and apparently functioning as transfer cells moving nutrients from the endosperm; Rudall and Dransfield (1989) speculate that nutrients are transferred from the pericarp to the embryo via the endosperm. The latter is liquid in the baccate fruits of the bamboo *Melocanna* (Ramanayake and Weerawardene 2003). Vivipary is also reported in *Dinochloa*; Dransfield (1981) suggests that this character is correlated with the presence of the fleshy fruit.

The point of attachment of the fruit to the plant is known, inaccurately, as the hilum. Since the term hilum normally refers to the point of attachment of the seed to the placenta, this use of the term in grasses is somewhat misleading, but is so widespread it will be followed here. It is in the

E.A. Kellogg, *Flowering Plants. Monocots*, The Families and Genera of Vascular Plants 13,
DOI 10.1007/978-3-319-15332-2_7, © Springer International Publishing Switzerland 2015

basal part of the pericarp, and corresponds to a region slightly below the placenta (Narayanaswami 1956). Other terms include "hilar orifice" (Harrington and Crocker 1923), "placental pad" (Rost et al. 1984), and "chalazal pad" (Fussell and Dwarte 1980). The latter is misleading as well since it is roughly opposite the micropylar end of the embryo, not the chalaza. The cells of the hilum may accumulate pigments in some species, such as *Cenchrus americanus* (= *Pennisetum glaucum*) (Fussell and Dwarte 1980) and *Sorghum bicolor* (Giles et al. 1975), in which the area becomes known as the "black region"; its formation indicates the end of carbon assimilation by the grain.

The shape and size of the fruit, embryo, and hilum are frequently used as taxonomic characters and can be diagnostic of large groups. For example, a punctate hilum is a putative synapomorphy for the tribe Poeae (Pooideae). The size of the embryo varies relative to the size of the caryopsis, and the caryopsis itself may be long and linear (e.g., some Stipeae) or nearly spherical (e.g., some Panicoideae). Shape and size of the fruit are under the control of different genetic programs (Gegas et al. 2010).

Dispersal

The dispersal unit in the grasses varies considerably, and reflects the many different plant tissues in which a disarticulation zone can form (Doust et al. 2014). In some taxa, the fruit is shed separately from the remainder of the spikelet. More commonly, the rachilla disarticulates at each node, so that the dispersal unit is the entire flower, including the lemma and palea, plus the attached bit of the rachilla internode. In other species, the rachilla disarticulates above the glumes but not between the flowers, so that the dispersal unit becomes all the flowers of the spikelet, often including a cluster of sterile lemmas at the distal end. In most Panicoideae, disarticulation is below the glumes so that the entire spikelet is shed, a characteristic that often provides a useful field identification character for the subfamily. In other taxa, disarticulation is at the base of the primary branch, which then is shed as a unit. In many Triticeae and Andropogoneae, the inflorescence axis breaks up at the nodes, dispersing spikelets or clusters of spikelets with an attached rachis internode. Finally, the entire inflorescence may be shed as a unit (e.g., some *Aegilops* and *Agrostis* species). Disarticulation – and hence dispersal – mode may vary among species of a single genus (e.g., *Muhlenbergia*), or may be characteristic of a genus or tribe.

Some species have more than one sort of propagule. For example, *Sphenopholis obtusata* disarticulates between the two flowers and also below the glumes so that it has one small diaspore, consisting of only the upper flower, and one large diaspore consisting of the lower flower plus the glumes (Rabinowitz and Rapp 1981).

Seed weight (and hence propagule weight) varies within an inflorescence and thus presumably affects dispersal ability as well as seedling survival. In *Eremopyrum distans*, seeds from the proximal flower of the spikelet are larger than those from distal flowers, have higher seedling survival, and form larger plants that produce more seeds (Wang et al. 2010). In *Leymus arenarius*, seeds are heavier from the upper parts of the inflorescence (the reverse of *Eremopyrum*), and in the proximal flowers of the spikelet (same as *Eremopyrum*) (Greipsson and Davy 1995).

Vegetative propagation is also common, although in most cases it results in very local dispersal. However, in some species that grow near or in water, fragmentation of the parent plant into separate ramets can lead to water dispersal. This has been reported for *Arundo donax* in South Africa, for example (Milton 2004). On the other hand, a study using microsatellite markers found little evidence that stands of *Phragmites australis* in a Canadian lake were propagated vegetatively; rather, most clumps were genetically distinct, pointing to dispersal and propagation by seed (Belzile et al. 2010).

The great diversity of grass dispersal units suggests that dispersal ability may have contributed to species diversity in the family. As evidence for this, Davidse (1987) notes that the pistillate spikelets or inflorescences of many dioecious species are highly modified with awns or burs that appear to assist in dispersal, whereas the staminate spikelets are generally unspecialized. However, as with most other aspects of grass ecology, dispersal has rarely been studied in a broad systematic context, nor is there a rich experimental literature. Instead the literature on grass dispersal is buried in studies of community ecology, range management, invasive species and community restoration; any one study is limited to the grasses of a particular, often limited, community. Most existing studies are observational and correlative, although there are some notable exceptions as outlined below. This summary is thus incomplete and the definitive review of the literature on grass dispersal has yet to be written.

Unassisted or wind dispersal

Dispersal in grasses is assumed to be predominantly passive, with seeds being dropped from the inflorescence ("unassisted dispersal") and/or spread by wind. This assumption is based on the generally small size of the dispersal units, lack of obvious morphological characteristics that might permit other modes of dispersal, or sometimes presence of structures that might reduce the terminal velocity of the propagule to keep it in the air column longer (Davidse 1987). For example, Jurado et al. (1991) classified 48 of 58 Australian grass species as wind-dispersed or unassisted.

Correlation between morphological categories and actual dispersal mechanism has rarely been tested, so the mechanism is usually inferred rather than demonstrated. For example, Thomson et al. (2010) found data on dispersal mechanism for only 15 % of eastern Australian grasses. Studies of grasses in Missouri and in Botswana found that propagule weight and morphology affected terminal velocity (measured as the speed with which a propagule fell from a specified height) and ability to be transported laterally (Ernst et al. 1992; Rabinowitz and Rapp 1981). In general, lighter propagules dispersed more widely, but distances were rarely more than a few meters at most (Cheplick 1998).

In one study, grasses were found to disperse soon after flowering, although some exceptions occurred (Rabinowitz and Rapp 1980). Thus, cool-season grasses (e.g., Pooideae) tend to disperse in summer, whereas warm-season grasses disperse in autumn. The exception was *Festuca paradoxa*, which dispersed throughout the growing season.

Dispersal in time: seed dormancy

The population dynamics of a plant species are controlled in part by the timing of germination and the ability of the seed to persist in the soil. In general, as a seed matures, the moisture content of the seed is reduced, particular sets of proteins accumulate, and the hormone abscisic acid (ABA) increases, as does sensitivity of the embryo to its effects; at the same time gibberellin content and sensitivity are low. In the grasses, this process occurs along with accumulation of starch in the endosperm. The details of seed maturation are best worked out in *Arabidopsis*, which lacks starchy endosperm and thus may not be comparable to the grasses; the controls of seed maturation and germination must be investigated independently in the grasses (Barrero et al. 2010; Gutierrez et al. 2007; Sabelli and Larkins 2009). (See also the section on Embryology above.)

While many grass seeds germinate immediately after they are shed, others require a period of after-ripening or are dormant. After-ripening is a short period of time in which the seed is unable to germinate, but germinability generally is gradually restored by a period of dry storage. Dormancy is the inability to germinate even when environmental conditions are otherwise favorable (Baskin and Baskin 1998; Simpson 1990). Dormancy is under genetic control, and varies among species and between plants within a species (Barrero et al. 2012; Baskin and Baskin 1998; Simpson 1990). Baskin and Baskin (1998) list a sample of 11 non-domesticated species in which fresh mature seeds will germinate immediately; these include representatives of the Pooideae, Chloridoideae and Panicoideae, indicating that lack of dormancy is phylogenetically widespread. A list of genera and species in which dormancy has been evaluated is provided by Simpson (1990); it would be interesting to update this list, which seems likely to expand.

Dormancy in the grasses is not related to embryo maturity, unlike the case in some other angiosperms (Baskin and Baskin 1998). In all grasses known, the embryo is mature when the seed is shed, and if the embryo is excised from the caryopsis it will almost always germinate (Simpson 1990). This observation thus points to the endosperm, aleurone, or pericarp as maintaining the dormant state.

The physical constraint of the pericarp appears unlikely to cause dormancy (Baskin and Baskin 1998; Simpson 1990). Whereas some angiosperms have seeds that must be scarified before germination occurs, this does not appear to be an absolute requirement in the grasses. However, puncturing the pericarp just above the embryo does enhance the percentage of germination in some species, apparently by permitting penetration of oxygen or additional water (Simpson 1990).

Dormancy is affected by the presence of lemmas, paleas and glumes, which are often part of the dispersal unit (Barrero et al. 2012; Baskin and Baskin 1998; Simpson 1990). These structures appear to limit access of water and light; when they are removed germination percentage is increased. In *Brachypodium*, for example, dormancy enforced by presence of the lemma and palea may last longer than dormancy of the naked caryopsis, but is eventually lost in darkness (Barrero et al. 2012). Although there has been some suggestion that the palea, lemma and glumes contain inhibitors to germination, there is little evidence for this.

The conditions that break dormancy vary widely and do not appear to correlate with phylogeny, hinting that the signaling pathways that control germination are exquisitely sensitive to modification by natural selection in response to particular environments. Presence or absence of light, and the quality of the light, are both major factors in germination (Simpson 1990). Both blue light, acting through the cryptochrome system, and red/far-red light, acting primarily through phytochrome B, regulate dormancy in grasses; both photoreceptors in turn regulate ABA (Gubler et al. 2008; Simpson 1990; Xu et al. 2009).

Moisture and temperature are also important for some species, but the particular combination of the two is species-specific. While some species (e.g., winter annuals such as *Avena fatua*) require high temperature and low moisture, others require exactly the opposite (Baskin and Baskin 1998). For most species, these requirements have never been dissected at the molecular level.

Exposure to smoke breaks dormancy in some grasses, an obvious adaptation of plants to fire-prone ecosystems. The effects of smoke interact with those of light, temperature and moisture content, so their effect on dormancy is not easy to predict (Turner et al. 2009).

Dormancy is affected by inflorescence architecture, suggesting a possible adaptive role for the considerable inflorescence diversity in the grasses. Dormancy and germinability vary within a plant, and are affected by position of the spikelet within the inflorescence and position of the flower within the spikelet (Simpson 1990). For example, in *Agrostis curtisii*, *Avenula marginata*, and *Pseudarrhenatherum longifolium*, position of the seed in the inflorescence influences dormancy, and interacts with high temperatures and plant ash to control germination (González-Rabanal et al. 1994).

The connection between inflorescence architecture and dormancy may be indirect and be mediated by seed provisioning. In *Eremopyrum distans*, seeds from the proximal flower of the spikelet had a higher proportion with physiological dormancy and lower seedling emergence the next spring (Wang et al. 2010). This pattern correlates with variation in seed weight mentioned above. Howard et al. (2012) show that dormancy is related to starch content of the seed, which is inversely related to sugar content. Higher sugar content led to reduced sensitivity of the embryo to ABA in barley, and hence less dormancy. However, the relationship between seed size and dormancy is not simple, and varies among species (Baskin and Baskin 1998).

Among domesticated species, selection has reduced dormancy (Barrero et al. 2010). In some cases this creates the problem of pre-harvest sprouting, in which the seeds germinate on the plant (Gubler et al. 2005). "Viviparous" mutants in maize generally result from defects in the ABA pathway, suggesting that the seed maturation program is disrupted. Loss of dormancy in crops is thus associated with a change in the hormonal environment of the developing seed.

Dispersal by animals

The dispersal units in many grasses have hooks or barbs on various parts, and these could be adaptations for dispersal by mammals or birds. Mediterranean annual grasses are reported to be transported by mammals in Israel (Shmida and Ellner 1983), and the pooid species *Agropyron cristatum* and *Stipa krylovii* were found to be transported easily in the fur of sheep and goats in Mongolia (Bläss et al. 2010). Diaspores of many species of grasses were found in the hair of bison in tallgrass prairie in North America; most of these did not have any obvious hooks or barbs that would attach them to the animals, although many did have hairs or awns whose function is unclear (Rosas et al. 2008). A careful developmental study of the hooks in *Pseudechinolaena* by Lucas (1979) showed that they are emergences of the back of the glumes, tipped by a single stiff macrohair.

Grass fruits will survive passage through the guts of mammals, and can be found germinating in dung. Somewhat surprisingly, the grasses reported to move in this way are predominantly pooids, whose caryopses are enclosed only in a membranous lemma and palea rather than any indurated covering. Introduced annuals of the genera *Briza, Bromus, Vulpia, Lolium* and *Poa* constitute a high proportion of seedlings sprouting from the dung of African ungulates (Milton 2004), and nearly half the seeds found in dung of bison in a tallgrass prairie were from grasses (Rosas et al. 2008). Cosyns et al. (2005) fed seeds of *Agrostis capillaris, Anthoxanthum odoratum* and *Poa pratensis* to sheep, cattle, rabbits, horses, and donkeys, and compared their germination to that of seeds placed directly on soil or on soil plus dung. Germinability for ingested seeds of all species was mostly less than 10 % of that of non-ingested seeds, but was more than zero. Seeds of *Eragrostis lehmanniana* can pass through the digestive tract of a sheep and still be viable (Fredrickson et al. 1997). While Bläss et al. (2010) contend that transport of seeds in the gut of grazers is only of minor importance, such transport could provide a mechanism for relatively long distance dispersal.

A few grasses have fleshy dispersal units. The pericarp is fleshy in a few species of bamboo (e.g., *Melocanna, Ferrocalamus*), whereas the fleshy structures in *Lasiacis* (Panicoideae) are glumes and sterile lemmas that become shiny black at maturity and accumulate oil in their epidermal layers (Davidse and Morton 1973). Spikelets of *Lasiacis* were collected from the stomachs of fruit-eating birds, indicating that the birds do consume them, and the seeds germinated, indicating that the seeds can be dispersed. It is unknown whether other panicoid grasses will survive passage through the gut of a bird, but this could provide a mechanism for long-distance dispersal.

Insect dispersal is rarely reported, but may simply be difficult to detect. Grasses with smooth caryopses are easily eaten by birds and rodents, but are harder for ants to grasp. Conversely, grasses in which the caryopsis is enclosed in a spikelet with awns or hairs are transported by ants in Arizona deserts (Pulliam and Brand 1975). Davidse (1987) lists 31 species in 11 genera of Panicoideae in which the rachilla contains oil, as indicated by staining with Sudan IV, suggesting that they may function as elaiosomes. *Streptostachys* has a thickened rachilla that also could be an elaiosome (Morrone and Zuloaga 1991).

Role of the awn

Awns are common in grasses and often form a prominent part of the spikelet. Awns most commonly occur on the lemma, but awned glumes are also found in some taxa. The position of the awn, whether on the lemma or glume, varies; it may form at the apex, or from a sinus, or on the abaxial side (generally called the dorsal side, although this usage contradicts the use of the word "dorsal" in other plant groups). It may be straight or twisted; generally if twisted the awn is hygroscopic, changing shape according to the amount of available moisture. While it is most common to have a single awn per lemma or glume, the lateral veins may be prolonged into awns so that the lemma or glume has three, five or even more awns.

Awns are thought to be adaptations for seed dispersal (Elbaum et al. 2007; Garnier and Dajoz 2001; Peart 1979, 1981, 1984; Peart and Clifford 1987). In addition, awns also affect seed provisioning (e.g., Li et al. 2006; Motzo and Giunta 2002), and possibly also drought stress (Abebe et al. 2010).

The role of awns in dispersal has been explored in a few species (Elbaum et al. 2007; Garnier and Dajoz 2001; Peart 1979, 1981, 1984; Peart and Clifford 1987). Wetting and drying of a hygroscopic awn causes the dispersal unit (often a flower or spikelet) to shift position and can even propel it several millimeters across the substrate. Hygroscopic awns in *Themeda triandra* move the spikelets along the soil and increase the chance that they will become buried (Sindel et al. 1993). The hygroscopic awn in *Macrochloa tenacissima* (= *Stipa tenacissima*) causes the diaspore to be buried in the soil or in crevices (Schöning et al. 2004). The presence of lateral awns, in addition to the central one, improves the effect of the hygroscopic awn in locating suitable microsites (Peart

1979). In *Avena fatua*, flowers within a spikelet mature from bottom to top (acropetally) and the hygroscopic action of the awn on the lowermost flower appears to force the flowers out of the spikelets. The result is dispersal of seeds of various sizes and at various stages of maturity (Raju and Ramaswamy 1983).

Presence or absence of an awn affects seed germination. In a series of experiments of six species of grasses with awns, Peart (1979) showed that twisting of the hygroscopic awn placed the anthecium into a microsite suitable for seed germination, such as a crack in the soil; removal of awns resulted in fewer seeds arriving in suitable microsites. He found an interaction between awn type (species) and soil type, with some awn morphologies more efficient than others in locating microsites in different sorts of soils. Schöning et al. (2004) show that seeds from diaspores with intact awns have higher germination than those with awns removed. In addition, awns make the seed less susceptible to predation by ants; although the awn breaks off eventually, once the diaspore is buried the ants are unable to find it. The experiments of Schöning et al. (2004) and Peart (1979) are highly informative and deserve to be repeated on a much larger set of species.

The length of the awn varies among closely related species, but within a species awn length is highly heritable, with minimal effect of the environment (Garnier and Dajoz 2001). In *Hyparrhenia diplandra* there was a positive relationship between the length of the twisted hygroscopic awn and both the proportion of buried seeds and mean seed burial depth (Garnier and Dajoz 2001).

Fire appears to be a selective force on awn morphology. Peart (1984) found that species lacking an awn were more likely to germinate after a fire. He postulated that the awned species tended to germinate on or near the surface and were thus destroyed by fire, whereas species without appendages were more likely to be buried. In contrast, Garnier and Dajoz (2001) found a positive relationship between awn length and burial depth, with longer awned species being more deeply buried and more likely to survive intense fires. In general, variability in fire occurrence and intensity also selects for variation in awn length (Garnier and Dajoz 2001).

Long twisted awns often occur in species in which the outside of the dispersal unit (lemma or glume) has long hairs and in which the callus is long and pubescent, a combination of characters described as the "burial syndrome" (Humphreys et al. 2011). Long hairs on the dispersal unit have an effect similar to that of hygroscopic awns in affecting orientation of the propagule relative to the soil (Peart 1979). Long hairs, whether on the propagule itself or only on the callus, also provide an anchor so that the radicle can penetrate the soil more effectively. In the absence of anchoring, the force of radicle expansion can push the anthecium out of the ground and can increase mortality (Peart 1981). Comparing one species each of *Aristida* and *Microlaena*, Peart (1981) found an interaction between presence of awns, presence of barbs on the callus, and soil type.

Although awns appear to provide a selective advantage in species in which they occur, awns have been lost many times in the evolution of grasses. In Danthonioideae, awns have been lost independently about 27 times, although the exact number depends on analytical assumptions (Humphreys et al. 2011). In that subfamily, loss of awns occurs preferentially in annual lineages, suggesting that perhaps awns do not confer an advantage in species with an annual life history. Awnless danthonioids tend to have glabrous lemmas and short glabrous calluses, both characters that imply that seeds are not buried. Conversely, awned danthonioids generally have longer calluses and tufts of hairs on the lemmas. Loss of the various components of the burial syndrome occurs in environments where active burial is less advantageous, and in which burial can be random.

Despite the ecological and agronomic importance of awns, the genetic control of their development has not been explored extensively. A recent study has identified one gene that controls cell division in the lemma of rice and thereby regulates awn formation (Luo et al. 2013). This same study found that awn length and grain length were positively correlated with each other, and inversely correlated with the number of grains per inflorescence, providing a genetic basis for the trade-off between fruit size and number.

Forcible dispersal

Forcible dispersal, or ballistochory, is not common in grasses but has been reported for *Sucrea* and *Raddia* (Olyreae, Bambuoisdeae) where the mode of dispersal may be synapomorphic (Sendulsky 1993; Oliveira et al. 2014). As the glumes dry out, cells on the adaxial side abruptly lose turgor; the glumes then roll up and eject the diaspore. The diaspore includes the caryopsis enclosed in an indurate and smoothly polished lemma and palea, which slips easily out of the glumes and is shot a distance of 1–2 m.

References

Abebe, T., Melmaiee, K., Berg, V., Wise, R.P. 2010. Drought response in the spikes of barley: gene expression in the lemma, palea, awn, and seed. Funct. Integr. Genomics 10: 191–205.

Barrero, J.M., Jacobsen, J., Gubler, F. 2010. Seed dormancy: Approaches for finding new genes in cereals. In: Pua, E.C., Davey, M.R. (eds.) Plant developmental biology - Biotechnological perspectives. Berlin: Springer. pp. 361–381

Barrero, J.M., Jacobsen, J.V., Talbot, M.J., White, R.G., Swain, S.M., Garvin, D.F. 2012. Grain dormancy and light quality effects on germination in the model grass *Brachypodium distachyon*. New Phytol. 193: 376–386.

Baskin, C.C., Baskin, J.M. 1998. Ecology of seed dormancy and germination in grasses. In: Cheplick, G.P. (ed.) Population biology of grasses. Cambridge: Cambridge University Press. pp. 30–83

Belzile, F., Labbé, J., LeBlanc, M.-C., Lavoie, C. 2010. Seeds contribute strongly to the spread of the invasive genotype of the common reed (*Phragmites australis*). Biol. Invasions 12: 2243–2250.

Bläss, C., Ronnenberg, K., Tackenberg, O., Hensen, I., Wesche, K. 2010. The relative importance of different seed dispersal modes in dry Mongolian rangelands. J. Arid Environ. 74: 991–997.

Cheplick, G.P. 1998. Seed dispersal and seedling establishment in grass populations. In: Cheplick, G.P. (ed.) Population biology of grasses. Cambridge: Cambridge University Press. pp. 84–105

Cosyns, E., Delporte, A., Lens, L., Hoffmann, M. 2005. Germination success of temperate grassland species after passage through ungulate and rabbit guts. J. Ecol 93: 353–361.

Davidse, G. 1987. Fruit dispersal in the Poaceae. In: Soderstrom, T.R., Hilu, K.W., Campbell, C.S., Barkworth, M.E. (eds.) Grass systematics and evolution. Washington, DC: Smithsonian Institution. pp. 143–155

Davidse, G., Morton, E. 1973. Bird-mediated fruit dispersal in the tropical grass genus *Lasiacis* (Gramineae: Paniceae). Biotropica 5: 162–167.

Doust, A.N., Mauro-Herrera, M., Francis, A.D., Shand, L. C. 2014. Morphological diversity and genetic regulation of inflorescence abscission zones in grasses. Amer. J. Bot. 101: 1759–1769.

Dransfield, S. 1981. The genus *Dinochloa* (Gramineae-Bambusoideae) in Sabah. Kew Bull. 36: 613–633.

Elbaum, R., Zaltzman, L., Burgert, I., Fratzl, P. 2007. The role of wheat awns in the seed dispersal unit. Science 316: 884–886.

Ernst, W.H.O., Veenendaal, E.M., Kebakile, M.M. 1992. Possibilities for dispersal in annual and perennial grasses in a savanna in Botswana. Vegetatio 102: 1–11.

Fredrickson, E.L., Estell, R.E., Havstad, K.M., Ksiksi, T., Van Tol, J., Remmenga, M.D. 1997. Effects of ruminant digestion on germination of Lehmann lovegrass seed. J. Range Manage. 50: 20–26.

Fussell, L.K., Dwarte, D.M. 1980. Structural changes of the grain associated with black region formation in *Pennisetum americanum*. J. Exp. Bot. 31: 645–654.

Garnier, L.K.M., Dajoz, I. 2001. Evolutionary significance of awn length variation in a clonal grass of fire-prone savannas. Ecology 82: 1720–1733.

Gegas, V.C., Nazari, A., Griffiths, S., Simmonds, J., Fish, L., Orford, S., Sayers, L., Doonan, J.H., Snape, J.W. 2010. A genetic framework for grain size and shape variation in wheat. Plant Cell 22: 1046–1056.

Giles, K.L., Bassett, H.C.M., Eastin, J.D. 1975. The structure and ontogeny of the hilum region in *Sorghum bicolor*. Aust. J. Bot. 23: 795–802.

González-Rabanal, F., Casal, M., Trabaud, L. 1994. Effects of high temperatures, ash and seed position in the inflorescence on the germination of three Spanish grasses. J. Veg. Sci. 5: 289–294.

Greipsson, S., Davy, A.J. 1995. Seed mass and germination behaviour in populations of the dune-building grass *Leymus arenarius*. Ann. Bot. 76: 493–501.

Gubler, F., Millar, A.A., Jacobsen, J.V. 2005. Dormancy release, ABA and pre-harvest sprouting. Curr. Opinion Pl. Biol. 8: 183–187.

Gubler, F., Hughes, T., Waterhouse, P., Jacobsen, J. 2008. Regulation of dormancy in barley by blue light and after-ripening: effects on abscisic acid and gibberellin metabolism. Plant Phys. 147: 886–896.

Gutierrez, L., Wuytswinkel, O.V., Castelain, M., Bellini, C. 2007. Combined networks regulating seed maturation. Trends Pl. Sci. 12: 1360–1385.

Harrington, G.T., Crocker, W. 1923. Structure, physical characteristics, and composition of the pericarp and integument of Johnson grass seed in relation to its physiology. J. Agric. Res. 23: 193–222.

Howard, T.P., Fahy, B., Craggs, A., Mumford, R., Leigh, F., Howell, P., Greenland, A., Smith, A.M. 2012. Barley mutants with low rates of endosperm starch synthesis have low grain dormancy and high susceptibility to preharvest sprouting. New Phytol. 194: 158–167.

Humphreys, A.M., Antonelli, A., Pirie, M.D., Linder, H.P. 2011. Ecology and evolution of the diaspore "burial syndrome". Evolution 65: 1163–1180.

Jurado, E., Westoby, M., Nelson, D. 1991. Diaspore weight, dispersal, growth form and perenniality of Central Australian Plants. J. Ecol. 79: 811–828.

Li, X., Wang, H., Li, H., Zhang, L., Teng, N., Lin, Q., Wang, J., Kuang, T., Li, Z., Li, B., Zhang, A., Lin, J. 2006. Awns play a dominant role in carbohydrate production during the grain-filling stages in wheat (*Triticum aestivum*). Physiol. Plant. 127: 701–709.

Lucas, M.A. 1979. Hooked spikelet emergences of *Pseudechinolaena* (Gramineae-Panicoideae). I. Comparative morphology and anatomy. Isleya 1: 115–139.

Luo, J., Liu, H., Zhou, T., Gu, B., Huang, X., Shangguan, Y., Zhu, J., Li, Y., Zhao, Y., Wang, Y., Zhao, Q., Wang, A., Wang, Z., Sang, T., Wang, Z., Han, B. 2013. *An-1* encodes a basic helix-loop-helix protein that regulates awn development, grain size, and grain number in rice. Plant Cell 25: 3360–3376.

Milton, S.J. 2004. Grasses as invasive alien plants in South Africa. S. Afr. J. Sci. 100: 69–75.

Morrone, O., Zuloaga, F. 1991. Revision de genero *Streptostachys* (Poaceae-Panicoideae), su posicion sistematica dentro de la tribu Paniceae. Ann. Missouri Bot. Gard. 78: 359–376.

Motzo, R., Giunta, F. 2002. Awnedness affects grain yield and kernel weight in near-isogenic lines of durum wheat. Aust. J. Agric. Res. 53: 1285–1293.

Narayanaswami, S. 1956. Structure and development of the caryopsis in some Indian millets. VI. *Setaria italica*. Bot. Gaz. 118: 112–122.

Oliveira, R.P., Clark, L.G., Schnadelbach, A.S., Monteiro, S.H.N., Borba, E.L., Longhi-Wagner, H.M., van den Berg, C. 2014. A molecular phylogeny of *Raddia* and its allies within the tribe Olyreae (Poaceae, Bambusoideae) based on non-coding plastid and nuclear spacers. Mol. Phylog. Evol. 78: 105–117.

Peart, M.H. 1979. Experiments on the biological significance of the morphology of seed-dispersal units in grasses. J. Ecol. 67: 843–863.

Peart, M.H. 1981. Further experiments on the biological significance of the morphology of seed-dispersal units in grasses. J. Ecol. 69: 425–436.

Peart, M.H. 1984. The effects of morphology, orientation and position of grass diaspores on seedling survival. J. Ecol. 72: 437–453.

Peart, M.H., Clifford, H.T. 1987. The influence of diaspore morphology and soil-surface properties on the distribution of grasses. J. Ecol. 75: 569–576.

Percival, J. 1921. The wheat plant: a monograph. New York: E. F. Dutton & Co.

Pulliam, H.R., Brand, M.R. 1975. The production and utilization of seeds in plains grassland of southeastern Arizona. Ecology 56: 1158–1166.

Rabinowitz, D., Rapp, J.K. 1980. Seed rain in a North American tall grass prairie. J. Appl. Ecol. 17: 793–802.

Rabinowitz, D., Rapp, J.K. 1981. Dispersal abilities of seven sparse and common grasses from a Missouri prairie. Amer. J. Bot. 68: 616–624.

Raju, M.V.S., Ramaswamy, S.N. 1983. Studies on the inflorescence of wild oats (*Avena fatua*). Can. J. Bot. 61: 74–78.

Ramanayake, S.M.S.D., Weerawardene, T.E. 2003. Flowering in a bamboo, *Melocanna baccifera* (Bambusoideae: Poaceae). Bot. J. Linn. Soc. 143: 287–291.

Rosas, C.A., Engle, D.M., Shaw, J.H., Palmer, M.W. 2008. Seed dispersal by *Bison bison* in a tallgrass prairie. J. Veg. Sci. 2008: 769–778.

Rost, T.L., Izaguirre de Artucio, P., Risley, E.B. 1984. Transfer cells in the placental pad and caryopsis coat of *Pappophorum subbulbosum* Arech. (Poaceae). Amer. J. Bot. 71: 948–957.

Rudall, P., Dransfield, S. 1989. Fruit structure and development in *Dinochloa* and *Ochlandra* (Gramineae-Bambusoideae). Ann. Bot. 63: 29–38.

Sabelli, P.A., Larkins, B.A. 2009. The development of endosperm in grasses. Plant Phys. 149: 14–26.

Schöning, C., Espadaler, X., Hensen, I., Roces, F. 2004. Seed predation of the tussock-grass *Stipa tenacissima* L. by ants (*Messor* spp.) in southeastern Spain: the adaptive value of trypanocarpy. J. Arid Environ. 56: 43–61.

Sendulsky, T. 1993. First report of ballistochory in Poaceae. Ann. Missouri Bot. Garden 80: 518–521.

Shmida, A., Ellner, S. 1983. Seed dispersal on pastoral grazers in open Mediterranean Chaparral, Israel. Israel J. Bot. 32: 147–159.

Simpson, G.M. 1990. Seed dormancy in grasses. Cambridge: Cambridge University Press.

Sindel, B.M., Davidson, S.J., Kilby, M.J., Groves, R.H. 1993. Germination and establishment of *Themeda triandra* (kangaroo grass) as affected by soil and seed characteristics. Aust. J. Bot. 41: 105–117.

Thomson, F.J., Moles, A.T., Auld, T.D., Ramp, D., Ren, S., Kingsford, R.T. 2010. Chasing the unknown: predicting seed dispersal mechanisms from plant traits. J. Ecol. 98: 1310–1318.

Turner, S.R., Merritt, D.J., Renton, M.S., Dixon, K.W. 2009. Seed moisture content affects afterripening and smoke responsiveness in three sympatric Australian native species from fire-prone environments. Aust. Ecol. 34: 866–877.

Wang, A.B., Tan, D.Y., Baskin, C.C., Baskin, J.M. 2010. Effect of seed position in spikelet on life history of *Eremopyrum distans* (Poaceae) from the cold desert of north-west China. Ann. Bot. 106: 95–105.

Xu, P., Xiang, Y., Zhu, H.L., Xu, H.B., Zhang, Z.Z., Zhang, C.Q., Zhang, L.X., Ma, Z.Q. 2009. Wheat cryptochromes: subcellular localization and involvement in photomorphogenesis and osmotic stress responses. Plant Phys. 149: 760–774.

Phytochemistry

Compared to many other angiosperms, the grasses are not noted for the diversity of their secondary compounds. However, as indicated below, this may simply reflect lack of investigation. A wide range of secondary compounds is reported, but in general the data come from a handful of taxa that are particularly important for grazing or field crops; the taxonomic distribution of most compounds is thus largely unknown. Secondary compounds may be produced by roots or by above ground tissue, and production may be constitutive, or be triggered by herbivore attack, or be produced in response to the presence of particular bacteria or fungi. Of the 2400 species in a database of plants with pest-control properties (Grainge and Ahmed 1988), about 50 are grasses; documentation of the chemicals produced by these plants and their effects on other organisms varies considerably in quality. Compounds produced in response to pathogens are sometimes called phytoalexins, a term that refers to their inducibility rather than their chemical composition. A number of grass species harbor fungal endophytes that produce secondary compounds that serve as plant defense; in this case the defense compound is in fact a product of the fungus rather than the plant. Finally, cell walls in the grasses have a distinctive chemical composition; whether it is adaptive in any way is unknown. The material presented here is not intended to be a comprehensive review, but will provide some hint of the phytochemical diversity to be explored in the family.

Much of the existing literature on secondary compounds in grasses reports on the composition of extracts, which are made up of the end products of biosynthetic pathways, and are necessarily biased toward compounds that are produced in relatively high amounts. Studies often rely on extraction of exudates or separation of compounds from whole tissue, followed by fractionation and analysis of compounds by HPLC, GC or other methods (e.g., Bais et al. 2006; Xuan et al. 2006). While such data provide good evidence of what the plant can produce under particular growth conditions, failure to find a compound in an extract does not mean that the plant cannot produce it. This inability to interpret negative data, combined with the patchy sampling, means that it is hard to generalize patterns to broad taxonomic groups.

More interesting evolutionary comparisons are possible in the cases where the biosynthetic pathway is known, or can be determined by genetic approaches, as described below for momilactones in rice. As more biosynthetic pathways are determined, it will be possible to try to locate orthologous genes in other taxa using whole genome sequences; this approach will ultimately provide a better picture of how phytochemical diversification has occurred in the family. In addition, by comparing normal plants with those that have a mutation in a single gene in the biosynthetic pathway, it is possible to isolate the effects of particular secondary compounds on the plant and the community of organisms around it. In general, however, these investigations have focused on crops and have yet to be extended to wild species.

Root exudates from grasses may have allelopathic properties (Duke 2007), and allelopathy has been reported in most major cereals (Belz 2007). Allelopathic effects are notoriously hard to demonstrate conclusively, because negative

E.A. Kellogg, *Flowering Plants. Monocots*, The Families and Genera of Vascular Plants 13,
DOI 10.1007/978-3-319-15332-2_8, © Springer International Publishing Switzerland 2015

plant-plant interactions can be caused by many other direct (e.g., competition) and indirect (e.g., effects on soil microbes) effects. In addition, such studies are necessarily biased toward compounds that are produced in relatively high quantities under particular growth conditions, or constitutively.

For at least two of the biosynthetic pathways described below – those for momilactones in rice and for avenacins in oats – the genes encoding the relevant enzymes are clustered in the genome (Field and Osbourn 2008; Qi et al. 2004; Xu et al. 2012). This is apparently the result of genomic rearrangement; the genes are not linked in other grasses. Such an arrangement has also been found for related genes in *Arabidopsis*, and may point to selection for co-regulation of components of the pathway (Field and Osbourn 2008). In some cases also the gene products are all anchored at a particular cellular location, forming metabolic units called metabolons; such localization appears to increase efficiency of biosynthesis and to reduce accumulation of potentially toxic intermediates.

Phenolics

Various secondary compounds are produced by the phenylpropanoid pathway, beginning with phenylalanine, and proceeding through steps catalyzed by phenyalanine ammonia lyase, cinnamate-4-hydroxylase, 4-coumarate: CoA ligase, and chalcone synthase. Some of the most widespread and best characterized of the products of this pathway are the flavonoids. Flavonoids are widespread plant pigments that are water soluble and generally stored in vacuoles. They exhibit highly diverse structures, and confer a variety of benefits on the plant, including disease resistance, deterrence of herbivory, and protection against UV damage (Grotewold 2006; Harborne and Williams 2000). The biosynthetic pathway of flavonoids is shared among seed plants (Tanaka et al. 2008). Like many plants, grasses produce a variety of flavonoids (Harborne and Williams 1987), the most common of which is tricin, found in over 90 % of grass species. Other flavonoids are apparently taxonomically restricted, although it is unclear whether these compounds really are specific, or whether they have simply not been detected in other taxa. For example, members of subfamily Pooideae are the most common grass producers of flavonols, the panicoid *Bothriochloa bladhii* is the only recorded source of 7,3′,4′-trihydroxyflavone, and the panicoids *Arthraxon hispidus* and *Miscanthus tinctorius* are the only species known to produce arthraxin, a derivative of luteolin (Harborne and Williams 1987). Rice seedlings may also release 5,7,4′-trihydroxy-3′,5′-dimethoxyflavone, and 3-isopropyl-5-acetoxycyclohexene-2-one-1 into the environment; these inhibit the growth of *Echinochloa crus-galli* and *Cyperus difformis*, and are particularly potent in combination (Kong et al. 2004).

Flavonoids are often found as glycosides, in which the flavonoid molecule is bound to a sugar; diversity comes from the kind and number of sugars and the nature of the bond between the sugar and the flavonoid (sometimes called the aglycone). Both *C*-glycosylflavones and *O*-glycosides are common in the grasses (Harborne and Williams 1987), as they are in other families. A leaf of barley or wheat may contain several dozen different *C*-glycosyl-flavones. Maysin is a *C*-glycosyl-flavone produced in the stigmas of maize, where it provides resistance to corn earworm (McMullen et al. 2009). In addition to the glycosides, some taxa, particularly in the PACMAD clade, produce flavonoid sulphates, but these are apparently not found (or not recorded) for members of the BEP clade.

Anthocyanins, a subgroup of flavonoids, are common in Poaceae, and probably occur in all species. The anthocyanins of maize are well known, producing brightly colored kernels and ears; study of these pigments and the enzymes that produce them has provided insight into fundamental aspects of gene structure and regulation, and famously led to the discovery of transposable elements (Comfort 2001). All grasses investigated produce the anthocyanin cyanidin and its derivatives (chiefly cyanidin 3-glucoside), whereas the distribution of delphinidin and peonidin in plant extracts is more sporadic (Escribano-Bailón et al. 2004; Fossen et al. 2002). Phlobaphenes, polymers of flavan-4-ols, form in the floral tissues of maize and sorghum and accumulate in the developing kernel. The biosynthetic pathway of these is becoming well known, and is shown to proceed via luteoforol or apiforol (Morohashi et al. 2012).

Other compounds produced by the phenylpropanoid pathway are the stilbenes. Sorghum seedlings produce the stilbene trans-piceid in response to fungal inoculation (Chong et al. 2009). In many plants, duplicate genes for chalcone synthase have been modified so they instead encode stilbene synthase, a protein that creates stilbene, a cyclic compound distinct from chalcone. Thus, stilbene synthase has been derived repeatedly from chalcone synthase (Chong et al. 2009), creating a novel branch in the phenylpropanoid pathway. In sorghum, stilbene synthase appears to be single copy (Paterson et al. 2009; Yu et al. 2005).

The cell wall composition of the grasses has attracted considerable attention because several members of the family (e.g., *Panicum virgatum*, *Miscanthus* x *giganteus*, and prairie grasses in general) have been suggested as possible sources of cellulosic biofuels. The accumulation of ferulic acid and coumaric acid in non-lignified cell walls is a characteristic shared with all other commelinid monocots such as palms, bananas and gingers. Primary walls include glucuronoarabinoxylans, a characteristic shared with Zingiberales and Commelinales. Presence of (1–3,1-4)-β-D-glucans is shared with the other members of the graminid clade.

Another major group of phenolics is the quinones, structurally similar to ubiquinone, an electron carrier in the photosynthetic electron transport chain. Sorgoleone is a *p*-benzoquinone produced in root hairs of *S. bicolor* and *S. halepense* (Czarnota et al. 2001, 2003). It is thought to inhibit growth of nearby plants by disrupting electron transport in the mitochondrion and the chloroplast by binding to specific components of the electron transport chains (Bais et al. 2006; Czarnota et al. 2001). It seems likely that benzoquinones are produced by other grasses as well.

Derivatives of Amino Acids (Tryptophan, Phenylalanine, Tyrosine)

Grasses produce benzoxazinones, which appear to be useful in defending the plant against bacteria, fungi, and insects (Sicker et al. 2000); the best known of these are DIBOA (2,4-dihydroxy-1,4-benzoxazin-3-one) and DIMBOA (produced by adding a methoxyl group to DIBOA). The compounds are produced as D-glucosides and are stored in the vacuole; when the cell is damaged, the glucose moiety is removed, releasing the toxic aglucone (Sicker et al. 2000). The ability to produce these compounds appears to be synapomorphic for the grasses or at least the members of the BEP plus PACMAD clades; the few dicots that produce them do so using non-orthologous proteins (Dick et al. 2012). The benzoxazinones are derived from indole, which ultimately comes from tryptophan, and their biosynthesis is catalyzed by a set of genes whose chromosomal locations vary among the grasses (Sue et al. 2011). The genes necessary for synthesizing benzoxazinones are missing in cultivated barley, but present in other species of *Hordeum* (Grün et al. 2005). DIBOA is exuded by the roots of rye, and appears to have some allelopathic activity (Sicker et al. 2000).

A variety of indole alkaloids has been reported from grasses in subfamily Pooideae, where they are of particular interest because of their toxicity to livestock. The indole alkaloids dimethyl tryptamine and tyramine-related compounds are common in *Phalaris*, where they cause toxicity to grazers, particularly sheep (Edgar 1994). Thus, breeding efforts in *Phalaris* have focused on trying to reduce the concentration of these compounds in the plants. Levels of the toxins vary according to cultivar, time of year, and moisture conditions (Culvenor et al. 2005). Perloline (a diazaphenanthrene alkaloid) and E/Z-thesenine-rhamnoside have been isolated from *Lolium perenne* (Grimmett and Waters 1943; Koulman et al. 2008).

While some indole alkaloids in Pooideae are undesirable because of their effects on mammals, others are beneficial because of their deterrent effects on insects. The indole alkaloid gramine is found in barley and in *Phalaris*, and has been investigated in the former species because it confers resistance to aphid attack. Barley cultivars differ in production of gramine, and production is also affected by specific aphids and by drought stress (Larsson et al. 2011). Some of the genes for gramine biosynthesis have been identified (Larsson et al. 2006).

Elymus (= *Elytrigia*) *repens* induces nearby plants to become more resistant to insect attack. It releases a complex mixture of substances from its roots, including carboline (6-hydroxy-1.2.3.4-

tetrahydro-b-carboline-3-carboxylic acid, an indole alkaloid). When barley plants are grown with *E. repens*, carboline appears to stimulate chemical changes in nearby barley plants to make them less palatable to aphids (Glinwood et al. 2003). The aphids do not respond to carboline itself, so must react to a chemical released by the barley plants. Thus, chemical interactions between two grasses affect herbivory indirectly.

In the pooid genus *Festuca*, the non-protein amino acid *m*-tyrosine is allelopathic, inhibiting germination and growth of other species of plants. The compound is thought to be derived from L-phenylalanine (Duke 2007). In the roots of *Festuca rubra* and *F. arizonica* *m*-tyrosine makes up 33–43 % of root extract by dry weight and is toxic to a variety of dicots and monocots, but not to fescues that produce the product (Bertin et al. 2007). *F. rubra* and *F. arizonica* are in separate clades in the genus *Festuca* with *F. rubra* in the Aulaxyper group, and *F. arizonica* classified in the *F. ovina* group (Darbyshire and Pavlick 2007; Inda et al. 2008). In contrast, three other members of the *F. ovina* group, *F. longifolia* (= *F. trachyphylla*), *F. ovina*, and *F. idahoensis*, did not produce *m*-tyrosine and indeed were sensitive to its effects on growth. Thus, *m*-tyrosine production apparently does not correlate with phylogeny. The only other species known to produce *m*-tyrosine is the dicot *Euphorbia myrsinitis*, but it is not clear how many other grasses have been tested.

Data are sparse on amino acid derivatives outside subfamily Pooideae. Rice (subfamily Ehrhartoideae) produces various compounds derived from tryptophan when attacked by larvae of the rice striped stem borer; these include tryptamine and serotonin, the latter a well-known animal hormone (Ishihara et al. 2008). Sorghum (subfamily Panicoideae) produces the cyanogenic glycoside dhurrin by modifying the amino acid tyrosine (Poulton 1990). The biosynthetic pathway is short, consisting of two cytochromes P450, a UDP-glycose-glycosyltransferase, and a NADPH-cytochrome P450 reductase, which together form a membrane-bound biosynthetic complex or metabolon (Jensen et al. 2011; Nielsen et al. 2008). The complex has been transferred to *Arabidopsis*, which then produces dhurrin, confirming the functional importance of the pathway (Tattersall et al. 2001). In the plant, dhurrin is sequestered in the vacuole in epidermal cells of the shoot, where it may constitute 30 % of the dry weight; tissue damage then releases the enzymes that break it down to produce HCN (Poulton 1990). Dhurrin may also be modified to a diglucoside, which accumulates in the apoplast and is excreted in guttation droplets (Selmar et al. 1996). Dhurrin is also reported in *Sorghastrum nutans* (Gorz et al. 1979), and in members of the *S. bicolor* complex plus *S. arundinaceum*, *S. halepense*, *S. propinquum*, and *S. versicolor* (Haskins and Gorz 1985).

Terpenoids

Grasses produce a wide variety of terpenoids. These compounds provide protection against fungal and bacterial pathogens, and some may also be allelopathic. In addition, volatile terpenoids affect herbivory either directly by deterring insect herbivores, or indirectly by attracting parasitoids that attack the herbivores in complex tritrophic interactions.

A major group of terpenoids is the diterpenoids, particularly the group known as the labdane-related diterpenes, which protect against fungal disease (Peters 2006, 2010). These are synthesized from (*E*, *E*, *E*)-geranylgeranyl diphosphate, also a precursor of gibberellin; their synthesis is characterized by two sequential cyclization reactions. Many of the genes encoding enzymes in this pathway have been duplicated and modified repeatedly over the course of plant evolution leading to several thousand compounds that are involved in plant defense (Peters 2006, 2010).

Structurally similar compounds may have different effects. For example, in rice, the labdane-related diterpenoid momilactones are allelopathic, whereas the diterpenoid phytocassanes are not. The proteins that synthesize momilactones are encoded by a set of genes that are clustered in the genome (Shimura et al. 2007; Xu et al. 2012). Rice plants lacking either of two of the biosynthetic genes were grown with lettuce, which is unusually sensitive to allelopathic compounds, as well as the common weed *Echinochloa crus-galli* (barnyard grass) (Xu et al. 2012). Plants unable to produce momilactone permitted much greater growth of lettuce and barnyard grass than did the momilactone producers; however, growth of the mutant rice plants was reduced, perhaps by accumulation of toxic intermediates (Xu et al. 2012). Like momilactones, phytocassanes are released from the roots of rice seedlings. However, these diterpenoids have no allelopathic effects on lettuce or Chinese cabbage (Toyomasu et al. 2008), suggesting that the compounds may be produced in response to other pathogens such as fungi (Hasegawa et al. 2010; Umemura et al. 2003).

Triterpenoids may be metabolized to produce saponins, which have antimicrobial activity. While this is most common in dicots, triterpenoid saponins - avenacins - are produced in epidermal cells of the root tip of species of *Avena* and in *Arrhenatherum elatius*, both in Aveninae (Osbourn 2003). (Oats also produce saponins from steroids, the avenosides, but these are produced in the leaves.) Avenacin confers resistance to the fungal pathogen *Gaeumannomyces graminis* var. *tritici* ("take-all") (Papadopoulou et al. 1999). The protein that produces avenacin, β-amyrin synthase, is produced by a gene that is a recent duplicate of a cycloartenol synthase-like gene, which is shared with wheat; apparently after duplication the protein function was slightly modified to produce avenacin (Qi et al. 2004). Avenacin producing genes are clustered in the oat genome, in a location that is not syntenic with the rice genome (Qi et al. 2004).

Many terpenoids are volatile. Some grasses release isoprene or monoterpenes to the atmosphere, but the significance of this is unknown, as is the taxonomic distribution of these emissions (Gibson 2009). Other volatile compounds are released when the plant is attacked by herbivores. While the volatile compounds attract parasitoids, they can also attract more herbivores, with different compounds having distinct effects on particular suites of insects. For example, rice releases (E)- β-caryophyllene constitutively; when the gene producing this compound is silenced, both planthoppers and their parasitoids are reduced, presumably because without the signal the plant becomes effectively invisible (Xiao et al. 2012). Rice also produces S-linalool when attacked by brown planthoppers; in transgenic plants that do not produce the compound, herbivory increased (Xiao et al. 2012). *Melinis minutiflora* (molasses grass) releases e-ocimene, 3–4,8-dimethyl-1,3,7-nonatriene, β-caryophyllene, humulene, and alpha-terpinolene; these compounds appear to repel stem borers that attack many grasses in Africa, as well as crops such as maize and sorghum (Khan et al. 2010). The compounds released constitutively by *M. minutiflora* resemble those released by maize after insect attack; thus, intercropping maize and *Melinis* provides protection against stem borers. While comparable experiments have not been done on grasses, recent work in Brassicaceae suggests that intraspecific variation in defense compounds is likely to be maintained by contrasting selective regimes in different environments throughout the range of a species (Prasad et al. 2012).

A few grasses, particularly in Andropogoneae, produce oils of commercial importance. *Chrysopogon* (= *Vetiveria*) *zizanioides* is the source of vetiver oil, which forms the basis of many western perfumes, including the classic Chanel No. 5 (Joy undated; Maffei 2002). Vetiver oil is produced in cells just outside the root endodermis, cells that are also colonized by bacteria. Vetiver plants grown without bacteria fail to produce oil, and presence of the bacteria correlates with expression of terpene synthase in the plant; the bacteria then metabolize the terpenes produced by the plant to produce the complex mix of compounds characteristic of vetiver oil (Del Giudice et al. 2008).

Cymbopogon species are characterized by their oil production and pleasant scent. Among other compounds, they produce oil that has a high percentage of geraniol, which is commercially valuable for flavoring and for perfume, but is also antimicrobial and may be useful for pest control (Chen and Viljoen 2010). *C. citratus* (lemon grass) produces a lemon-scented oil that is used in cooking; the major components are geranial and its isomer neral (Rodríguez Quintanilla et al. 2012). *C. martinii* (palmarosa) produces an oil that is 65 % geraniol (a monoterpene) and 20 % geranyl acetate (Prashar et al. 2003; Rodríguez Quintanilla et al. 2012), although the relative proportions of these changes during development; palmarosa oil inhibits growth of yeast (*Saccharomyces cerevisiae*) (Prashar et al. 2003). *C. nardus* produces mostly citronellal, geraniol and citronelol, and *C. flexuosus* produces geranial and neral (Rodríguez Quintanilla et al. 2012). Oil from *C. validus* (giant turpentine grass) has as its major constituents myrcene, (E)- β-ocimene, geraniol, linlol, and camphene (Chagonda et al. 2000).

Another member of Andropogoneae, *Elionurus elegans*, produces a variety of sesquiterpenes, the major one being campherenone, which constituted more than a third of the essential oils extracted from the roots and shoots (Mevy et al. 2002). *E. muticus* produces primarily geranial, neral, and geranyl acetate (Chagonda et al. 2000).

Chemicals Produced by Symbionts or Pathogens

In some grasses, toxins are produced by fungi or bacteria living on the plants. While some of these are well studied, others are only referenced briefly in the literature and thus their prevalence in natural ecosystems is unknown.

Fungi in the family Clavicipitaceae (Hypocreales, Ascomycota) are endophytes of grasses (White, Jr. et al. 2003). The most famous of these is ergot (*Claviceps purpurea*), which infects rye; alkaloids from the fungus affect the mammalian nervous system, and in humans cause convulsions and seizures, as well as inducing psychoactive effects. Other well-studied members of the family occur in the genus *Epichloë* (and its anamorph *Neotyphodium*) (White, Jr. et al. 2003). These infect many pooid grasses, including *Lolium cuneatum* (darnel), the "tares" of the gospel of Matthew, which is a weed of wheat fields. Other *Epichloë* infections are responsible for "sleepygrass", an infestation of *Achnatherum robustum* of western North America, which causes grazing horses to go to sleep for two or three days, and drunken horse grass (*Achnatherum inebrians*; Asian); the alkaloid responsible has been identified as lysergic acid amide. Indolediterpenes are also produced and are another cause of livestock poisoning, as found in dronkgras (*Melica decumbens*; South Africa) and huecú (*Poa huecu*; Argentina). Other hosts of clavicipatalean fungi include *Echinopogon*, and species of *Festuca* and *Poa*. The alkaloids have also been widely used medicinally.

Fungal hyphae extend throughout the above ground parts of the plant and infect the developing grain, where they form a thin layer outside the aleurone and inside the seed coat (Schardl et al. 2004). The fungus is thus transmitted maternally through the seed. When the fungus enters its sexual phase, it invades inflorescence tissue and produces conidia; the inflorescence aborts, a condition known as "choke".

Fungal infection is beneficial to the plant in that it reduces herbivory by insects (Schardl et al. 2004). Infected plants are more productive than their uninfected neighbors. For example, lolines (saturated pyrolizidines) produced by fungi infecting *Lolium cuneatum* (darnel) and *L. arundinaceum* (tall fescue) are apparently not toxic to mammals, but are as effective as nicotine in deterring insect feeding (Schardl et al. 2007).

Species of Clavicipitaceae that infect plants are derived from ancestors that were pathogens in animals (Spatafora et al. 2007). Thus, the ability to produce pathogenic alkaloids is ancestral in the fungi, and the grasses have been able to take advantage of this capacity in the establishment of the symbiosis. The shift to plants appears to have occurred only once. *Epichloë* species have been found in many Pooideae, including the early diverging *Brachyelytrum*. Although there is a general pattern of co-speciation of hosts and fungi, some host switching is observed (Schardl et al. 1997).

A number of grasses in Poeae in New Zealand, along with *Danthonia caespitosa*, can be infected with the bacterium *Clavibacter toxicus* (= *Corynebacterium* spp.), which produces corynetoxins (tunicaminyluracil) (Edgar 1994). The bacterium is spread by a nematode in the genus *Anguina*. Livestock that ingest infected leaves suffer neurological damage or death. While this sort of toxin production has been investigated only in an agricultural context, it suggests a possibility of complex interactions of vector, pathogen, and host ecology.

Phytosiderophores

A number of grasses secrete chelating agents, phytosiderophores, which enhance iron uptake in response to low iron concentrations in the soil (Kobayashi and Nishizawa 2012; Römheld and Marschner 1986). The phytosiderophore is secreted from the roots into the soil, where it binds Fe^{3+} in a 1:1 ratio. The complex is then taken up into the roots by a transporter (Murata et al. 2006; Schaaf et al. 2004). This mechanism is apparently restricted to grasses, and has been called Strategy II; other angiosperms have mechanisms that increase iron solubility in the soil, or Strategy I (Kobayashi and Nishizawa 2012; Römheld and Marschner 1986). The phytosiderophores include mugeneic acid, 2′-deoxymugineic acid, and various derivatives of these compounds, all of which are derived from methionine (Ma 2005; Ueno et al. 2007).

References

Bais, H.P., Weir, T.L., Perry, L.G., Gilroy, S., Vivanco, J.M. 2006. The role of root exudates in rhizosphere interactions with plants and other organisms. Ann. Rev. Plant Biol. 57: 233–266.

Belz, R.G. 2007. Allelopathy in crop/weed interactions. Pest Manag. Sci. 63: 308–326.

Bertin, C., Weston, L.A., Huang, T., Jander, G., Owens, T., Meinwald, J., Schroeder, F.C. 2007. Grass roots chemistry: *meta*-tyrosine, an herbicidal nonprotein amino acid. Proc. Natl. Acad. Sci. 104: 16964–16969.

Chagonda, L.S., Makanda, C., Chalchat, J.-C. 2000. The essential oils of wild and cultivated *Cymbopogon validus* (Stapf) Stapf ex Burtt Davy and *Elionurus muticus* (Spreng.) Kunth from Zimbabwe. Flavour Fragrance J. 15: 100–104.

Chen, W., Viljoen, A.M. 2010. Geraniol - a review of a commercially important fragrance material. South Afr. J. Bot. 76: 643–651.

Chong, J., Poutaraud, A., Hugueney, P. 2009. Metabolism and roles of stilbenes in plants. Plant Sci. 177: 143–155.

Comfort, N.C. 2001. The tangled field: Barbara McClintock's search for the patterns of genetic control. Cambridge, Massachusetts: Harvard University Press.

Culvenor, R.A., Reed, K.F.M., McDonald, S.E. 2005. Comparative levels of dimethyltryptamine- and tyramine-related alkaloid toxins in Australian cultivars and some wild populations of *Phalaris aquatica*. Aust. J. Agric. Res. 56: 1395–1403.

Czarnota, M.A., Paul, R.N., Dayan, F.E., Nimbal, C.I., Weston, L.A. 2001. Mode of action, localization of production, chemical nature, and activity of sorgoleone: a potent PSII inhibitor in *Sorghum* spp. root exudate. Weed Technol. 15: 813–825.

Czarnota, M.A., Paul, R.N., Weston, L.A., Duke, S.O. 2003. Anatomy of sorgoleone-secreting root hairs of *Sorghum* species. Int. J. Plant Sci. 164: 861–866.

Darbyshire, S.J., Pavlick, L.E. 2007. *Festuca* L. In: Barkworth, M.E., Capels, K.M., Long, S., Anderton, L.K., Piep, M.B. (eds.) Flora of North America north of Mexico. Vol. 24. Magnoliophyta: Commelinidae (in part): Poaceae, part 1. New York: Oxford University Press. pp. 389–443

Del Giudice, L., Massardo, D.R., Pontieri, P., Bertea, C.M., Mombello, D., Carata, E., Tredici, S.M., Tala, A., Mucciarelli, M., Groudeva, V.I., De Stefano, M., Vigliotta, G., Marrei, M.E., Alifano, P. 2008. The microbial community of vetiver root and its involvement into essential oil biogenesis. Environ. Microbiol. 10: 2824–2841.

Dick, R., Rattei, T., Haslbeck, M., Schwab, W., Gierl, A., Frey, M. 2012. Comparative analysis of benzoxazinoid biosynthesis in monocots and dicots: independent recruitment of stabilization and activation functions. Plant Cell 24: 915–928.

Duke, S.O. 2007. The emergence of grass root chemical ecology. Proc. Natl. Acad. Sci. 104: 16729–16730.

Edgar, J.A. 1994. Toxins in temperate grasses - implications and solutions. N. Z. J. Agric. Res. 37: 341–347.

Escribano-Bailón, M.T., Santos-Buelga, C., Rivas-Gonzalo, J.C. 2004. Anthocyanins in cereals. J. Chromatog. A 1054: 129–141.

Field, B., Osbourn, A.E. 2008. Metabolic diversification - independent assembly of operon-like gene clusters in different plants. Science 320: 543–547.

Fossen, T., Slimestad, R., Øvstedal, D.O., Andersen, Ø.M. 2002. Anthocyanins of grasses. Biochem. Syst. Ecol. 30: 855–864.

Gibson, D.J. 2009. Grasses and grassland ecology. New York: Oxford University Press.

Glinwood, R., Pettersson, J., Ahmed, E., Ninkovic, V., Birkett, M., Pickett, J. 2003. Change in acceptability of barley plants to aphids after exposure to allelochemicals from couch-grass (*Elytrigia repens*). J. Chem. Ecol. 29: 261–274.

Gorz, H.J., Haskins, F.A., Dam, R., Vogel, K.P. 1979. Dhurrin in *Sorghastrum nutans*. Phytochemistry 18: 2024.

Grainge, M., Ahmed, S. 1988. Handbook of plants with pest-control properties. New York: Wiley.

Grimmett, R.E., Waters, D.F. 1943. A fluorescent alkaloid in ryegrass (*Lolium perenne* L.) II. Extraction from fresh ryegrass and separation from other bases. N. Z. J. Sci. Technol. Sect. B 24: 151–155.

Grotewold, E. (ed.) 2006. The science of flavonoids. New York: Springer.

Grün, S., Frey, M., Gierl, A. 2005. Evolution of the indole alkaloid biosynthesis in the genus *Hordeum*: Distribution of gramine and DIBOA and isolation of the benzoxazinoid biosynthesis genes from *Hordeum lechleri*. Phytochemistry 66: 1264–1272.

Harborne, J.B., Williams, C.A. 1987. Flavonoid patterns of grasses. In: Soderstrom, T.R., Hilu, K.W., Campbell, C.S., Barkworth, M.E. (eds.) Grass systematics and evolution. Washington, DC: Smithsonian Institution Press. pp. 107–113

Harborne, J.B., Williams, C.A. 2000. Advances in flavonoid research since 1992. Phytochemistry 55: 481–504.

Hasegawa, M., Mitsuhara, I., Seo, S., Imai, T., Koga, J., Okada, K., Yamane, H., Ohashi, Y. 2010. Phytoalexin accumulation in the interaction between rice and the blast fungus. Mol. Plant-Microbe Interact. 23: 1000–1011.

Haskins, F.A., Gorz, H.J. 1985. Dhurrin and *p*-hydroxybenzaldehyde in seedlings of various *Sorghum* species. Phytochemistry 24: 597–598.

Inda, L.A., Segarra-Moragues, J.G., Müller, J., Peterson, P. M., Catalán, P. 2008. Dated historical biogeography of the temperate Loliinae (Poaceae, Pooideae) grasses in the northern and southern hemispheres. Mol. Phyl. Evol. 46: 932–957.

Ishihara, A., Hashimoto, Y., Miyagawa, H., Wakasa, K. 2008. Induction of serotonin accumulation by feeding of rice striped stem borer in rice leaves. Plant Signaling Behav. 3: 714–716.

Jensen, K., Osmani, S.A., Hamann, T., Naur, P., Møller, B. L. 2011. Homology modeling of the three membrane proteins of the dhurrin metabolon: catalytic sites, membrane surface association and protein-protein interactions. Phytochemistry 72: 2113–2123.

Joy, R.J. Undated. "Sunshine" Vetivergrass *Chrysopogon zizanioides* (L.) Roberty. USDA-NRCS Plant Guide.

Khan, Z.R., Midega, C.A.O., Bruce, T.J.A., Hooper, A.M., Pickett, J.A. 2010. Exploiting phytochemicals for developing a 'push-pull' crop protection strategy for cereal farmers in Africa. J. Exp. Bot. 61: 4185–4196.

Kobayashi, T., Nishizawa, N.K. 2012. Iron uptake, translocation, and regulation in higher plants. Ann. Rev. Plant Biol. 63: 131–152.

Kong, C., Liang, W., Xu, X., Hu, F., Wang, P., Jiang, Y. 2004. Release and activity of allelochemicals from allelopathic rice seedlings. J. Agric. Food Chem. 52: 2861–2865.

Koulman, A., Seeliger, C., Edwards, P.J.B., Fraser, K., Simpson, W., Johnson, L., Cao, M., Rasmussen, S., Lane, G.A. 2008. E/Z-Thesinine-O-4′-alpha-rhmanoside, pyrrolizidine conjugates produced by grasses (Poaceae). Phytochemistry 69: 1927–1932.

Larsson, K.A.E., Zetterlund, I., Delp, G., Jonsson, L.M.V. 2006. *N*-methyltransferase involved in gramine biosynthesis in barley: cloning and characterization. Phytochemistry 67: 2002–2008.

Larsson, K.A.E., Saheed, S.A., Gradin, T., Delp, G., Karpinska, B., Botha, C.E.J., Jonsson, L.M.V. 2011. Differential regulation of 3-aminomethylindole/N-methyl-3-aminomethylindole N-methyltransferase and gramine in barley by both biotic and abiotic stress. Plant Physiol. Biochem. 49: 96–102.

Ma, J.F. 2005. Plant root responses to three abundant soil minerals: silicon, aluminum and iron. Crit. Rev. Plant Sci. 24: 267–281.

Maffei, M. 2002. Introduction to the genus *Vetiveria*. In: Maffei, M. (ed.) The genus *Vetiveria*. London: Taylor & Francis.

McMullen, M.D., Frey, M., Degenhardt, J. 2009. Genetics and biochemistry of insect resistance in maize. In: Bennetzen, J.L., Hake, S.C. (eds.) Handbook of Maize: Its Biology. New York: Springer. pp. 271–289

Mevy, J.P., Bessiere, J.-M., Dherbomez, M., Viano, J. 2002. Composition and some biological activities of the essential oils from an African pasture grass: *Elionurus elegans* Kunth. J. Agric. Food Chem. 50: 4240–4243.

Morohashi, K., Casas, M.I., Ferreyra, L.F., Mejia-Guerra, M.K., Pourcel, L., Yilmaz, A., Feller, A., Carvalho, B., Emiliani, J., Rodriguez, E., Pellegrinet, S., McMullen, M., Casati, P., Grotewold, E. 2012. A genome-wide regulatory framework identifies maize *Pericarp Color1* controlled genes. Plant Cell 24: 2745–2764.

Murata, Y., Ma, J.F., Yamaji, N., Ueno, D., Nomoto, K., Iwashita, T. 2006. A specific transporter for iron(III)-phytosiderophore in barley roots. Plant J. 46: 563–572.

Nielsen, K.A., Tattersall, D.B., Jones, P.R., Møller, B.L. 2008. Metabolon formation in dhurrin biosynthesis. Phytochemistry 69: 88–98.

Osbourn, A.E. 2003. Saponins in cereals. Phytochemistry 62: 1–4.

Papadopoulou, K., Melton, R.E., Leggett, M., Daniels, M.J., Osbourn, A.E. 1999. Compromised disease resistance in saponin-deficient plants. Proc. Natl. Acad. Sci. 96: 12923–12928.

Paterson, A.H., Bowers, J.E., Bruggmann, R., Dubchak, I., Grimwood, J., Gundlach, H., Haberer, G., Hellsten, U., Mitros, T., Poliakov, A., Schmutz, J., Spannagl, M., Tang, H., Wang, X., Wicker, T., Bharti, A.K., Chapman, J., Feltus, F.A., Gowik, U., Grigoriev, I.V., Lyons, E., Maher, C.A., Martis, M., Narechania, A., Otillar, R.P., Penning, B.W., Salamov, A.A., Wang, Y., Zhang, L., Carpita, N.C., Freeling, M., Gingle, A.R., Hash, C.T., Keller, B., Klein, P., Kresovich, S., McCann, M.C., Ming, R., Peterson, D.G., Mehboob-ur-Rahman, Ware, D., Westhoff, P., Mayer, K.F.X., Messing, J., Rokhsar, D.S. 2009. The *Sorghum bicolor* genome and the diversification of grasses. Nature 457: 551–556.

Peters, R.J. 2006. Uncovering the complex metabolic network underlying diterpenoid phytoalexin biosynthesis in rice and other cereal crop plants. Phytochemistry 67: 2307–2317.

Peters, R.J. 2010. Two rings in them all: the labdane-related diterpenoids. Nat. Prod. Rep. 27: 1521–1530.

Poulton, J.E. 1990. Cyanogenesis in plants. Plant Phys. 94: 401–405.

Prasad, K.V.S.K., Song, B.-H., Olson-Manning, C., Anderson, J.T., Lee, C.-R., Schranz, M.E., Windsor, A.J., Clauss, M.J., Manzaneda, A.J., Naqvi, I., Reichelt, I. M., Gershenzon, J., Rupasinghe, S.G., Schuler, M.A., Mitchell-Olds, T. 2012. A gain-of-function polymorphism controlling complex traits and fitness in nature. Science 337: 1081–1084.

Prashar, A., Hili, P., Veness, R.G., Evans, C.S. 2003. Antimicrobial action of palmarosa oil (*Cymbopogon martinii*) on *Saccharomyces cerevisiae*. Phytochemistry 63: 569–575.

Qi, X., Bakht, S., Leggett, M., Maxwell, C., Melton, R., Osbourn, A. 2004. A gene cluster for secondary metabolism in oat: implications for the evolution of metabolic diversity in plants. Proc. Natl. Acad. Sci. USA 101: 8233–8238.

Rodríguez Quintanilla, R., Ruiz Nova, C., Arias Moyano, G., Castro Salazar, H., Martínez, J., Stashenko, E. 2012. Comparative study of the essential oil compositions of four *Cymbopogon* (Poaceae) species grown in Colombia. Bol. Latinamer. Caribe Plantas Medicin. Aromat. 11: 77–85.

Römheld, V., Marschner, H. 1986. Evidence for a specific uptake system for iron phytosiderophores in roots of grasses. Plant Phys. 80: 175–180.

Schaaf, G., Ludewig, U., Erenoglu, B.E., Mori, S., Kitahara, T., Wiren, N.V. 2004. *ZmYs1* functions as a proton-coupled symporter for phytosiderophore- and nicotianamine-chelated metals. J. Biol. Chem. 279: 9091–9096.

Schardl, C.L., Leuchtmann, A., Chung, K.-R., Penny, D., Siegel, M.R. 1997. Coevolution by common descent of fungal symbionts (*Epichloë* spp.) and grass hosts. Mol. Biol. Evol. 14: 133–143.

Schardl, C.L., Leuchtmann, A., Spiering, M.J. 2004. Symbioses of grasses with seedborne fungal endophytes. Ann. Rev. Plant Biol. 55: 315–340.

Schardl, C.L., Grossman, R.B., Nagabhyru, P., Faulkner, J. R., Mallik, U.P. 2007. Loline alkaloids: currencies of mutualism. Phytochemistry 68: 980–996.

Selmar, D., Irandoost, Z., Wray, V. 1996. Dhurrin-6′-glucoside, a cyanogenic diglucoside from *Sorghum bicolor*. Phytochemistry 43: 569–572.

Shimura, K., Okada, A., Okada, K., Jikumaru, Y., Ko, K.-W., Toyomasu, T., Sassa, T., Hasegawa, M., Kodama, O., Shibuya, N., Koga, J., Nojiri, H., Yamane, H. 2007. Identification of a biosynthetic gene cluster in rice for momilactones. J. Biol. Chem. 282: 34013–34018.

Sicker, D., Frey, M., Schulz, M., Gierl, A. 2000. Role of natural benzoxazinones in the survival strategy of plants. Int. Rev. Cytol. 198: 319–346.

Spatafora, J.W., Sung, G.-H., Sung, J.-M., Hywel-Jones, N. L., White, Jr., J.F. 2007. Phylogenetic evidence for an animal pathogen origin of ergot and the grass endophytes. Mol. Ecol. 16: 1701–1711.

Sue, M., Nakamura, C., Nomura, T. 2011 Dispersed benzoxazinone gene cluster: molecular characterization and chromosomal localization of glucoxyltransferase and glucosidase genes in wheat and rye. Plant Phys. 157: 985–997.

Tanaka, Y., Sasaki, N., Ohmiya, A. 2008. Biosynthesis of plant pigments: anthocyanins, betalains and carotenoids. Plant J. 54: 733–749.

Tattersall, D.B., Bak, S., Jones, P.R., Olsen, C.E., Nielsen, J. K., Hansen, M.L., Høj, P.B., Møller, B.L. 2001. Resistance to an herbivore through engineered cyanogenic glucoside synthesis. Science 293: 1826–1828.

Toyomasu, T., Kagahara, T., Okada, K., Koga, J., Hasegawa, M., Mitsuhashi, W., Sassa, T., Yamane, H. 2008. Diterpene phytoalexins are biosynthesized in and exuded from the roots of rice seedings. Biosci. Biotech. Biochem. 72: 562–567.

Ueno, D., Rombola, A.D., Iwashita, T., Nomoto, K., Ma, J. F. 2007. Identification of two novel phytosiderophores secreted by perennial grasses. New Phytol. 174: 304–310.

Umemura, K., Ogawa, N., Shimura, M., Koga, J., Usami, H., Kono, T. 2003. Possible role of phytocassane, rice phytoalexin, in disease resistance of rice against blast fungus *Magnaporthe grisea*. Biosci. Biotech. Biochem. 67: 899–902.

White, Jr., J.F., Bacon, C.W., Hywel-Jones, N.L., Spatafora, J.W. 2003. Historical perspecitives: human interactions with clavicipitalean fungi. In: White, Jr., J.F., Bacon, C.W., Hywel-Jones, N.L., Spatafora, J.W. (eds.) Clavicipitalean fungi: evolutionary biology, chemistry, biocontrol, and cultural impacts. New York: Marcel Dekker, Inc. pp. 1–14

Xiao, Y., Wang, Q., Erb, M., Turlings, T.C.J., Ge, L., Hu, L., Li, J., Han, X., Zhang, T., Lu, J., Zhang, G., Lou, Y. 2012. Specific herbivore-induced volatiles defend plants and determine insect community composition in the field. Ecol. Lett. 15: 1130–1139.

Xu, M., Galhano, R., Wiemann, P., Bueno, E., Tiernan, M., Wu, W., Chung, I.-M., Gershenzon, J., Tudzynski, B., Sesma, A., Peters, R.J. 2012. Genetic evidence for natural product-mediated plant-plant allelopathy in rice (*Oryza sativa*). New Phytol. 193: 570–575.

Xuan, T.D., Chung, I.M., Khanh, T.D., Tawata, S. 2006. Identification of phytotoxic substances from early growth of barnyard grass (*Echinochloa crusgalli*) root exudates. J. Chem. Ecol. 32: 895–906.

Yu, C.K.Y., Springob, K., Schmidt, J., Nicholson, R.L., Chu, I.K., Yip, W.K., Lo, C. 2005. A stilbene synthase gene (*SbSTS1*) is involved in host and nonhost defense responses in sorghum. Plant Phys. 138: 393–401.

Distribution, Habitats and Conservation

Distribution and Maintenance of Grasslands

Grasses are found on all continents, including Antarctica (Convey 2001), and exhibit remarkable ecological diversity. Species occur in deep shade of tropical forests, in full sun in deserts, from near the poles to the equator, and from sea level to high altitudes.

Not only are grasses widespread, but they are often ecologically dominant, giving their name to a major ecosystem type, grasslands. The prairies of North America and the vast grasslands of Africa, the annual grasslands of the Mediterranean and California, the steppes of Russia, the pampas of Argentina, and the great bamboo forests of South America and Asia all attest to the importance of this one family of plants.

The definition of grassland varies among authors (Gibson 2009). In general, grassland is a biome with predominantly herbaceous vegetation, in which grasses are ecological dominants, and in which trees and shrubs are not common. Some definitions (e.g., that of White et al. 2000) include areas such as tundra and shrub lands. It is common to include ecological forces in the definition, such that a grassland is "maintained by fire, grazing, drought and/or freezing temperatures" (White et al. 2000). Because of the variety of definitions of what constitutes grassland, estimates vary as to the percent of the Earth's land surface covered by grasses, but generally range between 30 and 40 % (Gibson 2009; White et al. 2000).

Grasslands (excluding bamboo forests) are one of the major biomes of the world, and were defined by Whitaker (1975) as areas of relatively low precipitation and high temperature. However, the conditions of temperature and precipitation that support grasslands also support forests; thus, these two axes are insufficient for determining where grasslands will occur (Bond 2008; Lehmann et al. 2011). Dynamic global vegetation models, which attempt to predict vegetation based on climate and soil characteristics, also fail to predict the extent of grasslands; instead such models suggest that forest should be far more common than it is (Bond et al. 2005). While these environmental parameters focus on the water use efficiency of grasses, such conditions also create an opportunity for fire. Bond et al. (2005) argue that grasslands in many parts of the world are maintained by fire; they cite multiple fire-exclusion studies in which grassland gradually shifts to forest or woodland. Lehmann et al. (2011) also find that rainfall seasonality is important in maintaining savannas.

Herbivory also affects the balance between grasses and woody vegetation, but its effects are context-dependent (Bond 2008). Whether grazing favors trees or grasses depends primarily on the forage preferences of the herbivores. Herbivory also interacts with fire. For example, the extent of tree cover in the Serengeti is primarily regulated by fire (Holdo et al. 2009). After eradication of rinderpest, a virus that caused high mortality in wildebeest, the wildebeest population expanded; this intensified grazing pressure and reduced grass cover, leading to reduction in available fuel. Fire frequency dropped and tree cover increased (Holdo et al. 2009).

The grasses became ecologically dominant during the Miocene, with the forest to grassland transition occurring at different times on different continents (Edwards et al. 2010; Strömberg

E.A. Kellogg, *Flowering Plants. Monocots*, The Families and Genera of Vascular Plants 13,
DOI 10.1007/978-3-319-15332-2_9, © Springer International Publishing Switzerland 2015

2011). The exact drivers of this transition are not clear, but likely included a combination of increasing aridity and stronger seasonality, which in turn led to altered fire regimes and increased herbivory; the latter two then may have led to transition to grassland. In all cases, the transition to C_4 from C_3 grassland came later, roughly between 8 and 3 million years ago (Edwards et al. 2010). The causes of the shift to C_4-dominated ecosystems are poorly understood; atmospheric CO_2 concentrations have been low enough to favor C_4 photosynthesis since the Oligocene, ca. 30 million years ago (mya), and there was no noticeable shift at 8 mya (Pagani et al. 1999, 2005); therefore CO_2 concentrations cannot explain C_4 grassland expansion. Changes in precipitation are also insufficient to explain C_4 grassland expansion. Models that introduce fire as a major parameter fare better in modeling current grassland distributions (Scheiter et al. 2012).

The rise of grasslands also affected global cycling of silica (Kidder and Gierlowski-Kordesch 2005). The dry weight of C_3 grasses can be 1–3 % silica, a number that may be higher for C_4 species (6–7 %) and for those in wetlands (10–15 %). As grasses die or are eaten, the silica is recycled within the grassland, or is released into runoff and transported to lakes and oceans, where it is available for incorporation into diatoms. Geological evidence for accumulation of diatomite apparently coincides with the earliest expansion of the grasslands in the Miocene (Kidder and Gierlowski-Kordesch 2005). Herbivory and subsequent passage through a mammalian or insect gut appears to increase the solubility of silica and thus enhance silica cycling through the ecosystem, providing a potential link between grassland formation, the rise of large mammalian herbivores, and accumulation of diatoms in aquatic environments (Vandevenne et al. 2013).

Distribution of Major Clades

Individual clades of grasses have distinct ecological preferences, and it is possible that the synapomorphies of many groups are physiological rather than simply morphological. For example, the North American and African grasslands are largely covered with members of the subfamilies Panicoideae and Chloridoideae, with their high efficiency C_4 photosynthesis and preference for warm temperatures.

While the grasses ancestrally occupied warm moist habitats, there have been two independent radiations into cool temperate zones, the Danthonioideae and the Pooideae (Edwards and Smith 2010). The Pooideae not only tolerate cool winters, but many have a specific requirement for vernalization to trigger flowering (Preston and Kellogg 2008). The ability of Pooideae to tolerate cold correlates with an increased number of genes whose expression is modified in response to low temperatures (Sandve and Fjellheim 2010; Sandve et al. 2008) and that show evidence of positive (adaptive) selection (Vigeland et al. 2013). Pooideae are the major members of Mediterranean, Californian, Russian, Argentine, and New Zealand grasslands. Danthonioideae have developed cold tolerance independently of the Pooideae. While an ability to tolerate cold apparently coincided with the origin of the subfamily, major clades within it (*Chionochloa*, *Danthonia*, and *Rytidosperma*) have developed even greater tolerance for lower temperatures (Humphreys and Linder 2013). The mechanisms of cold tolerance in this group are unknown.

Studies are just beginning to tease apart abiotic and biotic environmental correlates of species diversity in particular clades. In South Africa, the C_4 subfamilies Aristidoideae, Chloridoideae, and Panicoideae sort out differently along gradients of moisture, temperature, fire frequency, and grazing intensity (Visser et al. 2012). Aristidoideae are most diverse in hot, dry, disturbed or grazed areas, whereas Chloridoideae reach their highest diversity in areas that are hot but wet, with high grazing pressure but infrequent fire. Panicoideae are most diverse in areas that are warm and mesic, but more species of Paniceae occur with high grazing intensity whereas Andropogoneae are the opposite. Andropogoneae are also most diverse in areas with high fire frequency. It will be of considerable interest to extend this study to other parts of the world.

Species Diversity

Grasslands often harbor high numbers of species, and are commonly listed among the world's diversity hotspots. For example, the IUCN lists

234 Centres of Plant Diversity of which 40 are grasslands, and the World Wildlife Fund (WWF)-US Global 200 Programme identifies 136 terrestrial ecoregions as having outstanding diversity of which 35 are grasslands (Gibson 2009). Among 25 biodiversity hotspots identified by Myers and Mittermeier (2000) is the Cape Floristic Province in South Africa, an extraordinarily species-rich grassland, with over 1700 species of plants.

Despite their high numbers of species, most grasslands are characterized by a handful of dominants (Coupland 1992, 1993); in other words, most grasses in most ecosystems are relatively uncommon and fewer than 10 % of grass species make up most of the biomass in major grasslands. For example, in the North American tallgrass prairie, the dominant species are *Andropogon gerardii*, *Schizachyrium scoparium*, *Sorghastrum nutans*, and *Panicum virgatum*. All four happen to be C_4 taxa, but the less common species are also C_4. Thus, the photosynthetic pathway may be only one aspect of their dominance. Grassland dominants are significantly clustered phylogenetically (Edwards et al. 2010), indicating that some clades have characteristics that confer a selective advantage. At the same time the characters that make a grass dominant are not shared by all members of the family.

Grazing is known to affect species diversity in grasslands, but whether the effect is positive, negative or neutral depends on the size of the herbivore and the productivity of the particular plant community. Moderate grazing by large herbivores in productive communities increases diversity, at least in part by reducing the cover of dominant species; in communities with lower productivity, however, grazing generally reduces diversity (Bakker et al. 2006; Gibson 2009). Not surprisingly, over-grazing reduces the number of species and can lead to ecosystem change or even loss of plant cover entirely. The field of range management focuses on techniques that maximize forage quality and diversity on a sustainable basis. Nonetheless, as population pressure increases in many parts of the world, the temptation to put too many grazing animals on a piece of land will often win out.

Clayton and Renvoize (1986) note that grass genera rarely occur on more than one particular continent. This may point to a broad biogeographic pattern, but also may be influenced by taxonomic artifacts. For example, many monotypic or oligotypic segregate genera have been described in Africa by Stapf and Nees. This could indicate remarkable diversification on the African continent, or could represent the work of a couple of enthusiastic splitters.

Conservation and Response to Climate Change

Grasslands typically have excellent soil and thus are attractive sites for agriculture. They are also easily converted to other human-dominated landscapes such as housing developments and shopping malls. As such they are surprisingly endangered in many parts of the world. Of the original tall-grass prairies in North America, less than 10 % of the original acreage remains (Gibson 2009). While many of the grasslands in the British Isles are themselves the result of forest removal by humans, the native species are often replaced by introduced pasture grasses, which provide better forage for cattle (Gibson 2009). Despite their ongoing conversion to other uses, only a small fraction of world grasslands is protected, making them a highly endangered ecosystem. Hoekstra et al. (2005) estimate the percent of the world's biomes converted to other uses and divide this by the percent of the biome protected to develop a Conservation Risk Index. Temperate grasslands, savannas and shrublands had the highest risk index of any of the world's biomes, in part because the percentage of protected habitat was so low.

Grasslands are important buffers against increasing global CO_2. Worldwide, grasslands store almost as much carbon as forests (34 vs. 39 %) (White et al. 2000). Because of the deep roots of many perennial grasses, much of this storage is in the soil, rather than above ground as in forests.

Given the complex and poorly understood set of factors maintaining grasslands, the future of grasslands in the face of climate change is hard to predict (Bond and Midgley 2012). In many parts of the world, grasslands are being taken over by woody vegetation, a pattern that may be caused by changes in land use or increased atmospheric CO_2 or a combination of the two (Bond 2008). Changes in the amount and timing of precipitation also affect the presence and structure of grasslands. A particularly detailed description of the famous North American drought of the 1930s is provided by Gibson (2009), summarizing the work of Weaver (1954); the decade-long series of droughts altered the structure and dominance relationships of grasses in the prairies. In contrast to the studies of Bond (2008) and Bond and Midgley (2012), based on modeling in warm areas of the world, and Weaver's (1954) observations in North America, a long-term experiment in a low-productivity grassland in England found no significant change in community composition despite substantial changes in moisture and temperature (Grime et al. 2008). Plots in a community dominated by perennial grasses were treated by excluding rain, providing extra water, or by heating, over a 13-year period. Species abundances were largely unchanged. Thus, certain grasslands may be resistant to the changes in climate expected over the next century.

Increased heat and drought might favor grasses over trees, but changes to local fire regimes could have the opposite effect; increased atmospheric CO_2 by itself should favor tree growth, but the interactions with other factors are complex. Recent models of vegetation in Africa predict that the boundary between forest and savanna will shift abruptly in some areas, with savanna converting to forest; however, the timing and conditions under which such changes occur is likely to vary from site to site (Higgins and Scheiter 2012). In particular, the history of the site, the ambient rainfall and frequency of fire affect its sensitivity to elevated CO_2. While many factors can affect the ability of grasslands to provide ecosystem services such as carbon sequestration, carbon storage, grassland production of livestock, and water provision, species diversity is not important; Naidoo et al. (2008) note that there is no correlation between species diversity and any standard measure of ecosystem services.

Many of the characteristics that make grasses valuable crops also make some of them aggressive weeds, particularly in response to disturbance such as over-grazing. For example, over-grazing of the perennial grasslands in California in the 19th century created disturbed areas into which Mediterranean annuals could move. The California annual grassland of today is thus a stable ecosystem of introduced species. Restoration of the original grassland is unlikely, since this would require elimination of weeds of the same family (Young and Allen 1997). Many other grasses (e.g., *Imperata cylindrica, Sorghum halepense*) are noxious weeds that negatively affect agriculture throughout the world.

References

Bakker, E.S., Ritchie, M.E., Olff, H., Milchunas, D.G., Knops, J.M.H. 2006. Herbivore impact on grassland plant diversity depends on habitat productivity and herbivore size. Ecol. Lett. 9: 780–788.

Bond, W.J. 2008. What limits trees in C_4 grasslands and savannas? Ann. Rev. Ecol. Evol. Syst. 39: 641–659.

Bond, W.J., Midgley, G.F. 2012. Carbon dioxide and the uneasy interactions of trees and savannah grasses. Phil. Trans. R. Soc. London B 367: 601–612.

Bond, W.J., Woodward, F.I., Midgley, G.F. 2005. The global distribution of ecosystems in a world without fire. New Phytol. 165: 525–538.

Clayton, W.D., Renvoize, S.A. 1986. Genera graminum: grasses of the world. London: Her Majesty's Stationery Office.

Convey, P. 2001. Antarctic ecosystems. In: Levin, S.A. (ed.) Encyclopedia of biodiversity. San Diego: Academic Press. pp. 171–184

Coupland, R.T. 1992. Natural grasslands: Introduction and western hemisphere. Amsterdam: Elsevier.

Coupland, R.T. 1993. Natural grasslands: Eastern hemisphere and résumé. Amsterdam: Elsevier.

Edwards, E.J., Smith, S.A. 2010. Phylogenetic analyses reveal the shady history of C_4 grasses. Proc. Natl. Acad. Sci. USA 107: 2532–2537.

Edwards, E.J., Osborne, C.P., Strömberg, C.A.E., Smith, S. A., C_4 grasses consortium. 2010. The origins of C_4 grasslands: integrating evolutionary and ecosystem science. Science 328: 587–591.

Gibson, D.J. 2009. Grasses and grassland ecology. New York: Oxford University Press.

Grime, J.P., Fridley, J.D., Askew, A.P., Thompson, K., Hodgson, J.G., Bennett, C.R. 2008. Long-term resistance to simulated climate change in an infertile grassland. Proc. Natl. Acad. Sci. USA 105: 10028–10032.

Higgins, S.I., Scheiter, S. 2012. Atmospheric CO_2 forces abrupt vegetation shifts locally, but not globally. Nature 488: 209–212.

Hoekstra, J.M., Boucher, T.M., Ricketts, T.H., Roberts, C. 2005. Confronting a biome crisis: global disparities of habitat loss and protection. Ecol. Lett. 8: 23–29.

Holdo, R.M., Sinclair, A.R.E., Dobson, A.P., Metzger, K.L., Bolker, B.M., Ritchie, M.E., Holt, R.D. 2009. A disease-mediated trophic cascade in the Serengeti and its implications for ecosystem C. PLoS Biol. 7: e1000210.

Humphreys, A.M., Linder, H.P. 2013. Evidence for recent evolution of cold tolerance in grasses suggests current distribution is not limited by (low) temperature. New Phytol. 198: 1261–1273.

Kidder, D.L., Gierlowski-Kordesch, E.H. 2005. Impact of grassland radiation on the nonmarine silica cycle and Miocene diatomite. PALIOS 20: 198–206.

Lehmann, C.E.R., Archibald, S.A., Hoffmann, W.A., Bond, W.J. 2011. Deciphering the distribution of the savanna biome. New Phytol. 191: 197–209.

Myers, N., Mittermeier, R.A. 2000. Biodiversity hotspots for conservation priorities. Nature 403: 853–859.

Naidoo, R., Balmford, A., Costanza, R., Fisher, B., Green, R.E., Lehner, B., Malcolm, T.R., Ricketts, T.H. 2008. Global mapping of ecosystem services and conservation priorities. Proc. Natl. Acad. Sci. USA 105: 9495–9500.

Pagani, M., Freeman, K.H., Arthur, M.A. 1999. Late Miocene atmospheric CO_2 concentrations and the expansion of C_4 grasses. Science 285: 876–879.

Pagani, M., Zachos, J.C., Freeman, K.H., Tipple, B., Bohaty, S. 2005. Marked decline in atmospheric carbon dioxide concentrations during the Paleocene. Science 309: 600–603.

Preston, J.C., Kellogg, E.A. 2008. Discrete developmental roles for temperate cereal grass *AP1/FUL*-like genes in flowering competency and the transition to flowering. Plant Phys. 146: 265–276.

Sandve, S.R., Fjellheim, S. 2010. Did gene family expansions during the Eocene-Oligocene boundary climate cooling play a role in Pooideae adaptation to cool climates? Mol. Ecol. 19: 2075–2088.

Sandve, S.R., Rudi, H., Asp, T., Rognli, O.A. 2008. Tracking the evolution of a cold stress associated gene family in cold tolerant grasses. BMC Evol. Biol. 8: 245.

Scheiter, S., Higgins, S.I., Osborne, C.P., Bradshaw, C., Lunt, D., Ripley, B.S., Taylor, L.L., Beerling, D.J. 2012. Fire and fire-adapted vegetation promoted C_4 expansion in the late Miocene. New Phytol. 195: 653–666.

Strömberg, C.A.E. 2011. Evolution of grasses and grassland ecosystems. Ann. Rev. Earth Planet. Sci. 39: 517–544.

Vandevenne, F.I., Barão, A.L., Schoelynck, J., Smis, A., Ryken, N., Van Damme, S., Meire, P., Struyf, E. 2013. Grazers: biocatalysts of terrestrial silica cycling. Proc. Roy. Soc. B 280: 20132083.

Vigeland, M.D., Spannagl, M., Asp, T., Paina, C., Rudi, H., Rognli, O.-A., Fjellheim, S., Sandve, S.R. 2013. Evidence for adaptive evolution of low-temperature stress response genes in a Pooideae grass ancestor. New Phytol. 199: 1060–1068.

Visser, V., Woodward, F.I., Freckleton, R.P., Osborne, C. P. 2012. Environmental factors determining the phylogenetic structure of C_4 grass communities. J. Biogeog. 39: 232–246.

Weaver, J.E. 1954. North American prairie. Lincoln, Nebraska, USA: Johnsen Publishing.

Whitaker, R.H. 1975. Communities and ecosystems. New York: Macmillan.

White, R., Murray, S., Rohweder, M. 2000. Pilot analysis of global ecosystems (PAGE): Grassland ecosystems: World Resources Institute.

Young, J.A., Allen, F.L. 1997. Cheatgrass and range science: 1930–1950. J. Range Manage. 50: 530–535.

Reproductive Systems

Reproductive systems in the grasses are diverse. This section begins by describing the morphological variation in placement of the stamens and pistils, and then follows with discussion of outcrossing, self-pollination and apomixis.

Unisexual vs. Bisexual Flowers

Grasses may have unisexual or bisexual flowers. Because of the distribution of the unisexual versus bisexual condition in the immediate outgroups and early diverging taxa, attempts to determine the ancestral condition for the family are highly sensitive to analytical method and assumptions (Malcomber and Kellogg 2006). Among the graminid Poales, flowers of Flagellariaceae are bisexual, whereas those of Restionaceae, Centrolepidaceae and Anarthriaceae are unisexual. The two immediate outgroups of the grasses, Ecdeiocoleaceae and Joinvilleaceae, have unisexual and bisexual flowers, respectively. In the grasses, flowers in Anomochlooideae are bisexual, but those in Pharoideae are unisexual, and Puelioideae are bisexual. This distribution of character states indicates either that unisexual flowers have arisen repeatedly, or that reversion from unisexual to bisexual is fairly easy over evolutionary time, or both.

Virtually all possible arrangements of unisexual flowers on the plant are known in the grasses (Connor 1981, 1987). Monoecy is widespread, appearing in all Pharoideae, tribe Olyreae of Bambusoideae, many members of Chloridoideae and Panicoideae, and some Ehrhartoideae. Among monoecious grasses, sex expression may vary between flowers within a spikelet (e.g., *Lecomtella*, *Ixophorus*), between spikelets within an inflorescence (e.g., *Pharus*, *Hypogynium*, *Zizania*), or between inflorescences within a plant (e.g., *Zea*, many Olyreae).

Andromonoecy is common, particularly in the Panicoideae, in which spikelets are conventionally two-flowered, with the lower flower often producing stamens and the upper one being bisexual. Additional variations on this theme occur in the tribe Andropogoneae, with sex expression varying between the flowers in each of the two spikelets of the spikelet pair. For example, in many species of *Andropogon*, the sessile spikelet has a sterile lower flower and bisexual upper flower, while the pedicellate spikelet has a sterile lower flower and staminate upper one; this produces an inflorescence of bisexual and staminate flowers in a 1:1 ratio. In other Andropogoneae (e.g., *Elymandra*), the most proximal spikelet pairs on the inflorescence consist entirely of staminate flowers, thus increasing the ratio of staminate to bisexual flowers.

In contrast to andromonoecy, gynomonoecy is rare in the grasses as a whole. Connor (1981) reports only eight genera with this distribution of flower types. In two of these, the bambusoids *Diandrolyra* and *Piresia*, the apparently bisexual flowers are in fact staminate with the gynoecium non-functional; thus, the plants are actually monoecious. In the chloridoid genus *Munroa*, pistillate flowers are proximal to bisexual ones in the spikelets of most species, but in *M. squarrosa*, pistillate spikelets occur only on lateral inflorescence branches, whereas the terminal spikelet is bisexual (Anton and Hunziker 1978); spikelets of *M. argentina* are all bisexual. In *Centotheca*, *Coelachne* and *Heteranthecia*, the distal flower of the

E.A. Kellogg, *Flowering Plants. Monocots*, The Families and Genera of Vascular Plants 13,
DOI 10.1007/978-3-319-15332-2_10, © Springer International Publishing Switzerland 2015

spikelet is often pistillate whereas proximal flowers are bisexual.

Dioecy is not particularly common in the grasses. Although Connor (1987) and Connor et al. (2000) list 22 and 20 dioecious genera respectively, several are monotypic genera that are now placed in synonymy in *Distichlis*, a genus in which all species are dioecious, and in *Bouteloua*, a well supported clade that includes species with almost every conceivable form of dicliny (Kinney et al. 2007). The two dioecious species formerly segregated as *Neeragostis* are now placed in *Eragrostis*, in which they are the only dioecious members (Ingram and Doyle 2004). Following current taxonomy, dioecy thus occurs in 14 genera in three subfamilies: Chloridoideae (*Allolepis*, *Bouteloua*, *Cyclostachya*, *Distichlis*, *Eragrostis*, *Jouvea*, *Scleropogon*, *Sohnsia*), Panicoideae (*Gynerium*, *Pseudochaetochloa*, *Spinifex*, *Zygochloa*), and Pooideae (*Poa*, the "*Leucopoa* grade").

While dioecy is uncommon, gynodioecy (plants either entirely pistillate or with bisexual flowers) is even more rare. In *Bouteloua* and *Poa*, both dioecious and gynodioecious species are known. Gynodioecy also appears in *Austroderia*, *Chimaerochloa*, and *Cortaderia* in Danthonioideae. In *Poa* and the three danthonioid genera, some species seem to be entirely pistillate and apomictic, with staminate plants unknown. Plants that appear to be gynodioecious may in fact be functionally dioecious; morphologically bisexual plants in *Cortaderia selloana* in Argentina serve primarily as pollen donors, so are functionally staminate despite having gynoecia (Connor et al. 2000). Male sterility in *Cortaderia* is under the control of three loci, but these have not been investigated at the molecular level (Connor and Charlesworth 1989).

Unisexual flowers in the grasses always initiate both gynoecium and androecium; one or the other of these then aborts during development (Le Roux and Kellogg 1999). In staminate flowers of all panicoids investigated to date, the gynoecium develops until the nucellus is visible and carpel walls form a ridge surrounding it; at this stage cell death occurs in the subepidermal cells and growth ceases. Gynoecial abortion in panicoids is under the control of the proteins TASSELSEED1 (Acosta et al. 2009) and TASSELSEED2 (DeLong et al. 1993), which function in the jasmonic acid pathway (Yan et al. 2012). However, Malcomber and Kellogg (2006) find that TS2 is expressed throughout the plant including anthers, and is highly conserved in all grasses, suggesting that it may have been coopted for sex determination only in the panicoids. In addition, variation in the sequence of the protein does not correlate with the presence of unisexual flowers. Thus, its major role is unknown but certainly is more general than simply to cause organ abortion. Development of staminate flowers in *Bouteloua* (Chloridoideae) is histologically similar to that in Panicoideae, and gynoecial abortion also requires TS2 (Chandra and Huff 2010); interestingly, infection of staminate flowers with the fungus *Salmacisia* (= *Tilletia*) *buchloëana* suppresses TS2 expression and causes pistils to develop. In contrast, development of staminate flowers in *Zizania* (Ehrhartoideae) is distinct from the panicoid or chloridoid pattern. In *Zizania*, the stigmatic arms of the gynoecium develop, the ovule differentiates and integuments become visible, and only then does growth arrest, accompanied by deposition of a dark-staining substance (Zaitchik et al. 2000).

Sex expression is also affected by the hormone gibberellin. In maize, the gene known as *anther ear1* encodes ent-kaurene synthase, in the gibberellin pathway; when it is mutated the pistillate flowers of the ear become bisexual and at the same time the number of branches in the tassel is reduced (Bensen et al. 1995). *Bouteloua dactyloides* (buffalograss) is normally dioecious, but some plants are trimonoecious, producing staminate, pistillate and bisexual flowers. In such plants, sex expression can be manipulated by applying gibberellin, in which case inflorescences become entirely staminate (Quinn 2000). Inhibiting gibberellin with paclobutrazol creates inflorescences that are entirely pistillate.

Self-pollination and Self-incompatibility Systems

It is surprising, given the ecological dominance of the grasses, how few studies have been undertaken of their breeding systems. Fewer than 100 genera have been characterized as self-compatible or self-incompatible (Campbell et al. 1983; Connor 1979), so self-compatibility is unknown for the majority of the family. Self-compatibility is generally more common among annuals than perennials; because most cereal crops are domesticated annuals, it is not surprising that many of these are also self compatible (e.g., barley, wheat, foxtail millet) (Glémin and Bataillon 2009).

Of the grasses that have been studied, many are capable of both self-pollination and outcrossing. In the few such taxa that have been studied in detail, however, outcross pollen leads to higher seed set and germinability than self pollen (Fang et al. 2004; Kettenring et al. 2011). Even though barriers to self-pollination are rarely absolute, estimated rates of outcrossing are strongly bimodal, with self-incompatible species generally above 80 % outcrossing and self-compatible ones mostly well under 15 % (Connor 1987; Raduski et al. 2012).

The genetics of self-incompatibility have been worked out in most detail in *Secale cereale* and seven other Pooideae. Self-incompatibility in these species is controlled by the pollen (gametophytic) alleles of two loci, called S and Z (Hayman 1956; Heslop-Harrison 1982; Lundqvist 1954). Pollen growth will be blocked if the S and Z alleles of the pollen both match S and Z alleles in the stigma. This system has two interesting effects; first, partial compatibility is possible, and second, reciprocal crosses may have different outcomes. For example, if a plant with the genotype S1S1 Z1Z3 is used as the seed parent and a plant with S1S2 Z2Z3 is the pollen donor, the possible pollen genotypes will be S1Z2, S1Z3, S2Z2, and S2Z3. Of these, three will be compatible and one incompatible, resulting in 75 % of the pollen being compatible. The reciprocal cross will be 50 % compatible. Unlike single locus systems that occur in other families, the S-Z system will function even if gametes are unreduced and the plant becomes autopolyploid. In addition to the eight species in which the genetics are known, another eight have a gametophytic self-incompability system in which reciprocal crosses have different outcomes; it seems likely that these also have an S-Z system (Baumann et al. 2000).

Even though the basic genetics of the S-Z system was determined more than half a century ago, neither gene has been cloned and their molecular identity and mode of action remain unknown. Loss of self-incompatibility can occur by mutation of alleles that occur only in the pollen, suggesting that perhaps both the S and the Z locus are actually complex loci with at least two genes (Baumann et al. 2000). In addition, a third locus known as T (for third) has also been identified recently in three pooid grasses (reviewed in Yang et al. 2008). Comparative genome mapping is likely ultimately to identify the specific loci corresponding to S and Z; unfortunately the pooids in which the genetic work has been done have some of the largest genomes in the family so genomic approaches will be relatively slow.

Pollen tube growth has been described above under Pollination (section Pollen). The self-incompatibility response appears to require an interaction between the tip of the pollen tube and the stigma, so occurs after pollen germination (Heslop-Harrison and Heslop-Harrison 1987). Barriers to self-pollination may occur quite rapidly (within 90 seconds in *Secale cereale*; Heslop-Harrison 1982) or much later after the pollen tube has grown into the style – e.g., in *Alopecurus pratensis* (Shivanna et al. (1982) and *Andropogon* species (Norrmann and Keeler 2003).

Many grass species are protogynous, which presumably reduces self-pollination, but as with self-compatibility, the distribution of this character is poorly documented.

Cleistogamy

More cleistogamous species are reported for the grasses than for any other plant family (Campbell et al. 1983; Culley and Klooster 2007). Cleistogamy, defined as "self-fertilization in an enclosed flower", is widespread in the family, occurring in over 300 species and 90 genera distributed among most subfamilies (Campbell et al. 1983; Connor 1987; Culley and Klooster 2007). In many taxa, cleistogamy is caused by failure of the inflorescence to emerge from the sheath; flower development is otherwise normal but self-fertilization occurs because the anthers and stigmas are never exposed. In some cases, retention of the inflorescence in the sheath is caused by environmental factors (e.g., low temperatures, low soil moisture), and this is particularly true for the failure of the terminal inflorescence to emerge. However, failure of axillary inflorescences to emerge is often genetically fixed within a species.

In this case, the enclosed axillary inflorescence may be morphologically distinct from the exserted inflorescences elsewhere on the plant, and is known as a cleistogene (Chase 1918).

Two genera - *Amphicarpum* and *Eremitis* - produce spikelets underground from positively geotropic shoots, as well as above ground, from open panicles (Campbell et al. 1983). In *Amphicarpum* the seeds of subterranean spikelets are borne more than 3 cm below the surface (Cheplick and Quinn 1988), allowing them to survive the frequent fires of the New Jersey pine barrens where the species is native. Two species of *Chloris* and one of *Paspalum* are also reported to have subterranean spikelets.

Cleistogamous flowers have much smaller anthers than chasmogamous ones, resulting in lower pollen to ovule ratios (Campbell et al. 1983). In plants producing cleistogenes or subterranean spikelets, fruits of the cleistogamous spikelets are heavier than those of chasmogamous spikelets, although their relative success in germination and seedling establishment varies between species (Connor 1987).

Barriers to Interspecific Crossing

Crosses between species of grasses are often but not always unsuccessful. Knobloch (1968) published a useful compendium of crosses that have been attempted in the grasses. In most cases the offspring were sterile, but exceptions occurred, particularly among polyploid species. For example, Norrmann and Keeler (2003) report production of viable hybrids between *Andropogon gerardii* and *A. lateralis*, *A. gerardii* and *A. glaucophyllus*, and *A. hypogynus* and *A. lateralis*. The parents of these hybrids differ in chromosome number, but are all high polyploid ($2n = 60$ and above). In general, whether crosses - even very wide crosses - are successful or not depends on the particular combination of genomes in the parents and not necessarily on their phylogenetic proximity (Jones and Pasakinskiene 2005).

As noted under Pollination (section Pollen), pre-zygotic isolating mechanisms exist in the grasses but are not universal and where they exist may still be somewhat porous. The stigma is often not a barrier to pollen germination, but foreign pollen may be blocked in passage down the style. The mechanism behind this is unknown. In a series of papers, Evans and Kermicle (Evans and Kermicle 2001; Kermicle 2006; Kermicle and Evans 2005, 2010; Kermicle et al. 2006) have identified three genetic loci, *Gametophyte1* (*Ga1*), *Gametophyte2* (*Ga2*), and *teosinte crossing barrier1* (*tcb1*), which they call pollen-pistil cross-compatibility (PPCC) loci. Each locus has three alleles that are found in wild populations of *Zea mays* (teosinte). One allele of *tcb1* and all three alleles of *Ga1* and *Ga2* are also found in cultivated maize. One allele creates a block in the pistil but pollen growth is competent, the second allows pollen growth but does not confer a corresponding block in the pistil, and the third has neither effect. Incompatibility at any of the three loci will lead to markedly reduced success of pollination. Together the three loci reduce cross-pollination between maize and teosinte in fields where the two subspecies co-occur. These loci are thus "speciation loci", providing an early barrier to gene flow between two partially differentiated taxa. This genetic block to hybridization must have arisen quite rapidly since maize was domesticated only about 9,000 years ago or less.

Wheat (*Triticum aestivum*) can be crossed with rye (*Secale cereale*), but crossability between these two species depends on the genotype of the seed parent (wheat) at a locus known as *Kr1*; because wheat is a hexaploid, there are three *Kr* loci, on chromosomes 5A, 5B, and 5D (Manickavelu et al. 2009). A dominant *Kr* allele at any locus blocks the growth of rye pollen tubes at the base of the style or in the ovary wall, whereas recessive alleles at *Kr* have no effect (Lange and Wojciechowska 1976). Recessive *kr* alleles occur naturally in *T. aestivum*, suggesting that wide crosses could occur from time to time in nature.

The possibility of wide hybridization has also been investigated in sorghum. Maize pollen will germinate on the stigma of most accessions of *Sorghum bicolor*, but fails to grow more than a few hundred microns (Laurie and Bennett 1989). However, in a landrace known as *S. nervosum* Nr481, maize pollen will germinate and occasionally will reach the ovule. The ability to permit or block maize pollen tube growth is controlled by a single locus known as *Iap* (*Inhibition of alien pollen tubes*). When the seed parent is homozygous recessive at this locus, maize pollen tube growth is permitted (Laurie and Bennett 1989).

Price et al. (2006) extended this result by testing pollen from *Sorghum nitidum*, *S. macrospermum*, and *Sarga angusta* (= *Sorghum angustum*) on *S. bicolor* stigmas. In *S. bicolor* plants with the dominant *Iap* allele, foreign pollen germinated, but was then blocked at various points in the stigmas or style branches such that no fertilization occurred. In plants that were recessive at *Iap*, however, a few pollen tubes reached the ovary, although they grew more slowly than pollen from *S. bicolor*. Some viable embryos were obtained from all three interspecific crosses, and *S. bicolor* x *S. macrospermum* yielded germinable seed. Thus, pre-zygotic isolation between species is not absolute.

Post-zygotic blocks to hybridization may occur by disruption of either endosperm or embryo sac development. Crosses between *Sorghum bicolor* and *S. angustum* or *S. nitidum* produced viable embryos that had to be rescued because of endosperm failure (Price et al. 2006). In a series of interspecific crosses between *Oryza sativa* and other diploid species of *Oryza*, pollination and fertilization were normal (Ishikawa et al. 2011). However, subsequent development of the endosperm was disrupted, and the timing of cellularization was abnormal. Seeds were unviable not because of problems with embryo development but because of the endosperm. Ishikawa et al. (2011) suggest that the failure of endosperm development may be caused by inappropriate imprinting of parental genes.

Other post-zygotic mechanisms affect the embryo sac. Recently, Yang et al. (2012) have identified a set of three linked genes in rice that confer intersterility between *indica* and *japonica* lines. These function as a set of Dobzhansky-Muller interactors, such that particular combinations of alleles cause abortion of the embryo sac, apparently by affecting the endoplasmic reticulum. As with the incompatibility genes described above, these confer incompatibility only in particular combinations, whereas in other combinations of alleles embryo sac formation and ultimately seed production proceed normally.

Asexual Reproduction

Asexual reproduction is common in the grasses. Many species, particularly perennials, can reproduce vegetatively via rhizomes or stolons (see discussion on Vegetative Morphology and Anatomy). Rhizomatous spread may be relatively slow, but the tillers produced may be quite long-lived. Harberd (1961) famously estimated that a colony of *Festuca rubra* might have been over 400 years old, and perhaps as old as 1000 years. Similar dates have been estimated for other clonal grasses (de Witte and Stöcklin 2010). A more recent study with molecular markers suggested a minimum age for a plant of *Festuca idahoensis* as at least 60 years (Liston et al. 2003).

A few species – notably *Poa alpina*, *P. bulbosa*, *Deschampsia alpina*, and several species of *Festuca* – produce onion-like bulbils in place of spikelets (Gustafsson 1947). This phenomenon is sometimes called pseudovivipary to distinguish it from vivipary, a term that should be reserved for germination of the seed before being shed from the parent (see discussion under the section Fruit and seed). Formation of bulbils is often under environmental control (Heide 1989; Youngner 1960), and is common in cool, wet environments (Lee and Harmer 1980) although *P. bulbosa* has become a widespread weed even in dry areas. In *P. alpina*, plants that produced bulbils are more common at high elevations and have reduced fitness if transplanted to lower sites (Steiner et al. 2012). The bulbils may be dormant (as in *P. bulbosa*) or may germinate immediately (as in *F. viviparoidea*) (Lee and Harmer 1980).

Many grasses are capable of producing asexual seed (apomixis or agamospermy). Although some authors (e.g., Gustafsson 1947; Richards 2003) use the term apomixis to refer to all modes of asexual reproduction, it is more common to use apomixis to refer only to asexual seed production (Ozias-Akins 2006; van Dijk 2009), a terminology followed here.

Asexual seed production in the grasses can occur via several mechanisms, all of which can be viewed as heterochronic changes in the normal reproductive process, such that the normal developmental program is triggered at an abnormal time or in inappropriate cells (Grimanelli 2012; Nygren 1967). In sporophytic apomixis (adventitious embryony), a nucellar (maternal) cell develops directly into an embryo, bypassing gametophyte production entirely. In gametophytic apomixis (apospory or diplospory), development of the female gametophyte is modified.

In aposporous apomixis, an unreduced embryo sac forms directly from cells in the nucellus, such that each cell is diploid rather than haploid. The unreduced egg then proceeds to form an embryo directly, without fertilization. Meanwhile, the megaspore mother cell undergoes meiosis as usual, and begins normal embryo sac formation. However, the normal megagametophyte generally breaks down before maturing. Aposporous embryo sacs tend to form earlier than the normal one (e.g., Kellogg 1987); it is not clear if they compete for resources or space, or if other processes lead to breakdown of the normal embryo sac.

In diplosporous apomixis, the megaspore mother cell never undergoes meiosis at all, but instead divides and forms a single embryo sac by mitosis. Alternatively, meiosis begins but after the first division the resulting nuclei fuse together to form a restitution nucleus. In either case, the embryo sac is diploid and the egg develops without fertilization.

Apomixis is relatively common in the grasses, having been reported in as many as 60 non-hybrid genera (Czapik 2000; Ozias-Akins 2006), and has clearly arisen more than once in the family (van Dijk and Vijverberg 2005). It is well known and extensively studied in species of *Poa* (Pooideae), in *Cortaderia* (Danthonioideae) and in a number of panicoids including *Bothriochloa*, *Tripsacum*, *Paspalum*, *Brachiaria*, *Urochloa maxima* (= *Megathyrsus*), and *Cenchrus* (which now includes species formerly placed in *Pennisetum*). In the panicoid species, aposporous embryo sacs are four-, rather than eight-celled, and lack antipodals.

It is not clear whether sporophytic apomixis occurs in grasses. Asker and Jerling (1992) indicate only gametophytic apomixis, whereas Naumova (1993) lists eight species with reported adventitious embryony. Although Yu et al. (2000) report sporophytic apomixis in *Bothriochloa* and *Dichanthium*, they do not provide supporting evidence. Gametophytic apomixis is by far more common in the grasses, with apospory more common than diplospory (Nygren 1967).

In pseudogamous apomixis, embryo formation is parthenogenetic but fertilization is still needed for endosperm development. Endosperm in turn often requires an appropriate ratio of maternal to paternal genomes, generally two maternal to one paternal. Some grasses (e.g., *Tripsacum*) appear to tolerate deviation from the 2:1 ratio. Others, such as aposporous Panicoideae, achieve the ratio by production of a four-celled embryo sac with a single unreduced central cell. This has twice the chromosome complement of the reduced sperm that fertilizes it, resulting in a 2:1 ratio of maternal to paternal genomes (Ozias-Akins 2006).

Pseudogamy makes apomixis difficult to detect because simple bagging experiments will allow seed development, and exclusion of pollen will prevent seed set; such results do not distinguish agamospermy from self-compatibility. Demonstration of apomixis thus requires either observation of megagametophyte development, assessment of ploidy levels in tissues of developing ovules, or analysis of variation among the offspring of a single plant (Ozias-Akins 2006; Ozias-Akins and van Dijk 2007).

Non-pseudogamous apomixis is much less common than pseudogamous, and in grasses is known only in species of *Calamagrostis*, *Cortaderia* (including *Lamprothyrsus*), *Nardus* and *Poa* (Connor 1981). It is unclear how endosperm development in these species is able to proceed without fertilization.

Apomixis requires both formation of unreduced gametes and parthenogenesis, processes that are not always linked. Unreduced gametes occur frequently in grasses and may or may not be correlated with subsequent parthenogenesis. In addition, the tendency to produce unreduced gametes is not necessarily shared among closely related species. For example, Norrmann and Keeler (2003) report evidence for an unreduced egg cell from *Andropogon lateralis*, but found no evidence for unreduced gametes in over 1000 plants of *A. gerardii*.

The genetic basis of apomixis is not well understood, although significant progress has been made in recent years (Grimanelli 2012; Ozias-Akins and van Dijk 2007). Because apomixis would permit fixation of particular hybrid combinations it is of considerable interest to agriculture. Crosses of sexual and apomictic species have shown that apomixis is often dominant (Harlan and de Wet 1963; Ozias-Akins 2006). Genetic studies with molecular markers and genomic comparisons have been undertaken in five panicoid genera (*Cenchrus*, *Brachiaria*, and

Urochloa (= *Panicum maximum* = *Megathyrsus maximus*) of tribe Paniceae; *Paspalum* in Paspaleae; and *Tripsacum* in Andropogoneae), and one pooid (*Poa*, tribe Poeae). All but *Tripsacum* are aposporous. Ozias-Akins et al. (2003) have shown that apomixis is associated with a large block of DNA that is maintained in hemizygous condition in polyploid *Cenchrus squamulatus* (= *Pennisetum squamulatum*) and *Cenchrus ciliaris*. This large region does not recombine and contains regions of highly repetitive DNA. In addition, its chromosomal location appears to vary over evolutionary time. Although Akiyama et al. (2011) argue that the apospory-specific region originated only once in *Cenchrus* and was transferred to other species via hybridization, their data are also consistent with independent recruitment of related genomic regions. A similar structure has been found in *Paspalum simplex* (Grimanelli 2012) and in *Tripsacum dactyloides*. The cytogenetic results reported by Harlan and de Wet (1963) in *Bothriochloa* could also be interpreted as indicating an apospory-specific region. In *Cenchrus*, the gene controlling production of apomictic embryo sacs is tightly linked to that for parthenogenesis (Ozias-Akins 2006), whereas in *Poa* the two aspects of agamospermy can be separated genetically (Albertini et al. 2001).

Apomixis appears to be controlled in part by epigenetic mechanisms (Garcia-Aguilar et al. 2010; Grimanelli 2012). Based on data from maize and *Arabidopsis*, Grimanelli (2012) presents a model suggesting that reproductive development represents inactivation of transcription of a still unknown (but presumably fairly large) set of genes. Failure of inactivation leads to production of unreduced gametes. Regulation of parthenogenesis is less well understood, but it too may involve epigenetic mechanisms.

Because grass genomes are largely collinear, several authors have attempted to determine whether apomixis involves the same set of genes wherever it occurs. While the regions controlling apomixis appear to be similar among congeneric species of *Paspalum* (Pupilli et al. 2004), between genera the apomixis-controlling region appears not to be conserved (Ozias-Akins 2006). This makes identification of the controlling factors much more difficult, but is consistent with the observation that apomixis has originated many times independently in the family.

Apomixis is strongly associated with polyploidy. Because polyploidy itself is so common in Poaceae, it is not surprising that apomixis should also be common. Nonetheless, it is not certain that polyploidy is absolutely required. Some evidence from *Paspalum* and *Hieracium* (Asteraceae) suggests that formation of apomictic embryo sacs can occur in diploids, but its frequency becomes more common upon chromosome doubling (Ozias-Akins 2006).

References

Acosta, I.F., Laparra, H., Romero, S.P., Schmelz, E., Hamberg, M., Mottinger, J.P., Moreno, M.A., Dellaporta, S.L. 2009. *tasselseed1* is a lipoxygenase affecting jasmonic acid signaling in sex determination in maize. Science 323: 262–265.

Akiyama, Y., Goel, S., Coner, J.A., Hanna, W.W., Yamada-Akiyama, H., Ozias-Akins, P. 2011. Evolution of the apomixis transmitting chromosome in *Pennisetum*. BMC Evol. Biol. 11: 289.

Albertini, E., Porceddu, A., Ferranti, F., Reale, L., Barcaccia, G., Romano, B., Falcinelli, M. 2001. Apospory and parthenogenesis may be uncoupled in *Poa pratensis*: a cytological investigation. Sex. Plant Reprod. 14: 213–217.

Anton, A.M., Hunziker, A.T. 1978. El genero *Munroa* (Poaceae): sinopsis morfológica y taxonomica. Bol. Acad. Nac. Cienc., Córdoba 52: 229–252.

Asker, S.E., Jerling, L. 1992. Apomixis in plants. Boca Raton, Florida: CRC Press.

Baumann, U., Juttner, J., Bian, X., Langridge, P. 2000. Self-incompatibility in the grasses. Ann. Bot. 85 (supplement A): 203–209.

Bensen, R.J., Johal, G.S., Crane, V.C., Tossberg, J.T., Schnable, P.S., Meeley, R.B., Briggs, S.P. 1995. Cloning and characterization of the maize *An1* gene. Plant Cell 7: 75–84.

Campbell, C.S., Quinn, J.A., Cheplick, G.P., Bell, T.J. 1983. Cleistogamy in grasses. Ann. Rev. Ecol. Syst. 14: 411–441.

Chandra, A., Huff, D.R. 2010. A fungal parasite regulates a putative female-suppressor gene homologous to maize *Tasselseed2* and causes induced hermaphroditism in male buffalograss. Mol. Plant-Microbe Interact. 23: 239–250.

Chase, A. 1918. Axillary cleistogenes in some American grasses. Amer. J. Bot. 5: 254–258.

Cheplick, G.P., Quinn, J.A. 1988. Subterranean seed production and population responses to fire in *Amphicarpum purshii* (Gramineae). J. Ecol. 76: 263–273.

Connor, H.E. 1979. Breeding systems in the grasses: a survey. N. Z. J. Bot. 17: 547–574.

Connor, H.E. 1981. Evolution of reproductive systems in the Gramineae. Ann. Missouri Bot. Gard. 68: 48–74.

Connor, H.E. 1987. Reproductive biology in the grasses. In: Soderstrom, T.R., Hilu, K.W., Campbell, C.S., Barkworth, M.E. (eds.) Grass systematics and

evolution. Washington, DC: Smithsonian Institution. pp. 117–132
Connor, H.E., Charlesworth, D. 1989. Genetics of male sterility in gynodioecious *Cortaderia* (Gramineae). Heredity 63: 373–382.
Connor, H.E., Anton, A.M., Astegiano, M.E. 2000. Dioecism in grasses in Argentina. In: Jacobs, S.W.L., Everett, J. (eds.) Grasses: systematics and evolution. Melbourne: CSIRO. pp. 287–293
Culley, T.M., Klooster, M.R. 2007. The cleistogamous breeding system: a review of its frequency, evolution, and ecology in angiosperms. Bot. Rev. 73: 1–30.
Czapik, R. 2000. Apomixis in monocotyledons. In: Jacobs, S.W.L., Everett, J. (eds.) Grasses: systematics and evolution. Melbourne: CSIRO. pp. 316–321
DeLong, A., Calderon-Urrea, A., Dellaporta, S.L. 1993. Sex determination gene *TASSELSEED2* of maize encodes a short-chain alcohol dehydrogenase required for stage-specific floral organ abortion. Cell 74: 757–768.
de Witte, L.C., Stöcklin, J. 2010. Longevity of clonal plants: why it matters and how to measure it. Ann. Bot. 106: 859–870.
Evans, M.M.S., Kermicle, J.L. 2001. *Teosinte crossing barrier-1*, a locus governing hybridization of teosinte with maize. Theor. Appl. Genet. 103: 259–265.
Fang, X., Subudhi, P.K., Venuto, B.C., Harrison, S.A. 2004. Mode of pollination, pollen germination, and seed set in smooth cordgrass (*Spartina alterniflora*, Poaceae). Int. J. Plant Sci. 165: 395–401.
Garcia-Aguilar, M., Michaud, C., Leblanc, O., Grimanelli, D. 2010. Inactivation of a DNA methylation pathway in maize reproductive organs results in apomixis-like phenotypes. Plant Cell 22: 3249–3267.
Glémin, S., Bataillon, T. 2009. A comparative view of the evolution of grasses under domestication. New Phytol. 183: 273–290.
Grimanelli, D. 2012. Epigenetic regulation of reproductive development and the emergence of apomixis in angiosperms. Curr. Opinion Pl. Biol. 15: 57–62.
Gustafsson, A. 1947. Apomixis in higher plants. Lund, Sweden: Lunds Universitets Arsskrift. N.F.
Harberd, D.J. 1961. Observations on population structure and longevity of *Festuca rubra* L. New Phytol. 60: 184–206.
Harlan, J.R., de Wet, J.M.J. 1963. The compilospecies concept. Evolution 17: 497–501.
Hayman, D.L. 1956. The genetic control of self-incompatibility in *Phalaris coerulescens* Desf. (Poaceae). Aust. J. Biol. Sci. 9: 321–331.
Heide, O.M. 1989. Environmental control of flowering and viviparous proliferation in seminiferous and viviparous arctic populations of two *Poa* species. Arctic Alpine Res. 21: 305–315.
Heslop-Harrison, J. 1982. Pollen-stigma interaction and cross-incompatibility in the grasses. Science 215: 1358–1364.
Heslop-Harrison, J., Heslop-Harrison, Y. 1987. Pollen-stigma interaction in the grasses. In: Soderstrom, T. R., Hilu, K.W., Campbell, C.S., Barkworth, M.E. (eds.) Grass systematics and evolution. Washington, DC: Smithsonian Institution. pp. 133–142
Ingram, A.L., Doyle, J.J. 2004. Is *Eragrostis* monophyletic? Insights from nuclear and plastid sequence data. Syst. Bot. 29: 545–552.
Ishikawa, R., Ohnishi, T., Kinoshita, Y., Eiguchi, M., Kurata, N., Kinoshita, T. 2011. Rice interspecies hybrids show precocious or delayed developmental transitions in the endosperm without change to the rate of syncytial nuclear division. Plant J. 65: 798–806.
Jones, N., Pasakinskiene, I. 2005. Genome conflict in the Gramineae. New Phytol. 165: 391–410.
Kellogg, E.A. 1987. Apomixis in the *Poa secunda* complex. Amer. J. Bot. 74: 1431–1437.
Kermicle, J.L. 2006. A selfish gene governing pollen-pistil compatibility confers reproductive isolation between maize relatives. Genetics 172: 499–506.
Kermicle, J.L., Evans, M.M.S. 2005. Pollen-pistil barriers to crossing in maize and teosinte result from incongruity rather than active rejection. Sex. Plant Reprod. 18: 187–194.
Kermicle, J.L., Evans, M.M.S. 2010. The *Zea mays* sexual compatibility gene *ga2*: naturally occurring alleles, their distribution, and role in reproductive isolation. J. Hered. 101: 737–749.
Kermicle, J.L., Taba, S., Evans, M.M.S. 2006. The *Gametophyte-1* locus and reproductive isolation among *Zea mays* subspecies. Maydica 51: 219–225.
Kettenring, K.M., McCormick, M.K., Baron, H.M., Whigham, D.F. 2011. Mechanisms of *Phragmites australis* invasion: feedbacks among genetic diversity, nutrients, and sexual reproduction. J. Appl. Ecol. 48: 1305–1313.
Kinney, M.S., Columbus, J.T., Friar, E.A. 2007. Dicliny in *Bouteloua* (Poaceae: Chloridoideae): implications for the evolution of dioecy. Aliso 23: 605–614.
Knobloch, I.W. 1968. A checklist of crosses in the Gramineae. [E. Lansing].
Lange, W., Wojciechowska, B. 1976. The crossing of common wheat (*Triticum aestivum* L.) with cultivated rye (*Secale cereale* L.). I. Crossability, pollen grain germination and pollen tube growth. Euphytica 25: 609–620.
Laurie, D.A., Bennett, M.D. 1989. Genetic variation in sorghum for the inhibition of maize pollen tube growth. Ann. Bot. 64: 675–681.
Lee, J.A., Harmer, R. 1980. Vivipary, a reproductive strategy in response to environmental stress? Oikos 35: 254–265.
Le Roux, L.G., Kellogg, E.A. 1999. Floral development and the formation of unisexual spikelets in the Andropogoneae (Poaceae). Amer. J. Bot. 86: 354–366.
Liston, A., Wilson, B.L., Robinson, W., Doescher, P.S., Harris, N.R., Svejcar, T. 2003. The relative importance of sexual reproduction versus clonal spread in an aridland bunchgrass. Oecologia 137: 216–225.
Lundqvist, A. 1954. Studies of self-sterility in rye, *Secale cereale* L. Hereditas 40: 278–294.
Malcomber, S.T., Kellogg, E.A. 2006. Evolution of unisexual flowers in grasses (Poaceae) and the putative sex-determination gene, *TASSELSEED2* (*TS2*). New Phytol. 170: 885–899.
Manickavelu, A., Koba, T., Mishina, K., Sassa, H. 2009. Molecular characterization of crossability gene *Kr1* for intergeneric hybridization in *Triticum aestivum* (Poaceae: Triticeae). Plant Syst. Evol. 278: 125–311.
Naumova, T.N. 1993. Apomixis in angiosperms: nucellar and integumentary embryony. Boca Raton, Florida: CRC Press.

Norrmann, G.A., Keeler, K.H. 2003. Cytotypes of *Andropogon gerardii* Vitman (Poaceae): fertility and reproduction of aneuploids. Bot. J. Linn. Soc. 141: 95–103.

Nygren, A. 1967. Apomixis in the angiosperms. Encyclopedia of Plant Physiology (Handbuch der Pflanzenphysiologie) 18: 551–596.

Ozias-Akins, P. 2006. Apomixis: developmental characteristics and genetics. Crit. Rev. Plant Sci. 25: 199–214.

Ozias-Akins, P., van Dijk, P.J. 2007. Mendelian genetics of apomixis in plants. Ann. Rev. Genet. 41: 509–537.

Ozias-Akins, P., Akiyama, Y., Hanna, W.W. 2003. Molecular characterization of the genomic region linked with apomixis in *Pennisetum/Cenchrus*. Func. Integ. Genomics 3: 94–104.

Price, H.J., Hodnett, G.L., Burson, B.L., Dillon, S.L., Stelly, D.M., Rooney, W.L. 2006. Genotype dependent interspecific hybridization of *Sorghum bicolor*. Crop Sci. 46: 2617–2622.

Pupilli, F., Martinez, E.J., Busti, A., Calderini, O., Quarin, C.L., Arcioni, S. 2004. Comparative mapping reveals partial conservation of synteny at the apomixis locus in *Paspalum* spp. Mol. Genet. Genomics 270: 539–548.

Quinn, J.A. 2000. Adaptive plasticity in reproduction and reproductive systems of grasses. In: Jacobs, S.W.L., Everett, J. (eds.) Grasses: systematics and evolution. Melbourne: CSIRO. pp. 281–286

Raduski, A.R., Haney, E.B., Igic, B. 2012. The expression of self-incompatibility in angiosperms is bimodal. Evolution 66: 1275–1283.

Richards, A.J. 2003. Apomixis in flowering plants: an overview. Phil. Trans. R. Soc. London B 358: 1085–1093.

Shivanna, K.R., Heslop-Harrison, Y., Heslop-Harrison, J. 1982. The pollen-stigma interaction in the grasses. 3. Features of the self-incompatibility response. Acta Bot. Neerl. 31: 307–319.

Steiner, B.L., Armbruster, G.F.J., Scheepens, J.F., Stöcklin, J. 2012. Distribution of bulbil- and seed-producing plants of *Poa alpina* (Poaceae) and their growth and reproduction in common gardens suggest adaptation to different elevations. Amer. J. Bot. 99: 2035–2044.

van Dijk, P. 2009. Apomixis: basics for non-botanists. In: Schön, I., Martens, K., van Dijk, P. (eds.) Lost sex: The evolutionary biology of parthenogenesis. Dordrecht: Springer. pp. 47–62.

van Dijk, P.J., Vijverberg, K. 2005. The significance of apomixis in the evolution of the angiosperms: a reappraisal. In: Bakker, F.T., Chatrou, L.W., Gravendeel, B., Pelser, P.B. (eds.) Plant species-level systematics: new perspectives on pattern and process. Ruggell, Liechtenstein: A.R.G. Gantner Verlag. pp. 101–116

Yan, Y., Christensen, S., Isakeit, T., Engelberth, J., Meeley, R., Hayward, A., Neil Emery, R.J., Kolomiets, M.V. 2012. Disruption of *OPR7* and *OPR8* reveals the versatile functions of jasmonic acid in maize development and defense. Plant Cell 24: 1420–1436.

Yang, B., Thorogood, D., Armstead, I., Barth, S. 2008. How far are we from unravelling self-incompatibility in grasses? New Phytol. 179: 740–753.

Yang, J., Zhao, X., Cheng, K., Du, H., Ouyang, Y., Chen, J., Qiu, S., Huang, J., Jiang, Y., Jiang, L., Ding, J., Wang, J., Xu, C., Li, K., Zhang, Q. 2012. A killer-protector system regulates both hybrid sterility and segregation distortion in rice. Science 337: 1336–1340.

Youngner, V.B. 1960. Environmental control of initiation of the inflorescence, reproductive structures, and proliferations in *Poa bulbosa*. Amer. J. Bot. 47: 753–757.

Yu, P., Prakash, N., Whalley, R.D.B. 2000. Comparative reproductive biology of the vulnerable and common grasses in *Bothriochloa* and *Dichanthium*. In Jacobs, S.W.L., Everett, J. (eds.) Grasses: systematics and evolution. Melbourne: CSIRO. pp. 307–315

Zaitchik, B.F., le Roux, L.G., Kellogg, E.A. 2000. Development of male flowers in *Zizania aquatica* (North American wild-rice; Gramineae). Int. J. Plant Sci. 161: 345–351.

Fossil Record and Dates of Diversification

The fossil record of the grasses has been comprehensively reviewed by Thomasson (1987), Jacobs et al. (1999), and most recently by Strömberg (2011), and what follows is based heavily on those papers. Available fossils consist of pollen, macrofossils (leaf fragments, spikelets, fruits), and phytoliths (silica bodies).

As noted above, pollen of all members of Poales is similar in being spherical and monoporate. In grasses and other graminid Poales, the pore is surrounded by an annulus. In the past, pollen with this general morphology was placed in the form taxon *Monoporites annulatus* and often assigned to the grasses; however, the annulus is not synapomorphic and many specimens assigned to *M. annulatus* have been re-assigned to other families (Linder and Ferguson 1985). The diagnostic character for grasses is the lack of scrobiculi through the exine, a character that must be assessed with transmission electron microscopy. *M. annulatus* from 57 million years ago (mya) appears in northern South America and is likely to be Poaceae, whereas earlier Maastrichtian records (ca. 70 mya) may or may not represent grasses (Jacobs et al. 1999). Linder (1987) notes that the Maastrichtian grains "are rather inadequately illustrated by drawings" in van der Hammen (1954); more recent reports are also not fully diagnostic (Srivastava 2011). Once grass pollen appears in the record, it provides little further information on diversification of the grasses. Grass pollen is simply too uniform to assign to any particular major clade.

Grass macrofossils - particularly fruits, lemmas and glumes - dating from the Miocene (23–5.3 mya) have been found in North and South America, Europe, and Africa, showing that the family was widespread and diverse by that time (Jacobs et al. 1999; Strömberg 2011). Many of these fossils are well preserved and can be identified to modern genera or tribes on the basis of overall morphology and epidermal characters (e.g., Elias 1942; Thomasson 1985). The Miocene record clearly shows the expansion of grassland habitats. The earliest grass fossil with Kranz anatomy dates to the mid-Miocene, ca. 12.5 mya, in North America (Nambudiri et al. 1978; Whistler and Burbank 1992).

Earlier macrofossils, from the Eocene and Oligocene, are less common and also harder to assign taxonomically, as might be expected. A two-flowered grass spikelet has been found in Eocene deposits of North America, dated to about 55 mya (Crepet and Feldman 1991). Although the authors assigned this to Arundinoideae based on the broad concept of the subfamily prevalent at the time, the characters only allow assignment to the Bistigmatic (multi-flowered spikelet) clade; the spikelet does not match any currently known taxon. In the early Eocene of western Europe, the London clay flora (56–49 mya) is reported to contain grass leaves and inflorescences (Chandler 1964), but detailed descriptions of these are not available. The next earliest macrofossil in North America is from about 34 mya (upper Eocene) in the Florissant flora, assigned to *Stipa* or a close relative (MacGinitie 1953; Manchester 2001). Early grass macrofossils are not known from South America (Jacobs et al. 1999; Strömberg 2011), despite the early pollen records, nor are there macrofossils reported from Africa before the Miocene. Macrofossils have not been found in Australia (Jacobs et al. 1999; Strömberg 2011).

E.A. Kellogg, *Flowering Plants. Monocots*, The Families and Genera of Vascular Plants 13,
DOI 10.1007/978-3-319-15332-2_11, © Springer International Publishing Switzerland 2015

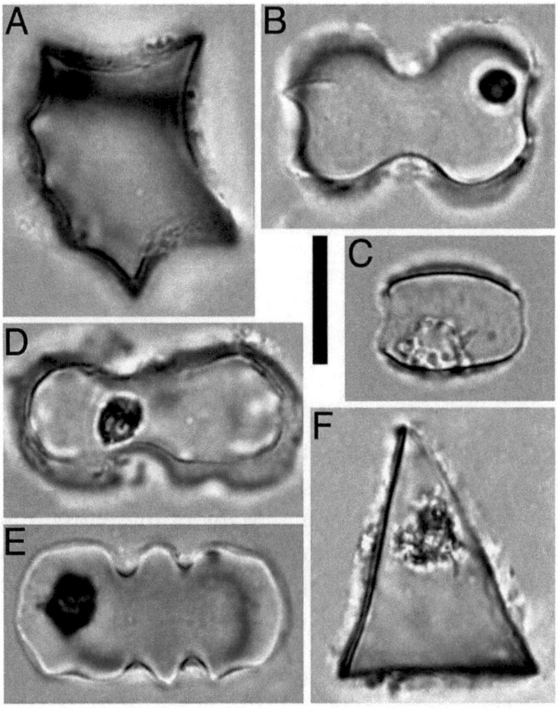

Fig. 8. Grass phytoliths from the Eocene to Miocene of North America. **A** Rondel, assigned to Bambusoideae, similar to *Chusquea*. **B** Two-lobed phytolith, similar to many in the PACMAD clade; compare to Fig. 2B. **C** Saddle-shaped, similar to modern Chloridoideae. **D** Two-lobed phytolith, similar to those in Stipeae. **E** Crenate phytolith, similar to those in Pooideae. **F** Tall, keel-shaped structure, unlike any known in modern grasses. (From Strömberg 2005; copyright 2005, National Academy of Sciences, U.S.A.)

The lack of early macrofossils may indicate absence of grasses, but could also reflect poor preservation. Pollen preserves poorly in semi-dry environments where the soil pH is above about 6.0 (Bryant, Jr. et al. 1994); this may affect the pollen record of dry grasslands in particular. (Note, however, that the pollen preservation study was done in the American Southwest, which is drier than many of the world's grasslands.) Likewise, macrofossils are rare in part because the open habitats of grasses are not conducive to preservation.

A handful of Cretaceous macrofossils have been suggested to represent grasses, although the evidence for this assignment is not strong. *Programinis laminatus* is a leaf fragment embedded in Burmese amber, and is thought to be early Cretaceous, upper Albian (97–110 mya) (Poinar, Jr. 2004, 2011). The rounded structures that are visible on the leaves are interpreted as silica bodies similar to those in the Pooideae, although this identification is far from certain. *P. burmitis* is a bracteate structure from the same deposit. It is interpreted as a spikelet, but lacks diagnostic characters that would make this interpretation definitive (Poinar, Jr. 2004, 2011). Because the Cretaceous date of these fossils makes them almost as old as the oldest angiosperm, accepting them as Pooideae would require a major revision of all dates of all land plants, and on that basis their identity as grasses is suspect.

The most widespread but arguably the hardest data to interpret come from phytoliths, or fossilized silica bodies (Fig. 8). If these are interpreted as current literature suggests, they will require revision not only of the history of the grasses, but also the history of all angiosperms. Phytoliths vary widely in shape among the grasses and thus provide much more taxonomic information than pollen grains do. They are far better preserved in the fossil record and thus provide more data than the sparse set of macrofossils. Silica bodies have been described for virtually all genera of grasses, beginning in the first part of the 20th century (Metcalfe 1960; Prat 1932; Watson and Dallwitz 1992 onward), but these descriptions were based on the shape in surface view (see also Description of the family, Vegetative morphology and anatomy: Leaves: Leaf epidermis). While certain shapes are more common in some subfamilies than others (e.g., cross shaped silica bodies in Chloridoideae), the shapes are rarely diagnostic. In an early phylogenetic analysis of the family, Kellogg and Campbell (1987) excluded silica bodies entirely because of unusually high homoplasy. Recent work on phytolith morphology, however, has shown that with analysis in three dimensions additional characters can be retrieved and phytoliths can in some cases be assigned reliably to particular taxa (Piperno 2006; Strömberg 2004). For example, most bambusoid silica bodies are saddle shaped or bilobate in surface view, but when removed from the leaf tissue and viewed from the side, they appear thick, with irregular points (Piperno and Pearsall 1998), and this morphotype appears to be diagnostic for the subfamily. Nonetheless, much phytolith data relies on the frequency of particular morphs in an assemblage, rather than a single synapomorphic form (Piperno 2006; Piperno and Pearsall 1998; Strömberg 2004).

The presumed earliest records of grass phytoliths are from India. Phytoliths found in a dinosaur coprolite suggest that dinosaurs consumed grass (Prasad et al. 2005), a suggestion at odds with all other fossil information. Morphological diversity of the phytoliths suggests that the BEP clade had already diversified by the date of the coprolite (65–67 mya). Additional phytolith data from similar fossil formations indicate presence of Ehrhartoideae phytoliths at about that date as well (Prasad et al. 2011), again a surprisingly early record. Because the Indian subcontinent had no connections to other landmasses at the time, accepting the date of the Indian phytoliths would force the conclusion that grass diversification must have either occurred in India, or have occurred before the breakup of Gondwana. Phytolith assemblages have also been described by Brea et al. (2009) and Zucol et al. (2010) for a mid- to late-Eocene formation in Patagonia (ca. 39–41 mya); these include a diverse set of phytoliths including some that could be assigned to chloridoids or panicoids.

Grass fossils have been used to calibrate the grass phylogeny and estimate the timing of major diversification events (Bouchenak-Khelladi et al. 2010, 2014; Christin et al. 2008, 2014; Jones et al. 2014; Vicentini et al. 2008). Pollen, macrofossils, and the majority of the phytolith data point to origin of the family about 80 mya, or in the upper Cretaceous. The divergence of the BEP and PACMAD clades (crown grasses) then is placed at 50 to 60 mya (Bouchenak-Khelladi et al 2010; Jones et al. 2014; Vicentini et al. 2008). Christin et al. (2014) also estimate the BEP-PACMAD split at 51.2 mya (±12.3) or 62.6 (±7.6) using two different dating programs, a set of nuclear genes, and calibrations using only fossils from non-Poaceae. These calibrations and dates all have the satisfying property of placing the origin of the PACMAD clade and hence the earliest appearance of C_4 photosynthesis about 32 mya in the Oligocene, corresponding to a drop in atmospheric CO_2.

In contrast to all other data, the Indian phytoliths would place the origin of grasses appreciably earlier, around 128 mya (Vicentini et al. 2008); this date corresponds to the earliest fossil evidence of monocots, which is about 125 mya (Herendeen and Crane 1995) and thus would require pushing the monocot-dicot split much earlier. Using this same calibration point for a dataset based on whole chloroplast genomes, Jones et al. (2014) placed the stem node of Poaceae at 180 mya. Christin et al. (2014), using nuclear genes, the phytolith calibration plus non-Poaceae fossils, placed the BEP-PACMAD divergence at 82.4 (±14.8) or 79.1 (±3.0) mya, depending on the dating program used. Dates based on such calibrations imply a major slowdown of evolutionary rates in the Poaceae. In addition, such dates would require revision of hypotheses regarding C_4 evolution because atmospheric CO_2 was about 1000 ppm at that time. If the older calibration and dates are correct, then lower atmospheric CO_2 cannot have been the driver of C_4 but rather it may have been a change in seasonality or aridity (Strömberg 2011). The older dates also contradict the suggestion of Ehleringer et al. (1997), which argues that C_4 photosynthesis can only be more efficient than C_3 when CO_2 concentrations are relatively low. The older date is also incompatible with estimates of the evolution of cold tolerance in Pooideae

(Sandve and Fjellheim 2010), which would have to have occurred when the climate was unusually warm. The older date for grass origins also affects estimates of evolutionary rates and dates of other parts of the angiosperm tree. Therefore, accepting the date and identity of the Indian phytoliths requires overturning virtually all other data and inferences about grass, and indeed monocot, evolution.

In summary, the age of the grasses and their rate of molecular evolution are unknown, although most lines of evidence point to the BEP-PACMAD split having occurred 50–60 mya in the Paleocene or early Eocene, and the grass stem node being dated to the upper Cretaceous. In this scenario, both the earliest C_4 origins and the evolution of pooid cold tolerance originated around the Eocene-Oligocene boundary. However, these dates, despite being supported by many disparate sorts of fossils and dating methods, are incompatible with phytolith data (Christin et al. 2014; Prasad et al. 2011; Strömberg 2011). If the phytolith data are accepted at face value, then the paleohistory of grasses must be revised by millions of years. Such revisions change estimates of molecular evolution in the grasses, from being much more rapid than most monocots to being much slower; such changes also affect estimates of dates for many important events in angiosperm history. In addition, apparent correlations between physiology and paleoclimate are broken down. Much thus rests on the reliable dating of the Indian phytoliths and the phylogenetic distribution of their distinctive shapes. Their use in dating assumes that they represent Oryzeae, which in turn assumes that costal rows of vertical bilobate scooped silica bodies occur only in that tribe. This assumption can only be addressed by broad surveys of silica body morphology in other Poaceae and graminid lineages.

References

Bouchenak-Khelladi, Y., Verboom, G.A., Savolainen, V., Hodkinson, T.R. 2010. Biogeography of the grasses (Poaceae): a phylogenetic approach to reveal evolutionary history in geographical space and geological time. Bot. J. Linn. Soc. 162: 543–557.

Bouchenak-Khelladi, Y., Muasya, A.M., Linder, H.P. 2014. A revised evolutionary history of Poales: evolution and diversification. Bot. J. Linn. Soc. 175: 4–16.

Brea, M., Zucol, A.F., Raigemborn, M.S., Matheos, S. 2009. Reconstrucción de paleocomunidades arbóreas mediante análisis fitolíticos en sedimentos del paleoceno superio-eoceno? (formación las flores), Chubut, Argentina. BAR Int. Series S1870 9: 91–108.

Bryant, Jr., V.M., Holloway, G.R., Jones, J.G., Carlson, D.L. 1994. Pollen preservation in alkaline soils of the American Southwest. In: Traverse, A. (ed.) Sedimentation of organic particles. Cambridge: Cambridge University Press. pp. 47–58

Chandler, M.E.J. 1964. The lower Tertiary flora of southern England IV. London: British Museum (Natural History).

Christin, P.-A., Besnard, G., Samaritani, E., Duvall, M.R., Hodkinson, T.R., Savolainen, V., Salamin, N. 2008. Oligocene CO_2 decline promoted C_4 photosynthesis in grasses. Curr. Biol. 18: 37–43.

Christin, P.-A., Spriggs, E., Osborne, C.P., Strömberg, C.A. E., Salamin, N., Edwards, E.J. 2014. Molecular dating, evolutionary rates, and the age of the grasses. Syst. Biol. 63: 153–165.

Crepet, W.L., Feldman, G.D. 1991. The earliest remains of grasses in the fossil record. Amer. J. Bot. 78: 1010–1014.

Ehleringer, J.R., Cerling, T.E., Helliker, B.R. 1997. C_4 photosynthesis, atmospheric CO_2, and climate. Oecologia 112: 285–299.

Elias, M.K. 1942. Tertiary prairie grasses and other herbs from the High Plains. Geol. Soc. Amer. Spec. Pap. 41: 1–176.

Herendeen, P.S., Crane, P.R. 1995. The fossil history of the monocotyledons. In: Rudall, P.J., Cribb, P.J., Cutler, D.F., Humphries, C.J. (eds.) Monocotyledons: systematics and evolution. Kew: Royal Botanic Gardens. pp. 1–21

Jacobs, B.F., Kingston, J.D., Jacobs, L.L. 1999. The origin of grass-dominated ecosystems. Ann. Missouri Bot. Gard. 86: 590–643.

Jones, S.S., Burke, S.V., Duvall, M.R. 2014. Phylogenomics, molecular evolution, and estimated ages of lineages from the deep phylogeny of Poaceae. Plant Syst. Evol. 300: 1421–1436.

Kellogg, E.A., Campbell, C.S. 1987. Phylogenetic analyses of the Gramineae. In: Soderstrom, T.R., Hilu, K.W., Campbell, C.S., Barkworth, M.E. (eds.) Grass systematics and evolution. Washington, DC: Smithsonian Institution Press. pp. 310–322

Linder, H.P. 1987. The evolutionary history of the Poales/ Restionales - a hypothesis. Kew Bull. 42: 297–318.

Linder, H.P., Ferguson, I.K. 1985. On the pollen morphology and phylogeny of the Restionales and Poales. Grana 24: 65–76.

MacGinitie, H.D. 1953. Fossil plants of Florissant Beds, Colorado. Publ. Carnegie Inst. Wash. 599: 1–198.

Manchester, S.R. 2001. Update on the megafossil flora of Florissant, Colorado. In: Evanoff, E., Gregory-Wodzicki, K.M., Johnson, K.R. (eds.) Fossil flora and stratigraphy of the Florissant Formation, Colorado. Denver, Colorado: Denver Museum of Nature and Science. pp. 137–161

Metcalfe, C.R. 1960. Anatomy of the monocotyledons. I. Gramineae. Oxford: Clarendon Press.

Nambudiri, E.M.V., Tidwell, W.D., Smith, B.N., Hebbert, N.P. 1978. A C_4 plant from the Pliocene. Nature 276: 816–817.

Piperno, D.R. 2006. Phytoliths: a comprehensive guide for archaeologists and paleoecologists. New York: Alta Mira.

Piperno, D.R., Pearsall, D.M. 1998. The silica bodies of tropical American grasses: morphology, taxonomy, and implications for grass systematics and fossil phytolith identification. Smithsonian Contr. Bot. 85: 1–40.

Poinar, Jr., G.O. 2004. *Programinis burmitis gen. et sp. nov.*, and *P. laminatus sp. nov.*, Early Cretaceous grass-like monocots in Burmese amber. Aust. Syst. Bot. 17: 497–504.

Poinar, Jr., G. 2011. Silica bodies in the Early Cretaceous *Programinis laminatus* (Angiospermae: Poales). Palaeodiversity 4: 1–6.

Prasad, V., Strömberg, C.A.E., Alimohammadian, H., Sahni, A. 2005. Dinosaur coprolites and the early evolution of grasses and grazers. Science 310: 1177–1180.

Prasad, V., Strömberg, C.A.E., Leaché, A.D., Samant, B., Patnaik, R., Tang, L., Mohabey, D.M., Ge, S., Sahni, A. 2011. Late Cretaceous origin of the rice tribe provides evidence for early diversification in Poaceae. Nature Comm. 2: 480.

Prat, H. 1932. L'épiderme des graminées: étude anatomique et systématique. Ann. Sci. Nat.: Botanique, Séries 10 14: 117–324.

Sandve, S.R., Fjellheim, S. 2010. Did gene family expansions during the Eocene-Oligocene boundary climate cooling play a role in Pooideae adaptation to cool climates? Mol. Ecol. 19: 2075–2088.

Srivastava, S.K. 2011. The occurrence of the fossil genus *Graminidites* in the Maastrichtian Scollard Formation, Alberta, Canada, and its palaeoecological and palaeogeographical significance. Bot. J. Linn. Soc. 167: 235–248.

Strömberg, C.A.E. 2004. Using phytolith assemblages to reconstruct the origin and spread of grass-dominated habitats in the great plains of North America during the late Eocene to early Miocene. Palaeogeog. Palaeoclimatol. Palaeoecol. 207: 239–275.

Strömberg, C.A.E. 2005. Decoupled taxonomic radiation and ecological expansion of open-habitat grasses in the Cenozoic of North America. Proc. Natl. Acad. Sci. USA 102: 11980–11984.

Strömberg, C.A.E. 2011. Evolution of grasses and grassland ecosystems. Ann. Rev. Earth Planet. Sci. 39: 517–544.

Thomasson, J.R. 1985. Miocene fossil grasses: possible adaptation in reproductive bracts (lemma and palea). Ann. Missouri Bot. Gard. 72: 843–851.

Thomasson, J.R. 1987. Fossil grasses: 1820–1986 and beyond. In: Soderstrom, T.R., Hilu, K.W., Campbell, C.S., Barkwroth, M.E. (eds.) Grass systematics and evolution. Washington DC: Smithsonian Institution Press. pp. 159–167

van der Hammen, T. 1954. El desarollo de la flora Colombiana en los periodos geologicos I. Maestrichtiano hasta terciario más inferior. Bol. Geolog. Bogotá 2: 49–106.

Vicentini, A., Barber, J.C., Giussani, L.M., Aliscioni, S.S., Kellogg, E.A. 2008. Multiple coincident origins of C_4 photosynthesis in the Mid- to Late Miocene. Global Change Biol. 14: 2963–2977.

Watson, L., Dallwitz, M.J. 1992 onward. The grass genera of the world: descriptions, illustrations, identification, and information retrieval; including synonyms, morphology, anatomy, physiology, phytochemistry, cytology, classification, pathogens, world and local distribution, and references, 23rd April 2010 edn. Wallingford: CAB International.

Whistler, D.P., Burbank, D.W. 1992. Miocene biostratigraphy and biochronology of the Dove Spring Formation, Mojave Desert, California, and characterization of the Clarendonian mammal age (late Miocene) in California. Geol. Soc. Amer. Bull. 104: 644–658.

Zucol, A.F., Brea, M., Bellosi, E.S. 2010. Phytolith studies in Gran Barranca (central Patagonia, Argentina): the middle-late Eocene. In: Madden, R.H., Carlini, A.A., Bucetich, M.G., Kay, R.F. (eds.) The paleontology of Gran Barranca. Cambridge: Cambridge University Press. pp. 317–340

Domestication

"... agriculture [is] something the grasses did to people as a way to conquer the trees." (Pollan 2001), p. xxi

Origins of Major Cereal Crops

Because the grasses are common plants, often ecologically dominant and with nutritious seeds, it is not surprising that they were harvested and later domesticated by the earliest agriculturalists. Archaeological evidence shows that wild grasses were used well before agriculture began (Bettinger et al. 2010; Kislev et al. 1992; Purugganan and Fuller 2009), and only gradually did early hunter-gatherer groups shift to farming. Fuller (2007) suggests a model, modified from Harris (1989), that describes gathering of wild food and cultivation of domesticated plants as end-points of a continuum, with intermediate stages of "semi-domestication" that spread over various lengths of time in various crops (e.g., Tanno and Willcox 2006). In general, this model shows that domestication was a long slow process, occurring over hundreds to thousands of years. As population genetic data have been applied more extensively to the question of domestication, and as more archaeological sites have been investigated, it has also become clear that many domesticated species were domesticated more than once (Weiss et al. 2006). The best studied of the cereals are the so-called Big Three - wheat, rice, and maize - plus barley, all of which have an appreciable archaeological record. However, additional data are accumulating on many other species that are somewhat less important for modern agriculture, but that have been important historically and regionally. These data permit some generalizations about the domestication process as well as the effects of selection on genes and genomes.

The ancestry of wheat is complex but well studied, involving several species with three distinct genomes, conventionally labeled A, B, and D. Diploid A genome wheats include *Triticum urartu* and *T. boeoticum*. *T. boeoticum* was domesticated in what is now southeast Turkey about 9000 years before present, to give rise to einkorn wheat (*T. monococcum*) (Peng et al. 2011). An A genome wheat similar to modern *T. urartu* crossed with a species of goatgrass similar to modern *Aegilops speltoides*, which provided the B genome, and formed a stable allotetraploid. Because *A. speltoides* is widespread and weedy it is not surprising that the two species would have come into contact. The resulting tetraploid wheats include *T. dicoccoides* (= *T. turgidum* ssp. *dicoccoides*), *T. dicoccum* (= *T. turgidum* ssp. *dicoccum*), and *T. araraticum*; the former two are known as wild emmer and emmer, respectively, with *T. dicoccum* being the domesticated form of *T. dicoccoides*. (Note that the taxonomy within *Triticum* is not stable, with each taxon being named variously at the species or subspecies level. In addition, not all authors recognize all species as distinct. Nomenclature here follows Peng et al. 2011.) Like einkorn, emmer wheat was domesticated in southeast Turkey or the southern Levant or both ca. 9500–9000 years before present; the wild ancestor, *T. dicoccoides* is fairly widespread in the region and authors disagree on the number of domestication events (Luo et al. 2007; Matsuoka 2011; Peng et al. 2011). *Triticum durum*, now used for making pasta, is derived from *T. dicoccum*. Other independent, domestication events of einkorn and emmer can be detected in the archaeological

E.A. Kellogg, *Flowering Plants. Monocots*, The Families and Genera of Vascular Plants 13,
DOI 10.1007/978-3-319-15332-2_12,

record but did not persist to modern times (Fuller 2007; Peng et al. 2011).

Although most of the morphological changes associated with domestication occurred in the diploid and tetraploid species einkorn and emmer, neither is a significant crop today. Instead they have been supplanted by the allohexaploid, *Triticum aestivum*. Soon after emmer domestication, cultivated emmer or a very similar tetraploid with A and B genomes crossed with a second species of goatgrass, *A. tauschii* (with the D genome), to produce *T. aestivum* or bread wheat (Kihara 1944; McFadden and Sears 1946). This hybridization probably occurred multiple times in the southwestern Caspian region of present-day Iran (Wang et al. 2013). The D genome is responsible for providing the genes that give wheat flour its many desirable properties for baking (Kaltsikes et al. 1968; Kerber and Tipples 1969; Nelson et al. 2006). Evidence for the hybridization events leading to bread wheat was first provided by painstaking cytogenetic studies; Kihara then resynthesized *T. aestivum* from its tetraploid and diploid ancestors (Kihara 1982; Kihara and Lilienfeld 1949). His results have since been confirmed by genome mapping (Devos et al. 2008; Gale and Devos 1998) and population genetics (Wang et al. 2013).

Rice was domesticated in Asia about 9000 years ago from *Oryza rufipogon*, but the precise number of origins in a matter of contention (Kellogg 2009; Vaughan et al. 2008); population genetic data have been equivocal. Recent evidence points to a single origin of *O. sativa japonica* in southern China followed by wide dispersal (Huang et al. 2012; Molina et al. 2011). Extensive human transport of grains over millennia plus continued introgression with wild rice (*Oryza rufipogon*) later led to the origin of *O. sativa indica* (Huang et al. 2012). *Oryza glaberrima* was domesticated independently in western Africa from *O. barthii*, no later than 300–200 BC (Sweeney and McCouch 2007; Vaughan et al. 2008). Although rice grains as old as 3500 years before present have been identified at sites in northeast Nigeria, they were apparently not domesticated (Sweeney and McCouch 2007).

Zea mays ssp. *mays* was domesticated in the Balsas River Basin in Mexico, about 9000 years before present (Matsuoka et al. 2002; Piperno et al. 2009; van Heerwaarden et al. 2011), from teosinte, *Zea mays* ssp. *parviglumis*. Early introgression into maize from another teosinte, *Zea mays* ssp. *mexicana* created some confusion about the precise locality of domestication, but this has been clarified by population genetic data on maize landraces (van Heerwaarden et al. 2011).

Domesticated barley (*Hordeum vulgare*) appears in archaeological remains about the same time as einkorn and emmer wheat. In addition to its center of domestication in the Fertile Crescent, a second domestication occurred to the east of the Zagros Mountains (Morrell and Clegg 2007; Saisho and Purugganan 2007). Even within the Fertile Crescent, there is evidence for at least two domestication events, based on the genes controlling rachis disarticulation. These two loci, *Brittlerachis1* (*Btr1*) and *Brittlerachis2* (*Btr2*) are linked on chromosme 3H; the alleles conferring rachis disarticulation are dominant. Loss of function at either locus creates a non-shattering phenotype. Barley cultivars that are mutant at *btr1* are phylogenetically distinct from plants that are mutant at *btr2* (Azhaguvel and Komatsuda 2007), based on a phylogeny of a marker linked to the *Btr* loci.

Foxtail millet (*Setaria italica*) was domesticated from green millet (*S. viridis*) in northwestern China about 7000–8000 years ago (Hunt et al. 2008; Le Thierry d'Ennequin et al. 2000; Li and Wu 1996). Although archaeological sites in the Caucasus hint at a possible second center of domestication, the evidence for this is not firm (Harlan 1992; Hunt et al. 2008). The world's oldest recorded noodles, dated at 4000 years before present were thought to have been made with foxtail millet, but recent information questions this claim and suggests that the noodles could have been made from wheat or barley, both of which were also cultivated in the region at the time (Ge et al. 2011).

Finger millet, *Eleusine coracana* ssp. *coracana*, is an allotetraploid native to Africa, domesticated from *E. coracana* ssp. *africana*. The date of domestication is uncertain. Grains of *E. coracana* have been found in Ethiopia from a site with a possible date of the third millennium BC, but the date has not been verified (Hilu et al. 1979). The crop was introduced to India from Africa, where it has developed several characteristics distinct from its African forebears (Dida et al.

2008; Hilu and de Wet 1976). One genome of *E. coracana* was clearly donated by a parent similar to the African species *E. indica*, but the other genome is unknown (Neves et al. 2005).

Common (Proso or broomcorn), millet (*Panicum miliaceum*) was domesticated in China in the same general area as foxtail millet but somewhat earlier, perhaps about 10,000 years before present (Lu et al. 2009a). The two species have similar grains, making interpretation of archaeological data difficult, but some details of the lemma epidermis differ between them (Lu et al. 2009b). *Panicum miliaceum* is an allopolyploid. One genome comes from a clade that includes the tetraploids *P. bergii* and *P. stramineum*, and the diploids *P. capillare* and *P. nepheliphilum*, whereas the other genome is connected to a group of polyploid species placed in *Panicum* sect. *Dichotomiflora* (Triplett et al. 2012). However, few Asian species have been included in any phylogenetic analysis to date, so the closest relatives and diploid progenitors of common millet remain unknown.

Pearl millet (*Cenchrus americanus* = *Pennisetum glaucum*) was domesticated in western Africa, with the oldest remains discovered in the Tilemsi Valley in present-day Mali, about 4500 years before present (Manning et al. 2011; Oumar et al. 2008).

In contrast to other cereal crops, the domestication history of *Sorghum bicolor* is poorly known. It appears to have originated in sub-Saharan Africa from a wild progenitor similar to *S. bicolor* ssp. *arundinaceum* (Zohary and Hopf 2000). Domestication is thought to have occurred in Ethiopia beginning 6000–5000 years before present, but the resultant landraces are highly diverse both morphologically and genetically (Dillon et al. 2007), reflecting the many disparate uses of the crop.

Sugarcane (*Saccharum officinarum*) originated in New Guinea, where it appears to have been derived from the wild species, *S. robustum* (Artschwager and Brandes 1958; Lebot 1999), presumably by selection for clones with high sugar and low fiber contents. Unlike the cereals, it is generally propagated vegetatively, leading to a set of widespread clones with low genetic diversity. The specific steps in domestication of this species are unknown; in general it is much less well studied than domestication of the major cereals. *S. officinarum* has hybridized repeatedly with *S. spontaneum*, a species that is widespread in Southeast Asia, New Guinea and nearby islands. In the last century, the two species were deliberately hybridized to give rise to most modern cultivars and breeding lines of sugarcane. *S. officinarum* has a chromosome number of $2n = 80$, and often produces unreduced gametes. When it crosses with *S. spontaneum* ($2n = 64$), in which gametes are generally haploid, the chromosome complement of the resulting hybrid is roughly 80 + 32 chromosomes. Loss of *S. spontaneum* chromosomes is common, giving rise to a range of aneuploid numbers (Piperidis et al. 2009).

Tef (*Eragrostis tef*) was domesticated in Ethiopia a little more than 2000 years before present, from an ancestor similar to *E. pilosa* (D'Andrea 2008; Ingram and Doyle 2003); both species are tetraploid. Among the cereals, tef is distinctive for its incredibly tiny grains, 0.8–1.5 mm long. This is only slightly longer than the size of grains in *E. pilosa* (0.5–1.1 mm), which makes distinguishing the two difficult (D'Andrea 2008). Selection appears to have focused on increasing inflorescence branching and producing additional tillers, rather than increasing size of the grains and density of the inflorescence. It is possible that this selection strategy results from concerns about lodging (D'Andrea 2008).

The domestication history of many minor grains, such as fonio (*Digitaria exilis*), Guinea millet (*Brachiaria deflexa*), Kodo millet (*Paspalum scrobiculatum*), barnyard millet (*Echinochloa crus-galli*), sauwi millet (*Panicum sonorum*), raishan (*Digitaria* sp.), and *Bromus mango* remains poorly studied. Lines of fonio and Guinea millet with larger seeds and non-shattering inflorescences are reported (National Research Council 1996). In the Chad Basin of northeastern Nigeria, it is possible to buy in local markets a mixture known as kreb, which consists of small grains from several genera of Paniceae, including *Echinochloa*, *Brachiaria*, and *Panicum* (Klee et al. 2000). *Bromus mango* was apparently cultivated in what is now Chile before the Spanish conquest (Planchuelo and Peterson 2000), but is now rare.

Modern oats and rye were domesticated in Europe, but were previously cultivated and possibly domesticated in several distinct sites in the Near East; these earlier domesticated strains have not persisted (Weiss et al. 2006). Rye (*Secale cereale*) was domesticated from wild plants of the same species, which occur in areas corresponding to present-day Turkey (Khush 1963; Sencer and Hawkes 1980; Zohary and Hopf 2000). In comparison to species of *Triticum* and *Hordeum*, it is more tolerant of cool temperatures and is generally found at higher altitudes (Burger and Ellstrand 2005). Domestication appears to have occurred more than once. A few, possibly domesticated, grains appear at a site known as Abu Hureyra, on the Euphrates in current Syria, dating from about 12,500 years before present, but these were mixed with wild grains (Weiss et al. 2006; Willcox 2005). Domesticated grains (plump, on tough rachises) have been found about 8600 years before present in south Central Anatolia, and ca. 4000 years BP in north central Anatolia (Weiss et al. 2006).

Cultivated oats first appear in the archaeological record in Europe in the second millennium BC (Zohary and Hopf 2000). Large caches of oats have been found in archaeological sites in the Near East. While these caches appear to lack the characteristic morphology of domesticated oats, the numbers of grains suggest that they were likely cultivated (Weiss et al. 2006). Zohary and Hopf (2000) speculate that oats were probably domesticated more or less incidentally. The hexaploid oats in the *sativa* complex are common and widespread weeds in fields of wheat and barley throughout the Mediterranean; it is thus likely that they were harvested along with other grains and were thus selected alongside them. Cultivated oat (*Avena sativa*) is a hexaploid, with genomes labeled A, C, and D. The A genome appears to have come from the maternal parent, as shown by sequences of chloroplast genes (Peng et al. 2010). Based on phylogenetic analysis, the maternal parent was similar to the diploid *A. wiestii*, a parent also shared by the tetraploid *A. murphyi*, and the hexaploids *A. sterilis* and *A. occidentalis*. Chloroplast data suggest that the hexaploid wild oats (*A. fatua*) were derived from an independent allopolyploidization event.

Most domesticated species are crossable with their wild ancestors, and the wild ancestors often persist as weeds in cultivated fields (Harlan 1992). For example, *Setaria italica* has been crossed with its progenitor, *S. viridis* (Wang et al. 1998). *Zea mays* ssp. *mays* (cultivated maize) can be crossed with its progenitor teosinte (*Zea mays* ssp. *parviglumis*) as long as maize is the female parent and teosinte the pollen parent. (See section on Reproductive systems above.) In addition, *Zea mays* ssp. *mays* has crossed with *Zea mays* ssp. *mexicana*, somewhat confusing efforts to disentangle the domestication history of the species (van Heerwaarden et al. 2011). Introgression has been demonstrated between finger millet (*Eleusine coracana* ssp. *coracana*) and its wild progenitor (*Eleusine coracana* ssp. *africana*) (Hilu and de Wet 1976), between pearl millet (*Cenchrus americanus* = *Pennisetum glaucum*) and its wild ancestor (*P. glaucum* ssp. *monodii*) (Mariac et al. 2006; Oumar et al. 2008), between wild (*Triticum dicoccoides*) and cultivated (*T. dicoccum*) emmer wheat (Luo et al. 2007), and between *Oryza sativa* and *O. rufipogon* (Huang et al. 2012).

Morphological Changes in Domestication

Domestication results in a set of well-documented genetic changes caused by the intense selection pressure created by growing in tilled soil, annual harrowing, annual harvesting and subsequent planting (Harlan 1992). One of the earliest traits to be modified during domestication is grain size (Fuller 2007; Purugganan and Fuller 2009). In part, this appears to be the result of unconscious selection for plants that produce vigorous seedlings and that can tolerate the increased planting depth of tilled soil (Purugganan and Fuller 2009). Changes in grain size occurred rapidly in cultivated wheat, barley and rye, with larger grains appearing in the archaeological record over a period of about 500 years (Fuller 2007). The change of grain size in rice is confounded with changes in shape, but in general the change in grain size appears to have been somewhat slower in rice (Fuller 2007).

Another early change is loss of disarticulation (shattering). (See section on Inflorescence structure: Disarticulation.) Whether this is selected for depends on the mode of harvest. If seeds are collected by shaking the plants and collecting the falling seeds in a basket, then shattering will be retained. This is seen in North American wild rice (*Zizania*), which has been harvested by Native Americans for centuries by moving through the stands of wild rice in boats and beating the plants so that the seeds fall in to the boat. In contrast, if harvesting is done with a sickle, the selection pressure is reversed to favor plants that retain their seed. A plant that lacks shattering is heavily, if not completely, dependent on humans for propagation and dispersal, so lack of shattering is often taken as diagnostic of a domesticated plant. Several of the genetic loci controlling shattering have been cloned in rice (Konishi et al. 2006; Li et al. 2006; Zhou et al. 2012), and one in wheat (Simons et al. 2006). Whether these same loci were selected in other domesticated grasses is unknown. The *Btr1* and *Btr2* loci of barley map to chromosome 3 (known as 3H, for chromosome 3 in *Hordeum*); there are two brittle rachis loci in emmer wheat, *Br1* and *Br2* that also map to the homoeologous chromosomes (3A and 3B) (Matsuoka 2011). Whether the *Hordeum* and wheat loci are orthologous is unknown and will await cloning of one or more of the genes (Li and Gill 2006).

Whereas development of larger grains occurred before loss of shattering in wheat, barley and rice (Fuller 2007; Purugganan and Fuller 2009), the order was reversed in pearl millet (Fuller 2007). While loss of shattering could in principle have occurred in less than 100 years, such rapid genetic change requires intense (albeit unconscious) selection imposed by harvesting semi-ripe grains with a sickle or by uprooting (Hillman and Davies 1990). Rapid selection also requires sowing seed in new sites each year to avoid interbreeding with unselected plants. The archaeological record, in contrast, indicates that the transition to non-shattering lines occurred over about 1500 years in wheat and 2000 years in barley, suggesting that wild populations continued to mix with domesticated ones and that selection for non-shattering grains was relatively weak (Fuller 2007).

Domestication also affects tillering and culm branching, although both the phenotype and the underlying genes vary among domesticated grasses. In the domestication of maize, tillering was suppressed so that cultivated maize generally has a single culm, in contrast to its highly branched wild ancestor, teosinte. The change in architecture was caused by selection on regulation of the gene *teosinte branched1* so that it is more strongly expressed in the axillary buds near the base of the plant, preventing bud outgrowth (Clark et al. 2006; Doebley et al. 1997; Zhou et al. 2011). Superficially similar phenotypic changes occurred during domestication of foxtail millet (*Setaria italica*) and pearl millet (*Cenchrus americanus* = *Pennisetum glaucum*), but quantitative genetic studies find that *tb1* was only a minor player in those transitions (Doust et al. 2004; Remigereau et al. 2011). In both maize and pearl millet, the axillary buds that would form tillers are well developed but fail to elongate (Remigereau et al. 2011). In foxtail millet, in contrast, axillary buds fail to initiate at all, consistent with a different locus of selection (Doust et al. 2004).

In pooid and ehrhartoid grasses, tillering is common, with each tiller ending in an inflorescence; in contrast culm branching is rare or absent. In domestication of crops in these subfamilies, the extent of tillering appears not to have been a target of selection (Doust 2007). In rice and wheat the extent of tillering does not differ in a consistent way between wild and domesticated plants (Remigereau et al. 2011). Wild finger millet (*Eleusine coracana* ssp. *africana*) develops more tillers than cultivated (*E. coracana* ssp. *coracana*), but cultivated plants have more axillary branches than wild ones (Dida et al. 2008).

The process of sowing and harvesting selects automatically for plants that germinate promptly and synchronously so dormancy and often day-length requirements tend to be lost in domesticated species (Harlan et al. 1973). In wild oats, einkorn and emmer wheat, spikelets generally have two flowers; the proximal one is larger and non-dormant whereas the distal one is dormant. Dormancy is lost during domestication. In addition to loss of dormancy (dispersal in time), domestication also often correlates with loss of dispersal structures, such as awns in wheat (Fuller 2007) and reduction or loss of awns in rice (Sweeney and McCouch 2007).

In cultivated rice, most lines have white or brownish pericarp, representing modification of the ancestral red color. Pericarp color is regulated by the Rc locus, a bHLH transcription factor (Sweeney et al. 2006). The Rc allele in cultivated rice is non-functional, having a 14 bp deletion that abolishes DNA binding. The allele appears to have arisen in a *japonica* rice and then spread via introgression throughout almost all cultivated lines (Sweeney et al. 2006).

In both maize and barley, human selection led to inflorescences that are not merely larger, but that have changes in architecture. Wild barley has three spikelets per node, only one of which is fertile. The clusters of three spikelets form on alternate sides of the inflorescence such that there is one row of fertile spikelets on either side; this is known as two-row barley. Some modern cultivars retain this trait, but more often selection has fixed mutations that cause the sterile spikelets to become fertile. This is six-row barley, with three fertile spikelets in clusters on alternate sides of the rachis. Fertility of the lateral spikelets is controlled by at least two loci. One, a homeodomain-leucine zipper (HD-ZIP I) transcription factor, *Vrs1*, has been mutated several times independently in the history of barley (Komatsuda et al. 2007). A second locus, *INT-C* is an ortholog of *Tb1*, the gene that affects tillering in maize (Ramsay et al. 2011). Mutations in *INT-C* create a phenotype that is intermediate between two-rowed and six-rowed barley; the lateral spikelets are partially fertile, and are better developed than those in two-rowed lines but smaller than those in six-rowed lines.

The most dramatic and most famous of the domesticated phenotypes is the cob or ear of maize (corn). This is a botanical monstrosity that bears little resemblance to the small pistillate inflorescence of the other subspecies of *Zea mays* and other species of *Zea*. The wild type inflorescence of teosinte has distichous spikelets. At maturity the outer glume plus the associated rachis internode form a rock-hard structure that encloses the grain, and the inflorescence rachis disarticulates at the nodes as in other Andropogoneae. The cultivated inflorescence (cob or ear), in contrast, has spikelets in many rows; the spikelets are naked, and at maturity the grain grows well beyond the flimsy glumes, lemmas and paleas. The rachis in maize is firm, with no vestige of disarticulation. This complex set of morphological changes is under the control of a single major locus, TEOSINTE GLUME ARCHITECTURE1, which is a transcription factor that normally regulates spikelet development (Preston et al. 2012; Wang et al. 2005). (See also Inflorescence structure: Spikelets, above.)

In wild wheat and barley, as well as in the so-called "hulled wheats" - einkorn, emmer, and the hexaploid *Triticum spelta* (spelt) - grains are dispersed with the glumes, lemma, and palea still attached. Grains require drying to be broken free from the surrounding floral bracts (Peng et al. 2011). Mutations in the genes Q and *Tenacious glumes* led to free-threshing varieties, in which the glumes break easily (Nalam et al. 2006, 2007; Simons et al. 2006). (See above, Inflorescence structure: Spikelets.) In addition to conferring the free threshing characteristic, the dominant (domesticated) Q allele makes the inflorescence shorter and broader, and the rachis tough and non-disarticulating.

Genetic and Genomic Changes in Domestication

The effects of domestication on the genome are the subject of much interest and active study. In general, domestication creates a population bottleneck, which leads to reduced genetic diversity (Doebley et al. 2006; Glémin and Bataillon 2009). In cultivated durum and bread wheat, diversity has been sharply reduced from wild strains, with only 16 % and 31 % of wild diversity respectively (Haudry et al. 2007). However, these estimates are based on a modest number of accessions and markers so should probably be expanded and verified. Foxtail millet retains about 45 % of the genetic diversity of its wild ancestor green millet (Wang et al. 2010), and pearl millet retains 67–74 % of the wild diversity (Clotault et al. 2012; Oumar et al. 2008). In contrast, landraces of maize harbor almost as much variation as is found in teosinte (83 %; Hufford et al. 2012).

Levels of diversity vary among the three genomes of hexaploid wheat (Akhunov et al. 2010). The A and B genomes, which were inherited from emmer wheat, are more diverse than the D genome, and single nucleotide polymorphisms (SNPs) in the A and B genomes are spread throughout the chromosomes. The simplest explanation for this pattern is past and continuing gene flow between wild and cultivated emmer before polyploidy with *Aegilops tauschii*, and continuing gene flow between wild emmer and cultivated wheat. The latter two form fertile pentaploid hybrids in nature (Kihara 1982). In contrast, SNP diversity in the D genome is very low, confirming previous studies suggesting only a handful of hybridization events leading to the hexaploid *T. aestivum* (Akhunov et al. 2010). In *T. aestivum*, loci related to domestication appear to be clustered in only about 16 regions of the genome

(Peng et al. 2011). However, because genes in wheat tend to be clustered in general, this clustering may reflect a peculiarity of the wheat genome, rather than a generalization about domestication.

Reduction of diversity is especially obvious in regions of the genome harboring genes that are directly selected. In such regions, genetic diversity in the gene under selection is reduced to near zero, as would be expected if a single allele is chosen for propagation. During the selection process, the region of the genome closely linked to the selected locus is effectively pulled along, so that an entire piece of the genome shows reduced diversity. Because selection effectively sweeps away variation in the selected region, such a pattern is taken as evidence of a "selective sweep". Now that it is possible to sequence portions of the genome of many members of a population, it is possible to scan the genome to look for regions of reduced variation and to infer past selection. Such a study in *Zea mays* estimated that 2–4 % of the genome had been selected during domestication and subsequent breeding (Wright et al. 2005), and similar work in rice finds evidence of 55 regions that have undergone selective sweeps (Huang et al. 2012).

Recent studies have been structured to disentangle the early genomic effects of domestication and the much later changes created during 20th century crop improvement. For example, Hufford et al. (2012) find that 1–2 % of the annotated genes in maize appear to have been selected during domestication, with a comparable number selected during improvement. While the two sets of loci overlap somewhat, in general domestication targeted different genomic regions than crop improvement did. This is not particularly surprising, since the process of domestication reduced variation at selected loci, so there was little for subsequent selection to act on. Comparison of gene expression between teosinte and maize finds that some of the same loci identified by population genetic studies exhibit differential expression patterns between the wild and cultivated lines, consistent with the hypothesis of modification through selection (Swanson-Wagner et al. 2012).

The process of maize breeding has led to the development of three "heterotic groups", known as Iowa Stiff Stalk Synthetic (SS), Non-Stiff Stalk (NS), and Iodent (IDT). When plants from any two of these lines are hybridized, heterosis leads to high-yielding plants that are the basis of current maize production (Duvick 2001). The heterotic groups have become increasingly differentiated from each other over time; the genome of each shows evidence of directional selection but no strong selective sweeps (van Heerwaarden et al. 2012).

With the increased availability of molecular markers and DNA sequence data, it has become possible to identify many of the genes that were fixed during the domestication process. Although the sample size remains small, most such genes appear to be transcriptional regulators, consistent with the hypothesis that modification of such genes is responsible for much morphological change (Doebley et al. 2006). "Domestication genes" include *tb1* and *tga1* in maize, *vrs1* in barley and *prog1* in rice, all of which control plant or inflorescence architecture, *sh1*, *sh4*, and *shat1* in rice plus Q in wheat, all of which control shattering, and Rc in rice, which affects pericarp color (Glémin and Bataillon 2009). These loci have been discussed above and in the sections on Inflorescence structure and on Fruit and seed. While these nine loci are all transcriptional regulators, GIF1 in rice is a cell-wall invertase, an enzyme that controls carbohydrate partitioning and hence grain filling (Wang et al. 2008).

In contrast, genes that have been selected during crop improvement (post-domestication) are much more diverse and include many genes encoding parts of biosynthetic pathways. This pattern also would be expected if crop improvement focuses on such processes as starch biosynthesis (Doebley et al. 2006). Crop improvement genes include those controlling kernel color and endosperm composition in maize (*c1*, *r1*, *sh2*, *su1*, *y1*) and in rice (*waxy*), plant height (*sd1* in rice, *rht* in wheat), flowering time in rice (*hd1*, *hd6*), and response to vernalization in wheat (*vrn1*, *vrn2*). Mutations in *waxy* affect starch biosynthesis; by preventing amylose production they lead to sticky rice grains that are used for special culinary purposes. Sticky rice varieties exhibit a broad region (260 kb) of reduced diversity around the *waxy* locus, indicating strong selection (Olsen et al. 2006). Kernel color in maize varies, with yellow corn being caused by a dominant allele at Y1, which encodes phytoene synthase. Breeders have selected for yellow corn because of its high nutritional value, leaving a signature of selection in the form of reduced nucleotide diversity extending about 600 kbp from the gene (Palaisa et al. 2006).

References

Akhunov, E.D., Akhunova, A.R., Anderson, O.D., Anderson, J.A., Blake, N., Clegg, M.T., Coleman-Derr, D., Conley, E.J., Crossman, C.C., Deal, K.R., Dubcovsky, J., Gill, B.S., Gu, Y.Q., Hadam, J., Heo, H., McGuire, E., Morrell, P.L., Qualset, C.O., Renfro, J., Tabanao, D., Talbert, L.E., Tian, C., Toleno, D.M., Warburton, M., You, F.M., Zhang, W., Dvorak, J. 2010. Nucleotide diversity maps reveal variation in diversity among wheat genomes and chromosomes. BMC Genomics 11: 702.

Artschwager, E., Brandes, E.W. 1958. Sugarcane (*Saccharum officinarum* L.): origin, classification, characteristics and descriptions of representative clones. Washington, DC: U.S. Department of Agriculture.

Azhaguvel, P., Komatsuda, T. 2007. A phylogenetic analysis based on nucleotide sequence of a marker linked to the *brittle rachis* locus indicates a diphyletic origin of barley. Ann. Bot. 100: 1009–1015.

Bettinger, R.L., Barton, L., Morgan, C. 2010. The origins of food production in North China: a different kind of agricultural revolution. Evol. Anthropol. 19: 9–21.

Burger, J.C., Ellstrand, N.C. 2005. Feral rye - evolutionary origins of a weed. In: Gressel, J. (ed.) Crop ferality and volunteerism. Boca Raton, Florida: CRC Press. pp. 175–192

Clark, R.M., Wagler, T.N., Quijada, P., Doebley, J. 2006. A distant upstream enhancer at the maize domestication gene *tb1* has pleiotropic effects on plant and inflorescent architecture. Nat. Genet. 38: 594–597.

Clotault, J., Thuillet, A.-C., Buiron, M., DeMita, S., Couderc, M., Haussmann, B.I.G., Mariac, C., Vigouroux, Y. 2012. Evolutionary history of pearl millet (*Pennisetum glaucum* [L.] R. Br.) and selection on flowering genes since its domestication. Mol. Biol. Evol. 29: 1199–1212.

D'Andrea, A.C. 2008. T'ef (*Eragrostis tef*) in ancient agricultural systems of highland Ethiopia. Econ. Bot. 62: 547–566.

Devos, K.M., Costa de Oliveira, A., Xu, X., Estill, J., Estep, M., Jogi, A., Morales, M., Pinheiro, J., SanMiguel, P., Bennetzen, J.L. 2008. Structure and organization of the wheat genome - the number of genes in the hexaploid wheat genome. In: Appels, R., Eastwood, R., Lagudah, E., Langridge, P., Mackay, M., McIntyre, L., Sharp, P. (eds.) Proceedings of the 11th International Wheat Genetics Symposium. Sydney, Australia: Sydney University Press. pp. 1–5

Dida, M.M., Wanyera, N., Harrison Dunn, M.L., Bennetzen, J.L., Devos, K.M. 2008. Population structure and diversity in finger millet (*Eleusine coracana*) germplasm. Trop. Plant Biol. 1: 131–141.

Dillon, S.L., Shapter, F.M., Henry, R.J., Cordeiro, G., Izauierdo, L., Lee, L.S. 2007. Domestication to crop improvement: genetic resources for *Sorghum* and *Saccharum* (Andropogoneae). Ann. Bot. 100: 975–989.

Doebley, J., Stec, A., Hubbard, L. 1997. The evolution of apical dominance in maize. Nature 386: 485–488.

Doebley, J.F., Gaut, B.S., Smith, B.D. 2006. The molecular genetics of crop domestication. Cell 127: 1309–1321.

Doust, A. 2007. Architectural evolution and its implications for domestication in grasses. Ann. Bot. 100: 941–950.

Doust, A.N., Devos, K.M., Gadberry, M., Gale, M.D., Kellogg, E.A. 2004. Genetic control of branching in foxtail millet. Proc. Natl. Acad. Sci. USA 101: 9045–9050.

Duvick, D.N. 2001. Biotechnology in the 1930s: the development of hybrid maize. Nature Rev. Genet. 2: 69–64.

Fuller, D.Q. 2007. Contrasting patterns in crop domestication and domestication rates: recent archaeobotanical insights from the Old World. Ann. Bot. 100: 903–924.

Gale, M.D., Devos, K.M. 1998. Plant comparative genetics after 10 years. Science 282: 656–659.

Ge, W., Liu, L., Chen, X., Jin, Z. 2011. Can noodles be made from millet? An experimental investigation of noodle manufacture together with starch grain analyses. Archaeometry 53: 194–204.

Glémin, S., Bataillon, T. 2009. A comparative view of the evolution of grasses under domestication. New Phytol. 183: 273–290.

Harlan, J.R. 1992. Crops and man. Madison, Wisconsin: American Society of Agronomy and Crop Science Society of America.

Harlan, J.R., de Wet, J.M.J., Price, E.G. 1973. Comparative evolution of cereals. Evolution 27: 311–325.

Harris, D.R. 1989. An evolutionary continuum of people-plant interaction. In: Harris, D.R., Hillman, G.C. (eds.) Foraging and farming: the evolution of plant exploitation. London: Routledge. pp. 11–26

Haudry, A., Cenci, A., Ravel, C., Bataillon, T., Brunel, D., Poncet, C., Hochu, I., Poirier, S., Santoni, S., Glémin, S., David, J. 2007. Grinding up wheat: a massive loss of nucleotide diversity since domestication. Mol. Biol. Evol. 24: 1506–1517.

Hillman, G.C., Davies, M.S. 1990. Domestication rates in wild-type wheats and barley under primitive cultivation. Biol. J. Linn. Soc. 39: 39–78.

Hilu, K.W., de Wet, J.M.J. 1976. Racial evolution in *Eleusine coracana* ssp. *coracana* (Finger Millet). Amer. J. Bot. 63: 1311–1318.

Hilu, K.W., de Wet, J.M.J., Harlan, J.R. 1979. Archaeobotanical studies of *Eleusine coracana* ssp. *coracana*. Amer. J. Bot. 66: 330–333.

Huang, X., Kurata, N., Wei, X., Wang, Z.-X., Wang, A., Zhao, Q., Zhao, Y., Liu, K., Lu, H., Li, W., Guo, Y., Lu, Y., Zhou, C., Fan, D., Weng, Q., Zhu, C., Huang, T., Zhang, L., Wang, Y., Feng, L., Furuumi, H., Kubo, T., Miyabayashi, T., Yuan, X., Xu, Q., Dong, G., Zhan, Q., Li, C., Fujiyama, A., Toyoda, A., Lu, T., Feng, Q., Qian, Q., Li, J., Han, B. 2012. A map of rice genome variation reveals the origin of cultivated rice. Nature 490: 497–501.

Hufford, M.B., Xu, X., van Heerwaarden, J., Pyhäjärvi, T., Chia, J.-M., Cartwright, R.A., Elshire, R.J., Glaubitz, J. C., Guill, K.E., Kaeppler, S.M., Lai, J., Morrell, P.L., Shannon, L.M., Song, C., Springer, N.M., Swanson-Wagner, R.A., Tiffin, P., Wang, J., Zhang, G., Doebley, J., McMullen, M.D., Ware, D., Buckler, E.S., Yang, S., Ross-Ibarra, J. 2012. Comparative population genomics of maize domestication and improvement. Nat. Genet. 44: 808–811.

Hunt, H.V., van der Linden, M., Liu, X., Motuzaite-Matuzeviciute, G., Colledge, S., Jones, M.K. 2008. Millets across Eurasia: chronology and context of early records of the genera *Panicum* and *Setaria* from archaeological sites in the Old World. Veget. Hist. Archaeobot. 17 (suppl.): S5-S18.

Ingram, A.L., Doyle, J.J. 2003. The origin and evolution of *Eragrostis tef* (Poaceae) and related polyploids: evidence from nuclear *waxy* and plastid *rps16*. Amer. J. Bot. 90: 116–122.

Kaltsikes, P.J., Evans, L.E., Bushuk, W. 1968. Durum-type wheat with high bread-making quality. Science 159: 211–213.

Kellogg, E.A. 2009. The evolutionary history of Ehrhartoideae, Oryzeae, and *Oryza*. RICE 2: 1–14.

Kerber, E.R., Tipples, K.H. 1969. Effects of the D genome on milling and baking properties of wheat. Can. J. Plant Sci. 49: 255–263.

Khush, G.S. 1963. Cytogenetic and evolutionary studies in *Secale* III. Cytogenetics of weedy ryes and origin of cultivated rye. Econ. Bot. 17: 60–71.

Kihara, H. 1944. Discovery of the DD-analyser, one of the ancestors of *Triticum vulgare*. Agric. Hortic. 19: 13–14.

Kihara, H. 1982. Wheat studies - retrospect and prospects. Amsterdam: Elsevier.

Kihara, H., Lilienfeld, F. 1949. A new synthesized 6x-wheat. Hereditas 35 (S1): 307–319.

Kislev, M.E., Nadel, D., Carmi, I. 1992. Epipalaeolithic (19,000 BP) cereal and fruit diet at Ohalo II, Sea of Galilee, Israel. Rev. Palaeobot. Palynol. 73: 161–166.

Klee, M., Zach, B., Neumann, K. 2000. Four thousand years of plant exploitation in the Chad Basin of northeast Nigeria I: the archaeobotany of Kursakata. Veget. Hist. Archaeobot. 9: 223–237.

Komatsuda, T., Pourkheirandish, M., He, C., Azhaguvel, P., Kanamori, H., Perovic, D., Stein, N., Graner, A., Wicker, T., Tagiri, A., Lundqvist, U., Fujimura, T., Matsuoka, M., Matsumoto, T., Yano, M. 2007. Six-rowed barley originated from a mutation in a homeodomain-leucine zipper I-class homeobox gene. Proc. Natl. Acad. Sci. USA 104: 1424–1429.

Konishi, S., Izawa, T., Lin, S.Y., Ebana, K., Fukuta, Y., Sasaki, T., Yano, M. 2006. An SNP caused loss of seed shattering during rice domestication. Science 312: 1392–1396.

Lebot, V. 1999. Biomolecular evidence for plant domestication in Sahul. Genet. Res. Crop Evol. 46: 619–628.

Le Thierry d'Ennequin, M., Panaud, O., Toupance, B., Sarr, A. 2000. Assessment of genetic relationships between *Setaria italica* and its wild relative *S. viridis* using AFLP markers. Theor. Appl. Genet. 100: 1061–1066.

Li, W., Gill, B.S. 2006. Multiple genetic pathways for seed shattering in the grasses. Func. Integ. Genomics 6: 300–309.

Li, Y., Wu, S. 1996. Traditional maintenance and multiplication of foxtail millet (*Setaria italica* (L.) P. Beauv.) landraces in China. Euphytica 87: 33–38.

Li, C., Zhou, A., Sang, T. 2006. Rice domestication by reducing shattering. Science 311: 1936–1939.

Lu, H., Zhang, J., Liu, K.-b., Wu, N., Li, Y., Zhou, K., Ye, M., Zhang, T., Zhang, H., Yang, X., Shen, L., Xu, D., Li, Q. 2009a. Earliest domestication of common millet (*Panicum miliaceum*) in East Asia extended to 10,000 years ago. Proc. Natl. Acad. Sci. USA 106: 7367–7372.

Lu, H., Zhang, J., Wu, N., Liu, K.-b., Xu, D., Li, Q. 2009b. Phytoliths analysis for the discrimination of foxtail millet (*Setaria italica*) and common millet (*Panicum miliaceum*). PLoS ONE 4: e4448.

Luo, M.-C., Yang, Z.-L., You, F.M., Kawahara, T. 2007. The structure of wild and domesticated emmer wheat populations, gene flow between them, and the site of emmer domestication. Theor. Appl. Genet. 114: 947–959.

Manning, K., Pelling, R., Higham, T., Schwenniger, J.-L., Fuller, D.Q. 2011. 4500-year old domesticated pearl millet (*Pennisetum glaucum*) from the Tilemsi Valley, Mali: new insights into an alternative cereal domestication pathway. J. Archaeol. Sci. 38: 312–322.

Mariac, C., Robert, T., Allinne, C., Remigereau, M.S., Luxereau, A., Tidjani, M., Seyni, O., Bezancon, G., Pham, J.L., Sarr, A. 2006. Genetic diversity and gene flow among pearl millet crop/weed complex: a case study. Theor. Appl. Genet. 113: 1003–1014.

Matsuoka, Y. 2011. Evolution of polyploid *Triticum* wheats under cultivation: the role of domestication, natural hybridization and allopolyploid speciation in their diversification. Plant Cell Physiol. 52: 750–764.

Matsuoka, Y., Vigouroux, Y., Goodman, M.M., Sanchez G., J., Buckler, E., Doebley, J. 2002. A single domestication for maize shown by multilocus microsatellite genotyping. Proc. Natl. Acad. Sci. USA 99: 6080–6084.

McFadden, E.S., Sears, E.R. 1946. The origin of *Triticum spelta* and its free-threshing hexaploid relatives. J. Hered. 37: 81–89.

Molina, J., Sikora, M., Garud, N., Flowers, J.M., Rubinstein, S., Reynolds, A., Huang, P., Jackson, S., Schaal, B.A., Bustamente, C.D., Boyko, A.R., Purugganan, M. D. 2011. Molecular evidence for a single evolutionary origin of domesticated rice. Proc. Natl. Acad. Sci. USA 108: 8351–8356.

Morrell, P.L., Clegg, M.T. 2007. Genetic evidence for a second domestication of barley (*Hordeum vulgare*) east of the Fertile Crescent. Proc. Natl. Acad. Sci. USA 104: 3289–3294.

Nalam, V.J., Vales, M.I., Watson, C.J.W., Kianian, S.F., Riera-Lizarazu, O. 2006. Map-based analysis of genes affecting the brittle rachis character in tetraploid wheat (*Triticum turgidum* L.). Theor. Appl. Genet. 112: 373–381.

Nalam, V.J., Vales, M.I., Watson, C.J.W., Johnson, E.B., Riera-Lizarazu, O. 2007. Map-based analysis of genetic loci on chromosome 2D that affect glume tenacity and threshability, components of the free-threshing habit in common wheat (*Triticum aestivum* L.). Theor. Appl. Genet. 116: 135–145.

National Research Council. 1996. Lost crops of Africa. Volume 1. Grains. Washington, DC: National Academy Press.

Nelson, J.C., Andreescu, C., Breseghello, F., Finney, P.L., Gualberto, D.G., Bergmann, C.J., Peña, R.J., Perretant, M.R., Leroy, P., Qualset, C.O., Sorrells, M.E.

2006. Quantitative trait locus analysis of wheat quality traits. Euphytica 149: 145–159.
Neves, S.S., Swire-Clark, G., Hilu, K.W., Baird, W.V. 2005. Phylogeny of *Eleusine* (Poaceae: Chloridoideae) based on nuclear ITS and plastid *trnT-trnF* sequences. Mol. Phyl. Evol. 35: 395–419.
Olsen, K.M., Caicedo, A.L., Polato, N., McClung, A., McCouch, S., Purugganan, M.D. 2006. Selection under domestication: evidence for a sweep in the rice *Waxy* genomic region. Genetics 173: 975–983.
Oumar, I., Mariac, C., Pham, J.-L., Vigoroux, Y. 2008. Phylogeny and origin of pearl millet (*Pennisetum glaucum* [L.] R. Br.) as revealed by microsatellite loci. Theor. Appl. Genet. 117: 489–497.
Palaisa, K., Morgante, M., Tingey, S., Rafalski, A. 2006. Long-range patterns of diversity and linkage disequilibrium surrounding the maize *Y1* gene are indicative of an asymmetric selective sweep. Proc. Natl. Acad. Sci. USA 101: 9885–9890.
Peng, Y.-Y., Wei, Y.-M., Baum, B.R., Jiang, Q.-T., Lan, X.-J., Dai, S.-F., Zheng, Y.-L. 2010. Phylogenetic investigation of *Avena* diploid species and the maternal genome donor of *Avena* polyploids. Taxon 59: 1472–1482.
Peng, J.H., Sun, D., Nevo, E. 2011. Domestication evolution, genetics and genomics in wheat. Mol. Breeding 28: 281–301.
Piperidis, G., Piperidis, N., D'Hont, A. 2009. Molecular cytogenetic investigation of chromosome composition and transmission in sugarcane. Mol. Genet. Genomics 284: 65–73.
Piperno, D.R., Ranere, A.J., Holst, I., Iriarte, J., Dickau, R. 2009. Starch grain and phytolith evidence for early ninth millenium B.P. maize from the Central Balsas River Valley, Mexico. Proc. Natl. Acad. Sci. USA 106: 5019–5024.
Planchuelo, A.M., Peterson, P.M. 2000. The species of *Bromus* (Poaceae: Bromeae) in South America. In: Jacobs, S.W.L., Everett, J. (eds.) Grasses: Systematics and evolution. Melbourne: CSIRO. pp. 89–101
Pollan, M. 2001. The botany of desire. New York: Random House.
Preston, J.C., Wang, H., Doebley, J., Kellogg, E.A. 2012. The role of *teosinte glume architecture* (*tga1*) in coordinated regulation and evolution of grass glumes and inflorescence axes. New Phytol. 193: 204–215.
Purugganan, M.D., Fuller, D.Q. 2009. The nature of selection during plant domestication. Nature 457: 843–848.
Ramsay, L., Comadran, J., Druka, A., Marshall, D.F., Thomas, W.T.B., Macaulay, M., MacKenzie, K., Simpson, C., Fuller, J., Bonar, N., Hayes, P.M., Lundqvist, U., Franckowiak, J.D., Close, T.J., Muehlbauer, G.J., Waugh, R. 2011. *INTERMEDIUM-C*, a modifier of lateral spikelet fertility in barley, is an ortholog of the maize domestication gene *TEOSINTE BRANCHED 1*. Nat. Genet. 43: 169–172.
Remigereau, M.-S., Lakis, G., Rekima, S., Leveugle, M., Fontaine, M.C., Langin, T., Sarr, A., Robert, T. 2011. Cereal domestication and evolution of branching: evidence for soft selection in the *Tb1* orthologue of pearl millet (*Pennisetum glaucum* [L.] R. Br.). PLoS ONE 6: e22404.
Saisho, D., Purugganan, M.D. 2007. Molecular phylogeography of domesticated barley traces expansion of agriculture in the Old World. Genetics 177: 1765–1776.
Sencer, H.A., Hawkes, J.G. 1980. On the origin of cultivated rye. Biol. J. Linn. Soc. 13: 299–313.
Simons, K.J., Fellers, J.P., Trick, H.N., Zhang, Z., Tai, Y.-S., Gill, B.S., Faris, J.D. 2006. Molecular characterization of the major wheat domestication gene *Q*. Genetics 172: 547–555.
Swanson-Wagner, R., Briskine, R., Schaefer, R., Hufford, M.B., Ross-Ibarra, J., Myers, C.L., Tiffin, P., Springer, N.M. 2012. Reshaping of the maize transcriptome by domestication. Proc. Natl. Acad. Sci. USA 109: 11878–11883.
Sweeney, M., McCouch, S. 2007. The complex history of the domestication of rice. Ann. Bot. 100: 951–957.
Sweeney, M.T., Thomson, M.J., Pfeil, B.E., McCouch, S. 2006. Caught red-handed: *Rc* encodes a basic helix-loop-helix protein conditioning red pericarp in rice. Plant Cell 18: 283–294.
Tanno, K., Willcox, G. 2006. How fast was wild wheat domesticated? Science 311: 1886.
Triplett, J.K., Wang, Y., Zhong, J., Kellogg, E.A. 2012. Five nuclear loci resolve the polyploid history of switchgrass (*Panicum virgatum* L.) and relatives. PLoS ONE 7: e38702.
van Heerwaarden, J., Doebley, J., Briggs, W.H., Glaubitz, J. C., Goodman, M.M., Sanchez Gonzalez, J.D.J., Ross-Ibarra, J. 2011. Genetic signals of origin, spread, and introgression in a large sample of maize landraces. Proc. Natl. Acad. Sci. USA 108: 1088–1092.
van Heerwaarden, J., Hufford, M.B., Ross-Ibarra, J. 2012. Historical genomics of North American maize. Proc. Natl. Acad. Sci. USA 109: 12420–12425.
Vaughan, D.A., Lu, B.-R., Tomooka, N. 2008. The evolving story of rice evolution. Plant Sci. 174: 394–408.
Wang, Z.M., Devos, K.M., Liu, C.J., Wang, R.Q., Gale, M. D. 1998. Construction of RFLP-based maps of foxtail millet, *Setaria italica* (L.) P. Beauv. Theor. Appl. Genet. 96: 31–36.
Wang, H., Nussbaum-Wagler, T., Li, B., Zhao, Q., Vigouroux, Y., Faller, M., Bomblies, K., Lukens, L., Doebley, J.F. 2005. The origin of the naked grains of maize. Nature 436: 714–719.
Wang, E., Wang, J., Zhu, X., Hao, W., Wang, L., Li, Q., Zhang, L., He, W., Lu, B., Lin, H., Ma, H., Zhang, G., He, Z. 2008. Control of rice grain-filling and yield by a gene with a potential signature of domestication. Nat. Genet. 40: 1370–1374.
Wang, C., Chen, J., Zhi, H., Yang, L., Li, W., Wang, Y., Li, H., Zhao, B., Chen, M., Diao, X. 2010. Population genetics of foxtail millet and its wild ancestor. BMC Genet. 11: 90.
Wang, J., Luo, M.-C., Chen, Z., You, F.M., Wei, Y., Zheng, Y., Dvorak, J. 2013. *Aegilops tauschii* single nucleotide polymorphisms shed light on the origins of wheat D-genome genetic diversity and pinpoint the geographic origin of hexaploid wheat. New Phytol. 198: 925–937.
Weiss, E., Kislev, M.E., Hartmann, A. 2006. Autonomous cultivation before domestication. Science 312: 1608–1610.

Willcox, G. 2005. The distribution, natural habitats and availability of wild cereals in relation to their domestication in the Near East: multiple events, multiple centres. Veget. Hist. Archaeobot. 14: 534–541.

Wright, S.I., Bi, I.V., Schroeder, S.G., Yamasaki, M., Doebley, J.F., McMullen, M.D., Gaut, B.S. 2005. The effects of artificial selection on the maize genome. Science 308: 1310–1314.

Zhou, L., Zhang, J., Yan, J., Song, R. 2011. Two transposable element insertions are causative mutations for the major domestication gene *teosinte branched1* in modern maize. Cell Res 21: 1267–1270.

Zhou, Y., Lu, D., Li, C., Luo, J., Zhu, B.-F., Zhu, J., Shangguan, Y., Wang, Z., Sang, T., Zhou, B., Han, B. 2012. Genetic control of seed shattering in rice by the APETALA2 transcription factor *SHATTERING ABORTION1*. Plant Cell 24: 1034–1048.

Zohary, D., Hopf, M. 2000. Domestication of plants in the Old World. Oxford: Oxford University Press.

Affinities

Poaceae are members of the commelinid clade of monocots (Barrett et al. 2012; Givnish et al. 2010; Stevens 2012). Commelinids share the unique characteristic of accumulating ferulic and coumaric acids in the walls of unlignified cells; these compounds makes the cell walls fluoresce under ultraviolet light. All members of the clade, including Poaceae, produce silica (SiO_2) bodies in their leaves. Stomata have two subsidiary cells with the long axis parallel to the stomate (paracytic stomata), or may have an additional two short subsidiaries at each end of the stomate, perpendicular to the long axis (tetracytic). Inflorescences are bracteate. Embryos of the commelinids are short and broad, but this character has been modified in the Poaceae, as described below.

The commelinid clade includes Dasypogonaceae, Arecales (palms, family Arecaceae), plus the orders Commelinales, Zingiberales (gingers and bananas), and Poales. The latter three orders probably form a monophyletic group, with Poales sister to a clade of Commelinales plus Zingiberales. However, analyses of whole plastome data sets find that monophyly depends on methods of data analysis and the particular genomic regions used (Barrett et al. 2012). Non-sequence-based characters support monophyly of the three orders, which share a unique chemical composition of the primary cell wall (glucurono-arabinoxylans), and stomata with parallel cell divisions, characters that are retained throughout the grasses. Members of Commelinales, Zingiberales, and Poales also produce abundant starchy endosperm, a characteristic that is thus quite ancient. Starchy endosperm is strongly associated with the Poaceae and their overwhelming economic importance, but in fact is much more widely distributed taxonomically. The clade of Commelinales, Zingiberales, and Poales may be about 120 million years old (Janssen and Bremer 2004), unless dating of the Indian phytoliths proves to be correct, in which case it would be substantially older (see discussion above under Fossil record and dates of diversification).

The order Poales includes 17 families all of which accumulate silica in the epidermis and have endosperm in which multiple nuclear divisions occur before cell walls are formed (nuclear endosperm) (Stevens 2012). The embryo is tiny in all but the grasses. The order is also defined by molecular characters and sequence data; the mitochondrial gene, *succinate dehydrogenase 3* (*sdh3*) is absent from all sampled families (Adams and Palmer 2003). Many members are wind-pollinated and many have reduced, greenish flowers borne in condensed bracteate inflorescences. The term spikelet is applied to the tiny spike-like inflorescences of many members of Cyperaceae, as well as Poaceae and other poalean families. However, "spikelets" are clearly not homologous throughout the order and the term is only useful within the context of a particular family. The crown group node for the order is estimated to date to about 113 mya (Janssen and Bremer 2004), or 109–106 mya (Leebens-Mack et al. 2005), or 109–108 mya (Magallón and Castillo 2009), or 99 mya (Magallón and Castillo 2009) (all studies ignoring the dates of the Indian phytoliths).

Within Poales is a clade comprising seven families (Anarthriaceae, Restionaceae, Centrolepidaceae, Flagellariaceae, Joinvilleaceae, Ecdeiocoleaceae, and Poaceae). In much of the literature before 2000, the name Poales was applied only to

DOI 10.1007/978-3-319-15332-2_13,

this group of families, and hence had a more restricted application than it does currently (Campbell and Kellogg 1987; Kellogg and Linder 1995). This clade is now sometimes called the Poales s.s., or the graminid clade (Angiosperm Phylogeny Group 2009). The group is clearly monophyletic, whether defined by molecular or morphological characters. It includes plants with rhizomes, and distichous (two-ranked) sheathing leaves; flowers are small, with inconspicuous tepals, and are often imperfect. In the anther, the endothecial wall has girdle-like thickenings; the pollen is monoporate and annulate, and has tiny channels known as scrobiculi penetrating the wall. In the gynoecium of most taxa, the stigmas are plumose, and the receptive cells are on multicellular branches. Each carpel has a single orthotropous, pendulous ovule, and reduction to a single ovule per fruit is common. The graminid Poales also share chemical characteristics, including the ability to produce flavones, a primary cell wall that includes (1-3,1-4)-β-D-glucans, and sieve tube plastids with cuneate crystals.

Although past literature has inferred that the entire graminid clade is wind pollinated, Givnish et al. (2010) infer that wind pollination has been independently derived in Restionaceae + Centrolepidaceae, Ecdeiocoleaceae, and in the grasses descended from the common ancestor of Puelioideae and the BEP + PACMAD clades. They suggest that the ancestral condition for the graminid Poales is animal pollination, as postulated for Flagellariaceae, Joinvilleaceae, Anomochlooideae and Pharoideae. However, this inference is based largely on the androecial and stigmatic morphology of the groups, as well as on nectar production in *Flagellaria*; animal pollination has not actually been observed.

Within the graminid clade Flagellariaceae, Joinvilleaceae, Ecdeiocoleaceae and Poaceae form a clade, with Flagellariaceae (consisting only of the genus *Flagellaria*) sister to the rest (Campbell and Kellogg 1987; Kellogg and Linder 1995; Givnish et al. 2010). These four families are characterized by leaves with obvious cross veins, and a ligule (Stevens 2012). Inflorescences are branched, with a pulvinus (adaxial swelling) at the base of each branch. Fruits are indehiscent. Epidermal cells in the root epidermis are paired, with root hairs forming from the distal cell of the pair. Joinvilleaceae, Ecdeiocoleaceae, and Poaceae share a unique set of nested inversions in the chloroplast genome (Doyle et al. 1992; Michelangeli et al. 2003).

The sister group of Poaceae is either Joinvilleaceae, based on morphological characters (Campbell and Kellogg 1987; Kellogg 2000; Kellogg and Linder 1995), Ecdeiocoleaceae, based on sequences of whole plastomes (Givnish et al. 2010), or Ecdeiocoleaceae plus Joinvilleaceae (M. McKain, unpublished). Joinvilleaceae include a single genus, *Joinvillea*, which is distributed in southeast Asia and the Pacific Islands. Ecdeiocoleaceae include two genera, *Ecdeiocolea* and *Georgeantha*, both native to southwestern Australia. Pollen characters and leaf epidermal characters conflict in terms of which family is sister to Poaceae. Both Poaceae and Ecdeiocoleaceae have lost scrobiculi in the pollen wall; this shared loss is a potential synapomorphy linking the two families. However, both Joinvilleaceae and Poaceae have alternating long and short cells in the leaf epidermis, a character used by Campbell and Kellogg (1987), Kellogg (2000), and Kellogg and Linder (1995) to suggest a sister taxon relationship of the two families. Mature plants of *Ecdeiocolea* lack leaves so the character cannot be scored in that genus; however, alternating long and short cells are not observed in juvenile *Ecdeiocolea* or mature *Georgeantha* (Givnish et al. 2010 citing Briggs, pers. comm.). Additional morphological comparisons of Joinvilleaceae, Ecdeiocoleaceae, and Poaceae are provided in the sections on inflorescence and floral morphology.

References

Adams, K.L., Palmer, J.D. 2003. Evolution of mitochondrial gene content: gene loss and transfer to the nucleus. Mol. Phyl. Evol. 29: 380–395.

Angiosperm Phylogeny Group 2009. An update of the Angiosperm Phylogeny Group classification for the orders and families of flowering plants: APG III. Bot. J. Linn. Soc. 161: 105–121.

Barrett, C.F., Davis, J.I., Leebens-Mack, J., Conran, J.G., Stevenson, D.W. 2012. Plastid genomes and deep relationships among the commelinid monocot angiosperms. Cladistics 29: 65–87.

Campbell, C.S., Kellogg, E.A. 1987. Sister group relationships of the Poaceae. In: Soderstrom, T.R., Hilu, K. W., Campbell, C.S., Barkworth, M.E. (eds.) Grass systematics and evolution. Washington, DC: Smithsonian Institution. pp. 217–224

Doyle, J.D., Davis, J.I., Soreng, R.J., Garvin, D., Anderson, M.J. 1992. Chloroplast DNA inversions and the origin of the grass family (Poaceae). Proc. Natl. Acad. Sci. 89: 7722–7726.

Givnish, T.J., Ames, M., McNeal, J.R., McKain, M.R., Steele, P.R., dePamphilis, C.W., Graham, S.W., Pires, J.C., Stevenson, D.W., Zomlefer, W.B., Briggs, B.G., Duvall, M.R., Moore, M.J., Heaney, J.M., Soltis, D.E., Soltis, P.S., Thiele, K., Leebens-Mack, J.H. 2010. Assembling the tree of monocotyledons: plastome sequence phylogeny and evolution of Poales. Ann. Missouri Bot. Gard. 97: 584–616.

Janssen, T., Bremer, K. 2004. The age of major monocot groups inferred from 800+ *rbc*L sequences. Bot. J. Linn. Soc. 146: 385–398.

Kellogg, E.A. 2000. The grasses: a case study in macroevolution. Ann. Rev. Ecol. Syst. 31: 217–238.

Kellogg, E.A., Linder, H.P. 1995. Phylogeny of Poales. In: Rudall, P.J., Cribb, P.J., Cutler, D.F., Humphries, C.J. (eds.) Monocotyledons: systematics and evolution. Kew: Royal Botanic Gardens. pp. 511–542

Leebens-Mack, J., Raubeson, L.A., Cui, L., Kuehl, J.V., Fourcade, M.H., Chumley, T.W., Boore, J.L., Jansen, R.K., de Pamphilis, C.W. 2005. Identifying the basal angiosperm node in chloroplast genome phylogenies: sampling one's way out of the Felsenstein zone. Mol. Biol. Evol. 22: 1948–1963.

Magallón, S., Castillo, A. 2009. Angiosperm diversification through time. Amer. J. Bot. 96: 349–365.

Michelangeli, F.A., Davis, J.I., Stevenson, D.W. 2003. Phylogenetic relationships among Poaceae and related families as inferred from morphology, inversions in the plastid genome, and sequence data from the mitochondrial and plastid genomes. Amer. J. Bot. 90: 93–106.

Stevens, P.F. 2012. Angiosperm phylogeny website. Version 12, July 2012 [and more or less continuously updated since]. 2001 onward.

Classification of the Poaceae

Subdivision of the Family

The Poaceae are one of the few large plant families that currently enjoy a fully phylogenetic classification, based on a phylogeny generated largely from DNA sequence data, but supported by extensive morphological data (Fig. 9). All named subfamilies, tribes, and subtribes are strictly monophyletic (i.e., including all the descendants of a given ancestor). Of the ca. 11,000 species in the family, all can be placed in one of the early-diverging subfamilies, or in the BEP or PACMAD clades, using either molecular or morphological data or both. Subfamily assignment is uncertain for fewer than ten species, and a large majority of species can be placed into tribes and subtribes. Over 1/3 of the species in the family now have some sort of molecular data available, and the remainder exhibit enough morphological synapomorphies to place them confidently in clades identified by combined molecules and morphology.

The subfamilial classification used here follows that of the GPWG (2001), updated by Sánchez-Ken et al. (2007) and further confirmed by the GPWG II (2012). The GPWG (2001) aimed to maximize nomenclatural stability. Although there are many possible ways to convert a phylogeny into a hierarchical classification, the GPWG (2001) chose the combination of groups that retained the most similarity with previous classifications. Recent studies have also defined the limits of many of the tribes of grasses; these are cited under the appropriate subfamily and most are widely accepted. I have chosen not to recognize tribes that are fully redundant with subfamilies (e.g., Phareae in Pharoideae). Delimitation of subtribes remains a work in progress, and few subtribes have been in use long enough to be widely accepted.

This treatment of the grasses relies heavily on the tremendous synthetic work of Clayton and Renvoize (1986), Watson and Dallwitz (1992 onward), and Clayton et al. (2006 onward), and would have been completely impossible without those earlier works. Although I have included as much recent information as possible, many generic limits recognized here are based on the previous work. This is particularly true for genera of tropical Asia and Africa, many of which have yet to be incorporated in the recent wave of molecular phylogenetic studies. The major innovation of this book is the phylogenetic approach. The work of Clayton and Renvoize (1986) and Clayton et al. (2006 onward) arises from the philosophy of the evolutionary or phenetic school of taxonomy, and Watson and Dallwitz (1992 onward) use an explicitly phenetic approach. More recent checklists use a phylogenetic approach (Simon 2007; Simon et al. 2011 onward; Soreng et al. 2012 and onward), updating the classification frequently as indicated by recent molecular studies. The classification used here is likewise phylogenetic, and is thus similar to those in the recent checklists.

Many aspects of the classification presented here are congruent with the Classification of World Grasses (CWG; Soreng et al. 2012 and onward). Both this classification and that of the CWG are based on the same underlying data and are designed to be phylogenetic. The classification presented here uses fewer ranks (e.g., supertribes are not used), and also avoids monogeneric tribes or subtribes whenever possible. Therefore, the

E.A. Kellogg, *Flowering Plants. Monocots*, The Families and Genera of Vascular Plants 13,
DOI 10.1007/978-3-319-15332-2_14,

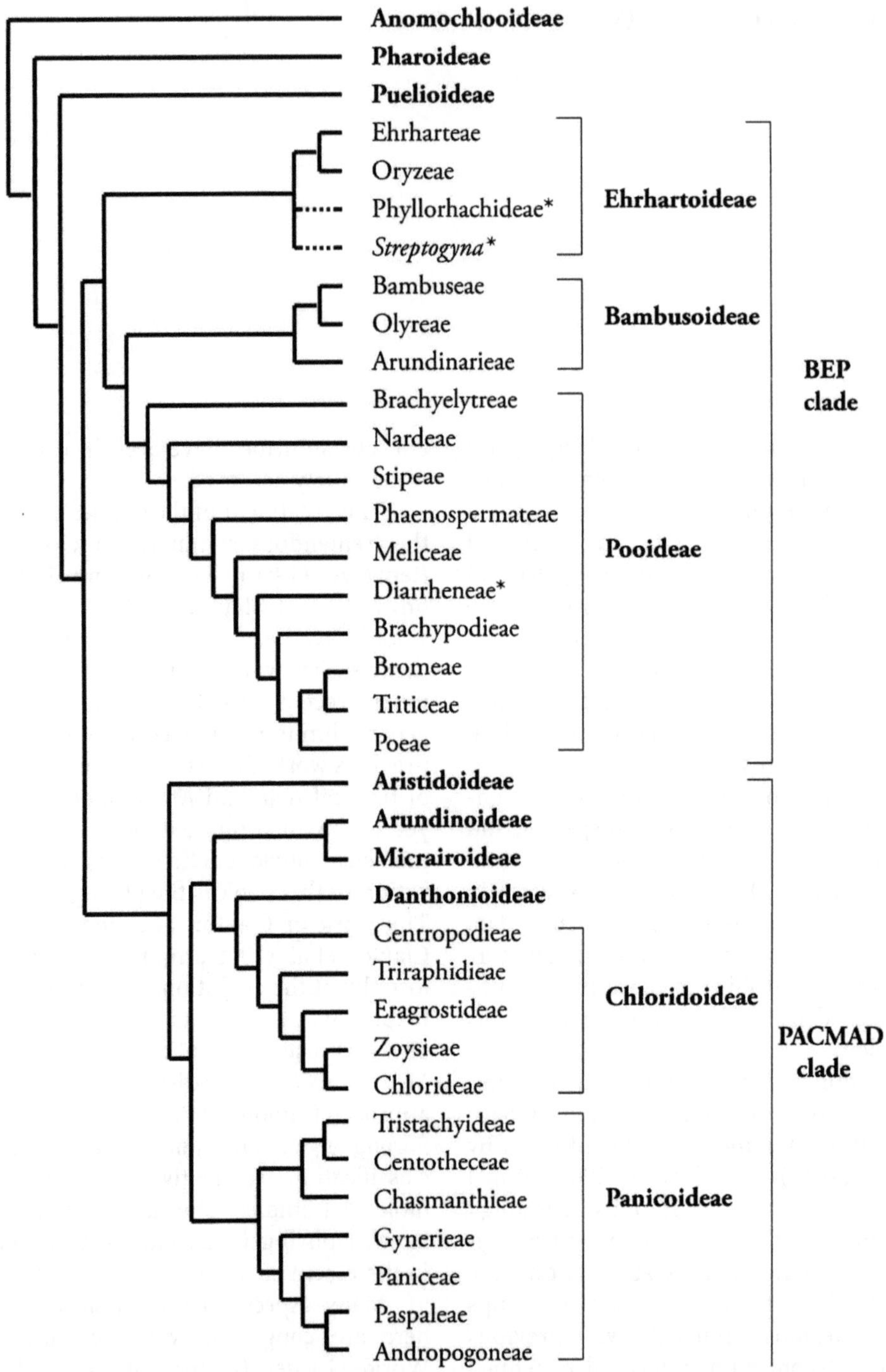

Fig. 9. Phylogeny of the tribes of Poaceae, based on GPWG II (2012). Subfamily names in bold. [1]denotes taxa not placed by GPWG II. *Streptogyna* is the only genus of the family unplaced in a subfamily, although it clearly belongs the BEP clade. *Dashed line* Uncertain position.

classification is less finely subdivided than that in the CWG (Soreng et al. 2012 and onward). In addition, the delimitation of subtribes sometimes differs, notably in the Pooideae and Chloridoideae. The structure of the phylogeny in those subfamilies forces a choice between narrowly defined subtribes that are morphologically diagnosable but contain only a single genus versus broader subtribes that provide information on relationships among genera but that have no morphological synapomorphy. The CWG has chosen the former option, creating a subtribal classification that provides little information about relationships. I have chosen the alternative, naming more inclusive clades as subtribes; the cost of this decision is that the included genera rarely share morphological characters. These differences reflect personal preferences in ranking and do not affect the underlying circumscription of groups.

The grass family includes about 30 % monotypic genera. These are almost always easily recognizable species, with clear morphological synapomorphies, and are generally segregates from larger genera. Being monotypic, their generic classification conveys no phylogenetic information, reflecting only degree of difference. Some of these genera are undoubtedly independent lineages, but others have been found to be derived from within well-known genera. I have returned the latter to their "parent" genera where molecular and/or morphological data support the combinations, but I have hesitated to make changes where shared derived characters (molecular or morphological) are lacking.

The phylogenetic approach used here does affect many familiar names of species. In an explicit comparison of names in GrassBase (Clayton et al. 2006 onward) with those in a recent phylogenetic checklist (Simon 2007; Simon et al. 2011 onward), generic names were changed for an estimated 10–20 % of species (Vorontsova and Simon 2012).

Despite the advantages of a phylogenetic classification, the grass classification illustrates one serious disadvantage. The groups shown by molecular data to be monophyletic often do not have obvious morphological characters by which they can be reliably diagnosed. The subfamily Pooideae is a case in point. Most species in the subfamily lack culm branches, lack microhairs, have parallel-sided subsidiary cells in the stomata, have laterally compressed spikelets, disarticulate above the glumes, have a chromosome complement based on seven pairs of large chromosomes, and have inflorescences in which the primary branches initiate in two ranks rather than a spiral. When these characters are mapped on the phylogeny, however, none actually appears at the origin of the subfamily, and thus none is synapomorphic or fully diagnostic. They become handy generalizations, but not infallible guides to history. The pattern tells us a great deal about how morphological evolution probably happened – a stepwise accretion of characteristics that in aggregate permitted a major radiation of several thousand species in temperate habitats. However, it is inconvenient for easy identification and generation of keys. In the keys throughout this book, therefore, taxa frequently have to be keyed multiple times, and the leads contain weasel words such as "generally" and "usually".

A second challenge for taxonomy in the grasses is the nature of the morphological characters themselves. The characters most useful for delimiting taxonomic groups are often difficult or impossible to use for identification. This dilemma first became obvious in the 1930s, when Avdulov showed that chromosome number – a totally useless field character – varied in a way that correlated well with some gross morphological characters. Additional studies on leaf anatomy were influential in refining the classification, and most recently molecular data have provided clear support for large groups based on other (micromorphological) characters. This illustrates the remarkable ability of natural selection to shape the gross morphology of the grasses, but also leads to difficulties in writing keys. Again the reader will have a chance to experience for him- or herself this double-edged sword. The key to the subfamilies below is a case in point.

Key to the Subfamilies

1. Pseudopetioles present 2
1. Leaves without pseudopetioles 11
2 (1). Conventional spikelets absent **I. Anomochlooideae**
2. Conventional spikelets present 3
3 (2). Leaf blades resupinate, major veins at an oblique angle to the midrib 4
3. Leaf blades not resupinate, major veins parallel to the midrib 5
4 (3). Stigmas three; leaf blades with microhairs; inflorescence branches and spikelets covered with hooked hairs **II. Pharoideae**
4. Stigmas two; leaf blades without microhairs; inflorescence branches and spikelets smooth **VI. Pooideae** (*Phaenosperma*)
5 (3). Culms woody, usually well over 1 m tall; leaves dimorphic, the culm leaves with prominent sheaths and blades reduced or absent, the foliage leaves with broad blades **V. Bambusoideae** (Bambuseae and Arundinarieae)
5. Culms herbaceous; if lignified and over 1 m tall, then the leaves not dimorphic 6

6 (5). Inflorescence axis flattened, leaf-like, with the spikelets borne on one side **IV. Ehrhartoideae** (Phyllorachideae)
6. Inflorescence axis not flattened 7
7 (6). Spikelets with exactly two flowers, the distal one bisexual and the proximal one staminate or sterile; rachilla extension lacking **VIII. Panicoideae** (Paspaleae, Paniceae, Andropogoneae)
7. Spikelets with 1 or more flowers; if exactly two, then not as above 8
8 (7). Stamens 6 **III. Puelioideae**
8. Stamens 2 or 3 (numerous in *Pariana*), not 6 9
9 (8). Spikelets with 1 flower, unisexual **V. Bambusoideae** (Olyreae)
9. Spikelets with more than 1 flower, generally bisexual 10
10 (9). Style branches and stigmas growing into long tangled tendrils that harden at maturity **8. *Streptogyna* (BEP clade Incertae sedis)**
10. Style branches and stigmas not as above **VIII. Panicoideae** (Chasmanthieae and Centotheceae)
11 (1). Spikelets with exactly two flowers; rachilla extension lacking 12
11. Spikelets with 1 to several flowers, if two then rachilla extension present 14
12 (11). Upper flower of spikelet bisexual, the proximal one staminate or sterile **VIII. Panicoideae** (Paspaleae, Paniceae, Andropogoneae)
12. Both flowers of spikelet fertile 13
13 (12). Plants several meters tall, reed-like; dioecious *Gynerium*
13. Plants less than 1 m tall, often tiny; flowers bisexual **XI. Micrairoideae**
14 (11). Ligule membranous 15
14. Ligule a fringe of hairs or a fringed membrane 16
15 (14).[1] Lodicules membranous; plants lacking two-celled microhairs; stomatal subsidiary cells with parallel sides; leaf veins widely spaced; plants C_3 **VI. Pooideae**
15. Lodicules fleshy; plants with two-celled microhairs; stomatal subsidiary cells dome shaped or triangular; leaf veins closely spaced; plants C_4 **X. Chloridoideae**
16 (14). Spikelets with one fertile flower 17
16. Spikelets with more than one flower 19
17 (16). Lemma terminating in three awns, their bases fused to form a column **VII. Aristidoideae**
17. Lemma awnless or with a single awn; if more than one awn then these not fused at their bases 18
18 (17). Fertile flower subtended by two sterile flowers, these often reduced to bract-like structures **IV. Ehrhartoideae**
18. Fertile flower without proximal sterile flowers **X. Chloridoideae**
19 (16). Glumes as long as the flowers or longer **IX. Danthonioideae**
19. Glumes generally shorter than the flowers 20
20 (19)[1]. Plants C_4, with veins separated by no more than two mesophyll cells **X. Chloridoideae**
20. Plants C_3, with veins separated by more than two mesophyll cells **XII. Arundinoideae** plus several Panicoideae (*Centotheca, Megastachya, Cyperochloa, Spartochloa, Chasmanthium*)

References

Clayton, W.D., Renvoize, S.A. 1986. Genera graminum: grasses of the world. London: Her Majesty's Stationery Office.

Clayton, W.D., Harman, K.T., Williamson, H. 2006 onward. GrassBase - the online world grass flora. http://www.kew.org/data/grasses-db.html

GPWG 2001. Grass Phylogeny Working Group. Phylogeny and subfamilial classification of the Poaceae. Ann. Missouri Bot. Gard. 88: 373–457.

GPWG II 2012. Grass Phylogeny Working Group II. New grass phylogeny resolves deep evolutionary relationships and discovers C_4 origins. New Phytol. 193: 304–312.

Sánchez-Ken, J.G., Clark, L.G., Kellogg, E.A., Kay, E.E. 2007. Reinstatement and emendation of subfamily Micrairoideae (Poaceae). Syst. Bot. 32: 71–80.

Simon, B.K. 2007. GrassWorld: Interactive key and information system of world grasses. Kew Bull. 62: 475–484.

Simon, B.K., Clayton, W.D., Harman, K.T., Vorontsova, M.S., Brake, L., Healy, D., Alfonso, Y. 2011 onward. GrassWorld. http://grassworld.myspecies.info/

Soreng, R.J., Davidse, G., Peterson, P.M., Zuloaga, F.O., Judziewicz, E.J., Filgueiras, T.S., Morrone, O., Romaschenko, K. 2012 and onward. A world-wide phylogenetic classification of Poaceae (Gramineae). http://www.tropicos.org/docs/meso/CLASSIFICATION%20OF%20world%20grasses%202013%20Oct%2029.htm

Vorontsova, M.S., Simon, B.K. 2012. Updating classifications to reflect monophyly: 10 to 20 percent of species names change in Poaceae. Taxon 61: 735–746.

Watson, L., Dallwitz, M.J. 1992 onward. The grass genera of the world: descriptions, illustrations, identification, and information retrieval; including synonyms, morphology, anatomy, physiology, phytochemistry, cytology, classification, pathogens, world and local distribution, and references, 23rd April 2010 edn. Wallingford: CAB International. http://delta-intkey.com/grass/

[1] Use of macromorphological characters to distinguish these subfamilies would require developing a key to the level of genera or small clades. In terms of overall plant structure and spikelet morphology, Chloridoideae are quite similar to several other subfamilies.

I. Subfamily Anomochlooideae Pilg. ex Potztal (1957)

Rhizomatous perennials, herbaceous. Ligule fringed or lacking. Leaves with a pulvinus at the distal end of the pseudopetiole, with broad blades. *Inflorescence made up of units that defy conventional descriptions, generally known as "spikelet equivalents," more or less equivalent to a single flower,* bisexual. The spikelet equivalents sessile, bracteate or not. *Additional bract-like structures arranged in unusual phyllotaxis, but probably representing highly modified perianth.* Caryopsis with a hard endosperm, with the hilum long but inconspicuous, the embryo with or without an epiblast, with a scutellar cleft, without a mesocotyl internode, embryonic leaf margins overlapping or not.

Tropical America. Two genera and four species.

Judziewicz and Soderstrom (1989) suggest that evidence for monophyly of the Anomochlooideae is provided by their remarkably large microhairs (75–150 μ long), which have a constriction part way up the basal cell, by their distinctive midrib structure, in which the midrib projects above the surface of the blade on both the ab- and adaxial sides, and by their silica bodies elongated perpendicular to the proximo-distal axis of the leaf. However, because they compared *Anomochloa* and *Streptochaeta* only to the Bambusoideae and select Oryzeae, it is not clear whether these characters are truly synapomorphic. The GPWG (2001) questioned whether any synapomorphies link the two genera, and raised the possibility that their sister group relationship in phylogenetic trees is an artifact. Nonetheless, subsequent phylogenetic analyses continue to place them as sister taxa (GPWG II 2012).

Although each genus in this subfamily has been placed in its own tribe, division of such a small subfamily seems unnecessary.

KEY TO THE GENERA OF ANOMOCHLOOIDEAE

1. Leaves basal, with long pseudopetioles; stamens surrounded by a fringe like structure outside of which are two bracts; bract closest to the floral organs leaf-like, with sheath and blade **1. *Anomochloa***
1. Leaves borne along the culm, with short pseudopetioles; bracts subtending the flower 11; bracts closest to the floral organs without obvious sheath and blade, but one with a long coiled awn **2. *Streptochaeta***

1. ***Anomochloa*** Brongniart Fig. 10

Anomochloa Brongniart, Ann. Sci. Nat. Bot. III, 16: 368 (1851).

Leaves broad, looking more like a member of the Zingiberales than a grass. Ligule a fringe of hairs. *Pulvinus at both ends of the pseudopetiole. Inflorescence complex, spike-like; each branch subtended by an enveloping bract. Flowers subtended by two additional bracts that are separated by an internode, the upper bract extended into a narrow curved leaf blade. Stamens 4, surrounded by a ring of cilia.* Stigma 1. Embryo with an epiblast, leaf margins meeting. $2n = 36$.

One sp., *A. marantoidea* Brongn., Bahia, Brazil.

DOI 10.1007/978-3-319-15332-2_15,

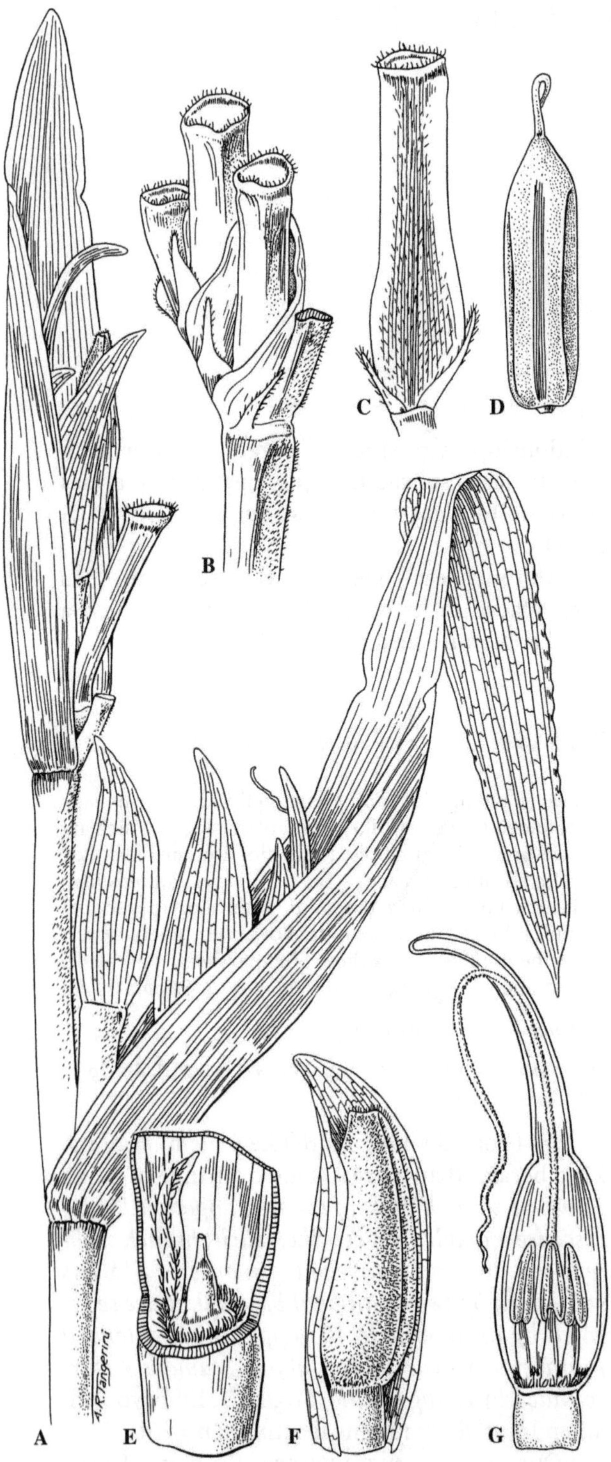

Fig. 10. Anomochlooideae. *Anomochloa marantoidea*. **A** Portion of flowering part of the plant, showing the lowest and next to lowest bracts subtending spikelet equivalents. **B** Base of inflorescence branch, viewed from the adaxial side, subtending bract removed. **C** Basal portion of primary axis, showing small, lower (*left*) and large, upper (*right*) bracteoles, the bract between bracteoles and axis removed. **D** Caryopsis, view from the adaxial side showing linear hilum. **E** Portion of flower; upper bract partially dissected away, showing ovary surrounded by ring of cilia and with anomalous bract found in one spikelet, stamens removed. **F** Detail of bracts surrounding the flower, deciduous lower bract and upper bract separated by an internode, the blade-like extension fallen from the latter. **G** Longitudinal section of flower, lower bract removed, with thickened internode supporting upper bract, ring of cilia, stamens, and gynoecium. (Reprinted from Judziewicz and Soderstrom 1989, courtesy of Smithsonian Institution; drawn by A.R. Tangerini)

This is one of the most curious of the grasses, based on its phylogenetic position and its odd morphology. It is rare and nearly extinct in the wild and difficult to maintain in cultivation. It has been the subject of several careful developmental and anatomical studies, which describe in detail the nature of the floral parts and subtending bracts (Judziewicz et al. 1999; Judziewicz and Soderstrom 1989; Sajo et al. 2012). Establishment of homology with spikelets of other grasses is difficult and likely to remain controversial.

2. ***Streptochaeta*** Nees Figs. 11a, 11b

Streptochaeta Nees, Fl. Bras. Enum. Pl. 2: 536 (1829).

Leaves with auricles; ligule absent. Inflorescence spike-like, with spirally arranged, contracted lateral branches, *each bearing 11 spirally arranged bracts with varying morphologies, the sixth bract ending in a long coiled awn. Awns becoming entangled at maturity*. Embryo without an epiblast, leaf margins overlapping. $2n = 22$.

Three spp., Central and South America.

The development and possible homologies of the "spikelet equivalents" are discussed in detail by Preston et al. (2009) and Sajo et al. (2008).

Spikelet Clade

All remaining subfamilies have flowers arranged in spikelets, tiny spike-like structures subtended by two bracts (the glumes) (GPWG 2001). See discussion under Inflorescence structure.

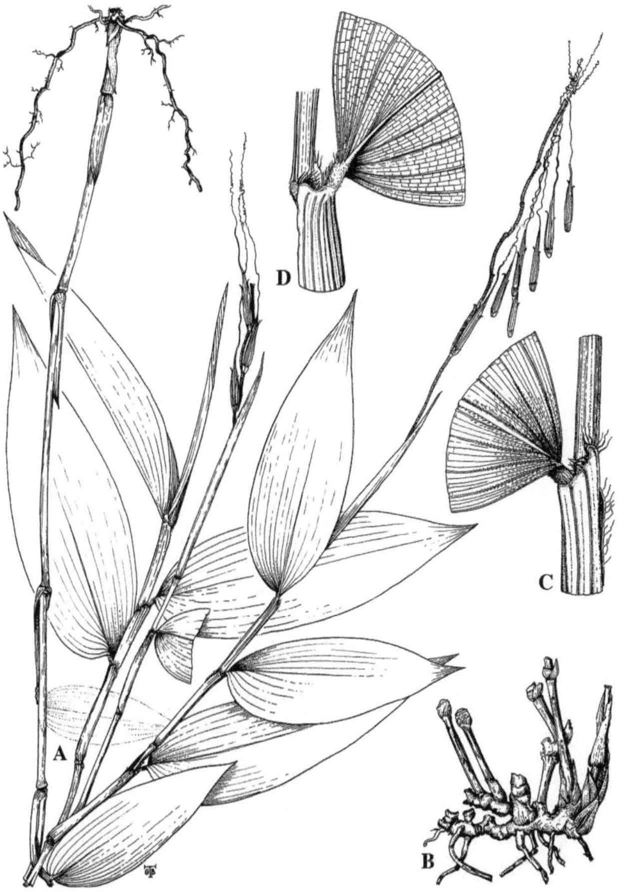

Fig. 11a. Anomochlooideae. *Streptochaeta spicata* subsp. *spicata*. **A** Habit, showing disarticulated spikelet equivalents hanging by awns from the apex of the rachis (note lateral branch of culm terminating in a secondary inflorescence). **B** Base of plant showing stout sympodial rhizomes. **C** Summit of leaf sheath and base of blade showing prominent sheath auricle. **D** Opposite side of summit of leaf sheath and base of blade with smaller sheath auricle and turgid pulvinus. (Reprinted from Judziewicz and Soderstrom 1989, courtesy of Smithsonian Institution; drawn by G.B. Threlkeld).

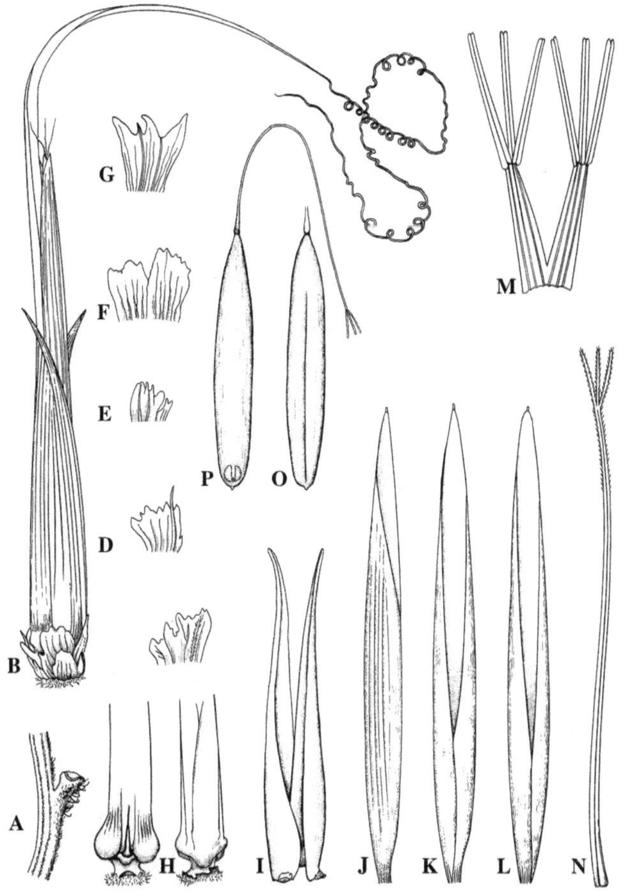

Fig. 11b. Anomochlooideae. *Streptochaeta spicata* subsp. *spicata*, showing details of spikelet equivalent. **A** Portion of rachis with pedicel of spikelet-equivalent. **B** Spikelet equivalent, showing long awn on bract VI. **C–G** Bracts I–V at the base of the spikelet equivalent, showing their variable venation and shape. **H** Saccate base of bract VI, showing the internode between bract V and bract VI. **I** Bracts VII and VIII. **J–L** Bracts X–XII. **M** Androecium showing short, elongating stamen tube. **N** Gynoecium. **O** Caryopsis, view from adaxial side, showing linear hilum. **P** Same, from abaxial side, showing persistent style and small, basal embryo. (Reprinted from Judziewicz and Soderstrom 1989, courtesy of Smithsonian Institution; drawn by G.B. Threlkeld)

References

GPWG 2001. Grass Phylogeny Working Group. Phylogeny and subfamilial classification of the Poaceae. Ann. Missouri Bot. Gard. 88: 373–457.

GPWG II 2012. Grass Phylogeny Working Group II. New grass phylogeny resolves deep evolutionary relationships and discovers C_4 origins. New Phytol. 193: 304–312.

Judziewicz, E.J., Soderstrom, T.R. 1989. Morphological, anatomical, and taxonomic studies in *Anomochloa* and *Streptochaeta* (Poaceae: Bambusoideae). Smithsonian Contr. Bot. 68: 1–52.

Judziewicz, E.J., Clark, L.G., Londoño, X., Stern, M.J. 1999. American bamboos. Washington, DC: Smithsonian Institution Press.

Preston, J.C., Christensen, A., Malcomber, S.T., Kellogg, E. A. 2009. MADS-box gene expression and implications for developmental origins of the grass spikelet. Amer. J. Bot. 96: 1419–1429.

Sajo, M.G., Longhi-Wagner, H.M., Rudall, P.J. 2008. Reproductive morphology of the early-divergent grass *Streptochaeta* and its bearing on the homologies of the grass spikelet. Plant Syst. Evol. 275: 245–255.

Sajo, M.G., Pabón-Mora, N., Jardim, J., Stevenson, D.W., Rudall, P.J. 2012. Homologies of the flower and inflorescence in the early-divergent grass *Anomochloa* (Poaceae). Amer. J. Bot. 99: 614–628.

II. Subfamily Pharoideae L.G. Clark & Judz. (1996)

Rhizomatous perennials, herbaceous. *Leaf blades resupinate*, with pseudopetiole; *veins running obliquely from midrib to margin.* Inflorescence branched, *with hooked hairs on the branches and the spikelets.* Spikelets unisexual, with one flower per spikelet, staminate and pistillate spikelets paired, the staminate one pedicellate and the pistillate one sessile. Glumes 2. Lemmas tubular or inflated. Lodicules generally absent, or present in staminate flowers only. Caryopsis with hard endosperm, with the hilum extending the full length of the caryopsis, the embryo with an epiblast, with or without a scutellar tail, without a mesocotyl internode, the embryonic leaf margins overlapping. *Inner bundle sheath of veins with multiple layers of cells. Intercostal epidermis with files of long fibers alternating with normal files of intercostal long cells.*

Three genera and 12 species.

Both *Scrotochloa* and *Pharus* have a fragile panicle, which could be a synapomorphy linking the two genera; however, utriculate female lemmas are present in *Leptaspis* and *Scrotochloa*, and could be a synapomorphy for that pair. This character conflict prevents determining which characters are synapomorphic. Virtually all molecular data on this subfamily come from *Pharus*, which has been studied extensively. Good comparative data on *Leptaspis* and *Scrotochloa* would be particularly valuable. The embryonic characters have only been sampled for *Pharus*, but are presumed to apply to the whole subfamily. The presence of epidermal fibers is unusual and may be unique in Poaceae (Soderstrom et al. 1987). The genus *Suddia* has been placed in this subfamily by Clayton and Renvoize (1986) and Watson and Dallwitz (1992 and onward). However, it appears not to have pharoid leaf anatomy, and may not have the resupinate leaf blades that characterize the rest of the subfamily (Judziewicz and Clark 2007). Following the suggestion of Judziewicz and Clark (2007), it is placed in Ehrhartoideae.

Key to the Genera of Pharoideae

1. Lemma margins fused, the resulting structure inflated 2
1. Lemma margins free; if fused, the structure not inflated **4. *Pharus***
2 (1). Ligule membranous, inflorescence disarticulating from plant **5. *Scrotochloa***
2. Ligule a fringed membrane, inflorescence rachis tough **3. *Leptaspis***

Genera of Pharoideae

3. ***Leptaspis*** R. Br.

Leptaspis R. Br., Prodr. Fl. Nov. Holl.: 211 (1810).

Ligule a fringed membrane. Inflorescence branches not disarticulating. *Female lemmas urceolate, the margins fused, becoming indurate.* Lodicules present or absent. $2n = 24$.

Three spp., Old World tropics.

4. ***Pharus*** P. Browne Fig. 12

Pharus P. Browne, Civ. Nat. Hist. Jamaica: 344 (1756).

Ligule membranous, a fringed membrane, or a fringe of hairs. *Inflorescence branches or the whole inflorescence breaking away from the plant.* Female lemmas cylindrical, in some species

DOI 10.1007/978-3-319-15332-2_16,

with a tiny beak; *covered with hooked hairs*. Lodicules sometimes present in staminate flowers. $2n = 24$.

Seven spp., Central and South America, West Indies.

The margins of the lemma are fused in one species of *Pharus*.

5. ***Scrotochloa*** Judziewicz

Scrotochloa Judziewicz, Phytologia 56: 299–304 (1984).

Ligule membranous, pubescent. *Inflorescence branched, falling entire. Female lemmas urceolate, the margins fused, becoming indurate, covered with hooked hairs.* Lodicules absent. Caryopsis with a groove on the side of the hilum. Invaginations of mesophyll cell walls obscure.

Two spp., Ceylon, SE Asia, New Guinea, Solomon Is.

Watson and Dallwitz (1992 onward) report that this genus has silica bodies with a raised prickly belt around the equator; this characteristic is not known elsewhere in the grasses.

BISTIGMATIC CLADE

Stigmas two. Orders of stigmatic branching two. Spikelets with multiple flowers. Anther walls with a middle layer that breaks down during development; the inner walls of endothecial cells becoming fibrous at maturity.

All remaining subfamilies are members of a large group in which the characters listed above are synapomorphic (GPWG 2001). Stigma number reverses to three in some species of *Puelia*, and in the Bambusoideae. Flower number per spikelet has reversed to one in multiple lineages.

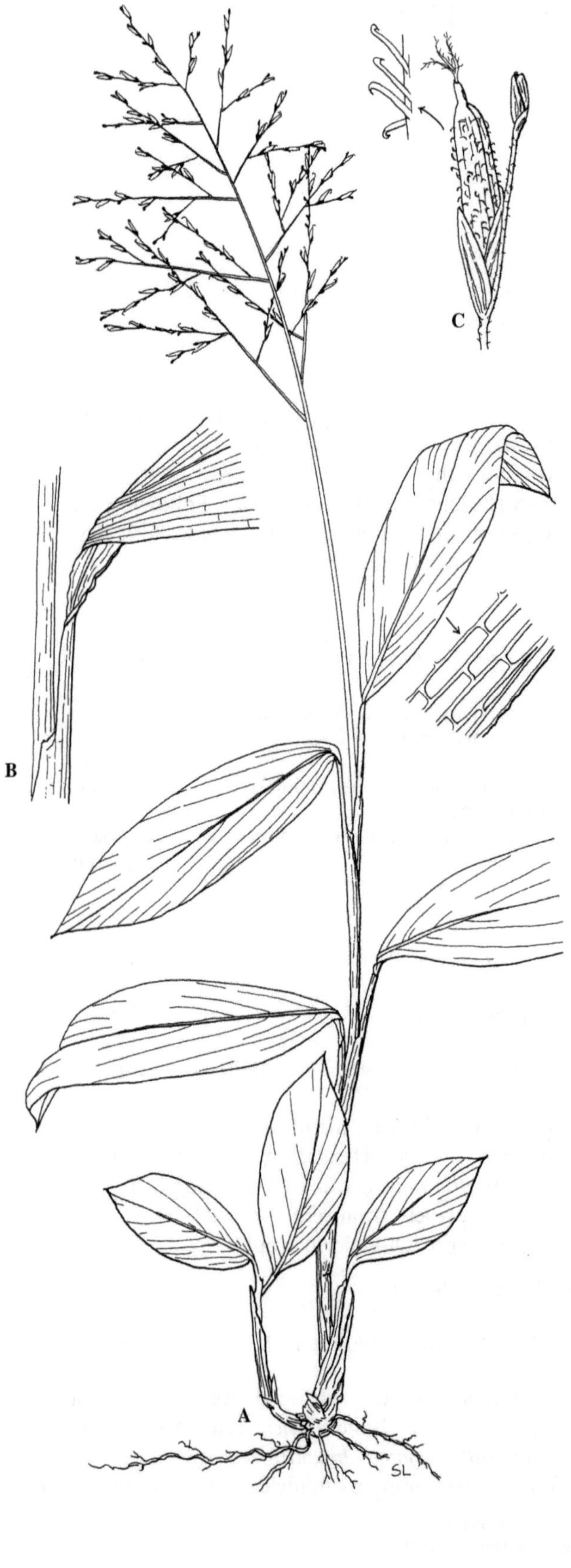

Fig. 12. Pharoideae. *Pharus glaber*. **A** Flowering plant. **B** Leaf base showing pseudopetiole. **C** Pistillate spikelet (*left*) with uncinate trichomes, and staminate spikelet (*right*). (From Barkworth et al. 2007, drawn by S. Long)

References

Barkworth, M.E., Capels, K.M., Long, S., Anderton, L.K., Piep, M.B. (eds.) 2007. Flora of North America north of Mexico. Vol. 24. Magnoliophyta: Commelinidae (in part): Poaceae, part 1. New York: Oxford University Press.

Clayton, W.D., Renvoize, S.A. 1986. Genera graminum: grasses of the world. London: Her Majesty's Stationery Office.

GPWG 2001. Grass Phylogeny Working Group. Phylogeny and subfamilial classification of the Poaceae. Ann. Missouri Bot. Gard. 88: 373–457.

Judziewicz, E.J., Clark, L.G. 2007. Classification and biogeography of New World grasses: Anomochlooideae, Pharoideae, Ehrhartoideae, and Bambusoideae. Aliso 23: 303–314.

Soderstrom, T.R., Ellis, R.P., Judziewicz, E.J. 1987. The Phareae and Streptogyneae (Poaceae) of Sri Lanka. Smithsonian Contr. Bot. 65: 1–27.

Watson, L., Dallwitz, M.J. 1992 onward. The grass genera of the world: descriptions, illustrations, identification, and information retrieval; including synonyms, morphology, anatomy, physiology, phytochemistry, cytology, classification, pathogens, world and local distribution, and references, 23rd April 2010 edn. Wallingford: CAB International. http://delta-intkey.com/grass/

III. Subfamily Puelioideae L.G. Clark et al. (2000)

Rhizomatous perennials, herbaceous, the culms hollow. Leaves with broad blades, with pseudopetioles, without auricles, the ligule a fringed membrane. Inflorescences branched or unbranched, sometimes with minute bracts subtending the branches. Spikelets with multiple flowers, *the proximal several staminate or sterile*, the distal ones pistillate or bisexual. Stigmas 2 or 3. Caryopsis with a long, linear hilum, the embryo small, its detailed structure unknown.

Two genera and eleven species; tropical Africa.

This subfamily is poorly known. Material is difficult to acquire and the plants are rarely if ever found in cultivation. Presence of proximal staminate or neuter flowers in the spikelet may be uniquely derived for this group, but the character appears in several other grass taxa, most notably in the Panicoideae (GPWG 2001). Although each genus has been placed in its own tribe, division of such a small subfamily seems unnecessary.

Key to the Genera

1. Abaxial ligule absent; spikelets with short rachilla internodes, more than six flowers, the distalmost 5–12 bisexual **6. *Guaduella***
1. Abaxial ligule present; spikelets with long rachilla internodes; flowers six or fewer, the distalmost one pistillate **7. *Puelia***

6. ***Guaduella*** Franch. Fig. 13

Guaduella Franch., Bull. Mens. Soc. Linn. Paris 1: 676 (1887).

Spikelets with two to four glumes; *distal 5 to 12 flowers bisexual*, proximal 1 to 3 staminate. Stigmas 2.

Six spp., western Africa.

7. ***Puelia*** Franch.

Puelia Franch., Bull Mens. Soc. Linn. Paris 1: 674 (1887).

Abaxial ligule present. Spikelets with 2 or 3 (up to 7) glumes; *distal flower pistillate*, proximal 3 to 6 flowers staminate. *Pistillate flower on a long internode and subtended by a broad outgrowth of the rachilla. Filaments of stamens fused.* Pericarp free from seed coat. $2n = 24$.

Five spp., western Africa.

BEP Plus PACMAD Clades

Leaves lacking a pseudopetiole. Lodicules 2, stamens 3. Mesophyll without fusoid cells.

These clades include all the remaining subfamilies and thus the overwhelming majority of the species of the family. The characters listed were suggested by the GPWG (2001) as possible synapomorphies for this clade, although all of them reverse elsewhere in the tree. In addition, the GPWG (2001) suggested that loss of invaginated cells walls might be synapomorphic. However, optimization of this character on recent trees is ambiguous. These characters will need to be optimized on the most recent grass phylogeny (GPWG II 2012).

BEP Clade

This clade has no known morphological synapomorphies.

The BEP clade was first identified by Clark et al. (1995) (as the BOP clade), and includes Bambusoideae, Ehrhartoideae, and Pooideae.

E.A. Kellogg, *Flowering Plants. Monocots*, The Families and Genera of Vascular Plants 13,
DOI 10.1007/978-3-319-15332-2_17,

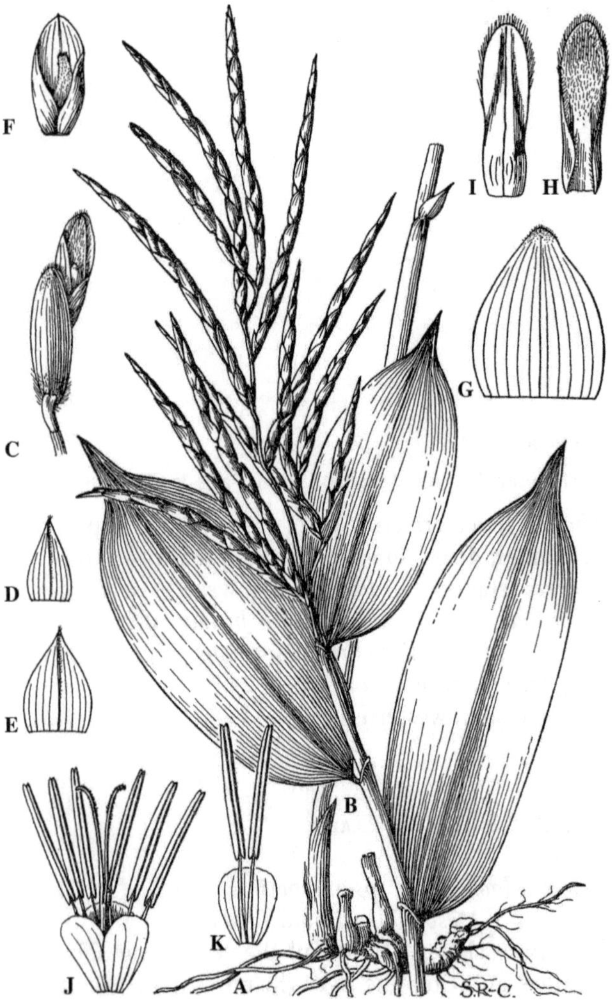

Fig. 13. Puelioideae. *Guaduella oblonga*. **A** Rhizome. **B** Flowering culm. **C** Portion of spikelet. **D** Lower glume. **E** Upper glume. **F** Lemma and palea. **G** Lemma. **H** and **I** Palea. **J** Flower. **K** Lodicule and two stamens. (From Clayton 1972, drawn by S. Ross-Craig)

Early phylogenetic analyses provided little support for the group (e.g., GPWG 2001; Kellogg and Linder 1995), but recent data show that the group is monophyletic (Bouchenak-Khelladi et al. 2008; Edwards and Smith 2010; GPWG II 2012; Vicentini et al. 2008). The amphi-Atlantic genus *Streptogyna* has been difficult to place, and its variable position in phylogenies has contributed to low statistical support for the BEP clade and for relationships within it. All possible relationships of the three subfamilies have been suggested, but increasing support is gathering for the topology [(Ehrhartoideae) (Bambusoideae, Pooideae)] (GPWG II 2012; Zhao et al. 2013). The position of *Streptogyna* remains ambiguous but it may be sister to Ehrhartoideae.

BEP clade incertae sedis:

8. ***Streptogyna*** P. Beauv. Fig. 14

Streptogyna P. Beauv., Ess. Agrost. 80 (1812).

Rhizomatous or caespitose perennials. Leaf blades with a pseudopetiole, deciduous, the ligule membranous, abaxial ligule present. Inflorescence narrow, branched. Spikelets with several flowers. Distal flowers reduced. Rachilla breaking up between flowers, *the internode developing into a hook*. Lemma awned. Lodicules 3. Stamens 2. Style branches and stigmas 2 or 3, *extending and twisting together at maturity, becoming hard*. Caryopsis with a long linear hilum. Mesophyll cells with or without invaginated cell walls, mesophyll with fusoid cells. $2n = 24$.

Two spp., tropical America, Africa, Sri Lanka, India.

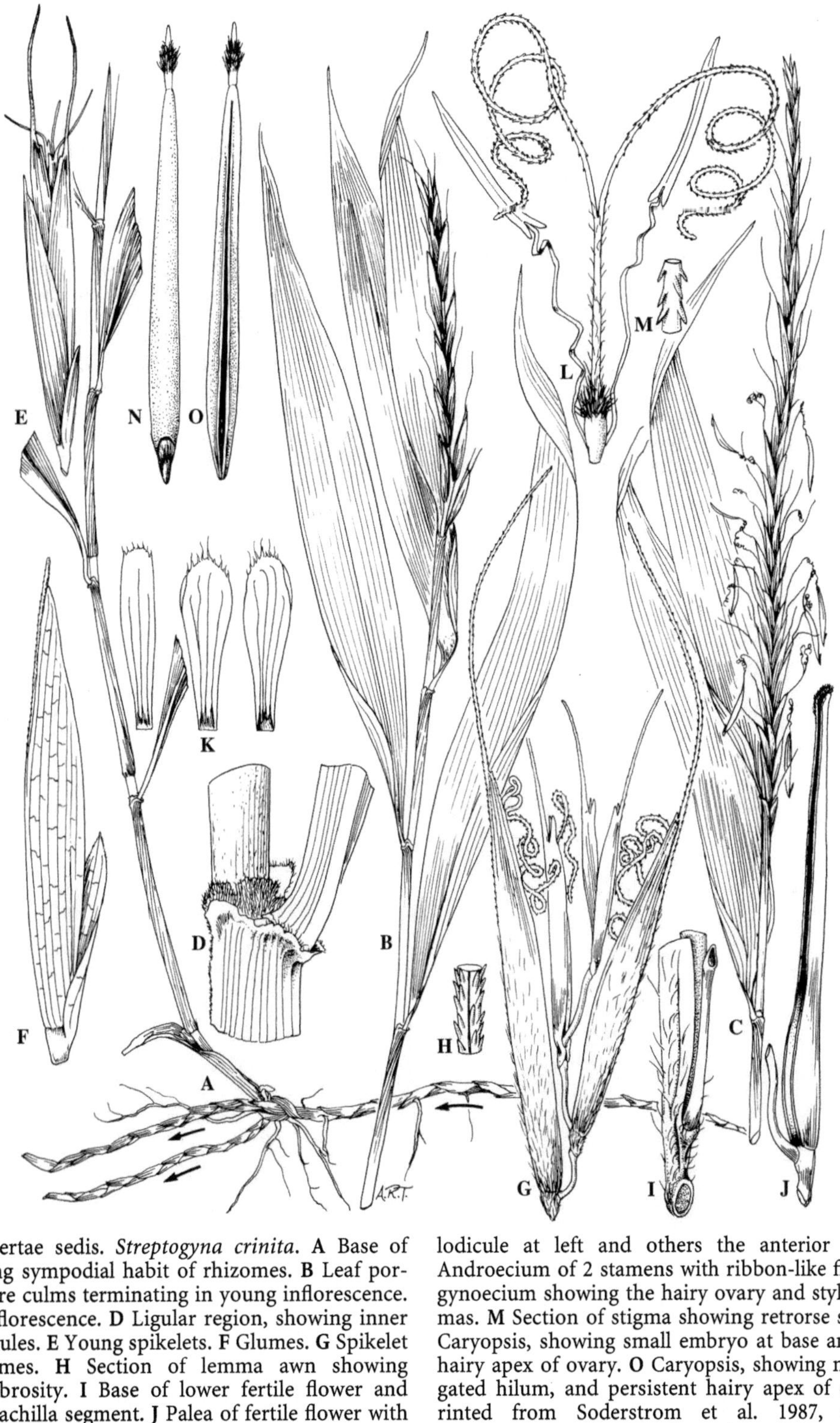

Fig. 14. Incertae sedis. *Streptogyna crinita*. **A** Base of plant showing sympodial habit of rhizomes. **B** Leaf portion of mature culms terminating in young inflorescence. **C** Mature inflorescence. **D** Ligular region, showing inner and outer ligules. **E** Young spikelets. **F** Glumes. **G** Spikelet without glumes. **H** Section of lemma awn showing antrorse scabrosity. **I** Base of lower fertile flower and succeeding rachilla segment. **J** Palea of fertile flower with succeeding rachilla segment. **K** Lodicules, the posterior lodicule at left and others the anterior lodicules. **L** Androecium of 2 stamens with ribbon-like filaments and gynoecium showing the hairy ovary and style and 2 stigmas. **M** Section of stigma showing retrorse scabrosity. **N** Caryopsis, showing small embryo at base and persistent hairy apex of ovary. **O** Caryopsis, showing narrow, elongated hilum, and persistent hairy apex of ovary. (Reprinted from Soderstrom et al. 1987, courtesy of Smithsonian Institution; drawn by A.R. Tangerini)

References

Bouchenak-Khelladi, Y., Salamin, N., Savolainen, V., Forest, F., van der Bank, M., Chase, M.W., Hodkinson, T. R. 2008. Large multi-gene phylogenetic trees of the grasses (Poaceae): progress towards complete tribal and generic taxon sampling. Mol. Phyl. Evol. 47: 488–505.

Clark, L.G., Zhang, W., Wendel, J.F. 1995. A phylogeny of the grass family (Poaceae) based on *ndhF* sequence data. Syst. Bot. 20: 436–460.

Clayton, W.D. 1972. Gramineae. In: Hepper, F.N. (ed.) Flora of West Tropical Africa, 2nd edn. Vol. 3, part 2. London: Crown Agents. p. 361

Edwards, E.J., Smith, S.A. 2010. Phylogenetic analyses reveal the shady history of C_4 grasses. Proc. Natl. Acad. Sci. USA 107: 2532–2537.

GPWG 2001. Grass Phylogeny Working Group. Phylogeny and subfamilial classification of the Poaceae. Ann. Missouri Bot. Gard. 88: 373–457.

GPWG II 2012. Grass Phylogeny Working Group II. New grass phylogeny resolves deep evolutionary relationships and discovers C_4 origins. New Phytol. 193: 304–312.

Kellogg, E.A., Linder, H.P. 1995. Phylogeny of Poales. In: Rudall, P.J., Cribb, P.J., Cutler, D.F., Humphries, C.J. (eds.) Monocotyledons: systematics and evolution. Kew: Royal Botanic Gardens. pp. 511–542

Soderstrom, T.R., Ellis, R.P., Judziewicz, E.J. 1987. The Phareae and Streptogyneae (Poaceae) of Sri Lanka. Smithsonian Contr. Bot. 65: 1–27.

Vicentini, A., Barber, J.C., Giussani, L.M., Aliscioni, S.S., Kellogg, E.A. 2008. Multiple coincident origins of C_4 photosynthesis in the Mid- to Late Miocene. Global Change Biol. 14: 2963–2977.

Zhao, L., Zhang, N., Ma, P.-F., Liu, Q., Li, D.-Z., Guo, Z.-H. 2013. Phylogenomic analyses of nuclear genes reveal the evolutionary relationship within the BEP clade and the evidence of positive selection in Poaceae. PLoS ONE 8: e64642.

IV. Subfamily Ehrhartoideae Link (1827)

Rhizomatous, stoloniferous, or caespitose perennials, herbaceous or with a woody caudex. Ligule membranous, a fringed membrane, or a fringe of hairs. Inflorescences branched or unbranched, the spikelets pedicellate. *Spikelets with glumes 0 or 2, with the two proximal flowers sterile and often reduced, only the distal flower fertile.* Flowers bisexual or unisexual. Lemma with or without an awn. Lodicules 2. Caryopsis with hard endosperm, with a long-linear hilum, the embryo with an epiblast, with a scutellar cleft, without a mesocotyl internode, with the embryonic leaf margins overlapping. Fusoid cells absent or present, mesophyll cells with or without invaginated cell walls, midrib simple or complex. Silica bodies in leaves transverse to the proximo-distal axis. Photosynthetic pathway C_3. Worldwide distribution.

Fifteen genera and 112 spp.

Recent literature on this subfamily is summarized by Kellogg (2009); little is known about Phyllorachideae, whereas Oryzeae have been studied in great depth (see below). The spikelet structure is unique to the subfamily, and has led to some misleading terminology in the literature. In some genera, particularly *Oryza*, the two glumes are so reduced that they are either ignored or called "rudimentary glumes", while the lemmas of the two sterile flowers are inaccurately called "empty glumes". Molecular genetic studies show that the sterile lemmas in fact express some genes that prevent them from developing like the glumes and other genes that prevent them from developing like the fertile lemma (Ren et al. 2013); thus, the sterile lemmas do share aspects of both lemma and glume identity and constitute a novel organ. The situation is further complicated by loss of one or more glumes and sterile flowers in some taxa. Depending on the species, the rachilla may be elongated. In species in which the glumes and sterile lemmas are missing (e.g., *Hygroryza*, *Chikusiochloa*), the one fertile flower then appears on a long stipe. Maturation of the spikelet is basipetal, and ectopic expression of some floral identity genes affects the sterile lemmas but not the true glumes (Komatsu et al. 2003; Malcomber and Kellogg 2004, 2006; Malcomber et al. 2006).

Silica bodies of this subfamily are distinctive and have been called "oryzoid" (Metcalfe 1960). They occur primarily over the veins, and are elongated transverse to the proximo-distal axis of the leaf. In addition, more conventional bilobate silica bodies occur parallel to the proximo-distal axis. Most members of the subfamily are perennial and this is likely the ancestral state, with the annual habit having arisen several times independently (Kellogg 2009). All Oryzeae and the early-diverging Ehrharteae are plants of moist habitats, suggesting that this is the ancestral environment. Oryzeae and Ehrharteae are thought to have diverged about 35 million years ago, during a period of global cooling and a reduction in atmospheric CO_2 (Vicentini et al. 2008), although data on phytolith assemblages in India suggest that the tribe might be twice as old (Prasad et al. 2011) (see section on Fossil record). The common ancestor of Ehrhartoideae may have been native to what is now the Old World with several independent dispersal events to North America (Kellogg 2009; Tang et al. 2010).

The chromosome base number of most species of Ehrhartoideae is $x = 12$, although *Zizania* is unusual in having $2n = 30$ or 34.

E.A. Kellogg, *Flowering Plants. Monocots*, The Families and Genera of Vascular Plants 13,
DOI 10.1007/978-3-319-15332-2_18, © Springer International Publishing Switzerland 2015

Key to the Genera

1. Inflorescence axis terete or angled, but not broad and leaf-like 2
1. Inflorescence axis broad, flattened, leaf-like, enclosing the primary branches and spikelets 14
2 (1). Plant scrambling; apices of some leaves extended into long tendrils **16. *Prosphytochloa***
2. Plant habit various, but not scrambling and leaf apices not tendril-like 3
3 (2). Spikelets with two well developed glumes, two sterile lemmas, and one bisexual flower **12. *Ehrharta***
3. Spikelets with glumes minute or absent, sterile lemmas absent or reduced to small subulate structures, and one bisexual or unisexual flower 4
4 (3). Spikelets with two sterile lemmas, these often subulate 5
4. Spikelets lacking sterile lemmas, thus consisting of a single flower with its lemma and palea 8
5 (4). Leaf blades with a pseudopetiole **14. *Maltebrunia***
5. Leaf blades without a pseudopetiole 6
6 (5). Distal part of lemma with an attenuate beak **20. *Rhynchoryza***
6. Distal part of lemma awned or not, but not with a beak 7
7 (6). Fertile lemma and palea hardened, with rough tuberculate outgrowths, keeled **15. *Oryza***
7. Fertile lemma and palea chartaceous or membranous, not keeled **19. *Potamophila***
8 (4). Fertile flower on a long stipe 9
8. Fertile flower not on a stipe 10
9 (8). Leaf blades elliptic and floating; plants of India and Sri Lanka **18. *Hygroryza***
9. Leaf blades linear; plants of southeast Asia **17. *Chikusichloa***
10 (8). Spikelets strongly laterally compressed, keeled; flowers bisexual **13. *Leersia***
10. Spikelets terete or only weakly compressed, not keeled; flowers unisexual 11
11 (10). Lemma unawned 12
11. Lemma awned 13
12 (11). Inflorescence branches verticillate; spikelets bisexual; plants of Sudan **9. *Suddia***
12. Inflorescence branches not verticillate; spikelets unisexual; plants of the New World **22. *Luziola***
13 (11). Fruit ovoid, an achene, the seed coat free from the pericarp; staminate and pistillate spikelets on the same inflorescence branches **23. *Zizaniopsis***
13. Fruit linear, a caryopsis, the seed coat fused to the pericarp; staminate and pistillate spikelets separate in the inflorescence **21. *Zizania***
14 (1). Staminate and pistillate spikelets on the same inflorescence branch **11. *Phyllorachis***
14. Staminate and pistillate spikelets in separate inflorescences **10. *Humbertochloa***

Ehrhartoideae Incertae Sedis

9. *Suddia* Renvoize

Suddia Renvoize, Kew Bull. 39: 455 (1984).

Plants 1–3 m tall, *with spongy culms*. Leaves up to 1 m long, the ligule membranous, with a pseudopetiole, *the blades sagittate. Inflorescence branches verticillate. Glumes lacking*. Mesophyll cells without invaginated cell walls. Lemma with prominent veins, the apex acute. Stamens 4.

One sp., *S. sagittifolia* Renvoize, wet sites in Sudan.

Suddia has been placed in Pharoideae by Clayton and Renvoize (1986) and Watson and Dallwitz (1992 and onward). However, it appears not to have pharoid leaf anatomy, and may not have the resupinate leaf blades that characterize the rest of the subfamily (Judziewicz and Clark 2007). Following the suggestion of Judziewicz and Clark (2007), it is placed incertae sedis in Ehrhartoideae pending more morphological and molecular data.

I. Tribe Phyllorachideae C.E. Hubbard (1939)

Monoecious perennials with hard culms. Leaf blade cordate to sagittate, with a pseudopetiole, the ligule a ciliolate membrane. *Inflorescence axis flattened, broad and leaf-like, enclosing the short primary branches; primary branches each bearing a condensed cluster of 1 to 4 spikelets.* Spikelets with *two flowers, the distal one unisexual, the proximal one sterile. Glumes distinct, the lower one subulate and the upper oblong.* Mesophyll cells with invaginated cell walls.

Two genera and three species.

Phyllorachideae are placed in Ehrhartoideae based on morphology plus a DNA sequence of the *rpl16* intron for *Humbertochloa greenwayii* (Zhang 2000). Additional information on this pair of genera would be valuable.

10. *Humbertochloa* A. Camus & Stapf

Humbertochloa A. Camus & Stapf, Bull. Soc. Bot. France 81: 467 (1934).

Inflorescences unisexual. Lower lemma of pistillate spikelet rounded; upper lemma tuberculate.

Two spp., Tanzania and Madagascar.

11. *Phyllorachis* Trimen. Fig. 15

Phyllorachis Trimen., J. Bot. 17: 353 (1879).

Primary branches of the inflorescence each with one pistillate spikelet and two or three staminate

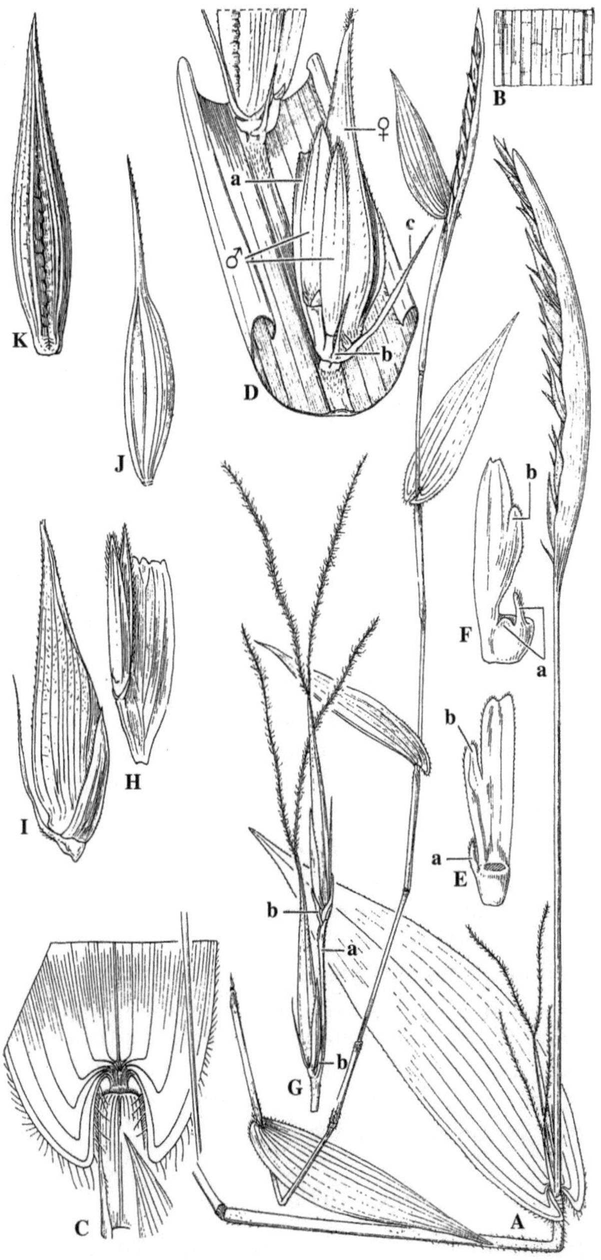

Fig. 15. Ehrhartoideae-Phyllorhachideae. *Phyllorhachis sagittata*. **A** Culm with axillary and terminal inflorescences. **B** Detail from underside of leaf blade showing venation. **C** Base of leaf blade. **D** Raceme showing male and female spikelets, secondary rachis (*a*), vestigial spikelet (*b*) and lower glume of female spikelet (*c*). **E, F** Secondary rachis, front and back views, showing vestigial spikelets at *a* and *b*. **G** Axillary inflorescence, showing primary (*a*) and secondary (*b*) rachis. **H** Staminate spikelet on secondary rachis. **I** Pistillate spikelet. **J** Upper flower of female spikelet. **K** Dorsal view of lower lemma from female spikelet. (From Clayton 1970)

spikelets. Lower lemma of pistillate spikelet grooved; upper lemma smooth. $2n = 24$.

One sp., *P. sagittata* Trimen, equatorial Africa.

II. Tribe Ehrharteae Nevski (1937)

Annuals or perennials, rhizomatous, stoloniferous, or caespitose, sometimes suffrutescent. Leaves with or without auricles, the ligule membranous or a fringed membrane. Inflorescence branched or not. Spikelets with one fertile flower, subtended by two empty lemmas plus two glumes. Stamens 2, 3, 4, or 6. *Epiblast absent.* Mesophyll cells with or without invaginated cell walls; fusoid cells absent. *Vascular system of the midrib simple.*

One genus and thirty-seven species.

12. ***Ehrharta*** Thunb. Fig. 16

Ehrharta Thunb., Vet. Acad. Handl. Stockholm 40: 217 (1779); phylog. (Verboom et al. 2003).
Tetrarrhena R. Br., Prodr. Fl. Nov. Holl. 209 (1810).
Microlaena R. Br., Prodr. Fl. Nov. Holl. 210 (1810).
Petriella Zotov, Trans. Roy. Soc. New Zealand 73: 235 (1943).
Zotovia Edgar & Connor, New Zealand J. Bot. 36: 569 (1998).

Characters of the tribe. $2n = 24, 48$.

Thirty-seven spp., southern Africa, Southeast Asia, New Zealand.

Recognition of this group as a single genus is supported by the work of Verboom et al. (2003), who found that *Microlaena* is polyphyletic, and that recognition of *Tetrarrhena* and *Zotovia* make *Ehrharta* paraphyletic. Stamen number is reduced to four in the clade that includes species formerly in *Tetrarrhena* plus *Ehrharta multinoda* (as *Microlaena polynoda*) and *E. stipoides* (as *M. stipoides*). Stamen number is reduced to three in the clade of *E. delicatula* and *E. triandra*, and independently to two in the clade that includes *E. colensoi* (= *Zotovia colensoi*), *E. thomsonii* (= *Z. thomsonii*), *E. diplax* (= *M. avenacea*), *E. tasmanica* (= *M. tasmanica*) and *E. avenacea*. The lack of fusoid cells and the simple midrib anatomy might be synapomorphic; under this interpretation, these two characters would also be independently derived in the Pooideae.

Ehrharta species growing in the dry winter-rainfall habitats of South Africa represent a relatively recent (possibly late Miocene) radiation

Fig. 16. Ehrhartoideae-Ehrharteae. *Ehrharta erecta.* A Habit. B Node with leaf sheath and ligule. C Spikelet. D Florets. (From Barkworth et al. 2007, drawn by H. Paźkirková)

(Verboom et al. 2003). The early-diverging members of the group occupied regions of year-round rainfall, estimated at 689–1,400 mm per year. *E. stipoides* is being investigated as a possible drought-tolerant crop for human consumption (Shapter et al. 2013).

III. Tribe Oryzeae Dumort. (1824)

Annuals or perennials. Leaf blades with or without a pseudopetiole, the ligule membranous. Spikelets with 1 to 3 flowers, one fertile, the other 0 to 2 sterile and reduced to lemmas. *Glumes reduced or absent. Vascular system of the midrib with at least two vascular bundles, often more, with bundles placed ad- and abaxially, and often with air spaces.*

Eleven genera and seventy species.

A robust and well-sampled molecular phylogeny for the tribe shows that all genera are monophyletic (Tang et al. 2010). Previous data linking *Prosphytochloa* and *Potamophila* were based on misidentification of the *Prosphytochloa* specimen. A new specimen of *Prosphytochloa* plus an accession of *Maltebrunia* shows that the two are sister taxa that together are sister to *Leersia*. Data on leaf anatomy relies on the comprehensive work of Tateoka (1963); vascular structure of the midrib in *Luziola* and *Hygroryza* is simple, similar to that in Ehrharteae.

All members of Oryzeae grow in damp areas, many occupy flooded sites, and still others (e.g., *Hygroryza*) are truly aquatic. Although this tribe scarcely needs dividing into subtribes, the names are available and correspond to well-supported clades, and so are provided here.

III.1. Subtribe Oryzinae Griseb. (1853)

This subtribe has no diagnostic morphological characters. The leaves may be auriculate or not. Spikelets are laterally compressed in all genera, but this character may be ancestral. Fusoid cells are lacking in the mesophyll of all genera except for *Maltebrunia*. Loss of fusoid cells appears to be a derived character here, although they were lost in parallel in *Ehrharta*.

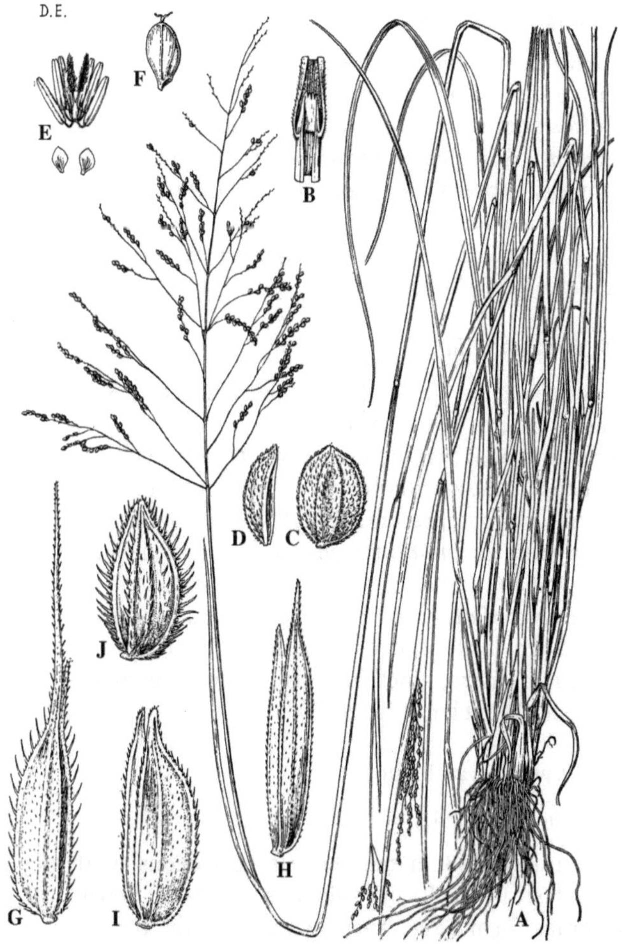

Fig. 17. Ehrhartoideae-Oryzeae-Oryziinae. *Leersia drepanothrix.* **A** Habit. **B** Ligule. **C** Spikelet. **D** Lemma. **E** Flower and lodicules separated. **F** Caryopsis. **G–J** Spikelets from other species: **G** *L. tisserantii.* **H** *L. friesii.* **I** *L. denudata.* **J** *L. hexandra.* (From Clayton 1970)

13. ***Leersia*** Sw. Fig. 17

Leersia Sw., Prodr. Veg. Ind. Occ. 21 (1788).

Annuals or perennials, caespitose or rhizomatous. Culm nodes spongy to solid. Inflorescence branches themselves branched. Spikelets with one flower, *sterile lemmas absent*. Fertile lemma asymmetrically keeled, awned or not, the margins inrolled. Stamens *1, 2, 3* or 6. *Scutellar cleft absent*. $2n = 24, 48, 60$.

Eighteen spp., tropical and warm temperate.

Loss of sterile lemmas is a synapomorphy for *Leersia* (Kellogg 2009). Parallel losses have occurred in some members of subtribe Zizaniinae. *Leersia oryzoides* is reported to take up large quantities of arsenic from the soil, making it potentially useful for phytoremediation (Ampiah-Bonney et al. 2007).

14. ***Maltebrunia*** Kunth

Maltebrunia Kunth, Rév. Gram. 1: 6 (1829).

Perennials, rhizomatous or not. *Leaf blade with a pseudopetiole*, with or without obvious cross veins. Inflorescence branches branched. Spikelets with two sterile and one fertile flowers. *Margins of the lemma clasping the palea.*

Four spp., Gabon, Tanzania, Madagascar.

15. ***Oryza*** L.

Oryza L., Sp. Pl.: 333 (1753); phylog.: Vaughan et al. (2005).
Porteresia Tateoka, Bull. Natl. Sci. Mus. 8: 406 (1965).

Annuals or perennials, caespitose or rhizomatous. Culm nodes spongy to solid. Ligule sometimes ciliolate. Inflorescence branches themselves branched. Spikelets with two sterile and one fertile flower, disarticulating below the sterile lemmas and above the true glumes. Fertile lemma coriaceous, asymmetrically keeled, the margins interlocking with the margins of the palea, awned or not. Epidermal cells of the lemma and palea *producing tuberculate outgrowths*. $2n = 24, 48$.

Twenty spp., tropical and subtropical worldwide.

The phylogeny of this group has been resolved by a large genomic data set (Zou et al. 2008). See also Kellogg (2009) for discussion.

16. ***Prosphytochloa*** Schweick.

Prosphytochloa Schweick., Züchter 31: 193 (1961).

Scandent perennials. *Apex of distal leaf blades extended into a long tendril.* Inflorescence branches branched. Spikelets with two sterile flowers and one fertile.

One sp., *P. prehensilis* (Nees) Schweick., South Africa.

III.2. Subtribe Zizaniinae Benth. (1881)

This subtribe lacks a clear diagnostic character, although some labile characters may be synapomorphic depending on their optimization on the phylogeny.

All members of the clade have branched inflorescence branches except for a few species of *Luziola*, and all except for *Luziola* have fusoid

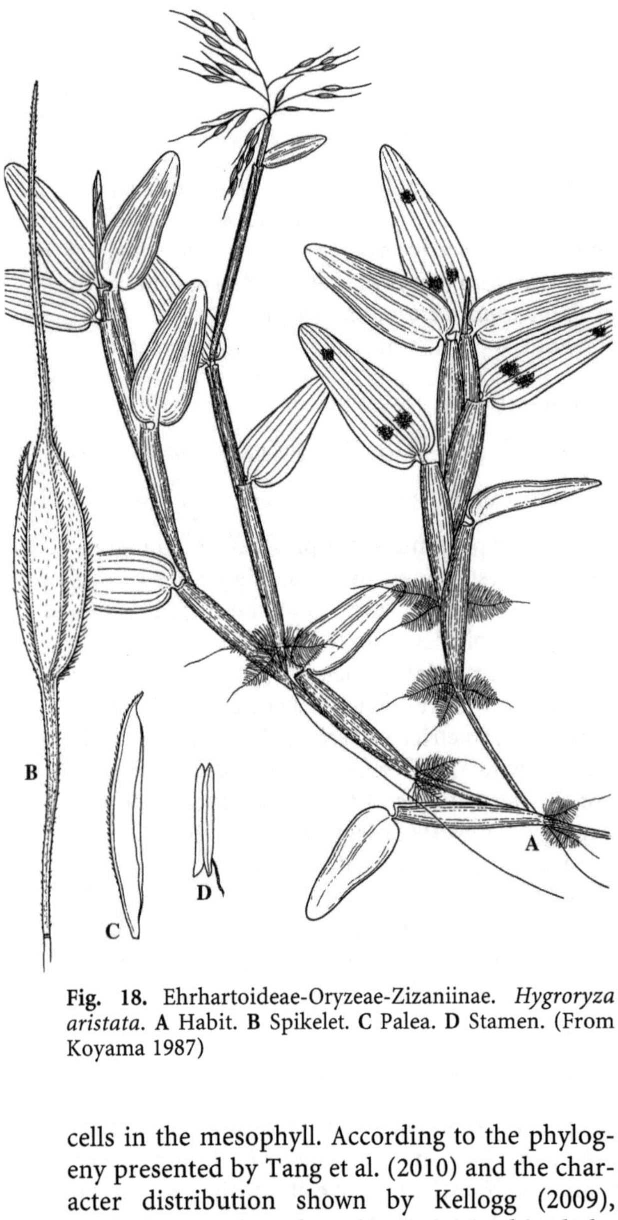

Fig. 18. Ehrhartoideae-Oryzeae-Zizaniinae. *Hygroryza aristata*. **A** Habit. **B** Spikelet. **C** Palea. **D** Stamen. (From Koyama 1987)

cells in the mesophyll. According to the phylogeny presented by Tang et al. (2010) and the character distribution shown by Kellogg (2009), sterile lemmas have been lost twice in this clade, and unisexual flowers have originated twice.

17. ***Chikusichloa*** Koidzumi

Chikusichloa Koidzumi, Bot. Mag. Tokyo 39: 23 (1925).

Caespitose perennials. *Ligule sometimes ciliolate. Spikelets on a long slender pedicel.* Glumes absent; sterile lemmas absent. *Stamen 1.* Mesophyll cells without invaginated cell walls. $2n = 24$.

Three spp., Southeast Asia to Japan.

The stamen number clearly represents a reduction from the ancestral number of 6.

18. ***Hygroryza*** Nees Fig. 18

Hygroryza Nees, Edinb. New Phil. J. 15: 380 (1833).

Prostrate stoloniferous perennials; aquatic. Culms spongy. *Leaf-blades floating, with inflated sheaths.* Spikelets laterally compressed. Sterile lemmas absent. *Flower on a long pedicel.* Lemma awned. $2n = 24$.

One sp., *H. aristata* (Retz.) Nees ex Wright & Arn., eastern Asia.

19. ***Potamophila*** R. Br.

Potamophila R. Br., Prodr. Fl. Nov. Holl: 211 (1810); phylog.: Abedinia et al. (1998).

Rhizomatous perennials. *Spikelets unisexual or bisexual, these intermixed in the inflorescence.* Sterile lemmas present. *Scutellar cleft absent, the embryonic leaf margins non-overlapping.* $2n = 24$.

One sp., *P. parviflora* R. Br., Australia.

P. parviflora may be a diplosporous apomict, based on lack of overall genetic diversity and irregular microgametogenesis and megagametogenesis (Wheeler et al. 2001).

20. ***Rhynchoryza*** Baillon

Rhynchoryza Baillon, Hist. Pl. 12: 291 (1893).

Perennials, 2–3 m tall. Leaves coriaceous, with auricles. Pedicels of spikelets cuneate. Sterile lemmas forming a short involucre at the base of the fertile flower. Fertile lemma with involute margins, *the apex extended to a long aerenchymatous beak and ultimately to an awn.*

One sp., *R. subulata* (Nees) Baill., Paraguay to Argentina.

21. ***Zizania*** L.

Zizania L., Sp. Pl.: 991 (1753); tax.: Terrell et al. (1997).

Aquatic annuals or perennials, caespitose, with or without stolons, 1–3 m tall. Culms spongy. Spikelets *unisexual, staminate spikelets on lower panicle branches, pistillate ones on upper branches.* Sterile lemmas absent. Lemmas awned. $2n = 30, 34$.

Five spp., Eastern Asia and North America.

Luziola + Zizaniopsis clade:

Sterile lemmas absent. *Flowers unisexual. Pericarp free from the seed coat, brittle.*

This origin of unisexual flowers appears to be independent from that in *Zizania.*

22. ***Luziola*** Juss.

Luziola Juss., Gen. Pl. 33 (1789); tax.: Martínez-y-Pérez et al. (2008).
Hydrochloa P. Beauv., Ess. Agrostogr. 135, 165, 182 (1812).

Caespitose perennials, with or without stolons. Leaves auriculate or not. Inflorescence branched or not, each branch bearing 6 or fewer spikelets. Spikelets *unisexual, often with the sexes in separate inflorescences*, the staminate inflorescence terminal and the pistillate axillary, or with staminate and pistillate spikelets in the same inflorescence. Spikelets terete or laterally compressed, neither the lemma nor the palea keeled. *Lodicules absent.* Stamens 6 *to 12. Fusoid cells lacking.*

Eleven spp., southern U.S.A. to Argentina.

The high stamen number in some species of the genus is derived.

23. ***Zizaniopsis*** Doell. & Asch.

Zizaniopsis Doell. & Asch. in Martius, Fl. Bras. 2, 2: 12 (1871).

Annuals or perennials, rhizomatous or not, 1–3 m tall. Culms spongy or solid. Inflorescence branches generally verticillate. Spikelets laterally compressed, unisexual, *both sexes occurring on the same inflorescence branch, staminate proximal and pistillate distal.* Lemma without a keel, awned. *Stamens 2*, stigmas 2. Caryopsis with a protrusion (rostrum). $2n = 24$.

Five spp., southern U.S.A. to Argentina.

References

Abedinia, M., Henry, R.J., Clark, S.C. 1998. Distribution and phylogeny of *Potamophila parviflora* R. Br. a wild relative of rice from eastern Australia. Genet. Res. Crop Evol. 45: 399–406.

Ampiah-Bonney, R.J., Tyson, J.F., Lanza, G.R. 2007. Phytoextraction of arsenic from soil by *Leersia oryzoides*. Int. J. Phytoremediation 9: 31–40.

Barkworth, M.E., Capels, K.M., Long, S., Anderton, L.K., Piep, M.B. (eds.) 2007. Flora of North America north of Mexico. Vol. 24. Magnoliophyta: Commelinidae (in part): Poaceae, part 1. New York: Oxford University Press.

Clayton, W.D. 1970. Gramineae (Part 1). In: Milne-Redhead, E., Polhill, R.M. (eds.) Flora of Tropical East Africa. London: Crown Agents for Oversea Governments and Administrations.

Clayton, W.D., Renvoize, S.A. 1986. Genera graminum: grasses of the world. London: Her Majesty's Stationery Office.

Judziewicz, E.J., Clark, L.G. 2007. Classification and biogeography of New World grasses: Anomochlooideae, Pharoideae, Ehrhartoideae, and Bambusoideae. Aliso 23: 303–314.

Kellogg, E.A. 2009. The evolutionary history of Ehrhartoideae, Oryzeae, and *Oryza*. RICE 2: 1–14.

Komatsu, M., Chujo, A., Nagato, Y., Shimamoto, K., Kyozuka, J. 2003. FRIZZY PANICLE is required to prevent the formation of axillary meristems and to establish floral meristem identity in rice spikelets. Development 130: 3841–3850.

Koyama, T. 1987. Grasses of Japan and its neighboring regions. Tokyo: Kodansha Ltd.

Malcomber, S.T., Kellogg, E.A. 2004. Heterogeneous expression patterns and separate roles of the SEPALLATA gene *LEAFY HULL STERILE* (*LHS1*) in grasses. Plant Cell 16: 1692–1706.

Malcomber, S.T., Kellogg, E.A. 2006. Evolution of unisexual flowers in grasses (Poaceae) and the putative sex-determination gene, *TASSELSEED2* (*TS2*). New Phytol. 170: 885–899.

Malcomber, S.T., Preston, J.C., Reinheimer, R., Kossuth, J., Kellogg, E.A. 2006. Developmental gene evolution and the origin of grass inflorescence diversity. Adv. Bot. Res. 44: 425–481.

Martínez-y-Pérez, J.L., Mejía-Saulés, T., Sosa, V. 2008. A taxonomic revision of *Luziola* (Poaceae: Oryzeae). Syst. Bot. 33: 702–718.

Metcalfe, C.R. 1960. Anatomy of the monocotyledons. I. Gramineae. Oxford: Clarendon Press.

Prasad, V., Strömberg, C.A.E., Leaché, A.D., Samant, B., Patnaik, R., Tang, L., Mohabey, D.M., Ge, S., Sahni, A. 2011. Late Cretaceous origin of the rice tribe provides evidence for early diversification in Poaceae. Nature Comm. 2: 480.

Ren, D., Li, Y., Zhao, F., Sang, X., Shi, J., Wang, N., Guo, S., Ling, Y., Zhang, C., Yang, Z., He, G. 2013. *MULTI-FLORET SPIKELET1*, which encodes an AP2/ERF protein, determines spikelet meristem fate and sterile lemma identity in rice. Plant Phys. 162: 872–884.

Shapter, F.M., Cross, M., Ablett, G., Malory, S., Chivers, I. H., King, G.J., Henry, R.J. 2013. High-throughput sequencing and mutagenesis to accelerate the domestication of *Microlaena stipoides* as a new food crop. PLoS ONE 8: e82641. doi: 10.1371/journal.pone.0082641

Tang, L., Zou, X., Achoundong, G., Potgieter, C., Second, G., Zhang, D., Ge, S. 2010. Phylogeny and biogeography of the rice tribe (Oryzeae): evidence from combined analysis of 20 chloroplast fragments. Mol. Phyl. Evol. 54: 266–277.

Tateoka, T. 1963. Notes on some grasses. XIII. Relationship between Oryzeae and Ehrharteae, with special reference to leaf anatomy and histology. Bot. Gaz. 124: 264–270.

Terrell, E.E., Peterson, P.M., Reveal, J.L., Duvall, M.R. 1997. Taxonomy of North American species of *Zizania* (Poaceae). Sida 17: 533–549.

Vaughan, D.A., Kadowaki, K., Kaga, A., Tomooka, N. 2005. On the phylogeny and biogeography of the genus *Oryza*. Breeding Sci. 55: P113-122.

Verboom, G.A., Linder, H.P., Stock, W.D. 2003. Phylogenetics of the grass genus *Ehrharta*: evidence for radiation in the summer-arid zone of the South African Cape. Evolution 57: 1008–1021.

Vicentini, A., Barber, J.C., Giussani, L.M., Aliscioni, S.S., Kellogg, E.A. 2008. Multiple coincident origins of C_4 photosynthesis in the Mid- to Late Miocene. Global Change Biol. 14: 2963–2977.

Watson, L., Dallwitz, M.J. 1992 onward. The grass genera of the world: descriptions, illustrations, identification, and information retrieval; including synonyms, morphology, anatomy, physiology, phytochemistry, cytology, classification, pathogens, world and local distribution, and references, 23rd April 2010 edn. Wallingford: CAB International. http://delta-intkey.com/grass/

Wheeler, M.A., Lee, L.S., Henry, R.J. 2001. The reproductive biology and genetic diversity of *Potamophila parviflora* R. Br. Genet. Res. Crop Evol. 48: 483–497.

Zhang, W. 2000. Phylogeny of the grass family (Poaceae) from *rpl16* intron sequence data. Mol. Phyl. Evol. 15: 135–146.

Zou, X.H., Zhang, F.M., Zhang, J.G., Zang, L.L., Tang, L., Wang, J., Sang, T., Ge, S. 2008. Analysis of 142 genes resolves the rapid diversification of the rice genus. Genome Biol. 9: R49.

V. Subfamily Bambusoideae Luerss. (1893)

This treatment of Bambusoideae includes extensive and detailed contributions from Lynn G. Clark (Iowa State University). All remaining errors and misinterpretations are mine.

Plants perennial. Culms woody or herbaceous, caespitose or rhizomatous, the nodes flat or swollen. Leaves pseudopetiolate, often with obvious cross veins, strongly dimorphic in woody taxa, with culm leaves distinctly different from leaves on lateral branches ("foliage leaves"), the culm leaves with prominent sheaths and reduced blades, the foliage leaves pseudopetiolate. Ligule membranous, with or without a fringe of hairs. Outer ligule present or absent. Inflorescences branched or unbranched, bearing spikelets or pseudospikelets. Glumes 1 to several, or lacking. Lodicules 3, with extensive vascular tissue. Stamens 3 or 6. Ovary with 2 or 3 stigmas. Hilum linear, as long as the caryopsis. Epiblast present, scutellar cleft present, mesocotyl internode absent, embryonic leaf margins overlapping. *Lamina of first seedling leaf absent.* Mesophyll with fusoid cells, *with arm cells with asymmetric invaginations in the cell walls.* C_3.

One hundred sixteen genera, 1441 spp.

Phylogenies of Bambusoideae consistently show strong support for three tribes, Bambuseae, Olyreae, and Arundinarieae, as recognized here (Bouchenak-Khelladi et al. 2008; Burke et al. 2012; Clark et al. 2007; Kelchner and Bamboo Phylogeny Group 2013; Kelchner and Clark 1997; Sungkaew et al. 2009; Zhang and Clark 2000). Bambuseae and Arundinarieae were long thought to be sisters based on the shared characters of woody culms, differentiated foliage and culm leaves, and gregarious flowering, all of which appear to be uniquely derived (Zhang and Clark 2000). However, molecular phylogenies based on nuclear genes indicate that the history of Bambusoideae is more complex and intriguing than previously thought (Triplett et al. 2014). Arundinarieae (the temperate woody bamboos) are tetraploids with genomes designated A and B. Bambuseae (the tropical woody bamboos) are tetraploid and hexaploid with genomes designated C, D and E. Olyreae are predominantly diploid with a genome designated H. Thus, the Arundinarieae and Bambuseae are the products of separate polyploidization events. Remarkably, however, the B genome of the temperate bamboos is most closely related to the C genome of the tropical ones; in other words, the temperate and tropical species have one genome donor in common. Chloroplast data place Olyreae sister to Bambuseae (Bouchenak-Khelladi et al. 2008; Burke et al. 2012; Kelchner and Bamboo Phylogeny Group 2013; Kelchner and Clark 1997; Sungkaew et al. 2009), although monophyly of the woody bamboos cannot be rejected statistically (Kelchner and Bamboo Phylogeny Group 2013). The data of Triplett et al. (2014) indicate that woodiness is derived within Bambusoideae, but it is unclear if it evolved once or if each independent polyploidization event correlates with an independent origin of woodiness.

Many traditionally recognized genera in Bambusoideae have proven to be polyphyletic with the addition of molecular data (see references below), indicating that the commonly used characters are in fact homoplasious and are not reliable indicators of relationship. Such characters include rhizome structure, presence or

E.A. Kellogg, *Flowering Plants. Monocots*, The Families and Genera of Vascular Plants 13,
DOI 10.1007/978-3-319-15332-2_19, © Springer International Publishing Switzerland 2015

absence of pseudospikelets, numbers and arrangement of culm buds and branches, and inflorescence architecture. In addition, homoploid hybridization has been documented and appears to be far more common than previously thought possible (Takahashi et al. 1994; Triplett and Clark 2010; Triplett et al. 2010, 2014). Several common "genera" appear to be simply early-generation (possibly F1) hybrids that combine the characters of their parent genera. If hybridization is common, then introgression, particularly of chloroplasts, may also occur frequently. In this scenario, the chloroplast phylogenies in the literature may thus be misleading about relationships (see, for example, Yang et al. 2013). An additional confounding factor is the slow rate of molecular evolution, presumably caused by the very long generation time of most bamboos. Thus, even very well sampled molecular phylogenies such as those of Triplett and Clark (2010) and Zeng et al. (2010) often provide only weak hints of phylogenetic history. Finally, all woody bamboos appear to be polyploid (e.g., Stebbins 1971, 1985; Triplett et al. 2014). Not surprisingly, the first genome sequence of a woody bamboo shows a whole genome duplication event, estimated to have occurred 7–12 mya (Peng et al. 2013), although it is unclear if the methods used account for the very slow mutation rate. More studies like that of Triplett et al. (2014) will be needed to address the effects of hybridization and polyploidy on bamboo evolution and hybridization.

A detailed account of bamboo morphology is provided by Stapleton (1997) and also on the web (http://www.bamboo-identification.co.uk/index/html and http://www.eeob.iastate.edu/research/bamboo/characters.html). In general, morphological work on the subfamily has been uneven, with some taxa receiving detailed and comprehensive study and others only being described in a fairly superficial way. Numerous good field characters – most notably the rhizome type and plant architecture, number of branches per midculm node, and color of hairs and leaf blades – are not readily available on herbarium specimens. An additional complication is the lack of flowering material. The bamboos are well known for cyclical flowering, with many species flowering at intervals of tens to over one hundred years (Janzen 1976; McClure 1966; Ramanayake and Weerawardene 2003), although sporadic (off-cycle) flowering is also known. This means that many collections are sterile, and taxonomists have had to develop ways to identify species without access to the often-diagnostic inflorescences and flowers. As noted by Wong (2004), "Attempts to identify bamboos using older accounts can turn out to be frustrating because they may not always include adequate descriptions of the most accessible parts of a bamboo plant . . .". This is something of an understatement.

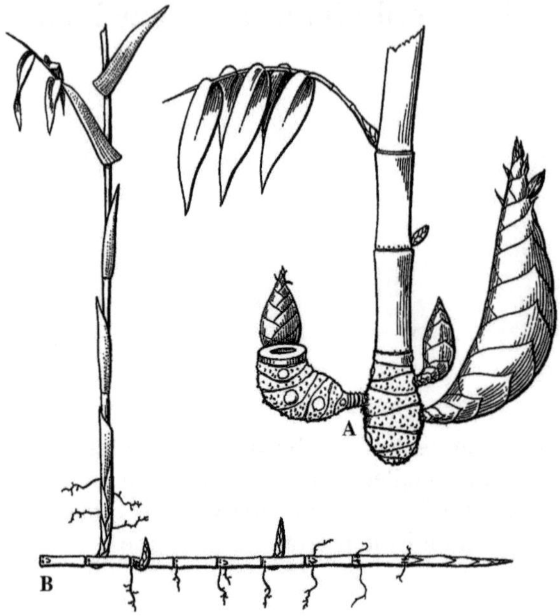

Fig. 19. Bambusoid rhizomes. **A** Pachymorph (sympodial) rhizome (*Bambusa beechyana*). **B** Leptomorph (monopodial) rhizome (*Arundinaria amabilis*). (From McClure 1966)

McClure (1966) drew attention to rhizome morphology in woody bamboos. In some taxa, the apical meristem of the rhizome is persistent and never turns upward to form a culm; all above ground culms are formed from meristems that are axillary on the rhizomes. Such rhizomes are known as leptomorph, or monopodial (Fig. 19B). In other taxa, the apical meristem of the rhizome eventually will form a culm, so that the growth of the plant is sympodial as is common in most grasses (Fig. 19A). Sympodial rhizomes are known as pachymorph, a term that reflects extensive variation in thickness from one internode to the next. As the rhizome initiates from a lateral meristem, the first few internodes are slender and lack axillary buds, although they may produce adventitious roots. These first internodes, which connect the newly formed rhizome to the parent axis, are known as the neck. In some cases, the neck may be one or occasionally several meters long, and will have many nodes. Subsequent internodes are noticeably wider in diameter, produce axillary buds and often numerous roots throughout the internode, and are asymmetric so that they curve upward toward the soil surface. After the meristem extends above ground and forms the culm, internode diameter is again reduced although not nearly to the extent seen in the neck. The result is a plant with curved swollen rhizomes, connected to their parent axes by narrow internodes.

Rhizome morphology affects the arrangement of culms. A plant with pachymorph rhizomes with short necks will be clump forming. Conversely, a plant with either leptomorph rhizomes or pachymorph rhizomes with long necks will produce more widely spaced culms.

The leaves of many bamboos bear fimbriae or oral setae, elongated structures up to several cm long, at the summit of the sheath of culm and/or foliage leaves, on the auricles, and sometimes around the ligule. Most commonly these fimbriae are slender, terete or slightly flattened, less than 0.1 mm wide, and look like particularly long hairs. A few taxa, however, have much more robust structures, 0.2–1.2 mm wide, occurring in the ligular area and often fused to the abaxial side of the ligule. In most of the literature (e.g., Judziewicz et al. 1999) both sorts of structures are called fimbriae, although other authors call both oral setae. Clark and Cortés R. (2004) note the distinction between the two, based largely on morphology but also position, and argue that the slender structures should be called fimbriae whereas the more robust ligular ones should be called oral setae. By this definition, in New World bamboos, oral setae occur only in Bambuseae subtribe Guaduinae, in *Eremocaulon*, *Olmeca* and *Otatea*. Clearly, this distinction needs to be considered for Old World bamboos as well. However, because most bamboo taxonomists use the terms fimbriae and oral setae interchangeably, in this account all such structures will be called fimbriae.

Bamboo inflorescences are described in detail in the section on Inflorescence structure. While many genera have conventional inflorescences with conventional grass spikelets like other grasses, others have inflorescences that are complex and difficult to describe using standard terminology. In this latter group, the distinction between vegetative and inflorescence structures appears to be blurred, as though the meristem goes through a slow transition from producing leaves to producing variously reduced and modified bracts, to ultimately producing grass flowers with lemmas and paleas. In such plants, it is common for inflorescence branches to be subtended by bracts and prophylls. In addition, there may be glume-like structures proximal to the flowers; these glume-like structures (so-called "gemmiparous bracts") have buds in their axils. Any bracts distal to the gemmiparous bracts and proximal to the flowers are then known as glumes. Thus, plants with gemmiparous bracts and no additional bracts below the flowers are described as lacking glumes, although this description assumes that a glume is necessarily a sterile structure. After the flowers are formed, the buds in the axils of the glume-like structures then grow out and reiterate the production of glumes and flowers, and the reiterated branchlets ("rachillas", although the homology is not clear) may themselves produce more branchlets and flowers. This complex, reiterated structure is known as a pseudospikelet. Inflorescences with pseudospikelets are known as iterauctant, in contrast to those with conventional spikelets, which are described as semelauctant. I have avoided these terms here, preferring to describe the morphology using more familiar – if more lengthy – descriptors. Presence or absence of pseudospikelets has never been analyzed in a phylogenetic context but the taxonomic distribution of the structure suggests that it is highly homoplasious, having been gained and/or lost many times in the history of the subfamily.

The tribal and subtribal classification presented here follows that of the Bamboo Phylogeny Group (BPG) (2012), which in turn is based on current phylogenetic studies (Kelchner and Bamboo Phylogeny Group 2013). The BPG (2012) presents a comprehensive table of these studies and their main conclusions, so only the most recent are referenced here. Information for New World genera of Bambusoideae relies heavily on Judziewicz et al. (1999), which should be consulted for more detailed information. The book is elegantly illustrated and the taxa are described much more completely than is attempted here. Information on Old World bamboos comes from various sources, but most notably the *Flora of China* (Li et al. 2006). Although I have attempted to provide parallel descriptions, information is lacking on many important characters, particularly for Old World taxa. This paucity of descriptive material combined with lack of specimens or incomplete specimens makes the bamboos the group of grasses for which morphological data are most urgently needed.

Tribes and Subtribes of Bambusoideae

IV. Tribe Arundinarieae
- Clade I. *Bergbambos*
- Clade II. *Oldeania*
- Clade III. *Chimonocalamus*
- Clade IV. *Shibataea* clade
- Clade V. *Phyllostachys* clade
- Clade VI. *Arundinaria* clade
- Clade VII. *Thamnocalamus*
- Clade VIII. "*Indocalamus*" *wilsonii*
- Clade IX. *Gaoligongshania*
- Clade X. *Indocalamus* s.s.
- Clade XI. "*Ampelocalamus*" *calcareus*
- Clade XII. *Kuruna*

V. Tribe Bambuseae
- Neotropical clade
  - V.1. Subtribe Chusqueinae
  - V.2. Subtribe Arthrostylidiinae
  - V.3. Subtribe Guaduinae
- Paleotropical clade
  - V.4. Subtribe Bambusinae
  - V.5. Subtribe Melocanninae
  - V.6. Subtribe Hickeliinae
  - V.7. Subtribe Racemobambosinae

VI. Tribe Olyreae
- VI.1. Subtribe Buergersiochloinae
- VI.2. Subtribe Parianinae
- VI.3. Subtribe Olyrinae

Key to the Genera of Bambusoideae

The following key should be considered provisional, especially as regards Old World taxa. Many of the genera are polyphyletic. Others are poorly known, with incomplete character information in the literature and few available specimens.

1. Culms herbaceous; abaxial ligule absent **Group I** (Olyreae)
1. Culms woody; abaxial ligule present 2
2 (1). Culms normally leafless, the plants clump forming and rush-like; internodes with cross partitions; near Rio de Janeiro, Brazil **56. *Glaziophyton***
2. Culms normally leafy, the plants clump forming or not; internodes lacking partitions; distribution various 3
3 (2). Culms bearing thorns and/or root spines at the nodes 4
3. Thorns and root spines absent 9
4 (3). Thorns formed from modified branches 5
4. Root spines formed from prop roots, especially at lower nodes 6
5 (4). Culms with white hairs below the nodes; culm leaves with blade margins continuous with the sheath margins; palea keels winged; plants of the New World **73. *Guadua***
5. Culms lacking white hairs at the nodes; culm leaves with the margins distinct from the sheath margins; palea keels lacking wings; plants of the Old World **97. *Bambusa***
6 (4). Plants of the New World; spikelets with one fertile flower **55. *Chusquea***
6. Plants of the Old World; spikelets with several fertile flowers 7
7 (6). Rhizomes leptomorph **34. *Chimonobambusa***
7. Rhizomes pachymorph 8
8 (7). Rhizomes with short necks, the culms clumped **26. *Chimonocalamus***
8. Rhizomes with long necks, the culms solitary **25. *Oldeania***
9 (3). Culms unbranched 10
9. Culms branched 11
10 (9). Inflorescence branches with subtending bracts; spikelets with a rachilla extension; plants of New Caledonia **79. *Greslania***
10. Inflorescence branches lacking subtending bracts; spikelets lacking a rachilla extension; plants of the New World **55. *Chusquea* subg. *Platonia*, subg. *Magnifoliae***
11 (9). Foliage leaves with an adaxially projecting midrib; plants of Madagascar, Africa, and Réunion [subtribe Hickeliinae] 12
11. Foliage leaves with the midrib not projecting on the adaxial side; distribution various, but generally not Madagascar or Réunion 19
12 (11). Young shoots covered with black hairs 13
12. Young shoots not covered with black hairs 14
13 (12). Inflorescences dense, capitate, the branches branched, lateral branches arranged in tight fans **112. *Cathariostachys***
13. Inflorescence unbranched, surrounded by a spathe, with 1 to 4 spikelets, the spatheate units clustered **113. *Decaryochloa***
14 (12). Culm leaf sheaths lacking auricles; glumes 1 to 5 15
14. Culm leaf sheaths with auricles; glumes 5 or 6 18
15 (14). Glumes 1 or 2 16
15. Glumes 3, 4 or 5 17
16 (14). Spikelets laterally compressed, keeled; style persistent in fruit **115. *Hitchcockella***
16. Spikelets terete, lacking keels; style not persistent **117. *Perrierbambus***
17 (15). Foliage leaves lacking auricles; ovary apex pubescent **114. *Hickelia***
17. Foliage leaves with prominent auricles; ovary glabrous **118. *Sirochloa***
18 (14). Lemma with few veins **116. *Nastus* s.s.**

18. Lemma with 29 to 41 veins **119. *Valiha***
19 (11). Branch complement of two or more subequal branches **Group II**
19. Branch complement with one branch only or one dominant branch and two to many smaller laterals **Group III**

Group I - Herbaceous bamboos

1. Pistillate lemma and palea leathery, but not hardened; lemma with an awn 5–30 mm long; plants of New Guinea **121. *Buergersiochloa***
1. Pistillate lemma and palea generally hardened; lemmas generally unawned, if awned then the awn less than 15 mm; plants of the New World 2
2 (1). Long slender fimbriae present at the top of the sheaths, near the sheath/blade junction; staminate spikelets in a whorl 3
2. Fimbriae absent; staminate spikelets not whorled 4
3 (2). Leaves with the top of the sheath variously decorated with translucent swellings ("lunar marks"); pistillate spikelets one per inflorescence node **123. *Pariana***
3. Leaves without lunar marks; pistillate spikelets one per inflorescence, terminal **122. *Eremitis***
4 (2). Upper leaf blade inverted, forming a canopy over the single terminal inflorescence 5
4. Leaf blades not inverted, not covering the inflorescence 6
5 (4). Bases of glumes of pistillate spikelets fused; glumes of staminate spikelets tiny, membranous **140. *Diandrolyra***
5. Bases of glumes of pistillate spikelets not fused; glumes of staminate spikelets absent **141. *Maclurolyra***
6 (4). Plants producing swollen corms at base of stems 7
6. Corms absent 9
7 (6). Inflorescences with several branches, each terminating in a pistillate spikelet, with one or two staminate spikelets below; plants of northern South America **139. *Rehia***
7. Inflorescences with only two branches, one with staminate spikelets and the other with pistillate; plants of Cuba 8
8 (7). Leaf blades with tapering bases; flowering culms longer than vegetative ones; pistillate flower longer than glumes; stamens 3 **137. *Mniochloa***
8. Leaf blades with truncate bases; flowering culms shorter than vegetative ones; pistillate flower shorter than glumes; stamens 2 **138. *Piresiella***
9 (6). Pistillate lemma awned 10
9. Pistillate lemma awnless, or if awned then the awn much shorter than the body of the lemma 11
10 (9). Awn up to 15 mm long, longer than the body of the lemma; leaf blades deciduous **127. *Ekmanochloa***
10. Awn <2 mm long, shorter than the lemma; leaf blades persistent on distal leaves **124. *Agnesia***
11 (9). Plants over 50 cm tall, often over 1 m; forming large clumps **130. "*Olyra*"**
11. Plants always less than 1 m tall, generally less than 50 cm 12
12 (11). Pistillate flower shaped like a helmet or tooth, laterally compressed **129. *Lithachne***
12. Pistillate flower symmetrical, generally ellipsoid or fusoid, compressed very little if at all 13
13 (12). Glumes of pistillate spikelets with thickened margins, extending into attenuate apices, these twisting at maturity 14
13. Glumes without thickened margins 15
14 (13). Sexes in the same inflorescence, pistillate spikelets above, staminate below **135. *Sucrea***
14. Sexes in separate inflorescences, the terminal inflorescence staminate and branched, the axillary ones pistillate and unbranched **136. *Raddia***
15 (13). Culms dimorphic on the same plant, with upright culms producing leaves with broad blades and decumbent culms producing leaves with blades reduced or lacking; inflorescences commonly but not exclusively borne on the decumbent culms **133. *Piresia***
15. Culms not dimorphic; inflorescences borne on leafy culms 16
16 (15). Inflorescence with one distal pistillate spikelet, the proximal spikelets staminate 17
16. Inflorescences various, often with pistillate and staminate spikelets mixed, but not with a single distal pistillate spikelet 19
17 (16). Spikelets clustered, with many male spikelets surrounding a single terminal female one; the cluster deciduous; 2 to 5 spikelet clusters on a zigzag axis **128. *Froesiochloa***
17. Spikelets not clustered 18
18 (17). Inflorescences produced from apical and axillary meristems **125. *Arberella***
18. Inflorescences produced only from apical meristems **134. *Reitzia***
19 (16). Disarticulation of female spikelet above the glumes **126. *Cryptochloa***
19. Disarticulation of female spikelet below the glumes 20
20 (19). Each inflorescence with few flowers; male and female flowers in separate inflorescences **132. *Raddiella***
20. Each inflorescence with many flowers; male and female flowers mixed in the same inflorescence **131. *Parodiolyra***

Group II - Branch complement of two or more subequal branches

1. Mid-culm branches apparently paired, with one main branch and one lateral, the two becoming subequal **40. *Phyllostachys***
1. Mid-culm branches more than two 2
2 (1). Mid-culm nodes with branches in a fan-shaped pattern 3
2. Mid-culm nodes with branching pattern not fan-shaped 6
3 (2). Culm internodes alternating in length between long and short; bud prophylls cordate; culm leaf blades erect; spikelets in clusters of 2 or 3 **65. *Didymogonyx***
3. Culm internodes not alternating in length; bud prophylls triangular; culm leaf blades reflexed; spikelets not clustered 4
4 (3). Foliage leaves dimorphic, the blades of those on the main culm much larger than the ones on the lateral branches; leaf sheaths with fused fimbriae creating a

"ruffled" margin; inflorescence branches branched **58. *Actinocladum***

4. Foliage leaves not dimorphic, leaf sheaths with fimbriae not fused, the margin not appearing ruffled; inflorescence generally unbranched 5

5 (4). Blades of culm leaves reflexed, the base narrower than the distal portion of the sheath; culms often with mottled and/or scabrous internodes **68. *Merostachys***

5. Blades of culm leaves erect, the base about the same width as the distal portion of the sheath; culm internodes generally not mottled or scabrous **70. *Rhipidocladum***

6 (2). Branches as many as 80 per node, in 2 or 3 rows, appearing verticillate **35. *Drepanostachyum***

6. Branches fewer than 80 per node, in 1 row or cluster, not verticillate 7

7 (6). First internode long, the upper ones very short and the leaves densely overlapping; branches only one per node, but appearing subequal because of very short internodes; plants of the Guyana Highlands **69. *Myriocladus***

7. First and second internodes roughly similar lengths, the upper ones not noticeably short with the leaves overlapping; branches more than one from the same node; plants not of the Guyana Highlands 8

8 (7). Mid-culm branches from multiple buds 9

8. Mid-culm branches from a single bud 14

9 (8). Culm leaf blade less than 1 cm long, narrow to acicular or nearly absent **34. *Chimonobambusa***

9. Culm leaf blade more than 1 cm long, flat 10

10. Buds at a single node obviously dimorphic or trimorphic; culm leaves lacking auricles or fimbriae **55. *Chusquea* subg. *Rettbergia*, Euchusquea clade**

10. Buds not clearly dimorphic or trimorphic, although sometimes varying somewhat in size; culm leaves generally with auricles and/or fimbriae 11

11 (10). Culms climbing or scrambling; flowers in pseudospikelets **91. *Holttumochloa***

11. Culms erect; flowers in conventional spikelets 12

12 (11). Shoots fragrant; plants of Asia **26. *Chimonocalamus***

12. Shoots not fragrant; plants of South America 13

13 (12). Plants <2 m tall; foliage leaves with white fimbriae **67. *Filgueirasia***

13. Plants over 2 m tall; foliage leaves with green or brown fimbriae **71. *Apoclada***

14 (8). Each lateral branch bearing only 1 or 2 leaves 15

14. Each lateral branch bearing more than 2 leaves 16

15 (14). Branches 4 or 5 per node at mid-culm; sheaths of foliage leaves contiguous with branches; terminal leaf lacking a sheath **30. *Shibataea***

15. Branches 7 to 12 per node at mid-culm; sheaths of foliage leaves not contiguous with branches; terminal leaf with a sheath **28. *Gelidocalamus***

16 (14). Culm nodes surrounded by a dense ring of spreading hairs **92. *Kinabaluchloa***

16. Culm nodes glabrous, or hairs not spreading 17

17 (16). Culm leaves with sheaths with a dark girdle between the node and the base of the sheath **93. *Dinochloa***

17. Culm leaves with sheaths lacking a dark girdle between the node and the base of the sheath 18

18 (17). Plants of the New World 19

18. Plants of the Old World 23

19 (18). Culms scandent, vine-like; pseudopetioles strongly reflexed; foliage leaves with one wide marginal sector appearing bright green, with no or reduced wax 20

19. Culms erect; pseudopetioles not reflexed; foliage leaves without a bright green marginal sector 21

20 (19). Flowers spindle-shaped, the lemmas sharp-pointed **62. *Atractantha***

20. Flowers not spindle-shaped, the lemmas blunt or acute but not sharp-pointed **61. *Athroostachys***

21 (19). Plants of the southeastern U.S.; rhizomes leptomorph **45. *Arundinaria***

21. Plants of Central or South America; rhizomes pachymorph 22

22 (21). Plants of Mexico, Central America and Colombia; rhizomes with a long neck **75. *Otatea***

22. Plants of southeastern Brazil; rhizomes with a short neck **57. *Cambajuva***

23 (18). Rhizomes leptomorph 24

23. Rhizomes pachymorph 31

24 (23). Flowers borne in conventional spikelets; spikelets generally pedicellate; inflorescence with bracts reduced or lacking 25

24. Flowers borne in pseudospikelets; pseudospikelets generally sessile; inflorescence with well-developed bracts 30

25 (24). Culm internodes flattened above the nodes **39. *Oligostachyum***

25. Culm internodes more or less terete, or slightly grooved above the nodes 26

26 (25). Branches 1–3 per node 27

26. Branches 3–7 per node 28

27 (26). Culm branches profusely branched; culm buds always open at front **42. *Sarocalamus***

27. Culm branches generally with few or no secondary branches; culm buds initially closed at front **29. *Pseudosasa***

28 (26). Stamens 6 **31. *Acidosasa***

28. Stamens 3 29

29 (28). Culm leaves with the sheath persistent, the blade reflexed **41. *Pleioblastus***

29. Culm leaves with the sheath deciduous, the blade erect **33. *Bashania***

30 (24). Stamens 6 **46. *Indosasa***

30. Stamens 3 **43. *Sinobambusa***

31 (23). Rhizomes with elongated necks 32

31. Rhizomes with short necks 35

32 (31). Inflorescence branches not subtended by bracts **44. *Yushania***

32. Inflorescence branches subtended by bracts 33

33 (32). Pericarp crustaceous **108. *Pseudostachyum***

33. Pericarp fleshy 34

34 (33). Inflorescence a dense head **110. *Stapletonia***

34. Inflorescence elongate, branching **105. *Melocanna***

35 (31). Style elongate, hollow 36

35. Style absent or short, or not hollow 39

36 (35). Palea with 2 awns; stamen filaments fused **106. *Neohouzeaua***
36. Palea unawned; stamen filaments free 37
37 (36). Culm leaf sheaths flat; blade of culm leaf rolled; pseudospikelets with several flowers **111. *Teinostachyum***
37. Culm leaf sheaths with 1 or 2 strong undulations toward the apex; blade of culm leaf flat; pseudospikelets with one fertile flower 38
38 (37). Glumes absent **109. *Schizostachyum***
38. Glumes 2 or 3, awned **103. *Cephalostachyum***
39 (35). Flowers borne in conventional spikelets 40
39. Flowers borne in pseudospikelets 45
40 (39). Spikelets with 2 to 8 fertile flowers; glumes 2 or 3 41
40. Spikelets with 1 fertile flower; glumes 2, 5 or 6 43
41 (40). Mid-culm branches 5 to 7 per node 42
41. Mid-culm branches many per node **90. *Vietnamosasa*** (keyed here provisionally assuming that the branch complement arises from a single bud)
42. (41). Culm leaves with a narrow girdle, ca. 1 mm wide; inflorescence one-sided, with bracts at all nodes **36. *Fargesia***
42. Culm leaves lacking a girdle; inflorescence open, lacking bracts **54. *Kuruna***
43 (40). Glumes 5 or 6 **81. Not "*Nastus*"** (southeast Asian material)
43. Glumes 2 44
44 (43). Culm leaves with sheaths persistent; plants of Africa **24. *Bergbambos***
44. Culm leaves with sheaths deciduous; plants of Thailand **89. *Temochloa***
45 (39). Stamens with filaments fused **87. *Pseudoxytenanthera***
45. Stamens with filaments free 46
46 (45). Flowers with palea longer than the lemma **86. *Pseudobambusa***
46. Flowers with palea shorter than the lemma 47
47 (46). Rachilla nodes not disarticulating **98. *Dendrocalamus***
47 Rachilla nodes disarticulating **97. *Bambusa***

Group III – Branch complement with one branch only or one dominant branch and two to many smaller laterals

1. Culm leaves not differentiated from foliage leaves; leaf sheaths densely overlapping **69. *Myriocladus***
1. Culm leaves morphologically distinct from foliage leaves; leaf sheaths generally not overlapping 2
2 (1). Mid-culm branches no more than one per node 3
2. Mid-culm branches more than one per node 22
3 (2). Culm internodes flattened or grooved **29. *Pseudosasa***
3. Culm internodes terete 4
4 (3). Plants of the New World 5
4. Plants of the Old World 10
5 (4). Foliage leaves with pseudopetioles reflexed, the blades thus appearing to droop; fimbriae curly **63. *Aulonemia***
5. Foliage leaves with blades erect; fimbriae, if present, not curly 6
6 (5). Culms with a band of white hairs below the nodes; flowers in pseudospikelets **73. *Guadua*** (non-thorny forms)
6. Culms lacking a band of white hairs below the nodes; flowers in conventional spikelets 7
7 (6). Culm leaves with sheaths persistent, indurate; lemma and palea fused to each other at the base **74. *Olmeca***
7. Culm leaves with sheaths deciduous (or if persistent not indurate); lemma and palea not fused 8
8 (7) Spikelets with 4 glumes and one flower **55. *Chusquea*** (few specimens of this genus will key here as most have multiple buds and branches)
8. Spikelets with 2 glumes and several flowers 9
9 (8). Foliage leaves with green or brown fimbriae, the blades with prominent tessellate venation; plants of southeastern Brazil **57. *Cambajuva***
9. Foliage leaves with white fimbriae, the blades without prominent tessellate venation; plants of cerrado, Brazil **67. *Filgueirasia*** (few specimens of this genus will key here as most have multiple buds and branches)
10 (4). Rhizomes leptomorph 11
10. Rhizomes pachymorph 14
11 (10). Plants less than 2 m tall 12
11. Plants generally over 2 m tall 13
12 (11). Culms and axis of inflorescence covered with white powdery wax; culm leaf sheaths longer than the internodes **48. *Sasamorpha***
12. Culm and axis lacking white powdery wax; culm leaf sheaths shorter than the internodes **47. *Sasa***
13 (11). Internodes with a ring of yellow-brown hairs below the node; fruit dry **38. "not *Indocalamus*"**
13. Internodes with a ring of white hairs below the node; fruit fleshy **27. *Ferrocalamus***
14 (10). Culms erect to arching 15
14. Culms scrambling 18
15 (14). Rhizome necks elongate, the culms solitary **44. *Yushania***
15. Rhizome necks short, the culms clumped 16
16 (15). Flowers in conventional spikelets; glumes 2; plants of New Caledonia **79. *Greslania***
16. Flowers in pseudospikelets; glumes several; plants of temperate to tropical Asia 17
17 (16). Inflorescence a cluster of pseudospikelets; pseudospikelets with 3 or 4 flowers **86. *Pseudobambusa***
17. Inflorescence elongate, branched; pseudospikelets with 1 flower **106. *Neohouzeaua***
18 (14). Flowers in conventional spikelets **51. *Gaoligongshania***
18. Flowers in pseudospikelets 19
19 (18). Culm nodes with a shelf-like extension; culm leaf sheaths with horn-like auricles **78. *Fimbribambusa***
19. Culm nodes lacking a shelf-like extension; culm leaf sheath with auricles not horn-like, or lacking 20
20. (19). Pseudospikelets with one flower **106. *Neohouzeaua***
20. Pseudospikelets with more than one flower 21
21 (20). Base of culm sheath smooth; plants of southern China and Vietnam **76. *Bonia***

21. Base of culm sheath wrinkled; plants of northern Australia **94. *Mullerochloa***
22 (2). Culms with multiple buds, usually dimorphic, per node **55. *Chusquea***
22. Culms with a single bud per node 23
23 (22). Culm leaves with a corky crest at the junction of the sheath and the girdle 24
23. Culm leaves lacking a corky crest at the junction of the sheath and the girdle 25
24 (23). Corky crest not bumpy; culms delicate; spikelets conventional, with 2 or 3 glumes and few to many flowers, the flowers not sharp-pointed **64. *Colanthelia***
24. Corky crest bumpy, tumor-like; pseudospikelets with one glume, one prophyll, and one flower, the flower slender, indurate, sharp-pointed, needle-like **62. *Atractantha***
25 (23). Blade of culm leaf a tiny acicular extension of the sheath, or short and awl-like 26
25. Blade of culm leaf broad and flat, not acicular, generally not an extension of the sheath 28
26 (25). Branches 2 or 3 per node; plants of Africa **83. *Oreobambos***
26. Branches 5 or more per node; plants of Asia 27
27. Secondary branches not pendulous **82. *Neomicrocalamus***
27. Secondary branches long and pendulous **49. *Thamnocalamus***
28 (25). Culms generally broadly spaced, rhizomes leptomorph, or pachymorph with elongate necks 29
28. Culms clumped, rhizomes pachymorph with short necks 35
29 (28). Foliage leaf blades with a strongly excentric midrib; plants of the New World **72. *Eremocaulon***
29. Foliage leaf blades with midrib central or only slightly off center; plants of the Old World 30
30 (29). Inflorescence branches subtended by well-developed bracts and prophylls; stamens 6 **46. *Indosasa***
30. Inflorescence branches with bracts and prophylls reduced or absent; stamens 3 31
31 (30). Culm internodes glabrous below the nodes 32
31. Culm internodes with a ring of hairs below the nodes 34
32 (31). Rhizomes pachymorph but with long necks **44. *Yushania***
32. Rhizomes leptomorph 33
33 (32). Ridge prominent above node; branch buds on a promontory; secondary branches profuse; culm buds open at the front **42. *Sarocalamus***
33. Ridge above node lacking; promontory lacking; secondary branches lacking; culm buds initially closed at the front **29. *Pseudosasa***
34 (31). Culm internodes with yellow brown hairs below the nodes; fruit dry **52. *Indocalamus***
34. Culm internodes with white hairs below the nodes; fruit fleshy **27. *Ferrocalamus***
35 (28). Midrib with a single vascular bundle; plants of the New World 36
35. Midrib with multiple vascular bundles; plants generally of the Old World 42
36 (35). Plants less than 2 m tall, erect 37
36. Plants tall, erect or scandent 38
37 (36). Foliage leaves with green or brown fimbriae, the blades with prominent tessellate venation; plants of southeastern Brazil **57. *Cambajuva***
37. Foliage leaves with white fimbriae, the blades without prominent tessellate venation; plants of cerrado, Brazil **67. *Filgueirasia***
38 (36). Flowers spindle-shaped, the lemmas sharp-pointed **62. *Atractantha***
38. Flowers not spindle-shaped, the lemmas blunt or acute but not sharp pointed 39
39 (38). Culm leaf blades erect 40
39. Culm leaf blades divergent or reflexed 41
40 (39). Culm leaves lacking a dark girdle at the base; inflorescence with conventional spikelets, the axis often zigzag; fruit dry; plants of Central and South America, but not extending to Bahia **60. *Arthrostylidium***
40. Culm leaves with a dark girdle at the base; inflorescence with pseudospikelets, the axis straight; fruit fleshy; plants of Bahia, Brazil **59. *Alvimia***
41 (39). Culm and foliage leaf sheaths with long straight fimbriae over 2 cm long; stamens 6 **66. *Elytrostachys***
41. Culm and foliage leaf sheaths with shorter, curly fimbriae; stamens 3 **63. *Aulonemia***
42 (35). Culm leaf sheaths with auricles 43
42. Culm leaf sheaths without auricles 73
43 (42). Auricles low, rim-like 44
43. Auricles various, but not rim-like 46
44 (43). Culms with a thick band of silvery white hairs below each node; plants of the Malay Peninsula **100. *Maclurochloa***
44. Culms lacking silvery white hairs; plants of various parts of Asia 45
45 (44). Flowers in conventional spikelets **120. *Racemobambos***
45. Flowers in pseudospikelets **99. *Gigantochloa***
46 (43). Culms erect 47
46. Culms scrambling 59
47 (46). Foliage leaves without fimbriae 48
47. Foliage leaves with fimbriae 53
48 (47). Stamens numerous; fruit fleshy; style elongate, hollow **107. *Ochlandra***
48. Stamens 6; fruit generally dry; style short, not hollow 49
49 (48). Flowers in conventional spikelets **81. Not "*Nastus*"** (southeast Asian material)
49. Flowers in pseudospikelets 50
50 (49). Palea divided at least 1/3 of its length **102. *Thyrsostachys***
50. Palea undivided, or merely notched 51
51 (50). Clusters of pseudospikelets in the inflorescence subtended by a two-keeled prophyll; rachilla generally disarticulating between the flowers **97. *Bambusa***
51. Clusters of pseudospikelets in the inflorescence subtended by a one-keeled prophyll; rachilla generally not disarticulating 52
52 (51). Stamens with filaments free **98. *Dendrocalamus***
52. Stamens with filaments fused **99. *Gigantochloa***
53 (47). Style elongate, hollow; ovary glabrous 54
53. Style short, not hollow; ovary generally pubescent 56
54 (53). Inflorescence densely capitate **104. *Davidsea***

54. Inflorescence elongate 55
55 (54). Pseudospikelets with one flower; palea awned **106. *Neohouzeaua***
55. Pseudospikelets with several flowers; palea lacking awns **111. *Teinostachyum***
56 (53). Stamens with filaments fused **87. *Pseudoxytenanthera***
56. Stamens with filaments free 57
57 (56). Pseudospikelets with 1 flower **77. *Cyrtochloa***
57. Pseudospikelets with 2 or more fertile flowers 58
58 (57). Culm leaves lacking a wrinkled zone at the base **97. *Bambusa***
58. Culm leaves with a wrinkled zone at the base **95. *Neololeba***
59 (46). Nodes crested with a narrow shelf-like projection 60
59. Nodes various, but not with a shelf-like projection 61
60 (59). Culm leaf with blade spreading, the sheath with stiff horn-like auricles; flowers in pseudospikelets, these in clusters of more than 25 per inflorescence node **78. *Fimbribambusa***
60. Culm leaf with blade erect, the sheath with rounded auricles; flowers in conventional spikelets, these in clusters of 2 or 3 per inflorescence node **88. *Temburongia***
61 (59). Culm leaf sheaths with a wrinkled base 62
61. Culm leaf sheaths with the base smooth 65
62 (61). Foliage leaf base strongly asymmetrical, with one side broad and rounded and the other narrow and straight; spikelets with one flower **77. *Cyrtochloa***
62. Foliage leaf base symmetrical; spikelets with more than one flower 63
63 (62). Culm leaf blades reflexed; spikelets with 2 flowers **80. *Melocalamus***
63. Culm leaf blades erect; spikelets with more than 2 flowers 64
64 (63). Leaf auricles leathery, crescent shaped; palea apex without hooked projections; lodicules 0; plants of Mindanao, Sulawesi, Maluku, New Guinea, Solomon Islands, Queensland **95. *Neololeba***
64. Leaf auricles papery, ovate to rounded; palea apex with two hooked projections; lodicules 3; plants of Peninsular Malaysia, Singapore **101. *Soejatmia***
65 (61). Inflorescence with conventional spikelets 66
65. Inflorescence with pseudospikelets 68
66 (65). Spikelets with one flower **81. Not "*Nastus*"** (southeast Asian material)
66. Spikelets with more than one flower 67
67 (66). Spikelets on long pedicels **32. *Ampelocalamus***
67. Spikelets sessile **120. *Racemobambos***
68 (65). Inflorescence a single dense cluster of pseudospikelets 69
68. Inflorescence elongate, with clusters of pseudospikelets at the nodes 70
69 (68). Blade of culm leaf reflexed **104. *Davidsea***
69. Blade of culm leaf erect **85. *Parabambusa***
70 (68). Style elongate; ovary glabrous 71
70. Style short; ovary often pubescent at least at the apex **87. *Pseudoxytenanthera***
71 (70). Palea longer than the lemma, the keels winged; style short, not hollow **96. *Sphaerobambos***
71. Palea shorter than the lemma, the keels without wings; style elongate, hollow 72
72 (71). Stamens with filaments fused; pericarp dry **106. *Neohouzeaua***
72. Stamens with filaments free; pericarp fleshy **111. *Teinostachyum***
73 (42). Inflorescence branches not subtended by bracts and prophylls; spikelets pedicellate 74
73. Inflorescence branches subtended by bracts and prophylls; spikelets sessile 75
74 (73). Rhizomes with elongated necks; spikelets with 2 or more fertile flowers **44. *Yushania***
74. Rhizomes with short necks; spikelets with 1 fertile flower **37. *Himalayacalamus***
75 (73). Palea divided at least 1/3 of its length **102. *Thyrsostachys***
75. Palea undivided, or merely notched 76
76 (75). Clusters of pseudospikelets in the inflorescence subtended by a two-keeled prophyll; rachilla generally disarticulating between the flowers 77
76. Clusters of pseudospikelets in the inflorescence subtended by a one-keeled prophyll; rachilla generally not disarticulating 78
77 (76). Culms erect; basal part of culm internode not swollen **97. *Bambusa***
77. Culms scrambling; basal part of culm internode swollen **94. *Mullerochloa***
78 (76). Stamens with filaments free **98. *Dendrocalamus***
78. Stamens with filaments fused 79
79 (78). Inflorescence with clusters of pseudospikelets at the nodes; culms often with yellow stripes; plants of India and southeast Asia **99. *Gigantochloa***
79. Inflorescence a capitate cluster of pseudospikelets at the end of each lateral branch; culms not yellow striped; plants of Africa **84. *Oxytenanthera***

TRIBES AND GENERA OF BAMBUSOIDEAE

IV. TRIBE ARUNDINARIEAE Asch. & Graebn. (1902)

Rhizomes generally leptomorph. Culms woody, generally hollow. *Branch development basipetal.* Foliage leaves with an outer ligule. Spikelets laterally compressed. $2n = 48$. Midrib complex.

Twenty-six genera, 533 species.

This tribe includes all the temperate woody bamboos, and is monophyletic in all molecular phylogenies (Bamboo Phylogeny Group 2012; Burke et al. 2012; Ma et al. 2014; Zhang et al.

2012, Triplett et al. 2014). Leptomorph rhizomes may be a synapomorphy for this clade, but if so then the character has apparently reversed in some taxa. Culms of members of Arundinarieae reach their full height without initiating branches and produce only culm leaves; later the culms lignify and produce branches and foliage leaves, beginning with the top of the plant and proceeding basipetally (Bamboo Phylogeny Group 2012).

Rather than recognize subtribes in Arundinarieae, I have followed the work of Triplett and Clark (2010), Zeng et al. (2010), and Zhang et al. (2012) and simply listed major clades. Even with large amounts of molecular data, relationships within the tribe are poorly resolved (Burke et al. 2012; Ma et al. 2014). Most genera are para- or polyphyletic, and generic limits are thus highly likely to change in the near future. Like rhizome morphology, characteristics of inflorescence structure are highly homoplasious on the molecular phylogeny as is leaf epidermal micromorphology (Zhang et al. 2014). Because these were the central characters on which bamboo classification was based, it is not surprising that generic limits are in extensive need of revision.

Another confounding factor is hybridization. While the irregular flowering of the bamboos was thought to minimize opportunities for hybridization, crossing has now been clearly documented; it appears that there are few reproductive barriers in this group other than differences in phenology (Triplett and Clark 2010; Zhang et al. 2012). The hybrid taxa tend to combine the morphological characteristics of the parents and have thus often been named as new genera. Hybrid genera include *Hibanobambusa*, a hybrid between *Sasa* and *Phyllostachys* (Triplett and Clark 2010; Triplett et al. 2014); *Pseudosasa japonica*, a hybrid between *Sasamorpha* and *Pleioblastus* (Triplett and Clark 2010; Triplett et al. 2014); *Sasaella*, a hybrid between *Sasa* and *Pleioblastus* (Triplett and Clark 2010, and references therein; Triplett et al. 2014); *Semiarundinaria* (including *Brachystachyum*), a hybrid between *Phyllostachys* and *Pleioblastus* (Triplett et al. 2010, 2014). These are not described here. *Pseudosasa japonica* is the type of *P.* subg. *Pseudosasa*, which also contains *P. owatarii*; nothing is known of the origin of the latter. *Pseudosasa* subg. *Sinicae* may be non-hybrid and is described below in Clade IV, the *Shibataea* clade.

Clade I. Bergbambos

24. ***Bergbambos*** Stapleton

Bergbambos Stapleton, Phytokeys 25: 99 (2013); Soderstrom and Ellis (1982), morphol., anat.

Culms erect or slightly arching. Rhizomes pachymorph, with short necks, with air canals. Branches subequal, 5 to 7, in a horizontal line at each node, from a single bud; nodes not prominent. Culm leaves with persistent sheaths, without auricles, the blade erect to reflexed. Foliage leaf blade with fimbriae. Inflorescence unbranched. Spikelets with short pedicels, with 1 fertile flower. Glumes 2. Lodicules 3. Stamens 3. Stigmas 3.

One sp., *B. tessellata* (Nees) Stapleton, mountains of South Africa.

This species is placed in an uncertain position in molecular analyses (Triplett and Clark 2010; Zeng et al. 2010; Zhang et al. 2012), and may be in a clade sister to all other Arundinarieae. It was formerly placed in *Thamnocalamus* but is clearly misclassified there. *Bergbambos* may be sister to the Sri Lankan *Kuruna* (Attigala et al. 2014), or to *Indocalamus wilsonii* (Zeng et al. 2010).

Clade II. Oldeania

25. ***Oldeania*** Stapleton

Oldeania Stapleton, Phytokeys 25: 100 (2013).

Culms 15 (−20) m tall, erect to nodding. Rhizomes pachymorph, with long necks. *Aerial roots common, particularly at lower nodes, spiny.* Branches 3 to 5, subequal, the culm slightly grooved above them, with a well-developed ridge above the node. Culm leaves with sheaths deciduous. Inflorescence branches branched, the branches subtended by bracts, remnants of sheaths or hairs. Spikelets with several flowers. Empty glumes 2, fertile glumes 4–8. Lodicules 3. Stamens 3. Stigmas 2.

Species one, *O. alpina* (K. Schum.) Stapleton, mountains of Africa.

O. alpina was originally described in *Yushania* but is clearly unrelated to the Asian members of that genus (Zeng et al. 2010). *Yushania ambositrensis* and possibly all Madagascan *Arundinaria* species may also belong to *Oldeania*, based on phylogenetic results from Triplett and Clark (2010).

Clade III. Chimonocalamus

26. ***Chimonocalamus*** J. R. Xue & T. P. Yi Fig. 20

Chimonocalamus J. R. Xue & T. P. Yi, Acta Bot. Yunnan 1: 75 (1979).

Culms erect, the *shoots fragrant*. Rhizomes short, pachymorph. *Aerial roots common, particularly at lower nodes, spiny.* Branches 3 to 5, in a horizontal line, subequal or with 3 dominant branches. Culm leaves with deciduous sheaths, auriculate or not, with fimbriae, the blade erect or reflexed. Foliage leaves without auricles, with or without fimbriae, without prominent cross veins. Inflorescence branches branched, without bracts. Spikelets with 4 to 12 flowers. Glumes 2. Lodicules 3. Stamens 3. Stigmas 2.

Eleven spp., southwest China, eastern Himalayas, Burma.

Chimonocalamus is generally monophyletic in molecular analyses (Triplett and Clark 2010; Yang et al. 2013; Zhang et al. 2012), but some species of *Chimonocalamus* have acquired chloroplasts from species in the *Phyllostachys* clade, indicating intergeneric hybridization (Yang et al. 2013). This probably explains the apparent polyphyly of the genus in the phylogeny of Zeng et al. (2010).

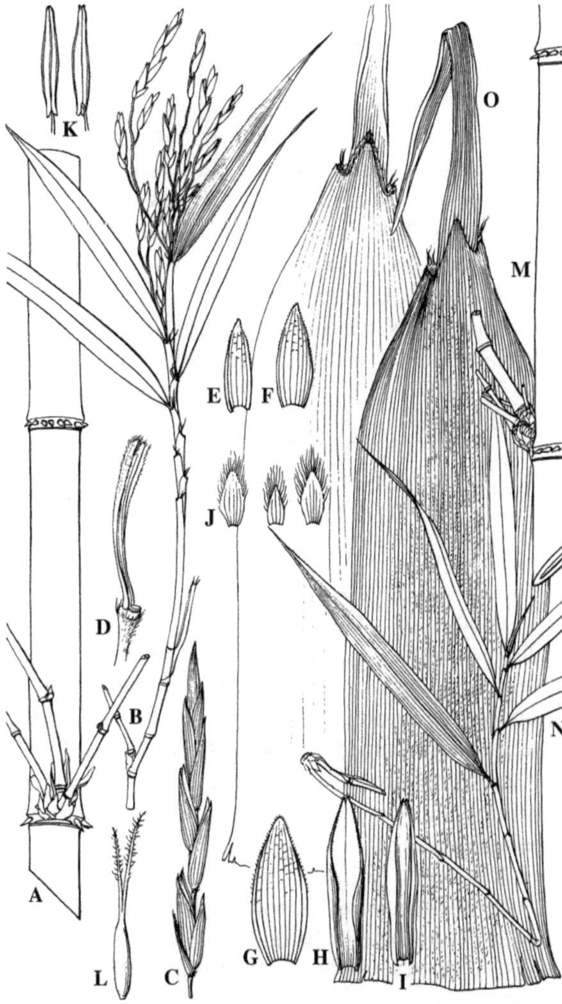

Fig. 20. Bambusoideae-Arundinarieae. **A–L** *Chimonocalamus delicatus.* **A** Portion of culm showing branching pattern and spiny adventitious roots. **B** Flowering branchlet. **C** Spikelet. **D** Portion of rachilla internode. **E** Lower glume. **F** Upper glume. **G** Lemma. **H** Abaxial view of palea. **I** Adaxial view of palea. **J** Lodicules. **K** Abaxial and lateral views of stamen. **L** Pistil. **M–O** *C. pallens.* **M** Portion of culm. **N** Leafy branchlet. **O** Abaxial and adaxial views of culm sheath. (From Wu et al. 2007, p. 133, with permission from the Missouri Botanical Garden Press, St. Louis, and Science Press, Beijing)

Clade IV. Shibataea clade

Rhizomes leptomorph. Lodicules 3.

Five genera, 37+ species.

The *Shibataea* clade is monophyletic in phylogenies based on plastid genes (Triplett and Clark 2010; Zeng et al. 2010) and could represent a recircumscribed Shibataeinae. The traditional Shibataeinae are polyphyletic. In addition to the genera listed below, three species of "*Indocalamus*" fall in this group (Zeng et al. 2010). The clade has no obvious morphological synapomorphy, and is not seen the nuclear gene (GBSSI) phylogeny of Zhang et al. (2012). However, the latter phylogeny is poorly supported and could be simply uninformative about the existence of the clade.

27. ***Ferrocalamus*** J. R. Xue & Keng f.

Ferrocalamus J. R. Xue & Keng f., J. Bamboo Res. 1: 3 (1982).

Culms erect or scrambling, plants shrubby. Internodes terete, *with white hairs below the nodes.*

Branches 1 per node, or 3 to 5 per node higher up on the plant. Culm leaves with persistent sheaths, with or without auricles and fimbriae, the blades reflexed. Foliage leaves with sheaths lacking auricles, with fimbriae, the blades with distinct transverse veins. Inflorescence branches branched, flexuous. Spikelets with 3 to 10 flowers. Glumes 2. Stamens 3. Style 1, stigmas 2. *Caryopsis with fleshy pericarp*. $2n = 48$.

Two spp., China.

This genus is monophyletic in the analyses of Triplett and Clark (2010) and Zeng et al. (2010).

28. **"Gelidocalamus"** T. H. Wen

"*Gelidocalamus*" T. H. Wen, J. Bamboo Res. 1: 21 (1982).

Culms erect, plants shrubby, caespitose. Branches 7 to 12 (to 20) per node, without additional orders of branching. Culm leaves with persistent sheaths, with or without auricles, the blades erect. *Foliage leaves generally 1 per branch*, apparently without auricles or fimbriae, the blades with distinct transverse veins. Inflorescence branches branched. Spikelets with 3 to 5 flowers. Glumes 2. Stamens 3. Stigmas 2. Fruit globose, beaked.

Nine spp., China.

"*Gelidocalamus*" is polyphyletic in molecular analyses (Zeng et al. 2010; Zhang et al. 2012), with some species perhaps related parts of the polyphyletic "*Indocalamus*".

29. **"Pseudosasa"** Makino ex Nakai

"*Pseudosasa*" Makino ex Nakai, J. Arnold Arbor. 6: 150 (1925), p.p.

Culms erect or nodding; internodes terete or grooved. Branches 1 to 3 per node. Culm leaves with sheaths deciduous to persistent, auricles and fimbriae present or absent, the blade erect to reflexed. Foliage leaves with sheaths persistent, auricles and fimbriae present or absent, the blades with distinct transverse veins. Inflorescence unbranched or branched, with the spikelets pedicellate on a central axis. Spikelets with 3 to 30 flowers. Glumes 2. Stamens 3. Stigmas 3.

Nineteen spp., China, Japan, Taiwan, Korea.

Triplett and Clark (2010) indicate that the Chinese species of *Pseudosasa* and *Sasa* form a monophyletic group that will probably be recognized as a new genus. "*Pseudosasa*" as conventionally circumscribed is polyphyletic (Zeng et al. 2010).

30. ***Shibataea*** Makino ex Nakai

Shibataea Makino ex Nakai, J. Jap. Bot. 9: 83 (1933).

Culms erect, the plants less than 1 m tall. Culms flattened above branches. Branches 3 to 5 per node, more or less equal, without additional orders of branching. Culm leaves with deciduous sheaths, lacking auricles, the blade slender, reflexed. *Foliage leaves with sheaths narrow, similar to petioles, contiguous with the branchlets, the ligule long and involute*, the blade with distinct transverse veins; *terminal leaf of each branch lacking a sheath*. Inflorescence a fascicle of pseudospikelets, each with a spathe-like prophyll and up to 3 bracts, with or without buds in their axils. Pseudospikelets with 2 to 7 flowers, only the lower 1 or 2 fertile. Stamens 3. Style 1, stigmas 3. $2n = 48$.

Seven spp., China and Japan.

This genus is monophyletic in molecular phylogenies (Triplett and Clark 2010; Zeng et al. 2010).

CLADE V. PHYLLOSTACHYS CLADE, sensu Zeng et al. (2010)

Lodicules 3. Stamens generally 3. Style 1.

Fourteen genera, 392 species.

In addition to the genera listed below, several species of "*Pseudosasa*" and "*Sasa*" fall in this clade. While the clade is retrieved in phylogenies based on plastid genes, it is not seen the nuclear gene (GBSSI) phylogeny of Zhang et al. (2012); however, the latter phylogeny is poorly supported and could be simply uninformative about the existence of the clade.

31. **"Acidosasa"** C. D. Chu & C. S. Chao ex Keng f.

"*Acidosasa*" C. D. Chu & C. S. Chao ex Keng f., J. Bamboo Res. 1: 31 (1982).
Metasasa W. T. Lin, Acta Phytotax. Sin. 26: 145 (1988).

Culms erect, shrubby to arborescent. Rhizomes leptomorph, elongate. Branches 3, or more at higher nodes. Culm leaves with deciduous sheaths, with or without auricles, the auricles bearing fimbriae, the blade erect or reflexed. Foliage leaves without auricles or fimbriae, the blades with distinct transverse veins. Inflorescence unbranched or branched, without bracts, the spikelets pedicellate on a central axis. Spikelets elongate, with several to many flowers. Glumes 2 to 4. *Stamens 6.* Stigmas 3.

Eleven spp., southern China, Vietnam.

This genus is polyphyletic in the analyses of Zeng et al. (2010). *A. purpurea* is supported as being sister to *Arundinaria gigantea* in the phylogenomic study of Burke et al. (2012), but other species of *Acidosasa* were not included.

32. "***Ampelocalamus***" S. L. Chen, T. H. Wen & G. Y. Sheng

"*Ampelocalamus*" S. L. Chen, T. H. Wen & G. Y. Sheng, Acta Phytotax. Sin. 19: 332 (1981); tax.: Stapleton (1994b).

Culms clumped, pendulous above. Rhizomes pachymorph. Branches several per node, 1 often dominant. Culm leaves with deciduous sheaths, auricles present, fimbriae generally present, the blade reflexed. Foliage leaves with sheaths with auricles, with prominent spreading fimbriae, the blades without obvious cross veins. Inflorescence pendulous, unbranched, or with sparse branched branches, with few or no bracts, the spikelets on long pedicels borne in clusters at the nodes. Spikelets with 2 to 7 fertile flowers plus 1 reduced flower distally. Glumes 2. Stigmas 2 or 3. Ovary glabrous, without an appendage.

Thirteen spp., central Himalayas to southern China.

This genus is polyphyletic in the analyses of Zeng et al. (2010) and Yang et al. (2013), although the latter study shows that most species form a clade, with only "*A*". *calcareus* originating independently; see Clade XI below.

33. "***Bashania***" Keng f. & T.P. Yi

"*Bashania*" Keng f. & T.P. Yi, J. Bamboo Res. 1: 171 (1982).

Culms erect. Rhizomes leptomorph. Branches 3 to 7 per node, from a single bud. Culm leaves deciduous, the sheaths without auricles, with fimbriae, the blade erect. Foliage leaves auriculate, without fimbriae, the blades tessellate with obvious cross veins. Inflorescence unbranched or with sparsely branched branches, the spikelets pedicellate on a central axis, subtended by reduced sheaths or rings of hairs. Spikelets with two glumes, the glumes mucronate, with 4 to 14 flowers. *Stamens sometimes up to 5.* Stigmas 2 to 3. Ovary glabrous.

Two spp., China.

"*Bashania*" is sometimes considered a subgenus of *Arundinaria*, but the two genera are clearly not related. "*Bashania*" is polyphyletic in the analyses of Zeng et al. (2010).

34. ***Chimonobambusa*** Makino

Chimonobambusa Makino, Bot. Mag. (Tokyo) 28: 153 (1914); tax.: Stapleton (1994b).
Oreocalamus Keng, Sunyatsenia 4: 146 (1940).
Qiongzhuea J. R. Xue & T. P. Yi, Acta Bot. Yunnan. 2: 91–92 (1980).
Menstruocalamus T. P. Yi, J. Bamboo Res. 11: 38 (1992).

Culms erect, plants shrubby. Rhizomes leptomorph, elongate. Culm internodes often grooved, *the basal nodes often with prop roots modified to form a ring of spines.* Branches 3 (–7), more or less equal. Culms leaves with deciduous sheaths, auricles tiny or absent, *the blade much reduced (less than 1 cm long), often simply an acicular point.* Foliage leaves with or without auricles and fimbriae. Inflorescence unbranched, or the branches branched, subtended by bracts with buds in their axils. Pseudospikelets with several to many flowers. Glumes 1 to 3, often with a bud in the axil of one. Stigmas 2 or 3. Fruit nut-like.

Thirty-seven spp., China and Japan.

The analyses of Triplett and Clark (2010) provide no evidence to accept or reject the monophyly of this genus. Zhang et al. (2012) and Peng et al. (2008) present molecular data to support synonymy of *Qiongzhuea* and *Menstruocalamus* under *Chimonobambusa.*

35. "***Drepanostachyum***" Keng f.

"*Drepanostachyum*" Keng f., J. Bamboo Res. 2: 15 (1983); tax., Stapleton (1994b).

Plants shrubby, caespitose. Rhizomes short, pachymorph. *Branches up to 80 per node, in 2 or 3 rows, all more or less equal.* Culm leaves with sheaths deciduous, *adaxially scabrous*, without auricles, with or without fimbriae, the blade reflexed. Foliage leaves lacking auricles or fimbriae, without transverse veins. Inflorescence branches branched, the branches subtended by reduced bracts or rings of hairs; basal internodes of the inflorescence short so that branches appear fascicled. Spikelets conventional, with 2 to 6 fertile flowers plus one reduced flower distally. Glumes 2. Stigmas 2. Ovary glabrous.

Ten spp., Himalayan regions of Bhutan, China, India, and Nepal.

"*Drepanostachyum*" is not monophyletic in the chloroplast phylogeny of Triplett and Clark (2010). The nuclear gene phylogeny of Zhang et al. (2012) suggests that it may be derived within "*Himalayacalamus*".

36. "*Fargesia*" Franchet

"*Fargesia*" Franchet, Bull. Mens. Soc. Linn. Paris 2: 1067 (1893).
Sinarundinaria Nakai, J. Jap. Bot. 11: 1 (1935).
Borinda Stapleton, Edinburgh J. Bot. 51(2): 284 (1994).

Culms erect. Rhizomes pachymorph, short. Branches 7 to numerous per node, becoming deflexed, subequal. Culm leaves with sheaths persistent or deciduous, leaving a short rim, with or without auricles, with or without fimbriae, the blade reflexed or erect. Foliage leaves with or without auricles, with or without fimbriae. Inflorescences unbranched, or the branches branched, generally subtended by spathe-like sheaths, the spikelets borne on one side of the axis or branch. Spikelets with several flowers. Glumes generally 2. Lemma apex mucronate to awned. *Style branches sometimes 2*, stigmas 2 or 3.

Ninety spp., east and southeast Asia.

"*Fargesia*" is not monophyletic in the chloroplast phylogenies of Triplett and Clark (2010) or Zeng et al. (2010). The inclusion of *Sinarundinaria* and *Borinda* in *Fargesia* follows the recommendation of the Bamboo Phylogeny Group (2012).

37. "*Himalayacalamus*" Keng f.

"*Himalayacalamus*" Keng f., J. Bamboo Res. 2: 23 (1983); tax., Stapleton (1994b).

Culms nodding or pendulous, plants shrubby to more or less arborescent. Rhizomes pachymorph, short. Branches up to 25 per node, the central one dominant, from a single bud. Culm leaves with sheaths deciduous, leaving a narrow rim, without auricles or fimbriae, the blade reflexed. Foliage leaves with sheaths with or without auricles and fimbriae, with transverse veins obscure or absent. Inflorescence unbranched or branched, the branches sometimes subtended by rings of hairs; basal internodes of the inflorescence short so that branches appear fascicled. Spikelets conventional, with 1 (or 2) fertile flowers proximally, 1 reduced flower distally. Glumes 2. Stigmas 2.

Eight spp., Bhutan, China, India, Nepal.

"*Himalayacalamus*" is not monophyletic in the chloroplast phylogeny of Triplett and Clark (2010). The nuclear gene phylogeny of Zhang et al. (2012) suggests that it may be paraphyletic, with "*Drepanostachyum*" derived from within it, but the tree topology is not well enough supported to defend combining the two.

38. Not "*Indocalamus*" Nakai

Not "*Indocalamus*" Nakai, J. Arnold Arbor. 6: 148 (1925).

Culms nodding, clumped, the plants shrubby, generally less than 2 (3) m tall. Rhizomes leptomorph. Internodes terete, with a dense yellow-brown ring of hairs below the nodes. Branches one per node, about the same thickness as the culm. Culm leaves with persistent sheaths, generally with auricles and fimbriae, the blade generally recurved. Foliage leaves with sheaths without auricles, with fimbriae, the blades with distinct transverse veins. Inflorescence unbranched or branched, the branches subtended by small bracts. Spikelets with 3 to more than 12 flowers. Glumes 2 or 3. Style 1 to nearly absent, stigmas 2 (3). $2n = 48$.

Twenty-one spp., one in Japan, all others in China.

"*Indocalamus*" is polyphyletic (Zeng et al. 2010; Zhang et al. 2012). The former authors included 21 of the 23 named species, which fall in virtually every clade of Arundinarieae. Most members of the "genus" belong here in the *Phyllostachys* clade, but the type (*I. sinicus* (Hance) Nakai) and *I. wilsonii* are in an uncertain position outside other major clades in the tribe. The description above refers to the majority of the species and will need to be emended as the genus is redefined.

39. "***Oligostachyum***" Z. P. Wang & G. H. Ye

"*Oligostachyum*" Z. P. Wang & G. H. Ye, J. Nanjing Univ., Nat. Sci. 1982: 95 (1982).
Clavinodum T. H. Wen, J. Bamboo Res. 3: 23 (1984).

Culms erect, plants shrubby to arborescent. Rhizomes leptomorph, elongate. Branches 3 (to 7) per node, subequal, from a single bud. Culm leaves with sheaths generally deciduous, with or without auricles and fimbriae, the blades erect or reflexed. Foliage leaves with or without auricles and fimbriae, with distinct transverse veins. Inflorescence terminal on lateral branches, with 2 or 3 (–6) spikelets pedicellate on a slender axis. Spikelets with several to many flowers, the proximal ones sterile. Glumes (1) 3 (5), forming a morphological transition to the proximal sterile lemmas. *Stamens sometimes 4 or 5.* Stigmas 3.

Fifteen spp., China.

"*Oligostachyum*" is not monophyletic in the analyses of Zeng et al. (2010).

40. "***Phyllostachys***" Siebold & Zucc.

"*Phyllostachys*" Siebold & Zucc., Abh. Math.-Phys. Cl. Königl. Bayer. Akad. Wiss. 3: 745, pl. 5, f. 3 (1843) nom. conserv.

Culms erect, the plants arborescent or shrubby. Rhizomes leptomorph, elongate. Culms flattened above branches. Branches 2 (to 3) at mid-culm nodes, unequal, with higher-order branches; internode at base of branch complement short, failing to elongate. Culm leaves with sheaths deciduous, with auricles and fimbriae, the blade erect or recurved. Foliage leaves with or without auricles and fimbriae, the blades with distinct transverse veins, generally pilose on the proximal portion of the abaxial side. Inflorescence of unbranched branches, arranged in bracteate fascicles, the most proximal bract often subtending a bud, the next 2 to 6 more distal bracts scale-like and the 2 to 7 distalmost spathe-like. Pseudospikelets with 2 to 6 fertile flowers plus one reduced flower distally. Glumes 0 or 1 (3). Stigmas (1 to) 3. Fruit with an abaxial groove. $2n = 48$.

Fifty-one spp., Asia.

The analyses of Triplett and Clark (2010) provide no evidence to accept or reject the monophyly of "*Phyllostachys*", but it is not monophyletic in the analyses of Zeng et al. (2010).

41. "***Pleioblastus***" Nakai

"*Pleioblastus*" Nakai, section *Amari* S.L. Chen & G.Y. Sheng, pro parte.
Nipponocalamus Nakai, J. Jap. Bot. 18: 350 (1942).
Polyanthus C. H. Hu, J. Bamboo Res. 10(3): 28–30 (1991).

Culms erect, sometimes less than 1 m tall. Rhizomes leptomorph. Internodes flattened or grooved. Branches 3 to 9 per node, subequal, from a single bud, producing extensive secondary branches. Culm leaves with sheaths persistent, with or without auricles, with or without fimbriae, the blade reflexed. Foliage leaves with or without auricles, with or without fimbriae, the ligules notably long. Inflorescence unbranched or with branched branches. Spikelets with 2 glumes. Stamens sometimes up to 5. Stigmas 2 or 3.

Ca. forty spp., Japan, China.

"*Pleioblastus*" in its conventional sense is not monophyletic (Zeng et al. 2010). The Japanese species form a well-supported clade (Zeng et al. 2010; J. Triplett, pers. comm.), but the relationship of these to the various Chinese species is less clear. At least some Chinese species likely belong in the genus, but others will need to be transferred elsewhere. J. Triplett (pers. comm.) notes that this genus includes species with rhizomes that are leptomorph (monopodial) but short, thick and tillering, as well as species with more conventional elongate leptomorph rhizomes. These two types of rhizome correspond to major phylogenetic subdivisions in the genus.

42. ***Sarocalamus*** Stapleton

Sarocalamus Stapleton, Novon 14: 346. (2004).

Culms erect to nodding, the internodes terete. Rhizomes leptomorph. Branches 1 to 3 per node, erect, branching profusely, basal internodes elongate. Culm leaves with sheaths deciduous or persistent, with or without fimbriae and auricles, the blade erect or reflexed. Foliage leaves with or without auricles and fimbriae. Inflorescence unbranched, or the branches branched. Spikelets with 4 to 10 flowers. Glumes 1 or 2. Stigmas 2 or 3.

Three spp., Bhutan, China, NE India, Nepal.

43. "*Sinobambusa*" Makino ex Nakai

"*Sinobambusa*" Makino ex Nakai, J. Arnold Arbor. 6: 152 (1925).

Culms erect to nodding, the plants arborescent or shrubby. Rhizomes leptomorph, elongated. Internodes grooved. Branches 3 (5 to 7), more or less equal. Culm leaves with deciduous sheaths, with auricles and fimbriae. Foliage leaves with or without auricles and fimbriae. Inflorescence bracteate, unbranched, bearing 1 to 3 pseudospikelets, each subtended by a prophyll and 2 or more bracts, the more distal bracts often bearing buds in their axils; buds developing into additional spikelets. Pseudospikelets with as many as 50 flowers. Stigmas 2 or 3.

Ten spp., China, Taiwan, Vietnam, introduced to Japan.

The distinction between "*Sinobambusa*" and "*Indosasa*" is based largely on stamen number, with fewer than 6 (generally 3) in the former versus 6 in the latter. "*Sinobambusa*" is not monophyletic in the analyses of Zeng et al. (2010).

44. *Yushania* Keng f.

Yushania Keng f., Acta Phytotax. Sin. 6: 355 (1957); tax.: Stapleton (1994c).
Burmabambus Keng f., J. Bamboo Res. 1(2): 173 (1982).
Butania Keng f., J. Bamboo Res. 1(2): 42 (1982).
Monospatha W.T. Lin, J. Bamboo Res. 13(4): 1–2 (1994).

Culms erect, plants shrubby; lower nodes sometimes developing root spines. Rhizomes pachymorph, *with elongated necks*. Branch number varying from 1 at most basal nodes to 9 to 11 (to 45) at upper nodes, more or less equal in size, or the central one larger, from a single bud. Culm leaves with sheaths persistent or deciduous, with or without auricles, with or without fimbriae, the blade erect or reflexed. Foliage leaves with sheaths with or without auricles, with or without fimbriae, with transverse veins. Inflorescence branches branched, flexuous, not subtended by bracts. Spikelets with several to many flowers, the distalmost flower reduced. Glumes 2. Stigmas 2.

Seventy-seven spp., Africa, Madagascar, East and southeast Asia.

Yushania is polyphyletic in the analyses of Triplett and Clark (2010), with the species from alpine regions in Africa (*Y. alpina*) and Madagascar (*Y. ambositrensis*) forming a clade distinct from the other species of *Yushania* sampled. However, if the African alpine bamboos (now in *Oldeania*) are excluded then the genus may be monophyletic. This description is based largely on Chinese material. The type species, *Y. niitakayamensis* also falls here, based on the phylogeny of Triplett and Clark (2010). The observation of root spines is from Wong (2004).

Clade VI. Arundinaria clade

Rhizomes leptomorph, elongate.

Five genera, 126 spp.

This clade has no obvious morphological synapomorphy.

Some species formerly known as *Pleioblastus* fall in this clade, but have not yet been placed in a new genus. Numbers of species are based on estimates from Triplett and Clark (2010).

45. *Arundinaria* Michx.

Arundinaria Michx., Fl. Bor.-Amer. 1: 73 (1803); phylog.: Triplett et al. (2010).

Culms erect, caespitose or not. Branches 1 to 3 per node, subequal, from a single bud, the branches often extensively rebranching. One to five internodes at the base of each branch complement failing to elongate. Culm leaves with sheaths persistent or deciduous, with auricles, the blades erect or reflexed. Foliage leaves with auricles and fimbriae, the blades with fine transverse veins. Inflorescence unbranched, or the branches branched, with the spikelets pedicellate. Spikelets with 6 to 12 flowers. Glumes 1 or 2. Stamens 3. Style reduced or absent, stigmas 3. $2n = 48$.

Three spp., southeastern US.

This circumscription of *Arundinaria* follows Triplett et al. (2010). *Arundinaria* subgenera *Bashania* and *Sarocalamus* fall in the *Phyllostachys* clade and are here recognized as distinct genera. Species previously classified as

Arundinaria in Sri Lanka are unrelated to *Arundinaria* s.s. and are now placed in the genus *Kuruna* (Attigala et al. 2014).

46. ***"Indosasa"*** McClure

"Indosasa" McClure, Lingnan Univ. Sci. Bull. 9: 28 (1940).

Culms erect to nodding. Branches 3 per node, more or less equal or the central one larger than the others, the internode grooved above the branches. Culm leaves with sheaths deciduous, with or without auricles and fimbriae, the blade erect or reflexed. Foliage leaves with or without auricles and fimbriae, the blades with distinct transverse veins. Inflorescence unbranched, bracteate, branches with prophylls. Pseudospikelets sessile, with many flowers. Glumes 2. Lodicules 3. Stamens 6. Style 1, stigmas 3. Fruit beaked.

Fifteen spp., China and Viet Nam.

"Indosasa" is not monophyletic in the analyses of Zeng et al. (2010).

47. ***Sasa*** Makino & Shibata, s.s.

Sasa Makino & Shibata, Bot. Mag. (Tokyo) 15: 18 (1901), s.s.

Plants shrubby, less than 2 m tall, culms erect or sometimes leaning. Internodes terete, nodes prominent. Branches 1 per node. Culm leaves with persistent sheaths, with or without auricles and fimbriae, the blades erect or reflexed. Foliage leaves with or without auricles and fimbriae, the blades with distinct transverse veins. Inflorescence branched or unbranched, with tiny bracts. Spikelets with 4 to 8 flowers. Glumes 2. Lodicules 3. Stamens 6. Style 1, stigmas 3.

Forty spp., Japan, Korea, eastern China and eastern Russia.

"Sasa" as conventionally circumscribed is polyphyletic (Triplett and Clark 2010; Zeng et al. 2010), with species falling in both the *Arundinaria* clade and the *Shibataea* clade. The type species, *S. veitchii* (Carrière) Rehder, falls into the *Sasa* subgroup in the *Arundinaria* clade (Triplett and Clark 2010), so this clade constitutes *Sasa* s.s. The placement of the named hybrid *Sasella* in this clade shows that the maternal parent of *Sasella* was a species of *Sasa*.

48. ***Sasamorpha*** Nakai

Sasamorpha Nakai, J. Fac. Sci. Hokkaido Univ. V, Bot. 26: 180 (1931).

Shrubby plants less than 1.5 m tall. *Culms and axis of inflorescence white powdery*. Internodes terete, nodes flat. Branches 1 per node. Culm leaves with persistent sheaths; auricles and fimbriae lacking. Foliage leaves with sheaths lacking auricles and fimbriae, with blades with distinct transverse veins. Inflorescence branched or unbranched, with tiny bracts. Spikelets with 4 to 8 flowers. Glumes 2. Lodicules 3. Stamens 6. Style 1, stigmas 3.

Five spp., China, Japan, Korea, Eastern Russia.

Sasamorpha was segregated from *Sasa* on the basis of the white powdery wax on the culms, and the non-auriculate culm leaves. Also the culm leaf sheaths are longer than the internodes, whereas they are shorter in *Sasa*.

CLADE VII. THAMNOCALAMUS Munro

49. ***Thamnocalamus*** Munro

Thamnocalamus Munro, Trans. Linn. Soc. London 26: 33 (1868); tax.: Stapleton (1994c); phylog.: Guo and Li (2004).

Plants shrubby, culms erect to nodding. Rhizomes pachymorph. Branches about 5 per node, with one somewhat larger than the others; branchlets long and pendulous. Culm leaves with sheaths deciduous, without auricles, with fimbriae, the blade awl-shaped, erect. Foliage leaf sheaths without auricles or fimbriae, the blades with distinct transverse veins. Inflorescence branches branched, the basal branches with bracts. Spikelets with 1 to many flowers, the distalmost flower reduced. Glumes 2. Stigmas 3.

Two spp., China and Himalayas.

Thamnocalamus s.l. is not monophyletic in the chloroplast phylogeny of Triplett and Clark (2010), but may form a clade when *Bergbambos* (Clade I) is excluded. Triplett and Clark (2010) place *Thamnocalamus* s.s in the *Phyllostachys* clade, whereas Zeng et al. (2010) place it in its own clade.

Clade VIII. "Indocalamus" wilsonii (Rendle) C. S. Chao & C. D. Chu

50. **"Indocalamus" wilsonii** (Rendle) C. S. Chao & C. D. Chu

"Indocalamus" wilsonii (Rendle) C. S. Chao & C. D. Chu, J. Nanjing Techn. Forest Prod. 1981: 43 (1981).

Culms less than 1 m tall. Culm leaves without auricles or fimbriae, the blade recurved. Foliage leaves with sheaths drying orange-red, without auricles or fimbriae, the blades wavy when dry, with obvious cross-veins. Inflorescence branches branched. Spikelets with 3 to 7 flowers. Glumes 2. Style reduced or absent, stigmas 3.

One sp., *I. wilsonii* (Rendle) C. S. Chao & C. D. Chu, China.

This species is sister to *Bergbambos*, but without support, in the analyses of Zeng et al. (2010). It is unrelated to other species of "*Indocalamus*", nearly all of which are members of the *Phyllostachys* clade.

Clade IX. Gaoligongshania D.Z. Li, Hsueh & N.H. Xia

51. ***Gaoligongshania*** D.Z. Li, Hsueh & N.H. Xia

Gaoligongshania D.Z. Li, Hsueh & N.H. Xia, Acta Phytotax. Sin. 33: 598 (1995).

Culms scrambling or epiphytic. Rhizomes pachymorph. Branches 1 per node, similar in thickness to the culm. Culm leaves with persistent leathery sheaths, with auricles, the auricles bearing fimbriae, the blade recurved. Foliage leaf sheaths with fimbriae, the blades with transverse veins. Inflorescence branches branched, sparse, without bracts. Spikelets pedicellate, with 4 to 10 flowers, the distal one reduced. Glumes 2. *Lemma with a long awn.* Lodicules 3. Stamens 3. Style 1, stigmas 3. Ovary stalked.

One sp., *G. megalothyrsa* (Hand.-Mazz.) D.Z. Li, J.R. Xue & N.H. Xia, China.

Clade X. Indocalamus Nakai, s.s.

52. ***Indocalamus*** Nakai

Indocalamus Nakai, s.s. , J. Arnold Arbor. 6: 148 (1925).

Culms 1–4 m tall, nodding, clumped. Internodes terete, glabrous. Rhizomes leptomorph. Branch generally 1 per node, nearly as thick as the culm. Culm leaves persistent, auricles absent, fimbriae present, blade erect to spreading. Foliage leaves without auricles, with fimbriae, the blades with obvious cross veins. Inflorescence branches branched. Spikelets with 3 or 4 flowers. Glumes 2. Stamen number unknown. Stigmas 2.

One sp., *I. sinicus* (Hance) Nakai, China.

Indocalamus sinicus is unrelated to other species of "*Indocalamus*", which fall in the *Phyllostachys* clade (Zeng et al. 2010). Because this is the type species, the remaining species will ultimately need to be reclassified.

Clade XI. "Ampelocalamus" calcareus

53. ***"Ampelocalamus" calcareus*** C.D. Chu & C.S. Chao

"Ampelocalamus" calcareus C.D. Chu & C.S. Chao, Acta Phytotaxonomica Sinica 21: 204–206.

"*A*". *calcareus* is in a position unrelated to other species of *Ampelocalamus* in the phylogenies of Yang et al. (2013) and Ma et al. (2014), and is thus placed in Clade XI of the Arundinarieae; in the latter phylogeny it is sister to all other species of the tribe. The species is poorly known and is not given a full description here, nor is it placed in the key pending more comprehensive analysis.

Clade XII. Kuruna Attigala, Kaththriarachchi & L.G. Clark

54. ***Kuruna*** Attigala, Kaththriarachchi & L.G. Clark

Kuruna Attigala, Kaththriarachchi & L.G. Clark, Phytotaxa 174: 199 (2014).

Culms erect to scandent, the internodes terete, flattened or sulcate. Rhizomes pachymorph, with short necks. Branches several per node, from a single bud. Culm leaves hispid abaxially with fimbriae, without auricles, the blades reflexed. Foliage leaf sheaths with fimbriae, erect. Inflorescences branched or unbranched, open, the branches lacking subtending bracts. Glumes 2, fertile flowers (1) 2 to 6. Palea with an acute apex or with two sharp points. Stigmas 2 or 3.

Five spp., Sri Lanka.

V. Tribe Bambuseae Kunth ex Dumort. (1829)

Culms woody. Rhizomes pachymorph. *Branch development acropetal or bidirectional.* Outer ligule (contraligule) present. Spikelets laterally compressed, disarticulating above the glumes. Midrib complex. $2n = (20)$ 40, (44), 46, 48, 70, 72.

Sixty-three genera, 784 spp.

All Bambuseae have pachymorph rhizomes, except for *Chusquea* sect. *Chusquea*, in which the rhizomes are amphimorph, presumably a derived character. As in Arundinarieae, culms of members of Bambuseae reach their full height without initiating branches and produce only culm leaves; later the culms lignify and produce branches and foliage leaves, beginning with the base or middle of the plant and proceeding acropetally or bidirectionally (Bamboo Phylogeny Group 2012). Height is strongly correlated with root pressure, which can refill embolized xylem vessels (Cao et al. 2012). Pseudospikelets are common among members of Bambuseae, but do not occur in all genera. The outer ligule also occurs in *Streptogyna* (BEP clade, incertae sedis). Phylogenetic analysis of the chloroplast genome shows that Bambuseae can be divided into a neotropical and a paleotropical group, each of which contains several well-supported subclades, recognized as subtribes (Kelchner and Bamboo Phylogeny Group 2013). Monophyly of the neotropical clade is not strongly supported, however. Phylogenies of nuclear genes also identify neotropical and paleotropical clades, with the C1 genome of the neotropical group most closely related to the B genome of Arundinarieae (Triplett et al. 2014).

Neotropical clade

V.1. Subtribe Chusqueinae Bews (1929)

Neurolepidinae Soderstr. & R. P. Ellis in Soderstr. et al., Grass Syst. Evol.: 238 (1987).

Culms erect to scandent or scrambling, sometimes hanging, shrubby or arborescent. Rhizomes pachymorph. Culms generally solid, unbranched, with one bud per node; prop roots sometimes forming root spines. Culm leaves present or

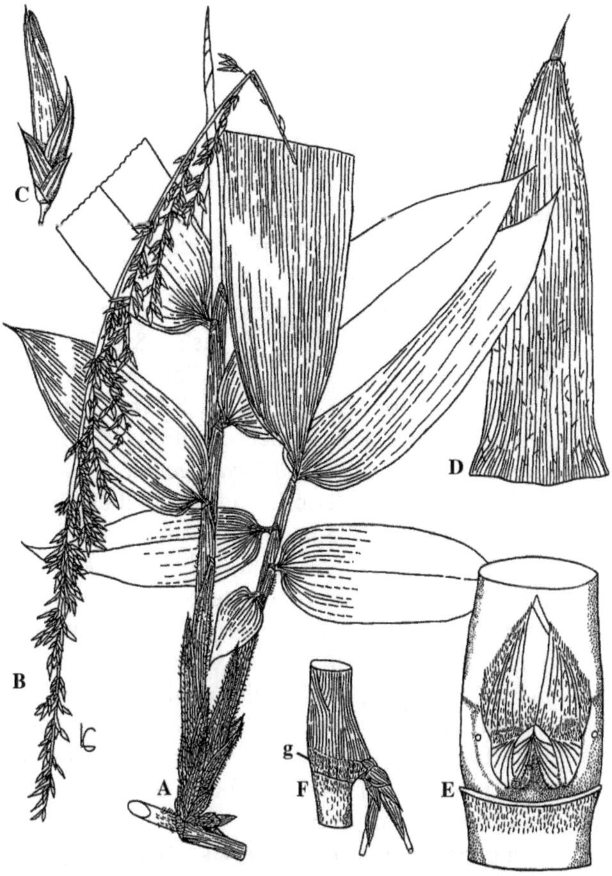

Fig. 21. Bambusoideae-Bambuseae-Chusqueinae. *Chusquea latifolia*. **A** Branch complement with foliage leaves. **B** Inflorescence. **C** Spikelet. **D** Culm leaf. **E** Bud complement. **F** Node with branches emerging infravaginally; *g* girdle, a persistent section of the leaf sheath encircling the culm. (From Clark 1989)

absent, the sheaths lacking fimbriae or auricles, the blades stiff and upright, rarely reflexed. Ligule with obvious veins. Foliage leaves lacking fimbriae or auricles. Inflorescence branches branched, sometimes capitate. *Spikelets with one flower, rachilla extension lacking. Glumes four*, the proximal two sometimes reduced, with or without awns. Lodicules 3. Stamens 3. Stigmas 2. *Stomatal subsidiary cells each with at least 2 papillae.*

One genus, 160 species.

55. ***Chusquea*** Kunth — Fig. 21

Chusquea Kunth, J. Phys. Chim. Hist. Nat. Arts 95: 151 (1822); phylog.: Kelchner and Clark (1997), Fisher et al. (2009, 2014).

Rettbergia Raddi, Agrostogr. Bras. 17–18, t. 1, f. 1 (1823).
Neurolepis Meisn., Pl. Vasc. Gen. 1: 426 (1843).
Swallenochloa McClure, Smithsonian Contr. Bot. 9: 106, f. 43–45 (1973).

Characters as for the subtribe. $2n = 20, 40, 44, 48$ (excluding counts listed as doubtful by Judziewicz et al. 1999).

One hundred sixty spp., Central and South America, West Indies.

Chusquea is made up of four major clades, *Chusquea* subgenus *Platonia* Fisher & L.G. Clark, *C.* subgenus *Magnifoliae* L.G. Clark & Fisher, *C.* subgenus *Rettbergia* (Raddi) L.G. Clark, and the Euchusquea clade, which includes subgenus *Chusquea* and subgenus *Swallenochloa*, both of which are polyphyletic (Fisher et al. 2009, 2014). The Euchusquea clade contains most of the species of the genus. Species in subgenera *Platonia* and *Magnifoliae* were formerly placed in the genus *Neurolepis*, but this "genus" is clearly paraphyletic; its former generic characters (lack of aerial branching, ligules with obvious veins, and a base chromosome number of $x = 12$) could possibly be synapomorphies for *Chusquea* s.l. The presence of multiple buds at each node, with the buds dimorphic or trimorphic, is synapomorphic for *Chusquea* s.s. (i.e., subgenus *Rettbergia* plus Euchusquea).

Arthrostylidiinae + Guaduinae

Arthrostylidiinae and Guaduinae are sister taxa, and share *refractive papillae on leaf epidermal cells* (Guala et al. 2000; Kelchner and Clark 1997; Ruiz-Sánchez et al. 2008; Sungkaew et al. 2009; Zhang and Clark 2000). Although refractive papillae on the leaf epidermis are likely synapomorphic for this pair of subtribes, they have originated apparently in parallel in Melocanninae (Bamboo Phylogeny Group 2012).

V.2. Subtribe Arthrostylidiinae Bews (1929)

Clump forming species with short thick (pachymorph) rhizomes. Branch buds one per node. Prophylls triangular. Foliage leaves with well-developed, curly or wavy fimbriae at the collar. *Wax distributed unequally over the foliage leaf blade surface, leaving one narrow marginal sector of the abaxial side with no or reduced wax.* Lodicules 3. Stamens 3. Stigmas 2. *Leaf mesophyll with sclerenchyma fibers in intercostal regions; midrib with a single vascular bundle.* $2n = 40$.

Fifteen genera, 172 spp.

The sclerenchyma fibers in the intercostal regions are unusual in Poaceae. Although the narrow waxless strip on the leaves is apparently synapomorphic for Arthrostylidiinae, the character is also found in other bamboos, particularly Melocanninae (Tyrell et al. 2012). Most, if not all, other members of Bambuseae have a complex midrib, consisting of several vascular bundles.

The phylogeny and considerable morphological diversity in Arthrostylidiinae are described in detail by Tyrell et al. (2012). The subtribe is morphologically diverse, with variation between genera in patterns of culm internode elongation, position of culm leaf blades (erect to reflexed), the shape of the prophylls (cordate or triangular), the number and patterning of branches at each node (one to several dominant ones, branches sometimes in a fan-shaped pattern), inflorescence architecture (branched or unbranched, capitate or not), presence of pseudospikelets or conventional spikelets, seed coat separable from the pericarp or not, and pericarp fleshy or not.

Glaziophyton + Cambajuva

Inflorescence branches erect, appressed to the rachis, *foliage leaves erect, strongly tessellate, fusoid cells in the leaf mesophyll lacking.*

Glaziophyton and *Cambajuva* form a clade in molecular phylogenies that is sister to all other members of the subtribe (Tyrell et al. 2012). Erect leaf blades also occur in some species of *Aulonemia* and *Myriocladus* but appear to be independently derived. Likewise tessellation of the abaxial surface of the leaf occurs in high-altitude species of both the latter genera; it is possible that this vein pattern is related to low temperature (Viana et al. 2013).

56. ***Glaziophyton*** Franch.

Glaziophyton Franch., J. Bot. (Morot) 3: 277 (1889).

Culms erect, *with chambered pith. First internode very long, subsequent ones very short.* Branches generally absent. Culm leaves rare, when present the sheath without auricles or fimbriae, the apex of the blade involute, forming a sharp point. Inflorescence branches branched, each subtended by a basal bract, and with a prophyll. Spikelets with 1 sterile flower and 2 fertile flowers. Glumes 2.

One sp., *G. mirabile* Franch., Rio de Janeiro, Brazil.

57. ***Cambajuva*** P. L. Viana, L. G. Clark & Filg.

Cambajuva P. L. Viana, L. G. Clark & Filg., Syst. Bot. 38: 98 (2013).

Culms erect, the internodes hollow. Culm leaves deciduous, the sheath apex with fimbriae, the blade erect. Branches 1 to 7 per node from a single bud, usually on a promontory. Foliage leaves with sheaths bearing erect undulate to crispate fimbriae, the blade apex sharp-pointed. Inflorescence branches branched, lacking subtending bracts and prophylls. Spikelets with 1 sterile flower, 2 to 4 fertile flowers, and 1 distal rudimentary flower. Glumes 2. Glumes and lemmas with awns. Lodicules 3. Anthers 3. Stigmas 2. *Leaf epidermis lacking refractive papillae.*

One sp., *C. ulei* (Hack.) P.L. Viana, L.G. Clark & Filg., southern Brazil.

Cambajuva was described to accommodate *Aulonemia ulei* (Hack.) McClure, which is morphologically and phylogenetically distinct from other species of *Aulonemia.*

Remaining species of the subtribe:
Reflexed pseudopetioles, the leaves thus appearing to droop. This character is reversed in *Filgueirasia.*

Tyrell et al. (2012) recover three clades among the remaining species: I. *Arthrostylidium, Rhipidocladum, Didymogonyx,* and *Elytrostachys*; II. *Aulonemia* and *Colanthelia,* plus two species of *Arthrostylidium*; III. *Actinocladum, Merostachys, Alvimia, Atractantha,* and *Filgueirasia.* Although these clades are moderately well supported by chloroplast sequence data, none has a diagnostic morphological character.

58. ***Actinocladum*** McClure ex Soderstr.

Actinocladum McClure ex Soderstr., Amer. J. Bot. 68: 1201 (1981).

Culms erect. Branch complements fan-shaped. Culm leaves with sheaths with ruffle-like fimbriae near their summit by the collar, the blades spreading to reflexed. Foliage leaves on terminal branches broader than those below and *often reflexed.* Inflorescence branches branched or not, the spikelets on long stalks. Spikelets with many flowers. Glumes 2, their apices acuminate. Lemma acuminate or with a short beak. Fruit with pericarp free from seed coat.

One sp., *A. verticillatum* (Nees) Soderstrom, Brazil and adjacent Bolivia.

Actinocladum is strongly supported by molecular data as sister to *Merostachys* (Tyrell et al. 2012). Both genera have caryopses in which the seed coat is free from the pericarp and fan-shaped clusters of branches. The latter character also appears independently in *Rhipidocladum* and some species of *Arthrostylidium.*

59. ***Alvimia*** C. E. Calderón ex Soderstr. & Londoño

Alvimia C. E. Calderón ex Soderstr. & Londoño, Amer. J. Bot. 75: 833 (1988).

Culms scrambling and hanging from vegetation. Branches several to 100 or more, 1 branch dominant, from a single bud on a promontory just above the node. Culm leaves with sheaths with a dark line at the base, with or without fimbriae, the blades erect. Foliage leaves with dark pseudopetioles. Inflorescence branches branched. Pseudospikelets long and narrow, with numerous flowers. Lemma with a narrow tip. *Stamens 2. Fruit fleshy,* falling with lemma, palea and rachis internode.

Three spp., Bahia, Brazil.

Alvimia is strongly supported by molecular data as sister to *Atractantha* (Tyrell et al. 2012).

60. **"*Arthrostylidium*"** Rupr.

"*Arthrostylidium*" Rupr., Bambuseae 27 (1839); tax.: Judziewicz and Clark (1993).

Culms scandent at maturity. Branches 3 to many, with one dominant, the branches from a single bud on a promontory just above the node. Culm leaf sheaths without fimbriae, the blades erect. Foliage leaf sheaths with fimbriae. Inflorescences unbranched, with a zigzag rachis in some species. Spikelets with 1 to several flowers. Glumes 1 to 3. Lemma with margins clasping the rachis.

Thirty-two spp., Central and South America.

The fimbriae on the foliage leaves may be inconspicuous. This genus is polyphyletic in the analyses of Tyrell et al. (2012).

61. ***Athroostachys*** Benth.

Athroostachys Benth., Gen. Pl. 3: 1208 (1883).

Culms scandent and vine-like. Branches several per node, all about the same diameter. Culm leaf sheaths persistent or deciduous, the blades reflexed, the sheath with a proximal dark-colored, thickened rim, with fimbriae at the apex, without auricles. Foliage leaf sheaths with fimbriae, without auricles. Inflorescence dense and capitate, the branches branched but not elongate. Spikelets with 1 flower plus a small rudimentary one. Glume 1, deciduous, the apex acuminate, awn-like.

One sp., *A. capitata* (Hook.) Benth., coastal Brazil.

Molecular data are unavailable for *Athroostachys*. Tyrell et al. (2012) speculate that it may be closely related to *Alvimia* and *Atractantha*.

62. ***Atractantha*** McClure

Atractantha McClure, Smithsonian Contr. Bot. 9: 42 (1973); tax.: Judziewicz (1992); Santos-Gonçalves et al. (2011).

Culms scandent, vine-like, or pendent. Branches three per node, rebranching, with one branch dominant, from a single bud on a promontory. Culm leaves deciduous, with a pseudopetiole, the sheaths with an obvious skirt of downward pointing fimbriae, with a bumpy, tumor-like, corky crest at the base of the sheath, the blades generally horizontal to reflexed. Inflorescence unbranched, or branched and capitate. Pseudospikelets with 1 glume, 1 prophyll, and 1 flower, *the flower slender, indurate, sharp-pointed, needle-like.*

Six spp., coastal Brazil, Amazonas, and southern Venezuela.

Atractantha is strongly supported by molecular data as sister to *Alvimia* (Tyrell et al. 2012). The culm leaf blades are small and erect in *A. radiata* McClure, an exception to the general state for *Atractantha*.

63. "***Aulonemia***" Goudot

"*Aulonemia*" Goudot, Ann. Sci. Nat., Bot. III, 5: 75 (1846).
Matudacalamus F. Maek., J. Jap. Bot. 36(10): 345 (1961).

Culms erect or scandent. Mid-culm nodes with one bud, producing one dominant branch nearly as large as the culm. Culm leaf sheaths with fimbriae, blades reflexed to spreading, deciduous. Foliage leaf sheaths often purple spotted. Inflorescence branches branched. Spikelets with 2 or 3 glumes and several flowers. Lemma with the apex obtuse, with a brief mucro or arista. $2n = 40$.

Forty spp., montane regions of Central and South America.

"*Aulonemia*" as traditionally circumscribed is polyphyletic. One species has been removed to *Cambajuva* (Viana et al. 2013), and two other species were transferred to *Olmeca* (Ruiz-Sánchez et al. 2011a). The remaining species form a paraphyletic group from which *Colanthelia* is derived (Tyrell et al. 2012). The *Aulonemia/Colanthelia* clade is strongly supported.

A few species of "*Aulonemia*", including the type, produce several smaller lateral branches in addition to the dominant one.

64. ***Colanthelia*** McClure & E. W. Sm.

Colanthelia McClure & E. W. Sm., Smithsonian Contr. Bot. 9: 77 (1973).

Culms erect or scandent. Branches several per node, with one branch dominant, divergent, from a single bud. Culm leaves often with a dark girdle between the node and the base of the sheath, with a corky crest at the junction of the girdle and the base of the sheath. Foliage leaves with sheaths sometimes auriculate, the blades deciduous. Inflorescence unbranched or with branched branches. Spikelets slender, with few to many flowers. Glumes 2 or 3. Stamens 3. Stigmas 2.

Seven spp., southern Brazil and northern Argentina.

See comments under "*Aulonemia*".

65. ***Didymogonyx*** (L.G. Clark & Londoño) C.D. Tyrrell, L.G. Clark & Londoño

Didymogonyx (L.G. Clark & Londoño) C.D. Tyrrell, L.G. Clark & Londoño, Mol. Phylog. Evol. 65: 146 (2012).

Rhipidocladum sect. *Didymogonyx* L. G. Clark & Londoño, Amer. J. Bot. 78: 1271 (1991).

Culms erect below, arching or scrambling above, with alternating long and short internodes, the long ones hollow, the short ones solid or pithy. *Prophylls cordate.* Culm leaves with sheaths with auricles and fimbriae. Foliage leaves with sheath fimbriate, the fimbriae with bases slightly flattened, auricles lacking. Inflorescence contracted, branched or unbranched. Spikelets in clusters of 2 or 3. Glumes 3 (−5), sterile lemmas 1 or 2, flowers 2 to 12. Mesophyll with sclerenchyma associated with the bulliform cells. Leaf epidermal long cells with papillae overarching the stomata.

Two spp., Colombia and Venezuela.

66. ***Elytrostachys*** McClure

Elytrostachys McClure, J. Wash. Acad. Sci. 32: 173 (1942).

Culms scandent, often drooping from trees. Branches several per node, with one branch dominant, the branches from a single bud. Culm leaves with sheaths with straight fimbriae, *the fimbriae 2–8 cm long.* Foliage leaf sheaths with or without auricles. Inflorescences on lateral shoots, dense and capitate. Pseudospikelets each subtended by a prophyll; each branch with 1 or 2 flowers subtended by 2 glumes. *Stamens 6.* Ovary apex pubescent.

Two spp., Central and South America.

67. ***Filgueirasia*** Guala

Filgueirasia Guala, Bamboo Sci. Cult. 17: 2 (2003).

Culms erect, <2 m tall. Mid-culm nodes producing 1 to 15 buds in a line above the node. Culm leaves with sheaths glabrous on both surfaces, the blades erect. Foliage leaves with white fimbriae, *the blades erect.* Inflorescence branches one-sided. Spikelets with 3 to 15 flowers. Fruit beaked.

Two spp., Brazilian cerrado.

68. ***Merostachys*** Spreng. Fig. 22

Merostachys Spreng., Syst. Veg. 1: 132, 249 (1824); tax., Sendulsky and Soderstrom (1984).

Culms scandent at maturity, often drooping from trees. Mid-culm branches in fan-shaped clusters, from a single bud. Culm leaves with sheaths bearing fimbriae. *Inflorescences unbranched, one-sided, the spikelets crowded, sessile or on short pedicels.* Spikelets with 1 flower, with a rachilla

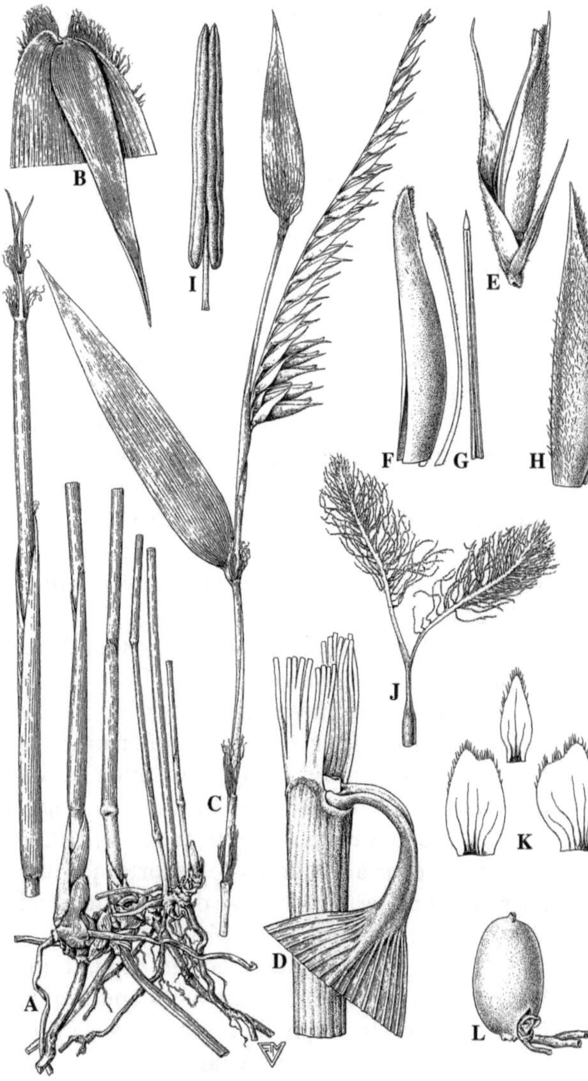

Fig. 22. Bambusoideae-Bambuseae-Arthrostylidiinae. *Merostachys* sp. (**A, L**) and *M. speciosa* (**B–K**). **A** Seedling plant, showing basal caespitose clump habit and pachymorph rhizomes, with young culm shoot at left. **B** Culm sheath from mid-culm node. **C** Flowering branch, terminating in secund spicate inflorescence. **D** Leaf sheath showing ligule, base of oral setae, petiole, and base of leaf blade. **E** A single spikelet of the inflorescence. **F** Palea. **G** Bristle-like prolongation of rachilla in two aspects. **H** Lemma. **I** Stamen. **J** Gynoecium. **K** Lodicule complement. **L** Seed after having produced a seedling plant. (From McClure 1973)

extension. Glumes 2, shorter than the flower. Lemma clasping the palea at the base only.

Forty-eight spp., Central and South America.

Merostachys is strongly supported by molecular data as sister to *Actinocladum* (Tyrell et al. 2012). Both genera have caryopses in which the seed coat is free from the pericarp and fan-shaped clusters of branches. The latter character also appears independently in *Rhipidocladum* and some species of *Arthrostylidium*. Species of *Merostachys* occasionally have two to several flowers per spikelet.

69. ***Myriocladus*** Swallen

Myriocladus Swallen, Fieldiana, Bot. 28: 34 (1951); rev.: Judziewicz (1998).

Culms erect, *the internodes often notably unequal in length*. Branches 1 per node, but appearing clustered as internodes fail to elongate. Prophylls cordate. Culm leaves not differentiated from foliage leaves. Leaves densely overlapping, with the sheaths thickened distally, the blades erect to spreading, deciduous. Inflorescence terminal, unbranched, or the branches branched. Spikelets with 1 to 10 flowers. Glumes generally 2, with or without awns.

Twelve spp., Guayana Highlands of Venezuela, Brazil and Guyana.

The pattern of internode elongation in *Myriocladus* varies. In two species, internodes are approximately equal in length. In other species the basal internode is long whereas the upper ones are scarcely elongate at all, and in still others long internodes are followed by short ones in an alternating pattern. Molecular data are not available for *Myriocladus*, so its placement is uncertain. Tyrell et al. (2012) speculate that it may belong in a clade with *Glaziophyton* and *Cambajuva*. The cordate prophylls are shared with *Didymogonyx*, but may not be synapomorphic.

70. **"*Rhipidocladum*"** McClure

"*Rhipidocladum*" McClure, Smithsonian Contr. Bot. 9: 101 (1973); tax., Clark and Londoño (1991).

Culms weak, erect below, scandent or clambering above. Mid-culm branches numerous, in a fan-shaped cluster, from a single bud. Culm leaves triangular, deciduous, the sheaths lacking auricles or fimbriae, the blade confluent with the sheath. Inflorescence unbranched, the spikelets often borne on one side. Spikelets with several flowers. Glumes 2 or 3 (to 5). Lemma apex obtuse to acute, awned or not.

Fifteen spp., Central to South America.

V.3. Subtribe Guaduinae Soderstr. & R. P. Ellis (1987)

Culm leaves often with the blade margins confluent with the sheath margins. Culm leaf sheath and foliage leaf sheath generally with fimbriae. *Both adaxial and abaxial surfaces of foliage leaf blades with stomata, the adaxial stomata with overarching papillae.* Lodicules 3. $2n = 46$ or 48.

Five genera, 44 species.

Species with elongated rhizome necks are common but not universal in this subtribe; they occur in *Eremocaulon*, *Guadua*, *Olmeca* and *Otatea*.

71. ***Apoclada*** McClure

Apoclada McClure, Fl. Il. Catarin. 1 (Gram.-Supl.): 57 (1967); rev., Guala II (1995).

Culms erect, clump forming. Mid-culm nodes with 2 to 5 buds. Branches as many as the number of buds, sometimes rebranching. Culm leaves with sheaths deciduous, without auricles, with fimbriae, with blades erect. Foliage leaves with fimbriae. Inflorescence with few branches. Spikelets with 7 to 28 flowers. Stamens 3. Stigmas 2. Ovary with an enlarged apex, glabrous.

One sp., *A. simplex* McClure & L.B. Smith, southeastern Brazil.

72. ***Eremocaulon*** Soderstr. & Londoño

Eremocaulon Soderstr. & Londoño, Amer. J. Bot. 74: 37 (1987); rev., Londoño and Clark (2002).
Criciuma Soderstr. & Londoño, Amer. J. Bot. 74: 35 (1987).

Plants strongly rhizomatous, producing culms at 1 to 2 m intervals. Branches more than 1 per node, the central one dominant. Culm leaves leathery, with obvious auricles, with fimbriae, the blade becoming reflexed. Foliage leaves with fimbriae, with or without auricles, the midrib strongly excentric. Inflorescence sparsely branched. Pseudospikelets elongate, with 1 to 13 (17) flowers, with 0 to 3 proximal bracts bearing buds in their axils, and 0 to 3 additional bracts (glumes) lacking buds. Stamens 6. Stigmas 2 or 3.

Four spp., Bahia, Brazil.

73. ***Guadua*** Kunth Fig. 23

Guadua Kunth, J. Phys. Chim. Hist. Nat. Arts 95: 150 (1822).

Culms arborescent to scandent, *usually with a band of white hairs below the nodes, and often above as well. Branches one per node, the secondary branches generally modified into thorns.* Culm leaves deciduous, with sheaths contiguous with the blades, fimbriae sometimes absent. Inflorescences of several pseudospikelets, each with 1 bract and 1 prophyll, 1 to several bracts with buds in their axils, zero to several glumes, and several flowers. *Palea keels winged.* Stamens 6. Stigmas 3. Fruit dry. $2n = 46, 48$.

Twenty-seven spp., Mexico, Central and South America.

Although most of the members of this genus have dry fruits, that of *G. sarcocarpa* Londoño is fleshy.

74. ***Olmeca*** Soderstr.

Olmeca Soderstr., Phytologia 51: 161 (1982); Ruiz-Sánchez et al. (2011a), rev., phylog.

Culms solitary, the rhizome necks elongate. *Nodes with one branch, the branch not itself branching*; branches at mid-culm with a diameter between 0.5 and 1.0 of that of the culm. Culm leaves with sheaths persistent, indurate, the blades erect or reflexed. Spikelets with 2 glumes and several flowers. Lemma glabrous or with long hairs at the apex. *Lemma and palea fused to each other at the base.* Stamens 3. Stigmas 2 or 3. *Fruit fleshy* or dry.

Five spp., southern Mexico, Honduras.

The rhizome necks are 3–9 m long in one clade (2 species), but 4–30 cm in the other (3 species) (Ruiz-Sánchez et al. 2011a). Soderstrom provided a description and diagnosis of *Olmeca* in 1981 (Amer. J. Bot. 68: 1362), but failed to designate a type. The name was subsequently validated in 1982 as noted above.

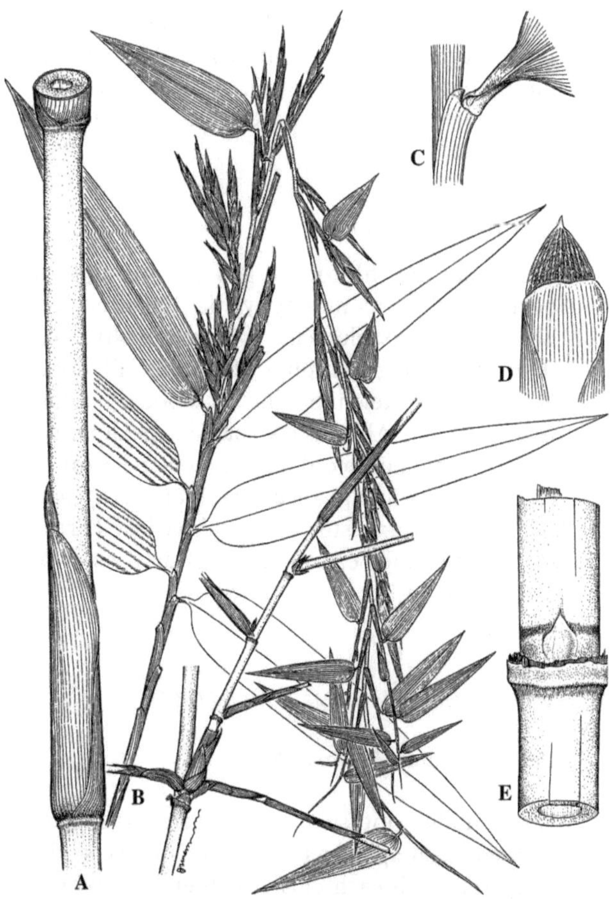

Fig. 23. Bambusoideae-Bambuseae-Guaduinae. *Guadua ciliata*. **A** Portion of stem showing culm leaf with erect blades. **B** Branches bearing pseudospikelets. **C** Apex of sheath of foliage leaf, with pseudopetiole and short ligule. **D** Culm leaf with triangular blade, blade margins continuous with the sheath. **E** Node with a single bud, with white hairs above and below the node, the remains of the leaf sheath forming a rim. (From Davidse et al. 2004, with permission from the Missouri Botanical Garden Press; drawn by Bruno Manara)

75. ***Otatea*** (McClure & E. W. Smith) C. Calderón & Soderstr.

Otatea (McClure & E. W. Smith) C. Calderón & Soderstr.; Ruiz-Sánchez et al. (2008, 2011b), rev., phylog.

Culms erect. Mid-culm nodes with one branch bud, *producing three principal ascending branches*, diameter of the branches at mid-culm less than 1/4 that of the culm. Culm and foliage leaves lacking auricles, with fimbriae (oral setae). Inflorescence branches branched. Spikelets with multiple flowers. Stamens 3. Stigmas 2 or 3.

Eight spp., Mexico, Central America and Colombia.

Paleotropical clade

Data from the chloroplast genome strongly support this clade as monophyletic, but there is no obvious morphological synapomorphy other than the biogeography (Bamboo Phylogeny Group 2012; Kelchner and Bamboo Phylogeny Group 2013). All species may be hexaploid or higher. The polyploid history of the group is complex, involving ancestors that were closely related to the ancestor of Arundinarieae (Triplett et al. 2014). Melocanninae are clearly sister to a clade of the other three subtribes.

V.4. Subtribe Bambusinae J. Presl (1830)

Dendrocalaminae Benth. (1881).

Culm nodes not swollen. Flowers arranged in conventional spikelets or pseudospikelets, these clustered, with each cluster subtended by bracts. Stamens 6. Style 1 or reduced. Stigmas 3. $2n = (48)\ 70, 72$.

Twenty-five genera, 264 spp.

This subtribe has no obvious morphological synapomorphy. The apex of the ovary of Bambusinae often has a different texture and color from the body of the ovary itself. The apical tissue is variously called an umbo, an appendage, or simply a fleshy outgrowth. The only detailed description of the structure is presented by Holttum (1956), who shows sketches of sections of ovaries and fruits of a few genera. He describes the outgrowth at the apex of the ovary as similar to a style base in *Bambusa* and *Gigantochloa*, whereas the structure is hollow in *Oxytenanthera*. I do not know of any investigation of its development, anatomy, or function.

Within Bambuseae, generic limits are most problematical in Bambusinae. The most comprehensive phylogeny to date is that of Goh et al. (2013), which finds evidence for monophyly of a group of climbing species and a complex history of apparent hybridization and/or lineage sorting in the genera *Bambusa*, *Dendrocalamus*, *Gigantochloa* and several smaller segregates. Intergeneric hybridization had already been suggested by Goh et al. (2010), and documented by Goh et al. (2011). A small population of hybrids between *Dendrocalamus pendulus* Ridl. and *Gigantochloa scortechinii* Gamble was found growing near both parental species, in an area that had been disturbed by clearing associated with highway construction (Wong and Low 2011). The hybrid plants flowered but were apparently sterile and there was no evidence of F2 plants.

76. ***Bonia*** Balansa

Bonia Balansa, J. Bot. (Morot) 4: 29 (1890); tax.: Xia (1996).
Monocladus L.C. Chia, H.L. Fung & Y.L. Yang, Acta Phytotax. Sin. 26(3): 212 (1988).

Culms scrambling. Branches 1 per node, about the same thickness as the culm. Culm leaves with sheaths persistent, with prominent auricles, the auricles bearing fimbriae or not, the blade erect or reflexed. Foliage leaves with or without falcate auricles, with fimbriae. Inflorescence unbranched, subtended by a bract and prophyll; distal to the prophyll are 2 or 3 bracts with buds in their axils, and distal to these a 1 cm rachilla internode. Pseudospikelets with 3 to 9 flowers; distal flowers under-developed. Lodicules 3. Ovary glabrous.

Five spp., southern China, Vietnam.

The bracts in this genus are sometimes called "glumes", but have buds in their axils and are thus gemmiparous bracts.

77. ***Cyrtochloa*** S. Dransf.

Cyrtochloa S. Dransf., Kew Bull. 53: 861 (1998).

Culms erect or scandent. Branches 3 to many per node, with 1 branch dominant. Culm leaves with sheaths persistent, wrinkled at the base, with auricles, the blades ovate or triangular and cordate. Foliage leaves with auricles absent or falcate, fimbriae sparse; the blades with strongly asymmetric bases. Inflorescence with capitate clusters of spikelets at the nodes, the clusters separated by internodes. Pseudospikelets with 1 fertile flower. Glumes 2 to several, the apex acute to acuminate, apex of the upper one sometimes mucronate. Lemma wrapped around the palea. Lodicules

2 or 3. Ovary with a long appendage, the apex pubescent. Caryopsis with the pericarp fleshy, or free from the seed coat and breakable.

Five spp., Tropical Asia.

78. ***Fimbribambusa*** Widjaja

Fimbribambusa Widjaja, Reinwardtia 11: 80 (1997).

Culms scrambling. *Nodes with a narrow shelf-like extension.* Branches 1 per node, or 5 to 10 per node with 1 dominant. Culm leaf sheaths with stiff linear, horn-like auricles, with fimbriae, the blades spreading to reflexed. Foliage leaf sheaths with auricles, with fimbriae. Inflorescences borne on leafless branches, the pseudospikelets in clusters of few to many at the nodes, the clusters subtended by 2 prophylls. Pseudospikelets with 2 or 3 flowers. Lodicules 2 or 3.

Two spp., East Java, New Guinea,

Fimbribambusa was segregated from *Bambusa* based on the shelf-like extension at the nodes, the lodicules being entire, and the ovary glabrous. *Fimbribambusa* is similar to *Temburongia* but the latter has only 2 or 3 spikelets per node.

79. ***Greslania*** Balansa

Greslania Balansa., Bull. Soc. Bot. France 19: 319 (1873).

Culms erect. Branches 1 per node or absent. Culm leaves similar to foliage leaves, the sheaths lacking auricles, fimbriae present or absent. Inflorescence branches extensively branched, with subtending bracts. Spikelets with 1 flower plus a rachilla extension. Glumes 2. Lodicules 3. Ovary glabrous. Fruit with pericarp easily removed from the seed coat.

Four spp., New Caledonia.

Notes on the morphology and phenology of two of the species are provided by Dransfield (2002).

80. ***Melocalamus*** Benth.

Melocalamus Benth., Gen. Pl. 3: 1212 (1883).

Culms scandent. Branches many per node, with one branch dominant. Culm leaves with sheaths persistent, with a wrinkled basal zone, with auricles, with or without fimbriae, with blades reflexed. Foliage leaf sheaths with or without auricles, with or without fimbriae. Inflorescence unbranched, bearing small clusters of pseudospikelets at the nodes. Pseudospikelets with 1 or 2 flowers, the proximal one sterile or fertile, the distal one fertile. Glumes 2, the glumes and lemmas orbicular to oblate. Lodicules 3. *Stigmas sometimes 2.* Ovary with or without a rounded projection at the apex, glabrous. Fruit spherical, the pericarp fleshy.

Five spp., Asia.

81. **Not "*Nastus*"** Juss.

Not "*Nastus*" Juss., Gen Pl. 34 (1789).

Culms erect or scandent. Lateral branches branched, 3 to many, subequal or one or two dominant, generally arising at the node, but in one species just below the node. Culm leaves with sheaths deciduous or persistent, covered with black hairs, with auricles and fimbriae. Foliage leaves with or without auricles, with or without fimbriae. Inflorescence unbranched or the branches branched. Spikelets conventional, with 1 fertile flower; rachilla extension present. Glumes 5 or 6. Lodicules 3. Ovary with or without an enlarged apex, the apex pubescent or not.

Two spp., Asia, Pacific.

Nastus s.s. belongs in Hickeliinae and is native to Madagascar and Réunion. The southeast Asian species fall here and are unrelated to the rest of the genus. The blades of *N. schlechteri* (Pilger) Holttum have obvious papillae on the adaxial surface, visible under low magnification.

82. ***Neomicrocalamus*** Keng f.

Neomicrocalamus Keng f., J. Bamboo Res. 2: 10 (1983); tax.: Stapleton (1994b).

Culms scandent. Branches many per node, the central one dominant. Culm leaves with sheaths deciduous, leaving a rim, auricles absent, *the blade a tiny acicular extension of the sheath.* Foliage leaves small, without auricles or fimbriae, *the apex acicular.* Inflorescence with primary branches only, the spikelets sessile, subtended by bracts and prophylls, but these not bearing axillary buds. Spikelets conventional, with several flowers, the distal ones reduced. Glumes 1 or lacking. Lodicules 3. Ovary apex pubescent.

Five spp., southern Asia.

Neomicrocalamus s.l. is sister to *Bonia* in the phylogeny of Yang et al. (2008).

83. ***Oreobambos*** K. Schum. Fig. 24

Oreobambos K. Schum., Notizbl. Königl. Bot. Gart. Berlin 1: 178 (1896).

Culms scandent or drooping. Branches 2 or 3 per node, more or less in a line, one branch dominant. Culm leaves without auricles, with fimbriae, the blade acicular, short. Foliage leaves without fimbriae, the blades without prominent cross veins. Inflorescence a fascicle of pseudospikelets, surrounded by bracts. Pseudospikelets with 2 flowers. Glume 1, with an obtuse apex. Lemma ovate, the margins convolute, the apex obtuse. Lodicules absent. Stamens 6. *Stigmas 1.* Ovary with a rounded projection at the apex, pubescent at the apex. Pericarp free from the seed coat, soft.

One sp., *O. buchwaldii* K. Schum., Tanzania, Uganda, Zimbabwe, Malawi, Zambia.

84. ***Oxytenanthera*** Munro

Oxytenanthera Munro, Trans. Linn. Soc. London 26: 126 (1868).

Culms erect. Branches many per node, in an irregular line, one branch dominant. Culm leaves without auricles, the blade erect on lower leaves, becoming reflexed on more apical ones. Foliage leaves with blades deciduous, the blades with faint cross-veins, with fimbriae. Inflorescence a capitate cluster of pseudospikelets at the ends of lateral branches. Pseudospikelets with 2 to 5 flowers, the lower 1 to 3 sterile. Glumes 2, hispidulous. Lemmas awned. *Lodicules absent. Stamens with filaments fused at the base.* Ovary with a long hollow apical extension, glabrous.

One sp., *O. abyssinica* (A. Rich.) Munro, Africa.

Molecular data (Yang et al. 2010) place this genus within *Dendrocalamus*, but without support. The ovary extension is sketched by Holttum (1956); it is hollow, with the hollow portion not contiguous with the ovary locule.

85. ***Parabambusa*** Widjaja

Parabambusa Widjaja, Reinwardtia 11: 121 (1997).
Pinga Widjaja, Reinwardtia 11: 123–124 (1997).

Culms scandent, sometimes producing aerial roots at the nodes. Branches 2 to 5, with 1 branch dominant, the dominant branch branching. Culm leaves with sheaths deciduous, leaving a persistent rim, with auricles and fimbriae, the blade erect. Foliage leaves with sheaths with erect auricles and fimbriae, the blades deciduous. Inflorescence a cluster of up to 10 pseudospikelets, the cluster subtended by a bract and a prophyll. Pseudospikelets with 5 to 12 flowers. Glumes 2. Lemma apex acuminate. *Lodicules absent.* Ovary glabrous.

Two spp., New Guinea.

86. ***Pseudobambusa*** T. Q. Nguyen

Pseudobambusa T. Q. Nguyen, Bot. Žurn. (Moscow & Leningrad) 76: 992 (1991).

Culms erect. Branches 1 to 4 per node, subequal. Culm leaves densely hairy, the sheaths with auricles and dense fimbriae. Foliage leaves with blades with 10 to 12 secondary veins. Inflorescence a cluster of pseudospikelets, subtended by bracts. Pseudospikelets with 3 or 4 flowers. Glumes several, the apex of the distal one acute and mucronate. Lemma apex acute and mucronate. *Palea longer than the lemma.* Lodicules 0 or 3. Ovary pubescent.

One sp., *P. schizostachyoides* (Kurz) T. Q. Nguyen, southern Burma and Vietnam.

87. ***Pseudoxytenanthera*** Soderstr. & R. P. Ellis

Pseudoxytenanthera Soderstr. & R. P. Ellis, Smithsonian Contr. Bot. 72: 52 (1988).

Culms erect below, scandent above. Branches many per node, the branches more or less equal or one branch dominant. Culm leaves with sheaths deciduous, leaving a persistent rim or girdle, with auricles, the blade horizontal or reflexed. Foliage leaves with auricles, with fimbriae. Inflorescence unbranched, with clusters of pseudospikelets at the nodes. Pseudospikelets

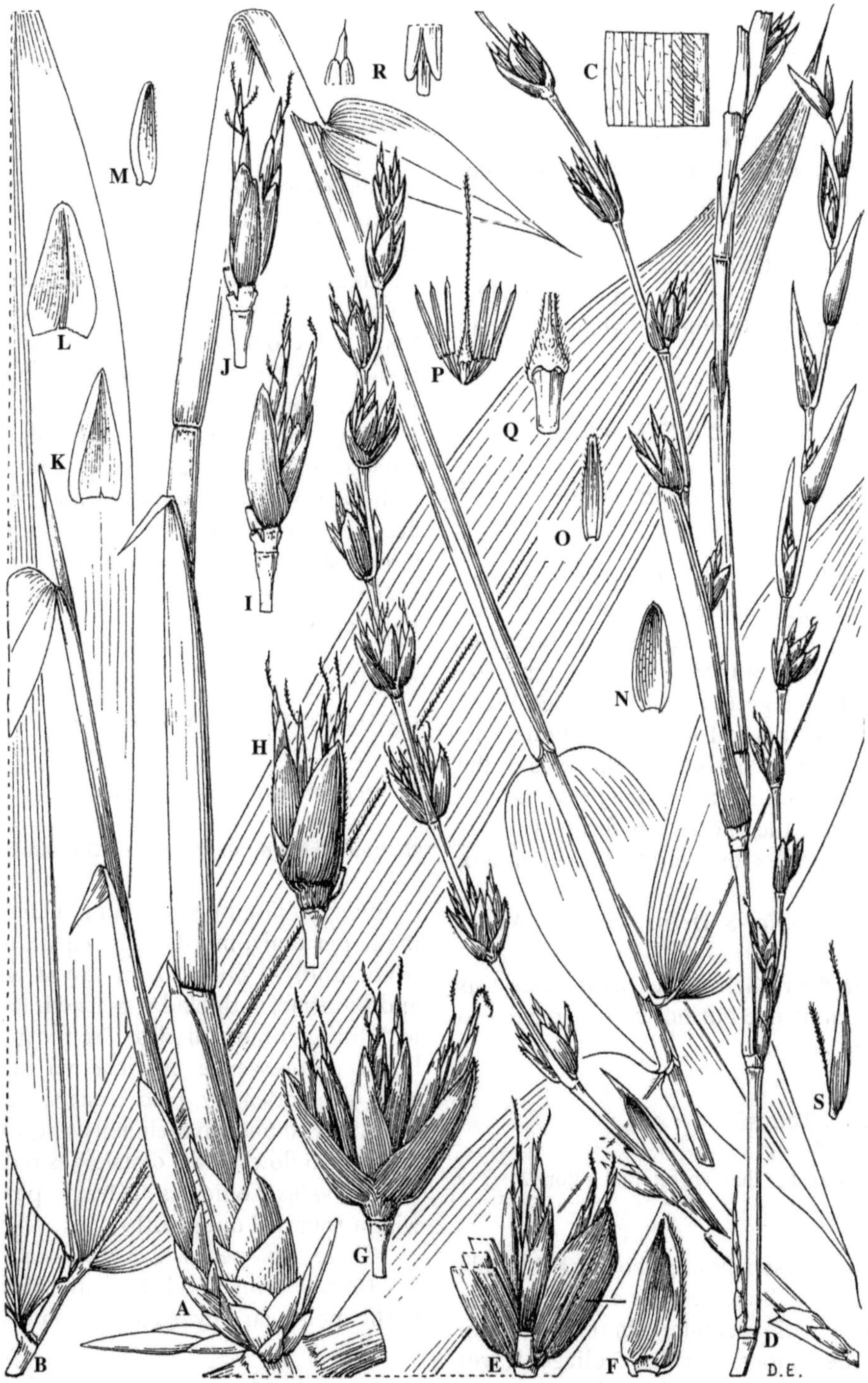

Fig. 24. Bambusoideae-Bambuseae-Bambusinae. *Oreobambos buchwaldii.* **A** Branch with foliage leaves. **B** Foliage leaves. **C** Detail of abaxial surface of leaf blade showing tessellate venation. **D** Flowering branches showing clusters of spikelets. **E** Spikelet cluster, viewed from the adaxial side. **F** Bract removed from spikelet cluster. **G** Spikelet cluster, viewed from the abaxial side. **H** Spikelet cluster, lateral view. **I** Spikelet cluster partly dissected. **J** Two spikelets after removal of bracts. **K, L** Bracts. **M** Upper glume. **N** Lemma. **O** Palea. **P** Gynoecium and androecium. **Q** Ovary. **R** Portions of anther. **S** Terminal flower showing rachilla extension. (From Clayton 1970)

with 1 to 3 flowers, the distal ones reduced. Glumes 2 to several, the upper glume mucronate. Lemma muticous or mucronate. *Lodicules absent. Stamens with filaments fused. Stigmas sometimes 1 or 2.* Ovary with or without a swollen apex, glabrous, or pubescent on the apex, or pubescent throughout. Pericarp sometimes free from the seed coat.

Twelve spp., southern India, Sri Lanka.

The number of species in *Pseudoxytenanthera* is based on Bamboo Phylogeny Group (2012).

88. ***Temburongia*** S. Dransf. & K. M. Wong

Temburongia S. Dransf. & K. M. Wong, Sandakania 7: 55 (1996).

Culms scandent. Nodes with a narrow shelf-like extension. Branches several, clumped, with one branch dominant, from a single bud. Culm leaves with small rounded auricles, without fimbriae, the blade erect. Foliage leaves with auricles. Inflorescence with clusters of 1 to 3 conventional spikelets, bracteate, without buds in the axils of the bracts. Spikelets with 3 flowers, the proximal and distal ones sterile, the middle one fertile. Glumes 2. Lodicules 3. Apex of ovary with an umbo, glabrous.

One sp., *T. simplex* S. Dransf. & K. M. Wong, Brunei.

Molecular data (Goh et al. 2013) place this species in Bambusinae but not closely related to other sampled genera.

89. ***Temochloa*** S. Dransf.

Temochloa S. Dransf., Thai Forest Bull., Bot. 28: 179 (2000).

Culms erect below, scrambling above, ca. 2 m tall; internodes slender. Branches 2 to 4 from a single bud, subequal, with secondary branches. Culm leaves with deciduous sheaths. Foliage leaves lacking auricles or fimbriae. Inflorescence unbranched, with 1 to 4 spikelets. Spikelets conventional, with 1 fertile proximal flower and 1 sterile distal flower. Glumes 2. Ovary glabrous.

One sp., *T. liliana* S. Dransf., Thailand.

90. ***Vietnamosasa*** T. Q. Nguyen

Vietnamosasa T. Q. Nguyen, Bot. Žurn. (Moscow & Leningrad) 75: 221 (1990); tax.: Haevermans et al. (2013); Xia (1996).

Culms generally less than 2 m tall. Branches many, dendroid. Culm leaves with auricles, the blade erect. Foliage leaves with the blade persistent or deciduous at the ligule. Inflorescence branches branched. Spikelets with 5 to 9 flowers plus additional reduced flowers distally. Glumes 2 to several. Lodicules 3. Ovary apex without an appendage.

Three spp., Vietnam.

Vietnamosasa is placed in Bambusinae by the phylogenetic analyses of Sungkaew et al. (2009). Although Watson and Dallwitz (1992 onward) describe the rhizomes as leptomorph, McClure (1966) shows them clearly as pachymorph, consistent with the placement of the genus here in Bambusinae.

Bambusinae: HK clade sensu Goh et al. (2013). *Holttumochloa* + *Kinabaluchloa*

91. ***Holttumochloa*** K. M. Wong

Holttumochloa K. M. Wong, Kew Bull. 48: 518 (1993).

Culms climbing or scrambling. Lateral branches many per node, more or less equal, unbranched, from multiple buds. Culm leaves with auricles and fimbriae, with blades reflexed. Foliage leaves with or without auricles. Pseudospikelets in a linear inflorescence or clustered at the nodes, each subtended by 1 to 3 sterile bracts, 1 or 2 bracts with buds in their axils, and 1 to 2 glumes, with 2 to 5 flowers, the distal ones reduced. *Palea with 10 veins.* Lodicules 3. Ovary apex pubescent, with a rounded outgrowth.

Three spp., peninsular Malaysia.

Holttumochloa and *Kinabaluchloa* are sisters in the molecular phylogeny of Goh et al. (2010).

92. ***Kinabaluchloa*** K. M. Wong

Kinabaluchloa K. M. Wong, Kew Bull. 48: 523 (1993).

Culms leaning or scrambling, with internodes up to 1.5 m long. Nodes ringed with spreading bristles. Branches many, more or less equal, from a single bud. Culm leaves with auricles. Foliage leaves with or without auricles, the ligule membranous or a fringed membrane. Inflorescence a cluster of pseudospikelets, subtended by a large bract. Individual pseudospikelets subtended by a bract as long as the lemma, followed by 1 or 2 bracts with buds in their axils, and 1 glume; flowers 1 to 2. *Palea with 10–16 veins*, the apex dentate. Lodicules 3. Ovary with or without a rounded projection at the tip, sometimes sparsely pubescent.

Two spp., Malay peninsula and Borneo.

Bambusinae: DMNS clade sensu Goh et al. (2013). *Dinochloa + Neololeba + Mullerochloa + Sphaerobambos*

Culms climbing or scrambling. Culm leaf sheath with a characteristic wrinkled zone at the base.

Chloroplast and nuclear DNA sequence data support the close relationship of the following four genera.

93. ***Dinochloa*** Büse

Dinochloa Büse, Pl. Jungh. 387 (1854), rev.: Dransfield (1981).

Culms *spiraling around tree trunks*. Branches 3 to 18 per node, from a single bud, more or less equal in size. Culm leaves with sheaths persistent or deciduous, *with a dark girdle between the node and the base of the sheath*, with or without auricles, the auricles bearing fimbriae, the blade constricted at the base or not, erect or deflexed. Foliage leaves with or without auricles, with fimbriae. Inflorescence elongate, up to 3 m long, the branches branched, each branch subtended by a sheath and a prophyll, the pseudospikelets in clusters separated by long internodes. Pseudospikelets with 1 fertile flower. Glumes 2 or 3. Lodicules 0 or 3. *Stigmas sometimes 2.* Ovary with or without an appendage, glabrous. Pericarp free from the seed coat, fleshy.

Thirty-one spp., mostly tropical Asia.

The fruit of *Dinochloa* is unusual in that the scutellum enlarges at the expense of the endosperm, and fills almost the entire fruit (Rudall and Dransfield 1989). This characteristic also appears in *Ochlandra*, *Melocalamus* and *Melocanna*. The base of each culm internode is swollen and the base of the sheath is wrinkled. This appears to function as a pulvinus, allowing the culm to twine around trees.

94. ***Mullerochloa*** K. M. Wong

Mullerochloa K. M. Wong, Blumea 50: 434–435, f. 1–9 (2005).

Culms with the base of the internodes swollen. Branches 1 per node, or 1 dominant and 2 secondary branches, from a single bud. Culm leaves with sheaths with auricles inconspicuous, the blades spreading to reflexed. Pseudospikelets in clusters at the nodes, bracteate, with 4 to 9 flowers. Palea longer than the lemma, densely pubescent adaxially. Lodicules 3. Stamens 4, filaments fused. Ovary glabrous, the apex not thickened.

One sp., *M. moreheadiana* (F. M. Bailey) K. M. Wong, North Queensland, Australia.

Sphaerobambos also has paleas longer than the lemmas, but unlike *Mullerochloa*, the palea keels are winged. The densely pubescent adaxial side of the palea is also seen in *Soejatmia*.

95. ***Neololeba*** Widjaja

Neololeba Widjaja, Reinwardtia 11: 112 (1997).

Culms branching only above several long proximal internodes. Branches 2 to 6, *very steeply ascending*, with one dominant, the dominant one branching again, the secondary branches unbranched. Culm leaves with light brown hairs, with sheaths persistent or deciduous, auricles present, fimbriae present or absent, the blade erect. Foliage leaves with or without auricles. Inflorescence a cluster of pseudospikelets, most sessile but one pedicellate. Pseudospikelets with 3 to 12 flowers, the distal flowers reduced in one species. Glumes 2 or 3, the apices long acuminate, mucronate. Lemma acuminate. *Lodicules absent.* Ovary with a rounded projection at the apex, pubescent at the apex or throughout.

Five spp., tropical Asia, Australia, Pacific.

96. ***Sphaerobambos*** S. Dransf.

Sphaerobambos S. Dransf., Kew Bull. 44: 428 (1989).

Culms with branches several to many per node, dendroid, with one branch dominant. Culm leaf sheaths with fimbriate auricles, the blade erect, sometimes becoming reflexed. Foliage leaves with fimbriate auricles. Inflorescence with clusters of a few pseudospikelets at the nodes. Pseudospikelets with 3 to 5 flowers. Glumes 2 to several. *Palea with narrow winged keels, the palea longer than the lemma at maturity*. Lodicules absent. Ovary with or without a solid swollen structure at the apex, glabrous.

Three spp., Borneo, the Philippines, Sulawesi. See comments under *Mullerochloa*.

BAMBUSINAE: BDG COMPLEX sensu Goh et al. (2010, 2013). ALL REMAINING GENERA.

Nuclear and chloroplast phylogenies are strongly incongruent for this group (Goh et al. 2013), suggesting either incomplete lineage sorting or hybridization. This could justify combining all taxa into a single large genus.

97. ***Bambusa*** Schreb. Fig. 25

Bambusa Schreb., Gen. Pl. 1: 236 (1789), nom. conserv.; phylog.: Goh et al. (2010).
Ischurochloa Büse, Pl. Jungh.: 389 (1854).
Leleba Rumph., Catal. Plantarum in Horto Botanico Bogoriensi 20 (1866).
Tetragonocalamus Nakai, J. Jap. Bot. 9: 86, 88–89 (1933).
Lingnania McClure, Lingnan Univ. Sci. Bull. 9: 34 (1940).
Dendrocalamopsis Q.H. Dai & X.L. Tao, Acta Phytotax. Sin. 20: 210 (1982).
Neosinocalamus Keng f., J. Bamboo Res. 2: 12 (1983).
Phuphanochloa Sungkaew & Teerawat., Kew Bull. 63: 669–671, f. 1 (2008).

Culms erect or scrambling. Branches many per node, sometimes with 1 to 3 dominant. Culm leaves with sheaths deciduous, with or without auricles, with marginal fimbriae, the blades erect. Foliage leaves without auricles, with or without fimbriae. Inflorescence with clusters of pseudospikelets at the nodes, with a 2-keeled prophyll. Pseudospikelets with 2 to many flowers, with a sterile or imperfect terminal flower; proximal 1 to several flowers sterile. Rachilla nodes disarticulating. Lodicules 2 or 3. Ovary borne on a stalk, sometimes with an enlarged apical portion, the apex pubescent.

One hundred species, tropical and subtropical Asia.

The inclusion of *Dendrocalamopsis* in *Bambusa* is supported by the molecular phylogeny of Yang et al. (2010). Although they could not rule out monophyly for *Bambusa s.s.*, Goh et al. (2010) find no support for the existing subgeneric classification.

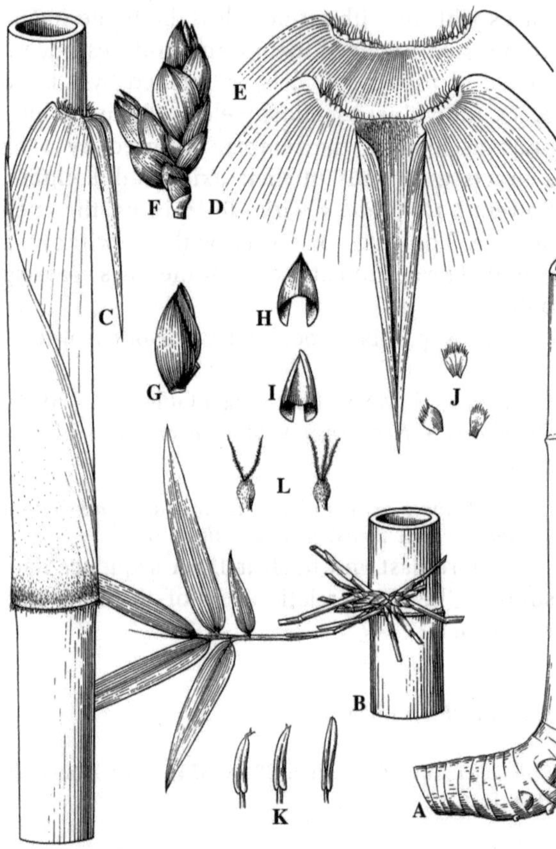

Fig. 25. Bambusoideae-Bambuseae-Bambusinae. *Bambusa chungii*. **A** Portion of culm base and rhizome. **B** Portion of culm, showing branching pattern. **C** Portion of young culm, showing culm sheath. **D** Abaxial view of apex of culm sheath. **E** Same, adaxial view. **F** Portion of inflorescence. **G** Floret. **H** Lemma. **I** Palea. **J** Lodicules. **K** Stamens. **L** Pistil. (From Wu et al. 2007, p. 23, with permission from the Missouri Botanical Garden Press, St. Louis, and Science Press, Beijing; drawn by Huang Yingqin)

98. ***Dendrocalamus*** Nees

Dendrocalamus Nees, Linnaea 9: 476 (1835); tax.: Stapleton (1994a).
Klemachloa R. Parker, Indian Forester 58: 7 (1932).
Sinocalamus s.s., McClure, Lingnan Univ. Sci. Bull. 9: 66 (1940).
Neosinocalamus Keng f., J. Bamboo Res. 2: 12 (1983).

Culms arborescent, caespitose, apex of the stem generally pendulous. Branches several to many at each node, none to 3 dominant. Culm leaves with sheaths deciduous, auricles absent or reduced, fimbriae present or absent, the blade reflexed or erect. Foliage leaves with or without short auricles, without fimbriae, sometimes lacking a pseudopetiole. Inflorescence generally unbranched, bearing clusters of pseudospikelets at the nodes. Pseudospikelets with 1 to 2 fertile flowers, sessile. Rachilla nodes not disarticulating. Lemma with a long mucro. Lodicules 0 to 3. Ovary on a stalk, the apex thick, pubescent. $2n = 70 + 2$.

Forty-one species, Asia.

In the phylogenies of H.-Q.Yang et al. (2008) and J.-B. Yang et al. (2010), species of *Dendrocalamus* form a clade within which members of *Gigantochloa* and *Oxytenanthera* are embedded. Unifying the three genera was suggested by Soderstrom and Ellis (1987), although Yang et al. (2010) argue that this conclusion is still premature. Yang et al. (2010) show that the type of *Sinocalamus* should be included in *Dendrocalamus*, but that other species assigned to *Sinocalamus* belong to *Bambusa*. The chromosome number report is based on Li et al. (2001).

99. ***Gigantochloa*** Kurz ex Munro

Gigantochloa Kurz ex Munro, Trans. Linn. Soc. London 26: 123 (1868).

Culms erect, densely caespitose. Internodes *often with yellow stripes*. Branches several per node, 1 dominant. Culm leaves with deciduous, densely hairy sheaths, auricles absent or reduced, the blade recurved or erect. Foliage leaves without auricles, with or without fimbriae. Inflorescence with clusters of pseudospikelets at the nodes. Pseudospikelets with 2 to 5 flowers, the terminal one reduced to a lemma. Empty bracts proximal to the glumes. Rachilla nodes not disarticulating. Lodicules absent. *Stamens with filaments fused*. Ovary on a stalk, the apex thick and hairy, *stigma 1*. $2n = 76$.

Thirty spp., India and southeast Asia.

Molecular data (Yang et al. 2010) place this genus in a clade with *Dendrocalamus*. It is possible that the two genera should be merged, but the phylogeny is not sufficiently well resolved to be convinced of this.

100. ***Maclurochloa*** K. M. Wong

Maclurochloa K. M. Wong, Kew Bull. 48: 528 (1993).

Culms climbing or scrambling, with a thick band of silvery white hairs below each node. Branches several per node, with 1 dominant, from a single bud. Culm leaves with sheaths bearing low rim-like auricles, with blades lanceolate. Inflorescence a cluster of pseudospikelets. Pseudospikelets with 1 to 2 flowers. Glumes 3 to 5, the distalmost 2 as large as the lemmas. Lodicules 3. Ovary with a rounded projection at the apex.

One sp., *M. montana* (Ridl.) K. M. Wong, Malay Peninsula.

101. ***Soejatmia*** K. M. Wong

Soejatmia K. M. Wong, Kew Bull. 48: 530 (1993).

Culms climbing or scrambling. Branches 3 per node. Culm leaves with sheaths with a wrinkled zone at the base, with auricles and fimbriae, the blade erect. Foliage leaves with fimbriae, with large, papery, ovate to rounded auricles. Inflorescence with clusters of pseudospikelets at the nodes. Pseudospikelets with 4 to 6 flowers. Glumes several. *Palea apex with two hooked projections*. Lodicules 3. Apex of ovary thickened, pubescent.

One sp., *S. ridleyi* (Gamble) Wong, Malay Peninsula and Singapore.

Clayton et al. (2006 onward) synonymize *Soejatmia* with *Kinabaluchloa*, but Goh et al. (2013) show that the two genera are unrelated with *Soejatmia* falling here in the large BDG complex.

102. ***Thyrsostachys*** Gamble

Thyrsostachys Gamble, Indian Forester 20: 1 (1894).

Culms erect, caespitose. Branches several to many per node, with 1 dominant. Culm leaves with persistent sheaths, without fimbriae, with inconspicuous auricles, the blade erect or recurved. Foliage leaves without auricles or fimbriae. Inflorescence elongate with branched branches. Pseudospikelets in clusters of 1 to 3 fertile ones and 1 or 2 sterile ones, the cluster subtended by a prophyll. Each pseudospikelet with a prophyll and 2 to 4 flowers. *Paleas of lower flowers deeply cleft*. Glumes 2 to 4. Lodicules 0 to 3. Ovary on a stalk, glabrous. $2n = 76$.

Two spp., Burma, Thailand, China.

V.5. Subtribe Melocanninae Benth. (1881)

Schizostachyidinae Soderstr. & R. P. Ellis in Soderstr. (1987).

Culms hollow. Foliage leaf blades generally with a narrow waxless stripe along the abaxial margin. Flowers borne in pseudospikelets, in clusters at the nodes, subtended by bracts, the spikelets generally with 2 glumes. Stamens 6. *Style elongate, hollow.* Stigmas generally 3. *Ovary glabrous.* Midrib with complex vasculature. *Leaf cross section asymmetrical around the midrib, with a set of prominent bulliform cells on one side only, creating a concave outline on one side of the midrib and convex on the other.* $2n = 72$.

Nine genera, 83 species.

Melocanninae are sister to all other paleotropical woody bamboos in the analyses of Sungkaew et al. (2009). Triplett et al. (2014) found a similar result albeit with a more limited sample of taxa. The asymmetric leaf cross section is described by Soderstrom and Ellis (1987) as an S-shaped midrib, a term that describes the shape when viewed in cross section.

103. ***Cephalostachyum*** Munro

Cephalostachyum Munro, Trans. Linn. Soc. London 26: 138 (1868); tax.: Stapleton (1994a).

Culms erect or scandent. Branches many per node, subequal. Culm leaf sheaths with 1 to 2 strong undulations near the apex, deciduous, with auricles and fimbriae, the blade reflexed. Foliage leaves with or without auricles and fimbriae, the blades without distinct cross veins. Inflorescence capitate, or an unbranched axis, with capitate clusters of pseudospikelets. Pseudospikelets with 1 fertile flower plus 1 or 2 distal reduced flowers. Glumes 2 or 3, awned. Lodicules 3. Stigmas 2 or 3. Ovary on a stalk. Pericarp free from the seed coat.

Twelve spp., southeast Asia.

Two Madagascan species formerly placed in *Cephalostachyum* belong in Hickeliinae and will need a new generic name (L.G. Clark, pers. comm.).

104. ***Davidsea*** Soderstr. & R. P. Ellis

Davidsea Soderstr. & R. P. Ellis, Smithsonian Contr. Bot. 72: 59 (1988).

Culms arching. Branches several per node, with 1 branch dominant, from a single bud. Culm leaves with sheaths deciduous, leaving a persistent girdle or rim, the blades horizontal or reflexed. Foliage leaves with sheaths with auricles, with fimbriae, the blade deciduous. Inflorescence a dense cluster of pseudospikelets, pseudospikelets with 1 to 3 flowers, the distal ones reduced. Glumes several. Lodicules 3. Stigmas 3.

One sp., *D. attenuata* (Thwaites) Soderstr. & R. P. Ellis, Sri Lanka.

105. ***Melocanna*** Trin.

Melocanna Trin., Neue Entdeck. Pflanzenk. 2: 43 (1820); tax.: Stapleton (1994a); flowering: Ramanayake and Weerawardene (2003).

Culms erect, arborescent. Rhizomes with elongated necks. Branches several to many per node, more or less equal. Culm leaves with persistent sheaths, with 1 to 2 strong undulations toward the apex, without auricles or fimbriae, the blade erect, at least proximally. Foliage leaves without auricles, with fimbriae, with or without obvious cross veins. Inflorescence elongate, the branches branched, bearing clusters of pseudospikelets. Pseudospikelets with 1 fertile flower plus several sterile flowers distally. Glumes 2 to 4, with buds in their axils. Paleas lacking keels. Lodicules 2. Stigmas 2 to 4. Pericarp fleshy, with an apical beak.

Two spp., eastern India to Burma.

106. ***Neohouzeaua*** A. Camus

Neohouzeaua A. Camus, Bull. Mus. Natl. Hist. Nat. 28: 100 (1922).

Culms erect or scandent. Branches 1 to several per node, more or less equal or one dominant. Culm leaves with persistent or deciduous sheaths, with auricles and fimbriae, the blade reflexed. Foliage leaves with or without auricles, with short fimbriae. Inflorescence elongate, branched. Pseudospikelets with 1 flower. Glumes 3 or 4 with buds in their axils. *Palea with 2 awns.* Lodicules 0 or 2. *Stamens with filaments fused.* Stigmas 3. Pericarp free from seed coat.

Seven spp., temperate to tropical Asia.

107. ***Ochlandra*** Thwaites

Ochlandra Thwaites, Enum. Pl. Zeyl. 376 (1864).

Culms erect. Branches several per node, in a line, with one dominant. Culm leaves with sheaths persistent or deciduous, with auricles and fimbriae, the blade reflexed. Foliage leaves with or without auricles, without fimbriae, the blades persistent or deciduous, the venation with or without cross veins. Inflorescence unbranched, with clusters of pseudospikelets at the nodes. Pseudospikelets more or less terete, with several sterile flowers proximally, and 1 fertile flower. Lodicules 1 or 3. *Stamens numerous*. Stigmas 3 to 9. Caryopsis with a fleshy pericarp.

Nine spp., southern India and Sri Lanka.

108. ***Pseudostachyum*** Munro

Pseudostachyum Munro, Trans. Linn. Soc. London 26: 141, pl. 4 (1868); tax.: Stapleton (1994a).

Culms erect, drooping at the apex. Rhizomes with elongated necks. Branches many per node, subequal. Culm leaves with the sheath deciduous, with small auricles, with fimbriae, the blade erect. Foliage leaf blades without auricles or fimbriae. Inflorescence branches branched, each branch subtended by a bract. Pseudospikelets pedicellate, with 1 flower. Glume 1, mucronate to long awned. Lodicules 3 to 5. Stigmas 2. Fruit globose; pericarp free from seed coat, brittle.

One sp., *P. polymorphum* Munro, Bhutan, China, NE India, Myanmar, Vietnam.

In the protologue, Munro comments that "perfect flowers are very rarely found in this species."

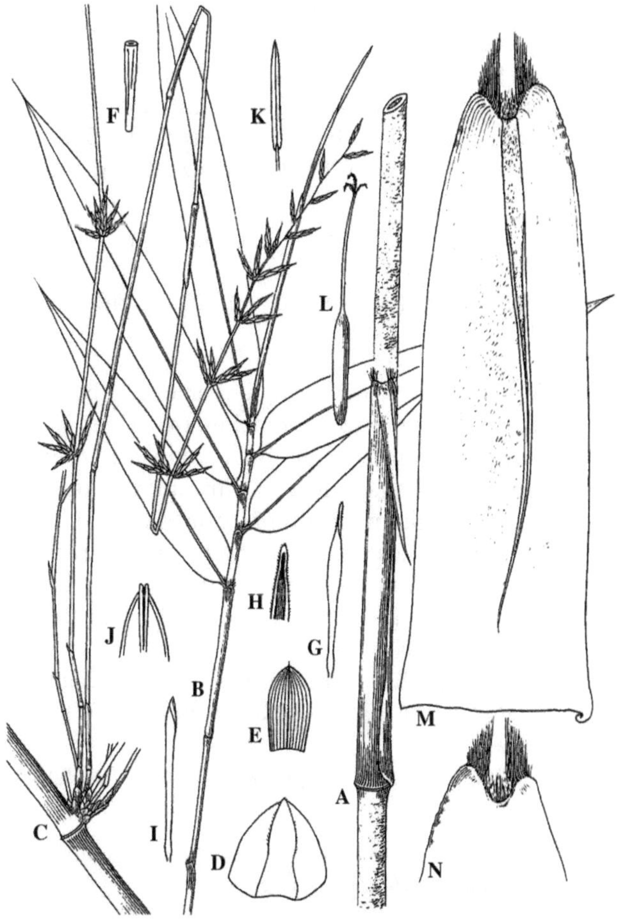

Fig. 26. Bambusoideae-Bambuseae-Melocanninae. *Schizostachyum dumetorum*. **A** Portion of culm and culm sheath, showing reflexed leaf blade. **B** Branch with foliage leaves. **C** Flowering branch. **D** Prophyll. **E** Bract. **F** Apical rachis joint. **G** Lemma. **H** Apex of lemma. **I** Palea. **J** Apex of palea. **K** Stamen. **L** Pistil. **M, N** *S. hainanense*. **M** Abaxial view of culm sheath. **N** Adaxial view of apical portion of culm sheath. (From Wu et al. 2007, p. 55, with permission from the Missouri Botanical Garden Press, St. Louis, and Science Press, Beijing)

109. ***Schizostachyum*** Nees Fig. 26

Schizostachyum Nees, Fl. Bras. Enum. Pl. 2: 535 (1829); phylog.: Yang et al. (2007).
Dendrochloa C. E. Parkinson, Indian Forester 59: 707 (1933).
Leptocanna L. C. Chia & H. L. Fung, Acta Phytotax. Sin. 19: 212–213 (1981).

Culms erect to scandent, or drooping at the apex. Branches many per node, subequal, from a single bud. Culm leaf sheaths with 1 to 2 strong undulations near the apex, deciduous, leaving a narrow rim, auricles inconspicuous, with fimbriae, the blade reflexed. Foliage leaves with or without auricles and fimbriae. Inflorescence an unbranched axis, each node with one or a few elongate pseudospikelets. Pseudospikelets with 1 to several bracts with buds in their axils, with up to 4 flowers but only the terminal 1 or 2 fertile. Glumes generally absent. Lodicules 0 to 3. Ovary on a stalk. Style 1, stigmas 3. Fruit fusiform, the style base persistent.

Fifty spp., Asia, Pacific islands.

Although most species have culm leaves with the blade reflexed, in *S. chinense* Rendel the culm sheath blade is erect. *Dendrochloa* is placed in synonymy here following the suggestion of Soderstrom and Ellis (1987), although the Bamboo Phylogeny Group (2012) recognize it as distinct.

110. ***Stapletonia*** P. Singh, S. S. Dash & P. Kumari

Stapletonia P. Singh, S. S. Dash & P. Kumari, Nelumbo 51: 241 (2009).

Culms erect below, scandent above. Rhizomes with elongate necks. Branches (4) 6 per node, subequal, from a single bud. Culm leaf sheaths deciduous. Foliage leaves with auricles bearing 1 cm long fimbriae. Inflorescence a large head. Pseudospikelets with 1 flower, with a rachilla extension. Glumes 6 to 8, the apex aristate. Ovary on a stalk. Style 1, stigmas 2. Fruit fleshy, ca. 5 cm across.

One sp., *S. arunachalensis* (H.B. Naithani) P. Singh, S.S. Dash & P. Kumari, Arunachal Pradesh, India.

Stapletonia is segregated from *Schizostachyum*.

111. ***Teinostachyum*** Munro

Teinostachyum Munro, Trans. Linn. Soc. London 26: 142, t. 3 (1868); tax.: Stapleton (1994a).

Culms erect below, scandent above, slender. Branches several per node, subequal or with one dominant. Culm leaves with the sheath margin thickened, fimbriae present, the blade rolled. Foliage leaf sheaths with fimbriae. Inflorescence unbranched, the pseudospikelets solitary or in clusters. *Pseudospikelets with several flowers separated by long rachilla internodes.* Glumes 2 to many. Lodicules 3. Stigmas 2 or 3. Fruit baccate, with a beak.

Two spp., Burma and India.

V.6. Subtribe Hickeliinae A. Camus (1924)

Nastinae Soderstr. & R. P. Ellis in Soderstr. (1987).

Rhizomes pachymorph. *Foliage leaf blades with adaxially projecting midribs.* Flowers borne in conventional spikelets. Spikelets with one fertile flower. Glumes 4 to 6. Stamens 6. Style 1. Stigmas 3.

Eight genera, 11 spp., Africa, Madagascar, and Réunion.

112. ***Cathariostachys*** S. Dransf.

Cathariostachys S. Dransf., Kew Bull. 53: 388–389 (1998).

Culms erect, the tips arching; *young shoots with black hairs.* Rhizome necks up to 4 m long. Branches several to many at each node, with 1 dominant, from a single bud. Culm leaf sheaths persistent. Inflorescences dense, capitate, the branches branched, *lateral branches arranged in tight fans.* Spikelets with 5 glumes. Lodicules absent. Stigmas 3. Fruit stalked.

Two spp., Madagascar.

One species is eaten by endemic lemurs.

113. ***Decaryochloa*** A. Camus

Decaryochloa A. Camus, Bull. Soc. Bot. France 93: 242 (1947); tax.: Dransfield (1997).

Culms scandent; *internodes covered with black hairs when young.* Rhizome necks elongate. Branches many in a clump. Culm leaf sheaths auriculate, the auricles adaxially pubescent, with fimbriae. Foliage leaves with erect auricles, with fimbriae. Inflorescence unbranched, surrounded by a spathe, with 1 to 4 spikelets, the spatheate units clustered. Glumes 3 or 4. Lodicules 3. Stigmas 3. Ovary umbonate, pubescent. Fruit fleshy.

One sp., *D. diadelpha* A. Camus, Madagascar.

The fimbriae on the auricles of the foliage leaves are 8–12 mm long, which according to the protologue "permettant de reconnaître cette espèce en l'absence d'épillets". In one specimen observed, the stamens are diadelphous, in two groups of three; in other specimens, however, the stamens are free.

114. ***Hickelia*** A. Camus Fig. 27

Hickelia A. Camus, Compt. Rend. Hebd. Séances Acad. Sci. 179: 479 (1924); rev.: Dransfield (1994).
Pseudocoix A. Camus, Compt. Rend. Hebd. Séances Acad. Sci. 179: 478 (1924).

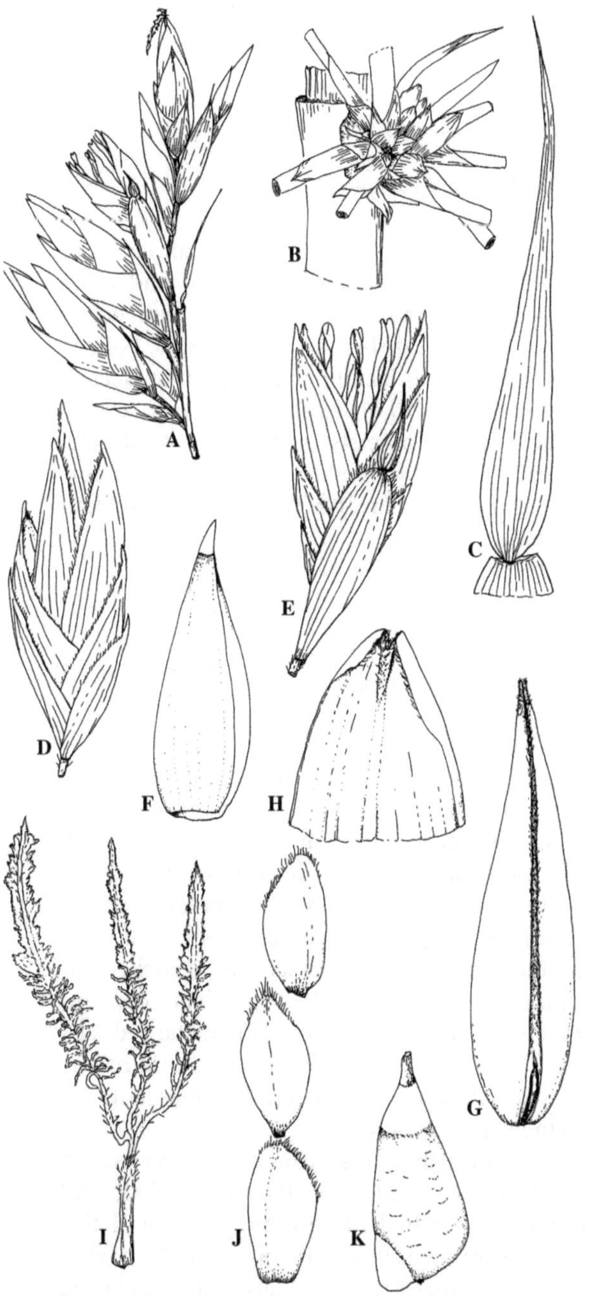

Fig. 27. Bambusoideae-Bambuseae-Hickeliinae. *Hickelia africana*. **A** Inflorescence. **B** Branches. **C** Culm sheath. **D** Spikelet before anthesis. **E** Anthetic spikelet. **F** Lemma. **G** Palea. **H** Apex of palea, viewed from inside. **I** Gynoecium. **J** Lodicules. **K** Caryopsis. (From Dransfield 1994, drawn by the author)

Culms erect, the upper part leaning. Rhizomes elongate or thickened. Branches many per node, one dominant, the others thinner, from a single bud. Culm leaves with sheaths lacking auricles, the blade erect initially, sometimes spreading at maturity. Foliage leaves lacking auricles. Inflorescence unbranched or the branches branched, if the latter then each branch subtended by a bract and prophyll. Glumes 3 to 5. Lodicules 3. Ovary umbonate, the apex pubescent, stigmas 3.

Four spp., Madagascar and eastern Africa.

115. ***Hitchcockella*** A. Camus

Hitchcockella A. Camus, Compt. Rend. Hebd. Séances Acad. Sci. 181: 253 (1925).

Culms scandent. Rhizomes unknown. Branches 3 to several. Inflorescence bracteate, terminal on lateral branches. Spikelets with 1 sterile flower proximal to the fertile one. Glumes 2. Lodicules 3. Stigmas 2. Ovary glabrous, rostrate.

One sp., *H. baronii* A. Camus, Madagascar.

Camus describes *Hitchcockella* as similar to *Perrierbambus*; *Hitchcockella* is distinguished from the latter genus by its strongly laterally compressed spikelets with keeled glumes and lemmas (vs. nearly terete and non-keeled in *Perrierbambus*) and its long style that persists in the fruit. The protologue provides little information on the leaves.

116. ***Nastus*** Juss., s.s.

Nastus Juss., s.s., Gen Pl. 34 (1789).

Culms arborescent or scandent. Lateral branches branched, 3 to many, subequal or one or two dominant, generally arising at the node. Culm sheath deciduous or not, auriculate. Foliage leaves with sheaths with or without auricles, with or without fimbriae. Inflorescence unbranched or the branches branched, surrounded by the subtending leaf sheath. Spikelets with rachilla extension absent. Glumes 5 or 6. Lodicules 3. Ovary with or without an enlarged apex, the apex pubescent or not.

Twenty spp., Réunion, Madagascar.

The type species of *Nastus*, *N. borbonicus* J.F. Gmel., is clearly placed here in Hickeliinae (L.G. Clark, pers. comm.), along with the Madagascan species of the genus. The southeast Asian species, however, belong in Bambusinae and need to be placed in a separate genus; available names are *Chloothamnus* Büse and *Oreiostachys* Gamble. These are based on Southeast Asian material and are not synonyms of *Nastus* s.s.

In some species, the nodal line is curved downward. This has led some authors to report the bud complement incorrectly as being below the node, whereas it is the nodal line itself whose position has shifted. This character also appears convergently in some species of *Chusquea*.

117. ***Perrierbambus*** A. Camus

Perrierbambus A. Camus, Bull. Soc. Bot. France 71: 697 (1924).

Culms erect. Branches numerous, thinner than the culm. Culm leaves without auricles or fimbriae. Foliage leaves with conspicuous transverse veins. Inflorescences terminal on each lateral branch, unbranched, with 1 to 3 spikelets hidden by the distalmost leaves of the branch. Glumes 1 or 2, their apices acuminate. Lodicules 3. Stigmas 2 or 3, *style branches 2 or 3, distinct to the base.* Ovary apex puberulent, extended to a narrow beak. Fruit with pericarp free from seed coat.

Two spp., Madagascar.

The beak of the ovary may actually correspond to a short style. The protologue notes "folia caduca", but it is unclear whether this refers to the culm leaves or foliage leaves.

118. ***Sirochloa*** S. Dransf.

Sirochloa S. Dransf., Kew Bull. 57: 965–966, f. 1 (2002).

Culms clumped, erect below, scrambling above. Branches several per node with one dominant, from a single bud. Culm leaves with sheaths with a rugose basal portion, "breaking up during branch development", without auricles, with blades spreading. Foliage leaves with auricles prominent, with fimbriae. Inflorescence terminal on lateral branches, unbranched, bracteate. Spikelets with a rachilla extension. Glumes 4. Lodicules 3. Stigmas 3. Ovary glabrous, style short or elongate. Fruit with pericarp free from seed coat, brittle.

One sp., *S. parvifolia* (Munro) S. Dransf., Madagascar.

119. ***Valiha*** S. Dransf.

Valiha S. Dransf., Kew Bull. 53: 380 (1998).

Culms erect, the tips arching. Rhizomes with necks elongate. Branches many per node, the central one dominant, from a single bud below the node. Culm leaf sheaths with auricles. Foliage leaves with sheaths lacking auricles. Inflorescence branches branched, bracteate, the bracts lacking axillary buds. Spikelets sometimes subterete. Glumes 5 or 6. *Lemma with 29 to 41 veins.* Lodicules 3. Stigmas 3. Fruit with pericarp free from seed coat.

Two spp., Madagascar.

V.7. Subtribe Racemobambosinae Stapleton (1994)

Culms generally scandent, over 3 m tall. Rhizomes pachymorph. Branches several to many, with one branch dominant, from a single bud. Culm leaves with sheaths deciduous, with small rim-like fimbriate auricles, the blades erect or reflexed. Foliage leaf sheaths generally with auricles and fimbriae. Inflorescence branched or unbranched ("racemose"). Spikelets conventional, with 3 to 8 flowers plus one distal reduced flower. Glumes 2 or 3. Stamens 6. Stigmas 3. Lodicules 3. Ovary apex generally pubescent.

One genus, 17 spp.

This subtribe was originally erected to include *Neomicrocalamus*, *Racemobambos*, and *Vietnamosasa*, which were thought to be distinguished from members of Bambusinae by having conventional spikelets. The analyses of Yang et al. (2008) place *Neomicrocalamus* plus two species formerly in *Racemobambos* with *Bonia* in Bambusinae. Specifically, *Racemobambos yunnanensis* is sister to *Melocalamus*, whereas *R. prainii* is sister to *Neomicrocalamus*, with which it has been combined. *Vietnamosasa* also is now placed in Bambusinae. Once these species are removed, *Racemobambos* appears to be monophyletic.

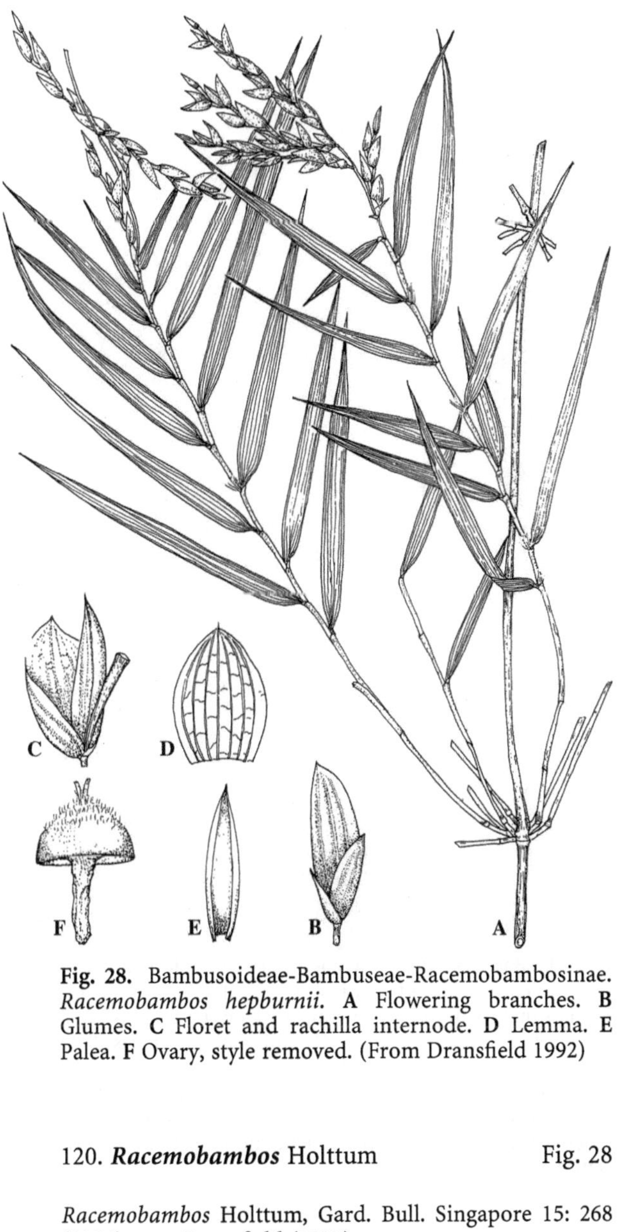

Fig. 28. Bambusoideae-Bambuseae-Racemobambosinae. *Racemobambos hepburnii*. **A** Flowering branches. **B** Glumes. **C** Floret and rachilla internode. **D** Lemma. **E** Palea. **F** Ovary, style removed. (From Dransfield 1992)

120. ***Racemobambos*** Holttum Fig. 28

Racemobambos Holttum, Gard. Bull. Singapore 15: 268 (1956); rev.: Dransfield (1992).

Characters as for the subtribe.

Seventeen spp., southeast Asia.

The dominant branch at a given node sometimes develops late, so that the minor branches appear to form a cluster.

VI. Tribe Olyreae Martinov (1820)

Buergersiochloeae S.T. Blake (1946).
Parianeae C.E. Hubb. (1934).

Culms herbaceous, rarely reed-like, caespitose. Culm leaves usually not differentiated from foliage leaves. Outer ligule absent. Leaf sheaths generally lacking auricles and fimbriae. Inflorescences generally axillary and terminal, sometimes borne on a leafless culm separate from the vegetative culm. *Plants monoecious*, generally with large pistillate spikelets and small staminate spikelets. Spikelets often dorsiventrally compressed, *with one flower, without a rachilla extension*. Pistillate spikelets with the pedicel apex generally wide and club-shaped; glumes 2, generally longer than the flower; lodicules 3; stigmas 2 (rarely three). Staminate spikelets generally lacking glumes; lemma and palea membranous; lodicules 0 or 3; stamens 2 or 3.

Twenty-one genera, 122 species.

We know little about the phylogeny of this group. *Buergersiochloa* has been supported as sister to the rest by both morphological (Kellogg and Watson 1993) and molecular (Bouchenak-Khelladi et al. 2008; Oliveira et al. 2014; Zhang and Clark 2000) data, and is the only genus that does not occur in the New World. The remaining genera share the synapomorphy of crenate and cross-shaped silica bodies, and all but *Pariana* and *Eremitis* have hardened lemmas and paleas in the pistillate spikelets. The latter two genera are clearly sisters, with strong morphological and molecular synapomorphies. *Sucrea* and *Raddia* are sisters in molecular phylogenies, a relationship that is consistent with their distinctive glume morphology (Oliveira et al. 2014; Zhang and Clark 2000). "*Olyra*" is apparently paraphyletic with *Lithachne* and *Arberella* derived within it (Oliveira et al. 2014), consistent with the suggestion of Judziewicz et al. (1999). As currently circumscribed, "*Olyra*" is defined as much by plant size as anything else.

The Bamboo Phylogeny Group (2012) note that there is no morphological synapomorphy for this tribe, although monoecy may be synapomorphic within the context of the subfamily. A major criterion for distinguishing genera of Olyreae is the distribution of the sexes in the inflorescence. This character may simply be highly labile over evolutionary time, and may not be a good indicator of relationship. However, testing this requires more phylogenetic, developmental, and genetic work.

The treatment here is based heavily on that of Judziewicz et al. (1999), which should be consulted for more extensive descriptions of genera, species, their geographic ranges, and their ecology.

VI.1. Subtribe Buergersiochloinae (S. T. Blake) L. G. Clark & Judz. (2007)

Culms dimorphic, the vegetative stems leafy, *flowering stems with leaf blades strongly reduced or entirely lacking.* Inflorescence branches branched, bearing pistillate spikelets distally, staminate proximally. Pistillate spikelet with glumes 2, shorter than the flower; lemma leathery but not hardened, pubescent, awned; staminodes 3; ovary with style 1, stigmas 2. Staminate spikelet with glumes lacking; lemma with a mucro or awn; stamens 2 or 3, the filaments becoming fused near anthesis. Lodicules 3.

121. ***Buergersiochloa*** Pilg. Fig. 29

Buergersiochloa Pilg., Bot. Jahrb. Syst. 52: 167 (1914).

Characters of the subtribe.
One sp., *B. bambusoides* Pilg., New Guinea.

The leaf epidermal structure and internal leaf anatomy are unusual and not obviously similar to other Bambusoideae (Soderstrom and Ellis 1987).

Remaining genera: *Leaf epidermis with cross-shaped and crenate silica bodies, the former over the vascular tissue (costal zone) and the latter over the mesophyll (intercostal zone).*

All species are restricted to the New World, except for "*Olyra*" *latifolia*, which also occurs in Africa where it is either native or naturalized.

VI.2. Subtribe Parianinae Hack. in Engler & Prantl (1887)

Plants caespitose, the clumps connected by rhizomes or stolons. Top of leaf sheath with thickened auricles bearing fimbriae. Inflorescence terminal only, unbranched. Pistillate spikelet terminal, *surrounded by whorls of staminate spikelets with flattened, hardened pedicels. Stamen filaments fused.*

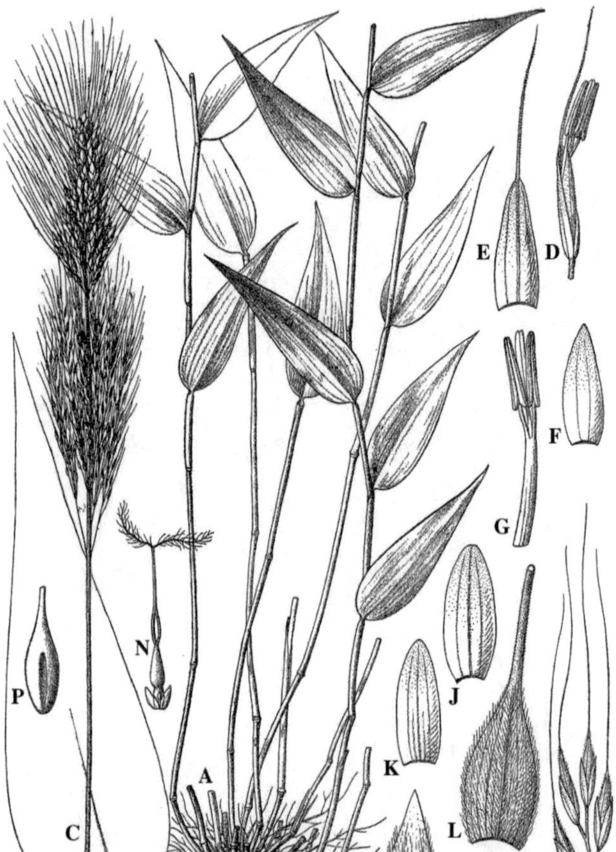

Fig. 29. Bambusoideae-Olyreae-Buergersiochloinae. *Buergersiochloa bambusoides.* **A** Habit. **B** Leaf. **C** Inflorescence. **D** Staminate spikelet. **E** Lemma of staminate spikelet. **F** Palea of staminate spikelet. **G** Androecium. **H** Pistillate spikelets. **J, K** Glumes of pistillate spikelets. **L** Lemma. **M** Palea. **N** Pistil, lodicules and staminodes. **O** Base of pistil. **P** Half-mature fruit. (From Pilger 1915, with permission from E.Schweizerbart'sche Verlagsbuchhandlung OHG, www.schweizerbart.de)

122. ***Eremitis*** Döll

Eremitis Döll, Fl. Bras. 2, 2: 338 (1877); tax.: Ferreira et al. (2013).

Inflorescences on erect leafy culms, on trailing culms with reduced leaves, and *on subterranean culms with inflated spathe-like leaves.* Pistillate spikelets with glumes with 1 vein. *Staminate spikelets on long fused pedicels in distal whorl,* shorter pedicels in lower whorls; lodicules 0 or similar to stamens; stamens 2. $2n \geq 60$, 67–94.

Three spp., Atlantic Brazil.

Two of the three species of *Eremitis* have papillae on the adaxial surfaces of their leaves; these papillae refract light and create blue iridescence, which may be an adaptation to shade (Ferreira et al. 2013).

123. ***Pariana*** Aubl. Fig. 30

Pariana Aubl., Hist. Pl. Guiane 2: 876, pl. 337 (1775).

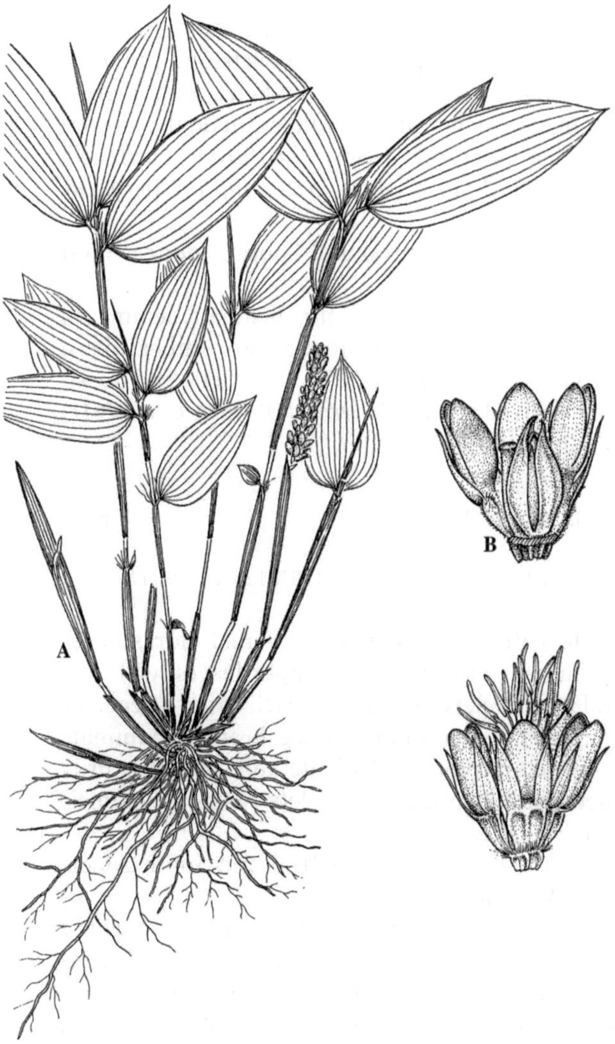

Fig. 30. Bambusoideae-Olyreae-Parianinae. *Pariana radiciflora*. **A** Whole plant. **B** Staminate (*above*) and pistillate (*below*) spikelets. (From Davidse et al. 2004, with permission from the Missouri Botanical Garden Press; drawn by Bruno Manara)

Leaves with the *top of the sheath variously decorated with translucent swellings ("lunar marks")*. Inflorescence cylindrical, *with a whorl of five staminate spikelets surrounding a single sessile pistillate spikelet and the associated rachis internode*. Pistillate spikelets with three *or more* lodicules. Staminate spikelets with pedicels often fused; *stamens numerous, showy*. $2n = 22$, 44, 38.

Thirty-five spp., Costa Rica and Trinidad, south through tropical South America.

Pariana is presumed to be insect pollinated (Soderstrom and Calderón 1971).

VI.3. Subtribe Olyrinae Kromb. (1875)

Pistillate spikelet with lemma and palea hardened. Staminate flowers with 3 stamens.

124. ***Agnesia*** Zuloaga & Judz.

Agnesia Zuloaga & Judz., Novon 3: 306 (1993).

Proximal leaves lacking blades, blades on leaves only above the middle of the culm. Inflorescences axillary, with one terminal pistillate spikelet and several proximal staminate spikelets, the latter on slender pedicels. Glumes of pistillate spikelet with long stiff awns; pistillate flowers covered with long hairs, awned.

One sp., *A. lancifolia* (Mez) Zuloaga & Judz., Peru, Colombia, Brazil.

125. ***Arberella*** Soderstr. & C.E. Calderón

Arberella Soderstr. & C.E. Calderón, Brittonia 31: 433 (1979).

Inflorescences with few flowers, borne at all nodes or only at upper and lower ones, with one terminal pistillate spikelet and several proximal staminate spikelets. Pistillate spikelets with glumes with many veins, with a prominent rachilla internode below the flower; lemma smooth, bearing hairs. $2n = 22$, 24.

Seven spp., Central and South America.

Arberella may be derived within "*Olyra*" (Oliveira et al. 2014).

126. ***Cryptochloa*** Swallen

Cryptochloa Swallen, Ann. Missouri Bot. Gard. 29: 317 (1942).

Inflorescences terminal and axillary, often partly hidden in leaf sheaths, the terminal inflorescence generally with only staminate spikelets, the axillary ones with spikelets of both sexes mixed. Pistillate spikelets with glumes with 5 veins, the rachilla internode below the flower generally thick and oil-containing, falling with the flower; lemma and palea becoming dark mottled at maturity. Staminate spikelets without glumes, with 3 or rarely with 2 stamens. $2n = 22$.

Eight spp., Central to South America.

The thickened oil-bearing rachilla internode is absent in *C. capillata* (Trin.) Soderstr. and *C. soderstromii* Davidse.

127. ***Ekmanochloa*** Hitchc.

Ekmanochloa Hitchc., Man. Grasses W. Ind. 374 (1936).

Vegetative culms with multiple foliage leaves, flowering culms with few leaves, these lacking blades. Inflorescence terminal, with 2 branches, one bearing staminate spikelets and the other, slightly longer, bearing pistillate spikelets. Pistillate spikelets elongate, narrow; *lemma awned, the awn slender, much longer than the body of the lemma*. Staminate spikelets elliptical, sometimes with 2 stamens. *Fusoid cells lacking.*

Two spp., eastern Cuba.

Ekmanochloa occurs only on serpentine soils (Zuloaga et al. 1993).

128. ***Froesiochloa*** G.A. Black

Froesiochloa G.A. Black, Bol. Tecn. Inst. Agron. N. 20: 29 (1950).

Leaf blades broad, with short petioles, *deciduous*. Inflorescence terminal, a cluster of spikelets, the terminal spikelet pistillate, proximal ones staminate, spirally arranged. Pistillate spikelet with the rachilla internode below the flower generally thick and oil-containing; lodicules absent. Staminate spikelets with *stamens with filaments fused.*

One sp., *F. boutelouoides* G.A. Black, French Guiana and adjacent Brazil.

The terminal inflorescence is showy, with the spikelets arranged so that they look like a single large spikelet rather than multiple small ones.

129. ***Lithachne*** P. Beauv.

Lithachne P. Beauv., Ess. Agrostogr. 135, 166, 168, t. 24, f. 2 (1812).

Plants generally less than 30 cm tall. Leaf blade with the base asymmetrical. Inflorescences terminal and axillary, the terminal inflorescence generally with only staminate spikelets, the axillary ones with distal pistillate spikelets and proximal staminate ones. Pistillate spikelet with glumes elongate, membranous, longer than the flower; rachilla internode below the flower hardened; *lemma and palea white (becoming mottled), the lemma gibbous*. $2n = 22$.

Four spp., Central and South America, West Indies.

The odd shape of the pistillate lemma in *Lithachne* makes the flower look rather like a large tooth. *Lithachne* is monophyletic and possibly derived within "*Olyra*" (Oliveira et al. 2014; Zhang and Clark 2000).

130. "***Olyra***" L. Fig. 31

Olyra L., Syst. Nat. (ed. 10), 2: 1253, 1261, 1379 (1759).

Culms often up to 3 m tall or more. Inflorescences terminal or axillary from upper leaves, with pistillate spikelets terminal in each inflorescence or on each major inflorescence branch, staminate spikelets proximal. Pistillate spikelets on sturdy pedicels. Staminate spikelets without glumes. $2n = 14, 22, 23, 40, 44$.

Twenty-four spp., Central and South America, West Indies.

Judziewicz et al. (1999) suggest that *Olyra* might be para- or polyphyletic, a hypothesis confirmed by the data of Oliveira et al. (2014). *Arberella* and *Lithachne* are derived within "*Olyra*", while the data on *Cryptochloa* are ambiguous. *Agnesia* could also be derived from within "*Olyra*" but data are not available to test this. Some species are visited and possibly pollinated

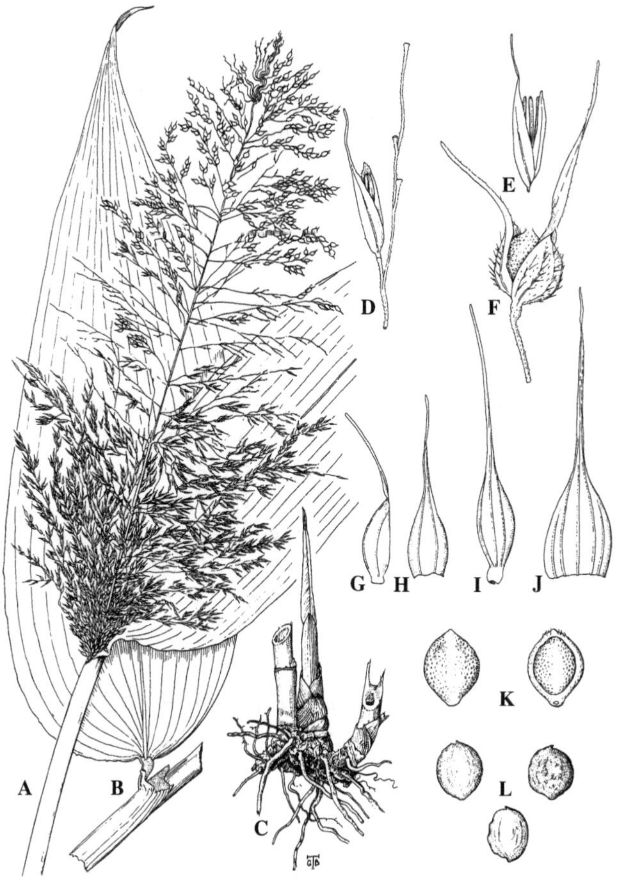

Fig. 31. Bambusoideae-Olyreae-Olyreae. *Olyra micrantha.* **A** Inflorescence. **B** Leaf blade, with pseudopetiole and ligule. **C** Base of the plant. **D** Portion of a branch of inflorescence with staminate spikelet. **E** Staminate spikelet. **F** Pistillate spikelet. **G** Lemma of staminate spikelet. **H** Palea of staminate spikelet, viewed from the adaxial side. **I** Lower glume of pistillate spikelet. **J** Upper glume of pistillate spikelet. **K** Lemma and palea of pistillate flower, *left* abaxial side, *right* adaxial side. **L** Three caryopses in side view. (Reprinted from Soderstrom and Zuloaga 1989, courtesy of Smithsonian Institution; drawn by G.B. Threlkeld)

by insects (Soderstrom and Calderón 1971). Although many species of "*Olyra*" range from 0.3–3 m in height, some species are reported to reach 5 or 6 m.

PARODIOLYRA + RADIELLA

Leaf blades with apex asymmetrical, apiculate.

Parodiolyra is paraphyletic with *Raddiella* derived from within it (Oliveira et al. 2014). The two genera should perhaps be combined.

131. *Parodiolyra* Soderstr. & Zuloaga

Parodiolyra Soderstr. & Zuloaga, Smithsonian Contr. Bot. 69: 64 (1989).

Culms scrambling. Inflorescence lax, the branches branched, distal branches with mixed staminate and pistillate spikelets, or pistillate spikelets only, the proximal branches with staminate spikelets only. Pistillate spikelets becoming black at maturity, falling with the glumes; rachis internode between glumes swollen. $2n = 36$.

Five spp., Costa Rica through tropical South America.

132. *Raddiella* Swallen

Raddiella Swallen, Bull. Torrey Bot. Club 75: 89 (1948).

Culms short, leaves <1.5 cm long in most species. Inflorescences terminal and axillary, with few spikelets, with staminate and pistillate spikelets mixed in the inflorescence, or terminal inflorescences wholly staminate, axillary wholly pistillate. Pistillate flower glabrous, pale but becoming dark at maturity in some species. $2n = 20$.

Eight spp., Panama, Trinidad, tropical South America.

133. *Piresia* Swallen

Piresia Swallen, Phytologia 11: 152 (1964); tax.: de Carvalho et al. (2012).

Culms dimorphic. Leaves of upright culms with broad blades; leaves of decumbent culms with blades reduced or lacking. Inflorescences terminal and axillary, most commonly on decumbent culms, but sometimes on upright culms as well. Pistillate and staminate spikelets morphologically distinct, or morphologically apparently bisexual but functionally unisexual, one distal pistillate spikelet and several proximal staminate spikelets near the apex of the inflorescence, sometimes additional staminate and pistillate spikelets intermixed proximally. Pistillate lemma densely pubescent. $2n = 22$.

Five spp., lowland tropics of South America.

134. ***Reitzia*** Swallen

Reitzia Swallen, Sellowia 7: 7 (1956).

Culms <30 cm tall. Inflorescence terminal, with several branches, each with a terminal pistillate spikelet and one or two proximal staminate spikelets. Pistillate flower shorter than the glumes, becoming darkly marbled at maturity.

One sp., *R. smithii* Swallen, southern Atlantic Brazil.

Sucrea + Raddia

Glumes of pistillate spikelets with thickened margins extending into long twisted apices.

Diaspores (anthecia) in these two genera are forcibly ejected from the glumes as the glumes dry and curl (Sendulsky 1993). This mode of dispersal, ballistochory, is otherwise unknown in the grasses.

135. ***Sucrea*** Soderstr.

Sucrea Soderstr., Brittonia 33: 200 (1981).

Plants with obvious prop roots. Leaves one or a few per culm, sheaths slightly inflated. Inflorescence branches branched, with pistillate spikelets mostly on distal branches, staminate generally on proximal branches. $2n = 22$.

Three spp., southeastern Brazil.

Sucrea may be paraphyletic with *Raddia* derived from within it (Oliveira et al. 2014).

136. ***Raddia*** Bertol.

Raddia Bertol., Opusc. Sci. 3: 410 (1819); tax.: Oliveira et al. (2008).

Leaves arranged along the culm with precise spacing, looking superficially similar to a large compound leaf. Inflorescences terminal and axillary, the terminal inflorescence with branches branched, bearing staminate spikelets, the axillary ones unbranched, with pistillate spikelets, or the branches branched and bearing staminate spikelets. $2n = 22$.

Nine spp., French Guiana, Brazil.

Molecular data from all nine species of *Raddia* show that the genus is monophyletic and either sister to or derived from within *Sucrea* (Oliveira et al. 2014).

Mniochloa + Piresiella + Rehia

Plants with corms.

137. ***Mniochloa*** Chase

Mniochloa Chase, Proc. Biol. Soc. Wash. 21: 185 (1908).

Culms dimorphic, vegetative culms leafy, flowering culms with leaves lacking blades. Inflorescence with 2 branches, one bearing pistillate spikelets, the other bearing staminate ones; spikelets on thickened pedicels. Pistillate spikelet with glumes longer than the flower. Staminate spikelets much smaller than pistillate ones.

One sp., *M. pulchella* (Griseb.) Chase, eastern Cuba; possibly extinct.

138. ***Piresiella*** Judz., Zuloaga & Morrone

Piresiella Judz., Zuloaga & Morrone, Ann. Missouri Bot. Gard. 80: 856 (1993).

Culms short, often only a few centimeters tall, dimorphic, vegetative culms leafy, flowering culms with leaves lacking blades, with slender stolons. Inflorescences with 2 branches, one bearing pistillate spikelets, the other bearing staminate ones. Pistillate spikelet with glumes longer than the flower, the lemma apiculate; stigmas 3. Staminate spikelets smaller than pistillate, lacking glumes; stamens 2.

One sp., *P. strephioides* (Griseb.) Judz., Zuloaga & Morrone, western Cuba.

139. ***Rehia*** Fijten

Rehia Fijten, Blumea 22: 416 (1975).

Culms short, caespitose. Inflorescence terminal, with several branches, each with a terminal pistillate spikelet and one or two proximal staminate spikelets. Pistillate flowers dark marbled color at maturity, glabrous except for a few hairs at the base, apex and margins of lemma. $2n = 20$.

One sp., *R. nervata* Fijten, northeastern Brazil and adjacent French Guiana, Surinam, and Guyana.

Diandrolyra + Maclurolyra

Blades of uppermost leaves inverted, covering the single terminal inflorescence like a canopy. Inflorescence terminal; *staminate and pistillate spikelets paired.* Pistillate flower with 3 staminodes.

140. ***Diandrolyra*** Stapf

Diandrolyra Stapf, Bull. Misc. Inform. Kew 1906: 204 (1906); tax.: de Oliveira and Clark (2009).

Culms dimorphic, the vegetative ones with multiple leaves, the flowering ones with one leaf covering the inflorescence. Pistillate spikelets with glumes with their bases fused; lemma and palea leathery. Staminate spikelets with tiny membranous glumes; *stamens 2*; pistillode with 3 stigmas. $2n = 18$.

Three spp., Atlantic Brazil.

Diandrolyra appears to be monophyletic (Oliveira et al. 2014).

141. ***Maclurolyra*** C.E. Calderón & Soderstr.

Maclurolyra C.E. Calderón & Soderstr., Smithsonian Contr. Bot. 11: 6 (1973).

Culms with 1 to 3 leaves. Inflorescence narrow, with staminate spikelets on long pedicels and pistillate spikelets on short ones. Pistillate spikelet with 2 glumes, the lemma fusoid, pubescent, with the margins inrolled over the palea. Staminate spikelet lacking glumes; stamens 3, staminodes 3. $2n = 22$.

One sp., *M. tecta* C.E. Calderón & Soderstr., Panama and adjacent Colombia.

References

Attigala, L., Triplett, J.K., Hashendra-Suvini, K., Clark, L.G. 2014. A new genus and a major temperate bamboo lineage of the Arundinarieae (Poaceae: Bambusoideae) from Sri Lanka based on a multi-locus plastid phylogeny. Phytotaxa 174: 187–205.

Bamboo Phylogeny Group 2012. An updated tribal and subtribal classification of the bamboos (Poaceae: Bambusoideae). J. Amer. Bamboo Soc. 24: 1–10.

Bouchenak-Khelladi, Y., Salamin, N., Savolainen, V., Forest, F., van der Bank, M., Chase, M.W., Hodkinson, T.R. 2008. Large multi-gene phylogenetic trees of the grasses (Poaceae): progress towards complete tribal and generic taxon sampling. Mol. Phyl. Evol. 47: 488–505.

Burke, S.V., Grennan, C.P., Duvall, M.R. 2012. Plastome sequences of two New World bamboos - *Arundinaria gigantea* and *Cryptochloa strictiflora* (Poaceae) - extend phylogenomic understanding of Bambusoideae. Amer. J. Bot. 99: 1951–1961.

Cao, K.-F., Yang, S.-J., Zhang, Y.-J., Brodribb, T.J. 2012. The maximum height of grasses is determined by roots. Ecol. Lett. 15: 666–672.

Clark, L.G. 1989. Systematics of *Chusquea* section *Swallenochloa*, section *Verticillatae*, section *Serpentes*, and section *Longifoliae* (Poaceae-Bambusoideae). Syst. Bot. Monographs 27: 1–127.

Clark, L.G., Cortés R., G. 2004. A new species of *Otatea* from Chiapas, Mexico. Bamboo Sci. Cult. 18: 1–6.

Clark, L., Londoño, X. 1991. A new species and new sections of *Rhipidocladum* (Poaceae: Bambusoideae). Amer. J. Bot. 78: 1260–1279.

Clark, L.G., Dransfield, S., Triplett, J., Sanchez-Ken, J.G. 2007. Phylogenetic relationships among the one-flowered, determinate genera of Bambuseae (Poaceae: Bambusoideae). Aliso 23: 315–332.

Clayton, W.D. 1970. Gramineae (Part 1). In: Milne-Redhead, E., Polhill, R.M. (eds.) Flora of Tropical East Africa. London: Crown Agents for Oversea Governments and Administrations.

Clayton, W.D., Harman, K.T., Williamson, H. 2006 onward. GrassBase - the online world grass flora. http://www.kew.org/data/grasses-db.html

Davidse, G., Judziewicz, E.J., Zuloaga, F.O. 2004. Poaceae. In: Berry, P.E., Yatskievych, K., Holst, B.K. (eds.) Flora of the Venezuelan Guyana. St. Louis, Missouri: Missouri Botanical Garden Press.

de Carvalho, M.L.S., Dórea, M.C., Pimenta, K.M., de Oliveira, R.P. 2012. *Piresia palmula*: a new species of herbaceous bamboo (Poaceae, Olyreae) endemic to the Atlantic rainforest, southern Bahia, Brazil. Syst. Bot. 37: 134–138.

de Oliveira, R.P., Clark, L.G. 2009. A tiny new Brazilian species of *Diandrolyra* (Poaceae, Bambusoideae, Olyreae), with notes on the systematics of the genus. Novon 19: 209–214.

Dransfield, S. 1981. The genus *Dinochloa* (Gramineae-Bambusoideae) in Sabah. Kew Bull. 36: 613–633.

Dransfield, S. 1992. The genus *Racemobambos* (Gramineae - Bambusoideae). Kew Bull. 37: 661–679.

Dransfield, S. 1994. The genus *Hickelia* (Gramineae: Bambusoideae). Kew Bull. 49: 429–443.

Dransfield, S. 1997. Notes on the genus *Decaryochloa* (Gramineae - Bambusoideae) from Madagascar. Kew Bull. 52: 593–600.

Dransfield, S. 2002. *Greslania circinata* and *Greslania rivularis* (Poaceae-Bambusoideae) from New Caledonia. Bamboo Sci. Cult. 16: 1–8.

Ferreira, F.M., Dórea, M.C., Leite, K.R.B., Oliveira, R.P. 2013. *Eremitis afimbriata* and *E. magnifica* (Poaceae, Bambusoideae, Olyreae): two remarkable new species from Brazil and a first record of blue iridescence in bamboo leaves. Phytotaxa 84(1): 31–45.

Fisher, A.E., Triplett, J.K., Ho, C.-S., Schiller, A.D., Oltrogge, K.A., Schroder, E.S., Kelchner, S.A., Clark, L.G. 2009. Paraphyly in the bamboo subtribe Chusqueinae (Poacae: Bambusoideae) and a revised infrageneric classification for *Chusquea*. Syst. Bot. 34: 673–683.

Fisher, A.E., Clark, L.G., Kelchner, S.A. 2014. Molecular phylogeny estimation of the bamboo genus *Chusquea* (Poaceae: Bambusoideae: Bambuseae) and description of two new subgenera. Syst. Bot. 39: 829–844.

Goh, W.L., Chandran, S., Lin, R.-S., Xia, N.-H., Wong, K. M. 2010. Phylogenetic relationships among Southeast Asian climbing bamboos (Poaceae: Bambusoideae) and the *Bambusa* complex. Biochem. Syst. Ecol. 38: 764–773.

Goh, W.L., Chandran, S., Kamiya, K., Wong, K.M. 2011. A natural hybrid between *Dendrocalamus pendulus* and *Gigantochloa scortechinii* (Poaceae: Bambusoideae: Babuseae) in Peninsular Malaysia. Gard. Bull. Singapore 62: 223–238.

Goh, W.L., Chandran, S., Franklin, D.C., Isagi, Y., Koshy, K.C., Sungkaew, S., Yang, H.Q., Xia, N.H., Wong, K. M. 2013. Multi-gene region phylogenetic analyses suggest reticulate evolution and a clade of Australian origin among paleotropical woody bamboos (Poaceae: Bambusoideae: Bambuseae). Plant Syst. Evol. 299: 239–257.

Guala II, G.F. 1995. A cladistic analysis and revision of the genus *Apoclada* (Poaceae: Bambusoideae: Bambusodae). Syst. Bot. 20: 207–223.

Guala, G.F., Bogler, D., Sadle, J., Francisco-Ortega, J. 2000. Molecular evidence for polyphyly in the genus *Apoclada* (Poaaceae: Bambusoideae). Bamboo Sci. Culture 14: 15–20.

Guo, Z.-H., Li, D.-Z. 2004. Phylogenetics of the *Thamnocalamus* group and its allies (Gramineae: Bambusoideae): inference from the sequences of GBSSI gene and ITS spacer. Mol. Phyl. Evol. 30: 1–12.

Haevermans, T., Nguyen, B.L., Gurgand, J., Haevermans, A., Dransfield, S., Diep, M.H. 2013. Clearing up *Vietnamosasa* (Poaceae, Bambusoideae): typification and nomenclature of a distinctive paleotropical bamboo genus. Phytotaxa 137: 57–60.

Holttum, R.E. 1956. The classification of bamboos. Phytomorphology 6: 73–90.

Janzen, D. 1976. Why bamboos wait so long to flower. Ann. Rev. Ecol. Syst. 7: 347–391.

Judziewicz, E. 1992. A revision of *Atractantha* (Poaceae: Bambusoideae: Bambuseae). Ann. Missouri Bot. Gard. 79: 160–183.

Judziewicz, E.J. 1998. A revision of *Myriocladus* (Poaceae: Bambusoideae: Bambuseae). Brittonia 50: 430–446.

Judziewicz, E.J., Clark, L.G. 1993. The South American species of *Arthrostylidium* (Poaceae: Bambusoideae: Bambuseae). Syst. Bot. 18: 80–99.

Judziewicz, E.J., Clark, L.G., Londoño, X., Stern, M.J. 1999. American bamboos. Washington, DC: Smithsonian Institution Press.

Kelchner, S.A., Bamboo Phylogeny Group 2013. Higher level phylogenetic relationships within the bamboos (Poaceae: Bambusoideae) based on five plastid markers. Mol. Phyl. Evol. 67: 404–413.

Kelchner, S.A., Clark, L.G. 1997. Molecular evolution and phylogenetic utility of the chloroplast *rpl16* intron in *Chusquea* and the Bambusoideae (Poaceae). Mol. Phyl. Evol. 8: 385–397.

Kellogg, E.A., Watson, L. 1993. Phylogenetic studies of a large data set. I. Bambusoideae, Pooideae, and Andropogoneae (Gramineae). Bot. Rev. 59: 273–343.

Li, X.-L., Lin, R.-S., Fung, H.-L., Qi, Z.-X., Song, W.-Q., Chen, R.-Y. 2001. Chromosome numbers of some bamboos native in or introduced to China. Acta Phytotax. Sinica 39: 433–442.

Li, D., Wang, Z., Wang, C., Zhu, Z., Xia, N., Jia, L., Guo, Z., Yang, G., Stapleton, C. 2006. I. Bambuseae. In: Wu, Z.-Y., Raven, P.H., Hong, D.-Y. (eds.) Flora of China, vol. 22 (Poaceae). Beijing: Science Press. pp. 7–180

Londoño, X., Clark, L.G. 2002. A revision of the Brazilian bamboo genus *Eremocaulon* (Poaceae: Bambuseae: Guaduinae). Syst. Bot. 27: 703–721.

Ma, P.-F., Zhang, Y.-X., Zeng, C.-X., Guo, Z.-H., Li, D.-Z. 2014. Chloroplast phylogenomic analyses resolve deep-level relationships of an intractable bamboo tribe Arundinarieae (Poaceae). Syst. Biol. 63: 933–950.

McClure, F.A. 1966. The bamboos: a fresh perspective. Cambridge, Massachusetts: Harvard University Press.

McClure, F.A. 1973. Genera of bamboos native to the New World (Gramineae: Bambusoideae). Smithsonian Contr. Bot. 9: 1–148.

Oliveira, R.P., Borba, E.L., Longhi-Wagner, H.M. 2008. Morphometrics of herbaceous bamboos of the *Raddia brasiliensis* complex (Poaceae - Bambusoideae): implications for the taxonomy of the genus and new species from Brazil. Plant Syst. Evol. 270: 159–182.

Oliveira, R.P., Clark, L.G., Schnadelbach, A.S., Monteiro, S.H.N., Borba, E.L., Longhi-Wagner, H.M., van den Berg, C. 2014. A molecular phylogeny of *Raddia* and its allies within the tribe Olyreae (Poaceae, Bambusoideae) based on non-coding plastid and nuclear spacers. Mol. Phylog. Evol. 78: 105–117.

Peng, S., Yang, H.-Q., Li, D.-Z. 2008. Highly heterogeneous generic delimitation within the temperate bamboo clade (Poaceae: Bambusoideae): evidence fom *GBSSI* and ITS sequences. Taxon 57: 799–810.

Peng, Z., Lu, Y., Li, L., Zhao, Q., Feng, Q., Gao, Z., Lu, H., Hu, T., Yao, N., Liu, K., Li, Y., Fan, D., Guo, Y., Li, W., Lu, Y., Weng, Q., Zhou, C., Zhang, L., Huang, T., Zhao, Y., Zhu, C., Liu, X., Yang, X., Wang, T., Miao, K., Zhuang, C., Cao, X., Tang, W., Liu, G., Liu, Y., Chen, J., Liu, Z., Yuan, L., Liu, Z., Huang, X., Lu, T., Fei, B., Ning, Z., Han, B., Jiang, Z. 2013. The draft genome of the fast-growing non-timber forest species moso bamboo (*Phyllostachys heterocycla*). Nat. Genet. 45: 456–461.

Pilger, R. 1915. Neue und weniger bekannte Gramineen aus Papuasien. Bot. Jahrb. 52: 167–176.

Ramanayake, S.M.S.D., Weerawardene, T.E. 2003. Flowering in a bamboo, *Melocanna baccifera* (Bambusoideae: Poaceae). Bot. J. Linn. Soc. 143: 287–291.

Rudall, P., Dransfield, S. 1989. Fruit structure and development in *Dinochloa* and *Ochlandra* (Gramineae-Bambusoideae). Ann. Bot. 63: 29–38.

Ruiz-Sánchez, E., Sosa, V., Mejía-Saules, M.T. 2008. Phylogenetics of *Otatea* inferred from morphology and chloroplast DNA sequence data and recircumscription of Guaduiinae (Poaceae: Bambusoideae). Syst. Bot. 33: 277–283.

Ruiz-Sánchez, E., Sosa, V., Mejía-Saules, M.T. 2011a. Molecular phylogenetics of the Mesoamerican bamboo *Olmeca* (Poaceae, Bambuseae): implications for taxonomy. Taxon 60: 89–98.

Ruiz-Sánchez, E., Sosa, V., Mejía-Saules, M.T., Londoño, X., Clark, L.G. 2011b. A taxonomic revision of *Otatea* (Poaceae: Bambusoideae: Bambuseae) including four new species. Syst. Bot. 36: 314–336.

Santos-Gonçalves, A.P., Filgueiras, T.S., Clark, L.G. 2011. *Atractantha shepherdiana*, a new species of woody bamboo (Poaceae: Bambusoideae: Bambuseae) from Brazil. Syst. Bot. 36: 310–313.

Sendulsky, T. 1993. First report of ballistochory in Poaceae. Ann. Missouri Bot. Gard. 80: 518–521.

Sendulsky, T., Soderstrom, T.R. 1984. Revision of the South American genus *Otachyrium* (Poaceae: Panicoideae). Smithsonian Contr. Bot. 57: 1–24.

Soderstrom, T.R., Calderón, C.E. 1971. Insect pollination in tropical rain forest grasses. Biotropica 3: 1–16.

Soderstrom, T.R., Ellis, R.P. 1982. Taxonomic status of the endemic South African bamboo, *Thamnocalamus tessellatus*. Bothalia 14: 53–67.

Soderstrom, T.R., Ellis, R.P. 1987. The position of bamboo genera and allies in a system of grass classification. In: Soderstrom, T.R., Hilu, K.W., Campbell, C.S., Barkworth, M.E. (eds.) Grass systematics and evolution. Washington, DC: Smithsonian Institution Press. pp. 225–238

Soderstrom, T.R., Zuloaga, F.O. 1989. A revision of the genus *Olyra* and the new segregate genus *Parodiolyra* (Poaceae, Bambusoideae, Olyreae). Smithsonian Contr. Bot. 69: 1–79.

Stapleton, C.M.A. 1994a. The bamboos of Nepal and Bhutan Part I: *Bambusa*, *Dendrocalamus*, *Melocanna*, *Cephalostachyum*, *Teinostachyum*, and *Pseudostachyum* (Gramineae: Poaceae, Bambusoideae). Edinburgh J. Bot. 51: 1–32.

Stapleton, C.M.A. 1994b. The bamboos of Nepal and Bhutan Part III: *Drepanostachyum*, *Himalayacalamus*, *Ampelocalamus*, *Neomicrocalamus* and *Chimonobambusa* (Gramineae: Poaceae, Bambusoideae). Edinburgh J. Bot. 51: 301–330.

Stapleton, C.M.A. 1994c. The bamboos of Nepal and Bhutan. Part II: *Arundinaria*, *Thamnocalamus*, *Borinda*, and *Yushania* (Gramineae: Poaceae, Bambusoideae). Edinburgh J. Bot. 51: 275–295.

Stapleton, C.M.A. 1997. The morphology of woody bamboos. In: Chapman, G.P. (ed.) The bamboos. New York: Academic Press. pp. 251–267

Stebbins, G.L. 1971. Chromosomal evolution in higher plants. London: Edward Arnold.

Stebbins, G.L. 1985. Polyploidy, hybridization and the invasion of new habitats. Ann. Missouri Bot. Gard. 72: 824–832.

Sungkaew, S., Stapleton, C.M.A., Salamin, N., Hodkinson, T.R. 2009. Non-monophyly of the woody bamboos (Bambuseae: Poaceae): a multi-gene region phylogenetic analysis of Bambusoideae s.s. J. Plant Res. 122: 95–108.

Takahashi, K., Watano, Y., Shimizu, T. 1994. Allozyme evidence for intersectional and intergeneric hybridization in the genus *Sasa* and its related genera (Poaceae: Bambusoideae). J. Phytogeog. Tax. 42: 49–60.

Triplett, J.K., Clark, L.G. 2010. Phylogeny of the temperate bamboos (Poaceae: Bambusoideae: Bambuseae) with an emphasis on *Arundinaria* and allies. Syst. Bot. 35: 102–120.

Triplett, J.K., Oltrogge, K., Clark, L.G. 2010. Phylogenetic relationships within *Arundinaria* (Poaceae: Bambusoideae) in North America. Amer. J. Bot. 97: 471–492.

Triplett, J.K., Clark, L.G., Fisher, A.E., Wen, J. 2014. Independent allopolyploidization events preceded speciation in the temperate and tropical woody bamboos. New Phytol. 204: 66–73. doi: 10.1111/nph.12988

Tyrell, C.D., Santos-Gonçalves, A.P., Londoño, X., Clark, L.G. 2012. Molecular phylogeny of the arthrostylidioid bamboos (Poaceae: Bambusoideae: Bambuseae: Arthrostylidiinae) and new genus *Didymogonyx*. Mol. Phyl. Evol. 65: 136–148.

Viana, P.L., Filgueiras, T.S., Clark, L.G. 2013. *Cambajuva* (Poaceae: Bambusoideae: Bambuseae: Arthrostylidiineae), a new woody bamboo genus from Southern Brazil. Syst. Bot. 38: 97–103.

Watson, L., Dallwitz, M.J. 1992 onward. The grass genera of the world: descriptions, illustrations, identification, and information retrieval; including synonyms, morphology, anatomy, physiology, phytochemistry, cytology, classification, pathogens, world and local distribution, and references, 23rd April 2010 edn. Wallingford: CAB International. http://delta-intkey.com/grass/

Wong, K.M. 2004. Bamboo: the amazing grass. Selangor, Malaysia and Kuala Lumpur, Malaysia: International Plant Genetic Resources Institute (IPGRI) and University of Malaya.

Wong, K.M., Low, Y.W. 2011. Hybrid zone characteristics of the intergeneric hybrid bamboo x *Gigantocalamus malpenensis* (Poaceae: Bambusoideae) in Peninsular Malaysia. Gard. Bull. Singapore 63: 375–383.

Wu, Z., Raven, P.H., Hong, D. (eds.) 2007. Flora of China Illustrations. Beijing: Science Press.

Xia, N.H. 1996. A study of the genus *Bonia* (Gramineae: Bambusoideae). Kew Bull. 51: 565–569.

Yang, H.-Q., Peng, S., Li, D.-Z. 2007. Generic delimitations of *Schizostachyum* and its allies (Gramineae: Bambusoideae) inferred from GBSSI and *trnL-F* sequence phylogenies. Taxon 56: 45–54.

Yang, H.-Q., Yang, J.-B., Peng, Z.-H., Gao, J., Yang, Y.-M., Peng, S., Li, D.-Z. 2008. A molecular phylogenetic and fruit evolutionary analysis of the major groups of the paleotropical woody bamboos (Gramineae: Bambusoideae) based on nuclear ITS, GBSSI gene and plastid *trn*L-F DNA sequences. Mol. Phyl. Evol. 48: 809–824.

Yang, J.-B., Yang, H.-Q., Li, D.-Z., Wong, K.-M., Yang, Y.-M. 2010. Phylogeny of *Bambusa* and its allies (Poaceae: Bambusoideae) inferred from nuclear *GBSSI* gene and plastid *psbA-trnH*, *rpl32-trnL* and *rps16* intron DNA sequences. Taxon 59: 1102–11110.

Yang, H.-M., Zhang, Y.-X., Yang, J.-B., Li, D.-Z. 2013. The monophyly of *Chimonocalamus* and conflicting gene trees in Arundinarieae (Poaceae: Bambusoideae) inferred from four plastid and two nuclear markers. Mol. Phyl. Evol. 68: 340–356.

Zeng, C.-X., Zhang, Y.-X., Triplett, J.K., Yang, J.-B., Li, D.-Z. 2010. Large multi-locus plastid phylogeny of the tribe Arundinarieae (Poaceae: Bambusoideae) reveals ten major lineages and low rate of molecular divergence. Mol. Phyl. Evol. 56: 821–839.

Zhang, W.P., Clark, L.G. 2000. Phylogeny and classification of the Bambusoideae (Poaceae). In: Jacobs, S.W. L., Everett, J. (eds.) Grasses: systematics and evolution. Melbourne: CSIRO.

Zhang, Y.-X., Zeng, C.-X., Li, D.-Z. 2012. Complex evolution in Arundinarieae (Poaceae: Bambusoideae): Incongruence between plastid and nuclear GBSSI gene phylogenies. Mol. Phyl. Evol. 63: 777–797.

Zhang, Y.-X., Zeng, C.-X., Li, D.-Z. 2014. Scanning electron microscopy of the leaf epidermis in Arundinarieae (Poaceae: Bambusoideae): evolutionary implications of selected micromorphological features. Bot. J. Linn. Soc. 176: 46–65.

Zuloaga, F.O., Morrone, O., Judziewicz, E.J. 1993. Endemic herbaceous bamboo genera of Cuba. Ann. Missouri Bot. Gard. 80: 846–861.

VI. Subfamily Pooideae Benth. (1861)

Festucoideae Link (1827).

Annuals or perennials, caespitose, rhizomatous or stoloniferous. Culms hollow. Ligule membranous, sometimes with a ciliolate margin but not long ciliate. Leaf blades lacking pseudopetioles. Leaf sheaths lacking auricles. Inflorescence unbranched, or of primary branches only, or the branches branched. Spikelets with 1 to many flowers, reduced flowers, if present, distal to the fertile ones. Rachilla generally breaking up between the flowers, i.e., disarticulation above the persistent glumes. Lemma awned or not. Palea with 2 veins, often keeled. Lodicules 2, membranous, not vascularized. Stamens 3. Stigmas 2, *borne on style branches that are inserted separately into the ovary apex.* Fruit with a linear hilum. Embryo small, with an epiblast, lacking an internode between the coleoptilar and scutellar vein, the embryonic leaf margins overlapping. C_3. Fusoid cells and arm cells absent. Leaf epidermal papillae absent.

One hundred seventy-seven genera, 3850 spp.

Pooideae are monophyletic in all molecular phylogenies to date. The character "loss of stylar fusion" appears to be synapomorphic (GPWG 2001), although anatomical interpretation of this character is unclear. It seems most likely that it represents a change in the zone of growth of the developing carpels so that fusion occurs below the separate lobes rather than above. In this case, the style "branches" might properly be called stylodia, but this term is not used regularly in the grass literature so I use the term style branches here.

Comprehensive phylogenetic studies of this subfamily have appeared only in the last few years, even though this group is the most economically and ecologically important of all temperate grasses. These studies have substantially revised our understanding of the history and classification of this group. In all molecular phylogenies, *Brachyelytrum* is sister to the rest of the subfamily, with Nardeae (*Nardus* plus *Lygeum*) the next diverging branch (Davis and Soreng 2007; GPWG 2001; GPWG II 2012; Romaschenko et al. 2012; Schneider et al. 2009, and many others). The relative positions of Meliceae, Phaenospermateae, and Stipeae differ among analyses and likely reflect differences in taxon sampling and molecular markers. For example, the GPWG II (2012), using three chloroplast loci, finds Stipeae, Phaenospermateae and Meliceae as successive sister groups to the remainder of the pooids (Diarrheneae not included), whereas Schneider et al. (2009, 2011) find the order to be Meliceae, Phaenospermateae, Stipeae, Diarrheneae and the rest. Other authors find other arrangements, which are often poorly resolved or weakly supported. For the moment then, the order of branching is unknown, which affects efforts to optimize characters on this part of the phylogeny.

After divergence of the tribes mentioned above, most studies agree that *Brachypodium* is sister to the rest of the subfamily (e.g., GPWG II 2012). After the divergence of *Brachypodium*, the remaining species are marked by a sharp increase in genome size (Bennetzen and Kellogg 1997; Kellogg and Bennetzen 2004). *Littledalea*, *Bromus* and Triticeae form a clade that is sister to the Poeae s.l. (Schneider et al. 2009, 2011). The latter group includes the former Aveneae as well as several smaller tribes, which are clearly nonmonophyletic and hence not recognized here.

DOI 10.1007/978-3-319-15332-2_20, © Springer International Publishing Switzerland 2015

Poeae s.l. has been supported as monophyletic in all molecular analyses.

Hybridization is common in Pooideae, as noted years ago by Stebbins (1950), and in a phylogenetic context by Kellogg and Watson (1993); hybridization is often associated with polyploidy. Many taxa exhibit morphological characters that combine characteristics of the putative parental taxa, leading to many named hybrid genera (e.g., x*Elyhordeum* Mansf. ex Tsitsin & K. A. Petrova, x*Pucciphippsia* Tzvelev). Additional evidence for hybridization has come from cytogenetics and isozymes, but the most compelling data have come from DNA sequences. Discrepancies between phylogenies produced by nuclear and chloroplast genes are increasingly demonstrating reticulation. In addition, single copy nuclear genes are clearly effective in resolving complex reticulate phylogenies (see, for example, studies on *Elymus*; Fan et al. 2013; Hu et al. 2013; Mason-Gamer et al. 2002, 2010a, 2010b), and these will need to be applied to problems throughout the subfamily. Several large polyploid genera appear to be polyphyletic (e.g., "*Calamagrostis*"; Saarela et al. 2010); the morphological characters associated with such genera are clearly homoplasious, creating difficulties with diagnosis.

Generic limits are a problem throughout Pooideae, perhaps more so than in any other subfamily of grasses except the bamboos. Some large genera (e.g., *Festuca*, *Poa*) are paraphyletic as traditionally circumscribed, with monotypic genera with one or two unusual characters having been segregated from the main genus. Monophyly can be restored by placing the segregates into synonymy, although this sometimes means that no obvious morphological character remains to identify the sensu lato genus.

All three phenomena – paraphyly, polyphyly, and frequent intergeneric hybridization – suggest that the subfamily taxonomy would be clarified by recognizing fewer, more broadly defined genera. While the classification presented here is a step in that direction, it certainly does not go far enough. As phylogenies are produced with more extensive species sampling, including the many monospecific genera, and as morphological studies become more detailed, the limits of the sensu lato genera are likely to become clearer.

Tribes and Subtribes of Pooideae

VII. Tribe Brachyelytreae
VIII. Tribe Nardeae
IX. Tribe Phaenospermateae
X. Tribe Stipeae
XI. Tribe Meliceae
XII. Tribe Diarrheneae
XIII. Tribe Brachypodieae
XIV. Tribe Bromeae
XV. Tribe Triticeae
XVI. Tribe Poeae
- Incertae sedis
- XVI.1 Subtribe Torreyochloinae
- XVI.2. Subtribe Aveninae
- XVI.3. Subtribe Phalaridinae
- XVI.4. Subtribe Anthoxanthinae
- XVI.5. Subtribe Agrostidinae
- XVI.6. Subtribe Scolochloinae
- XVI.7. Subtribe Airinae
- XVI.8. Subtribe Ammochloinae
- XVI.9. Subtribe Cynosurinae
- XVI.10. Subtribe Loliinae
- XVI.11. Subtribe Dactylidinae
- XVI.12. Subtribe Parapholiinae
- XVI.13. Subtribe Sesleriinae
- XVI.14. Subtribe Coleanthinae
- XVI.15. Subtribe Poinae

Key to the Genera of Pooideae

1. Base of inflorescence surrounded by a trumpet-shaped bract originating from the basal inflorescence node; plants of the Mediterranean **302. *Cornucopiae***
1. Inflorescence without a trumpet-shaped bract; plant distribution various 2
2 (1). Inflorescence composed of a single spikelet, barely extending above the leaves **162. *Aciachne***
2. Inflorescence with more than 1 spikelet 3
3. Inflorescence unbranched, with spikelets sessile on the rachis, or with short pedicels **Group I**
3. Inflorescence branched 4
4 (3). Spikelets without awns **Group II**
4. Spikelets with awns 5
5 (4). Awns straight **Group III**
5. Awns curved, twisted, or geniculate **Group IV**

Group I – Unbranched inflorescences

1. Glumes and/or lemmas with awns 2
1. Spikelets entirely without awns 40
2 (1). Spikelets on short pedicels 3
2. Spikelets sessile 12
3 (2). Flowers 1 per spikelet 4
3. Flowers more than 1 per spikelet 5
4 (3). Glumes large, enclosing the flower; awn from back of lemma; Afghanistan to western China **148. *Duthiea***

4. First glume missing, second glume minute; awn an extension of the lemma; eastern Asia and eastern North America **142. *Brachyelytrum***
5 (3). Lemma with 5 to 9 awns **194. *Bromus* (sect. *Boissiera*)**
5. Lemma with 1 awn 6
6 (5). Distal portion of glumes hyaline; apex of upper glume generally ragged; upper lemma margin hyaline; plants of wet sites **188. *Pleuropogon***
6. Glumes and lemmas membranous or firm, but not hyaline; plants of mesic to dry sites 7
7 (6). Lemma awned from sinus or back 8
7. Lemma awn terminal 10
8 (7). Spikelet with 2 flowers; lemma with 11 rib-like veins **186. *Lycochloa***
8. Spikelet with 4 or more flowers; lemma with few veins, these not rib-like 9
9 (8). Ligule 7 mm long; lodicules 2; western China and Himalayas **150. *Pseudodanthonia***
9. Ligule <3 mm long; lodicules 3; Balkans and Caucasus **147. *Danthoniastrum***
10 (7). Spikelets with 3 flowers, the lower 2 sterile; awn of upper flower ca. 2 cm; plants of China and Japan **184. *Brylkinia***
10. Spikelets with 2 to many flowers, but never with the 2 lower ones sterile; if 3 flowers, then awn of upper flower much less than 2 cm 11
11 (10). Spikelets with 2 or 3 flowers; delicate annuals with leaves generally less than 1 mm wide **266. *Festuca***
11. Spikelets with more than 5 flowers; perennials or sturdy annuals with leaves much wider than 1 mm **193. *Brachypodium***
12 (2). Spikelets 1 per node 13
12. Spikelets more than 1 per node, at least in the middle of the inflorescence 33
13 (12). Spikelets with glumes minute or absent 14
13. At least the first glume present, well developed 16
14 (13). Plants annual **266. *Festuca* (former *Psilurus*)**
14. Plants perennial, caespitose 15
15 (14). Spikelets borne on one side of the inflorescence axis; Europe **143. *Nardus***
15. Spikelets borne on opposite sides of the inflorescence axis; New Zealand **212. *Stenostachys***
16 (13). Lemma awned from the back or from a sinus 17
16. Lemma awnless or with an apical awn 19
17 (16). Plants annual, generally less than 30 cm tall 18
17. Plants perennial, generally well over 30 cm tall **149. *Metcalfia***
18 (17). Spikelets not sunken, borne on distinct pedicels **226. *Trisetaria***
18. Spikelets sunken in the inflorescence axis **258. *Scribneria***
19 (16). Spikelet edgewise to rachis **267. *Lolium***
19. Spikelet broadside to rachis 20
20 (19). Lemmas and often glumes strongly keeled 21
20. Neither lemmas nor glumes strongly keeled 23
21 (20). Margins of glumes and distal margins of lemmas with long silky hairs **200. *Dasypyrum***
21. Margins of glumes and lemmas not long hairy 22
22 (21). Glumes subulate, lemmas lanceolate **211. *Secale***
22. Glumes lanceolate or oblong, lemmas lanceolate, ovate, or oblong **215. *Triticum***
23 (20). Plants annual 24
23. Plants perennial 26
24 (23). Glumes with 1 vein **268. *Loliolum***
24. Glumes with >1 vein, these often prominent 25
25 (24). Glumes oblong, elliptic, or ovate; awned or not, their apices truncate **196. *Aegilops***
25. Glumes linear, extending to an attenuate apex **199. *Crithopsis***
26 (23). Spikelets strongly divergent from the rachis, or even somewhat reflexed **198. *Australopyrum***
26. Spikelets appressed to rachis, not reflexed 27
27 (26). Glumes subulate to narrowly lanceolate, 1-veined at mid-length 28
27. Glumes never subulate, variously lanceolate, rectangular, ovate or obovate, generally with 3 or more veins at mid-length 29
28 (27). Inflorescence erect; plants of cool temperate regions but not New Zealand **207. *Leymus***
28. Inflorescence nodding or drooping; plants of New Zealand **212. *Stenostachys***
29 (27). Spikelets about twice the length of inflorescence internodes or more, the spikelets thus overlapping in the inflorescence 30
29. Spikelets shorter than inflorescence internodes, the inflorescence thus appearing sparse 32
30 (29). Leaves >0.5 mm wide when dry **201. *Elymus* p.p.**
30. Leaves 0.5 mm wide when dry 31
31 (30). Spikelets with 3 to 5 flowers **203. *Festucopsis***
31. Spikelets with 5 to 10 flowers **208. *Peridictyon***
32 (29). Rachis tough **210. *Pseudoroegneria***
32. Rachis fragile **273. *Agropyropsis***
33 (12). Plants annual 34
33. Plants perennial 37
34 (33). Spikelets embedded in hollows in the rachis **258. *Scribneria***
34. Rachis terete, spikelets not embedded 35
35 (34). Lemmas bearing long tuberculate hairs **205. *Heteranthelium***
35. Lemmas glabrous or scabrous, but not with long hairs 36
36 (35). Awns many times longer than the lemmas, extending at least twice the length of the inflorescence, tangled **213. *Taeniatherum***
36. Awns up to twice as long as lemmas, extending no more than twice the length of the inflorescence and often much less, straight **206. *Hordeum***
37 (33). Spikelets three per node, the central 1 sessile and the lateral 2 pedicellate; spikelets with 1 flower **206. *Hordeum***
37. All spikelets at the node sessile, each with 2 or more flowers, although the distal ones sometimes reduced and sterile 38
38 (37). Rachis disarticulating, rachilla tough; spikelets with 1 or 2 flowers **209. *Psathyrostachys***
38. Rachis tough, rachilla disarticulating; spikelets generally with >2 flowers 39
39 (38). Glumes with 1 vein; leaf blades stiff, often bluish with wax **207. *Leymus***
39. Glumes with 3 veins; leaf blades lax, green **201. *Elymus***
40 (1) Spikelets 1 or 2 per inflorescence 41
40. Spikelets more than 2 per inflorescence, the inflorescences extending well above the leaves 42

41 (40). Inflorescences not or scarcely extending above the leaves; leaves terete, their apices subulate, plants cushion-forming; plants of the high Andes **162. *Aciachne***
41. Inflorescences borne well above the leaves; spikelet solitary, terminal, covered with dense golden or white hairs the length of the spikelet; spikelet subtended by a spathe; leaves flat; Mediterranean **144. *Lygeum***
42 (40). Spikelets partially embedded in hollows in the rachis 43
42. Rachis terete, spikelets not embedded 47
43 (42). Spikelets on one side of the inflorescence axis, each with 4 to 6 flowers **266. *Festuca* (former *Narduroides*)**
43. Spikelets on opposite sides of the inflorescence axis, each with 1 or 2 flowers 44
44 (43). Culm internodes solid; lower glume absent on all spikelets except the terminal one **276. *Hainardia***
44. Culm internodes hollow; lower glume present on all spikelets 45
45 (44). Leaf sheath with falcate auricles **204. *Henrardia***
45. Leaf sheath lacking auricles 46
46 (45). Inflorescence axis not disarticulating; spikelets with 2 flowers **310. *Pholiurus***
46. Inflorescence axis disarticulating; spikelets with 1 flower **277. *Parapholis***
47 (42). Plants annual 48
47. Plants perennial 53
48 (47). Inflorescence pectinate, the spikelets densely packed and stiffly divaricate **202. *Eremopyrum***
48. Inflorescence not pectinate, the spikelets overlapping only slightly if at all and not divaricate 49
49 (48). Lemma tuberculate **264. *Castellia***
49. Lemma smooth 50
50 (49). Spikelets with 2 or more flowers; lemmas membranous 51
50. Spikelets with 1 flower; lemmas translucent 52
51 (50). Spikelets edgewise to the rachis, the lower glume lacking **267. *Lolium***
51. Spikelets broadside to the rachis, both glumes present **266. *Festuca* (former *Micropyrum*)**
52 (50). Leaf sheaths open **310. *Poa***
52. Leaf sheaths closed **280. *Mibora***
53 (47). Spikelets >1 per node, at least in the middle of the inflorescence 54
53. Spikelets 1 per node 55
54 (53). Plants rhizomatous; rachilla disarticulating, rachis tough **207. *Leymus*** (some *Elymus* may key here as well)
54. Plants caespitose; rachilla tough, rachis disarticulating **209. *Psathyrostachys***
55 (53). Spikelets densely packed in the inflorescence, the internodes much shorter than the spikelets 56
55. Spikelets overlapping but largely erect, no more than about twice the length of the internodes 58
56 (55). Glumes and lemmas firm, lanceolate **197. *Agropyron***
56. Glumes and lemmas membranous to hyaline, ovate 57
57 (56). Inflorescence with 2 bracts at the base **282. *Sesleria***
57. Inflorescence without basal bracts **281. *Oreochloa***
58 (55). Glumes stiff, obtuse to truncate **214. *Thinopyrum***
58. Glumes and lemmas acute to acuminate, sometimes mucronate or awn-tipped, flexible 59
59 (58). Spikelets borne edgewise to the rachis, or if broadside to it, then the base corm-like and glumes membranous **267. *Lolium***
59. Spikelets borne broadside to the rachis, the base not corm-like; glumes as firm as or firmer than the lemmas 60
60 (59). Spikelets scarcely overlapping along the inflorescence axis, the apex of one just reaching the base of the next **210. *Pseudoroegneria***
60. Spikelets overlapping along the inflorescence axis, the apex of one reaching about the mid-point of the next **201. *Elymus***

Group II – Inflorescence axis branched, spikelets without awns

1. Glumes and lemmas broader than long, appearing inflated 2
1. Glumes and lemmas longer than broad, variously lanceolate or oblong, but not appearing inflated 4
2 (1). Glumes longer than lemmas; flowers 1 per spikelet **299. *Beckmannia***
2. Glumes shorter than lemmas; flowers more than 1 per spikelet 3
3 (2). Lemma apex abruptly narrowed to an acute or acuminate tip **231. *Chascolytrum***
3. Lemma apex broad, obtuse or very broadly acute **230. *Briza***
4 (1). Plants with 1 flower per spikelet 5
4. Plants with more than 1 flower per spikelet 21
5 (4). Lemma coriaceous, noticeably harder than the glumes 6
5. Lemma similar in texture to the glumes, or hyaline 8
6 (5). Lemma glabrous, shining, dorsiventrally compressed; most species north temperate 7
6. Lemma puberulent, at least near the apex, laterally compressed; plants of Argentina and Chile **232. *Relchela***
7 (6). Inflorescence an open panicle; spikelets borne on the distal half of the branches **306. *Milium***
7. Inflorescence a dense contracted panicle; branches spikelet-bearing to the base **227. *Phalaris***
8 (5). Inflorescence dense, cylindrical, much longer than broad **308. *Phleum***
8. Inflorescence open, much branched, most spikelets pedicellate 9
9 (8). Disarticulation below the glumes 10
9. Disarticulation above the glumes 12
10 (9). Spikelets laterally compressed **301. *Cinna***
10. Spikelets globose 11
11 (10). Leaf blades with pseudopetioles **145. *Phaenosperma***
11. Leaf blades without pseudopetioles **238. *Cyathopus***
12 (9). Spikelets dorsiventrally compressed; plants annual **293. *Zingeria***
12. Spikelets laterally compressed; plants annual or perennial 13

13 (12). Glumes shorter than the flower 14
13. Glumes longer than the flower 19
14 (13). Palea with 1 vein 15
14. Palea with 2 veins or keels 16
15 (14). Plants erect, rhizomatous; Arctic **314. *Arctagrostis***
15. Plants decumbent; New Zealand **312. *Simplicia***
16 (14). Plants 0.5–1 m tall; leaf sheaths open **296. *Aniselytron***
16. Plants much less than 0.5 m tall; leaf sheaths often closed >½ their length 17
17 (16). Glumes less than 1/3 the length of the flower, the lower glume sometimes absent; plants of the Arctic, Argentine Andes **288. *Phippsia***
17. Glumes well developed; plants neither arctic nor Andean 18
18 (17). Plants caespitose, or if rhizomatous then fruit lacking a beak **292. *Colpodium***
18. Plants rhizomatous; fruit with a small beak **287. *Paracolpodium***
19 (13). Palea minute or absent **233. *Agrostis***
19. Palea well developed, at least ½ the length of the lemma 20
20 (19). Callus hairs up to 2/3 the length of the lemma; Australia **243. "*Lachnagrostis*"**
20. Callus hairs short, much less then 2/3 the length of the lemma; western hemisphere **245. *Podagrostis***
21 (4). First glume greater than ½ the length of the first lemma 22
21. First glume ½ the length of the first lemma or less 49
22 (21). Plants annual 23
22. Plants perennial 28
23 (22). Inflorescence branches disarticulating at their bases, as well as at the bases of the pedicels and flowers **274. *Cutandia***
23. Inflorescence branches not disarticulating 24
24 (23). Inflorescence with spikelets on short pedicels or short sturdy primary branches **275. *Desmazeria***
24. Plants slender, the inflorescence much branched, with spikelets at the end of slender pedicels 25
25 (24). Glumes slightly shorter than the spikelet 26
25. Glumes longer than the spikelet 27
26 (25). Lemma pubescent, callus glabrous; flowers more than 2 in most spikelets **310. *Poa***
26. Lemma glabrous, callus pubescent; flowers exactly 2 per spikelet **256. *Molineriella***
27 (25). Flowers separated by an internode **248. *Antinoria***
27. Flowers scarcely separated **229. *Airopsis***
28 (22). Glumes notably dissimilar in shape, the first slender and linear, the second broadly oblanceolate to flabellate **225. *Sphenopholis***
28. Glumes similar in shape 29
29 (28). Lemma indurate, much firmer than the glumes; puberulent; plants of Argentina and Chile **232. *Relchela***
29. Lemma similar in texture to the glumes; glabrous to pubescent; distribution of plants various 30
30 (29). Lemmas with hairs on callus 31
30. Lemma callus glabrous 41
31 (30). Spikelets dorsiventrally compressed **231. *Chascolytrum***
31. Spikelets laterally compressed 32
32 (31). Lemma apex truncate, ragged; plants of SE Australia **265. *Dryopoa***
32. Lemma apex entire or bidentate, but not truncate and ragged; plants of various localities 33
33 (32). Leaf sheaths closed for more than ½ their length 34
33. Leaf sheaths open, or closed less than ½ their length 35
34 (33). Inflorescence branches spreading **286. *Hyalopoa***
34. Inflorescence branches upright **303. *Dupontia***
35 (33). Leaf blades with a prow-shaped apex, generally with a translucent line on either side of the mid-vein; inflorescence branches generally smooth **310. *Poa***
35. Leaf blades with apices acute or acuminate, not prow-shaped, lacking a translucent line on either side of the mid-vein; inflorescence branches generally scabrous 36
36 (35). Rachilla glabrous or scabrous 37
36. Rachilla pubescent 38
37 (36). Leaf blades with prominent cross veins; plants of Queensland and New South Wales **318. *Sylvipoa***
37. Leaf blades lacking prominent cross veins; plants of various localities **226. *Trisetaria***
38 (36). Leaf blades less than 1 mm wide **300. *Bellardiochloa***
38. Leaf blades more than 1 mm wide 39
39. Plants with sturdy rhizomes **299. *Arctopoa***
39. Plants generally caespitose, if rhizomatous then the rhizomes short 40
40. Apex of lemma minutely bifid, with an abaxial awn or mucro **226. *Trisetaria***
40. Apex of lemma entire, unawned **222. *Graphephorum***
41 (30). Sheath margins fused; rachilla ending in a club-like cluster of lemmas **187. *Melica***
41. Sheath margins fused less than ½ their length; rachilla terminating in a reduced flower, but not a cluster 42
42 (41). Callus broader than long, forming a smoothly rounded ridge at the base of the lemma 43
42. Callus about the same width as length, not obviously smoothly rounded 45
43 (42). Plants dioecious **263. *Leucopoa***
43. Plants bisexual 44
44 (43). Fruit free from the palea **266. *Festuca* s.s.**
44. Fruit adhering to the palea **267. *Lolium* s.l.**
45 (42). Lemma and glume margins ciliate 46
45. Lemma and glume margins not ciliate (although lemma often pubescent elsewhere) 47
46 (45). Rachilla internodes densely pubescent; plants of Australia **317. *Saxipoa***
46. Rachilla internodes glabrous, scabrous, or puberulent; plants of the Arctic **298. *Arctopoa***
47 (45). Leaf blades with a translucent line on either side of the midvein; blade tip prow-like; lemma apex entire **310. *Poa***

47. Leaf blades lacking a translucent line; blade tip acute to acuminate but not curved like the prow of a boat; lemma apex often minutely bifid 48

48 (47). Basal leaves often with sheaths persisting and forming bulb-like structures; callus glabrous **284. *Catabrosella***

48. Basal leaves not forming bulbs; callus pubescent **226. *Trisetaria***

49 (21). Lemma tuberculate **264. *Castellia***

49. Lemma smooth, scabrous or pubescent, but not tuberculate 50

50 (49). Leaf sheath margins fused more than ½ their length 51

50. Leaf sheath margins free for at least ½ their length 52

51 (50). Flowers 2 to 20 per spikelet; lemma with (5) 7 (11) veins **185. *Glyceria***

51. Flowers 1 to 3 per spikelet; lemma with 3 veins **283. *Catabrosa***

52 (50). Second glume and lemmas much >1 cm long, the glumes papery; leaf sheath with prominent auricles **195. *Littledalea***

52. Second glume and lemmas <1 cm long; leaf sheaths not auriculate 53

53 (52). Inflorescence one-sided; base of style branches prominent and persistent in fruit **291. *Sclerochloa***

53. Inflorescence not one-sided; base of style branches not persistent 54

54 (53). First glume minute, hyaline **272. *Sphenopus***

54. First glume well developed 55

55 (54). Lemmas keeled, their apices generally acute, the veins easily visible, converging toward the apex 56

55. Lemmas rounded on the back, their apices generally obtuse to truncate, the veins obscure, where visible then more or less parallel 59

56 (55). Plants generally >1 m tall **265. *Dryopoa***

56. Plant generally <1 m tall 57

57 (56). Leaf blades narrow, thread-like or bristle-like **300. *Bellardiochloa***

57. Leaf blades not thread-like or bristle-like 58

58 (57). Lemma apex obtuse; apex of leaf blades not prow-like **289. *Pseudosclerochloa***

58. Lemma apex acute; apex of leaf blades prow-like **310. *Poa***

59 (55). Plants mat-forming; spikelets with 1 or 2 flowers; plants of New Zealand **312. *Simplicia***

59. Plants erect, not mat-forming; spikelets generally with more than 2 flowers; plants elsewhere 60

60 (59). Lemma veins generally 7 to 9, rarely 5 **218. *Torreyochloa***

60. Lemma veins generally 5 (3–7) 61

61 (60). Culms decumbent but not rhizomatous; plants of saline or alkaline habitats **290. *Puccinellia***

61. Plant rhizomatous or stoloniferous; plants not of saline or alkaline habitats **292. *Colpodium***

Group III – Inflorescence branched, spikelets awned, the awn straight

1. Margins of leaf sheaths fused for most of their length 2

1. Margins of leaf fused less than ½ their length, if at all 3

2 (1). Leaf sheath forming a tube, the point of fusion of the margins not distinct **194. *Bromus***

2. Leaf sheath margins fused by a hyaline membrane **189. *Schizachne***

3 (1). Spikelets in pairs, with the outer and more obvious member of the pair made up solely of sterile lemmas, these well separated on the rachilla and stiffly divaricate **261. *Cynosurus***

3. Spikelets not paired, lacking sterile spikelets with divaricate lemmas 4

4 (3). Inflorescence capitate, more or less cylindrical 5

4. Inflorescence sparse or dense, if the latter then clearly longer than broad 8

5 (4). Lemmas stiff, coriaceous to indurate, divided into 5 to 7 stiff acuminate lobes **279. *Echinaria***

5. Lemmas membranous to coriaceous, ending in a mucro or single awn 6

6 (5). Plants perennial; Australia, New Guinea, New Zealand **240. *Echinopogon***

6. Plants annual, often minute; Mediterranean and Middle East 7

7 (6). Spikelets with 4 to 14 flowers **260. *Ammochloa***

7. Spikelets with 1 flower **311. *Rhizocephalus***

8 (4). Spikelets with one flower 9

8. Spikelets with more than one flower 23

9 (8). Tiny annuals; spikelets in clusters, appearing more or less verticillate; glumes absent, lemmas translucent **285. *Coleanthus***

9. Plants perennial, or if annual, glumes present and spikelets not in clusters 10

10 (9). Plants perennial, strongly rhizomatous; beaches in temperate regions **234. *Ammophila***

10. Plants annual or caespitose perennials; habitats various, but not beaches 11

11 (10). Disarticulation below the glumes; glumes and lemmas strongly compressed and keeled, the glumes often partially fused near their bases **295. *Alopecurus***

11. Disarticulation above the glumes; glumes and lemmas terete, if somewhat laterally compressed then not strongly so, not keeled, glume bases not fused 12

12 (11). Glumes awned; plants annual **233. *Agrostis* (former *Chaetopogon*)**

12. Glumes lacking awns; plants annual or perennial 13

13 (12). Callus hairs absent or less than 1/3 the length of the lemma 14

13. Callus hairs 1/3 the length of the lemma or more 21

14 (13). Lemma with margins fused proximally 15

14. Lemma margins not fused 16

15 (14). Basal leaves with blades lacking or <2 cm long **170. *Patis***

15. Basal leaves with blades present, generally >2 cm **166. *Piptatheropsis***

16 (14). Glumes shorter than the flower **165. *Ortachne***

16. Glumes as long as or longer than the flower 17

17 (16). Awn arising from the abaxial side of the lemma **245. *Podagrostis***

17. Awn apical or arising from the sinus between two lemma teeth 18

18 (17). Spikelets cylindrical 19

18. Spikelets laterally compressed 20

19 (18). Plants caespitose but <2 m tall, not forming massive tussocks; plants of the New World **163. *Anatherostipa***
19. Plants to 2.5 m tall, forming massive tussocks; plants of Eurasia **154. "Neotrinia"**
20 (18). Plants over 1 m tall; lemma awn 4–9 mm long; stamen 1 **174. *Anemanthele***
20. Plants much less than 0.4 m tall; lemma awn less than 4 mm long **294. *Agrostopoa***
21 (13). Callus hairs longer than the lemma **237. "*Calamagrostis*"**
21. Callus hairs from 1/3 the length of the lemma to nearly as long, but not longer 22
22 (21). Lemma stiff, hardened, scabrous **239. *Deyeuxia***
22. Lemma membranous, soft **243. *Lachnagrostis***
23 (8). Inflorescence a dense, spike-like panicle, with two sterile bracts at the base **282. *Sesleria***
23. Inflorescence open or contracted, but not spike-like, lacking basal bracts 24
24 (23). Spikelets borne in dense clusters at the ends of secondary branches 25
24. Spikelets not notably clustered, borne evenly along branches 26
25 (24). Spikelet clusters one-sided; all spikelets in a cluster fertile; plants perennial **270. *Dactylis***
25. Spikelet clusters not obviously one-sided; proximal spikelets in a cluster sterile; plants annual **271. *Lamarckia***
26 (24). Plants annual 27
26. Plants perennial 31
27 (26). Lemma awn terminal 28
27. Lemma awned from the sinus or from the abaxial side 29
28 (27). Lemmas with 3 veins; inflorescence branches disarticulating **274. *Cutandia***
28. Lemmas with 5 veins; inflorescence branches persistent **266. *Festuca***
29 (27). Lemma awned from the abaxial side; spikelets with 2 flowers **257. *Periballia***
29. Lemma awned from a short sinus; spikelets generally with more than 2 flowers 30
30 (29). Inflorescence with the proximal branches sterile **307. *Nephelochloa***
30. Inflorescence with all branches spikelet-bearing **278. *Vulpiella***
31 (26). Awn from the back of the lemma, with a crown of hairs at the middle, the tip flattened and spathulate; glumes hyaline, vase-shaped, noticeably longer than the flowers **253. *Corynephorus***
31. Awn terminal or from the back of the lemma, but never with a crown of hairs or with a flattened tip; glume texture various, but glumes not vase-shaped; glumes as long as or slightly shorter than the flowers 32
32 (31) Culm internodes solid **155. *Ampelodesmos***
32. Culm internodes hollow 33
33 (32). Lemmas broader than long **231. *Chascolytrum* (former *Erianthecium*)**
33. Lemmas longer than broad 34
34 (33). Callus hairs more than 2/3 the length of the flower **243. *Lachnagrostis***
34. Callus hairs less than 2/3 the length of the flower 35
35 (34). Awns of adjacent flowers coiling around each other; callus elongate **190. *Streblochaete***
35. Awns not coiling around each other; callus short 36
36 (35). Lemma awn terminal 37
36. Lemma awn abaxial or from a sinus between two teeth 40
37 (36). Lemmas with three prominent green nerves; grain with a prominent indurate apical structure ("beak") **192. *Diarrhena***
37. Lemmas with five or more nerves, these not prominent; grain lacking a beak 38
38 (37). Plants of tropical Africa and Madagascar; leaf blades with obvious transverse veins **269. *Pseudobromus***
38. Plants of temperate regions worldwide, not tropical; leaf blades lacking obvious transverse veins 39
39 (38). Glumes and lemmas keeled; plants dioecious **263. *Leucopoa***
39. Glumes and lemmas rounded on the back; plants bisexual **266. *Festuca***
40 (36). Plants with stout rhizomes; apex of leaf blade sharp pointed **316. *Nicoraepoa***
40. Plants generally caespitose, if rhizomatous then the apex of the leaf blade not sharp pointed 41
41 (40). Palea not tightly clasped by the lemma, clearly visible behind the lemma at maturity **225. *Trisetaria***
41. Palea tightly clasped by the lemma, the flower thus opening only at anthesis, but otherwise closed 42
42 (41). Spikelets with 2 or 3 flowers, 5–7 mm long; plants of North America **222. *Graphephorum***
42. Spikelets with 3 to 5 flowers, 7–14 mm long; plants of eastern Australia **315. *Hookerochloa***

Group IV – Inflorescence branched, spikelets awned, the awn curved, twisted or geniculate

1. Plants sweetly scented with coumarin; spikelets with more than one flower, but only the distal one bisexual; proximal flowers sterile or staminate **228. *Anthoxanthum***
1. Plants not scented; spikelets with one flower, or if more than one, then proximal flowers bisexual 2
2 (1). Disarticulation below the glumes; spikelets dispersed as a unit 3
2. Disarticulation above the glumes; flowers dispersed individually 4
3 (2). Tufted annuals; plants of the southern U.S. and Mexico **305. *Limnodea***
3. Perennials with short rhizomes; plants of Siberia **304. *Limnas***
4 (2). Inflorescence dense, lanceolate; glumes linear and covered with long hairs so that the entire inflorescence appears furry **224. *Lagurus***
4. Inflorescence open or compact; glumes generally glabrous, if pubescent then the hairs not so dense as to make the entire inflorescence appear furry 5
5 (4). Spikelets with one fertile flower 6
5. Spikelets with more than one fertile flower 53
6 (5). Glumes shorter than the flower 7

6. Glumes as long as or longer than the flower 14
7 (6). Rachilla extended beyond the distalmost flower, ending blindly or in a rudiment 8
7. Rachilla extension lacking 11
8 (7). Lemmas deeply two-lobed, with a line of long hairs around the base of the lobes, awned from the sinus; China and Central Asia **152. *Stephanachne***
8. Lemma apex entire or notched, lacking a line of long hairs; plants of Australia and the Pacific 9
9 (8). Lemma apex entire, acuminate, awned from the apex; plants of Juan Fernández Island, possibly extinct **217. *Podophorus***
9. Lemma apex notched or ragged, not acuminate, awned from the back; plants of Australia and New Guinea 10
10 (9). Lemma apex notched, the awn abaxial, hooked; callus hairs short **235. *Ancistragrostis***
10. Lemma apex ragged, the abaxial awn straight or geniculate; callus hairs at least 1/3 the length of the flower **239. *Deyeuxia***
11 (7). Palea with two lobes **191. *Triniochloa***
11. Palea not lobed 12
12 (11). Lemma apex with a cylindrical beak, with retrorse spines; plants of the Himalayas **159. *Orthoraphium***
12. Lemma not beaked; plants of Central and South America 13
13 (12). Awn clearly divided into proximal and distal segments, once or twice geniculate; lemma much more firm than the glumes **182. "*Jarava*"**
13. Awn not clearly divided into segments; lemma not hardened **165. *Ortachne***
14 (6). Lemma with at least three awns, a central one that is apical or abaxial, and two or four slender lateral ones 15
14. Lemma with only one awn 17
15 (14). Glumes ca. 4 cm long; major lemma awn from the sinus; plants of southeast Australia **146. *Anisopogon***
15. Glumes less than 1 cm long; major lemma awn abaxial; plants of South America or the Mediterranean 16
16 (15). Palea absent or nearly so; plants of South America **236. *Bromidium***
16. Palea present; plants Mediterranean, introduced elsewhere **247. *Triplachne***
17 (14). Lemma hyaline 18
17. Lemma membranous to coriaceous 20
18 (17). Glumes mucronate or awned **246. *Polypogon***
18. Glumes neither mucronate nor awned 19
19 (18). Spikelets vase-shaped, the glumes inflated around the fruit and narrowed above, then flaring distally; palea present **241. *Gastridium***
19. Spikelets ovate to oblong or lanceolate, but not vase-shaped; palea absent **233. *Agrostis***
20 (17). Lemma awn abaxial, inserted at or below the midpoint of the lemma 21
20. Lemma awn terminal or from a sinus 24
21 (20). Lemma crustaceous at maturity **242. *Hypseochloa***
21. Lemma not crustaceous, although sometimes hardened 22
22 (21). Callus hairs generally longer than the flower **237. "*Calamagrostis*"**
22. Callus hairs shorter than the flower 23
23 (22). Palea as long as the lemma **239. *Deyeuxia***
23. Palea shorter than the lemma **243. *Lachnagrostis***
24 (20). Lemma awn subapical or from a sinus 25
24. Lemma awn terminal 41
25 (24). Plants annual **297. *Apera***
25. Plants perennial 26
26 (25). Vegetative prophylls extending out of the leaf sheaths 27
26. Vegetative prophylls shorter than the leaf sheaths 28
27 (26). Inflorescence open, the branches spreading; awns twice geniculate **176. *Celtica***
27. Inflorescence narrow, the branches upright; awns once geniculate **153. *Macrochloa***
28 (26). Callus glabrous 29
28. Callus pubescent 31
29 (28). Plants rhizomatous; glume apices obtuse; plants of Mongolia **157. *Psammochloa***
29. Plants caespitose; glume apices acute to acuminate; plants of various localities 30
30 (29). Spikelets dorsiventrally compressed; awn deciduous **177. *Oloptum***
30. Spikelets laterally compressed; awn persistent **294. *Agrostopoa***
31 (28). Lemma glabrous or scabrous 32
31. Lemma pubescent 33
32 (31). Rachilla extension present; callus hairs at least 1/3 the length of the lemma **239. *Deyeuxia***
32. Rachilla extension absent; callus hairs much less than 1/3 the length of the lemma **174. *Anemanthele***
33 (31). Awn covered with hairs over its entire length, the hairs 1 mm long or more **160. *Ptilagrostis***
33. Awn glabrous, or pubescent only in the proximal part 34
34 (33). Lemma margins fused proximally **166. *Piptatheropsis***
34. Lemma margins free 35
35 (33). Palea much shorter than the lemma, glabrous **182. "*Jarava*"**
35. Palea about the same length as the lemma or longer, often pubescent 36
36 (35). Ligules densely pubescent; proximal segment of the awn with hairs 4–9 mm long **167. *Pappostipa***
36. Ligules glabrous or sparsely pubescent; proximal segment of awn with hairs short or lacking 37
37 (36). Lemma margins overlapping for most or all of their length at maturity 38
37. Lemma margins not overlapping 39
38 (37). Plants annual **178. *Stipellula***
38. Plants perennial **173. *Achnatherum* s.s., 179. "Timouria", and 172. "Stillmania" (= "*Achnatherum*" *stillmanii*)**
39 (37). Awns deciduous **171. *Piptatherum***
39. Awns persistent 40
40 (39). Lemma lobes <3 mm long; stigmas 2 **161. *Trikeraia***
40. Lemma lobes 3–6 mm long; stigmas often 3 **151. *Sinochasea***
41 (24). Palea deeply grooved, with lemma margins inrolled into the groove; palea apex extending beyond the lemma margins at the lemma apex **168. *Piptochaetium***
41. Palea and lemma margins flat, the latter overlapping or not; palea apex not extending beyond the lemma apex 42

42 (41). Flag leaf blades ca. 1 cm long or less **156. *Oryzopsis***
42. Flag leaf blades >1 cm 43
43 (42). Spikelets dorsiventrally compressed, the palea about the same length as the lemma and fully visible **171. *Piptatherum***
43. Spikelets terete, the palea generally shorter than the lemma, partially or wholly surrounded by it 44
44 (43). Apex of lemma with a circular rim or "crown", the awn originating within the rim 45
44. Apex of lemma acute to acuminate, without a circular rim 47
45 (44). Lemma margins overlapping at maturity, completely concealing the palea **183. *Nassella***
45. Lemma margins only partially overlapping, the palea partially visible at maturity 46
46 (45). Flowers large, 7.5–25 mm long **164. *Hesperostipa***
46. Flowers generally less than 7 mm long **180. "Eriocoma" group**
47 (44). Callus of flower obtuse, blunt 48
47. Callus of flower acute, often sharp-pointed 51
48 (47). Vegetative prophylls ending in two awns, 2–5 cm long **163. *Anatherostipa***
48. Vegetative prophylls with two teeth, but not awned 49
49 (48). Spikelets laterally compressed; lemma chartaceous, keeled **294. *Agrostopoa***
49. Spikelets terete; lemma hardened 50
50 (49). Plants with knotty rhizomes; leaf blades with the apices sharp-pointed **181. *Amelichloa***
50. Plants lacking knotty rhizomes; leaf blades with apices acuminate but not with sharp points **158. *Stipa***
51 (47). Spikelets generally >1 cm long, awns generally >10 cm long, plumose throughout **158. *Stipa***
51. Spikelets <1 cm long, awns <10 cm long, if plumose then only on the basal segment 52
52 (51). Plants not branching from the upper nodes **182. "*Jarava*"**
52. Plants branching from the upper nodes, the branches stiff **175. *Austrostipa***
53 (5). Disarticulation below the glumes, the spikelet falling entire **255. *Holcus***
53. Disarticulation above the glumes, flowers falling individually 54
54 (53). Glumes aristate to long awned 55
54. Glumes not awned 56
55 (54). Lemma keeled; lodicules 2; plants of Juan Fernandez Island **216. *Megalachne***
55. Lemma rounded on the back; lodicules 3; plants of Mexico **149. *Metcalfia***
56 (54). Lemma apex divided into four slender aristae above the large abaxial awn **244. *Pentapogon***
56. Lemma apex often bilobed, but not with four aristae 57
57 (56). Callus with hairs 2/3 the length of the lemma **243. *Lachnagrostis***
57. Callus glabrous or pubescent with hairs less than ½ the length of the lemma 58
58 (57). Plants annual 59
58. Plants perennial 62
59 (58). Second lemma with a twisted abaxial awn, other lemmas with a terminal long slender straight awn; first lemma and glumes remaining on the plant at maturity **313. *Ventenata***
59. Curved or twisted abaxial awn on all lemmas; all flowers falling off at maturity 60
60 (59). Glumes about the same length as the lowest lemma, but generally not extending beyond the flowers **226. *Trisetaria***
60. Glumes extending beyond the flowers, enclosing them 61
61 (60). Spikelets generally much >1 cm long; ovary pubescent **231. *Avena***
61. Spikelets generally <0.5 cm long; ovary glabrous **250. *Aira***
62 (58). First glume as long as or longer than the first lemma 63
62. First glume about ½ the length of the first lemma or less 71
63 (62). Flowers more than two per spikelet 64
63. Flowers exactly two per spikelet 67
64 (63). Awn from a sinus between two lobes, the lobes more than ¼ the length of the lemma 65
64. Awn abaxial, the lemma apex entire or with very short lobes 66
65 (64). Ligule membranous, lacking cilia; plants of tropical Asia **150. *Pseudodanthonia***
65. Ligule ciliate, ragged; plants primarily of temperate habitats, Eurasia **223. *Helictotrichon***
66 (64). Awn from about ½ way down the back of the lemma; rachilla pubescent on one side only **252. *Avenula***
66. Awn subapical; rachilla pubescent on all sides **226. *Trisetaria***
67 (63). Lower flower staminate, with a prominent twisted awn; upper flower bisexual, with a fragile straight awn **220. *Arrhenatherum***
67. Both flowers with twisted awns, both bisexual 68
68 (67). Panicle narrow, the branches upright; stamens 2 **226. *Trisetaria* (former *Peyritschia*)**
68. Panicle open, generally lax; stamens 3 69
69 (68). Ligule acute **254. *Deschampsia***
69. Ligule obtuse to truncate 70
70 (69). Ligule obtuse **251. *Avenella***
70. Ligule truncate with irregular teeth **259. *Vahlodea***
71 (62). Leaves narrow, folded, and with a long sharp tip **266. *Festuca* (former *Dielsiochloa*)**
71. Leaves broad, flat, not aristate 72
72 (71). Callus glabrous **187. *Melica***
72. Callus pubescent **218. *Amphibromus***

Tribes and Genera of Pooideae

VII. Tribe Brachyelytreae Ohwi (1941)

Perennials with short rhizomes. Leaf blades broad, lanceolate. Inflorescence branches branched. Spikelets terete to dorsiventrally compressed, with one flower, with a rachilla

Fig. 32. Pooideae-Brachyelytreae. *Brachyelytrum erectum.* **A** Flowering plant. **B** Adaxial leaf surface. **C** Flower. (From Barkworth et al. 2007, drawn by L.A. Vorobik)

extension. Lower glume minute or absent, upper glume much shorter than the flower. Callus of flower pubescent. Lemma apex entire, awned, the awn longer than the lemma. Fruit fusiform, grooved, pubescent, beaked. Embryo with or without a scutellar cleft. Microhairs lacking.

142. ***Brachyelytrum*** P. Beauv. Fig. 32

Brachyelytrum P. Beauv., Ess. Agrostogr. 39, 155, pl. 9, f. 2 (1812); rev.: Saarela et al. (2003).

Characters as for the tribe. $2n = 22$.

Three spp., eastern Asia, eastern North America.

Brachyelytrum was previously placed in the Bambusoideae on the basis of its embryo structure (Campbell et al. 1986), but is clearly sister to all other Pooideae. The presence of an epiblast, and the lack of an internode between the vascular trace to the coleoptile and that for the scutellum are shared with all other Pooideae. The character overlapping embryonic leaf margins is plesiomorphic. Campbell et al. (1986) show a clear demarcation between the scutellum and the coleorhiza, but others report that this is lacking (Reeder 1957); although Campbell et al. (1986) suggest that this character may be polymorphic, it seems equally possible that the observation of Reeder (1957) was in error. Campbell et al. (1986) report parallel-sided subsidiary cells, a character that is otherwise synapomorphic for the clade consisting of Brachypodieae, Bromeae, Triticeae, and Poeae.

Remainder of the subfamily:
Inflorescence phyllotaxis initially two-ranked, even if subsequent branching makes them appear spiral (Kellogg et al. 2013). *Spikelets laterally compressed. Embryo without a scutellar cleft, embryonic leaf margins non-overlapping.*

VIII. Tribe Nardeae W. D. J. Koch (1837)

Lygeeae J. Presl (1846).

Perennials. Leaves slender, leathery. Rachilla extension absent. Lower glume reduced to a rim, the upper glume absent or minute. *Lodicules absent; stigma one.*

Here the two genera are combined in a single tribe, following Döring et al. (2007).

143. ***Nardus*** L. Fig. 33

Nardus L., Sp. Pl. 1: 53–54 (1753).

Plants caespitose. Inflorescence unbranched, ending in a barren tip, the spikelets borne on one side, in two rows. Spikelet *dorsiventrally compressed*, with 1 flower. Upper glume minute. Lemma apex acute, awned. $2n = 26$.

One sp., *N. stricta* L., Europe.

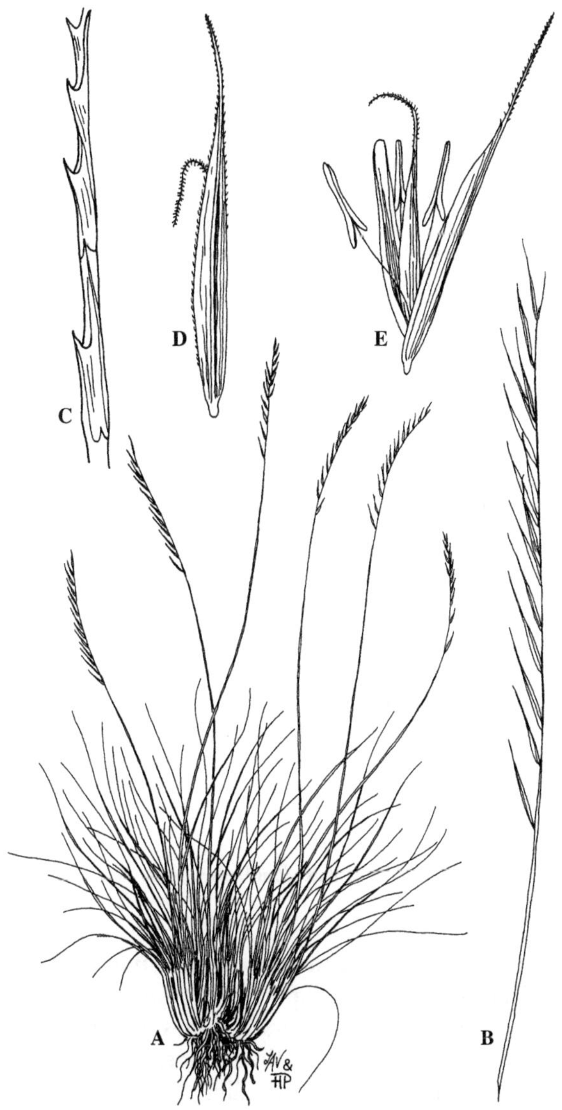

Fig. 33. Pooideae-Nardeae. *Nardus stricta*. **A** Habit. **B** Inflorescence. **C** Portion of the inflorescence axis, the spikelets fallen but the lower glumes retained. **D** Floret, style exserted. **E** Anthetic floret, palea to the left, lemma at right. (From Cope and Gray 2009)

144. ***Lygeum*** Loefl. ex L.

Lygeum Loefl. ex L., Gen. Pl. (ed. 5) 27, (522) (1754).

Plants with short rhizomes. Inflorescence with one spikelet, subtended by a leaf-like spathe. Spikelet laterally compressed, with 2 to 3 flowers, their lemmas fused to form a cup or urn and paleas fused to form a septum. Upper glume absent. $2n = 40$.

One sp., *L. spartum* L., Mediterranean.

Remaining tribes and genera:
Microhairs lacking.

Microhairs occur in a few Stipeae, where they appear to be secondarily derived; otherwise all remaining Pooideae lack them. If Stipeae diverged before Phaenospermateae, as suggested by GPWG II (2012), then loss of microhairs could have occurred after the divergence of Stipeae.

IX. Tribe Phaenospermateae Renvoize & Clayton (1985)

Duthieeae M. Röser & J. Schneider (2011).

Perennials. Ligule membranous. Lodicules 0, 2 or 3.

Eight genera, 14 spp.

Membership of this subtribe is gradually becoming clearer as sampling has improved in molecular phylogenetic studies. The GPWG (2001) found a sister relationship between *Anisopogon* and *Phaenosperma*, Davis and Soreng (2007) confirmed the addition of *Sinochasea* and *Duthiea* to the group, and Döring et al. (2007) found a clade including all four plus *Danthoniastrum*. The thorough study of Schneider et al. (2011) added the remaining genera listed below and discussed their placement, morphology and cytology. Schneider et al. (2011) argue for separating Phaenospermateae and Duthieeae because of the very different spikelet morphology of Duthieeae, and because the combined tribe lacks an obvious morphological synapomorphy. Like Soreng et al. (2012 onward), I combine them here to avoid a monogeneric Phaenospermateae. In addition the synapomorphies of Duthieae s.s. (i.e., without *Phaenosperma*) are not particularly distinctive within the context of Pooideae as a whole.

145. ***Phaenosperma*** Munro ex Benth.

Phaenosperma Munro ex Benth., J. Linn. Soc. Bot. 19: 59 (1881).
Euthryptochloa Cope, Kew Bull. 42(3): 707 (1987).

Plants 1–3 m tall. Leaf blade pseudopetiolate and resupinate, with primary veins extending at an acute angle from the midrib. Inflorescence branches branched. Spikelets *spherical* to somewhat laterally or dorsiventrally compressed, with

1 flower; *disarticulating below the glumes*; rachilla extension lacking. Glumes shorter than the flower, hyaline. Lemma apex entire, obtuse, awnless. Lodicules 3. Stigmas 2. Ovary glabrous. *Style bases persistent in fruit as a small beak; pericarp separable from seed.* $2n = 24$.

One sp., *P. globosa* Munro ex Benth., Assam to Japan.

The resupinate leaf blades are similar to those of *Pharus*. *Phaenosperma* was previously placed in Bambusoideae on the basis of embryo characters, but it shares with the Pooideae the presence of an epiblast and a short mesocotyl internode. The scutellar cleft and overlapping embryonic leaf margins are plesiomorphic so not indicative of relationship. The lack of microhairs and fusoid cells, and ample molecular data put *Phaenosperma* in Pooideae. Watson and Dallwitz (1992 onward) report that *Phaenosperma* has ruminate endosperm, a condition that is otherwise unknown in the grasses.

Remaining genera of the tribe:
Disarticulation above the glumes. *Rachilla extended beyond the distalmost flower. Lemma apex with two elongate lobes or teeth, awned from the sinus, the awn twisted.*

Disarticulation above the glumes is likely to be plesiomorphic, but the pattern of gain and loss of this character among the early-diverging tribes of Poeae is uncertain.

146. *Anisopogon* R. Br.

Anisopogon R. Br., Prodr. 176 (1810).

Leaf blades stiff and sharp-pointed. Ligule with a ciliate margin. Inflorescence branches branched. Spikelets with one flower. Glumes longer than the flower, their apices obtuse. Callus of flower sharply pointed, pubescent. Lemma wrapped around and entirely concealing the palea, pubescent, the lobes awn-like. Lodicules 3. Stigmas 2 or 3. Ovary pubescent. *Pericarp separable from the seed.*

Two spp., Southeastern Australia.

The single-flowered spikelet and sharp-pointed callus led previous authors to place this genus in Stipeae, but molecular data clearly place it here in Phaenospermateae (Schneider et al. 2009, 2011).

147. *Danthoniastrum* (J. Holub) J. Holub

Danthoniastrum (J. Holub) J. Holub, Folia Geobot. Phytotax. 5: 435 (1970).

Inflorescence unbranched or the branches occasionally branched, generally with fewer than 5 spikelets in the entire inflorescence. Spikelets with 4 to 9 flowers. Callus of flower with dense hairs. Glumes shorter than or as long as the flowers, their apices attenuate, unawned. Lodicules 3. Stigmas 2 or 3. Ovary pubescent at the apex or all over. Hilum linear. $2n = 24$.

Two spp., Balkans and Caucasus.

The two species of *Danthoniastrum* were considered a subgenus of *Helictotrichon* by J. Holub (*H.* subg. *Danthoniastrum* Holub), whereas Clayton et al. (2006 onward) and Clayton and Renvoize (1986) placed them in synonymy under *Metcalfia*. Neither placement is supported by the molecular phylogeny (Schneider et al. 2009). The leaf blades of *D. compactum* disarticulate from the sheath, and the anatomy of the disarticulation zone is described by Röser and Heklau (2011).

148. *Duthiea* Hack. Fig. 34

Duthiea Hack., Verh. K.K. Zool.-Bot. Ges. Wien 45: 200 (1896).

Inflorescence unbranched, *with some pedicels subtended by bracts*. Spikelets sometimes nearly terete, with 1 to 5 flowers. Glumes as long as or longer than the flowers, their apices acute. Callus of flower pubescent. Lodicules 0, 2 or 3. Ovary pubescent. Stigmas 2 or 3. *Base of stylar branches persistent in the fruit, forming a beak.* $2n = 14$.

Three spp., mountainous areas of Afghanistan to western China.

The bracts subtending the pedicels of the spikelets are generally interpreted as being the vestiges of sterile spikelets. The persistent base of the stylar branches also appears in *Phaenosperma* and in *Sclerochloa* (Puccinellinae).

149. *Metcalfia* Conert

Metcalfia Conert, Willdenowia 2: 417 (1960).

Inflorescence branches branched. Spikelets with 3 or 4 flowers. Glumes as long as the flowers, their

Fig. 34. Pooideae-Phaenospermateae. *Duthiea oligostachya*. **A** Habit. **B** Spikelet. **C, D** Glumes. **E** Lemma. **F** Palea. **G** Pistil. (From Aitchison 1882)

apices acute, short-awned. Callus of flower pubescent. Lodicules 3. *Stamens 2*. Stigmas 2. Ovary pubescent all over. Hilum linear.

One sp., *M. mexicana* (Scribn.) Conert, Mexico.

See comments under *Danthoniastrum*.

150. ***Pseudodanthonia*** Bor & C. E. Hubb.

Pseudodanthonia Bor & C. E. Hubb., Kew Bull. 12: 425 (1958).

Leaf blades leathery. Inflorescence unbranched or with branched branches. Spikelet with 4 to 8 flowers plus additional reduced flowers distally. Glumes as long as or longer than the flowers, their apices attenuate. Callus of flower pubescent. Lodicules 2. Stigmas 3. Ovary pubescent.

One sp., *P. himalaica* (Hook. f.) Bor & C. E. Hubb., montane areas of western China and NW Himalaya.

151. ***Sinochasea*** Keng

Sinochasea Keng, J. Wash. Acad. Sci. 48: 115, f. 1 (1958).

Plants with short rhizomes. Leaves with blades stiff. Inflorescence branches branched. Spikelet with one flower; *rachilla extension minute or absent on some spikelets*. Glumes longer than the flower, their apices acuminate. Callus of flower pubescent. Lemma pubescent. Lodicules 2 or 3. Stigmas 2 or 3. Ovary glabrous.

One sp., *S. trigyna* Keng, temperate Asia.

Sinochasea is the sister genus to *Stephanachne* in the molecular study of Schneider et al. (2011).

152. ***Stephanachne*** Keng

Stephanachne Keng, Contr. Biol. Lab. Chin. Assoc. Advancem. Sci., Sect. Bot. 9: 134 (1934).

Inflorescence unbranched. Spikelets with one flower. Glumes longer than the flower, their apices acute to acuminate. Callus of flower pubescent. *Lemma with a tuft of long hairs near the apex*, the hairs nearly as long as the slender lobes. Lodicules 2 or 3. Stamens 1 or 3. Stigmas 2. Ovary glabrous.

Three spp., Central Asia and western China.

X. Tribe Stipeae Dumort. (1824)

Perennials, generally caespitose but a few rhizomatous, most with stiff slender leaves. Inflorescence branches branched. Spikelets with one flower, without a rachilla extension. Glumes generally longer than the flower. Lemma generally wrapped around and often concealing the palea, pubescent, the apex entire or with two teeth; awned from the apex, the awn straight or twisted, often once or twice geniculate, scabrid to plumose. Lodicules 3. Anthers 3. Stigmas 2 (3, 4). Embryo less than 1/3 the length of the fruit.

Twenty-eight genera, 530 spp.

The one-flowered spikelet and lack of a rachilla extension may be synapomorphic for this tribe, although both characters appear elsewhere in the Pooideae. Stipeae include *Ampelodesmos*, which has spikelets with several flowers, which would then represent a reversal of the tribal synapomorphy. While all genera have pubescent lemmas, the trichomes may be in lines or distributed evenly.

Generic limits within this tribe are problematical and in a state of flux (Barkworth et al. 2008; Romaschenko et al. 2008, 2010, 2011, 2012). Traditional taxonomy separated plants with short roundish spikelets from those with narrow elongate spikelets, but this distinction is wholly artificial and does not indicate relationship (Romaschenko et al. 2011). The phylogenetic analyses of Romaschenko et al. (2010, 2011, 2012) place *Macrochloa* sister to all other members of the tribe, which is then divided into four clades. Clade I includes an Old World subclade (*Stipa* s.s., *Ampelodesmos*, *Psammochloa*, *Oryzopsis* s.s., *Trikeraia*, *Orthoraphium*, and *Ptilagrostis*, plus the New World *Oryzopsis* s.s.), and a New World subclade (*Aciachne*, *Anatherostipa*, *Hesperostipa*, *Lorenzochloa*, *Ortachne*, *Piptatheropsis*, *Piptochaetium*, and *Pappostipa*). Clade II includes only *Patis*, and Clade III only *Piptatherum* s.s., now restricted to include only Old World species, plus "*Achnatherum*" *stillmanii*. Clade IV, the "achnatheroid grasses", includes *Anemanthele*, *Austrostipa*, *Oloptum*, *Timouria*, "*Achnatherum*" (including "*Piptatherum*" sect. *Virescentia*), and the Major American Clade (MAC), which includes the "Eriocoma group" of *Achnatherum*, plus "*Jarava*", *Amelichloa*, and *Nasella*. Clade IV is distinguished by the epidermal morphology of the lemmas. In the ancestral lemma epidermal morphology, the long cells have sinuous walls, a condition that the authors described as "saw-like". In the derived condition, which appears in Clade IV, as well as in *Pappostipa* of Clade I, the cells are straight-walled and isodiametric or may be broader than long; this is the "maize-like" pattern because of its apparent similarity to the kernels on a maize ear. "*Achnatherum*" will ultimately be divided into several genera, as indicated below.

Putative hybrid genera include *Achnella* Barkworth, a hybrid between "*Achnatherum*" and *Nasella*.

153. *Macrochloa* Kunth

Macrochloa Kunth, Révis. Gramin. 1: 58 (1829); anat., tax.: Vázquez and Barkworth (2004).
Stipa subgen. *Macrochloa* (Kunth) Steud., Syn. Pl. Glum. 1: 132 (1854).

Plants up to 2 m tall; prophylls prominent, 1–2 cm long, pubescent. *Leaf blades hard, convolute, disarticulating from the sheath; ligules velutinous.* Ligule a line of hairs. Glumes longer than the flower, their apices attenuate. Callus of flower sharp. Lemma thick to hardened, pubescent, the margins not overlapping, the apex with two teeth, awned from the sinus, the awn twisted, once-geniculate, the proximal portion plumose. Lodicules 3. Hilum as long as the fruit. *Midvein of leaf with a prominent adaxial rib, in cross section with a single vascular bundle at the top of the rib and several pairs of bundles below it.* $2n = 24, 66, 72$.

Two spp., Mediterranean.

This genus is sister to all other Stipeae s.s. (Romaschenko et al. 2010, 2012). Vázquez and Barkworth (2004) draw attention to the long velutinous vegetative prophylls of this species, which are distinctive in the context of the Stipeae. The leaf blades of *M. tenacissima* (Loefl. ex L.) Kunth disarticulate from the sheath. The anatomy of the disarticulation zone is described by Röser and Heklau (2011).

Remainder of Stipeae:
Clade I, Eurasian subclade

Lemma lobes generally well developed. Long cells of the lemma epidermis with sinuous walls.

The sinuous walls on the long cells of the lemma epidermis represent the plesiomorphic condition. All genera are Eurasian except for the North American *Oryzopsis*.

154. "Neotrinia"

("Neotrinia group" of "*Achnatherum*")

Plants robust, to 2.5 m tall, forming massive tussocks. Sheaths of basal leaves persistent, fibrous. Glumes longer than the flower, their apices acute, the lower with 1 vein, the upper with 3 veins. Callus of flower obtuse, pubescent. Lemma densely pubescent, membranous, with two lobes, the awn straight, deciduous. Palea extending slightly beyond the lemma apex. $2n = 42, 48$.

One sp., "*Achnatherum*" *splendens* (Trin.) Nevski, Eurasia.

This species is unrelated to *Achnatherum* s.s. but is strongly supported as sister to *Psammochloa villosa* (Hamasha et al. 2012; Romaschenko et al. 2011, 2012).

155. *Ampelodesmos* Link

Ampelodesmos Link, Hort. Berol. 1: 136 (1827).

Plants robust, to 3.5 m tall, rhizomatous. *Culm solid. Leaf bases becoming hard and curved. Spikelets with 2 to 6 flowers*, with a rachilla extension, the rachilla pubescent. Glumes shorter than the flowers, their apices attenuate. Lemma apex with two teeth, awned or mucronate from the sinus. Awn twisted, not geniculate. *Lodicules 3*, vascularized. Ovary apex pubescent. $2n = 48$.

One sp., *A. mauritanicus* (Poir.) T. Durand & Schinz, Mediterranean.

The spikelets with multiple flowers have been used to exclude this species from Stipeae, but molecular data are unequivocal linking it with *Psammochloa* and "*Neotrinia*" (Romaschenko et al. 2012). The spikelet number thus appears to be a reversal.

156. *Oryzopsis* Michx.

Oryzopsis Michx., Fl. Bor.-Amer. 1: 51 (1803).

Inflorescence branches branched. Spikelets terete to laterally compressed. Glumes shorter than or equal to the flower, not hardened, with 6 to 10 veins, their apices mucronate. Callus of flower pubescent with a ring of hairs ≥1 mm long. Lemma leathery, pubescent, the margins overlapping and slightly involute, the apex lobed, awned from the sinus, the awn straight, deciduous. $2n = 46$.

One sp., *O. asperifolia* Michx., North America.

157. *Psammochloa* Hitchc.

Psammochloa Hitchc., J. Wash. Acad. Sci. 17: 140 (1927).

Plant robust, rhizomatous. Leaf blades flat. Glumes about the same length as the flower, their apices obtuse. Callus of flower obtuse, glabrous. Lemma not enclosing the palea, the apex with two short teeth, awned from the sinus, the awn bent, deciduous. *Palea with 5 to 7 veins*, similar to lemma, acute. Lodicules 3.

One sp., *P. villosa* (Trin.) Bor, Mongolia.

Psammochloa is clearly part of Stipeae, and sister to "*Neotrinia*" (Romaschenko et al. 2012; Döring et al. 2007).

158. *Stipa* L., s.s.

Stipa L., Sp. Pl. 1: 78–79 (1753); tax.: Gonzalo et al. (2012, 2013); Vázquez Pardo and Gutiérrez Esteban (2011).

Plants caespitose. Leaves often with blades convolute or involute. Glumes acuminate, much longer than the flower, with 3 to 7 veins. Callus of flower long, sharp, with the tip oblique or slightly curved to a sharp point, generally with long hairs. Lemma with the margins overlapping, the apex lacking lobes, the awn once or twice geniculate, the column twisted. Palea as long as the lemma or slightly longer, the apex somewhat wrinkled. Lodicules 3. Stigmas 2 or 3 (4). $2n = 44$.

One hundred ten spp., Eurasia and North Africa.

Stipa s.l. has been dismembered over the last decade, but remains a large genus. *Stipa* s.s. now includes sections *Barbatae*, *Leiostipa*, *Pseudoptilagrostis*, *Regelia*, *Stipa*, and *Smirnowia*. In its strict sense, the genus is monophyletic (Hamasha et al. 2012; Romaschenko et al. 2012), but subdivisions within the genus are not consistently supported by available molecular data.

Orthoraphium + *Ptilagrostis* + *Trikeraia*

These three genera form a clade (Romaschenko et al. 2010, 2012). Hamasha et al. (2012) identified a similar group, although with two species currently classified in *Stipa* plus two from "*Achnatherum*".

159. *Orthoraphium* Nees

Orthoraphium Nees, Proc. Linn. Soc. Lond. 1: 94 (1841).

Plants caespitose. Glumes shorter than the flower, their apices acute to obtuse. Callus of flower obtuse, pubescent. Lemma papery, margins overlapping only at the apex, the apex with a cylindrical beak, *with stout retrorse spines*, awned, the awn straight, persistent. Lodicules 3.

One sp., *O. roylei* Nees, Himalayas above 2700 m altitude.

160. ***Ptilagrostis*** Griseb., s.s. Fig. 35

Ptilagrostis Griseb., Fl. Ross. 4: 447 (1852).

Plants caespitose. *Leaf blades filiform*, convolute. Glumes longer than the flower, their apices obtuse to acute. Callus of flower blunt, pubescent. Lemma papery, the margins not overlapping, the apex with two teeth, awned from the sinus, the awn once- or twice-geniculate, covered with hairs for its entire length. Palea pubescent. Lodicules 3. Ovary glabrous. Stigmas 2. $2n = 22$.

Eight spp., mountains of Asia.

Ptilagrostis is probably monophyletic as long as "*P*". *kingii* is excluded (Romaschenko et al. 2010).

161. ***Trikeraia*** Bor

Trikeraia Bor, Kew Bull. 9: 555, f. s.n. (1954).

Plants robust, rhizomatous. Leaf blades rolled. Glumes longer than the flower, their apices acuminate. Spikelets weakly dorsiventrally compressed. Callus of flower obtuse, pubescent. Lemma covering the edges of the palea, with two short slender lobes flanking a weakly geniculate awn, the awn persistent. Lodicules 2 or 3.

Three spp., Tibet and Himalayas.

Clade I, New World subclade

Indurate lemmas.

This clade has been identified by several molecular phylogenetic studies, although support varies (Cialdella et al. 2010; Romaschenko et al. 2010, 2012). Cialdella et al. (2010) suggest that the presence of indurate lemmas may be synapomorphic for this clade, although the character reverses in *Anatherostipa* and *Ortachne*.

162. ***Aciachne*** Benth.

Aciachne Benth., Hooker's Icon. Pl. 14, t. 1362 (1881).

Fig. 35. Pooideae-Stipeae. *Ptilagrostis pelliotii*. **A** Habit. **B** Ligule of basal leaf. **C** Ligule of culm leaf. **D** Spikelet. **E** Glumes. **F** Flower. **G** Anther. (From Wu et al. 2007, p. 279, with permission from the Missouri Botanical Garden Press, St. Louis, and Science Press, Beijing)

Cushion plants. Leaf blades narrow and needle-like, with a sharp point. Inflorescence no taller than leaves, with only 1 or 2 spikelets. Glumes, lemma and palea indurate. Flower cylindrical, somewhat gibbous. Lemma with a long point.

Three spp., Costa Rica to Argentina and Chile.

Romaschenko et al. (2010, 2012) find *Aciachne* to be monophyletic in their molecular phylogenetic analysis, whereas Cialdella et al. (2010) suggest that it may be poly- or paraphyletic.

163. ***Anatherostipa*** (Hack. ex Kuntze) Peñailillo

Anatherostipa (Hack. ex Kuntze) Peñailillo, Gayana, Bot. 53: 277 (1996).
Stipa sect. *Anatherostipa* Hack. ex Kuntze, Revis. Gen. Pl. 3(3): 372 (1898).
Stipa sect. *Obtusae* Parodi, Revista Argent. Agron. 17(3): 201 (1950), invalid.
Lorenzochloa Reeder & C. Reeder, Bol. Soc. Argent. Bot. 11(4): 239 (1969).

Leaf blades narrow, filiform, convolute. Glumes papery, as long as or longer than the flower. Callus of flower blunt. Lemma margins not inrolled, not covering the palea, the apex awned, the awn straight or curved or reduced to a mucro. Lodicules 2 or 3. Hilum linear.

Nine spp., New World.

Lorenzochloa is placed within *Anatherostipa* by molecular data (Romaschenko et al. 2010, 2012). Plastid sequences hint that *Anatherostipa* may be paraphyletic, with *Aciachne* derived from within it, but the phylogeny is not well supported (Romaschenko et al. 2012). The two genera are kept separate here pending additional data.

164. ***Hesperostipa*** (M. K. Elias) Barkworth

Hesperostipa (M. K. Elias) Barkworth, Phytologia 74: 15 (1993).

Glumes extending to a long slender tip. Flower cylindrical, narrow, *unusually large, 7.5–25 mm long*, the callus sharp-pointed, with dense hairs. Lemma hard, the margins flat and slightly overlapping, *the apex without lobes, with a papillose, ciliate crown*, awned, the awn twice geniculate. Palea pubescent. Fruit fusiform. $2n = 38, 44, 46$.

Five spp., North America and Mexico.

Lemma epidermal morphology of *Hesperostipa* is similar to that of the fossil *Berriochloa*, from the late Miocene (Thomasson 1978, 1985).

165. ***Ortachne*** Nees ex Steud.

Ortachne Nees ex Steud., Syn. Pl. Glumac. 1: 121 (1854).

Leaves filiform. Spikelets cylindrical. Glumes shorter than the flower. Callus of flower blunt, pubescent. Lemma not hardened, awned, the awn not clearly differentiated into a base and tip. Lodicules 2 or 3.

Two spp., Costa Rica to montane South America.

166. ***Piptatheropsis*** Romasch., P. M. Peterson & Soreng

Piptatheropsis Romasch., P. M. Peterson & Soreng, Taxon 60: 1712 (2011).

Spikelets dorsiventrally compressed or terete. Glumes as long as or longer than the flower, with 1 to 3 veins. Callus of flower blunt, glabrous or pubescent. Lemma leathery, pubescent, lobed at the apex, with margins fused below, the awn persistent and straight or deciduous and twisted. Lodicules 3. Stigmas 2. Hilum linear. $2n = 20, 22, 24$.

Five spp., North America.

Romaschenko et al. (2010) describe the fused portion of the lemma margins as a "small fleshy wart".

167. ***Pappostipa*** (Speg.) Romasch., P. M. Peterson & Soreng

Pappostipa (Speg.) Romasch., P. M. Peterson & Soreng, J. Bot. Res. Inst. Texas 2: 181 (2008).

Leaves with apices acute, sometimes sharp; ligules densely pubescent. Glumes longer than the flower. Flower terete, fusiform, the callus obtuse to truncate, pilose. Lemma with long hairs at the apex, with two short lobes, lacking a crown-like rim; awn once geniculate, plumose. Palea about the same length as the lemma or shorter. Lodicules 3.

Thirty-one spp., South America, SW United States and northern Mexico.

Members of this genus have lemma epidermal cells that are straight-walled and nearly isodiametric or broader than long (Romaschenko et al. 2008). This lemma epidermal pattern appears in parallel in Clade IV. Discrepancies between gene trees led Romaschenko et al. (2012) to suggest that *Pappostipa* may be the product of ancient allopolyploidy.

168. ***Piptochaetium*** J. Presl

Piptochaetium J. Presl, Reliq. Haenk. 1: 222 (1830), nom. conserv.; Cialdella et al. (2007, 2010), phylog., key.

Glumes longer than the flower. Callus of flower blunt or sharp, glabrous or pubescent. *Lemma apex forming a firm crown-like rim*, awned, the awn deciduous or persistent, twice geniculate and twisted. *Palea grooved, extending beyond the lemma; lemma margins involute and fitting into the palea groove*. Lodicules 2 or 3. Stigmas 2. $2n = 28, 42$.

Thirty-five spp., temperate North and South America.

169. **"*Ptilagrostis*" *kingii*** (Bol.) Barkworth

"*Ptilagrostis*" *kingii* (Bol.) Barkworth

Leaves with blades convolute, their apices stiff. Glumes slightly longer than the flower, their apices rounded to acute. Callus of flower blunt, pubescent. Lemma membranous, with margins not overlapping, the awn once or twice geniculate, persistent, scabrous. Lodicules 3. Stigmas 2. $2n = 22$.

One sp., "*P.*" *kingii* (Bol.) Barkworth, Sierra Nevada, California.

Although Romaschenko et al. (2011, 2012) place "*P.*" *kingii* sister to *Piptatheropsis*, support for this is weak. There is little morphological similarity between the two.

Clade II

170. ***Patis*** Ohwi

Patis Ohwi, Actta Phytotax. Geobot. 11: 181 (1942).

Plants caespitose or short rhizomatous. Basal leaves *with blades lacking or less than 2 cm long*; cauline leaves with blades >10 cm long and >5 mm wide. Spikelets dorsiventrally compressed or terete. Glumes shorter than to much longer than the flower, with transverse veins. Callus of flower obtuse. Lemma with margins overlapping, fused below, the awn persistent, straight. Lodicules 3. Stigmas 2. $2n = 46, 48$.

Three spp., eastern Asia and eastern North America.

Patis was recircumscribed by Romaschenko et al. (2011) and now includes species placed by previous authors in *Oryzopsis* or *Piptatherum*. It is not closely related to any other genus in the tribe (Romaschenko et al. 2012). The short rhizomes may be uniquely derived here.

Clade III

171. ***Piptatherum*** P. Beauv., s.s.

Piptatherum P. Beauv., s.s., Ess. Agrostogr. 17, 173 (1812).

Plants caespitose or rhizomatous. *Spikelets dorsiventrally compressed*. Glumes with 3 to 9 veins, longer than the flower. Callus of flower blunt, flattened, glabrous. Lemma leathery, the margins not fused, the awn straight, deciduous. Palea visible proximally, not enclosed by the lemma at the base. Lodicules 3. Stigmas 2, exserted laterally. Hilum linear. $2n = 24$.

Thirty-two spp., Eurasian.

As traditionally circumscribed, *Piptatherum* is polyphyletic (Romaschenko et al. 2010). The delimitation and description followed here are based on Romaschenko et al. (2011), who provide a detailed description of the taxonomic history of the group and the reasons behind the problematical classifications. *Piptatherum* s.s. excludes the former sections *Miliacea* and *Virescentia*, which both belong in the achnatheroid grasses on the basis of their basally fused lemma margins, lemma epidermal micromorphology, and DNA sequences. Hamasha et al. (2012) also identify a monophyletic *Piptatherum* clade.

172. **"Stillmania"**

"*Achnatherum*" *stillmanii* (Bol.) Barkworth

Plants 0.6–1.5 m tall, with short rhizomes. Glumes longer than the flower, each with 1 vein. Callus of flower blunt. Lemma pubescent, the margins overlapping for most or all of their length, the apex with distinct slender lobes, awned, the awn once or twice geniculate. Palea about as long as or longer than the lemma, the keels extended to about the length of the lemma lobes.

One sp., *Achnatherum stillmanii* (Bol.) Barkworth, California.

This species is placed sister to *Piptatherum* s. s. by the analyses of Romaschenko et al. (2012).

Clade IV "achnatheroid grasses" sensu Romaschenko et al. (2012)

Lemma epidermis with cells generally isodiametric or broader than long.

Clade IVA. Core achnatheroid subclade

Glumes longer than the flower.

173. ***Achnatherum*** P. Beauv. s.s.

Achnatherum P. Beauv. s.s., Ess. Agrostogr. 19, 146, pl. 6, f. 7 (1812).
Aristella Bertol., Fl. Ital. [Bertoloni] I. 690 (1833).
Piptatherum sect. *Virescentia* Roshev. ex Freitag, Notes Royal Bot. Gard., Edinburgh 33: 363 (1975).

Plants caespitose. Spikelets fusiform. Glumes with apices acute to acuminate. Callus of flower acute to obtuse, pubescent. Lemma with margins not overlapping, the apex without a ciliate crown-like rim; awn persistent, not strongly twisted, straight, flexuous or once geniculate. $2n = 28, 36$.

Twenty-one spp., Asia.

"*Achnatherum*" in the sense of *Flora of China* (Wu and Phillips 2006) is polyphyletic (Romaschenko et al. 2010). *Achnatherum* s.s. includes only sections *Achnatherum*, *Achnatheropsis*, and *Aristella* (Romaschenko et al. 2011, 2012). Hamasha et al. (2012) suggest recognizing sect. *Aristella* as a separate genus, but acknowledge that it may equally well be accommodated within *Achnatherum* s.s., as done here.

174. ***Anemanthele*** Veldkamp

Anemanthele Veldkamp, Acta Bot. Neerl. 34: 107 (1985).

Plants over 1 m tall. Spikelets laterally compressed. Glumes with apices acute. Callus of flower short, obtuse, pubescent. Lemma margins convolute, not covering the palea, the awn straight, deciduous. *Stamen 1*. $2n = 40–44$.

One sp., *A. lessoniana* (Steud.) Veldkamp, New Zealand.

175. ***Austrostipa*** S. W. L. Jacobs & J. Everett

Austrostipa S. W. L. Jacobs & J. Everett, Telopea 6: 582 (1996); tax., micromorphology: Bustam (2010).

Plants often with knotty bases; culms branching from the upper nodes. Glumes with apices acute to acuminate, rarely muticous or mucronate. Callus of flower long, sharp-pointed, pubescent. Lemma tough, *dark at maturity, often with brownish pubescence*, the margins generally convolute, rarely involute, the apex sometimes with two minute lobes, awned, the awn once or twice geniculate. Lodicules 2 or 3.

Sixty-three spp., Australasia, several introduced elsewhere.

176. ***Celtica*** F. M. Vázquez & Barkworth

Celtica F. M. Vázquez & Barkworth, Bot. J. Linn. Soc. 144: 491 (2004).

Plants to 2.5 m tall. Glumes with apices acuminate or caudate. Callus of flower sharp-pointed, pubescent. Lemma leathery, the margins flat, overlapping, the apex with two teeth, awned from the sinus, the awn twisted and twice geniculate. *Palea veins extended into awns*. Lodicules 3. Stigmas 2. Hilum as long as the fruit. $2n = 96$.

One sp., *C. gigantea* (Link) F. M. Vázquez & Barkworth, Iberian Peninsula and northern Africa, cultivated elsewhere.

Vázquez and Barkworth (2004) note that *Celtica* has unusually long prophylls (14–15 cm) that terminate in two equally long teeth. This character is rarely reported in species descriptions of grasses, but may be diagnostic. The high chromosome number of this species points to a possible allopolyploid origin. This species is cultivated, as *Stipa gigantea* Link.

177. ***Oloptum*** M. Röser & H. R. Hamasha

Oloptum M. Röser & H. R. Hamasha, Pl. Syst. Evol. (2012).

Plants 1–1.5 m tall. Ligule membranous. Leaf blades stiff, scaberulous. Spikelets dorsiventrally compressed. Glumes with apices acuminate. Callus of flower glabrous. Lemma coriaceous, dark brown, shiny, with central vein prominent, the basal margins fused, the awn flexuous, deciduous. Anther apices minutely pubescent. $2n = 24$.

One sp., *Oloptum miliaceum* (L.) Röser & H. R. Hamasha, worldwide.

The genus *Oloptum* was erected to accommodate the taxa of the Miliacea group of "*Piptatherum*", which is clearly unrelated to *Piptatherum* s.s. (Hamasha et al. 2012; Romaschenko et al. 2011, 2012).

178. ***Stipellula*** M. Röser

Stipellula M. Röser, Schlechtendalia 4: 91–93 (2012).
Stipella (Tzvelev) M. Röser & H. R. Hamasha, nom. illegit.
Stipa sect. *Stipella* Tzvelev, Novosti Sist. Vyssh. Rast. 11: 15 (1974).

Annuals. Glumes with apices acuminate. Callus of flower sharp-pointed. Lemma hard, constricted below the apex, with lines of hairs, with margins overlapping, the apex awned; awn twice geniculate, twisted, the proximal portions pubescent, the distal portion glabrous. Palea much shorter than the lemma. Lodicules 2. $2n = 36$.

Two spp., Mediterranean.

Stipellula is clearly unrelated to *Stipa* s.s. The two species belong in the achnatheroid grasses on the basis of lemma epidermal pattern and molecular data (Romaschenko et al. 2011).

Clade IVB

179. ***Timouria*** Roshev.

Timouria Roshev., Fl. Aziatsk. Ross. 2: 173, pl. 12 (1916).

Leaf blades setaceous. Glumes with 3 veins. Callus of flower short, obtuse. Lemma short (3–4 mm long), densely pubescent, the margins overlapping, the apex lacking a ciliate crown, with apical lobes, the awn deciduous.

Five spp., Asia.

See discussion in Romaschenko et al. (2011) for species included here.

Clade IVC. Major American Clade

Plants of the New World.

This clade is strongly supported by chloroplast DNA sequences but not by nuclear ribosomal DNA (Cialdella et al. 2014; Romaschenko et al. 2012). Relationships within the clade are unclear as are the delimitation of groups within it. One possibility would be to include all members of the clade in a single large genus.

180. **"Eriocoma group"**

Spikelets fusiform or obovoid. Lemma with margins only partially overlapping, the apex lacking lobes, with a ciliate crown, the awn persistent or deciduous. Palea short.

Twenty-nine spp., New World.

The "Eriocoma group" includes species formerly placed in *Achnatherum*; these form a clade separate from *Achnatherum* s.s. (Romaschenko et al. 2012).

181. ***Amelichloa*** Arriaga & Barkworth

Amelichloa Arriaga & Barkworth, Sida 22: 146 (2006).

Plants with knotty rhizomes. Leaf blades stiff, inrolled, sharp-pointed. Ligule ciliate. Cleistogamous spikelets often present in upper leaf sheaths. Spikelets terete. Glumes longer than the flower, their apices acute to acuminate. Callus of flower pubescent, blunt. Lemma pubescent, awned, the awn twisted. Palea pubescent. Lodicules 3.

Five spp., South America, Mexico.

It is unclear whether *Amelichloa* is sister to *Nasella* or embedded within it. The genus is tentatively recognized here, following Romaschenko et al. (2012), although data presented by Cialdella et al. (2014) argue for combining the two genera.

182. **"*Jarava*"** Ruiz & Pav.

"*Jarava*" Ruiz & Pav., Fl. Peruv. Prodr. 2 (1794).

Leaf blades convolute, their apices pointed. Glumes shorter than to longer than the flower. Callus of flower with long white hairs. Lemma with margins not overlapping at maturity, the apex minutely bilobed, bearing long hairs, the awn terminal, deciduous or persistent, once or twice geniculate. Palea much shorter than the lemma, glabrous. Lodicules 2 or 3.

Thirty spp., South America.

Monophyly of "*Jarava*" is uncertain; it may be paraphyletic with *Nassella* and *Amelichloa* derived from within it (Romaschenko et al. 2012). In addition, a group of at least seven species of "*Achnatherum*", the "Pseudoeriocoma group", apparently fall within "*Jarava*" s.l.

183. ***Nassella*** (Trin.) E. Desv.

Nassella (Trin.) E. Desv., Fl. Chil. 6: 263 (1854); tax.: Barkworth (1990); phylog.: Cialdella et al. (2014).

Spikelets obovoid or terete. Glumes longer than the flower, their apices awned or awnless. Callus of flower blunt or sharp, glabrous or pubescent. Lemma hardened, *the margins convolute and completely wrapped around the palea and fruit, the apex bearing a crown-like rim*, the awn terminal, deciduous or persistent, twice geniculate. *Palea reduced, without veins*. Lodicules 2 or 3. Anthers 1 or 3. $2n = 28, 34, 36, 40, 42, 64, 70, 82, 88$.

One hundred seventeen spp., throughout the Western Hemisphere.

Barkworth (1990) notes that *Nassella* has an unusually long epiblast on the embryo, a character that it shares with *Hesperostipa*, some species of *Achnatherum* and two of *Piptochaetium*. Romaschenko et al. (2012) observe that the lemma epidermis of *Nassella* is unique, with short cells lacking silica bodies and "long" cells wider than long, forming a ladder-like pattern. Cialdella et al. (2014) show that members of *Nasella* fall into two clades, one of which is largely restricted to high elevations.

XI. Tribe Meliceae Rchb. (1828)

Brylkinieae Tateoka (1960).

Annuals or more often perennials. *Sheath margins fused, often joined by a hyaline membrane*. Inflorescence branched or unbranched. Glumes generally shorter than the adjacent lemmas. Lemma membranous below, papery distally, *veins generally not converging at the apex. Lodicules thick, truncate, fused for part of all of their length*. Ovary glabrous. Stigmas and style branches 2, free to the base, *the bases persistent*.

Eight genera, 163 spp.

The sister taxon relationship between *Brylkinia caudata* (formerly placed in its own tribe) and Meliceae has been shown by Schneider et al. (2009, 2011). The closed sheath margins are synapomorphic. The lodicules of *Brylkinia* are thick and fused proximally, although they are membranous and free distally. Previous reports suggesting that they are entirely free are in error.

184. ***Brylkinia*** F. Schmidt

Brylkinia F. Schmidt, Mém. Acad. Imp. Sci. Saint Pétersbourg VII, 12: 199 (1868).

Rhizomatous perennials. Inflorescence unbranched. Spikelets *with two sterile flowers proximally, and one bisexual flower distally, disarticulating at the base of the pedicel*. Lower glume acuminate, the upper narrowing to an attenuate apex. Lemma of sterile flower acuminate; lemma of bisexual flower with two tiny hyaline lobes, awned from the sinus, the awn straight, as long as or longer than the lemma. Lodicules membranous and free distally, thick and fused proximally. $2n = 40$.

One sp., *B. caudata* (Munro) F. Schmidt, China and Japan.

Remaining genera of the tribe:
Lodicules lacking a distal membranous portion. Spikelets ending in a reduced flower or flower-like cluster of lemmas.

185. ***Glyceria*** R. Br. Fig. 36

Glyceria R. Br., Prodr.: 179 (1810), nom. conserv.; tax., phylog., Whipple et al. (2007).

Perennials, rarely annual, caespitose, rhizomatous. Inflorescence branches branched. Spikelets with 2 to 20 flowers, laterally compressed or subterete. Glumes shorter than the spikelet, their apices obtuse to acute. Lemmas with 5 to 11 veins, *these generally prominent and not converging at the apex. Stamens 2* or 3. $2n = 20, 40, 60$.

Forty-eight spp., wet sites in temperate regions worldwide.

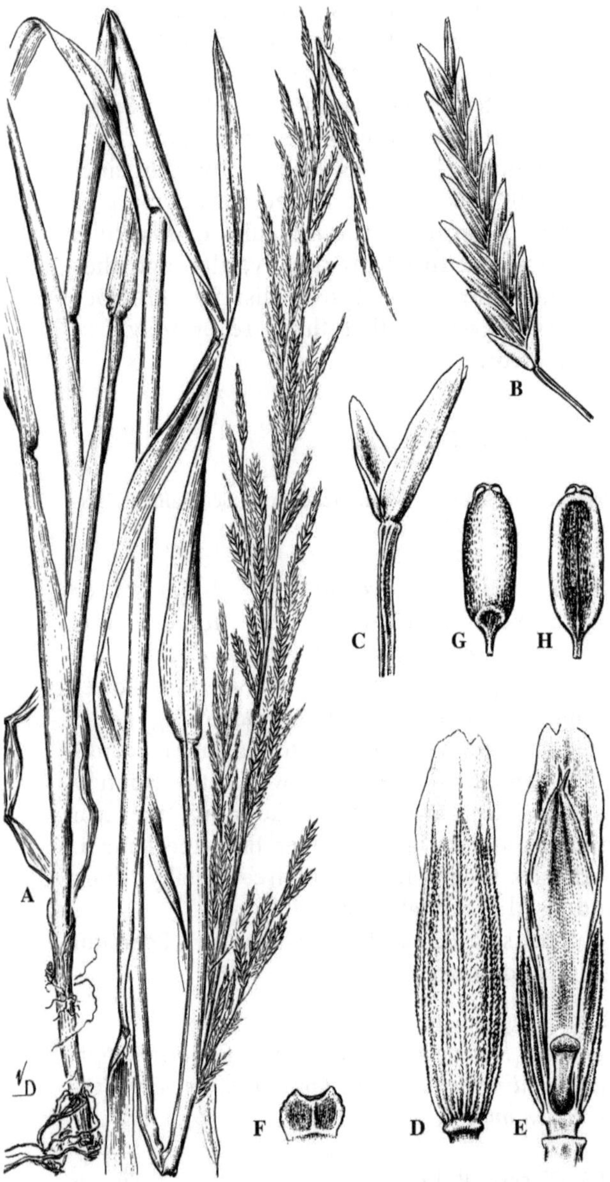

Fig. 36. Pooideae-Meliceae. *Glyceria multiflora*. **A** Culm with inflorescence. **B** Spikelet. **C** Glumes. **D** Lemma. **E** Palea and rachilla. **F** Fused lodicules. **G** Caryopsis, showing embryo. **H** Caryopsis, showing hilum. (From Zuloaga et al. 2012)

186. ***Lycochloa*** Samuelsson

Lycochloa Samuelsson, Ark. Bot. 25(8): 4 (1933).

Perennials. Inflorescence unbranched. Spikelets laterally compressed, with 2 flowers. Glumes shorter than the spikelet, papery, their apices acuminate. Callus of flower densely pubescent. Lemma with 11 rib-like veins, the apex with two teeth, with an abaxial awn.

One sp., *L. avenacea* Samuelsson, Syria.

187. ***Melica*** L.

Melica L., Sp. Pl. 1: 66–67 (1753); Hempel (2011), rev.

Perennials, caespitose or with short rhizomes, sometimes with the lowest culm internode enlarged and bulb-like. Inflorescence branches branched. Spikelets laterally or dorsiventrally compressed, with 1 to 8 flowers, disarticulating above or below the glumes, the rachilla terminating in a club-like cluster of lemmas. Glumes shorter or longer than the spikelet, membranous to papery, *the apices translucent*, obtuse to acuminate. Callus of flower glabrous. Lemma muticous to awned. $2n = 18, 36$.

Ninety spp., worldwide, temperate areas.

Disarticulation below the glumes occurs only in species with sharply bent pedicels.

188. ***Pleuropogon*** R. Br.

Pleuropogon R. Br., Chlor. Melvill.: 31–32, pl. D (1823).

Annuals or perennials, caespitose or rhizomatous. Inflorescence unbranched. Spikelets laterally compressed, with 5 to 20 flowers. Glumes shorter than the flowers, hyaline, their apices ragged. Callus of flower glabrous. Lemma apex ragged or toothed, sometimes awned, the awn terminal or from a sinus. Palea keels awned. $2n = 16, 32, 36, 40, 42$.

Six spp., wet areas of the Arctic, Pacific Coast of North America south to California.

189. ***Schizachne*** Hack.

Schizachne Hack., Repert. Spec. Nov. Regni Veg. 7: 322 (1909).

Caespitose perennials. Inflorescence unbranched, or with branched branches, sparse. Spikelets laterally compressed, with 3 to 6 flowers. Glumes shorter than the spikelet, papery proximally, hyaline distally, their apices acute. Callus of flower densely pubescent. Lemma apex bilobed, awned from the sinus, the awn straight. $2n = 20$.

One sp., *S. purpurascens* (Torr.) Swallen, North America, north temperate Asia.

The leaf sheath margins are fused by a hyaline-membrane that appears to disintegrate eventually.

190. ***Streblochaete*** Hochst. ex Pilg.

Streblochaete Hochst. ex Pilg., Bot. Jahrb. Syst. 37, Beibl. 85: 61 (1906).

Perennials. Inflorescence branches branched. Spikelets laterally compressed, with 2 to 6 flowers. Glumes shorter than the spikelet, their apices acute. Callus of flower sharp-pointed, pubescent. Lemma with 7 veins, with two apical teeth and an abaxial awn; awns of adjacent flowers coiling around each other. *Lodicules not fused.* $2n = 20$.

One sp., *S. longiaristata* (A. Rich.) Pilg., tropical Africa, Réunion, southeast Asia.

Streblochaete is strongly supported as being part of Meliceae, but its precise position within the tribe is not certain (Schneider et al. 2011). The lack of fusion in the locidules is probably a reversal, but might prove to be plesiomorphic.

191. ***Triniochloa*** Hitchc.

Triniochloa Hitchc., Contr. U.S. Natl. Herb. 17: 303 (1913).

Perennials, caespitose or rhizomatous. Inflorescence branches branched. *Spikelets with one flower.* Callus of flower pubescent. Glumes shorter than or as long as the spikelet, their apices acute to acuminate. Lemma with two slender lobes, awned; the awn abaxial or from the sinus. Palea with two slender lobes. $2n = 32$.

Six spp., Mexico, Central and South America.

Triniochloa is sister to *Glyceria* (Schneider et al. 2011), indicating that the one-flowered spikelet is derived and synapomorphic here.

XII. TRIBE DIARRHENEAE C.S. Campb. (1985)

Perennials, rhizomatous or loosely clumped; culm nodes concealed by leaf sheaths. *Leaf blades glossy, narrowing toward the ligule, the midvein usually excentric.* Inflorescence branches branched. Spikelets laterally compressed, with 2 to 7 flowers, the distal ones reduced and sterile. Glumes shorter than the flowers, their apices acute. *Callus of flower glabrous, rounded.* Lemma with three prominent veins, the apex acute, sometimes mucronate. Stamens generally 2. Fruit apex with an indurate beak, pericarp free from seed. $2n = 38, 60$.

Fig. 37. Pooideae-Diarrheneae. *Diarrhena obovata.* **A** Plant. **B** Spikelet. **C** Flower. **D** Caryopsis. (From Barkworth et al. 2007, drawn by H. Paździrková)

One genus, five spp.

The embryo of *Diarrhena* has been described as lacking (Reeder 1957) or having (Macfarlane and Watson 1980) a scutellar cleft, and embryonic leaf margins meeting (Reeder 1957) or overlapping (Macfarlane and Watson 1980). The genus may be polymorphic, or some of the observations may be erroneous.

192. ***Diarrhena*** P. Beauv. Fig. 37

Diarrhena P. Beauv., Ess. Agrostogr.: 142, 160, 162 (1812), nom. conserv.; rev.: Brandenburg et al. (1991b).
Neomolinia Honda, J. Fac. Sci. Univ. Tokyo, Sect. 3, Bot. 3: 110 (1930).

Characters as for the tribe.

Five spp., eastern Asia, eastern North America.

The East Asian species form a monophyletic group (Schneider et al. 2009) and are sometimes segregated as *Neomolinia*, but the clade is clearly monophyletic and there seems little reason to divide this tiny genus into two.

Remaining genera of the subfamily:
Parallel-sided subsidiary cells.

XIII. Tribe Brachypodieae Harz (1880)

Annuals or perennials, caespitose or rhizomatous. *Inflorescence unbranched.* Spikelets laterally compressed, with (3) 5 to 24 flowers, the distal flowers reduced. Glumes shorter than the flowers, their apices obtuse to acute or acuminate. Callus of flower glabrous. Lemma veins not converging toward the apex, the apex obtuse to acute, awned or not. Ovary apex pubescent. $2n = 10, 14, 16, 18, 20, 28, 30$.

One genus, 16 spp.

193. ***Brachypodium*** P. Beauv. Fig. 38

Brachypodium P. Beauv., Ess. Agrostogr.: 100, 15, pl. 19, f. 35 (1812); phylog.; Catalán and Olmstead (2000), Catalán et al. (2012).

Characters as for the tribe.

Sixteen spp., temperate regions of Eurasia, Mexico, Central and South America.

The annual species *Brachypodium distachyon* has become a valuable model system for genomic and genetic research. Its tiny stature and small genome make it a more tractable study system than others in Triticeae and Poeae (Mur et al. 2011). Plants previously classified as *B. distachyon* have 10, 20 or 30 pairs of chromosomes; while these had been treated as chromosomal races, they now appear to constitute three separate species (Catalán et al. 2012).

Remaining genera of the subfamily:
Lemma with veins converging distally. *Chromosome base number 7; chromosomes large. Genomes large.*

See data and discussion of genomic synapomorphies in Bennetzen and Kellogg (1997) and Kellogg and Bennetzen (2004).

Bromeae + Triticeae
Apex of ovary pubescent.

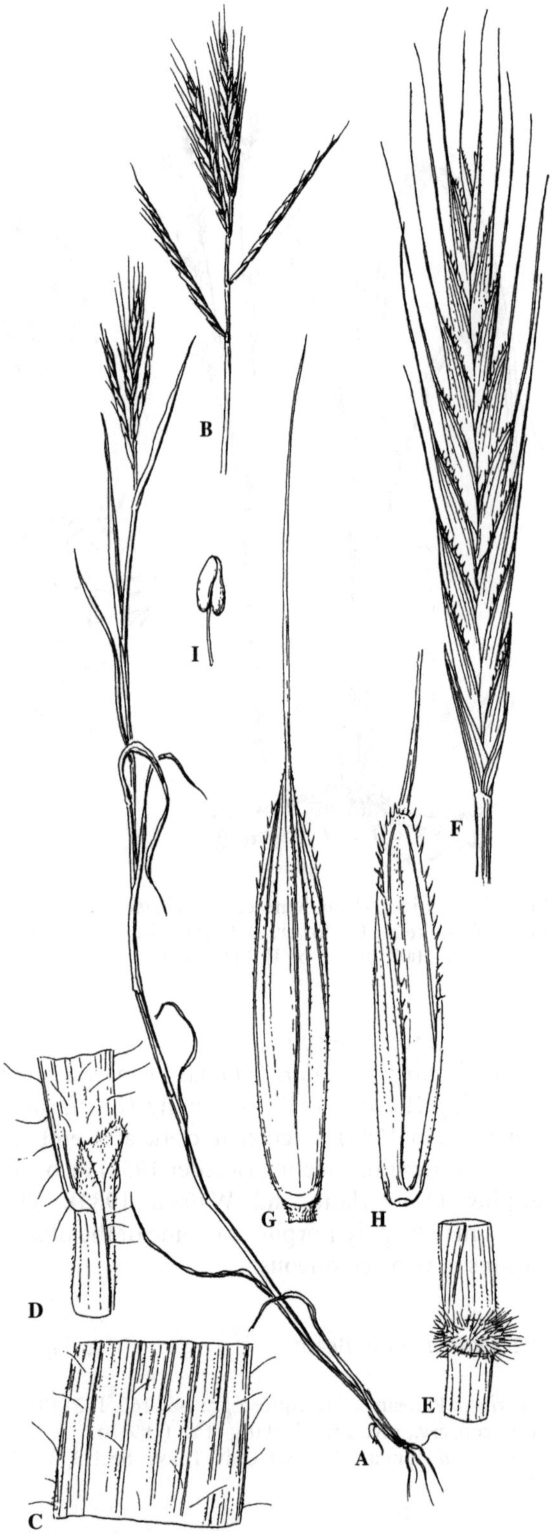

Fig. 38. Pooideae-Brachypodieae. *Brachypodium distachyon*. **A** Plant. **B** Inflorescence. **C** Adaxial leaf blade surface. **D** Leaf sheath, blade, and ligule. **E** Culm node. **F** Spikelet. **G** Flower, abaxial side. **H** Flower, adaxial side. **I** Stamen. (From Barkworth et al. 2007, drawn by C. Roché and A. Miller)

XIV. Tribe Bromeae Dumort. (1824)

Annuals or perennials, caespitose, sometimes rhizomatous. Inflorescence unbranched, or with branches branched. Spikelets generally laterally compressed (sometimes terete), with 2 or more flowers, the distal ones progressively reduced; rachilla extension present. Glumes shorter than the flowers. Callus of flower glabrous. Lemma apex entire or with two teeth or lobes, awned from the sinus or abaxially just below the apex, the awn generally straight.

Two genera, one hundred sixty-two species.

Inclusion of *Littledalea* here renders Bromeae paraphyletic (Davis and Soreng 2007; Schneider et al. 2009). Ultimately either *Littledalea* should be placed in its own tribe, or Bromeae as recognized here should be incorporated in Triticeae as recommended by Schneider et al. (2009). The former solution would be better for nomenclatural stability, as well as making distinctive tribes. Bromeae as described here have no synapomorphies.

194. ***Bromus*** L. Fig. 39

Bromus L., Sp. Pl. 1: 76–78 (1753); phylog.: Saarela et al. (2007).
Boissiera Hochst. ex Steud., Nomencl. Bot. (ed. 2) 1: 213 (1840).

Annuals or perennials, caespitose or rhizomatous. *Leaf sheath margins fused.* Inflorescence branches branched or unbranched. Spikelets with 3 to 30 flowers. Glume apices acute, sometimes mucronate. Lemma entire to bidentate, awnless to more often awned, the awn from a sinus or abaxial just below the apex. *Style branches originating on the abaxial side of the ovary, the ovary extending adaxial to the style branches and ending in two lobes.* $2n = 14, 28, 42, 56, 70, 84$.

One hundred sixty-one spp., temperate regions worldwide.

Boissiera is clearly derived within *Bromus* (GPWG II 2012) and is thus placed in synonymy here.

195. ***Littledalea*** Hemsl. Fig. 40

Littledalea Hemsl., Hooker's Icon. Pl. 25: t. 2472 (1896).

Perennials, caespitose or rhizomatous. Leaf sheath with prominent auricles. Inflorescence with branched branches. Spikelets with 2 to 11

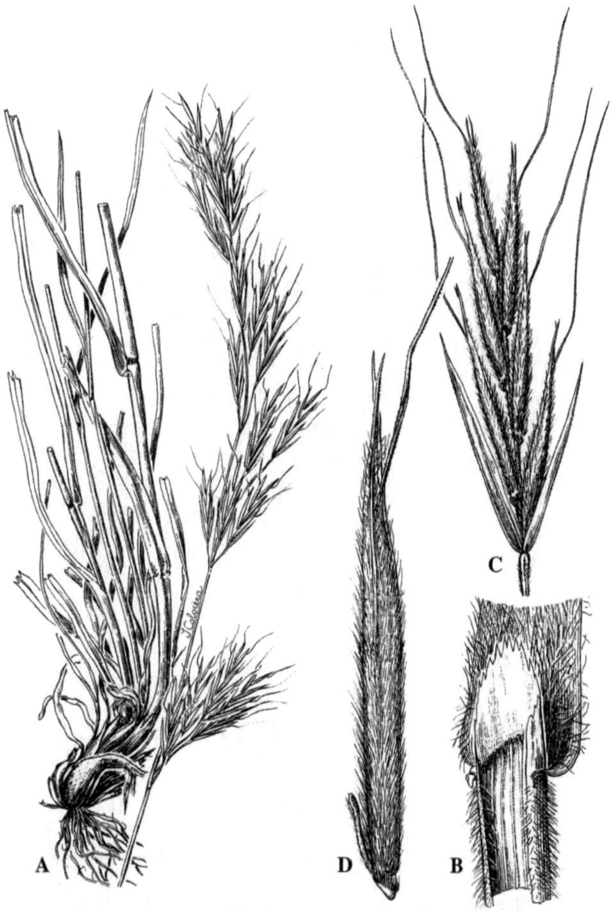

Fig. 39. Pooideae-Bromeae. *Bromus berteroanus.* **A** Plant. **B** Ligule region. **C** Spikelet. **D** Anthecium. (From Zuloaga et al. 2012, drawn by J. Coliera)

flowers. Glumes papery, shorter than the flowers, their apices obtuse to acute. Lemmas papery, over 1 cm long, the apex obtuse or truncate, ragged, awnless or mucronate.

Four spp., Central Asia.

Photographs of the ovary of *L. alaica* (Korsh.) Petrov ex Nevski in Schneider et al. (2009) show clearly that it lacks the ovary appendage characteristic of *Bromus.*

XV. Tribe Triticeae Dumort. (1824)

Annuals or perennials, caespitose or rhizomatous. Leaf sheaths open, often with conspicuous extensions (auricles) from the apex of the sheath and surrounding the stem. *Inflorescence unbranched,* with spikelets crosswise to the rachis. Spikelets 1 to 5 per node, generally laterally compressed. *Lodicules ciliate.* Embryo about 1/3 the length of the fruit; *starch grains simple.*

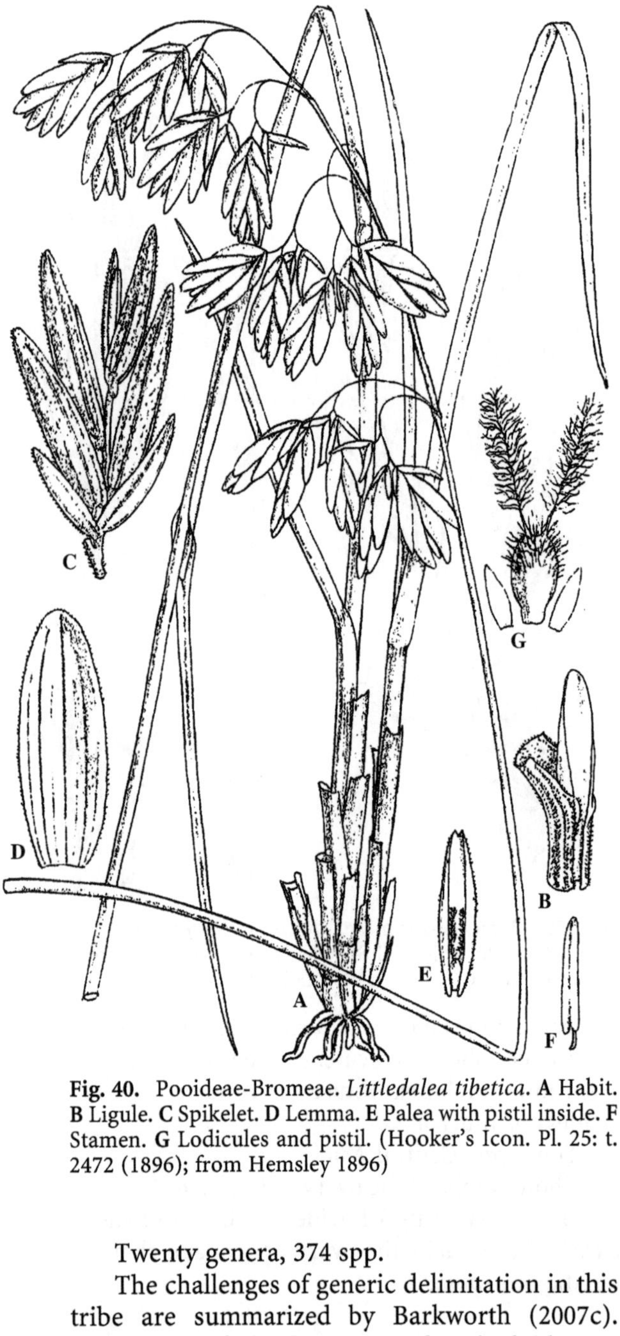

Fig. 40. Pooideae-Bromeae. *Littledalea tibetica*. **A** Habit. **B** Ligule. **C** Spikelet. **D** Lemma. **E** Palea with pistil inside. **F** Stamen. **G** Lodicules and pistil. (Hooker's Icon. Pl. 25: t. 2472 (1896); from Hemsley 1896)

Twenty genera, 374 spp.

The challenges of generic delimitation in this tribe are summarized by Barkworth (2007c). Intergeneric hybridization and polyploidy are common. Extensive cytogenetic studies have assessed genomic groups within which chromosomes pair freely and between which chromosome pairing is reduced or absent (Dewey 1984; Löve 1984). Löve (1984) and Barkworth and Dewey (1985) have suggested that each distinct genome or distinct combination of genomes should be the basis of a genus. While this approach has been influential it is rarely followed blindly. Generic limits here largely follow those laid out by Barkworth (2007c), except that I take a broader view of *Elymus*, which here includes most polyploids with an St genome from *Pseudoroegneria* irrespective of which other genomes are present. The central problem is created by placing *Hordeum*, *Triticum*, and *Secale* in separate genera; were they not such important crops, their names could easily be changed and the entire tribe could be placed in a single genus as repeatedly suggested by Stebbins (Stebbins 1956; Stebbins and Snyder 1956; Stebbins, Jr. and Walters 1949). However, maintaining nomenclatural stability for the major crops forces inherent ambiguity in the classification of the rest of the tribe. The evolutionary history of the group is clearly highly reticulate and any attempt to force a reticulum into a taxonomic hierarchy becomes arbitrary (Kellogg et al. 1996). The website at http://herbarium.usu.edu/Triticeae/ provides useful species descriptions and other information relevant to the tribe.

Named intergeneric hybrids include x*Triticosecale* Wittm. ex A. Camus, x*Pseudelymus* Barkworth & D. R. Dewey, x*Elyhordeum* Mansf. ex Tsitsin & K. A. Petrova, x*Elyleymus* B. R. Baum, x*Pascoleymus* (B. Boivin) Barkworth, and x*Leydeum* Barkworth. These are sterile and morphologically intermediate between their parental genera, and are not described here.

196. *Aegilops* L.

Aegilops L., Sp. Pl. 2: 1050 (1753); tax.: Kimber and Feldman (1987), van Slageren (1994).
Ambylopyrum (Jaub. & Spach) Eig, Agric. Rec. P.Z.E. Inst. Agric. Nat. Hist. (Tel Aviv) 2: 199 (1929).

Annuals. Leaves with falcate auricles. Inflorescence axis disarticulating at the nodes or not, or shed as a unit. Spikelets one per node, laterally compressed to *terete, often partially embedded in the rachis*, with 2 to 7 flowers. Glumes often with prominent veins, their apices truncate, toothed or awned, or awned only on the distal spikelets. Lemma apex toothed, awnless or with 1 to 3 awns. $2n = 14, 28, 42, 48$.

Twenty-eight spp., Europe, Mideast, North America.

Aegilops includes plants with genomes designated B, C, D, S, T, U, M, and N.

197. ***Agropyron*** Gaertn.

Agropyron Gaertn., Novi Comment. Acad. Sci. Imp. Petrop. 14: 539 (1770).

Perennials, caespitose or rhizomatous. Leaf sheaths with auricles. Inflorescence axis not disarticulating. Spikelets one per node, generally more than 3 times as long as the internodes, laterally compressed, with 3 to 16 flowers. Glumes shorter than the flowers, *asymmetrically keeled*, their apices acute, awned or not. Lemma *asymmetrically keeled*, the apex acute to awned. $2n = 14, 28, 42$.

Fifteen spp., Eurasia, introduced in North America.

Agropyron includes plants with the genome designated P.

The name *Agropyron* was at one time applied to any perennial member of the Triticeae with one spikelet per node, and thus included species now placed in *Pseudoroegneria*, *Thinopyrum*, *Elymus*, *Eremopyrum*, and *Pascopyrum*. This use of the name still appears in some treatments (e.g., Clayton et al. 2006 onward), but such a circumscription renders *Agropyron* polyphyletic. As described here, *Agropyron* includes only perennial species with asymmetrically keeled glumes and lemmas (the crested wheatgrasses); the estimated number of species follows Chen and Zhu (2006).

198. ***Australopyrum*** (Tzvelev) Á. Löve

Australopyrum (Tzvelev) Á. Löve, Feddes Rep. 95: 442 (1984).

Perennials, caespitose. Inflorescence with one spikelet per node, not disarticulating, the *spikelets broadly spreading or reflexed, much longer than the internodes*. Spikelets *disarticulating below the glumes*, with 3 to 9 flowers. Glumes subulate. Lemma with short awns. $2n = 14$.

Five spp., Australia, New Guinea, New Zealand.

Australopyrum includes plants with the W genome.

199. ***Crithopsis*** Jaub. & Spach

Crithopsis Jaub. & Spach, Ill. Pl. Orient. 4: 30 (1851); tax.: Frederiksen (1993).

Annuals, less than 30 cm tall. Leaf sheath with falcate auricles. Inflorescence disarticulating just above the nodes. Spikelets two per node, dorsiventrally compressed, with 1 fertile flower and 1 reduced flower distally. Glumes longer than the flowers, linear, their apices attenuate. Lemma apex attenuate, awned, the awns several times the length of the lemma. $2n = 14$.

One sp., *C. delileana* (Schult.) Roshev., Morocco to Afghanistan, but most common in the eastern Mediterranean.

Crithopsis includes plants with the K genome. Attempts to cross *C. delileana* with other Triticeae have been unsuccessful (Frederiksen 1993).

200. ***Dasypyrum*** (Coss. & Durieu) T. Durand

Dasypyrum (Coss. & Durieu) T. Durand, Index Gen. Phan.: 504 (1888); tax.: Frederiksen (1991a).
Haynaldia Schur, Verh. K.K. Zool.-Bot. Ges. Wien 16: 37 (1866).

Annuals or perennials, caespitose or rhizomatous. Inflorescence with one spikelet per node, disarticulating at the nodes. Spikelets laterally compressed, with 2 flowers plus 1 or 2 reduced flowers distally. Glumes shorter than the flowers, 2-keeled, apices tapering into long awns, margins with long silky hairs. Lemma apex acuminate, awned, the awn much longer than the lemma. $2n = 14$.

Two spp., Mediterranean.

Dasypyrum includes plants with the V genome.

201. ***Elymus*** L.

Elymus L., Sp. Pl. 1: 83 (1753); phylog.: Fan et al. (2013); Hu et al. (2013); Mason-Gamer et al. (2002, 2010a, 2010b); Petersen and Seberg (2008); Sun et al. (2008).
Hystrix Moench., Methodus: 294 (1794).
Elytrigia Desv., Nouv. Bull. Sci. Soc. Philom. Paris 2: 190 (1810).
Sitanion Raf., J. Phys. Chim. Hist. Nat. Arts 89: 103 (1819).
Roegneria K. Koch, Linnaea 21: 413 (1848).
Anthosachne Steud., Syn. Pl. Glumac. 1: 237 (1855).
Campeiostachys Drobow, Fl. Uzbekistan. 1: 300, 540 (1941).
Pascopyrum Á. Löve, Taxon 29: 547 (1980).
Kengyilia C. Yen & J.L. Yang, Canad. J. Bot. 68: 1897 (1990).
Douglasdeweya C. Yen, J.L. Yang & B. R. Baum, Canad. J. Bot. 83: 416 (2005).
Connorochloa Barkworth, S.W.L. Jacobs & H. Q. Zhang, Breed. Sci. 59: 685–686 (2009).

Perennials, caespitose, rhizomatous or stoloniferous. Auricles often present. Inflorescence with 1 to 3 (5) spikelets per node, generally not disarticulating at the nodes. Spikelets with 1 to 12 flowers, the distal ones often reduced. Glumes linear or awl-like, awned or not, sometimes absent or reduced. Lemma apex acute, entire or with two tiny teeth, awned from the apex or sinus, or awnless. $2n = 28, 42, 56$.

One hundred fifty spp., north temperate regions worldwide.

Elymus is defined by its genomic constitution as including only polyploids; these all include at least one diploid ancestor related to *Pseudoroegneria* (the St genome). Some taxa also include a genome from *Hordeum* sect. *Critesion* (the H genome), whereas others include the P genome from *Agropyron*, and/or the Y genome, which has not yet been found in any diploid. Species with the genomic constitution StY have sometimes been segregated as *Roegneria* but are retained here in *Elymus*. *Elymus smithii* is an allooctoploid with the St, H, Ns, and Xm genomes (Dewey 1975); it has been segregated into its own genus (*Pascopyrum*) on the basis of its novel genomic constitution, but is scarcely distinct morphologically. No comprehensive taxonomic treatment exists for this genus.

202. ***Eremopyrum*** (Ledeb.) Jaub. & Spach

Eremopyrum (Ledeb.) Jaub. & Spach, Ann. Sci. Nat., Bot. III, 14: 360 (1851); tax.: Frederiksen (1991b).

Annuals. Leaf sheath with inconspicuous auricles. Inflorescence with one spikelet per node, disarticulating at the nodes or not, internodes flat, much shorter than the spikelets, the spikelets broadly spreading. Spikelets laterally compressed, with 2 to 5 flowers, sometimes with additional reduced flowers distally. Glumes shorter than the flowers, their apices acute to acuminate, muticous or with a short awn. Lemma apex muticous to awned. *Palea keels sometimes extended into teeth.* $2n = 14, 28$.

Four spp., North America, central Asia.

Eremopyrum includes plants with the F and Xe genomes.

203. ***Festucopsis*** Melderis

Festucopsis Melderis, Bot. J. Linn. Soc. 76: 317 (1978).

Caespitose perennials. Leaf sheaths lacking auricles, leaf blades filiform. Inflorescence with one spikelet per node, not disarticulating. Spikelets with 3 to 5 flowers. Glumes with acute apices, unawned. Proximal lemma of spikelet with a short awn, distal lemma lacking an awn. $2n = 14$.

One sp., *Festucopsis serpentini* (C.E. Hubb.) Melderis, Albania.

Festucopsis and *Peridictyon* are sister taxa in molecular phylogenies (Petersen and Seberg 2008), but are distinguished by the presence or absence of awns, the disarticulation of the rachis, and the number of flowers in each spikelet.

204. ***Henrardia*** C. E. Hubb.

Henrardia C. E. Hubb., Blumea, Suppl. 3: 15 (1946); tax.: Frederiksen (1993).

Annuals, less than 35 cm tall. Leaf sheath with falcate auricles. Inflorescence disarticulating below the nodes. Spikelets one per node, dorsiventrally compressed, *embedded in the inflorescence axis*, with 1 or 2 flowers. *Glumes placed side by side*, their margins overlapping, slightly longer than the flowers, their apices obtuse to acute. Lemma apex acute. $2n = 14$.

Two spp., southwest and central Asia.

Henrardia includes diploid plants with the O genome. *H. persica* will not cross with *Aegilops* species, *Elymus tsukushiense*, *Eremopyrum bonaepartis*, *Heteranthelium*, or *Triticum boeoticum*, but forms sterile hybrids with *Eremopyrum orientale* (Sakamoto 1972). A few seeds were formed in a cross with *Taeniatherum* (Frederiksen 1993).

205. ***Heteranthelium*** Hochst. in Kotschy

Heteranthelium Hochst. in Kotschy (ed. Hohenacker), Pl. Aleppo exsiccatae 130a (1843); tax.: Frederiksen (1993).

Annuals. Leaf sheath and blade densely pubescent, the sheath with falcate auricles. Inflorescence with one spikelet per node, fertile and sterile spikelets interspersed, *disarticulating at some nodes, so that the dispersal unit includes both fertile and sterile spikelets*; distal inflorescence nodes with spikelets reduced to awns. Spikelets laterally compressed, with 1 or 2 fertile flowers and a rachilla ending in a cluster of awns. Glumes longer than the flowers, slender, awl-shaped, densely pubescent. *Lemma with long tuberculate hairs*, the apex with a long awn. $2n = 14$.

One sp., *H. piliferum* (Banks & Sol.) Hochst., Turkey to Pakistan.

Heteranthelium includes plants with the Q genome. *H. piliferum* will produce sterile hybrids with *Eremopyrum bonaepartis* (both diploid and

Fig. 41. Pooideae-Hordeeae. *Hordeum brevisubulatum* subsp. *turkestanicum*. **A** Habit. **B** Spikelet. **C** Middle flower. **D** Lateral flower. (From Wu et al. 2007, with permission from the Missouri Botanical Garden Press, St. Louis, and Science Press, Beijing)

tetraploid) and *Hordeum depressum* (a tetraploid) but not with *E. triticeum* (diploid), *Henrardia persica* (diploid) or *Elymus tsukushiense* (hexaploid) (Frederiksen 1993; Sakamoto 1974).

206. ***Hordeum*** L. Fig. 41

Hordeum L., Sp. Pl. 1: 84–85 (1753); tax.: Blattner (2009).
Critesion Raf., J. Phys. Chim. 89: 103 (1819).

Annuals or perennials, caespitose, rarely rhizomatous. Leaf sheath with or without auricles. Inflorescence axis generally disarticulating at the nodes; spikelets in groups of three at each node, with the central spikelet fertile and sessile, the lateral two generally sterile and pedicellate. Spikelets dorsiventrally compressed, with 1 flower. Glumes shorter or more often longer than the spikelet, generally subulate to linear, their apices long attenuate or extending into an awn. Rachilla extension generally present on central spikelet. Lemma apex attenuate, generally with a long apical awn, the awn straight or curved. $2n = 14, 28, 42$.

Forty-three spp., temperate and subtropical regions of Eurasia, North and South America, Africa, introduced elsewhere.

Hordeum vulgare, barley, was domesticated in the Fertile Crescent about 8500 years ago (Harlan and Zohary 1966), with a possible second center of domestication in the Zagros Mountains in present day Iran (Morrell and Clegg 2007). Other species of the genus are weedy. *Hordeum* includes plants with genomes designated H, I, X, and Y. Twenty species are diploid and the remainder are polyploid, so the history of the genus is reticulate (Blattner 2006, 2009). Species with the H genome have hybridized in the past with *Pseudorogeneria* (St genome; see below) to form allopolyploids classified here as *Elymus* (Mason-Gamer et al. 2010a, 2010b).

207. ***Leymus*** Hochst.

Leymus Hochst., Flora 31: 118 (1848); phylog.: Petersen and Seberg (2008).
Hordelymus (Jess.) Harz, Landw. Samenk. 2: 1147 (1885).
Eremium Seberg & Linde-Laursen, Syst. Bot. 21: 10, f. 4 (1996).

Perennials, rhizomatous or sometimes caespitose. Leaf blades stiff, leaf sheath with auricles. Inflorescence with 1 to 8 spikelets per node, occasionally branched, not disarticulating. Spikelets sessile or on pedicels up to 2 mm long, laterally compressed, with 2 to 12 flowers, the distal flower(s) reduced. Glumes with apices acute, acuminate, awl-like, or tapering to a short awn. Lemma apex acute, awned or awnless, the awn generally short but sometimes as long as the lemma, straight. $2n = 28, 42, 56$.

Fifty-four spp., temperate regions of the northern hemisphere.

Leymus includes polyploid plants with the Ns genome, derived from *Psathyrostachys*. A second genome may be present but the data are ambiguous (Bödvarsdottir and Anamthawat-Jónsson 2003; Wang and Jensen 1994). *Hordelymus* is included here (rather than in *Elymus*) because of the presence of an N genome and apparent lack of an St genome (Petersen and Seberg 2008; Zhang and Sun 2010).

208. ***Peridictyon*** Seberg, Fred. & Baden

Peridictyon Seberg, Fred. & Baden, Willdenowia 21: 96 (1991).

Caespitose perennials. Leaf sheaths with auricles, leaf blades filiform. Inflorescence with one spikelet per node, disarticulating at the nodes. Spikelets with 5 to 10 flowers. Glumes shorter than the flowers, their apices acute, with short awns. Lemma apex awned. $2n = 14$.

One sp., *Peridictyon sanctum* (Janka) Seberg, Fred. & Baden, Northern Greece and southwestern Bulgaria.

See comments under *Festucopsis*.

209. ***Psathyrostachys*** Nevski

Psathyrostachys Nevski, Trudy Bot. Inst. Akad. Nauk S.S.S. R., Ser. 1, Fl. Sist. Vyssh. Rast. 2: 57 (1934); tax.: Baden (1991).

Perennials, caespitose, rhizomatous, or stoloniferous. Leaf sheath with or without auricles. Inflorescence axis disarticulating at the nodes. Spikelets borne in groups of two or three on the rachis, each with 1 or 2 (3) flowers. Glumes as long as the flowers or longer, both abaxial to the rest of the spikelet, subulate. Lemma apex entire or with minute lobes, awned from the apex or from the sinus, the awns straight. $2n = 14, 28$.

Nine spp., northern Asia and the Caucasus, south to Turkey and Afghanistan.

This genus includes diploids and tetraploids bearing the N genome.

210. ***Pseudoroegneria*** (Nevski) Á. Löve

Pseudoroegneria (Nevski) Á. Löve, Taxon 29: 168 (1980).

Perennials, caespitose or rhizomatous. Leaf sheaths with auricles. Inflorescence with one spikelet per node, the rachis not disarticulating, with spikelets appressed to the axis, about the same length as the internodes or only slightly longer. Spikelets laterally compressed, with 4 to 9 flowers. Glumes shorter than the flowers, their apices acute to obtuse or truncate. Lemma apex acute, awned or unawned, the awn straight and erect or bent and divergent. $2n = 14, 28$.

Fifteen spp., Eurasia, Asia, North America.

Pseudoroegneria includes plants with the St genome. It is clearly one progenitor of most of the polyploids in *Elymus* s.l. (Mason-Gamer et al. 2002, 2010a, 2010b) and *Thinopyrum* (Liu and Wang 1993; Mahelka et al. 2011). Not surprisingly, because it has so many polyploid derivatives, the exact delimitation of *Pseudoroegneria* and its descendants has been controversial, with some species placed in *Elytrigia* and others in *Elymus*.

211. ***Secale*** L.

Secale L., Sp. Pl. 1: 84 (1753); tax.: Frederiksen and Petersen (1998).

Annuals or perennials. Leaf sheath with falcate auricles. Inflorescence with one spikelet per node, disarticulating at the nodes in wild species, not disarticulating in cultivated ones. Spikelets laterally compressed, with 2 flowers. Glumes subulate, keeled, shorter than the flowers, their apices extending into a terminal awn. Lemma keeled, acuminate, awned. $2n = 14$.

Eight spp., Mediterranean and western Asia, introduced and often cultivated elsewhere.

Secale includes plants with the R genome.

212. ***Stenostachys*** Turcz.

Stenostachys Turcz., Bull. Soc. Imp. Naturalistes Moscou 35: 330 (1862); phylog.: Petersen et al. (2011).

Stoloniferous perennials. Leaves with auricles minute. Inflorescence *drooping or nodding*, not disarticulating at the nodes. Spikelets one per node. Glumes shorter than the flower or absent, awl-shaped or lanceolate, their apices attenuate. Lemma with a short awn. $2n = 28$.

Four spp., New Zealand.

Stenostachys includes polyploid plants with the H and W genomes.

213. ***Taeniatherum*** Nevski

Taeniatherum Nevski, Trudy Sredne-Aziatsk. Gosud. Univ., Ser. 8b, Bot. 17: 38 (1934); tax.: Frederiksen (1986).

Annuals. Leaf sheath with falcate auricles. Inflorescence not disarticulating at the nodes. Spikelets 2 (rarely 3) per node, dorsiventrally compressed, with 1 flower plus 1 additional reduced flower distally. Glumes placed side by side, longer than the flowers, awl-like, *their bases fused.* Lemma apex acuminate, awned, the awn many times longer than the lemma; *awns becoming entangled at maturity.* $2n = 14$.

One sp., *T. caput-medusae* (L.) Nevski, Mediterranean and mid-East to Pakistan; introduced elsewhere.

The entangled awns are responsible for the Latin and common names of this grass, Medusa-head. *Taeniatherum* includes plants with the Ta genome.

214. ***Thinopyrum*** Á. Löve

Thinopyrum Á. Löve, Taxon 29: 351 (1980).
Lophopyrum Á. Löve, Taxon 29: 351 (1980).

Perennials, caespitose or rhizomatous. Auricles present or absent. Inflorescence generally with one spikelet per node, not disarticulating at the nodes. Spikelets appressed to ascending, 1 to 3 times as long as the internodes, with 3 to 12 flowers. Glumes thick, stiff, keeled, the apices truncate to acute, sharp-pointed but awnless. Lemma thick, stiff, the apex truncate to acute, mucronate or with short awns. $2n = 14, 42, 43, 56, 69, 70$.

Ten spp., Mediterranean to western Asia, introduced elsewhere.

Thinopyrum includes plants with the E genome; the polyploids also incorporate the St genome from *Pseudoroegneria*, the P genome from *Agropyron*, and maybe the L genome of *Festucopsis*.

215. ***Triticum*** L.

Triticum L., Sp. Pl. 1: 85–87 (1753).

Annuals. Leaf sheaths with auricles. Inflorescence with one spikelet per node, disarticulating at the nodes, the spikelets generally falling with the internode below but sometimes with the adjacent internode, or the entire inflorescence shed as a unit. Spikelets with 2 to 9 flowers, the distal ones often sterile. Glumes keeled, the apex with a tooth or awn. Lemma keeled, the proximal ones awned, the distal ones awned or not. $2n = 14, 28, 42$.

Eighteen spp., western and central Asia, cultivated and escaped in temperate regions worldwide.

Triticum includes diploid plants with the A genome and polyploids with the A, B, D, and G genomes; the B and D genomes are derived from species of *Aegilops*. *T. aestivum* L. (bread wheat) is a hexaploid species that is one of the most widely cultivated grasses in the world. *T. durum* Desf. is tetraploid, and is the major wheat used for pasta, semolina and bulghur.

XVI. Tribe Poeae R. Br. (1814), s.l

Annuals or perennials. Leaf sheaths generally lacking auricles. Inflorescence branches branched in most species. Spikelets laterally compressed, generally disarticulating above the glumes. *Hilum punctate; endosperm with lipid.*

One hundred-six genera, 2578 spp.

This tribe includes most of the species in the subfamily, following the circumscription of Soreng et al. (2007, 2012 and onward). Synapomorphies of the tribe include a punctate hilum and lipid in the endosperm (Soreng et al. 2007), although these characters reverse in some taxa. Abaxial awn attachment, absence of cilia on the lodicules, lack of ovary pubescence, and presence of compound starch grains may also be derived here, although they are homoplasious and thus their position on the phylogeny depends on the method of optimization.

The classification of Soreng et al. (2007), continually updated by Soreng et al. (2012 and onward), is used as the basis of the classification presented here, although some differences are noted below. Soreng et al. (2012 and onward) divide Poeae into 19 subtribes, of which 15 are recognized here. Within tribe Poeae s.l., Gillespie et al. (2008) used both nuclear (ITS) and plastid (*trnT-trnL-trnF*) sequences to assess relationships within and among genera in subtribe Poinae, extending work done by the same group of investigators (Soreng et al. 2007) proposing a subtribe phylogeny for tribe Poeae. Sequences of the plastid genome fall into two clades, which are noted here; however, these clades may not reflect the evolution of the nuclear genome.

Molecular phylogenetic studies consistently identify a clade within Poeae that has been called PPAM, an acronym for Poinae, Puccinelliinae, Alopecurinae and Miliinae (all tribes sensu Soreng et al. 2007); ITS data also place Cinninae in this group. Relationships among the five former subtribes vary depending on the molecular marker used, but they seem to form a coherent group (albeit without an obvious morphological synapomorphy or even a diagnostic combination of characters), with Puccinelliinae (now recognized as Coleanthinae) sister to the rest. Accordingly, here I place the entire PAM clade plus Cinninae in subtribe Poinae.

Named hybrid genera include x*Agropogon* P. Fourn., x*Arctodupontia* Tzvelev, x*Dupoa* J. Cay. &

Darbysh., and x*Pucciphippsia* Tzvelev. These are not described here.

Incertae sedis

216. *Megalachne* Steud.

Megalachne Steud., Syn. Pl. Glumac. 1: 237 (1854).

Perennials. Spikelets with 3 to 6 flowers plus additional reduced flowers distally. Glumes about as long as flowers, their apices acuminate to attenuate, with aristae or long awns. Callus of flower pubescent. Lemma keeled, the apex acuminate, awned from the apex, the awn long and curved. Ovary apex pubescent.

Two spp., Juan Fernández Island.

217. *Podophorus* Phil.

Podophorus Phil., Bot. Zeitung (Berlin) 14: 648 (1856).

Perennials <40 cm tall. Spikelets with 1 flower plus 1 reduced flower distally. Glumes shorter than or as long as the flowers, their apices acuminate but unawned. Callus of flower pubescent. Lemma rounded, the apex acuminate, awned, the awn bent. Ovary apex pubescent.

One sp., *P. bromoides* Phil., Juan Fernández Island.

Only a handful of specimens exist of this species, which has not been collected since the late 19th century (Baeza et al. 2002). It is presumed to be extinct.

Group with Aveneae-type plastid sequences

This clade is retrieved in all molecular studies of the plastid (e.g., Davis and Soreng 2007; Gillespie et al. 2008; Soreng and Davis 2000; Soreng et al. 2007), but appears to have no morphological synapomorphies. The group is supported poorly or not at all by analyses of nuclear gene sequences (Gillespie et al. 2008; Quintanar et al. 2007; Saarela et al. 2010), with the position of Sesleriinae being particularly different between the two sorts of analyses.

XVI.1. *Subtribe Torreyochloinae Soreng & J. I. Davis (2003)*

Aquatic or semi-aquatic perennials. Glumes shorter than the flowers, unawned. Lemma awned or unawned. Hilum 1/3 to 1/2 the length of the fruit.

Fig. 42. Pooideae-Poeae-Torreyochloinae. *Amphibromus scabrivalvis*. **A** Plant. **B** Spikelet. **C** Lemma. **D** Palea. **E** Palea, lodicules and anthers. **F** Lodicules, callus, and base of the palea. (From Zuloaga et al. 2012)

Two genera, 16 species.

This subtribe is placed sister to the rest of the *Avena*-plastid-type clade in chloroplast molecular phylogenies (Davis and Soreng 2007; Soreng and Davis 2000).

218. *Amphibromus* Nees Fig. 42

Amphibromus Nees, London J. Bot. 2: 420 (1843); tax.: Jacobs and Lapinpuro (1986).

Perennials, caespitose, rhizomatous or stoloniferous, often with cleistogamous inflorescences enclosed by the lower leaf sheaths. Spikelets with 3 to 10 flowers; rachilla extension present or absent. Glumes with apices obtuse to acute. Callus of flower pubescent. Lemma awned from the abaxial side, the awn twisted and geniculate, spreading or recurved, lemma apex with 2 to 4 teeth or lobes, the lobes obtuse or with aristae. Ovary glabrous. $2n = 42$.

Twelve spp., Australia, New Zealand, South America.

Amphibromus is morphologically similar to *Helictotrichon*, but with a glabrous ovary; the two genera are not related (Saarela et al. 2010).

219. ***Torreyochloa*** G. L. Church

Torreyochloa G. L. Church, Amer. J. Bot. 36: 163 (1949); tax.: Church (1952).

Perennials, rhizomatous. Spikelets with 2 to 8 flowers; rachilla extension present. Glumes with apices acute. Callus of flower glabrous. Rachilla glabrous. Lemma with 7 to 9 prominent veins, the veins parallel and not converging toward the apex, lemma apex truncate to acute, sometimes with a ragged edge. Ovary pubescent. $2n = 14$.

Four spp., North America, Northeast Asia.

XVI.2. SUBTRIBE AVENINAE J. Presl (1830)

Annuals or perennials. Lemma awnless, mucronate, or with an abaxial awn, the awn geniculate. Hilum short or linear, the latter state presumably a reversal.

Seven genera, 324 species.

A detailed phylogeny of this subtribe is presented by Saarela et al. (2010).

220. ***Arrhenatherum*** P. Beauv.

Arrhenatherum P. Beauv., Ess. Agrostogr. 55: 152–153 (1812). *Pseudarrhenatherum* Rouy, Bull. Soc. Bot. France 68:401 (1921).

Perennials, caespitose or rhizomatous, the base of the plant sometimes forming corms. Spikelets with 2 flowers, the proximal one staminate, the distal one pistillate or bisexual. Glumes as long as the flowers, their apices acute to acuminate. Callus of flower pubescent. Lemma of proximal flower with the apex acute, awned, the awn twisted, geniculate; lemma of distal flower with the apex acute, entire or bidentate, with or without a small awn. Ovary pubescent. *Hilum long-linear.* $2n = 14, 28, 42$.

Eight spp., Europe, mid-East.

Clayton and Renvoize (1986) argue that *Arrhenatherum* is simply a segregate from *Helictotrichon* and division between the two might be unwarranted. Döring et al. (2007) present data indicating that *Arrhenatherum* could be linked with *Pseudarrhenatherum*, and the two could perhaps be included in *Helictotrichon*. Saarela et al. (2010) resolve *Arrhenatherum* as monophyletic, not forming a clade with *Helictotrichon*. *Arrhenatherum* is thus recognized here.

221. ***Avena*** L.

Avena L., Sp. Pl. 1: 79–81 (1753); tax.: Baum (1977).

Annuals. Spikelets with 1 to 6 flowers plus a rachilla extension. Glumes generally enclosing the flowers, but sometimes shorter, their apices obtuse to acuminate. Callus of flower glabrous or pubescent. Lemma apex with two teeth or slender setae, awned, the awn curved or twisted, generally geniculate. Ovary pubescent. *Hilum linear.* $2n = 14, 28, 42$.

Twenty-four spp., Europe, Mediterranean areas, northern Africa.

Oats, *Avena sativa* L., are an important grain, particularly in temperate regions of the world. The species of *Avena* form a complex series of polyploids; although there is one phylogeny of these (Peng et al. 2010), many questions remain.

222. ***Graphephorum*** Desv.

Graphephorum Desv., Nouv. Bull. Sci. Soc. Philom. Paris 2: 189 (1810); tax.: Finot et al. (2005).

Caespitose perennials. Spikelets with 2 or 3 flowers; *rachilla with dense hairs.* Glumes shorter than or as long as the spikelet, their apices acute to acuminate. Lemma apex obtuse to acute, muticous or with an awn just below the apex.

Two spp., North and Central America.

Plastid data place *Sphenopholis* and *Graphephorum* as sisters and the two as sister to *Trisetum* s.l., whereas ITS data put *Graphephorum* within *Trisetum* s.l. and *Sphenopholis* sister to the *Graphephorum*/*Trisetum* clade (Quintanar et al. 2007, 2010). In the absence of a more consistent phylogenetic signal, I keep *Sphenopholis* and *Graphephorum* separate here.

223. ***Helictotrichon*** Besser

Helictotrichon Besser, Mant. 3: 526, in obs. (1827); tax.: Lange (1995).

Caespitose perennials. Ligule ciliate, ragged. Spikelets with 2 to 8 flowers plus additional reduced flowers distally; rachilla pubescent. Glumes as long as or longer than the flowers, their apices acute to acuminate, sometimes mucronate or awned. Callus of flower acute, with straight hairs. Lemma apex with two teeth, awned, the awn twisted and geniculate. Ovary pubescent. *Hilum long-linear.* $2n = 14, 28, 42$.

One hundred two spp., temperate regions of Eurasia.

Based on molecular data, *Helictotrichon* in its traditional circumscription is paraphyletic, with subg. *Pratavenastrum* Holub more closely related to *Deschampsia* than to subg. *Helictotrichon* or subg. *Tricholemma* M. Röser (Grebenstein et al. 1998). The three subgenera are all strongly supported as monophyletic. Therefore, *Helictotrichon* is restricted to the latter two sections, and subg. *Pratavenastrum* is placed in *Avenula*; see comments under the latter genus.

224. ***Lagurus*** L.

Lagurus L., Sp. Pl. 1: 81 (1753).

Tufted annuals. Inflorescence dense, ovoid. Spikelets with 1 flower and a pubescent rachilla extension. Glumes longer than the flower, *covered with wooly hairs*, their apices acuminate, awned, *the awns covered with hairs.* Lemma pubescent, the apex with two teeth or slender setae, awned, the awn twisted. $2n = 14$.

One sp., *L. ovatus* L., Mediterranean but widely cultivated.

The wooly hairs on the glumes make the entire inflorescence look furry, giving rise to the common name "rabbit's foot".

225. ***Sphenopholis*** Scribn.

Sphenopholis Scribn., Rhodora 8: 142 (1906), tax.: Erdman (1965); Finot et al. (2004).

Perennials, rarely annual. Spikelets with 2 or 3 flowers, *disarticulating below the glumes*, with a rachilla extension. Glumes shorter than or as long as the flowers; lower glume slender, the apex acute, *upper glume broader above than below, broadly oblanceolate to flabellate, appearing wedge shaped in lateral view*, the apex obtuse to acute. Callus of flower glabrous. Lemma apex obtuse to acute, with a small mucro or a twisted awn. *Fruit with liquid endosperm.* $2n = 14$.

Eight spp., North America and Mexico.

See comments under *Graphephorum*. *Sphenopholis* is distinguished from other members of this subtribe by having spikelets that disarticulate below the glumes.

Fig. 43. Pooideae-Poeae-Aveninae. *Trisetum barbinode* Trin. **A** Plant. **B** Spikelet. **C** Lowermost flower. (From Zuloaga et al. 2012)

226. ***Trisetaria*** Forssk. Fig. 43

Trisetaria Forssk., Fl. Aegypt.-Arab. 60: 27 (1775); tax.: Finot et al. (2004, 2005).
Trisetum Pers., Syn. Pl. 1: 97 (1805), nom. conserv.
Koeleria Pers., Syn. Pl. 1: 97 (1805).

Gaudinia P. Beauv., Ess. Agrostogr. 95, 164 (1812).
Rostraria Trin., Fund. Agrost. 149 (1820).
Avellinia Parl., Pl. Nov. 59 (1842).
Peyritschia E. Fourn., Mexic. Pl. 2: 109 (1886).
x*Trisetokoeleria* Tzvelev, Novosti Sist. Vyssh. Rast. 7: 73 (1970 [1971]).
Leptophyllochloa C. E. Calderón, Fl. Patagónica 3: 69 (1978).
Parafestuca E.B. Alexeev, Bjull. Moskovsk. Obač. Isp. Prir., Otd. Biol. 90: 107–108 (1985).

Perennials (most species) or annuals, caespitose, rhizomatous or stoloniferous. Inflorescences unbranched (in some annuals) or the branches branched. Spikelets with 2 to 7 (11) flowers; rachilla extension present, glabrous or pubescent, sometimes ending in a reduced flower. Glumes shorter than to longer than the flowers, their apices obtuse to attenuate, generally mucronate but rarely awned. Callus of flower generally pubescent. Lemma apex entire or with two teeth, awnless or with a mucro or awn from the abaxial side. Ovary glabrous or pubescent. Fruit sometimes with liquid endosperm. $2n = 14, 28, 42$.

One hundred seventy-nine spp., temperate regions worldwide.

Neither *Trisetum* nor *Koeleria* is monophyletic in any molecular phylogeny to date, but together they comprise a clade as long as several smaller genera are included as well (Grebenstein et al. 1998; Quintanar et al. 2007, 2010; Saarela et al. 2010; Soreng et al. 2007). Species of *Rostraria*, *Gaudinia*, *Avellinia*, *Peyritschia*, and *Parafestuca* are intermixed with those of *Trisetum* and *Koeleria* (Quintanar et al. 2007; Saarela et al. 2010) so are placed in synonymy here.

Among these smaller genera is *Trisetaria*, a name that has priority over *Trisetum*. The annual species of *Trisetaria* are clearly embedded in *Trisetum*, as noted by Quintanar et al. (2010). Although *Trisetum* is a *nomen conservandum*, it is not conserved against *Trisetaria*. The nomenclatural solutions are either to conserve *Trisetum* against *Trisetaria*, or to transfer all *Trisetum* species to the older genus; a few such transfers have already been done (Paunero Ruiz 1950).

As delimited here, *Trisetaria* corresponds to the subtribe Koeleriinae (Quintanar et al. 2010), except that *Graphephorum* and *Sphenopholis* are kept separate pending additional data. Saarela et al. (2010) identify two well-supported clades within *Trisetum* s.l., an Old World group and a New World one; the New World clade includes several species of *Calamagrostis* (subtribe Agrostidinae). In the future these clades may be recognized as sister genera or subgenera. No molecular data are available for *Leptophyllochloa*; it is placed here based on morphology. Species formerly placed in *Gaudinia* are distinctive in that the inflorescence is unbranched and the rachis disarticulates just above the spikelet.

XVI.3. Subtribe Phalaridinae Fr. (1835)

Annuals or perennials, caespitose or rhizomatous, in some perennials the lowest internode enlarged and corm-like. Inflorescence dense, capitate or spike-like. Spikelets bisexual, or in clusters with the proximal ones staminate and the terminal pistillate or bisexual; with 1 to 3 (4) flowers; disarticulation above the glumes, or beneath the distalmost spikelet of the cluster, or the clusters shed as a unit. Glumes longer than the flowers, keeled, their apices truncate to attenuate. *Proximal two flowers sterile, reduced to awl-shaped or flap-like lemmas.* Lemma and palea of distal bisexual flower polished and hard, the lemma apex generally acute to acuminate or beaked. *Hilum long-linear.*

One genus, 22 species.

227. ***Phalaris*** L. Fig. 44

Phalaris L., Sp. Pl. 1: 54–55 (1753); tax., Baldini (1995); phylog.: Voshell et al. (2011).

Characters as for the subtribe. $2n = 12, 14, 28, 29, 42$.

Twenty-two spp., north temperate and temperate regions of South America.

Chromosome numbers reported for *P. arundinacea* include the aneuploid counts 27, 29, 31, and 35 (Barkworth 2007b).

XVI.4. Subtribe Anthoxanthinae A. Gray (1856)

Annuals or perennials, caespitose or rhizomatous. Inflorescence *sometimes unbranched*, or the branches branched. Spikelets with 3 flowers, *the 2 proximal ones staminate or sterile, if sterile then consisting solely of a lemma*; rachilla

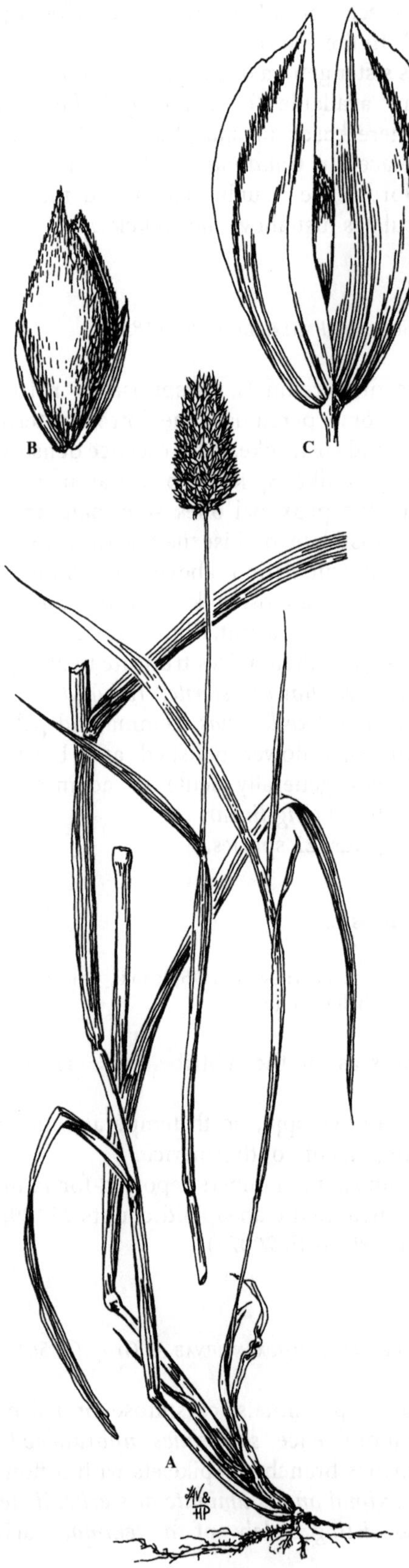

extension absent. Glumes as long as or longer than the flowers, their apices acute to acuminate, mucronate. Callus glabrous or hairy. Lemma of sterile flowers longer than that of the upper one, with long hairs, the apex bilobed, unawned or awned from the abaxial side; lemma of bisexual flower hardened, shiny, with margins enclosing the palea, the apex notched, awnless. *Lodicules absent* or 2. Stamens 2 or 3.

One genus, 52 species.

228. ***Anthoxanthum*** L. Fig. 45

Anthoxanthum L., Sp. Pl. 1: 28 (1753); tax.: Pereira et al. (2007); phylog.: Pimentel et al. (2013).
Hierochloe R. Br., Prodr.: 208 (1810), nom. conserv.

Characters as for the subtribe. $2n = 10, 20, 28, 42, 56$.

Fifty-two spp., temperate and arctic regions throughout the world, widely introduced for hay.

Species in this genus *produce coumarin* (presumed to be a synapomorphy), which makes them sweetly scented and accounts for the pleasant odor of some hay. Species formerly in *Hierochloe* were and are used by Native Americans to make baskets and decorations. The genus *Hierochloe* has been separated from *Anthoxanthum* on the basis of the proximal flowers of the spikelet, which are staminate in the former and sterile in the latter. However, Schouten and Veldkamp (1985) suggest that the two intergrade in parts of their range, and the genera are treated as a single genus by Allred and Barkworth (2007). Merging the two is also supported by molecular data (Quintanar et al. 2010; Saarela et al. 2010). While most reported chromosome numbers are euploid, those for *A. monticola* are cited as 56, 58, 63, 66, 72 (Allred and Barkworth 2007).

XVI.5. Subtribe Agrostidinae Fr. (1835)

Subtribe Brizinae Tzvelev, Bot. Žurn. (Moscow & Leningrad) 53: 310 (1968).

Fig. 44. Pooideae-Poeae-Phalaridinae. *Phalaris canariensis*. **A** Habit. **B** Sterile flowers and a fertile one. **C** Spikelet. (From Barkworth et al. 2007, drawn by L.A. Vorobik and H. Paźdirková)

Lemma awnless or with an abaxial awn, the awn often geniculate. Palea generally hyaline, the margins often wider than the space between the veins.

Twenty-one genera, 681 species.

Saarela et al. (2010) show that Brizinae and Agrostidinae are intermixed and argue against recognizing both. The two are combined here.

229. ***Airopsis*** Desv.

Airopsis Desv., J. Bot. (Desvaux) 1: 200 (1809).

Annuals. *Spikelets nearly spherical, with 2 flowers.* Glumes longer than the flowers, their apices obtuse. Lemma orbicular, the apex obtuse, with three broad lobes. $2n = 8$.

One sp., *A. tenella* Coss. & Durand, Mediterranean.

Briza s.s. and *Airopsis* form a weakly supported clade in the analyses of Saarela et al. (2010).

230. ***Briza*** L.

Briza L., Sp. Pl. 1: 70–71 (1753); phylog.: Essi et al. (2008).

Annuals or perennials, caespitose. Spikelets oval to elliptic in outline, sometimes as broad as long, with 3 to 12 (15) flowers that diverge strongly from the rachilla; rachilla extension absent. Glumes shorter than the flowers, their apices obtuse. Callus of flower glabrous. Lemma inflated, the base often cordate, the margins translucent, the apex unawned. $2n = 10, 14, 28$.

Five spp., Mediterranean, introduced and sometimes cultivated elsewhere.

231. ***Chascolytrum*** Desv.

Chascolytrum Desv., Nouv. Bull. Sci. Soc. Philom. Paris 2: 190 (1810); tax.: Essi et al. (2011).
Calotheca Desv., Nouv. Bull Sci. Soc. Philom. Paris 2: 190 (1810).
Rhombolytrum Link, Hort. Berol. 2: 296 (1833).
Poidium Nees, Intr. Nat. Syst. Bot.: 450 (1836).
Erianthecium Parodi, Notas Mus. La Plata, Bot. 8: 75 (1943).

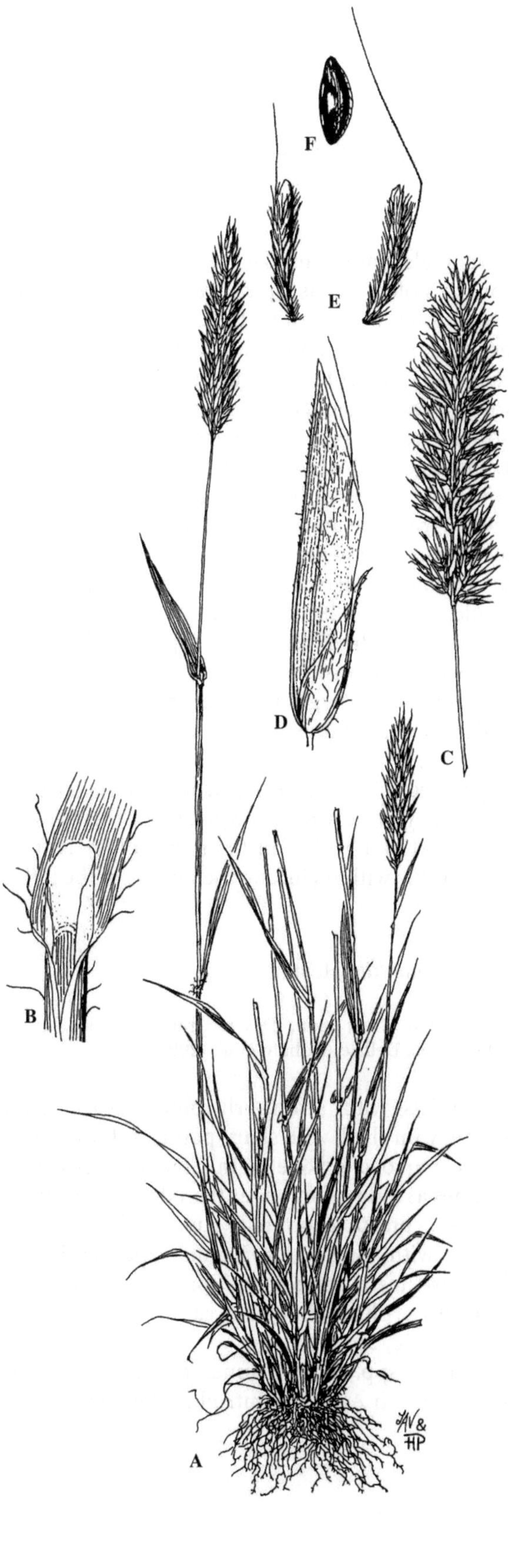

Fig. 45. Pooideae-Poeae-Anthoxanthinae. *Anthoxanthum odoratum*. **A** Plant. **B** Ligule region. **C** Inflorescence. **D** Spikelet. **E** Sterile flowers. **F** Caryopsis. (From Barkworth et al. 2007, drawn by L.A. Vorobik and H. Paźdirková)

Microbriza Parodi ex Nicora & Rúgolo, Darwiniana 23: 292 (1981).
Gymnachne Parodi, Notas Mus. La Plata, Bot. 3: 29 (1938).

Perennials, with culms sometimes forming corms. Inflorescence branched, condensed. Spikelets with 2 to 4 (to 10) flowers, ovate or orbicular; flowers barely separated on the rachilla. Glumes shorter than the flowers, often keeled, their apices acute to obtuse. Callus hairy or not. Lemmas broader than long, contracted distally to an acute or acuminate apex, or with 2 teeth or lobes, mucronate or with a short awn from the sinus. Rachilla disarticulating. $2n = 28$.

Twenty-eight spp., South America.

Clayton et al. (2006 onward) and Clayton and Renvoize (1986) consider *Chascolytrum* to be synonymous with *Briza*, but molecular data (Davis and Soreng 2007; Essi et al. 2008) indicate that the two genera are distinct. Essi et al. (2008) show convincingly that the two species formerly placed in *Microbriza* (= *B. poaemorpha* and *B. brachychaete*) are not sisters in either chloroplast or nuclear gene trees. *Erianthecium* and *Rhombolytrum* (represented by *Gymnachne koelerioides*) also fall in this clade. *Erianthecium* was segregated from *Chascolytrum* because it forms bulbs at the base of the culm, and the lemmas are awned.

232. *Relchela* Steud.

Relchela Steud., Syn. Pl. Glumac. 1: 101 (1855).

Stoloniferous perennials. Spikelets with 1 or 2 flowers; rachilla extension present or absent. Glumes as long as or longer than the flowers, their apices acute. *Lemma hard*, the margins inrolled and largely covering the palea, the apex obtuse. *Ovary apically pubescent.*

One sp., *R. panicoides* Steud., Argentina and Chile.

Remaining genera of the subtribe:
Spikelets with 1 flower.
Named hybrids include x*Agrocalamagrostis* Asch. & Graebn., x*Agropogon* P. Fourn., and x*Ammocalamagrostis* P. Fourn. These have not been verified by molecular data, but their existence shows the intergrading nature of morphological variation in this group.

233. *Agrostis* L.

Agrostis L., Sp. Pl. 1: 61–63 (1753).
Chaetopogon Janch., Eur. Gatt. Farn. Bl.-Pfl. (ed. 2): 33 (1913).

Perennials, rarely annuals, caespitose, rhizomatous or stoloniferous. Spikelets generally less than 7 mm long, without a rachilla extension. Glumes generally longer than the flower, with one vein, their apices acute to acuminate, muticous or rarely awned. Callus of flower glabrous or with straight hairs less than ½ the length of the lemma. Lemma generally hyaline or thinly membranous, with 3 or 5 veins, the apex truncate or toothed, awned from the apex or more often abaxially, the awn straight or geniculate. Palea generally *minute or lacking*. Fruit with endosperm hard, soft or liquid. $2n = 14, 28, 35, 42, 56$.

Two hundred sixty-four spp., worldwide in temperate regions.

Agrostis intergrades morphologically with *Calamagrostis*, but Saarela et al. (2010) find that most species of *Agrostis* form a reasonably well-supported clade, distinct from *Calamagrostis*. *Chaetopogon* falls within *Agrostis* in the trees of Saarela et al. (2010) and the morphology is fully compatible with including the former in the latter genus.

234. *Ammophila* Host

Ammophila Host, Icon. Descr. Gram. Austriac. 4: 24 (1809); tax.: Delisle-Oldham et al. (2008).

Perennials with strong rhizomes. Leaf blades strongly inrolled, with sharp points. Inflorescence dense, narrow. Spikelet with a rachilla extension. Glumes as long as or longer than the flowers, their apices acute to acuminate, unawned. Callus of flower with long hairs. Lemma apex entire or with two teeth, mucronate, the mucro attached abaxially. Hilum about 2/3 as long as the fruit. $2n = 14, 28, 56$.

Three spp., ocean dunes of Europe, North Africa, North America. Introduced elsewhere.

Hybridization with *Calamagrostis* has been documented by Reznicek and Judziewicz (1996). The hybrids are morphologically intermediate between the two parents and apparently largely sterile.

235. ***Ancistragrostis*** S.T. Blake

Ancistragrostis S.T. Blake, Blumea, Suppl. 3: 56 (1946).

Tiny perennials, <10 cm tall. Leaves filiform. Inflorescence branches branched. Spikelets with a rachilla extension, the rachilla extension with long hairs. Glumes shorter than the spikelet, translucent, their apices acute. Lemma apex notched, awned from the abaxial side; the *awn hooked.*

One sp., *A. uncinioides* S.T. Blake, New Guinea and Australia.

236. ***Bromidium*** Nees & Meyen

Bromidium Nees & Meyen, Gramineae 22–23 (1841); tax.: Rúgolo de Agrasar (1982).

Annuals or perennials, caespitose or rhizomatous. Inflorescence dense. Spikelets without a rachilla extension. Glumes longer than the flower, the apices unawned. Callus of flower with hairs less than 1/3 the length of the lemma, evenly distributed or in two tufts. *Lemma apex with four teeth or awns, plus a central awn from the abaxial side*, the awn twisted, geniculate.

Five spp., South America.

Bromidium is sometimes included in *Agrostis* but is kept separate here because of the distinctive long awns on the lemma. No molecular data are available.

237. ***"Calamagrostis"*** Adans.

"*Calamagrostis*" Adans., Fam. Pl. 2: 31, 530 (1763).

Perennials, caespitose or rhizomatous. Spikelets occasionally with 2 flowers; rachilla extension present, glabrous. Glumes longer than the flower, generally with one vein, their apices acute to attenuate, muticous or rarely awned. Callus of flower pubescent, the hairs generally longer than the flower. Lemma *generally more firm than glumes*, generally with 5 veins, the apex generally with two lobes or irregular teeth, awned from the apex or from the abaxial side, the awn straight or bent.

Ninety-eight spp., worldwide in temperate regions.

This genus is polyphyletic in the analyses of Saarela et al. (2010), who indicate that *Deyeuxia* should be excluded from "*Calamagrostis*", a conclusion followed here. Some New World species currently classified as "*Calamagrostis*" appear to belong instead to *Trisetaria*, q.v. (as *Trisetum* s.l., Saarela et al. 2010). The number of species belonging to *Calamagrostis* s.s. is uncertain; the number here is based on the number in GrassBase (Clayton et al. 2006 onward) excluding *Deyeuxia*, but Lu (2006) suggests that there are only 20 species. I have found no indication that any species of "*Calamagrostis*" is diploid; all appear to be tetraploid or higher-order polyploid. Polyploidy and apomixis are well documented, suggesting that "*Calamagrostis*" may be a genus entirely of allopolyploid origin, analogous to *Elymus*, so it is not surprising that the evolutionary history and taxonomy are complex.

238. ***Cyathopus*** Stapf

Cyathopus Stapf, Hooker's Icon. Pl. 24: t. 2395 (1895).

Tufted perennials. *Inflorescence branches numerous at most nodes.* Spikelets without a rachilla extension, *disarticulating below the glumes.* Glumes slightly longer than the flower, their apices extended into long tails. Lemma apex acute, awnless.

One sp., *C. sikkimensis* Stapf, Sikkim, Bhutan, China.

Clayton et al. (2006 onward) indicate that the primary inflorescence branches are whorled, but this is unlikely. In other supposedly "whorled" pooids, the primary branch complexes are on either side of the inflorescence suggesting that the branches initiate in a distichous phyllotaxis as in all other pooids.

239. ***Deyeuxia*** Clarion ex P. Beauv. Fig. 46

Deyeuxia Clarion ex P. Beauv., Ess. Agrostogr. 43 (1812); tax.: Phillips and Chen (2003).
Dichelachne Endl., Prodr. Fl. Ins. Norf.: 20 (1833).

Perennials, caespitose or rhizomatous, rarely annual. Spikelets usually with a slender rachilla extension. Glumes somewhat shorter or longer than the flower, their apices acute to attenuate. Callus of flower pubescent, the hairs from 1/3 the length of the flower to about the same length. Lemma apex ragged, with minute teeth, awned from the abaxial side, the awn straight or geniculate. Stamens sometimes reduced to 2 or 1. $2n =$ 28, 42, 56, 70.

Two hundred-seven spp., temperate regions worldwide, tropical mountains.

This genus appears to be entirely polyploid; no diploids are listed either in the *Flora of China* treatment (Lu et al. 2006) or in the Index to Plant Chromosome Numbers. Species formerly recognized as *Dichelachne* are reported to have $2n = 70$ (Beuzenberg and Hair 1983; Murray et al. 2005).

240. ***Echinopogon*** P. Beauv.

Echinopogon P. Beauv., Ess. Agrostogr.: 42, 148, 161 (1812).

Perennials, caespitose or rhizomatous. Spikelets with a rachilla extension. Glumes as long as or longer than the flower, their apices obtuse to attenuate. Callus of flower pubescent. Lemma apex entire or with two teeth or lobes, awned from the apex or the abaxial side, the awn stiff, the lobes also awned in some species. $2n = 42$.

Seven spp., New Guinea, Australia, and New Zealand.

241. ***Gastridium*** P. Beauv.

Gastridium P. Beauv., Ess. Agrostogr.: 21, 164 (1812).

Annuals. Spikelets vase-shaped, with or without a rachilla extension. *Glumes inflated around the fruit and narrowed above, then flaring distally, hardened and enlarged proximally*, membranous distally, their apices acuminate, unawned. Callus of flower glabrous. Lemma apex with small teeth, muticous or with an abaxial awn, the awn geniculate; *awned and unawned lemmas in the same inflorescence*. $2n = 14$.

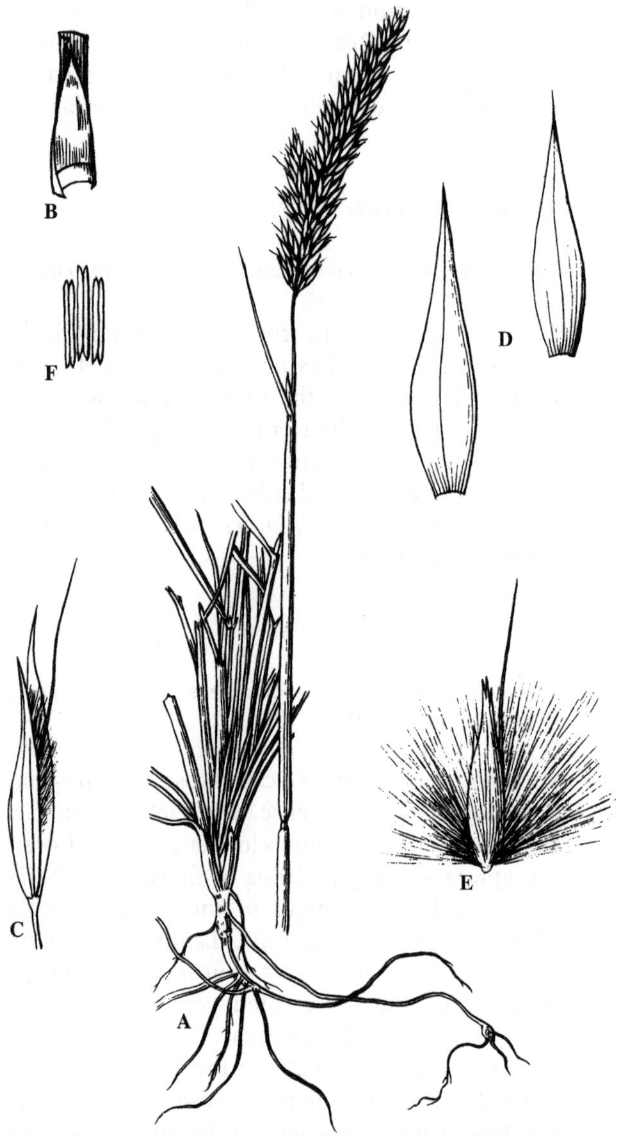

Fig. 46. Pooideae-Poeae-Agrostidinae. *Deyeuxia zangxiensis*. **A** Habit. **B** Ligule. **C** Spikelet. **D** Glumes. **E** Flower. **F** Anthers. (From Wu et al. 2007, with permission from the Missouri Botanical Garden Press, St. Louis, and Science Press, Beijing)

Two spp., Europe, North Africa, mid-East.

Saarela et al. (2010) place *Gastridium* sister to *Triplachne*.

242. ***Hypseochloa*** C. E. Hubb.

Hypseochloa C. E. Hubb., Bull. Misc. Inform. Kew 1936: 300, f. 1 (1936).

Annuals. Spikelets with a rachilla extension. Glumes longer than the flower, their apices acuminate. Callus of flower pubescent. Lemma *becoming indurate and crustaceous at maturity with involute margins*, the apex with two teeth, with an abaxial awn, the awn twisted and geniculate.

Two spp., Mount Cameroon and Tanzania.

Hubbard segregated *Hypseochloa* from *Agrostis* and *Aira* on the basis of its 5-veined glumes.

243. ***"Lachnagrostis"*** Trin.

"Lachnagrostis" Trin., Fund. Agrost.: 128, pl. 10 (1820); tax.: Jacobs (2001).

Annuals or perennials, caespitose or rhizomatous. Inflorescence often shed and dispersed as a whole. Spikelets sometimes with 2 flowers; with a rachilla extension. Glumes longer than the flower, their apices acuminate, unawned. Callus of flower pubescent, the hairs up to about 2/3 the length of the lemma. Lemma with minute teeth at the apex, unawned or with an abaxial awn, the awn straight or geniculate. $2n = 56$.

Twenty spp., Australasia.

"Lachnagrostis" is polyphyletic in the analyses of Saarela et al. (2010) with some species related to *Polypogon* and others to *Agrostis.* Jacobs (2001) distinguishes *"Lachnagrostis"* from *Agrostis* by its much better developed palea. He also compares the lemma epidermal pattern of *"Lachnagrostis"* to that in *Deyeuxia* and *Agrostis.*

244. ***Pentapogon*** R. Br.

Pentapogon R. Br., Prodr.: 173 (1810); tax.: MacFarlane and Hearn (2007).

Perennials. Spikelets without a rachilla extension. Glumes as long as the flowers, *with 1 vein*, their apices acuminate, awned. Callus of flower pubescent. *Lemma with margins convolute, covering the palea, with four apical aristae and a fifth abaxial.*

One sp., *P. quadrifidus* (Labill.) Baill., SE Australia.

245. ***Podagrostis*** (Griseb.) Scribn. & Merr.

Podagrostis (Griseb.) Scribn. & Merr., Contr. U.S. Natl. Herb. 13: 58 (1910).

Perennials, caespitose or rhizomatous. Spikelets with or without a rachilla extension. Glumes longer than the flower, their apices acute to acuminate. Callus of flower glabrous or with short hairs. Lemma with the apex truncate to acute, awned or unawned, the awn abaxial. $2n = 14$.

Six spp., New World.

Podagrostis is distinguished from *Agrostis* by the relatively long palea, and from *"Calamagrostis"* and *"Lachnagrostis"* by the short callus hairs.

246. ***Polypogon*** Desf.

Polypogon Desf., Fl. Atlant. 1: 66 (1798).

Annuals or perennials. Spikelets without a rachilla extension, disarticulating midway up the pedicel. Glumes longer than the flower, their apices entire or emarginate or bilobed, mucronate or awned from the sinus or apex, the awn flexuous. Lemma apex entire to lobed, muticous to awned, the awn generally just below the apex or from the sinus. $2n = 14, 28, 35, 42, 56,$

Twenty-six spp., worldwide in warm temperate regions and on tropical mountains.

Some species currently classified in *"Lachnagrostis"* should perhaps be placed in or near *Polypogon* (Saarela et al. 2010).

247. ***Triplachne*** Link

Triplachne Link, Hort. Berol. 2: 241 (1833).

Annuals. Glumes longer than the flower, with one vein, their apices acute. Callus of flower pubescent. *Lemma with lateral awns on either side of a central, abaxial awn.* $2n = 14$.

One sp., *T. nitens* (Guss.) Link, Mediterranean.

Saarela et al. (2010) find *Triplachne* to be strongly supported as sister to *Gastridium.*

Fig. 47. Pooideae-Poeae-Scolochloinae. *Scolochloa festucacea.* **A** Habit. **B** Spikelet. **C** Flower. (From Barkworth et al. 2007, drawn by L.A. Vorobik)

Group with Poeae-type plastid sequences

This clade is retrieved in all molecular studies of the plastid (e.g., Davis and Soreng 2007; Gillespie et al. 2008; Soreng and Davis 2000; Soreng et al. 2007), but appears to have no morphological synapomorphies.

Incertae sedis in Poeae-type plastid clade:
248. ***Antinoria*** Parl.

Antinoria Parl., Fl. Palerm. 1: 92 (1845).

Annuals or perennials. Spikelets with 2 flowers, without a rachilla extension. Glumes longer than the flower, their apices obtuse. Lemma apex truncate, unawned, with three teeth. $2n = 18$.

Two spp., Mediterranean.

Antinoria is similar to *Airopsis*, but the flowers are separated by a short internode. This genus is sister to Loliinae in the analyses of Quintanar et al. (2007) and to the Parapholiinae/Cynosurinae clade in Inda et al. (2008); both placements are based solely on an ITS sequence for *A. agrostidea* (DC) Parl.

XVI.6. Subtribe Scolochloinae Tzvelev (1987)

Perennials up to 2 m tall, with fleshy rhizomes. Spikelets with 3 or 4 flowers; rachilla extension present. Glumes shorter than or as long as the flowers, their apices acute to acuminate. Callus of flower pubescent, with stiff hairs. Lemma apex with 3 lobes or teeth, unawned. Ovary apex pubescent.

249. ***Scolochloa*** Link Fig. 47

Scolochloa Link, Hort. Berol. 1: 136 (1827).

Characters as for the subtribe. $2n = 28$.

One sp., *S. festucacea* (Willd.) Link, north temperate regions.

Scolochloa looks superficially like *Festuca*.

XVI.7. Subtribe Airinae Fr. (1835)

Annuals or perennials. Spikelets generally with 2 flowers. Lemma generally with an abaxial awn, the awn geniculate.

Ten genera, 97 spp.

250. ***Aira*** L.

Aira L., Sp. Pl. 1: 63–66 (1753).

Delicate tufted annuals. Spikelets generally <5 mm long, lacking a rachilla extension. Glumes longer than the flowers, their apices obtuse to acute, unawned. Callus of flower pubescent. Lemma apex bilobed. $2n = 14, 28$.

Eight spp., Europe, Mediterranean, Iran and weedy elsewhere in the world.

Aira includes plants that are distinctively delicate, with capillary inflorescence branches and pedicels.

251. ***Avenella*** (Bluff & Fingerh.) Drejer

Avenella (Bluff & Fingerh.) Drejer, Fl. Excurs. Hafn.: 32 (1837); tax.: Chiapella and Zuloaga (2010).

Caespitose perennials, with or without rhizomes. Spikelets with a pubescent rachilla extension. Glumes as long as the flowers, their apices acute. Callus of flower pubescent. Lemma apex truncate or toothed, the awn hair-like. $2n = 28$.

One sp., *A. flexuosa* (L.) Drejer, widespread in temperate and cold regions worldwide, including Antarctic islands.

Clayton and Renvoize (1986) synonymize this genus with *Deschampsia*, whereas Chiapella (2007) presents molecular phylogenetic data supporting the distinction of the two and a close relationship instead with *Aira*.

252. ***Avenula*** (Dumort.) Dumort.

Avenula (Dumort.) Dumort., Bull. Soc. Roy. Bot. Belgique 7: 68 (1868); cytogenetics: Winterfeld (2006); tax.: Lange (1995); phylog.: Winterfeld et al. (2012).
Helictotrichon subg. *Pratavenastrum* (Vierh.) Holub, Philipp Maxm. Opiz Bedeut. Pfl.-tax.: 125 (1958).

Perennials, caespitose or stoloniferous. Sheaths open or closed. Spikelets with 2 to 7 flowers; *rachilla glabrous when adjacent to the palea, pubescent elsewhere.* Glumes as long as or longer than the flowers. Callus of flower pubescent. Lemma apex obtuse, with two teeth, awned, the awn from about ½ way down the abaxial side. *Endosperm liquid.* $2n = 14, 28$.

Thirty spp., Eurasia, North Africa, North America.

Avenula is similar to *Helictotrichon*, but the latter has ligules that are truncate to rounded, leaf blades that are ribbed, rachillas that are pubescent on all sides, a long hilum, and solid endosperm. *Avenula pubescens* falls in conflicting positions in phylogenetic trees, and may be an intergeneric hybrid (Gillespie et al. 2008; Soreng and Davis 2000). It is probably more closely related to Poinae than to *Avenula* s.s., but is retained here pending a better placement. The closed sheaths are unusual for this subtribe.

253. ***Corynephorus*** P. Beauv.

Corynephorus P. Beauv., Ess. Agrostogr.: 90, 159 (1812), nom. conserv.

Annuals or perennials. Spikelets with a rachilla extension. Glumes hyaline, vase-shaped, longer than the flowers. Callus of flower pubescent. Lemma apex acute, awned from the abaxial side just above the base, *the awn articulated in the middle, twisted below, club-shaped above, with hairs at the connection between the two portions.* $2n = 14$.

Five spp., coastal dunes of Europe, Mediterranean, mid-East.

The molecular analyses of Hoffmann et al. (2013) place *Corynephorus* in Poinae, but on a very long branch. Because this placement contradicts other information on the genus, it is retained here pending additional data.

254. ***Deschampsia*** P. Beauv. Fig. 48

Deschampsia P. Beauv., Ess. Agrostogr.: 91–92, pl. 18, f. 3 (1812); tax.: Chiapella and Zuloaga (2010).

Annuals or perennials, caespitose. Spikelets generally with a pubescent rachilla extension. Glumes generally longer than the spikelet, their apices acute to acuminate, muticous or mucronate. Callus of flower pubescent. Lemma apex truncate or toothed, the abaxial awn hair-like, straight or geniculate and twisted. $2n = 26, 52$.

Forty spp., temperate areas worldwide.

Barkworth (2007a) cites various aneuploid chromosome counts for several species but notes that the voucher specimens have not been checked. These counts are not listed here.

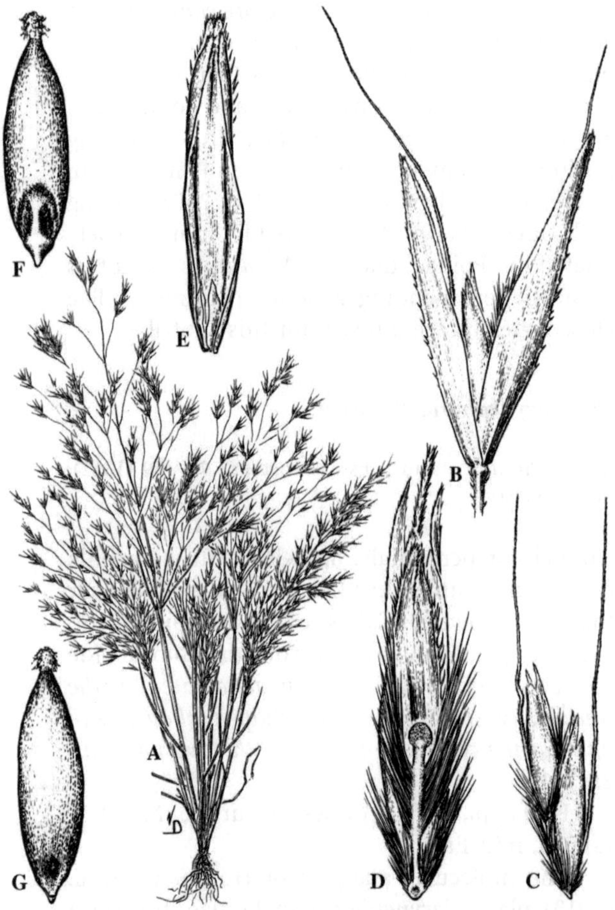

Fig. 48. Pooideae-Poeae-Airinae. *Deschampsia antarctica*. **A** Plant. **B** Spikelet, lateral view. **C** Flowers. **D** Base of the flower, viewed from the side of the palea and rachilla. **E** Palea, internal face, and lodicules. **F** Caryopsis, scutellar view. **G** Same, hilar view. (From Zuloaga et al. 2012)

255. *Holcus* L.

Holcus L., Sp. Pl. 2: 1047 (1753), nom. conserv.

Perennials, rarely annual, caespitose or rhizomatous. Spikelets disarticulating below the glumes, with the proximal flower(s) bisexual and the distal one staminate. Glumes as long as or longer than the flowers, keeled, their apices obtuse to acuminate, unawned. Callus of flower glabrous or pubescent. Lemma firm, shiny, the apex acute or bidentate, the proximal lemma in the spikelet unawned, the distal one with an abaxial awn. $2n = 14, 28$ (35, 42, 49).

Eight spp., Europe, north Africa, mid-East. See comments under *Vahlodea*.

256. *Molineriella* Rouy

Molineriella Rouy, Fl. France 14: 102 (1913).

Annuals. Spikelets without a rachilla extension. Glumes shorter than or about the same length as the flowers, their apices obtuse to acute. Callus of flower pubescent. Lemma apex obtuse, *unawned*. $2n = 8$.

One sp., *M. minuta* (L.) Rouy, Mediterranean, Australasia.

257. *Periballia* Trin.

Periballia Trin., Fund. Agrost.: 133 (1820).

Annuals. Spikelets without a rachilla extension. Glumes shorter than or about the same length as the flowers, their apices acute. Callus of flower pubescent. Lemma awned from the abaxial side, *the awn straight*, or the apex acute, mucronate. $2n = 8, 14, 18$.

Two spp., Mediterranean.

258. *Scribneria* Hack.

Scribneria Hack., Bot. Gaz. 11: 105 (1886).

Tufted annuals. Leaf blades filiform. *Inflorescence unbranched, the spikelets distichous and sunken in the rachis*, generally one per node, but sometimes up to 4; rachis not disarticulating. Spikelets with 1 flower; rachilla extension present. Glumes longer than the flower, coriaceous, unawned. Callus of flower pubescent. Lemma apex with two teeth, awned from the sinus. *Stamen 1*. $2n = 26$.

One sp., *S. bolanderi* (Thurb.) Hack., western U.S.

Scribneria is placed here on the basis of the data of Schneider et al. (2012).

259. *Vahlodea* Fr.

Vahlodea Fr., Bot. Not. 1842: 141, 178 (1842); tax.: Chiapella and Zuloaga (2010).

Caespitose perennials, *the leaves mostly cauline*. Spikelets with a rachilla extension. Glumes as long as or longer than the flowers, their apices acute to acuminate. Callus of flower pubescent, the hairs about ½ the length of the lemma. Lemma apex with minute irregular teeth, the awn hair-like, twisted, geniculate. *Hilum linear.* $2n = 14$.

One sp., *V. atropurpurea* (Wahlenb.) Fr. ex Hartm., temperate regions throughout the world.

Clayton and Renvoize (1986) synonymize *Vahlodea* with *Deschampsia*, whereas molecular phylogenetic data (Chiapella 2007; Döring et al. 2007; Schneider et al. 2009) support the distinction of the two, and place *Vahlodea* sister to *Holcus* with strong support. In *Vahlodea* the leaves are mostly cauline, whereas they are largely basal in *Deschampsia*. The rachilla extension is less than 0.5 mm long in *Vahlodea* and is glabrous, whereas it is generally more than 0.5 mm long in *Deschampsia* and is pubescent.

XVI.8. Subtribe Ammochloinae Tzvelev (1976)

Annuals, often tiny. Inflorescence capitate, surrounded by leaf sheaths in some species. Spikelets with 4 to 14 flowers plus additional reduced flowers distally or a barren rachilla extension. Glumes shorter than the flowers, sometimes each with an asymmetrical wing on the keel, their apices obtuse to acute, muticous to mucronate. Lemma with broad margins, the apex entire or with two teeth, mucronate or awned, the awn from the apex or sinus. *Lodicules absent. Base of style branches persistent in fruit, forming a beak.*

260. ***Ammochloa*** Boiss. Fig. 49

Ammochloa Boiss., Diagn. Pl. Orient. ser. 1. 2(13): 51 (1854).

Characters as for the subtribe.

Three spp., Mediterranean and mid-East.

In *A. pungens* the proximal 1 or 2 spikelets are sterile.

XVI.9. Subtribe Cynosurinae Fr. (1835)

Annuals or perennials, caespitose or rhizomatous. Inflorescence *unbranched* or the branches branched, *with paired spikelets, the proximal*

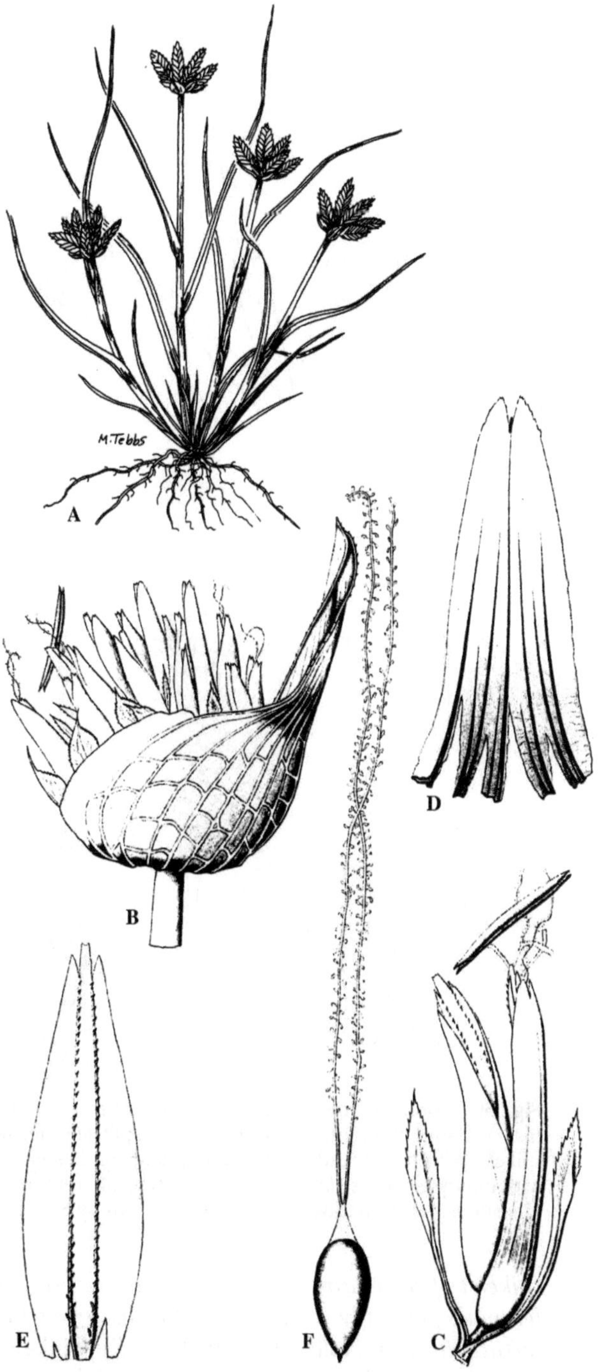

Fig. 49. Pooideae-Poeae-Ammochloinae. *Ammochloa*. **A** *A. palaestina*. Habit. (From Boulos 2005). **B–F** *Ammochloa involucrata*. **B** Capitulum of spikelets subtended by an involucre. **C** Spikelet with two flowers. **D** Lower glume viewed from the abaxial side. **E** Upper glume viewed from the abaxial side. **F** Caryopsis with stylopodium and stigmas. (From Murbeck 1899)

Fig. 50. Pooideae-Poeae-Cynosurinae. *Cynosurus echinatus*. **A** Habit. **B** Inflorescence. **C** Pair of spikelets, fertile spikelet with two spread glumes (*right*), and sterile spikelet with multiple glumes (*left*). **D** Lemma, at two different magnifications. (From Barkworth et al. 2007, drawn by S. Long)

spikelet of each pair made up of a cluster of sterile lemmas, generally abaxial to and obscuring the fertile distal spikelet. Proximal sterile spikelet with 6 to 18 flowers, the glumes and lemmas narrow, sometimes awned. Distal fertile spikelet with 1 to 5 flowers; rachilla extension present, glabrous. Glumes shorter than the flowers, their apices acute to attenuate, sometimes awned. Callus of flower glabrous. Lemma apex entire or with two teeth, unawned or awned from the apex or sinus. $2n = 14$.

261. ***Cynosurus*** L. Fig. 50

Cynosurus L., Sp. Pl. 1: 72–73 (1753).

Characters as for the subtribe.

Ten spp., Europe and Mediterranean.

XVI.10. Subtribe Loliinae Dumort. (1829)

Annuals or perennials. Leaf sheaths with or without auricles. Spikelets with more than one flower. Glumes generally shorter than the adjacent lemma.

Eight genera, 543 spp.

Loliinae are dominated by the large genus *Festuca*, the circumscription of which varies extensively in the literature. Molecular phylogenies show that *Festuca* as traditionally defined is paraphyletic, with many smaller genera derived from within it (Catalán et al. 2004; Inda et al. 2008; Torrecilla and Catalán 2002; Torrecilla et al. 2004). Catalán et al. (2004) identified a large clade that they called the "the fine-leaved fescues" and another "the broad-leaved fescues"; these clades include most, but not all species conventionally recognized as *Festuca*. The classification adopted here is influenced by the phylogeny, recognizing the "fine-leaved fescues" as *Festuca* s.s. This clade includes *Vulpia*, which has long been known to be related to *Festuca*, as well as several other smaller genera. The "broad-leaved fescues" include *Lolium*, also well known as a close relative of species formerly classified as *Festuca*; this clade thus becomes *Lolium* s.l. The remaining species of *Festuca* fall into one of two poorly supported grades, here listed as the "intermediate grade" and the "*Leucopoa* grade." I have not attempted to dismember these into subgroups. There are several reasonably well-supported subclades within each of the grades, but all require additional taxon sampling and sequence data. A more radical treatment than that adopted here would be to place the entire subtribe in a single genus, although even with this solution the genus would lack any obvious diagnostic characters.

262. **Incertae sedis – "intermediate grade" of Loliinae**

Festuca sect. *Subulatae* Tzvelev, Ukrain. Bot. Rev. 56(9): 1253 (1971).

Festuca sect. *Subuliflorae* (E.B. Alexeev) Darbysh., Phytologia 82(2): 77 (1997).
Festuca sect. *Amphigenes* (Janka) Tzvelev, Bot. Žurn. (Kiev) 56(9): 1253 (1971).

Catalán et al. (2004) and Inda et al. (2008) identify a grade that they called "intermediate", which is a set of poorly supported lineages sister to *Festuca* s.s. (the fine-leaved fescues). Their analysis includes representatives of the three sections listed above, plus *Castellia*. I have not attempted a diagnostic description of members of this grade.

263. **"Leucopoa grade"**

"Leucopoa grade"
Leucopoa Griseb., Fl. Ross. 4: 383 (1852).
Drymochloa Holub., Folia Geobot. Phytotax. 19: 96 (1984).
Festuca subg. *Drymanthele* V.I. Krecz. & Bobrov, Fl. URSS 2: 532 (1934).
Festuca sect. *Montanae* Hack., Monogr. Festuc. Eur. 195 (1882).
Festuca subg. *Obtusae* E.B. Alexeev, Novosti Sist. Vyssh. Rast. 17: 45 (1980).
Festuca sect. *Scariosae* Hack., Monogr. Festuc. Eur.: 193 (1882).
Festuca sect. *Pseudoscariosa* Krivot., Bot. Mater. Gerb. Bot. Inst. Komarov.Akad. Nauk. SSSR 20: 61 (1960).

Perennials, generally caespitose, rhizomatous or not; plants dioecious or bisexual. Leaves relatively broad and flat, with sclerenchyma girders adaxial and abaxial to the veins. Auricles absent. Spikelets with 2 to 10 flowers. Glumes shorter than the flowers, their apices acute to acuminate, unawned. Callus of flower generally wider than long, glabrous. Lemma apex acute to attenuate, mucronate to awned. Apex of ovary densely pubescent, rarely glabrous. $2n = 28, 42, 56$.

Thirty-seven spp., temperate regions.

Lolium s.l. is derived from within a largely unresolved grade including *Leucopoa* and several sections formerly assigned to *Festuca* (Catalán et al. 2004; Inda et al. 2008). I am calling this simply "the *Leucopoa* grade", pending further study. *Festuca* subg. *Montanae* presumably falls here, based on its leaf anatomy, although its species have not been placed phylogenetically. The description offered here is only approximate.

264. ***Castellia*** Tineo

Castellia Tineo, Pl. Rar. Sicil. 2: 17 (1846).

Annuals. Leaf sheaths with falcate auricles. *Inflorescence unbranched, or of unbranched branches borne along a central axis.* Spikelets broadside to the rachis, with 6 to 12 flowers plus additional reduced flowers distally. Glumes with apices obtuse to acute. *Lemma tuberculate*, the apex obtuse to acute, unawned. $2n = 14$.

One sp., *C. tuberculosa* (Moris) Bor, Canary Is., Mediterranean, Sudan, Somalia, Pakistan.

Catalán et al. (2004) and Inda et al. (2008) place *Castellia* in a position between the broad-leaved and fine-leaved fescues.

265. ***Dryopoa*** Vickery

Dryopoa Vickery, Contr. New South Wales Natl. Herb. 3: 195 (1963).

Perennials. Spikelets with 3 to 7 flowers plus additional reduced flowers distally. Glumes with apices acute. Lemmas with prominent raised veins, the apex truncate, ragged, unawned. Ovary apex pubescent.

One sp., *D. dives* (F. Muell.) Vickery, SE Australia.

While chloroplast data place *Dryopoa* near or in Loliinae (Gillespie et al. 2008), data from the nuclear internal transcribed spacer (ITS) and external transcribed spacer (ETS) of the ribosomal RNA place it sister to Agrostidinae (Birch et al. 2014).

266. ***Festuca*** L. Fig. 51

Festuca L., Sp. Pl. 1: 73–76 (1753).
Wangenheimia Moench, Methodus: 200 (1794).
Vulpia C. C. Gmel., Fl. Bad. 1: 8 (1805); rev.: Cotton and Stace (1967); Lonard and Gould (1974).
Micropyrum (Gaudin) Link, Linnaea 17: 397 (1844).
Ctenopsis De Not., Index Sem. Hort. Genuensis: 26 (1847).
Narduroides Rouy, Fl. France 14: 301 (1913).
Hellerochloa Rauschert, Taxon 31: 561 (1982).
Dielsiochloa Pilg., Bot. Jahrb. Syst. 73: 99 (1943).

Annuals (some very tiny) or perennials, caespitose or rhizomatous. Leaves generally slender. Inflorescence unbranched or with branched branches. Spikelets with 1 to 11 (17) flowers, or pseudoviviparous in a few species. Glumes shorter than the flowers, their apices acuminate, awned or unawned. Callus of flowers often wider than long, pubescent or glabrous. Lemma apex acute to acuminate, entire or sometimes minutely bidentate, mucronate to awned. Apex of ovary

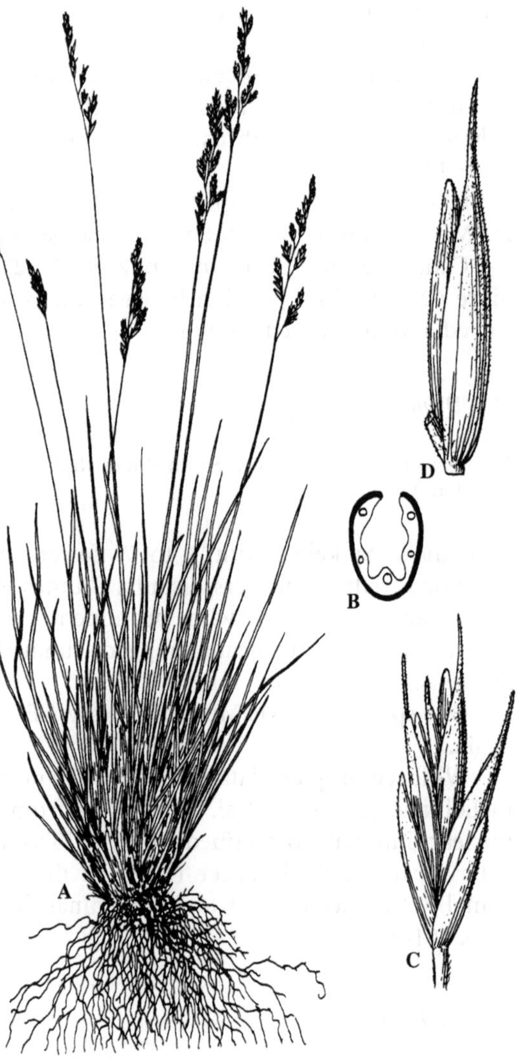

Fig. 51. Pooideae-Poeae-Loliinae. *Festuca ovina* L. **A** Habit. **B** Cross section of leaf blade. **C** Spikelet. **D** Flower. (From Wu et al. 2007, with permission from the Missouri Botanical Garden Press, St. Louis, and Science Press, Beijing)

glabrous or pubescent. Stamens 1 or 3. $2n = 14$, 28, 42, 44, 49, 50, 56, 63, 70.

Ca. 430 spp., temperate regions of North America and Eurasia.

Festuca as circumscribed here corresponds to the "fine-leaved fescues" (Catalán et al. 2004; Inda et al. 2008; Torrecilla and Catalán 2002; Torrecilla et al. 2004), including taxa formerly classified as *Festuca* subgenus *Festuca* sects. *Festuca*, *Aulaxyper* Dumort., *Eskia* Willk., and *Dimorphae* Joch. Müll. & Catalán. In addition the group contains several annual species each of which has some distinctive morphological characteristic that led it to be placed in its own genus at one time. These genera are clearly all derived from within *Festuca* and the distinctive characters are phylogenetically uninformative autapomorphies. While the group recognized here as *Festuca* is clearly monophyletic, it lacks any obvious morphological synapomorphies. Darbyshire and Pavlick (2007) describe patterns of leaf cross-sectional anatomy that may be diagnostic for *Festuca* as circumscribed here, although comparable data are not available for the segregate genera, which need to be checked.

267. *Lolium* L.

Lolium L., Sp. Pl. 1: 83 (1753); tax.: Charmet et al. (1996); Darbyshire (1993).
Schedonorus P. Beauv., Ess. Agrostogr.: 99, 162, 177 (1812).
Festuca subg. *Schedonorus* (P. Beauv.) Peterm., Deutschl. Fl.: 643 (1849).
Micropyropsis Romero Zarco & Cabezudo, Lagascalia 11: 95 (1983).
Festuca sect. *Subbulbosae* Nyman ex Hack., Bot. Centralbl. 8: 413 (1881), p.p.

Annuals or perennials, caespitose, sometimes rhizomatous. Leaf sheath with or without falcate auricles. Inflorescence unbranched, or with branched branches; if unbranched then spikelets edgewise to the rachis. Spikelets with 2 to 22 flowers. Glumes generally shorter than the flowers, their apices acute, unawned; *in unbranched inflorescences the lower glume lacking except on the terminal spikelet, upper glume on side away from rachis.* Callus of flower glabrous or with sparse hairs. Lemma apex acute, awned or unawned, the awn straight. Apex of ovary glabrous. Hilum linear. $2n = 14$, 28, 42, 56, 63, 70.

Twenty-six spp., temperate regions of Eurasia; widely introduced as a lawn and forage grass.

Inclusion of *Schedonorus* in *Lolium* is supported by morphological and molecular data (Catalán et al. 2004); as noted by Darbyshire (2007), the two "could appropriately be treated as congeneric subgenera . . .". The characters shared by *Lolium* and *Schedonorus* and not with *Festuca* s.s. include the falcate leaf auricles, flat, wide leaf blades, ovaries that are glabrous at the apex, stylar attachment abaxial, below the apex of the ovary, and fruit that adheres firmly to the palea. Hybridization between species of *Lolium* s.s. and *S. arundinaceus* (= *Festuca arundinacea*) is common, further supporting the close relationship between the two.

268. ***Loliolum*** V. I. Krecz. & Brobov.

Loliolum V.I. Krecz. & Bobrov, Fl. URSS 2: 544, 766 (1934).

Annuals. *Inflorescence unbranched, the spikelets all borne on one side.* Spikelets wedge-shaped, with 3 to 7 flowers plus additional reduced flowers distally. Glumes longer than the flowers, their apices attenuate, mucronate. Lemma apex notched to truncate to acute, muticous or with a short awn.

One sp., *L. subulatum* (Banks & Sol.) Eig., mid-East to central Asia.

No molecular data are available for *L. subulatum*, and it may be misplaced here. It has also been placed in *Festuca* and *Nardurus*, the latter now a synonym of *Festuca*.

269. ***Pseudobromus*** K. Schum.

Pseudobromus K. Schum., Pflanzenw. Ost-Afrikas C: 108 (1895).

Perennials. *Leaf blades with transverse veins.* Spikelets with 1 to 4 flowers, the rachilla terminating in a rudiment. Glumes shorter than the flowers, their apices acuminate. Callus of flower puberulous. Lemma awned, the awn from the apex or just below. Palea longer than the lemma.

Six spp., tropical Africa and Madagascar.

No molecular data are available for *Pseudobromus*; it is placed here based on morphology.

XVI.11. Subtribe Dactylidinae *Stapf (1898)*

Leaf sheaths closed for part of their length. Spikelets borne in dense clusters.

Dactylidinae are strongly supported as monophyletic by molecular data (Catalán et al. 2004; Inda et al. 2008).

Fig. 52. Pooideae-Poeae-Dactylidinae. *Dactylis glomerata.* **A** Plant. **B** Ligule area. **C** Open panicle. **D** Closed panicle. **E** Spikelet. **F** Flower. (From Barkworth et al. 2007, drawn by L.A. Vorobik and K. Klitz)

270. ***Dactylis*** L. Fig. 52

Dactylis L., Sp. Pl. 1: 71 (1753); tax.: Stewart and Ellison (2011).

Perennials, caespitose or short rhizomatous. *Spikelet clusters borne on one side of the inflorescence branches, the branches naked proximally.* Spikelets with 2 to 6 flowers; rachilla extension lacking. Glumes shorter than the flowers, their apices acute, mucronate or awn-tipped. Callus of flower glabrous. Lemma apex acuminate, with a short awn. Hilum round. $2n = 14, 21, 27–31, 42$.

Five spp., temperate Eurasia, widely introduced elsewhere as a forage grass.

271. ***Lamarckia*** Moench

Lamarckia Moench, Methodus: 201 (1794).

Annuals. Inflorescence with all branches borne on one side, *secondary inflorescence branches ending in clusters of spikelets, bent and disarticulating below the clusters. Proximal spikelets within each cluster sterile, with 5 to 10 flowers, distal spikelet fertile*, with 2 flowers, the proximal flower bisexual, on an elongate rachilla internode, the distal flower reduced to an awned rudiment. Glumes as long as or longer than the flower, their apices acuminate, mucronate to awned. Callus of flower glabrous. Lemma apex with two minute teeth, awned from the abaxial side. $2n = 14$.

One sp., *L. aurea* (L.) Moench, Mediterranean and Middle East.

XVI.12. Subtribe Parapholiinae Caro (1982)

Annuals or perennials. Inflorescences branched or unbranched, often disarticulating at the nodes. Spikelets with one to many flowers. Glumes shorter or longer than the flowers.

Seven genera, 25 species.

Many members of Parapholiinae were formerly placed in the tribe Hainardieae (Clayton and Renvoize 1986), a group that has since been shown to be polyphyletic (Schneider et al. 2012). *Narduroides* is now included in *Festuca* s.s., *Pholiurus* is in Poinae, whereas *Scribneria* is in Airinae, unrelated to the others. The remainder form Parapholiinae. The subtribe includes the named hybrid x*Hainardiopholis* Castrov.

272. ***Sphenopus*** Trin.

Sphenopus Trin., Fund. Agrost.: 135 (1820).

Annuals. Leaf blades filiform. Inflorescence branches branched. Spikelets with 2 to 6 flowers plus additional reduced flowers distally; *pedicels notably wider toward their apices*. Glumes less than ½ the length of the flowers, hyaline, their apices obtuse or the upper one acute. Lemma apex truncate or obtuse, muticous to mucronate. $2n = 12, 14, 28$.

Two spp., Mediterranean and mid-East.

Sphenopus is sister to the rest of the subtribe in molecular phylogenies (Inda et al. 2008; Soreng et al. 2007).

Remaining genera of the subtribe:
Inflorescence generally unbranched, often disarticulating at the nodes.

273. ***Agropyropsis*** (Trab.) A. Camus

Agropyropsis (Trab.) A. Camus, Bull. Soc. Bot. France 82: 11 (1935).

Perennials. Inflorescence unbranched, disarticulating at the nodes. Spikelets distichous, with 3 to 6 flowers plus additional reduced flowers distally. Glumes borne side by side on the abaxial side of the spikelet, shorter than the flowers, their apices obtuse. Callus of flower glabrous. Lemma apex obtuse, unawned. Ovary with a conspicuous apical appendage.

One sp., *A. lolium* (Balansa ex Coss. & Durieu) A. Camus, Algeria.

Schneider et al. (2012) argue that *Agropyropsis* should be placed in Loliinae, based on morphology. No molecular data are available, so the genus is retained here for the time being.

274. ***Cutandia*** Willk.

Cutandia Willk., Bot. Zeitung (Berlin) 18: 130 (1860).

Annuals. Inflorescence branches branched, *disarticulating at the base of the branches, pedicels or flowers*. Spikelets with 2 to 9 flowers. Glumes shorter than the flowers, their apices acute to rounded, awned or not. Callus of flower glabrous. Lemma keeled, with 3 conspicuous veins, the apex rounded to bifid to acuminate and entire, awned or awnless. $2n = 14$.

Six spp., Mediterranean and western Asia, one species introduced to North America.

Suggestions for placing this genus near *Vulpia* (Catalán et al. 2004; Torrecilla et al. 2004) were erroneous (Inda et al. 2008). It is clearly placed here, sister to *Desmazeria*.

Fig. 53. Pooideae-Poeae-Parapholinae. *Hainardia cylindrica*. **A** Plant. **B** Spikelet, with rachis internode (*right*) and upper glume (*left*) enclosing the flower (tip visible). **C** Flower concealed by upper glume. (From Barkworth et al. 2007, drawn by L.A. Vorobik and H. Paździrková)

275. *Desmazeria* Dumort.

Desmazeria Dumort., Comment. Bot.: 26 (1822).
Catapodium Link, Hort. Berol. 1: 44 (1827).
Scleropoa Griseb., Spic. Fl. Rumel. 2: 431 (1846).

Annuals. Inflorescences branched or unbranched, one-sided, not disarticulating at the nodes. Spikelets with 4 to 25 flowers, the distal ones reduced. Glumes shorter than the flowers, their apices obtuse to acute. Callus of flower glabrous. Lemma glabrous or pubescent with capitate hairs, the apex obtuse to acute, with or without a mucro. $2n = 14$.

Eight spp., Mediterranean.

276. *Hainardia* Greuter Fig. 53

Hainardia Greuter, Boisseria 13: 178 (1967).

Annuals. *Culm internodes solid.* Inflorescence unbranched, disarticulating at the nodes. Spikelets embedded in the inflorescence internodes, with 1 or 2 flowers; rachilla extension present. *Lower glume absent except on the terminal spikelet*, upper glume longer than the flowers, leathery, ribbed, the apex acute. Lemma membranous, the apex acuminate, unawned. Stamens 1 or 3. Fruit with an apical appendage. $2n = 26$.

One sp., *H. cylindrica* (Willd.) Greuter, coastal areas of the Mediterranean.

Hainardia is strongly supported as sister to *Parapholis* (Catalán et al. 2004; Schneider et al. 2012). The solid culm internodes for this species are unusual in the Pooideae. The name *Monerma* is sometimes misapplied to this genus; the latter is a superfluous name for the chloridoid genus *Lepturus*.

277. *Parapholis* C. E. Hubb.

Parapholis C. E. Hubb., Blumea, Suppl. 3: 14 (1946); tax.: Runemark (1962).

Annuals, tufted. Inflorescence unbranched, disarticulating at the nodes. Spikelets embedded in the inflorescence internodes, with one flower; rachilla extension present or absent. Glumes borne side by side on the abaxial side of the spikelet, generally longer than the flower, coriaceous, their apices acute to acuminate. Lemma translucent, the apex obtuse to acute. $2n = 14$, 18, 38.

Six spp., coastal areas of Europe, Mediterranean, and the mid-East.

278. ***Vulpiella*** (Batt. & Trab.) Burollet

Vulpiella (Batt. & Trab.) Burollet, Ann. Serv. Bot. Tunisie 4: 68 (1927).

Annuals. Spikelets with 5 to 18 flowers plus additional reduced flowers distally. Glumes shorter than the flowers, their apices acute, mucronate. Lemma apex with two teeth, awned from the sinus.

One sp., *V. stipoides* (L.) Maire, Western Mediterranean.

XVI.13. Subtribe Sesleriinae Parl. (1845)

Miborinae Asch. & Graebn., Syn. Mitteleur. Fl. 2: 118 (1899).

Annuals or perennials, most less than 40 cm tall. *Leaf sheaths with fused margins.* Inflorescence dense, one-sided or capitate.

Four genera, 33 spp.

This subtribe is placed within Aveninae in ITS analyses, whereas it clearly has a plastid similar to that of Poinae (Quintanar et al. 2007; Soreng et al. 2007; Gillespie et al. 2008). Although the discrepancy between the ITS and plastid data suggests a hybrid origin for the tribe, the low chromosome numbers indicate that polyploidization may not be involved. Both Quintanar et al. (2007) and Saarela et al. (2010) show that *Mibora* is derived within Sesleriinae. A few species of *Sesleria* reach 60 or 70 cm tall, but most members of this subtribe are tiny.

279. ***Echinaria*** Desf.

Echinaria Desf., Fl. Atlant. 2: 385 (1799).

Annuals, less than 25 cm tall. *Inflorescence capitate, prickly.* Spikelets with 2 to 4 flowers plus additional reduced flowers distally. Glumes shorter than the flowers, their apices obtuse, lower glume with two keels, upper glume with a short mucro. Callus of flower pubescent. *Lemma leathery, with 5 to 7 veins extending beyond the apex as narrow lobes or awns. Palea with two awns.* $2n = 14$, 18.

One sp., *E. capitata* (L.) Desf., Mediterranean and mid-East.

280. ***Mibora*** Adans.

Mibora Adans., Fam. Pl. 495 (1763).

Tiny annuals, less than 15 cm tall. Inflorescence unbranched, the spikelets imbricate, borne on short pedicels on one side of the rachis. Spikelet with one flower; rachilla extension absent. Glumes longer than the flower, their apices truncate or obtuse. Callus of flower glabrous. Lemma translucent, pubescent, the apex truncate and ragged. *Palea pubescent. Lodicules absent.* Style 1, stigmas 2. $2n = 14$.

Two spp., Western Europe and North Africa.

281. ***Oreochloa*** Link

Oreochloa Link, Hort. Berol. 1: 44 (1827).

Perennials, less than 40 cm tall. Inflorescence unbranched, one-sided, capitate. Spikelets with 3 to 7 flowers plus additional reduced flowers apically. Glumes hyaline, shorter than the flowers, their apices acute. Callus of flower glabrous. Lemmas hyaline, their apices acute, mucronate.

Four spp., southern Europe.

282. ***Sesleria*** Scop. Fig. 54

Sesleria Scop., Fl. Carniol.: 189 (1760); cytol.,: Lysak and Dolezel (1998).

Perennials, caespitose, rhizomatous or stoloniferous. Inflorescence spike-like or capitate, *often with two bracts or scales at the base.* Spikelets with 2 to 5 flowers; rachilla extension present. Glumes shorter than the flowers, their apices acute to acuminate, muticous or awned. Callus of flower glabrous, broadly rounded. Lemma apex entire or with 3 to 5 teeth, the teeth generally awned. Apex of ovary pubescent. $2n = 28$, 56.

Twenty-six spp., Europe.

The bracts at the base of the inflorescence are generally interpreted as sterile spikelets (Clayton and Renvoize 1986).

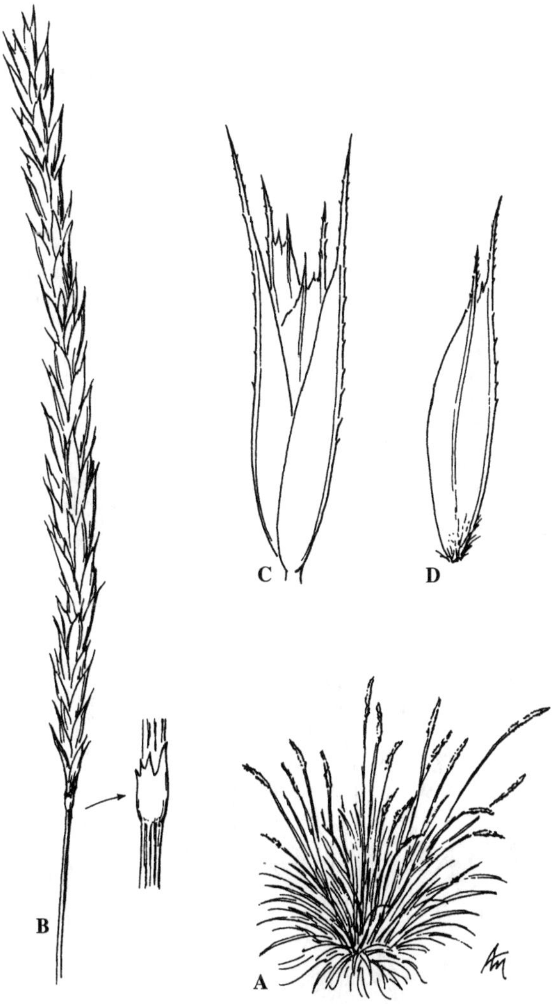

Fig. 54. Pooideae-Poeae-Seslerinae. *Sesleria autumnalis*. **A** Habit. **B** Inflorescence. **C** Spikelet. **D** Lemma. (From Barkworth et al. 2007, drawn by A. Miller)

XVI.14. Subtribe Coleanthinae *Rouy (1913)*

Subtribe Puccinelliinae Soreng & J. I. Davis, Contr. U.S. Natl. Herb. 48: 721 (2003).

Annuals or perennials, tufted or rhizomatous. Leaf sheaths often partially to completely closed. Spikelets with one to several flowers; awns absent. Glumes often with obtuse apices, *the veins prominent or obscure, approximately parallel and not reaching the apex.* Lemma apex obtuse to acute, often scarious, *veins approximately parallel and not reaching the apex.*

Twelve genera, 156 spp.

Named hybrid: x *Pucciphippsia* Tzvelev

283. ***Catabrosa*** P. Beauv.

Catabrosa P. Beauv., Ess. Agrostogr.: 97 (1812); tax.: Soreng and Fish (2011).

Perennials, sometimes stoloniferous. Leaves with sheath margins fused most of their length. Spikelets with 1 to 3 flowers; rachilla extension present. Glumes shorter than the flowers, their apices truncate to obtuse. Callus of flower glabrous. Lemma with 3 prominent veins, the apex rounded to truncate. $2n = 20, 30$.

Three spp., north temperate regions and Chile.

284. ***Catabrosella*** (Tzvelev) Tzvelev

Catabrosella (Tzvelev) Tzvelev, Bot. Žurn. (Moscow & Leningrad) 50: 1320 (1965).

Caespitose perennials, basal leaves often with sheaths persisting and forming bulb-like structures. Leaves with sheath margins fused 1/6 to 1/4 of their length. Spikelets with 2 to 4 flowers. Glumes much shorter than the flowers, the apices obtuse to subacute. Callus of flower glabrous or pubescent. Lemma apex obtuse. $2n = 10$.

Seven spp., Europe to the Himalayas.

285. ***Coleanthus*** Seidl — Fig. 55

Coleanthus Seidl, Syst. Veg. 2: 11, 276 (1817).

Tiny tufted annuals, mostly less than 7 cm tall. Leaves with sheath margins fused most of their length. Inflorescence dense, with spikelets in capitate clusters, surrounded by the upper sheath. Spikelets with one flower, without rachilla extension. *Glumes absent.* Callus of flower glabrous. Lemma translucent, the apex acute, awned. *Lodicules absent. Stamens 2.* $2n = 14$.

One sp., *C. subtilis* (Tratt.) Seidl, Central Europe, northern Asia, and North America.

286. ***Hyalopoa*** (Tzvelev) Tzvelev

Hyalopoa (Tzvelev) Tzvelev, Bot. Žurn. (Moscow & Leningrad) 50: 1320 (1965).

Rhizomatous perennials. Leaves with sheath margins fused 1/3 to 3/4 of their length. Spikelets with

2 to 4 flowers. Glumes shorter than flowers. Callus of the flower pubescent. Lemma weakly keeled, the apex obtuse to acute, sometimes with a sharp point. $2n = 28, 42$.

Four spp., Central Asia.

Although Clayton and Renvoize (1986) place this in the synonymy of *Colpodium*, Kim et al. (2009) and Hoffmann et al. (2013) suggest it is more closely related to *Catabrosa*, *Paracolpodium*, and *Catabrosella*.

287. *Paracolpodium* (Tzvelev) Tzvelev

Paracolpodium (Tzvelev) Tzvelev, Bot. Žurn. (Moscow & Leningrad) 50: 1320 (1965).

Rhizomatous perennials. Leaves with sheath margins fused ¼ to almost their entire length. Spikelets with one flower, rarely with a reduced flower distally. Glumes almost as long as the flower. Callus of flower glabrous or pubescent. Lemma apex acute, the veins not reaching the apex. Fruit with a small beak. $2n = 28, 42$.

Four spp., Caucasus to Himalayas.

Kim et al. (2009) show that *Paracolpodium* is phylogenetically distinct from *Colpodium*, and may be sister to *Catabrosella*.

288. *Phippsia* (Trin.) R. Br.

Phippsia (Trin.) R. Br., Chlor. Melvill.: 27 (1823); tax.: Aares et al. (2000).

Tiny perennials, caespitose or mat-forming. Sheaths of culm leaves closed at least ½ their length. Spikelets with one flower; rachilla extension absent. Glumes less than 1/3 the length of the flower, *without veins*, their apices obtuse to acute; *lower glume sometimes absent*. Callus of flower glabrous. Lemma translucent to membranous, the apex truncate to acute. *Stamens 1 or 2. Fruit extending beyond the flower at maturity*.

Three spp., Arctic, Argentine Andes. $2n = 28$.

Phippsia will hybridize with *Puccinellia* (Steen et al. 2004). The species of *Phippsia* appear to be allopolyploids but the parents are unknown (Aares et al. 2000).

Fig. 55. Pooideae-Poeae-Coleanthinae. *Coleanthus subtilis*. **A** Habit. **B** Single flowering shoot. **C** Spikelet with protruding caryopsis. **D** Lemma. **E** Palea. **F** Stamens. **G** Pistil. (From Wu et al. 2007, with permission from the Missouri Botanical Garden Press, St. Louis, and Science Press, Beijing)

289. *Pseudosclerochloa* Tzvelev

Pseudosclerochloa Tzvelev, Bot. Žurn. (Moscow & Leningrad) 89: 840 (2004).

Annuals. Leaves with sheaths open. Spikelets with 2 to 7 flowers, the distalmost ones reduced. Glumes shorter than the spikelet, their apices obtuse to acute. Callus of flower pubescent. Lemma distally keeled, the apex obtuse.

Two spp., western Europe, China.

These species have been segregated from *Puccinellia* on the basis of their keeled lemmas.

290. ***Puccinellia*** Parl.

Puccinellia Parl., Fl. Ital. 1: 366 (1848), nom. conserv.; tax.: Consaul et al. (2008a, 2008b).

Annuals or perennials, caespitose, occasionally stoloniferous. Leaves with sheaths closed 1/4 to 1/3 of their length. Spikelets with 2 to 10 flowers. Glumes shorter than the spikelet, their apices notched or truncate to acute or acuminate. Callus of flower glabrous or pubescent. Lemma generally with 5 veins, *the lateral veins obscure,* lemma apex entire or ragged, notched or truncate to acute, unawned. $2n = 14, 21, 28, 42, 56, 77$.

One hundred eleven spp., temperate regions in salty soil.

Puccinella has lemmas rounded on the back, which distinguished it from most species of *Poa* in which the lemmas are generally keeled. It also has obtuse lemma apices and rounded hila (the latter synapomorphic for Poeae) that distinguish if from *Festuca,* which has generally acute or acuminate lemma apices and elongate hila. *Puccinella* has a 3-veined upper glume, which distinguishes it from *Glyceria* in which the upper glume has 1 vein. Morphological similarity and documented hybridization between *Puccinellia* and *Phippsia* have led to the suggestion that the two be combined (Steen et al. 2004); Löve and Löve (1976) provide some but not all of the necessary combinations. *Puccinellia* may be sister to *Sclerochloa* (Hoffmann et al. 2013).

291. ***Sclerochloa*** P. Beauv.

Sclerochloa P. Beauv., Ess. Agrostogr.: 97, 177 (1812).

Annuals, tufted. Leaves with sheaths closed ¼ to ½ of their length. *Inflorescence unbranched and one-sided,* sometimes the branches branched. Spikelets with 2 to 7 flowers, *proximal rachilla internode noticeably thicker than the distal ones.* Glumes shorter than the flowers, with hyaline margins, their apices obtuse to emarginate. Callus of flower glabrous. Lemma hardened, with 7 to 9 prominent veins, the apex obtuse. *Base of style branches persistent in fruit, forming a beak.* $2n = 14$.

Two spp., southern Europe and mid-East, weedy elsewhere.

Beaked fruits also appear in *Duthiea,* but have been acquired independently. *S. dura* (L.) P. Beauv. is widely introduced and weedy.

COLPODIUM + ZINGERIA

Chromosome base number $x = 2$.

Colpodium and *Zingeria* are sisters in molecular phylogenetic studies (Gillespie et al. 2008; Kim et al. 2009; Schneider et al. 2009). These two genera were once placed in Miliinae (Soreng et al. 2007), but more recent data clearly place them here in Coleanthinae (Hoffmann et al. 2013; Schneider et al. 2009, 2012; Soreng et al. 2012 and onward).

292. ***Colpodium*** Trin.

Colpodium Trin., Fund. Agrost.: 119, pl. 7 (1820); tax.: Soreng and Fish (2011).

Perennials, <30 cm tall, caespitose, rhizomatous or stoloniferous. Spikelets with 1 to 4 flowers, with or without additional reduced flowers distally. Glumes shorter or longer than the flowers, their apices obtuse to acute. Callus of flower glabrous or with cobweb-like hairs. Lemma glabrous or pubescent, the apex truncate to acute, awnless (mucronate in one species). $2n = 4, 8, 20$.

Thirteen spp., central Asia to eastern Siberia, south to mountainous areas of Africa.

293. ***Zingeria*** P. A. Smirn.

Zingeria P. A. Smirn., Bjull. Moskovsk. Obač. Isp. Prir., Otd. Biol. 51(2): 67 (1946).

Delicate annuals. *Spikelet dorsiventrally compressed,* with one flower, without a rachilla extension. Glumes shorter or longer than the spikelets, their apices obtuse to acute, upper glume apex with a sharp point. Lemma apex obtuse, awnless. $2n = 4, 8, 12$.

Five spp., the Caucasus, Anatolia, northern Iraq and Iran.

The hexaploid species *Z. kochii* (Mez) Tzvel. is an allopolyploid with one genome donated by *Colpodium versicolor* (Kotseruba et al. 2010).

XVI.15. Subtribe Poinae Dumort. (1829)

Miliinae Dumort. (1829).
Alopecurinae Dumort. (1829).
Cinninae Caruel (1892).

Annuals or perennials. Leaf sheaths of upper culm leaves often closed up to ¼ of their length. Flower callus generally pubescent. Lemma generally keeled, the apex with or without an awn, the awn generally terminal. Palea generally membranous, green. Ovary glabrous.

Twenty-six genera, 636 spp.

This subtribe corresponds to the PAM clade of Gillespie et al. (2008). The entire clade is well supported (Gillespie et al. 2008; Hoffmann et al. 2013), but recognition of the subtribes Alopecurinae, Cinninae, and Miliinae renders Poinae paraphyletic so they are placed in synonymy here. Named hybrids include x*Dupoa* J. Cay. & Darbysh.

294. ***Agrostopoa*** Davidse, Soreng & P. M. Peterson

Agrostopoa Davidse, Soreng & P. M. Peterson, Novon 19: 33 (2009).

Annuals or perennials, caespitose. Spikelets with one flower, without a rachilla extension. Glumes about the same length as the flower, their apices acute to acuminate. Callus of flower glabrous. Lemma apex entire or with two teeth, mucronate or awned from the sinus, the awn straight or twisted.

Three spp., Colombia.

Davidse et al. (2008) observe that members of this genus are somewhat similar to Old World species of *Colpodium*.

295. ***Alopecurus*** L.

Alopecurus L., Sp. Pl. 1: 60–61 (1753).

Annuals or perennials, caespitose or rhizomatous. Inflorescence densely contracted, cylindrical. Spikelets with one flower; rachilla extension lacking; *disarticulation below the glumes.* Glumes as long as or longer than the flower, *strongly compressed and keeled*, often partially fused near their bases, their apices truncate to acute, muticous to awned. Callus of flower glabrous. Lemma keeled, the margins often partially fused, the apex truncate to acute, awned from the abaxial side, the awn straight, curved or geniculate. *Lodicules absent. Stamens 2* or 3. Style 1. $2n = 14, 26, 28, 30, 42, 98, 100, 105, 112, 117, 119, 120$.

Fifty-two spp., north temperate regions and South America.

296. ***Aniselytron*** Merr.

Aniselytron Merr., Philipp. J. Sci. 5: 328 (1910); anatomy: Ma et al. (2005).

Perennials, the culms solitary or tufted, 0.5–1 m tall. Leaf sheaths open. Spikelets with one flower, with a rachilla extension. Glumes shorter than the spikelet, the margins hyaline, lower glume apex truncate to acute, upper glume apex acute to acuminate. Lemma leathery, the apex entire or with two teeth, acute to acuminate, awnless. $2n = 42$.

Two spp., Asia.

Aniselytron – or at least *A. treutleri* (Kuntze) Soják – is likely to be of hybrid origin. One parent comes from the Poinae (minus *Poa*) clade, whereas the other is related to *Poa* sect. *Sylvestres* (Gillespie et al. 2008, 2010).

297. ***Apera*** Adans.

Apera Adans., Fam. Pl. 2: 495 (1763).

Annuals. Spikelet with one flower, with a rachilla extension. Glumes slightly shorter or longer than the flowers, their apices acute to acuminate. Callus of flower glabrous or with sparse hairs. Lemma apex acute, awned, the awn originating just below the lemma apex, generally several times as long as the lemma. $2n = 14, 28$.

Five spp., Europe through the Caucasus to Afghanistan.

Apera may be closely related to *Nephelochloa* and *Bellardiochloa*, as suggested by DNA sequence data (Hoffmann et al. 2013).

298. ***Arctopoa*** (Griseb.) Prob.

Arctopoa (Griseb.) Prob., Novosti Sist. Vyssh. Rast. 11: 49 (1974).

Rhizomatous perennials. Spikelets with 2 to 6 (8) flowers, the rachilla pubescent. Glumes acute to acuminate. Callus of flower glabrous or with a ring of short hairs. Lemma keeled, the apex acute. $2n = 28, 29\pm, 42, 62$.

Five spp., North America and Asia.

Some authors recognize this as *Poa* subgenus *Arctopoa*, but it is most likely to be intergeneric hybrid. The plastid is similar to *Poa* sect. *Sylvestres*, but the nuclear gene (ITS) sequences are similar to *Cinna* (Gillespie et al. 2008, 2010). This is the opposite of what is found for *Aniselytron*. The sterile hybrid between *A. eminens* and *Dupontia fisheri* is known as x*Dupoa* J. Cay. & Darbysh.

299. ***Beckmannia*** Host

Beckmannia Host, Icon. Descr. Gram. Austriac.: 3: 5 (1805).

Tufted annuals or rhizomatous perennials. *Primary inflorescence branches unbranched, borne on an elongated central axis, the spikelets borne on one side, imbricate. Spikelets orbicular or circular and slightly flattened*, with 1 or 2 flowers; rachilla extension absent; *disarticulation below the glumes. Glumes gibbous*, shorter than or as long as the flowers, their apices obtuse to acute. Callus of flower glabrous. Lemma apex muticous. $2n = 14$.

Two spp., north temperate regions.

Beckmannia is sister to *Pholiurus* in the study of Hoffmann et al. (2013), but the two genera are quite different morphologically.

300. ***Bellardiochloa*** Chiov.

Bellardiochloa Chiov., Stud. Veg. Piemonte: 60 (1929).

Densely tufted perennials. Leaf blades 1 mm wide; bristle-like. Spikelets with 2 to 8 flowers, the rachilla pubescent. Glumes shorter than or about as long as the flowers, their apices acuminate. Callus of flower pubescent. Lemma with the apex entire or with two minute teeth, acute, unawned or with a short mucro from the sinus. $2n = 14, 28$.

Four spp., southern Europe and Middle East.

In molecular analyses, *Bellardiochloa* is sister to *Nephelochloa* (Schneider et al. 2012), or to *Nephelochloa* plus *Apera* (Hoffmann et al. 2013).

301. ***Cinna*** L.

Cinna L., Sp. Pl. 1: 5 (1753); tax.: Brandenburg et al. (1991a).

Perennials, caespitose or rhizomatous. Spikelets with one flower, *the rachilla internode notably elongate below the flower; rachilla extension present; disarticulation below the glumes.* Glumes as long as the flowers or longer, their apices acute. Callus of flower glabrous. Lemma apex acute, muticous to awned, the awn subapical. *Stamens 1 or 2. Endosperm liquid.* Fruit often with a beak. $2n = 28$.

Four spp., north temperate regions, Mexico, Central America and northern South America.

302. ***Cornucopiae*** L.

Cornucopiae L., Sp. Pl. 1: 54 (1753).

Annuals, *with branched culms. Upper leaf sheaths surrounding 1 or more head-like inflorescences; stem below the inflorescence expanded into a broad cup.* Spikelets with one flower, without a rachilla extension. Glumes as long as or longer than the flowers, keeled, partially fused along their margins, the apices truncate to acute, muticous to mucronate. Lemma keeled, the margins fused, the apex truncate, awned from the abaxial side, the awn straight or twisted. *Palea minute or lacking. Lodicules absent.*

Two spp., eastern Mediterranean region.

The branched culms are unusual among pooid grasses.

303. ***Dupontia*** R. Br.

Dupontia R. Br., Chlor. Melvill.: 32–33 (1823).
Arctophila (Rupr.) Rupr. ex Andersson, Pl. Scand. Gram. x: 48 (1852).

Rhizomatous perennials. Leaf sheaths closed for more than ½ their length. Spikelets with 2 to 7 flowers, with a rachilla extension or the rachilla ending in a vestigial flower. Glumes shorter than or as long as the flower, their apices acute to obtuse. Callus of flower glabrous or pubescent. Lemma keeled, apex obtuse to acute or acuminate, awnless or with a short awn tip. $2n = 42$, 44, 63, 66, 84, 88.

Two spp. and a named hybrid, Arctic.

Specimens of the putative hybrid between *Arctophila fulva* and *Dupontia fisheri* (x*Arctodupontia scleroclada* (Rupr.) Tzvelev) are misidentified material of *Arctophila fulva*, except for the type, whose identity is uncertain (Brysting et al. 2003).

Dupontia fisheri and *Arctophila fulva* are strongly supported as sisters in the molecular phylogenies of Gillespie et al. (2010), and ITS sequences from the two are intermixed in the work of Hoffmann et al. (2013). Morphologically they differ principally in the relative length of the glumes. This close relationship, their morphological similarity and their ability to hybridize supports their combination.

304. ***Limnas*** Trin.

Limnas Trin., Fund. Agrost. 116, t. 6 (1820).

Perennials with short rhizomes. Inflorescence densely contracted. Spikelets with one flower, without a rachilla extension, *disarticulating below the glumes.* Glumes as long as the flower, their apices acute. Lemma apex acute, awned from the abaxial side, the awn twisted. *Stamens 2.*

Three spp., Siberia.

305. ***Limnodea*** L.H. Dewey

Limnodea L.H. Dewey, Contr. U.S. Natl. Herb. 2: 518 (1894).

Tufted annuals. *Internodes solid.* Spikelets with one flower, with a rachilla extension, *disarticulating below the glumes.* Glumes as long as the flower, coriaceous, their apices acute. Callus of flower glabrous. Lemma apex entire or with two teeth, awned from the abaxial side, the awn twisted and geniculate. $2n = 14$.

One sp., *L. arkansana* (Nutt.) L. H. Dewey, southern U.S. and Mexico.

Limnodea was included in *Cinna* by Tucker (1996), whereas Brandenburg and Thieret (2000) keep them separate. The difference in length of the rachilla internode below the lemma (long in *Cinna*, short in *Limnodea*), and the beaked fruit in *Cinna* differentiate the two genera. No molecular data are available for *Limnodea.*

306. ***Milium*** L.

Milium L., Sp. Pl. 1: 61 (1753).

Annuals or perennials, caespitose or rhizomatous. *Spikelets dorsiventrally compressed*, with one flower, without a rachilla extension. Glumes as long as or longer than the flowers, their apices obtuse to acute. Callus of flower glabrous. *Lemma and palea hardened, smooth, shiny*, the lemma apex obtuse. $2n = 14$, 18, 28.

Five spp., north temperate Eurasia and eastern North America.

The flowers of *Milium* are superficially similar to those in the Panicoideae, but there is only one flower per spikelet.

307. ***Nephelochloa*** Boiss.

Nephelochloa Boiss., Diagn. Pl. Or. 1: 72 (1844).

Annuals, less than 25 cm tall. Inflorescence branches appearing to be in whorls, *the proximal one or two sets of branches generally sterile.* Spikelets with 3 to 6 flowers, rachilla internode with a ring of hairs. Glumes shorter than the flowers, their apices acute. Callus of flower pubescent. Lemma apex with two lobes, awned from the sinus. *Stamens 2.*

One sp., *N. orientalis* Boiss., Turkey.

Mill (1985) describes the lodicules of this species as being fused and thickened at the base, an unusual character for the tribe. *Nephelochloa* is sister to *Bellardiochloa* in molecular analyses (Schneider et al. 2012).

308. ***Phleum*** L.

Phleum L., Sp. Pl. 1: 59–60 (1753).
Pseudophleum Dogan, Notes Roy. Bot. Gard. Edinburgh 40: 75 (1982).

Annuals or perennials, caespitose or rhizomatous. Inflorescence dense, cylindrical, spike-like. Spikelets with one flower; rachilla extension

Fig. 56. Pooideae-Poeae-Poinae. *Poa lilloi*. **A** Habit, with perfect flowers. **B** Habit, with pistillate flowers. **C, D** Ligule. **E** Spikelet. **F** Pistillate flower with large stamens. **G** Basal perfect flower. **H** Perfect flower with 3 stamens and lodicules. **I** Detail of the lodicules. (From Zuloaga et al. 2012)

present or absent. Glumes shorter or more commonly longer than the flower, strongly compressed, keeled, *keels extending into stiff awns*, glume apices truncate. Callus of flower glabrous. Lemma white, the apex acute, muticous or with a short awn. $2n = 14, 28, 42$.

Sixteen spp., north temperate regions and South America.

P. pratense L. is widely cultivated as a pasture grass. *Pseudophleum* was segregated from *Phleum* because the glumes are shorter than the flower, but there is no other reason to separate it.

309. ***Pholiurus*** Host ex Trin.

Pholiurus Host ex Trin., Fund. Agrost.: 131 (1820).

Annuals or perennials. Inflorescence unbranched, the rachis not disarticulating. Spikelets on opposite sides of the rachis, with 2 flowers; rachilla extension absent. Glumes borne side by side on the abaxial side of the spikelet, longer than the flower, coriaceous, their apices obtuse. Lemma apex acute, unawned. $2n = 14$.

One sp., *P. pannonicus* (Host) Trin., eastern Europe and central Asia.

Pholiurus has been placed here, rather than in Parapholiinae, based on the work of Schneider et al. (2012) and Hoffmann et al. (2013).

310. ***Poa*** L. Fig. 56

Poa L., Sp. Pl. 1: 67–70 (1753); tax.: Soreng et al. (2010); phylog.: Gillespie et al. (2008), Birch et al. (2014).
Anthochloa Nees & Meyen, Reise Erde: 2: 14 (1834).
Dissanthelium Trin., Linnaea 10: 305 (1836).
Aphanelytrum (Hack.) Hack., Oesterr. Bot. Z. 52: 12 (1902).
Libyella Pamp., Bull. Soc. Bot. Ital. 1925: 151 (1925).
Eremopoa Roshev., Fl. URSS 2: 429, 756 (1934).
Lindbergella Bor, Svensk. Bot. Tidskr. 63: 368 (1969).
Austrofestuca (Tzvelev) E.B. Alexeev, Bjull. Moskovsk. Obač. Isp. Prir., Otd. Biol. 81: 55 (1976).
Parodiochloa C. E. Hubb., Bull. Brit. Mus. (Nat. Hist.), Bot. 8: 395 (1981).
Tovarochloa T. D. Macfarl. & But, Brittonia 34: 478 (1982).
Neuropoa Clayton, Kew Bull. 40: 728 (1985).
Tzvelevia E. B. Alexeev, Bjull. Moskovsk. Obač. Isp. Prir., Otd. Biol. 90: 103 (1985).
Oreopoa H. Scholz & Parolly, Willdenowia 34: 146, f. 1 (2004).

Perennials or annuals, caespitose, rhizomatous, or stoloniferous; plants bisexual, monoecious, dioecious, or gynodioecious. Spikelets with (1) 2 to 6 (13) flowers; rachilla extension present or absent. Glumes generally shorter than the flowers, rarely longer, their apices obtuse to acuminate, sometimes muticous, rarely mucronate. Callus of flower glabrous, or with long cobweb-like hairs, or with a tuft of hairs. Lemma with a strong keel in many species, rounded in others, rarely flabellate, the apex obtuse to acuminate, generally awnless. $2n = 14, 28, 42, 56, 70$, as well as virtually all aneuploid numbers up to and including 147.

500 spp., worldwide in cool temperate regions.

In its current circumscription, *Poa* is the largest genus of grasses. Five subgenera are recognized by Soreng et al. (2010): *Ochlopoa*, *Poa*, *Pseudopoa*, *Stenopoa*, and *Sylvestres*. Of these, subgenus *Poa* is by far the largest and is divided into two supersections, *Homalopoa* (called HAMBADD by Gillespie et al. 2007) and *Poa* (called POM by Gillespie et al. 2007).

The work of Gillespie et al. (2007, 2008, 2009, 2010) and Soreng et al. (2010, 2011) has gone a long way to resolving phylogenetic issues within *Poa* and its many segregate genera. The synonymy presented here largely follows that of Gillespie et al. (2008). *Libyella* and *Lindbergella* are placed in synonymy based on morphology (R. Soreng and L. Gillespie, pers. comm.) and molecular data (Hoffmann et al. 2013), although the DNA sequences are quite short and other loci may need to be explored. *Anthochloa* was segregated from *Poa* on the basis of its flabellate lemma, and *Aphanelytrum* because of its awl-shaped lemma. *Oreopoa* was segregated based on its convolute leaf blades and somewhat rounded lemmas, but these characters also appear in some species of *Poa*, and the genus was placed in synonymy with *Poa* by Soreng et al. (2012 and onward); the molecular data on which the placement is based are unpublished. *Austrofestuca* was segregated based on having five or more veins on the lemma, but it clearly belongs in *Poa* (Jacobs et al. 2008). In addition, Gillespie et al. (2008) and Refulio-Rodriguez et al. (2012) provide convincing evidence for placing *Dissanthelium* and *Tovarochloa* in *Poa*. The former genus had been segregated (and even placed in a different tribe) because its glumes are longer than the flowers, while the latter was recognized on the basis of its tiny stature and capitate inflorescence. However, both are clearly derived from ancestors that must have looked like conventional *Poa*. It is not clear why *Eremopoa* or *Neuropoa* were segregated. *Libyella* was segregated based on its unbranched inflorescence and lack of lodicules.

Species of *Poa* are often polyploid, and hybridization and apomixis are common (e.g., Huff 2010; Rodionov et al. 2010; Soreng et al. 2010). The morphological characters that separate the species are subtle and generally quantitative. Despite the enormous strides made in disentangling the phylogeny of this genus, much work remains to be done, particularly on the evolution of the many polyploid complexes. The genus is the result of rapid and recent diversification, with much of the speciation occurring in the last 2 million years (Hoffmann et al. 2013).

311. ***Rhizocephalus*** Boiss.

Rhizocephalus Boiss., Diagn. Pl. Orient. 1: 68 (1844).

Tiny annuals, less than 2 cm tall. Inflorescence capitate, *surrounded by a leaf sheath*. Spikelets with one flower, without a rachilla extension. Glumes shorter than the flower, *with fused margins*, their apices muticous. Lemma with 5 veins, pubescent, *with club-shaped hairs*, the apex attenuate, with a mucro. *Lodicules absent. Stamens 2. Caryopsis with an apical beak.* $2n = 14$.

One sp., *R. orientalis* Boiss., eastern Mediterranean region and Iran.

312. ***Simplicia*** Kirk

Simplicia Kirk, Trans. Proc. New Zealand Inst. 29: 497 (1897); tax.: Smissen et al. (2011).

Decumbent, mat-forming perennials. Spikelets with 1 or 2 flowers. Glumes shorter than the flowers, the distal one less than ½ the length of the flower; glume apices acute. Lemma apex acute, with or without a mucro. Palea with one vein. Stamens 2 or 3. $2n = 28$.

Two spp., New Zealand.

Simplicia is morphologically similar to "*Calamagrostis*", but has a 3-veined lemma, and is similar to *Poa* but has a 1-keeled palea.

313. ***Ventenata*** Koeler

Ventenata Koeler, Descr. Gram.: 272 (1802), nom. conserv.
Gaudiniopsis (Boiss.) Eig., Repert. Spec. Nov. Regni Veg. 26: 74 (1929), nom. invalid.

Tufted annuals. Spikelets with 2 to 10 flowers, with reduced flowers distally or with a rachilla extension; *the most proximal flower remaining on the plant after the other flowers have been shed, eventually falling with the glumes.* Glumes shorter than or as long as the flowers, their apices acute to acuminate. Callus of flower pubescent.

Proximal lemmas unawned or with a straight terminal awn; distal lemmas with a twisted abaxial awn; in some species, additional awns developed on the lemma lobes. $2n = 14$.

Thirteen spp., Eurasia.

Gillespie et al. (2008) place *Gaudiniopsis* in Poinae, but keep it distinct from *Ventenata*, citing data from Döring et al. (2007). However, the latter authors do not mention *Gaudiniopsis*, and no molecular data are available for it. The five species of *Gaudiniopsis* are synonymized here with *Ventenata* following (Clayton and Renvoize (1986), pending further information.

HSAQN clade

The following four genera form a strongly supported clade, called HSAQN, in the study of Gillespie et al. (2009). The clade is sister to *Cinna*, and HSAQN plus *Cinna* is sister to *Aniselytron*. The HSAQN species share no obvious morphological characters, although they are generally similar to *Poa* from which they have all been segregated. All except for *Arctagrostis* are native to the Southern Hemisphere.

314. ***Arctagrostis*** Griseb.

Arctagrostis Griseb., Fl. Ross. 4: 434 (1852).

Rhizomatous perennials. Spikelets with one flower; rachilla extension present or the rachilla ending in a vestigial flower. Glumes shorter than the flower, their apices acute to obtuse. Callus of flower glabrous. Lemma translucent distally, keeled, the apex obtuse to acute, awnless. Palea with one vein, folded. Hilum less than 1/3 the length of the fruit. $2n = 28, 42, 56$.

Two spp., Arctic.

315. ***Hookerochloa*** E. B. Alexeev

Hookerochloa E. B. Alexeev, Bjull. Moskovsk. Obač. Isp. Prir., Otd. Biol. 90: 106 (1985).
Festucella E. B. Alexeev, Byull. Moskovsk. Obshch. Isp. Prir. Otd. Biol. 90: 104 (1985).

Caespitose perennials. Spikelets with 3 to 5 flowers. Glumes shorter than the flowers, their apices acute, *margins ciliolate. Rachilla internodes densely pubescent.* Callus of flower pubescent. Lemma apex entire or with 2 teeth, acute or truncate, awned, the awn subapical or from the sinus. Palea tightly clasped by the lemma.

Two spp., Australia, New Zealand.

Festucella was synonymized with *Hookerochloa* by Jacobs et al. (2008).

316. ***Nicoraepoa*** Soreng & L. J. Gillespie

Nicoraepoa Soreng & L. J. Gillespie, Ann. Missouri Bot. Gard. 94: 842 (2007).

Dioecious, gynodioecious or hermaphrodite perennials, with stout rhizomes. *Ligule pubescent abaxially. Leaf blade with sharp-pointed apex;* adaxial surface with prominent ridges. Spikelets with 2 to 5 flowers. Callus of flower glabrous or with short hairs. Lemmas acute, mucronate or with a short awn from a shallow sinus.

Six spp., Argentina, Chile, Islas Malvinas.

317. ***Saxipoa*** Soreng, L. J. Gillespie & S. W. L. Jacobs

Saxipoa Soreng, L. J. Gillespie & S. W. L. Jacobs, Austral. Syst. Bot. 22: 406–407, f. 1C, 3 (2009).

Caespitose perennials. Spikelets with 2 to 4 flowers. Glumes shorter than the flowers, margins scarious, their apices acute to obtuse. Callus of flower with a few straight hairs. *Rachilla densely pubescent.* Lemma margin ciliate, the apex acute, sometimes with a tiny subapical mucro.

One sp., *S. saxicola* (R. Br.) Soreng, L. J. Gillespie & S. W. L. Jacobs, Australia.

Saxipoa is similar to *Hookerochloa* in having leaf sheaths open almost their entire length, and the rachilla densely pubescent; the two also share minor characters of the lodicules and lemma margins. Their habitat is quite different however, and the two are not sisters in the analyses of Gillespie et al. (2009).

318. ***Sylvipoa*** Soreng, L. J. Gillespie & S. W. L. Jacobs

Sylvipoa Soreng, L. J. Gillespie & S. W. L. Jacobs, Austral. Syst. Bot. 22: 404, f. 1D (2009).

Caespitose perennials. *Leaves with sheaths open to the base, blades with obvious cross veins.* Spikelets with 2 to 4 (7) flowers. Glumes shorter than the flowers, with scarious margins, their apices acute. Callus of flower with minute hairs. Lemma strongly keeled, the apex obtuse to acute, unawned.

One sp., *S. queenslandica* (C. E. Hubb.) Soreng, L. J. Gillespie & S. W. L. Jacobs, Queensland and New South Wales.

References

Aares, E., Nurminiemi, M., Brochmann, C. 2000. Incongruent phylogeographics in spite of similar morphology, ecology, and distribution: *Phippsia algida* and *P. concinna* (Poaceae) in the North Atlantic region. Plant Syst. Evol. 220: 241–261.

Aitchison, J.E.T. 1882. On the flora of the Kuram Valley, etc., Afghanistan - Part II. Bot. J. Linn. Soc. 19: 139–200.

Allred, K.W., Barkworth, M.E. 2007. *Anthoxanthum* L. In: Barkworth, M.E., Capels, K.M., Long, S., Anderton, L. K., Piep, M.B. (eds.) Flora of North America north of Mexico. Vol. 24. Magnoliophyta: Commelinidae (in part): Poaceae, part 1. New York: Oxford University Press. pp. 758–764

Baden, C. 1991. A taxonomic revision of *Psathyrostachys* (Poaceae). Nordic J. Bot. 11: 3–26.

Baeza, C.M., Stuessy, T.F., Marticorena, C. 2002. Notes on the Poaceae of the Robinson Crusoe (Juan Fernández) Islands, Chile. Brittonia 54: 154–163.

Baldini, R.M. 1995. Revision of the genus *Phalaris* L. (Gramineae). Webbia 49: 265–329.

Barkworth, M.E. 1990. *Nassella* (Gramineae, Stipeae): revised interpretation and nomenclatural changes. Taxon 39: 597–615.

Barkworth, M.E. 2007a. *Deschampsia* P. Beauv. In: Barkworth, M.E., Capels, K.M., Long, S., Anderton, L.K., Piep, M.B. (eds.) Flora of North America north of Mexico. Vol. 24. Magnoliophyta: Commelinidae (in part): Poaceae, part 1. New York: Oxford University Press. pp. 624–633

Barkworth, M.E. 2007b. *Phalaris* L. In: Barkworth, M.E., Capels, K.M., Long, S., Anderton, L.K., Piep, M.B. (eds.) Flora of North America north of Mexico. Vol. 24. Magnoliophyta: Commelinidae (in part): Poaceae, part 1. New York: Oxford University Press. pp. 764–773

Barkworth, M.E. 2007c. Triticeae. In: Barkworth, M.E., Capels, K.M., Long, S., Anderton, L.K., Piep, M.B. (eds.) Flora of North America north of Mexico. Vol. 24. Magnoliophyta: Commelinidae (in part): Poaceae, part 1. New York: Oxford University Press. pp. 238–240

Barkworth, M.E., Dewey, D.R. 1985. Genomically based genera in the perennial Triticeae of North America: identification and membership. Amer. J. Bot. 72: 767–776.

Barkworth, M.E., Capels, K.M., Long, S., Anderton, L.K., Piep, M.B. (eds.) 2007. Flora of North America north of Mexico. Vol. 24. Magnoliophyta: Commelinidae (in part): Poaceae, part 1. New York: Oxford University Press.

Barkworth, M.E., Arriaga, M.O., Smith, J.F., Jacobs, S.W. L., Valdés-Reyna, J., Bushman, B.S. 2008. Molecules and morphology in South American Stipeae (Poaceae). Syst. Bot. 33: 719–731.

Baum, B.R. 1977. Oats: wild and cultivated; a monograph of the genus *Avena* L. (Poaceae). Biosyst. Res. Inst. Monograph 14: 1–463.

Bennetzen, J.L., Kellogg, E.A. 1997. Do plants have a one-way ticket to genomic obesity? Plant Cell 9: 1509–1514.

Beuzenberg, E.J., Hair, J.B. 1983. Contributions to a chromosome atlas of the New Zealand flora - 25 miscellaneous species. N. Z. J. Bot. 21: 13–20.

Birch, J.L., Cantrill, D.J., Walsh, N.G., Murphy, D.J. 2014. Phylogenetic investigation and divergence dating of *Poa* (Poaceae, tribe Poeae) in the Australasian region. Bot. J. Linn. Soc. 175: 523–552.

Blattner, F.R. 2006. Multiple intercontinental dispersals shaped the distribution area of *Hordeum* (Poaceae). New Phytol. 169: 603–614.

Blattner, F.R. 2009. Progress in phylogenetic analysis and a new infrageneric classification of the barley genus *Hordeum* (Poaceae: Triticeae). Breeding Sci. 59: 471–480.

Bödvarsdottir, S.K., Anamthawat-Jónsson, K. 2003. Isolation, characterization, and analysis of *Leymus*-specific DNA sequences. Genome 46: 673–682.

Boulos, L. 2005. Flora of Egypt, vol. 4. Cairo, Egypt: Al Hadara Publishing.

Brandenburg, D.M., Thieret, J.W. 2000. *Cinna* and *Limnodea* (Poaceae): not congeneric. Sida 19: 195–200.

Brandenburg, D.M., Blackwell, W.H., Thieret, J.W. 1991a. Revision of the genus *Cinna* (Poaceae). Sida 14: 581–596.

Brandenburg, D.M., Estes, J.R., Collins, S.L. 1991b. A revision of *Diarrhena* (Poaceae) in the United States. Bull. Torrey Bot. Club 118: 128–136.

Brysting, A.K., Aiken, S.G., Lefkovitch, L.P., Boles, R.L. 2003. *Dupontia* (Poaceae) in North America. Can. J. Bot. 81: 769–779.

Bustam, B.M. 2010. Systematic studies of Australian stipoid grasses (*Austrostipa*) based on micro-morphological and molecular characteristics. Biodiversitas 11: 9–14.

Campbell, C.S., Garwood, P.E., Specht, L.P. 1986. Bambusoid affinities of the north temperate genus *Brachyelytrum* (Gramineae). Bull. Torrey Bot. Club 113: 135–141.

Catalán, P., Olmstead, R.G. 2000. Phylogenetic reconstruction of the genus *Brachypodium* P. Beauv. (Poaceae) from combined sequences of chloroplast *ndhF* gene and nuclear ITS. Plant Syst. Evol. 220: 1–19.

Catalán, P., Torrecilla, P., López Rodríguez, J.A., Olmstead, R.G. 2004. Phylogeny of the festucoid grasses of subtribe Loliinae and allies (Poeae, Pooideae) inferred from ITS and *trn*L-F sequences. Mol. Phyl. Evol. 31: 517–541.

Catalán, P., Müller, J., Hasterok, R., Jenkins, G., Mur, L.A. J., Langdon, T., Betekhtin, A., Siwinska, D., Pimentel, M., López-Alvarez, D. 2012. Evolution and taxonomic split of the model grass *Brachypodium distachyon*. Ann. Bot. 109: 385–405.

Charmet, G., Balfourier, F., Chatard, V. 1996. Taxonomic relationships and interspecific hybridization in the genus *Lolium* (grasses). Genet. Res. Crop Evol. 43: 319–327.

Chen, S.-L., Zhu, G.-H. 2006. *Agropyron*. In: Wu, Z.-Y., Raven, P.H., Hong, D.-Y. (eds.) Flora of China, vol. 22 (Poaceae). Beijing: Science Press. pp. 439–441.

Chiapella, J. 2007. A molecular phylogenetic study of *Deschampsia* (Poaceae: Aveneae) inferred from nuclear ITS and plastid *trn*L sequence data: support for the recognition of *Avenella* and *Vahlodea*. Taxon 56: 55–64.

Chiapella, J., Zuloaga, F.O. 2010. A revision of *Deschampsia*, *Avenella*, and *Vahlodea* (Poaceae, Poeae, Airinae) in South America. Ann. Missouri Bot. Gard. 97: 141–162.

Church, G.L. 1952. The genus *Torreyochloa*. Rhodora 54: 197–200.

Cialdella, A.M., Giussani, L.M., Aagesen, L., Zuloaga, F.O., Morrone, O. 2007. A phylogeny of *Piptochaetium* (Poaceae: Pooideae: Stipeae) and related genera based on a combined analysis including *trnL-F*, *rpl16*, and morphology. Syst. Bot. 32: 545–559.

Cialdella, A.M., Salariato, D.L., Aagesen, L., Giussani, L. M., Zuloaga, F.O., Morrone, O. 2010. Phylogeny of New World Stipeae (Poaceae): an evaluation of the monophyly of *Aciachne* and *Amelichloa*. Cladistics 26: 563–578.

Cialdella, A.M., Sede, S.M., Romaschenko, K., Peterson, P. M., Soreng, R.J., Zuloaga, F.O., Morrone, O. 2014. Phylogeny of *Nassella* (Stipeae, Pooideae, Poaceae). Syst. Bot. 39: 814–828.

Clayton, W.D., Renvoize, S.A. 1986. Genera graminum: grasses of the world. London: Her Majesty's Stationery Office.

Clayton, W.D., Harman, K.T., Williamson, H. 2006 onward. GrassBase - the online world grass flora. http://www.kew.org/data/grasses-db.html

Consaul, L.L., Gillespie, L.J., Waterway, M.J. 2008a. Systematics of North American arctic diploid *Puccinellia* (Poaceae): morphology, DNA content, and AFLP markers. Syst. Bot. 33: 251–261.

Consaul, L.L., Gillespie, L.J., Waterway, M.J. 2008b. Systematics of three North American polyploid arctic alkali grasses (*Puccinellia*, Poaceae) morphology, ploidy, and AFLP markers. Botany 86: 916–937.

Cope, T., Gray, A. 2009. Grasses of the British Isles. London: Botanical Society of the British Isles.

Cotton, R., Stace, C.A. 1967. Taxonomy of the genus *Vulpia* (Gramineae): I. Chromosome numbers and geographical distribution of the Old World species. Genetica 46: 235–255.

Darbyshire, S.J. 1993. Realignment of *Festuca* subgenus *Schedonorus* with the genus *Lolium* (Poaceae). Novon 3: 239–243.

Darbyshire, S.J. 2007. *Schedonorus* P. Beauv. In: Barkworth, M.E., Capels, K.M., Long, S., Anderton, L.K., Piep, M.B. (eds.) Flora of North America north of Mexico. Vol. 24. Magnoliophyta: Commelinidae (in part): Poaceae, part 1. New York: Oxford University Press. pp. 445–448.

Darbyshire, S.J., Pavlick, L.E. 2007. *Festuca* L. In: Barkworth, M.E., Capels, K.M., Long, S., Anderton, L.K., Piep, M.B. (eds.) Flora of North America north of Mexico. Vol. 24. Magnoliophyta: Commelinidae (in part): Poaceae, part 1. New York: Oxford University Press. pp. 389–443.

Davidse, G., Soreng, R.J., Peterson, P.M. 2008. *Agrostopoa* (Poaceae, Pooideae, Poeae, Poinae), a new genus with three species from Colombia. Novon 19: 32–40.

Davis, J.I., Soreng, R.J. 2007. A preliminary phylogenetic analysis of the grass subfamily Pooideae (Poaceae), with attention to structural features of the plastid and nuclear genomes, including an intron loss in GBSSI. Aliso 23: 335–348.

Delisle-Oldham, M.B., Oldham, M.J., Catling, P.M. 2008. Taxonomic recognition of *Ammophila champlainensis* and morphological variation in northeastern North American *Ammophila* (Poaceae). Rhodora 110: 129–156.

Dewey, D.R. 1975. The origin of *Agropyron smithii*. Amer. J. Bot. 62: 524–530.

Dewey, D.R. 1984. The genomic system of classification as a guide to intergeneric hybridization in the perennial Triticeae. In: Gustafson, J.P. (ed.) Gene manipulation in plant improvement. New York: Plenum Press. pp. 209–279.

Döring, E., Schneider, J., Hilu, K.W., Röser, M. 2007. Phylogenetic relationships in the Aveneae/Poeae complex (Pooideae, Poaceae). Kew Bull. 62: 407–424.

Erdman, K.S. 1965. Taxonomy of the genus *Sphenopholis* (Gramineae). Iowa State Coll. J. Sci. 39: 289–336.

Essi, L., Longhi-Wagner, H.M., Teixeira de Souza-Chies, T. 2008. Phylogenetic analysis of the *Briza* complex (Poaceae). Mol. Phyl. Evol. 47: 1018–1029.

Essi, L., Longhi-Wagner, H.M., Chies, T.T.S. 2011. New combinations within the *Briza* complex (Poaceae, Pooideae, Poeae). Novon 21: 326–330.

Fan, X., Sha, L.-N., Dong, Z.Z., Zhang, H.-Q., Zhang, H.-Y., Kang, H.-Y., Wang, Y., Wang, X.-L., Zhang, L., Ding, C.-B., Yang, R.-W., Zheng, Y.-L., Zhou, Y.-H. 2013. Phylogenetic relationships and Y genome origin in *Elymus* L. sensu lato (Triticeae: Poaceae) based on single-copy nuclear *Acc1* and *Pgk1* gene sequences. Mol. Phyl. Evol. 69: 919–928.

Finot, V.L., Peterson, P.M., Soreng, R.J., Zuloaga, F.O. 2004. A revision of *Trisetum*, *Peyritschia*, and *Sphenopholis* (Poaceae: Pooideae: Aveninae) in Mexico and Central America. Ann. Missouri Bot. Gard. 91: 1–30.

Finot, V.L., Peterson, P.M., Soreng, R.J., Zuloaga, F.O. 2005. A revision of *Trisetum* and *Graphephorum* (Poaceae: Pooideae: Aveninae) in North America north of México. Sida 21: 1419–1453.

Frederiksen, S. 1986. Revision of *Taeniatherum* (Poaceae). Nordic J. Bot. 6: 389–397.

Frederiksen, S. 1991a. Taxonomic studies in *Dasypyrum* (Poaceae). Nordic J. Bot. 11: 135–142.

Frederiksen, S. 1991b. Taxonomic studies in *Eremopyrum* (Poaceae). Nordic J. Bot. 11: 271–285.

Frederiksen, S. 1993. Taxonomic studies in some annual genera of the Triticeae. Nordic J. Bot. 13: 481–493.

Frederiksen, S., Petersen, G. 1998. A taxonomic revision of *Secale* L. (Triticeae, Poaceae). Nordic J. Bot. 18: 399–420.

Gillespie, L.J., Archambault, A., Soreng, R.J. 2007. Phylogeny of *Poa* (Poaceae) based on *trnT-trnF* sequence data: major clades and basal relationships. Aliso 23: 420–434.

Gillespie, L.J., Soreng, R.J., Bull, R.D., Jacobs, S.W.L., Refulio-Rodriguez, N.F. 2008. Phylogenetic relationships in subtribe Poinae (Poaceae, Poeae) based on nuclear ITS and plastid *trnT-trnL-trnF* sequences. Botany 86: 938–967.

Gillespie, L.J., Soreng, R.J., Jacobs, S.W.L. 2009. Phylogenetic relationships of Australian *Poa* (Poaceae: Poinae), including molecular evidence for two new genera, *Saxipoa* and *Sylvipoa*. Aust. Syst. Bot. 22: 413–436.

Gillespie, L.J., Soreng, R.J., Paradis, M., Bull, R.D. 2010. Phylogeny and reticulation in subtribe Poinae and related subtribes (Poaceae) based on nrITS, ETS, and *trnTLF* data. In: Seberg, O., Petersen, G., Barfod, A.S., Davis, J.I. (eds.) Diversity, phylogeny and evolution in the monocotyledons. Aarhus, Denmark: Aarhus University Press. pp. 589-617

Gonzalo, R., Aedo, C., Nickrent, D.L., Garcia, M.A. 2012. A numerical taxonomic investigation of *Stipa* sect. *Smirnovia* and *S.* sect. *Subsmirnovia* (Poaceae). Syst. Bot. 37: 655–670.

Gonzalo, R., Aedo, C., García, M.A. 2013. Taxonomic revision of the Eurasian *Stipa* subsections *Stipa* and *Tirsae* (Poaceae). Syst. Bot. 38: 344–378.

GPWG 2001. Grass Phylogeny Working Group. Phylogeny and subfamilial classification of the Poaceae. Ann. Missouri Bot. Gard. 88: 373–457.

GPWG II 2012. Grass Phylogeny Working Group II. New grass phylogeny resolves deep evolutionary relationships and discovers C_4 origins. New Phytol. 193: 304–312.

Grebenstein, B., Röser, M., Sauer, W., Hemleben, V. 1998. Molecular phylogenetic relationships in Aveneae (Poaceae) species and other grasses as inferred from ITS1 and ITS2 rDNA. Plant Syst. Evol. 213: 233–250.

Hamasha, H.R., von Hagen, K.B., Röser, M. 2012. *Stipa* (Poaceae) and allies in the Old World: molecular phylogenetics realigns genus circumscription and gives evidence on the origin of American and Australian lineages. Plant Syst. Evol. 298: 351–367.

Harlan, J.R., Zohary, D. 1966. Distribution of wild wheats and barley. Science 153: 1074–1080.

Hempel, W. 2011. Revision und Phylogenie der Arten der Gattung *Melica* L. (Poaceae) in Eurasien und Nordafrika. Feddes Repert. 122: 1-253.

Hemsley, W.B. 1896. *Littledalea tibetica*, Hemsl. Hooker's Icon. Plant. 25: t. 2472.

Hoffmann, M.H., Schneider, J., Hase, P., Röser, M. 2013. Rapid and recent world-wide diversification of bluegrasses (*Poa*, Poaceae) and related genera. PLoS ONE 8: e60061.

Hu, Q., Yan, C., Sun, G. 2013. Phylogenetic analysis revealed reticulate evolution of allotetraploid *Elymus ciliaris*. Mol. Phyl. Evol. 69: 805–813.

Huff, D.R. 2010. Bluegrasses. In: Boller, B., Posselt, U.K., Veronesi, F. (eds.) Fodder crops and amenity grasses. New York: Springer.

Inda, L.A., Segarra-Moragues, J.G., Müller, J., Peterson, P. M., Catalán, P. 2008. Dated historical biogeography of the temperate Loliinae (Poaceae, Pooideae) grasses in the northern and southern hemispheres. Mol. Phyl. Evol. 46: 932–957.

Jacobs, S.W.L. 2001. The genus *Lachnagrostis* (Gramineae) in Australia. Telopea 9: 439–448.

Jacobs, S.W.L., Lapinpuro, L. 1986. The Australian species of *Amphibromus* (Poaceae). Telopea 2: 715–729.

Jacobs, S.W.L., Gillespie, L.J., Soreng, R.J. 2008. New combinations in *Hookerochloa* and *Poa* (Gramineae). Telopea 12: 273–278.

Kellogg, E.A., Bennetzen, J.L. 2004. The evolution of nuclear genome structure in seed plants. Amer. J. Bot. 91: 1709–1725.

Kellogg, E.A., Watson, L. 1993. Phylogenetic studies of a large data set. I. Bambusoideae, Pooideae, and Andropogoneae (Gramineae). Bot. Rev. 59: 273–343.

Kellogg, E.A., Appels, R., Mason-Gamer, R.J. 1996. When genes tell different stories: the diploid genera of Triticeae (Gramineae). Syst. Bot. 21: 321–347.

Kellogg, E.A., Camara, P.E.A.S., Rudall, P.J., Ladd, P., Malcomber, S.T., Whipple, C.J., Doust, A.N. 2013. Early inflorescence development in the grasses (Poaceae). Frontiers Plant Sci. 4. doi: 10.3389/fpls.2013.00250

Kim, E.S., Bolsheva, N.L., Samatadze, T.E., Nosov, N.N., Nosova, I.V., Zelenin, A.V., Punina, E.E., Muravenko, O.V., Rodionov, A.V. 2009. The unique genome of two-chromosome grasses *Zingeria* and *Colpodium*, its origin, and evolution. Russian J. Genet. 45: 1329–1337.

Kimber, G., Feldman, M. 1987. Wild wheat: an introduction. Columbia, Missouri: College of Agriculture, University of Missouri-Columbia.

Kotseruba, V., Pistrick, K., Blattner, F.R., Kumke, K., Weiss, O., Rutten, T., Fuchs, J., Endo, T., Nasuda, S., Ghukasyan, A., Houben, A. 2010. The evolution of the hexaploid grass *Zingeria kochii* (Mez) Tzvel. ($2n=12$) was accompanied by complex hybridization and uniparental loss of ribosomal DNA. Mol. Phyl. Evol. 56: 146–155.

Lange, D. 1995. Untersuchungen zur Systematik und Taxonomie der Gattung *Helictotrichon* Besser ex J. A. Schultes & J. H. Schultes (Poaceae) in Südosteuropa und Vorderasien. Bibl. Bot. 144: 1–238.

Liu, Z.W., Wang, R.R.C. 1993. Genome analysis of *Elytrigia caespitosa*, *Lophopyrum nodosum*, *Pseudoroegneria geniculata* ssp. *scythica*, and *Thinopyrum intermedium* (Triticeae: Gramineae). Genome 36: 102–111.

Lonard, R.I., Gould, F.W. 1974. The North American species of *Vulpia* (Gramineae). Madroño 22: 217–230.

Löve, A. 1984. Conspectus of the Triticeae. Feddes Repert. 95: 425–521.

Löve, A., Löve, D. 1976. Nomenclatural notes on arctic plants. Bot. Not. 128: 497–523.

Lu, S. 2006. *Calamagrostis*. In: Wu, Z., Raven, P.H., Hong, D. (eds.) Flora of China. Beijing: Science Press. pp. 359–363

Lu, S., Chen, W., Phillips, S.M. 2006. *Deyeuxia*. In: Wu, Z., Raven, P.H., Hong, D. (eds.) Flora of China: Poaceae. Beijing: Science Press. pp. 348–359

Lysak, M.A., Dolezel, J. 1998. Estimation of nuclear DNA content in *Sesleria* (Poaceae). Caryologia 51: 123–132.

Ma, H.-Y., Peng, H., Li, D.-Z. 2005. Taxonomic significance of leaf anatomy of *Aniselytron* (Poaceae) as an evidence to support its generic validity against *Calamagrostis* s.l. J. Plant Res. 118: 401–414.

MacFarlane, T.D., Hearn, R.W. 2007. Occurrence and status of *Pentapogon quadrifidus* (Poaceae) in Western Australia. Nuytsia 16: 469–471.

Macfarlane, T.D., Watson, L. 1980. The circumscription of Poaceae subfamily Pooideae, with notes on some controversial genera. Taxon 29: 645–666.

Mahelka, V., Kopecky, D., Pastova, L. 2011. On the genome constitution and evolution of intermediate wheatgrass (*Thinopyrum intermedium*: Poaceae, Triticeae). BMC Evol. Biol. 11: 127.

Mason-Gamer, R.J., Orme, N.L., Anderson, C.M. 2002. Phylogenetic analysis of North american *Elymus* and the monogenomic Triticeae (Poaceae) using three chloroplast DNA data sets. Genome 45: 991–1002.

Mason-Gamer, R.J., Burns, M.M., Naum, M. 2010a. Phylogenetic relationships and reticulation among Asian *Elymus* allotetraploids: analyses of three nuclear gene trees. Mol. Phyl. Evol. 54: 10–22.

Mason-Gamer, R.J., Burns, M.M., Naum, M. 2010b. Reticulate evolutionary history of a complex group of grasses: phylogeny of *Elymus* StStHH allotetraploids based on three nuclear genes. PLoS ONE 5: e10989.

Mill, R.R. 1985. *Nephelochloa* Boiss. In: Davis, P.D. (ed.) Flora of Turkey. Edinburgh: Edinburgh University Press. pp. 492–493

Morrell, P.L., Clegg, M.T. 2007. Genetic evidence for a second domestication of barley (*Hordeum vulgare*) east of the Fertile Crescent. Proc. Natl. Acad. Sci. USA 104: 3289–3294.

Mur, L.A., Allainguillaume, J., Catalán, P., Hasterok, R., Jenkins, G., Lesniewska, K., Thomas, I., Vogel, J. 2011. Exploiting the *Brachypodium* tool box in cereal and grass research. New Phytol. 191: 334–347.

Murbeck, S. 1899. Contributions à la connaissance de la flore du nord-ouest de l'Afrique. III. Plumbaginaceae-Graminaceae. Lund: E. Malmström.

Murray, B.G., de Lange, P.J., Ferguson, A.R. 2005. Nuclear DNA variation, chromosome numbers and polyploidy in the endemic and indigenous grass flora of New Zealand. Ann. Bot. 96: 1293–1305.

Paunero Ruiz, E. 1950. Las especies españolas de género *Trisetaria* Forssk. Anales Jard. Bot. Madrid: 503–582.

Peng, Y.-Y., Wei, Y.-M., Baum, B.R., Jiang, Q.-T., Lan, X.-J., Dai, S.-F., Zheng, Y.-L. 2010. Phylogenetic investigation of *Avena* diploid species and the maternal genome donor of *Avena* polyploids. Taxon 59: 1472–1482.

Pereira, M.P., Pérez, G.E., Balbuena, E.S. 2007. European sweet vernal grasses (*Anthoxanthum*: Poaceae, Pooideae, Aveneae): a morphometric taxonomical approach. Syst. Bot. 32: 43–59.

Petersen, G., Seberg, O. 2008. Phylogenetic relationships of allotetraploid *Hordelymus europaeus* (L.) Harz (Poaceae: Triticeae). Plant Syst. Evol. 273: 87–95.

Petersen, G., Seberg, O., Salomon, B. 2011. The origin of the H, St, W, and Y genomes in allotetraploid species of *Elymus* L. and *Stenostachys* Turcz. (Poaceae: Triticeae). Plant Syst. Evol. 291: 197–210.

Phillips, S.M., Chen, W.-L. 2003. Notes on grasses (Poaceae) for the Flora of China, I: *Deyeuxia*. Novon 13: 318–321.

Pimentel, M., Sahuquillo, E., Torrecilla, Z., Popp, M., Catalán, P., Brochmann, C. 2013. Hybridization and long-distance colonization at different time scales: towards resolution of long-term controversies in the sweet vernal grasses (*Anthoxanthum*). Ann. Bot. 112: 1015–1030.

Quintanar, A., Castroviejo, S., Catalán, P. 2007. Phylogeny of the tribe Aveneae (Pooideae, Poaceae) inferred from plastid *trnT-F* and nuclear ITS sequences. Amer. J. Bot. 94: 1554–1569.

Quintanar, A., Catalán, P., Castroviejo, S. 2010. A review of the systematics and phylogenetics of the Koeleriinae (Poaceae: Poeae). In: Seberg, O., Petersen, G., Barfod, A.S., Davis, J.I. (eds.) Diversity, phylogeny, and evolution in the monocotyledons. Aarhus, Denmark: Aarhus University Press. pp. 539–556

Reeder, J.R. 1957. The embryo in grass systematics. Amer. J. Bot. 44: 756–768.

Refulio-Rodriguez, N.F., Columbus, J.T., Gillespie, L.J., Peterson, P.M., Soreng, R.J. 2012. Molecular phylogeny of *Dissanthelium* (Poaceae: Pooideae) and its taxonomic implications. Syst. Bot. 37: 122–133.

Reznicek, A.A., Judziewicz, E.J. 1996. A new hybrid species, x*Calammophila don-hensonii* (*Ammophila breviligulata* x *Calamagrostis canadensis*, Poaceae) from Grand Island, Michigan. Michigan Bot. 35: 35–50.

Rodionov, A.V., Nosov, N.N., Kim, E.S., Machs, E.M., Punina, E.O., Probatova, N.S. 2010. The origin of polyploid genomes of bluegrasses *Poa* L. and gene flow between Northern Pacific and Sub-Antarctic Islands. Russian J. Genet. 46: 1407–1416.

Romaschenko, K., Peterson, P.M., Soreng, R.J., Garcia-Jacas, N., Futorna, O., Susanna, A. 2008. Molecular phylogenetic analysis of the American Stipeae (Poaceae) resolves *Jarava* sensu lato polyphyletic: evidence for a new genus, *Pappostipa*. J. Bot. Res. Inst. Texas 2: 165–192.

Romaschenko, K., Peterson, P.M., Soreng, R.J., Garcia-Jacas, M., Susanna, A. 2010. Phylogenetics of Stipeae (Poaceae: Pooideae) based on plastid and nuclear DNA sequences. In: Seberg, O., Petersen, G., Barfod, A.S., Davis, J.I. (eds.) Diversity, phylogeny, and evolution in the monocotyledons. Aarhus, Denmark: Aarhus University Press.

Romaschenko, K., Peterson, P.M., Soreng, R.J., Futorna, O., Susanna, A. 2011. Phylogenetics of *Piptatherum* s. l. (Poaceae: Stipeae): evidence for a new genus, *Piptatheropsis*, and resurrection of *Patis*. Taxon 60: 1703–1716.

Romaschenko, K., Peterson, P.M., Soreng, R.J., Garcia-Jacas, N., Futorna, O., Susanna, A. 2012. Systematics and evolution of the needle grasses (Poaceae: Pooideae: Stipeae) based on analysis of multiple chloroplast loci, ITS, and lemma micromorphology. Taxon 61: 18–44.

Röser, M., Heklau, H. 2011. Abscission of leaf laminas - an unnoticed factor in tussock grass formation. Flora 206: 32–37.

Rúgolo de Agrasar, Z.E. 1982. Revalidación del género *Bromidium* Nees et Meyen emend. Pilger (Gramineae). Darwiniana 24: 187–216.

Runemark, H. 1962. A revision of *Parapholis* and *Monerma* (Gramineae) in the Mediterranean. Bot. Not. 115: 1–17.

Saarela, J.M., Peterson, P.M., Soreng, R.J., Chapman, R.E. 2003. A taxonomic revision of the eastern North

American and eastern Asian disjunct genus *Brachyelytrum* (Poaceae): evidence from morphology, phytogeography and AFLPs. Syst. Bot. 28: 674–692.

Saarela, J.M., Peterson, P.M., Keane, R.M., Cayouette, J., Graham, S.W. 2007. Molecular phylogenetics of *Bromus* (Poaceae: Pooideae) based on chloroplast and nuclear DNA sequence data. Aliso 23: 450–467.

Saarela, J.M., Liu, Q., Peterson, P.M., Soreng, R.J., Paszko, B. 2010. Phylogenetics of the grass 'Aveneae-type plastid DNA clade' (Poaceae: Pooideae, Poeae) based on plastid and nuclear ribosomal DNA sequence data. In: Seberg, O., Petersen, G., Barfod, A.S., Davis, J.I. (eds.) Diversity, phylogeny, and evolution in the monocotyledons. Aarhus, Denmark: Aarhus University Press. pp. 557–587

Sakamoto, S. 1972. Intergeneric hybridization between *Eremopyrum orientale* and *Henrardia persica*, an example of polyploid species formation. Heredity 1972: 109–115.

Sakamoto, S. 1974. Intergeneric hybridization among three species of *Heteranthelium*, *Eremopyrum* and *Hordeum*, and its significance for the generic relationships within the tribe Triticeae. New Phytol. 73: 341–350.

Schneider, J., Döring, E., Hilu, K.W., Röser, M. 2009. Phylogenetic structure of the grass subfamily Pooideae based on comparison of plastid *matK* gene-3′*trnK* exon and nuclear ITS sequences. Taxon 58: 405–424.

Schneider, J., Winterfeld, G., Hoffmann, M.H., Röser, M. R. 2011. Duthieeae, a new tribe of grasses (Poaceae) identified among the early diverging lineages of subfamily Pooideae: molecular phylogenetics, morphological delineation, cytogenetics and biogeography. Syst. Biodiv. 9: 27–44.

Schneider, J., Winterfeld, G., Röser, M. 2012. Polyphyly of the grass tribe Hainardieae (Poaceae: Pooideae): identification of its different lineages based on molecular phylogenetics, including morphological and cytogenetic characteristics. Organisms Diversity Evol. 12: 113–132.

Schouten, Y., Veldkamp, J.F. 1985. A revision of *Anthoxanthum* including *Hierochloë* (Gramineae) in Malaysia and Thailand. Blumea 30: 319–351.

Smissen, R.D., de Lange, P.J., Thorsen, M.J., Ogle, C.C. 2011. Species delimitation and genetic variation in the rare New Zealand endemic grass genus *Simplicia*. N. Z. J. Bot. 49: 187–199.

Soreng, R.J., Davis, J.I. 2000. Phylogenetic structure in Poaceae subfamily Pooideae as inferred from molecular and morphological characters: misclassification vs. reticulation. In: Jacobs, S.W.L., Everett, J.E. (eds.) Grasses: systematics and evolution. Collingwood, Victoria, Australia: CSIRO Publishing. pp. 61–74

Soreng, R.J., Fish, L. 2011. *Catabrosa* versus *Colpodium* (Poaceae: Poeae) in southern Africa, with a key to these genera and their species in Africa. Kew Bull. 66: 1–10.

Soreng, R.J., Davis, J.I., Voionmaa, M.A. 2007. A phylogenetic analysis of Poaceae tribe Poeae sensu lato based on morphological characters and sequence data from three plastid-encoded genes: evidence for reticulation and a new classification for the tribe. Kew Bull. 62: 425–454.

Soreng, R.J., Bull, R.D., Gillespie, L.J. 2010. Phylogeny and reticulation in *Poa* based on plastid *trnTLF* and nrITS sequences with attention to diploids. In: Seberg, O., Petersen, G., Barfod, A.S., Davis, J.I. (eds.) Diversity, phylogeny, and evolution in the monocotyledons. Aarhus: Aarhus University Press. pp. 619–643

Soreng, R.J., Gillespie, L.J., Jacobs, S.W.L. 2011. *Saxipoa* and *Sylvipoa* - two new genera and a new classification for Australian *Poa* (Poaceae: Poinae). Aust. Syst. Bot. 22: 401–412.

Soreng, R.J., Davidse, G., Peterson, P.M., Zuloaga, F.O., Judziewicz, E.J., Filgueiras, T.S., Morrone, O., Romaschenko, K. 2012 and onward. A world-wide phylogenetic classification of Poaceae (Gramineae). http://www.tropicos.org/projectwebportal.aspx?pagename=ClassificationNWG&projectid=10

Stebbins, G.L. 1950. Variation and evolution in plants. New York and London: Columbia University Press.

Stebbins, G.L. 1956. Taxonomy and the evolution of genera, with special reference to the family Gramineae. Evolution 10: 235–245.

Stebbins, G.L., Snyder, L.A. 1956. Artifical and natural hybrids in the Gramineae, tribe Hordeae. IX. Hybrids between western and eastern North American species. Amer. J. Bot. 43: 305–312.

Stebbins, Jr., G.L., Walters, M.S. 1949. Artifical and natural hybrids in the Gramineae, tribe Hordeae. III. Hybrids involving *Elymus condensatus* and *E. triticoides*. Amer. J. Bot. 36: 291–301.

Steen, N.W., Elven, R., Nordal, I. 2004. Hybrid origin of the arctic x *Pucciphippsia vacillans* (Poaceae): evidence from Svalbard plants. Plant Syst. Evol. 245: 215–238.

Stewart, A.V., Ellison, N.W. 2011. Chapter 5. *Dactylis*. In: Kole, C. (ed.) Wild crop relatives: Genomic and breeding resources. Berlin Heidelberg: Springer-Verlag. pp. 73–87

Sun, G., Ni, Y., Daley, T. 2008. Molecular phylogeny of *RPB2* gene reveals multiple origin, geographic differentiation of H genome, and the relationship of the Y genome to other genomes in *Elymus* species. Mol. Phyl. Evol. 46: 897–907.

Thomasson, J.R. 1978. Epidermal patterns of the lemma in some fossil and living grasses and their phylogenetic significance. Science 199: 975–977.

Thomasson, J.R. 1985. Miocene fossil grasses: possible adaptation in reproductive bracts (lemma and palea) Ann. Missouri Bot. Gard. 72: 843–851.

Torrecilla, P., Catalán, P. 2002. Phylogeny of broad-leaved and fine-leaved *Festuca* lineages (Poaceae) based on nuclear ITS sequences. Syst. Bot. 27: 241–251.

Torrecilla, P., López-Rodríguez, J.-A., Catalan, P. 2004. Phylogenetic relationships of *Vulpia* and related genera (Poeae, Poaceae) based on analysis of ITS and *trn*L-F sequences. Ann. Missouri Bot. Gard. 91: 124–158.

Tucker, G.C. 1996. The genera of Poöideae (Gramineae) in the southeastern United States. Harvard Papers Bot. 9: 11–90.

van Slageren, M.W. 1994. Wild wheats: a monograph of *Aegilops* L. and *Amblyopyrum* (Jaub.& Spach) Eig.

Wageningen, Netherlands: Wageningen Agricultural University and ICARDA.

Vázquez, F.M., Barkworth, M.E. 2004. Resurrection and emendation of *Macrochloa* (Gramineae: Stipeae). Bot. J. Linn. Soc. 144: 483–495.

Vázquez Pardo, F.M., Gutiérrez Esteban, M. 2011. Classification of species of *Stipa* with awns having plumose distal segments. Telopea 13: 155–176.

Voshell, S.M., Baldini, R.M., Kumar, R., Tatalovich, H., Hilu, K.W. 2011. Canary grasses (*Phalaris*, Poaceae): Molecular phylogenetics, polyploidy and floret evolution. Taxon 60: 1306–1316.

Wang, R.R.-C., Jensen, K.B. 1994. Absence of J genome in *Leymus* species (Poaceae: Triticeae): evidence for DNA hybridization and meiotic pairing. Genome 37: 231–235.

Watson, L., Dallwitz, M.J. 1992 onward. The grass genera of the world: descriptions, illustrations, identification, and information retrieval; including synonyms, morphology, anatomy, physiology, phytochemistry, cytology, classification, pathogens, world and local distribution, and references, 23rd April 2010 edn. Wallingford: CAB International. http://delta-intkey.com/grass/

Whipple, I.G., Barkworth, M.E., Bushman, B.S. 2007. Molecular insights into the taxonomy of *Glyceria* (Poaceae: Meliceae) in North America. Amer. J. Bot. 94: 551–557.

Winterfeld, G. 2006. Molekular-cytogenetische untersuchungen an hafer-gräsern (Aveneae) und anderen Poaceae. Stapfia 86: 1–170.

Winterfeld, G., Schneider, J., Perner, K., Röser, M. 2012. Origin of highly polyploid species: different pathways of auto- and allopolyploidy in 12-18X species of *Avenula*. Int. J. Plant Sci. 173: 399–411.

Wu, Z., Phillips, S.M. 2006. Tribe Stipeae. In: Wu, Z., Raven, P.H., Hong, D. (eds.) Flora of China. Beijing: Science Press. pp. 188–212

Wu, Z., Raven, P.H., Hong, D. (eds.) 2007. Flora of China Illustrations. Beijing: Science Press.

Zhang, X., Sun, G. 2010. *RPB2* sequences reveal a close phylogenetic relationship between tetraploid *Hordelymus* and diploid *Hordeum* species in Triticeae (Poaceae). Biochem. Syst. Ecol. 38: 789–795.

Zuloaga, F.O., Rúgolo de Agrasar, Z.E., Anton, A.M. (eds.) 2012. Flora vascular de la República Argentina, Vol. 3, Tomos 1 y 2. Córdoba, Argentina: Gráficamente Ediciones.

PACMAD Clade

Embryonic mesocotyl internode elongate. Plants often C_4.

The PACMAD clade appeared as a strongly supported group in some of the earliest molecular phylogenies of the grasses, and remains a well-documented clade (Davis and Soreng 1993; GPWG 2001; GPWG II 2012). Perhaps surprisingly for a clade with such firm molecular support, there are no obvious morphological synapomorphies, with the possible exception of the elongate mesocotyl internode in the embryo. All 24+ origins of C_4 photosynthesis occur in this clade. While the photosynthetic pathway is not universal, it is derived wherever it occurs. Thus, the propensity to evolve C_4 may indeed be a synapomorphy. While the subfamilies within the clade are all clearly monophyletic, relationships among them have been unclear and highly sensitive to sampling. Addition of a large amount of data and a broad sample of species has shown that Aristidoideae are sister to all other members of the clade (GPWG II 2012) (Fig. 9).

VII. Subfamily Aristidoideae Caro (1982)

Annuals or perennials, herbaceous. Ligule a fringe of hairs, or a fringed membrane. Leaf blades generally narrow. Inflorescence branches branched. Spikelets cylindrical or laterally compressed, with one flower and no rachilla extension. Glumes generally longer than and enclosing the flower. Disarticulation above the glumes. Callus of flower short or long, blunt or sharp-pointed, pubescent. Lemma terete, leathery, the margins enfolding the palea, with a germination flap at the base, *with three awns*, these separate or *fused at the base*. Palea less than half the length of the lemma. Style branches 2. Hilum linear. Embryo with epiblast absent, scutellar cleft present or absent, embryonic leaf margins meeting.

Three genera, 365 spp.

Placement of the three genera now in Aristidoideae has been a long-standing problem in grass taxonomy. They are superficially similar to Stipeae, in that the spikelets are often fusiform in shape, and the callus is often sharp-pointed. However, the presence of microhairs on the epidermis, the embryo characters, and the C_4 photosynthesis in *Aristida* and *Stipagrostis* exclude Aristidoideae from Pooideae. Once DNA sequences became available for *Aristida* and *Stipagrostis*, it was clear that Aristidoideae formed a separate lineage within the PACMAD clade (Barker et al. 1995; Clark et al. 1995; GPWG 2001), and the subfamily now appears as sister to all other subfamilies in the clade (GPWG II 2012). The subfamily itself is clearly monophyletic, and each of the three included genera is also monophyletic (Cerros-Tlatilpa et al. 2011).

C_4 photosynthesis has been derived twice within Aristidoideae, once in *Aristida* and once in *Stipagrostis*. *Stipagrostis* is sister to the C_3 genus *Sartidia*, and the C_3 species *Aristida longifolia* is sister to all remaining species of *Aristida* (Cerros-Tlatilpa and Columbus 2009; GPWG II 2012). As in all C_4 species, *Stipagrostis* and *Aristida* have co-opted an existing isoform of PEP-carboxylase for carbon fixation. However, they have co-opted different paralogues, consistent with the observation of independent origins of C_4 (Christin and Besnard 2009). Leaf anatomy and gene expression differ between the two as

E.A. Kellogg, *Flowering Plants. Monocots*, The Families and Genera of Vascular Plants 13,
DOI 10.1007/978-3-319-15332-2_21,

well. All Aristidoideae, whether C_3 or C_4, have two sheaths around the vascular bundles (Cerros-Tlatilpa and Columbus 2009; Hattersley and Watson 1992). The inner sheath of plants in the *Sartidia*/*Stipagrostis* clade and in *A. longifolia* consists of thick-walled cells lacking chloroplasts (a mestome sheath), whereas the outer sheath is parenchymatous and contains chloroplasts (Cerros-Tlatilpa and Columbus 2009). However, in C_4 *Aristida* both sheaths are parenchymatous. Decarboxylating enzymes appear only in the inner sheath (Sinha and Kellogg 1996; Voznesenskaya et al. 2005b), pointing to that sheath as the site of carbon reduction. Rubisco is expressed in both sheaths, suggesting that it may capture carbon leaked out of the inner sheath.

Key to the Genera of Aristidoideae

1. Awns plumose — 321. ***Stipagrostis***
1. Awns glabrous — 2
2 (1). Glumes with 3 veins or more; leaves with C_3 anatomy — 320. ***Sartidia***
2. Glumes with 1 vein; leaves of most plants with unusual C_4 leaf anatomy — 319. ***Aristida***

Genera of Aristidoideae

319. *Aristida* L.

Aristida L., Sp. Pl.: 82 (1753); phylog.: Cerros-Tlatilpa et al. (2011).

Annuals or perennials, mostly caespitose. Leaf apex often sharp-pointed. Ligule a fringe of hairs, or a fringed membrane. Glumes generally with one vein, their apices often mucronate or awned. Lemma awns often fused to form a twisted column. C_4, NADP-ME subtype, except for *A. longifolia*. $2n = 22, 24, 36, 44, 48, 66$.

Three hundred four spp., warm areas throughout the world.

Cerros-Tlatilpa et al. (2011) find no evidence to support previously recognized subgenera in *Aristida*. Instead they identify several large clades that primarily reflect geography.

320. *Sartidia* de Winter

Sartidia de Winter, Kirkia 3: 137 (1963); tax.: Balkwill et al. (2011); phylog.: Cerros-Tlatilpa et al. (2011).

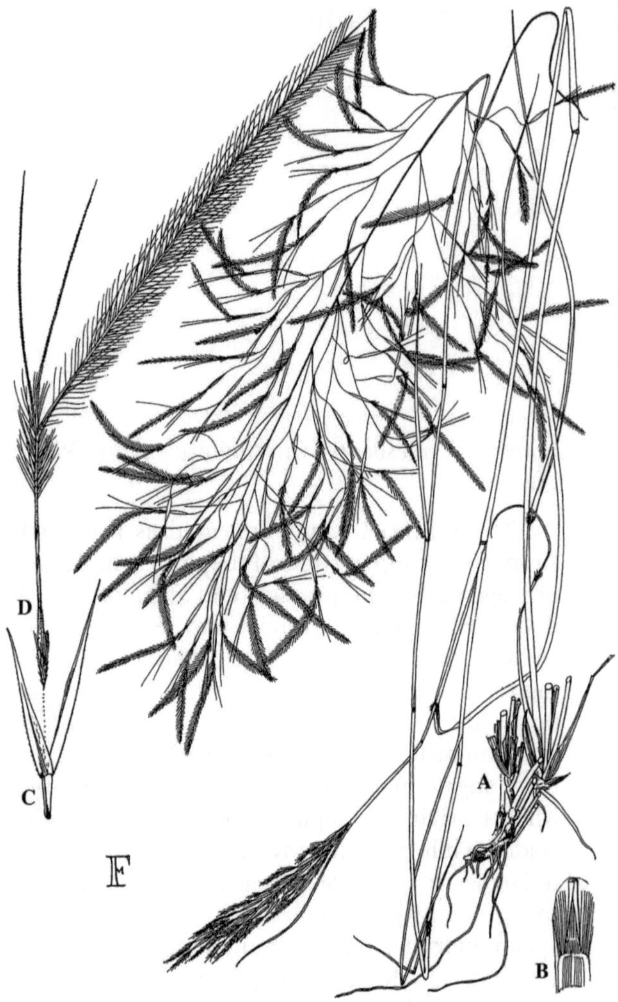

Fig. 57. Aristidoideae. *Stipagrostis hirtigluma* subsp. *patula*. **A** Habit. **B** Ligule. **C** Glumes. **D** Flower. (From Launert 1971)

Perennials. Ligule a fringe of hairs. Glumes with 3 or 5 veins, their apices acute to long attenuate or with an awn tip. C_3. $2n = 22$.

Five spp., Zambia, Angola, Congo, northeastern South Africa, Madagascar.

S. dewinteri occurs on serpentine soils in South Africa (Balkwill et al. 2011).

321. *Stipagrostis* Nees — Fig. 57

Stipagrostis Nees, Linnaea 7: 290 (1832); phylog.: Cerros-Tlatilpa et al. (2011).

Annuals or perennials, caespitose, rhizomatous or stoloniferous, sometimes becoming somewhat woody at the base. Leaf blades inrolled, the apex often sharp-pointed. Ligule a fringe of hairs. Glumes with 1 to 9 veins, their apices emarginate to long attenuate, muticous to awned. Lemma awns deciduous, *one or more plumose.* C_4, NADP-ME subtype. $2n = 22, 44$.

Fifty-six spp., Africa to central and western Asia.

Evidence that *Stipagrostis* uses NADP-ME as a decarboxylating enzyme is provided by Voznesenskaya et al. (2005a).

References

Balkwill, K., Campbell-Young, G.J., Fish, L., Munday, J., Frean, M.L., Stalmans, M. 2011. A new species of *Sartidia* (Gramineae), endemic to ultramafic soils. South Afr. J. Bot. 77: 598–607.

Barker, N.P., Linder, H.P., Harley, E.H. 1995. Polyphyly of Arundinoideae (Poaceae): evidence from *rbcL* sequence data. Syst. Bot. 20: 423–435.

Cerros-Tlatilpa, R., Columbus, J.T. 2009. C_3 photosynthesis in *Aristida longifolia*: implication for photosynthetic diversification in Aristidoideae (Poaceae). Amer. J. Bot. 96: 1379–1387.

Cerros-Tlatilpa, R., Columbus, J.T., Barker, N.P. 2011. Phylogenetic relationships of *Aristida* and relatives (Poaceae, Aristidoideae) based on noncoding chloroplast (*trn*L-F, *rpl*16) and nuclear (ITS) DNA sequences. Amer. J. Bot. 98: 1868–1886.

Christin, P.-A., Besnard, G. 2009. Two independent C_4 origins in Aristidoideae (Poaceae) revealed by the recruitment of distinct phosphoenolpyruvate carboxylase genes. Amer. J. Bot. 96: 2234–2239.

Clark, L.G., Zhang, W., Wendel, J.F. 1995. A phylogeny of the grass family (Poaceae) based on *ndhF* sequence data. Syst. Bot. 20: 436–460.

Davis, J.I., Soreng, R.J. 1993. Phylogenetic structure in the grass family (Poaceae) as inferred from chloroplast DNA restriction site variation. Amer. J. Bot. 80: 1444–1454.

GPWG 2001. Grass Phylogeny Working Group. Phylogeny and subfamilial classification of the Poaceae. Ann. Missouri Bot. Gard. 88: 373–457.

GPWG II 2012. Grass Phylogeny Working Group II. New grass phylogeny resolves deep evolutionary relationships and discovers C_4 origins. New Phytol. 193: 304–312.

Hattersley, P.W., Watson, L. 1992. Diversification of photosynthesis. In: Chapman, G.P. (ed.) Grass evolution and domestication. Cambridge: Cambridge University Press. pp. 38–116

Launert, E. 1971. Gramineae. Flora Zambesiaca. Kew, Richmond, Surrey: Royal Botanic Gardens.

Sinha, N.R., Kellogg, E.A. 1996. Parallelism and diversity in multiple origins of C_4 photosynthesis in the grass family. Amer. J. Bot. 83: 1458–1570.

Voznesenskaya, E.V., Chuong, S.D.X., Kiirats, O., Franceschi, V.R., Edwards, G.E. 2005a. Evidence that C_4 species in genus *Stipagrostis*, family Poaceae, are NADP-malic enzyme subtype with nonclassical type of Kranz anatomy (Stipagrostoid). Plant Sci. 168: 731–739.

Voznesenskaya, E.V., Chuong, S.D.X., Koteyeva, N.K., Edwards, G.E., Franceschi, V.R. 2005b. Functional compartmentation of C_4 photosynthesis in the triple layered chlorenchyma of *Aristida* (Poaceae). Func. Plant Biol. 32: 67–77.

VIII. Subfamily Panicoideae Link (1827)

Plants annual or perennial, caespitose, rhizomatous or stoloniferous; monoecious, dioecious, andromonoecious or bisexual. Culms often branching. Leaf blade with or without pseudopetiole; ligule membranous, a fringed membrane, or a fringe of hairs. Inflorescences unbranched, or with the primary branches unbranched, or the primary branches themselves branched to form higher-order branches; branch complexes sometimes present in the axils of upper culm leaves. Spikelets laterally or dorsiventrally compressed, disarticulating above or below the glumes. Glumes two, membranous, hyaline or indurate, with or without awns. Lemmas and paleas membranous, hyaline or indurate, with or without awns. Paleas with 0 to 2 keels. Lodicules 2, fleshy. Stamens 3, stigmas 2. Caryopsis with *hilum punctate.* Embryo with a long mesocotyl, with a scutellar cleft and with *the embryonic leaf margins overlapping*; epiblast present, but lost in Panicoideae s.s. Stomatal subsidiary cells triangular or domed (a plesiomorphy for the family); microhair apical cells generally much longer than wide ("the panicoid type"), but this character variable. Midribs simple or complex. Ancestrally C_3.

Two hundred-twelve genera, 3316 species.

In its traditional, strict sense, Panicoideae include the tribes Andropogoneae, Paspaleae, and Paniceae; this clade is referred to here as Panicoideae s.s. Recent molecular phylogenetic analyses, however, have led to the current broad circumscription of the subfamily, called here Panicoideae s.l., which also includes the tribes Gynerieae, Chasmanthieae, Centotheceae, and Tristachyideae (Morrone et al. 2012; Sánchez-Ken and Clark 2010) (Fig. 58). Both Panicoideae s.s. and s.l. are consistently monophyletic in all phylogenetic analyses of all sets of characters (see section on Subdivision of the family). In contrast, the former subfamily "Centothecoideae" forms a paraphyletic assemblage from which Panicoideae s.s. are derived. Accordingly, here I recognize Panicoideae in its broad sense. Eriachneae and Isachneae were once included as small tribes in Panicoideae (Clayton and Renvoize 1986), but are clearly unrelated to the panicoids and are placed in Micrairoideae (Sánchez-Ken et al. 2007).

Chromosome numbers are generally consistent within major clades and are synapomorphic for some. Gynerieae, Chasmanthieae, Centotheceae, and Tristachyideae have chromosome base numbers of 11, 12, 12, and 10–12, respectively. Andropogoneae and Paspaleae share a base chromosome number of 10, while Paniceae have a base chromosome number of 9 (Giussani et al. 2001). Both $x = 9$ and $x = 10$ are probably derived and synapomorphic.

C_4 photosynthesis has originated 22 to 24 times within Panicoideae s.l. (Christin et al. 2007; Edwards and Smith 2010; Giussani et al. 2001; GPWG II 2012; Vicentini et al. 2008). Gynerieae, Chasmanthieae and Centotheceae are entirely C_3, whereas Andropogoneae are wholly C_4, and Paniceae, Paspaleae, and Tristachyideae each contain a mix of C_3 and C_4 lineages. All known C_4 subtypes are represented, but most C_4 species are NADP-ME, with a single bundle sheath surrounding the veins. The NAD-ME and PCK subtypes are synapomorphic for Panicinae and Melinidinae, respectively, both monophyletic subtribes of Paniceae.

Spikelet structure has been noted since the time of Robert Brown (1810, 1814) to be a defining characteristic of Panicoideae s.s. Spikelets in

E.A. Kellogg, *Flowering Plants. Monocots*, The Families and Genera of Vascular Plants 13,
DOI 10.1007/978-3-319-15332-2_22, © Springer International Publishing Switzerland 2015

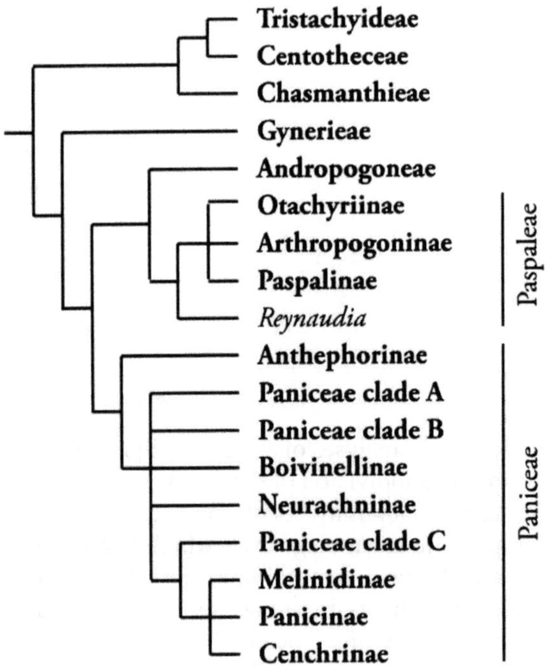

Fig. 58. Relationships of the subtribes and tribes of Panicoideae. Tree topology based on figures in Morrone et al. (2012).

Panicoideae s.s. are two-flowered, with the upper flower generally hermaphrodite and the lower one staminate or sterile. Developmental studies show that the upper flower develops ahead of the lower one, and exhibits a characteristic pattern of expression of genes that control floral organ identity (Doust and Kellogg 2002; Malcomber and Kellogg 2004; Reinheimer et al. 2006). Staminate flowers in all species studied initiate both a pistil and stamens (Le Roux and Kellogg 1999; Reinheimer et al. 2005; Zaitchik et al. 2000). The gynoecium develops a clear nucellus, surrounded by an emerging ridge of tissue that would normally form the ovary wall, but then the pistil ceases development. Cessation of growth correlates with loss of nuclei and cytoplasm in subepidermal cells, apparently a pattern of controlled cell death. In Andropogoneae, the dead cells are in the center of the gynoecium, whereas in the one species of *Panicum* studied, cell death appeared near the base of the ovary.

Pairing of spikelets, with one sessile or on a short pedicel and the other on a longer pedicel, is usually cited as a character useful for recognizing Andropogoneae. However, this character is much more widespread in the subfamily (Kellogg 2000; Reinheimer et al. 2005; Zanotti et al. 2010). Developmentally, paired spikelets in all Panicoideae form from a single primordium that appears to divide later in development, giving rise to one terminal spikelet and one lateral (Bommert et al. 2005; Kellogg 2000; Zanotti et al. 2010).

Tribes and Subtribes of Panicoideae

XVII. Tribe Steyermarkochloeae
XVIII. Tribe Tristachyideae
XIX. Tribe Chasmanthieae
XX. Tribe Gynerieae
XXI. Tribe Centotheceae
Panicoideae s.s.
Panicoideae s.s. incertae sedis
XXII. Tribe Andropogoneae
  XXII.1. Subtribe Arundinellinae
  Andropogoneae s.s.
    Andropogoneae s.s. incertae sedis
    XXII.2. Subtribe Tripsacinae
    XXII.3. Subtribe Rottboelliinae
    XXII.4. Subtribe Ischaeminae
    XXII.5. Subtribe Saccharinae
    XXII.6. Subtribe Germainiinae
    XXII.7. Subtribe Andropogoninae
XXIII. Tribe Paspaleae
  XXIII.1. Subtribe Arthropogoninae
  XXIII.2. Subtribe Otachyriinae
  XXIII.3. Subtribe Paspalinae
XXIV. Tribe Paniceae
  Incertae sedis
  XXIV.1. Subtribe Anthephorinae
  XXIV.2. Subtribe Boivinellinae
  XXIV.3. Subtribe Neurachninae
  XXIV.4. Subtribe Cenchrinae
  XXIV.5. Subtribe Melinidinae
  XXIV.6. Subtribe Panicinae

Key to the Genera of Panicoideae

1. Leaf blades absent from inflorescence-bearing culms or from all culms 2
1. Leaf blades present 4
2 (1). Plants annual; leaf blades absent from upper leaves, present on lower ones; axillary branch complexes present; upper glume gibbous; plants of India **363. *Pogonachne***
2. Plants perennial; blades absent from entire culms; axillary branch complexes absent; upper glume not gibbous; plants not occurring in India 3
3 (2). Spikelets laterally compressed, with 4 to 9 flowers, bisexual; plants of southwest Australia **339. *Spartochloa***
3. Spikelets dorsiventrally compressed to terete, with 2 or 3 flowers, most unisexual; plants of northern South America **322. *Steyermarkochloa***

4 (1). Upper surface of leaves with sinuous lamellae; plants aquatic, central Africa **342. *Hydrothauma***
4. Upper surface of leaves flat or ribbed, but not with sinuous lamellae; habitat various 5
5 (4). Inflorescence a dense spherical or hemispherical head 6
5. Inflorescence variously elongate, open or contracted, but not a dense head 12
6 (5). Spikelets with both flowers bisexual; plants of India **323. *Chandrasekharania***
6. Spikelets with only one flower bisexual; plants of Australia or, if in India, then only on beaches 7
7 (6). Plants annual **471. *Thedachloa***
7. Plants perennial 8
8 (7). Lower glume absent, if present then thin, less than ½ the length of the spikelet, the apex obtuse to acute; leaf blades disarticulating **497. *Cleistochloa***
8. Lower glume stiff, as long as the spikelet, the apex acuminate; leaf blades not disarticulating 9
9 (8). Plants with unisexual flowers, dioecious; forming clumps connected by long rhizomes; glumes membranous 10
9. Plants with bisexual flowers; caespitose or stoloniferous; glumes coriaceous to indurate 11
10 (9). Pistillate spikelets subtended by papery bracts, these acute to acuminate but not with long stiff tips; staminate spikelets 1 per inflorescence branch; plants of interior Australia **522. *Zygochloa***
10. Pistillate spikelets subtended by bracts with long stiff tips; staminate spikelets several per inflorescence branch; plants of coastal dunes **521. *Spinifex***
11 (9). Lower glume with a transverse line of hairs above a hyaline patch; glume apices truncate to obtuse **499. *Thyridolepis***
11. Lower glume lacking a transverse line of hairs and lacking a hyaline patch; glume apices acuminate **498. *Neurachne***
12 (5). Spikelets with one to many flowers, if two flowered then the reduced flower distal to the bisexual one 13
12. Spikelets with exactly two flowers, the distal one fertile and generally bisexual, the proximal one generally staminate or reduced to a sterile lemma 22
13 (12). Leaf blades with a pseudopetiole 14
13. Base of leaf blades not pseudopetiolate 17
14 (13). Sterile lemmas awned, the awns strongly recurved **333. *Lophatherum***
14. Sterile lemmas unawned, or if awned then the awns straight 15
15 (14). Flowers unisexual, the proximal flower in the spikelet pistillate with a gibbous lemma **335. *Zeugites***
15. At least some of the flowers in the spikelet bisexual, none gibbous 16
16 (15). Lemmas awned; spikelets with paleas free from the rachilla **332. *Chevalierella***
16. Lemmas without awns; spikelets with palea keels fused to the rachilla **334. *Orthoclada***
17 (13). Spikelets with only one flower, sterile flowers lacking **347. *Garnotia***
17. Spikelets with more than one flower, although some flowers reduced 18
18 (17). Plants annual 19
18. Plants perennial 20
19 (18). Base of leaf blades clasping the stem; spikelets with 8 to 20 flowers; lemma without reflexed bristles **340. *Megastachya***
19. Base of leaf blades narrow, not clasping; spikelets with 1 to 4 flowers; lemma with reflexed bristles **341. *Centotheca***
20 (18). Plants monoecious, to 3 m tall, forming an elevated tussock, the base of the plant covered by remnants of leaf sheaths and aerial roots; plants of central Brazil **330. *Arundoclaytonia***
20. Plants with flowers bisexual, generally less than 3 cm tall, not forming elevated tussocks, not occurring in Brazil 21
21 (20). All flowers fertile but the distal ones reduced; palea straight; stamens 3 **338. *Cyperochloa***
21. Proximal 1 to 2 flowers sterile; palea gibbous, arched; stamens 1 **331. *Chasmanthium***
22 (12). All or most spikelets associated with one or more bristle-like sterile branches 23
22. Inflorescences without bristle-like sterile branches; if bristle-like branches present, then associated with only some of the spikelets, usually just the terminal ones on the lateral branches 32
23 (22). Lateral branches disarticulating below any bristles and falling off as a unit 24
23. Lateral branches not falling off as a unit, spikelets disarticulating below the glumes but above any associated bristles 27
24 (23). Leaves pseudopetiolate, the blades cordate or sagittate **516. *Streptolophus***
24. Leaves lacking pseudopetioles, the blades narrow, linear 25
25 (24). Spikelets associated with one bristle or a fan of bristles, if the latter then the bristles not enclosing the spikelets; plants of Australia **508. *Paractaenum***
25. Spikelets subtended and surrounded by a cluster of bristles, these often forming a cup or involucre, but not a flattened fan; plants widespread 26
26 (25). Spikelets bisexual; inflorescence branches contracted creating a bottle-brush appearance **501. *Cenchrus***
26. Spikelets unisexual, the plants dioecious; inflorescence branches elongate, with 2 to 5 spikelets **509. *Pseudochaetochloa***
27 (23). Spikelets and bristles strictly paired, the two borne on a common pedicel 28
27. Spikelets and bristles not in strict pairs, spikelets either associated with multiple bristles, or some spikelets lacking bristles entirely 30
28 (27). Lateral branches of inflorescence elongate, with numerous spikelet/bristle pairs; plants of Central America, cultivated elsewhere **505. *Ixophorus***
28. Inflorescence slender, spike-like, lateral branches very short if developed at all; plants of the West Indies, Pacific, South America or the Old World 29
29 (28). Apex of first lemma rounded; plants of Australia **502. *Chamaeraphis***
29. Apex of first lemma acuminate; plants of Old and New World tropics **507. *Paratheria***
30 (27). First glume and lemma saccate, the sacs collapsing upon drying to form two abaxial protuberances; plants of Mexico **512. *Setariopsis***

30. First glume and lemma not saccate, not collapsing on drying; worldwide distribution 31
31 (30). Spikelets with the upper flower bisexual, the lower flower staminate or sterile **511. *Setaria***
31. Spikelets with both flowers bisexual **503. *Dissochondrus***
32 (22). Primary branches of inflorescence broad, leaf-like, often drooping to surround the spikelets; spikelets in clusters on the abaxial side of the branches **478. *Chlorocalymma***
32. Primary branches of inflorescence slender, if flattened not drooping to surround the spikelets, spikelets not clustered abaxially 33
33 (32). Inflorescence unbranched above the uppermost spathe or leaf **Group I** (includes plants with paired spikelets borne on otherwise unbranched axes)
33. Inflorescence with two or more branches above the uppermost spathe or leaf 34
34 (33). Inflorescence axis and/or branches disarticulating at the nodes; spikelets generally in pairs, less frequently single or in sets of 3 or more; dispersal unit including one or more spikelets and their associated internodes **Group II**
34. Inflorescence axis tough; spikelets borne singly or in pairs; dispersal units rarely including the inflorescence axis 35
35 (34). Primary branches of inflorescences not themselves branched **Group III**
35. Primary branches of inflorescence branched, although branches sometimes very short **Group IV**

Group I - Inflorescence unbranched

1. Glumes membranous, less firm than or about as firm as the lemmas, one or both often shorter than the spikelet 2
1. Glumes coriaceous to indurate, much firmer than the lemmas, generally longer than the spikelet 12
2 (1). Inflorescence axes surrounded by spathe-like bracts 3
2. Inflorescence axes not subtended by spathe-like bracts 4
3 (2). Inflorescence axes in clusters at nodes, each axis ending in a sharp point; all spikelets bisexual; plants of dry inland sites **519. *Xerochloa***
3. Inflorescence axes not clustered, ending in a spikelet; proximal spikelets bisexual, distal ones staminate; plants of coastal dunes **528. *Thuarea***
4 (2). Spikelets laterally compressed 5
4. Spikelets dorsiventrally compressed 8
5 (4). Lemmas awned **326. *Jansenella***
5. Lemmas lacking awns 6
6 (5). Base of leaf blades cordate; glumes with tuberculate-based hairs **457. *Echinolaena***
6. Base of leaf blades narrow; glumes glabrous, or if pubescent then the hairs not tuberculate-based 7
7 (6). Culms generally over 25 cm tall; plants caespitose, rarely stoloniferous; plants of Central and South America **446. *Mesosetum***
7. Culms less than 25 cm tall; plants stoloniferous, restricted to New South Wales **500. *Alexfloydia***
8 (4). Spikelets partially embedded in hollows in the broad flat inflorescence axis **514. *Stenotaphrum***
8. Spikelets not embedded in the inflorescence axis 9
9 (8). Both glumes present **523. *Chaetium***
9. Lower glume absent or minute 10
10 (9). Lemma with margins flat, overlapping and partially covering the palea; lower lemma awned **513. *Snowdenia***
10. Lemma with margins inrolled, covering only the margins of the palea; lower lemma unawned 11
11 (10). Lower lemma adaxial **453. *Axonopus***
11. Lower lemma abaxial **462. *Paspalum***
12 (1). Spikelets not obviously paired 13
12. Spikelets in sessile-pedicellate pairs, although the pedicellate spikelet sometimes represented only by the pedicel 16
13 (12). Spikelets all pedicellate, strongly laterally compressed **389. *Dimeria***
13. Spikelets all sessile, dorsiventrally compressed 14
14 (13). Lower glume smooth, not pitted or sculptured **369. *Oxyrhachis***
14. Lower glume sculptured or pitted 15
15 (14). Spikelets numerous on each inflorescence branch **386. *Ophiuros***
15. Spikelets few (often <3) with a long naked peduncle below **387. *Thaumastochloa***
16 (12). Lower glume of sessile spikelet hard, shiny, and variously sculptured, with pits, hooks, warts, or ridges 17
16. Lower glume of sessile spikelet smooth and shiny or green and leaf-life, with or without prominent veins or grooves, but not sculptured 22
17 (16). Pedicel free from rachis internode 18
17. Pedicel fused partially to wholly to rachis internode 19
18 (17). Spikelets unawned **383. *Rottboellia***
18. Spikelets with the pedicellate spikelet awned; awns generally also on the upper glume of the sessile spikelet **370. *Rhytachne***
19 (17). Lower glume of sessile spikelet awned **376. *Glyphochloa***
19. Glumes unawned 20
20 (19). Sessile spikelets two per node, on either side of a slender flattened pedicel bearing a tiny flap at its summit representing the pedicellate spikelet **381. *Mnesithea***
20. Sessile spikelets one per node; pedicellate spikelet well developed with obvious glumes 21
21 (20). Lower glume of sessile spikelet suborbicular **377. *Hackelochloa***
21. Lower glume of sessile spikelet broadly lanceolate-ovate **385. *Heteropholis***
22 (16). Lemmas lacking awns; lower flower of sessile spikelet often with lemma, palea, and stamens 23
22. Lemmas of one or both spikelets awned; lower flower of sessile spikelet often reduced 34
23 (22). Distal spikelet pairs of the inflorescence staminate, proximal ones pistillate **351. *Chionachne*** (note: if *Coix* is interpreted as having an unbranched inflorescence it will key here as well, but can be easily distinguished by the hard bead-like sheaths surrounding the pistillate spikelets)

23. Spikelet pairs similar in sex expression throughout the inflorescence 24
24 (23). Lower glume of pedicellate spikelet extended into a long slender, tail-like awn, becoming recurved after anthesis **371. *Urelytrum***
24. Lower glume of pedicellate spikelet not awned, or the awn <5 mm long 25
25 (24). Sessile spikelets two per node 26
25. Sessile spikelets one per node 28
26 (25). Glumes acuminate **361. *Lasiurus***
26. Glumes obtuse 27
27 (26). Lower glume of sessile spikelet with two keels, these expanding to narrow wings near the apex **383. *Rottboellia***
27. Lower glume of sessile spikelet lacking keels **381. *Mnesithea***
28 (25). Lower glume with a constriction about half way up, with an obtuse apex; plants of India **380. *Manisuris***
28. Lower glume without a constriction; plants of Old World tropics 29
29 (28). Upper glume of sessile spikelet laterally compressed **350. *Chasmopodium***
29. Upper glume of sessile spikelet dorsiventrally compressed 30
30 (29). Lower glume of sessile spikelet more or less oblong, with a long extended apex **384. *Hemarthria***
30. Lower glume with an acute to obtuse apex, not extended 31
31 (30). Upper part of plant with axillary branch complexes, producing complex compound inflorescences **383. *Rottboellia***
31. Upper part of plant without axillary branch complexes, inflorescence terminal on the culm 32
32 (31). Pedicellate spikelets rudimentary, or reduced only to an awn, or lacking, leaving a barren pedicel; nodes of inflorescence transverse **375. *Eremochloa***
32. Sessile and pedicellate spikelets morphologically similar, the pedicellate well developed, the sessile with non-winged glumes; nodes of inflorescence oblique 33
33 (32). Upper part of inflorescence internode not surrounding the base of sessile spikelet above; lower glume of sessile spikelet with a distinctly bifid tip **355. *Elionurus***
33. Upper part of inflorescence internode forming a crater or cup, surrounding the base of sessile spikelet above; lower glume of sessile spikelet entire **379. *Loxodera***
34 (22). Spikelets <5 mm long; glume apices obtuse to truncate 35
34. Spikelets generally over 5 mm long; glume apices generally acute to acuminate, rarely truncate 36
35 (34). Upper glume awned, sessile spikelets one per node **405. *Pogonatherum***
35. Upper glume awnless, sessile spikelets two per node **364. *Polytrias***
36 (34). Base of spikelet-bearing branch with four unawned spikelets, sessile or short pedicellate, arranged in a whorl, surrounding the single awned spikelet, another two unawned spikelets terminating the branch 37
36. Awned and unawned spikelets not arranged as above 38
37 (36). Awn covered with short hairs **423. *Themeda***
37. Awn glabrous **421. *Iseilema***
38 (36). Pedicellate spikelet scarcely developed, forming a tiny point at the apex of the pedicel, or lacking entirely 39
38. Pedicellate spikelet developed, staminate or sterile 43
39 (38). Inflorescence partially or almost wholly enclosed by the subtending leaf sheath or spathe; upper glume with a slender awn 40
39. Inflorescence not completely enclosed by leaves; upper glume of sessile spikelet awned or not 41
40 (39). Plants annual; inflorescence reduced to a set of three spikelets; plants of northern Australia **353. *Clausospicula***
40. Plants perennial; inflorescence with more than three spikelets; plants of tropical Asia **359. *Kerriochloa***
41 (39). Inflorescence internodes as broad as the spikelet **390. *Ischaemum***
41. Inflorescence internodes slender 42
42 (41). Base of sessile spikelet sunken into a crater-like node; inflorescence axis with internodes clavate **433. *Schizachyrium***
42. Base of sessile spikelet somewhat flattened, connected apparently laterally to a filiform inflorescence axis **413. *Anadelphia***
43 (38). Lemma of pedicellate spikelet awned, that of sessile spikelet unawned **412. *Trachypogon***
43. Lemma of pedicellate spikelet unawned, that of sessile spikelet awned 44
44 (43). Proximal spikelet pairs without awns, both sessile and pedicellate spikelets of these pairs staminate 45
44. Proximal spikelet pairs with awns, the sessile spikelets of these pairs bisexual, the pedicellate ones staminate or sterile 47
45 (44). Callus rounded **422. *Pseudodichanthium***
45. Callus sharp, barbed 46
46 (45). Pedicellate spikelets smaller than the sessile ones; lower glume of sessile spikelet not deeply grooved, the spikelet terete **420. *Heteropogon***
46. Pedicellate spikelets larger than sessile ones; lower glume of sessile spikelet deeply grooved, the groove translucent **410. *Agenium***
47 (44). Upper glume of sessile spikelet with a slender awn 48
47. Upper glume of sessile spikelet lacking an awn 52
48 (47). Inflorescence completely enclosed by a reddish brown spathe **415. *Monocymbium***
48. Inflorescence exserted, not completely enclosed by a spathe 49
49 (48). Pedicellate spikelets bisexual, similar to the sessile ones; lower glume of the pedicellate spikelet about the same width and color as that of the sessile spikelet **362. *Microstegium***
49. Pedicellate spikelets staminate or sterile, clearly different from the sessile ones; lower glume of the pedicellate spikelet often a different size or color from that of the sessile spikelet 50
50 (49). Lower glume of pedicellate spikelet firm to hard and shiny, the margins enclosing the upper glume **413. *Anadelphia***

50. Lower glume of pedicellate spikelet broad, flat, firm to papery but not hard and shiny, not enclosing the upper glume 51
51 (50). Lower glume of pedicellate spikelet well developed, membranous, with prominent green veins converging toward the apex **365. *Sehima***
51. Lower glume of pedicellate spikelet papery, purplish, without prominent veins **430. *Diectomis***
52 (47). Rachis internodes and pedicels broadly club-shaped; sessile spikelet inserted into a crater-like node; glumes of pedicellate spikelets with prominent wings 53
52. Rachis internodes and pedicels about the same diameter throughout their length; sessile spikelet not inserted into a crater-like node; glumes without wings 54
53 (52). Inflorescence exserted, not enclosed in a spathe; lower glume of sessile spikelet two-keeled with strong central groove; plants of Africa **388. *Andropterum***
53. Inflorescence subtended and partially enclosed by a broad spathe; lower glume of sessile spikelet neither two-keeled nor grooved; plants of tropical Asia **348. *Apluda***
54 (52). Lower glume with a prominent circular pit; spikelets strongly dorsiventrally compressed; plants of Africa **419. *Eremopogon***
54. Lower glume not pitted; spikelets more or less round in cross section 55
55 (54). Plants perennial; plants of Africa and South America **358. *Hypogynium***
55. Plants annual; plants of India and Southeast Asia 56
56 (55). Plants trailing, <0.8 m tall, rooting from the nodes; pedicellate spikelet staminate **429. *Pseudanthistiria***
56. Plants upright, 0.5–2 m tall, with prop roots; pedicellate spikelet rudimentary **368. *Triplopogon***

Group II – Inflorescence branched; inflorescence or branch axes disarticulating at the nodes

1. Spikelets without awns 2
1. Spikelets bearing awns from lemmas, glumes or both 17
2 (1). Primary inflorescence branches broad and leaf-like; spikelets in clusters of 2 to 6 on the abaxial side of the branches **482. *Trachys***
2. Primary inflorescence branches narrow, not leaf like; spikelets single or paired but not in abaxial clusters 3
3 (2). Glumes with prominent silicified hooks on the margin, and often over the back of the lemma as well **378. *Jardinea***
3. Glumes smooth, without prominent hooks 4
4 (3). Plants strictly monoecious, the pistillate spikelets often distinctly different from the staminate 5
4. Plants andromonoecious, with the upper flower of each sessile spikelet bisexual, the lower staminate or reduced; sex expression of the pedicellate spikelet various 8
5 (4). Staminate and pistillate spikelets in separate inflorescences, the staminate inflorescence branched, the pistillate unbranched **374. *Zea***
5. Pistillate spikelets proximal to the staminate, in the same inflorescence 6
6 (5). Pistillate spikelet surrounded by a hard round bead-like sheath **354. *Coix***
6. Pistillate spikelet with hardened glumes, but not surrounded by a bead-like sheath 7
7 (6). Pedicellate spikelets developed in all parts of the inflorescence; glume of pistillate spikelet flattened above; plants of India and Southeast Asia **351. *Chionachne***
7. Pedicellate spikelets not developing in the lower (pistillate) portion of the inflorescence; glume of pistillate spikelet rounded; plants of the New World **373. *Tripsacum***
8 (4). Inflorescence branches themselves branched, the proximal part of the branches bare 9
8. Inflorescence branches themselves unbranched; spikelets borne the full length of the branch 12
9 (8). Plants annual: spikelets solitary **537. *Arthragrostis***
9. Plant perennial; spikelets paired 10
10 (9). Ultimate inflorescence branches with more than 10 spikelet pairs **402. *Lasiorhachis***
10. Ultimate inflorescence branches with fewer than 7 spikelet pairs 11
11 (10). Callus of spikelets with tiny hairs; sessile spikelet truly sessile **396. *Sorghum***
11. Callus of spikelets with hairs about half the length of the spikelet; "sessile" spikelet on a short stalk **399. *Eriochrysis***
12 (8). Spikelets borne singly **517. *Uranthoecium***
12. Spikelets borne in pairs, one sessile and one pedicellate 13
13 (12). Spikelets <5 mm long 14
13. Spikelets 10 mm long or more 15
14 (13). Inflorescence with two (to six) branches; sessile spikelets overlapping; sessile and pedicellate spikelets morphologically different, the sessile with lower glumes truncate **407. *Germainia***
14. Inflorescences with many branches; "sessile" spikelets actually short pedicellate, not overlapping on the branches; sessile and pedicellate spikelets morphologically similar, with glumes acute to acuminate **403. *Miscanthus***
15 (13). Lower glume of sessile spikelet with a long-flattened tail, several times longer than the spikelet **372. *Vossia***
15. Lower glume of sessile spikelet often long acuminate, but the apical portion no longer than the spikelet 16
16 (15). Lower glume of sessile spikelet not winged **382. *Phacelurus***
16. Lower glume of sessile spikelet with two winged keels **383. *Rottboellia* (former *Thyrsia*)**
17 (1). Primary inflorescence branches unbranched, digitate, two to many on inflorescence axis, generally spikelet-bearing to near the base 18
17. Primary inflorescence branches themselves branched, with multiple orders of branching; spikelets often restricted to the distal half of the branch 44
18 (17). Glumes sculptured; inflorescence internodes club-shaped **367. *Thelepogon***
18. Glumes smooth or with hooked prickles, but not sculptured; inflorescence internodes slender 19

19 (18). Plants stoloniferous, culms weak, rooting from the nodes; leaves broadly lanceolate, <6 cm long, but >1 cm wide, often clasping the stem 20
19. Plants caespitose or rhizomatous, culms robust; leaves linear, generally <1 cm wide, if lanceolate then much longer than 6 cm, generally not clasping the stem 22
20 (19). Lemma awn originating from below the midpoint of the abaxial side **349. *Arthraxon***
20. Lemma awn originating from a short notch near the apex 21
21 (20). Rachis internodes slender; stamens generally 2; glumes unawned **393. *Leptatherum***
21. Rachis internodes clavate; stamens 2 or 3; upper glume often awned **362. *Microstegium***
22 (19). "Sessile" spikelets actually on a short pedicel 23
22. Sessile spikelets truly sessile, without any subtending pedicel 24
23 (22). Inflorescence contracted, superficially spike-like, covered in long silky white hairs **401. *Imperata***
23. Inflorescence of spreading digitate branches, glabrous or pubescent but not with long silky hairs **390. *Ischaemum***
24 (22). Ultimate inflorescence branches exactly two, closely subtended and often surrounded by a spathe-like leaf 25
24. Ultimate inflorescence branches two to many, if exactly two then spathe-like leaf, if present, not surrounding the branches 29
25 (24). Upper glume deeply three lobed, the central lobe awned **409. *Bhidea***
25. Upper glume apex entire, if bifid only shallowly so, generally without an awn 26
26 (25). Lower glume of sessile spikelet with a deep groove; apex of inflorescence (between the branches) extended into a flattened appendage at least as long as spikelets **417. *Hyperthelia***
26. Lower glume without a groove, inflorescence branch without a flattened extension 27
27 (26). Base of spikelet pair inserted into a crater-like extension of the internode below; ultimate inflorescence branches generally reflexed at maturity; proximal, non-spikelet-bearing portions of branches generally more or less equal in length **418. *Cymbopogon***
27. Base of spikelet pair not inserted into a crater-like extension of the internode below; ultimate inflorescence branches usually not reflexed; proximal, non-spikelet-bearing portion of branches generally unequal in length 28
28 (27). Lower glume of sessile spikelet without a line of hairs; glumes of pedicellate spikelets slender, tapering; fertile lemma with two teeth, the awn originating between them; plants of Africa and Asia **431. *Hyparrhenia***
28. Lower glume of sessile spikelet with a line of hairs; glumes of pedicellate spikelets broad, acute to obtuse; fertile lemma apex entire, not two-toothed, awned from the apex; plants of Australia **398. *Spathia***
29 (24). Lower glume of sessile spikelet with an awn 30
29. Lower glume of sessile spikelet unawned 31
30 (29). Lower glume of sessile spikelet without a groove; callus of sessile spikelet not protruding beyond the rachis axis **414. *Elymandra***
30. Lower glume of sessile spikelet with a deep groove; callus of sessile spikelet obliquely attached to the rachis, and extending beyond it **428. *Parahyparrhenia***
31 (29). Inflorescence branches more than five; inflorescence axis often elongate 32
31. Inflorescence branches generally 2 or 3, up to 5, digitate or nearly so 34
32 (31). Pedicellate spikelet similar to sessile one in size and sex expression; both spikelets of the pair with callus hairs as long as the spikelet **404. *Saccharum***
32. Pedicellate spikelet differing from the sessile one, staminate or more often sterile, generally lacking an awn; spikelets generally with callus hairs shorter than the spikelet 33
33 (32). Pedicels and rachis with prominent marginal veins or keels and hyaline central region **425. *Bothriochloa***
33. Pedicels and rachis round in cross section, without prominent veins or keels **394. *Pseudosorghum***
34 (31). Sessile and pedicellate spikelets similar in morphology, sex expression and awn development 35
34. Sessile spikelets and pedicellate spikelets different; one of the pair bisexual and awned, the other staminate or sterile and unawned 38
35 (34). Glumes stiff, truncate, with three teeth; inflorescence branches two, overlapping **408. *Lophopogon***
35. Glumes tapering, if truncate, then not stiff; inflorescence branches generally more than two 36
36 (35). Upper glume with a slender awn; lower glume with hairs in tufts **357. *Eulaliopsis***
36. Upper glume unawned or muticous; hairs on glumes not tufted 37
37 (36). Callus rounded; spikelets compressed dorsiventrally **400. *Eulalia***
37. Callus sharp-pointed; spikelets round in cross section **411. *Homozeugos***
38 (34). Inflorescence branches paired, one spikelet-bearing to near the base, the other with a long naked portion (over 1.5 cm); lowermost spikelets on each branch staminate, in a whorl of four **416. *Exotheca***
38. Inflorescence branches paired or up to 5, both spikelet-bearing to near the base; lowermost spikelets not in a whorl of four 39
39 (38). Lower glume of sessile spikelet with two raised margins (keels), with a hollow or groove between them 40
39. Lower glume of sessile spikelet rounded on the back, not grooved or hollowed 42
40 (39). Callus of sessile spikelet rounded, not extended below the base of the glumes **432. *Andropogon***
40. Callus of sessile spikelet elongate, ending in a sharp point 41
41 (40). Pedicels the same color as the rachis; sessile spikelet inserted into a deep crater-like node; plants of Africa **427. *Diheteropogon***
41. Pedicels deeply pigmented; sessile spikelet not inserted into a crater-like node; plants of South America **410. *Agenium***
42 (39). Lower glume with a truncate apex, winged in some species; pedicellate spikelet reduced to a pedicel, not obvious, tucked behind and partially fused to the lower glume **406. *Apocopis***

42. Lower glume with a rounded to acute apex, without wings; pedicellate spikelet clearly present 43
43 (42). Inflorescence nodding; plants annual **356. *Euclasta***
43. Inflorescence upright; plants generally perennial **426. *Dichanthium***
44 (17). Spikelets in groups of three with one sessile and two pedicellate, laterally compressed **352. *Chrysopogon***
44. Spikelets solitary or paired, if in groups of three then these only at the ends of the branches that also bear pairs of spikelets; spikelets generally dorsiventrally compressed 45
45 (44). Pedicels and rachis internodes with a translucent median line **424. *Capillipedium***
45. Pedicels and rachis internodes without a translucent median line 46
46 (45). Pedicels ending blindly, without a spikelet 47
46. Pedicels terminated with a spikelet, although this sometimes rudimentary 49
47 (46). Upper glume keeled abaxially; plants of Southeast Asia **391. *Asthenochloa***
47. Upper glume not keeled; plants of Africa, SE Asia, and the Americas 48
48 (47). Sessile spikelet rounded in cross section; apex with two teeth or keels **397. *Sarga***
48. Sessile spikelet dorsiventrally compressed; apex tapering, not toothed **395. *Sorghastrum***
49 (46). Spikelets unpaired **392. *Cleistachne***
49. Spikelets in sessile-pedicellate pairs 50
50 (49). Sessile and pedicellate spikelets similar in shape, size and sex expression 51
50. Sessile and pedicellate spikelets differing in size, shape and sex expression; in some species the two superficially similar, but the pedicellate spikelet staminate and lacking an awn 53
51 (50). Inflorescence not hairy; glumes with prominent veins **366. *Spodiopogon***
51. Inflorescence and spikelets covered with long dense hairs; glumes without prominent veins 52
52 (51). Inflorescence hairs reddish-brown or golden-brown; primary branches short, dense **399. *Eriochrysis***
52. Inflorescence hairs silvery; primary branches spreading, diffuse **404. *Saccharum***
53 (50). Nodes with stiff spreading white hairs; awns more than 2 cm long; callus of sessile spikelet sharp-pointed in many species **397. *Sarga***
53. Nodes glabrous or pubescent, but without stiff white hairs; awns no more than 1.5 cm long; callus of sessile spikelet blunt 52
54 (53). Sessile spikelet well over 5 mm long **396. *Sorghum***
54. Sessile spikelet 5 mm long or less **360. *Lakshmia***

Group III – Inflorescence branches unbranched, not disarticulating

1. Spikelets surrounded by a set of stiff flattened, glume-like bracts 2
1. Spikelets without a cluster of flattened bracts 3
2 (1). Upper lemma awned **477. *Chaetopoa***
2. Upper lemma unawned **476. *Anthephora***
3 (1). Primary branches of the inflorescence ending in a sterile point or flexuous bristle 4
3. Primary branches of the inflorescence terminated by a spikelet 8
4 (3). Primary branches of the inflorescence very short, partially embedded in the broad flat inflorescence axis **514. *Stenotaphrum***
4. Primary branches of the inflorescence not embedded in the inflorescence axis 5
5 (4). Plants monoecious; spikelets distantly spaced along the axis, the lower ones pistillate and smaller than the upper, staminate ones; plants of wet sites in northern Australia **504. *Hygrochloa***
5. Plants with bisexual flowers; habitat various 6
6 (5). Spikelets on slender pedicels often longer than the spikelet itself; underdeveloped spikelets common; lower lemma deeply grooved **468. *Holcolemma***
6. Spikelets on short pedicels, nearly sessile; lower lemma without a groove 7
7 (6). Upper glume mucronate to awned **510. *Pseudoraphis***
7. Upper glume not mucronate or awned **511. *Setaria***
8 (3). Lower glume minute (ca. 0.1 the length of the spikelet) or absent 9
8. Lower glume well developed 17
9 (8). Spikelets with a globose swelling at the base **525. *Eriochloa***
9. Spikelets lacking a globose swelling at the base 10
10 (9). Upper lemma membranous to cartilaginous; margins broad, overlapping, lying flat on the palea 11
10. Upper lemma indurate; margins narrow, not overlapping, inrolled on the edges of the palea 13
11 (10). Spikelets lacking awns **479. *Digitaria***
11. Spikelets awned on the lower lemma 12
12 (11). Spikelets in pairs **515. *Stereochlaena***
12. Spikelets solitary **513. *Snowdenia***
13 (10). Inflorescence branches capillary; callus of spikelet with dense white hairs; plants of Madagascar **530. *Yvesia***
13. Inflorescence branches not capillary; callus of spikelet glabrous or with short hairs; plants of various locations 14
14 (13). All spikelets with the lower lemma facing the branch axis (adaxial) **453. *Axonopus***
14. All spikelets with the lower lemma facing away from the branch axis (abaxial), or spikelet orientation variable 15
15 (14). Spikelets hemispherical or orbicular in outline, often flat on one side and convex on the other; primary inflorescence branches often flattened or winged **462. *Paspalum***
15. Spikelets ovate to lanceolate, generally convex on both sides; primary inflorescence branches terete or angular, not flattened or winged 16
16 (15). Upper glume and lower lemma pubescent **452. *Acostia***
16. Upper glume and lower lemma glabrous to puberulous **454. *Anthaenantiopsis* s.l.**
17 (8). Spikelets bearing hooked hairs on one or both glumes **496. *Pseudechinolaena***
17. Spikelets lacking hooked hairs 18
18 (17). One or both glumes clearly awned 19

18. Glumes unawned, although the apex attenuate in some taxa 25
19 (18). Both glumes with long awns; spikelets paired 20
19. Only one glume with an awn; spikelets solitary or paired 21
20 (19). Plant erect, caespitose; upper lemma awned **523. *Chaetium***
20. Plants decumbent, trailing; upper lemma awnless **492. *Oplismenus***
21 (19). Awn from lower glume only, upper glume and both lemmas with at most a short mucro; plants of Madagascar and Africa 22
21. Awn from upper glume and lower lemma, lower glume awnless 23
22 (21). Upper lemma and palea cartilaginous, not hyaline; spikelets paired, similar in sex expression **495. *Poecilostachys***
22. Upper lemma and palea hyaline; spikelets paired with the sessile one staminate or sterile **486. *Cyphochlaena***
23 (21). Awn of lower lemma coiled **465. *Acritochaete***
23. Awn of lower lemma straight 24
24 (23). Upper palea reflexed; ligule often absent **467. *Echinochloa***
24. Upper palea erect; ligule present, a fringed membrane; **438. *Oplismenopsis***
25 (18). Upper lemma with an awn extending beyond the apex of the glumes 26
25. Upper lemma awnless; if awned then the awn minute, less than 1.2 mm long, not extending beyond the apex of the glumes 30
26 (25). Spikelets strongly laterally compressed **389. *Dimeria***
26. Spikelets dorsiventrally compressed or scarcely compressed 27
27 (25). Spikelets paired, the members of the pair notably different in morphology and sex expression 28
27. Spikelets all alike in morphology and sex expression 29
28 (27). Pedicellate spikelet bisexual, the sessile one staminate or sterile **412. *Trachypogon***
28. Pedicellate spikelet sterile or rarely staminate, notably smaller than the bisexual sessile spikelet **396. *Sorghum***
29 (27). Spikelets borne in clusters **484. *Alloteropsis***
29. Spikelets borne in pairs, one nearly sessile and the other clearly pedicellate **403. *Miscanthus***
30 (25). Spikelets laterally compressed **457. *Echinolaena***
30. Spikelets dorsiventrally compressed 31
31 (30). Upper lemma with a prominent beak; ligule generally absent **467. *Echinochloa***
31. Upper lemma not beaked; ligule present 32
32 (31). Rachilla internode elongate between the lower and upper flowers 33
32. Rachilla internode between the lower and upper flowers contracted 34
33 (32). Stoloniferous perennials; upper glume and lower lemma without net-like venation; plants of Central and South America **461. *Ocellochloa***
33. Annuals; upper glume and lower lemma with prominent net-like venation; plants of eastern Africa **524. *Eccoptocarpha***
34 (32). Upper glume and lower lemma with a transverse line of hairs; culms often woody **531. *Leucophrys***
34. Upper glume and lower lemma with hairs not in lines or glabrous; culms generally not woody 35
35 (34). Upper palea winged near the apex, the wings visible at maturity **459. *Hopia***
35. Upper palea not winged 36
36 (35). Apex of upper lemma with a tiny mucro **533. *Moorochloa***
36. Apex of upper lemma acute to obtuse, not mucronate 37
37 (36). Leaf blades rigid, sharp-pointed **445. *Keratochlaena***
37. Leaf blades membranous, lacking shart points 38
38 (37). Upper lemma and palea indurate 39
38. Upper lemma and palea cartilaginous to coriaceous 41
39 (38). Spikelet with thickened rachilla; upper lemma and palea covered by bicellular microhairs and simple papillae **464. *Streptostachys***
39. Spikelet with slender rachilla; upper lemma and palea without extensive bicellular microhairs; papillae simple or compound 40
40 (39). Spikelets borne on the abaxial side of the branch, solitary or paired **529. *Urochloa***
40. Spikelets borne on both sides of the branch, solitary **535. *Panicum* s.s.** (only a few *Panicum* species will key here; most will fall in Group IV)
41 (38). Lemma margins flat, largely concealing the palea 42
41. Lemma margins involute, the palea generally exposed 44
42 (41). Lower glume as long as the spikelet **480. *Megaloprotachne***
42. Lower glume shorter than the spikelet 43
43 (42). Ligule a fringed membrane **444. *Triscenia***
43. Ligule membranous **481. *Tarigidia***
44 (41). Proximal flower of spikelet staminate, with a palea; plants of Central and South America **454. *Anthaenantiopsis* s.l.**
44. Proximal flower of spikelet sterile, lacking a palea; plants of Africa and Australia **488. *Entolasia***

Group IV – Primary branches of inflorescence branched, although branches sometimes very short

1. Plants tall (up to 10 m), reed-like; leaf blades disarticulating 2
1. Plants less than 2 m tall, usually much less than 1 m; leaf blades generally persistent 3
2 (1). Plants with flowers unisexual, dioecious; spikelets disarticulating above the glumes; upper glume caudate, recurved **336. *Gynerium***
2. Plants with flowers bisexual; spikelets disarticulating below the glumes; upper glume apex erect **337. *Thysanolaena***
3 (1). Lemmas awned 4
3. Lemmas lacking awns, although awns sometimes present on the glumes 13
4 (3). Lemma awned from the apex, apex entire 5
4. Lemma awned from the sinus of a two lobed lemma 7
5 (4). Glumes lacking awns, although the upper glume apex often caudate **346. *Arundinella***

5. Glumes with awns about 10 times longer than the glume itself 6
6 (5). Callus of flower glabrous **436. *Canastra***
6. Callus of flower pubescent **435. *Arthropogon***
7 (4). Spikelets in groups of three, either all pedicellate or all sessile; glumes firm, often with pustule-based hairs **329. *Tristachya***
7. Spikelets solitary or paired, if the latter then on pedicels of unequal lengths; glumes various 8
8 (7). Callus of spikelet with two teeth, or a single sharp point **327. *Loudetia***
8. Callus of spikelet blunt 9
9 (8). Lower glume absent or a minute hair-covered flap **532. *Melinis***
9. Lower glume present, although sometimes only about 1/3 the length of the lower lemma 10
10 (9) Palea absent; plants of Cuba **345. *Reynaudia***
10. Palea present; plants of Africa and western Asia 11
11 (10). Keels of upper palea lacking wings **328. *Trichopteryx***
11. Keels of upper palea winged 12
12 (11). Pedicels of spikelets fused to the rachis; lower palea becoming thick and hard **325. *Gilgiochloa***
12. Pedicels of spikelets free from the rachis; lower palea membranous **324. *Danthoniopsis***
13 (3). Inflorescence branches verticillate, 10–60 per node **440. *Stephostachys***
13. Inflorescence not markedly verticillate 14
14 (13). Leaves stiff, coriaceous, without a clear distinction between sheath and blade; ligule often lacking 15
14. Leaf blades stiff or lax, but clearly distinct from the sheath; ligule present 17
15 (14). Plants rhizomatous; apex of leaf blades obtuse; rachilla extension present **490. *Mayariochloa***
15. Plants caespitose; apex of leaf blades sharp-pointed; rachilla extension absent 16
16 (15). Panicle open, spreading **434. *Apochloa***
16. Panicle dense, contracted **463. *Renvoizea***
17 (14). Culms woody, leaning or scrambling; upper lemma apex blunt, bearing short white hairs, with a concave depression **489. *Lasiacis***
17. Culms herbaceous, erect; upper lemma apex various, but not both concave and with short white hairs 18
18 (17). Lower palea enlarged at maturity, hardened, winged and enclosing upper flower **450. *Otachyrium***
18. Lower palea neither enlarged, hardened, nor winged 19
19 (18). One or both glumes gibbous, sac-like 20
19. Glumes not gibbous 22
20 (19). Lower glume absent **441. *Cyphonanthus***
20. Lower glume present 21
21 (20). Leaf blades with a sharp tip; spikelets clearly falcate in side view; lower lemma grooved, translucent **472. *Thyridachne***
21. Leaf blades without a sharp tip; spikelets not falcate; lower lemma not grooved **474. *Sacciolepis***
22 (19). Glumes extended into a long acuminate tail, otherwise unawned **344. *Oryzidium***
22. Glumes awnless, not extended to a long acuminate tail 23
23 (22). Upper lemma with a thickened crest at the tip 24
23. Upper lemma without a thickened crest 29
24 (23). Spikelets laterally compressed 25
24. Spikelets dorsiventrally compressed 27
25 (24). Spikelet obovate, asymmetrical, the upper lemma gibbous, enlarged abaxially **487. *Cyrtococcum***
25. Spikelet lanceolate to ovate, symmetrical, the lemma not gibbous 26
26 (25). Leaf blades broadly ovate to elliptical, narrowed toward the base, with a fringe of hairs on the abaxial side of the collar (contraligule); spikelet callus pubescent **491. *Microcalamus***
26. Leaf blades linear or lanceolate, not narrowed to the base; no hairs on the abaxial side of the collar; spikelet callus glabrous **483. *Acroceras***
27 (24). Upper lemma and palea leathery; plants of Australia **469. *Homopholis***
27. Upper lemma and palea indurate; widespread 28
28 (27). Spikelets borne all around the inflorescence branches; hairs on apex of upper lemma flattened **527. *Rupichloa***
28. Spikelets borne on one side of the inflorescence branches; hairs on apex of upper lemma cylindrical **529. *Urochloa***
29 (23). Lower glume minute or absent 30
29. Lower glume present 33
30 (29). Spikelets glabrous; plants with rhizomes ending in cleistogamous spikelets **485. *Amphicarpum***
30. Spikelets sparsely or densely pubescent; plants not producing cleistogamous spikelets 31
31 (30). Upper lemma cartilaginous **448. *Anthaenantia***
31. Upper lemma indurate 32
32 (31). Inflorescence open, the branches spreading; plants of Central America **455. *Aakia***
32. Inflorescence narrow, the branches upright; plants of South America **454. *Anthaenantiopsis***
33 (29). Spikelets attached obliquely to the pedicel 34
33. Spikelets attached upright on the pedicel 35
35 (33). Spikelets glabrous **443. *Coleataenia***
35. Spikelets with long hairs **447. *Tatianyx***
35 (33). Inflorescence narrow, dense, highly branched but the branches short, producing a bottle-brush appearance **449. *Hymenachne***
35. Inflorescence open; if narrow then the branching sparse, not brush-like 36
36 (35). Leaves with base of blades rounded to cordate, clasping the stem and/or pseudopetiolate 37
36. Leaves with base of blades narrow, neither cordate nor pseudopetiolate 46
37 (36). Spikelets minute, <2 mm long; plants of Africa **470. *Hylebates***
37. Spikelets over 2 mm long; plants of the New World 38
38 (37). Upper lemma transversely rugose **473. "*Panicum*" sects. Monticolae, Verrucosa, Ovalifoliae**
38. Upper lemma smooth, generally shiny 39
39 (38). Margins of upper lemma lying flat on the palea **437. *Homolepis***
39. Margins of upper lemma involute on the palea 40
40 (49). Upper lemma with a tiny mucro **494. *Parodiophyllochloa***
40. Upper lemma not mucronate 41

41 (40). Spikelets borne abaxially on the inflorescence branches, often crowded on short pedicels 42
41. Spikelets borne on both sides of the inflorescence branches, generally spread out on long pedicels 44
42 (41). Upper lemma and palea membranous; plants of Central and South America, the Caribbean and Australia **449. *Hymenachne***
42. Upper lemma and palea indurate; plants of southeastern North America 43
43 (42). Glumes approximately equal in length **439. *Phanopyrum***
43. Lower glume less than 1/2 the length of the upper one **451. *Rugoloa***
44 (41). Basal leaves generally forming a rosette, morphologically distinct from cauline leaves; spikelets obovate **466. *Dichanthelium***
44. Basal leaves not rosette forming, not morphologically distinct from the cauline leaves; spikelets generally globose or elliptic 45
45 (44). Leaf blades lacking a pseudopetiole, persistent **475. *Trichanthecium***
45. Leaf blades with a pseudopetiole, disarticulating from the sheath **458. *Gerritea***
46 (36). Both glumes less than 1/2 the length of the spikelet **493. *Ottochloa***
46. One or both glumes as long as the spikelet 47
47 (46). Glumes with tubercle-based hairs, lower lemma deeply sulcate **518. *Whiteochloa***
47. Glumes lacking tubercle-based hairs, lower lemma rounded 48
48 (47). Abaxial ligule (contraligule) present, fringed; plants of Madagascar **343. *Lecomtella***
48. Abaxial ligule absent; plants of various regions 49
49 (48). Plants with bulbous corms at the base, or with stout thick rhizomes with short internodes **520. *Zuloagaea***
49. Plants without corms 50
50 (49). Glumes covered with prominent bumps or warts **473. "*Panicum*" sect. *Verrucosa***
50. Glumes without warts 51
51 (50). Spikelets laterally compressed 52
51. Spikelets dorsiventrally compressed 54
52 (51). Spikelets paired; callus of upper flower expanded into two wings or scars on the base of the lemma **460. *Ichnanthus***
52. Spikelets solitary; callus of upper flower not expanded 53
53 (52). Lower glume present; lemma of lower flower awnless **534. *Tricholaena***
53. Lower glume tiny or absent; lemma of lower flower awned **532. *Melinis***
54 (51). Rachilla thickened above the upper glume and below the lower flower **442. *Oncorachis***
54. Rachilla not thickened above the upper glume and below the lower flower 55
55 (54). Upper lemma indurate, generally shiny, more firm than the glumes 56
55. Upper lemma coriaceous to membranous, about as firm as the glumes or less firm 59
56 (55). Upper lemma rugose, the apex acute to mucronate **529. *Urochloa***
56. Upper lemma smooth to slightly rough, but not strongly rugose, the apex obtuse 57
57 (56). Plants aquatic, the culms spongy, sheaths with aerenchyma **536. *Louisiella***
57. Plants terrestrial, the culms solid or hollow but not spongy; sheaths lacking aerenchyma 58
58 (57). Ligule a fringed membrane **535. *Panicum* s.s.**
58. Ligule an unfringed membrane **456. *Osvaldoa***
59 (55). Inflorescence branches deciduous; upper flower on an extended rachilla internode; plants of Australia **537. *Arthragrostis***
59. Inflorescence branches persistent; rachilla internode below the upper flower not extended; plants of the New World **454. *Anthaenantiopsis***

TRIBES AND GENERA OF PANICOIDEAE

XVII. TRIBE STEYERMARKOCHLOEAE Davidse & R.P. Ellis (1984)

Caespitose perennials; monoecious. Vegetative and flowering culms morphologically distinct, the former with only one leaf, the latter with multiple leaves that lack blades. Leaf sheaths hard and longer than culm internodes. Ligule absent. Inflorescence branches branched, short and appressed to the main axis of the inflorescence; staminate spikelets below, pistillate above. Spikelets dorsiventrally compressed to terete. Staminate spikelets with two fertile flowers and one sterile one distally. Pistillate spikelets with one fertile flower, with sterile flowers proximal and distal. Glumes shorter than the spikelet. Lemma apex truncate or obtuse. C_3.

One genus, one species.

Steyermarkochloeae are treated here as monogeneric, following the work of Sánchez-Ken and Clark (2010) and Morrone et al. (2012). Both authors exclude *Arundoclaytonia*, based on phylogenetic data. *Steyermarkochloa* is clearly a member of Panicoideae, although its precise placement remains unclear; it is sister to Tristachyideae in the combined morphological/ molecular analysis of Morrone et al. (2012), but support is lacking.

322. *Steyermarkochloa* Davidse & R. P. Ellis Fig. 59

Steyermarkochloa Davidse & R. P. Ellis, Ann. Missouri Bot. Gard. 71: 995 (1984).

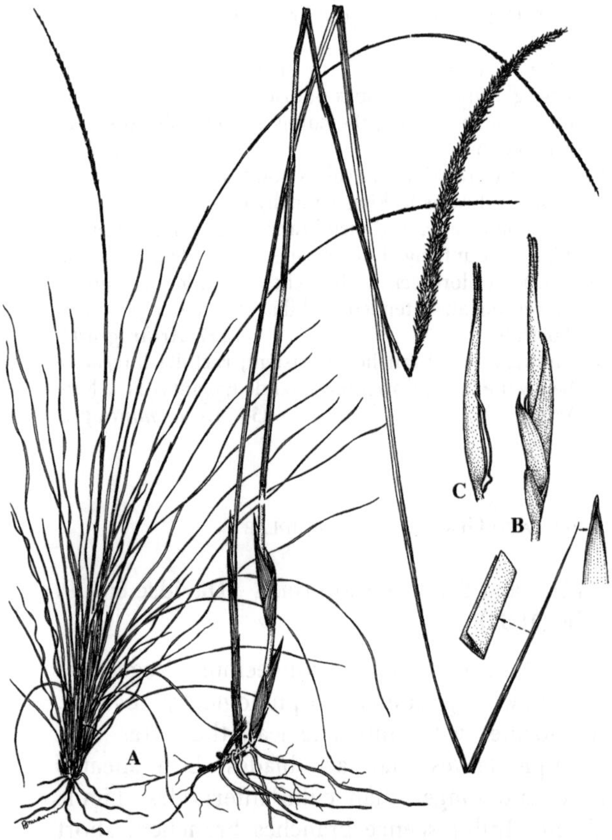

Fig. 59. Panicoideae-Steyermarkochloeae. *Steyermarkochloa angustifolia*. **A** Flowering plants. **B** Spikelet. **C** Flower. (From Davidse et al. 2004, with permission from the Missouri Botanical Garden Press; drawn by Bruno Manara)

Characters of the tribe.

One sp., *S. angustifolia* (Spreng.) Judz., northern South America.

XVIII. Tribe Tristachyideae Sánchez-Ken & L.G. Clark (2010)

Annuals or perennials with erect culms; caespitose, stoloniferous or short rhizomatous. Leaves without pseudopetioles, the ligule a line of hairs or membranous. Inflorescence unbranched to highly branched, with the spikelets in groups of two or three, *the spikelet cluster falling as a unit.* Spikelets laterally compressed or not, with two flowers, disarticulating above the glumes. Glumes 2, membranous, often a different color from the lemmas. Lemma of lower flower similar to the upper glume. Upper flower bisexual, the lemma membranous to coriaceous, not indurate, *often with tufts of hair*, with two teeth or lobes and awned from the sinus. Ancestrally C_3.

Seven genera, 89 species.

Most of the genera included here are members of the former Arundinelleae. However, that tribe is clearly an artificial assemblage representing at least two distinct lineages, *Arundinella* versus everything else (Li and Phipps 1973). *Arundinella* itself is sister to Andropogoneae and is therefore included in the latter tribe. *Chandrasekharania*, *Jansenella*, *Gilgiochloa*, and *Trichopteryx* are placed in Tristachyideae provisionally, based on the similarity of their lemmas to those of other genera included here. However, *Chandrasekharania* and *Jansenella* share some characters, including photosynthetic pathway, with Micrairoideae and may ultimately be better placed there.

323. *Chandrasekharania* V. J. Nair, Ramachandran, Sreekumar

Chandrasekharania V. J. Nair, Ramachandran, Sreekumar, Proc. Indian Acad. Sci. 91: 79 (1982).

Annuals. Ligule membranous. Inflorescence capitate. Spikelets with two *bisexual flowers*. Glumes awned. Lemma with a short awn originating between two teeth. C_3.

One sp., *C. keralensis* V.J. Nair, V.S. Ramach. & Sreek., India.

No molecular data are available for this genus. It is possible that it is misplaced here, given its C_3 leaf anatomy and spikelets with a pair of hermaphrodite flowers. Sánchez-Ken and Clark (2010) suggest that *Chandrasekharania* should be retained in Arundinelleae but without strong evidence. The genus could be a better fit for Micrairoideae.

324. *Danthoniopsis* Stapf

Danthoniopsis Stapf, Hooker's Icon. Pl. 31: pl. 3075 (1916).

Perennials or annuals. Ligule a fringe of hairs. Leaf blades with or without pseudopetioles. Spikelets laterally compressed, in pairs or groups of three, purplish. Callus of flower blunt, pubescent. Upper lemma glabrous or with 2 to 8 tufts of hairs, with two lobes, the margins clasping the palea. Upper palea with winged keels, expanded above into a club-shaped swelling. C_4, with a single bundle sheath. $2n = 24$.

Sixteen spp., Africa, east to Pakistan.

325. ***Gilgiochloa*** Pilg.

Gilgiochloa Pilg., Bot. Jahrb. Syst. 51: 415 (1914).

Annuals. Leaf margins cartilaginous. Ligule a fringe of hairs. Inflorescence narrow and spike-like, the *pedicels largely fused to the rachis*. Spikelets laterally compressed, disarticulating above the glumes. Glumes awned. Callus of flower pubescent. *Lower palea becoming thick and hard*. Upper lemma with tufts of hair at the margins and at the base of the awn, with two long narrow lobes; the awn flattened and geniculate. *Upper palea keels winged*. C_4, with a single bundle sheath.

One sp., *G. indurata* Pilg., central Africa.

326. ***Jansenella*** Bor

Jansenella Bor, Kew Bull. 10: 96 (1955).

Mat-forming annuals. Ligule membranous. Spikelets in pairs, but the inflorescence otherwise unbranched. Spikelets laterally compressed, disarticulating above the glumes. *Lower lemma with an awn*, upper lemma with two tufts of hair and two long narrow lobes. *Upper palea papillose pubescent, with winged keels*. C_3. $2n = 20$.

One sp., *J. griffithiana* (Müll. Stuttg.) Bor, India.

Türpe (1970) provides a detailed description of the leaf anatomy of this species in comparison to *Arundinella* and *Danthoniopsis*. Sánchez-Ken and Clark (2010) suggest that *Jansenella* should be retained in Arundinelleae but without strong evidence. The genus may be a better fit for Micrairoideae.

327. ***Loudetia*** Hochst. ex Steud.

Loudetia Hochst. ex Steud., Syn. Pl. Glumac. 1: 238 (1854).

Perennials. Ligule a fringe of hairs. Inflorescence branches branched or unbranched, the branches spreading or upright; spikelets borne singly or in pairs. *Callus of the spikelet with two teeth or with a single oblique sharp point*. Upper lemma glabrous to pubescent, with two teeth. *Upper palea keels thick. Stamens 2*. C_4, with one bundle sheath; isolated bundle sheath cells sometimes in place of minor veins.

Twenty-five spp., Africa, Madagascar and South America.

328. ***Trichopteryx*** Nees

Trichopteryx Nees, Nat. Syst. Bot.: 449 (1836).

Trailing annuals or perennials. Ligule a fringe of hairs. Inflorescence branches branched, spreading or upright. Spikelets laterally compressed or not compressed, disarticulating above the glumes. Callus of spikelet blunt, pubescent. Upper lemma with tufts of hair at the margins, with two long narrow lobes. C_4, with one bundle sheath; isolated bundle sheath cells sometimes in place of minor veins.

Five spp., Africa and Madagascar.

329. ***Tristachya*** Nees Fig. 60

Tristachya Nees, Fl. Bras. Enum. Pl. 2: 458 (1829).
Loudetiopsis Conert, Bot. Jahrb. Syst. 77: 277 (1957).
Dilophotriche (C. E. Hubb.) Jacq.-Fél., J. Agric. Trop. Bot. Appl. 7: 407 (1960).
Zonotriche (C.E. Hubb.) J.B. Phipps, Kirkia 4: 113 (1964).

Perennials, or a few species annual, caespitose or rhizomatous. Ligule a fringe of hairs. Spikelets in groups of three. Callus of spikelet blunt to sharp-pointed. Lower lemma with 3 to 9 veins. *Stamens 2* or 3. C_4, with one bundle sheath. $2n = 20$, 24, 40.

Forty spp., Central and southern Africa, Central and South America, Madagascar.

In contrast to Andropogoneae, in which spikelets are in sessile and pedicellate pairs or triplets, in *Tristachya* all three spikelets of a group are pedicellate or (in some African species) all sessile. In some species, the common branch supporting the cluster of three is sharply bent at maturity, with a point of disarticulation in the middle of the bend. The African species have pustule-based macrohairs on the glumes, the pustule itself black and contrasting with the light-colored hairs and glume surface. These

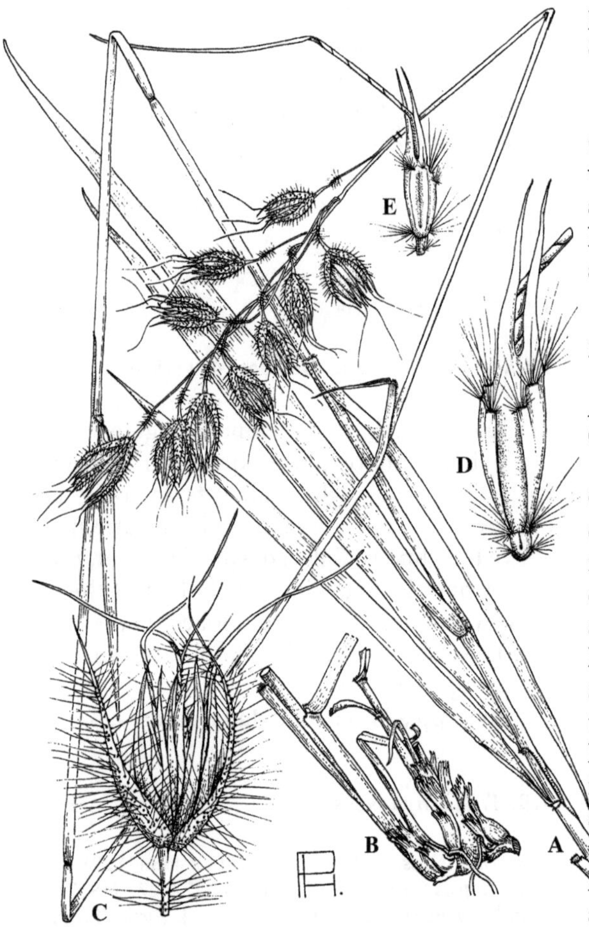

Fig. 60. Panicoideae-Tristachyideae. *Tristachya decora*. **A** Habit. **B** Rootstock. **C** Triad of spikelets. **D** Upper lemma, viewed from the adaxial side. **E** Upper lemma, viewed from the abaxial side. (From Clayton et al. 1974, drawn by P. Halliday)

distinctive characters of the inflorescence have led to recognition of several small segregate genera, but the group as a whole appears to be easily accommodated in a single genus.

XIX. Tribe Chasmanthieae W. V. Br. & B. N. Smith ex Sánchez-Ken & L. G. Clark (2010)

Zeugiteae Sánchez-Ken & L. G. Clark (2010).

Perennials, caespitose, rhizomatous, or stoloniferous. Leaf blades broad, with or without pseudopetioles. Inflorescence branches branched. Spikelets with one to many flowers, laterally compressed in most species, often strongly so. Stamens 1, 2 or 3. C_3.

Six genera, 25 species.

Chasmanthieae plus Zeugiteae form a clade (GPWG II 2012; Sánchez-Ken and Clark 2010), with reasonably good support in the GPWG II analysis; accordingly, the two are combined here. One species of *Zeugites* is annual, presumably a derived condition.

330. ***Arundoclaytonia*** Davidse & R. P. Ellis

Arundoclaytonia Davidse & R. P. Ellis, Ann. Missouri Bot. Gard. 74: 479 (1987).

Caespitose perennials, to 3 m tall, with an unusual elevated tussock, the base of the plant covered by remnants of leaf sheaths and aerial roots; monoecious. Leaf insertion reported to be spiral. Ligule a fringed membrane. Staminate and pistillate spikelets in separate capitate clusters, each cluster surrounded by bracts. Staminate spikelets with 3 to 9 flowers, glumes generally with cross veins; lemmas mucronate. Pistillate spikelets with one fertile flower, the sterile flowers both proximal and distal; fertile lemma pubescent, with 9 to 11 veins and cross veins. Palea of the pistillate flower longer than the lemma, curved, thickened and smooth below, convolute. C_3.

One sp., *A. dissimilis* Davidse & R. P. Ellis, central Brazil.

A partial sequence of *ndhF* places *Arundoclaytonia* near *Chasmanthium* (Sánchez-Ken and Clark 2007); when the same sequence is combined with morphological data, the species falls within *Chasmanthium* (Morrone et al. 2012), a position that needs to be confirmed. *Arundoclaytonia* was included with *Steyermarkochloa* in the tribe Steyermarkochloeae, but the two genera do not appear to be closely related (Morrone et al. 2012).

331. ***Chasmanthium*** Link

Chasmanthium Link, Hort. Berol. 1: 159 (1827).
Bromuniola Stapf & C.E. Hubb., Bull. Misc. Inform. Kew 1926: 366 (1926).
Gouldochloa Valdés-Reyna, Morden & S.L. Hatch, Syst. Bot. 11: 112 (1986).

Perennials. Culms hollow or solid. Ligule a fringed membrane. Primary branches of the inflorescence branched or unbranched. Spikelets with 2 to 10 flowers, the proximal 1 or 2 sterile. *Paleas gibbous*, the keels winged. *Stamens 1* (3). $2n = 24, 48$.

Seven spp., eastern USA, Mexico, tropical Africa.

Molecular data clearly unite *Chasmanthium* with *Bromuniola* and *Gouldochloa* (Sánchez-Ken and Clark 2007, 2010).

Remaining genera of the tribe:

Leaves with a pseudopetiole. Spikelets disarticulating below the glumes.

332. ***Chevalierella*** A. Camus

Chevalierella A. Camus, Rev. Bot. Appl. Agric. Trop. 13: 421 (1933).

Perennials, 1–2 m tall. Culms hollow. Leaf blades broad. Ligule a ciliolate membrane. Primary branches of the inflorescence unbranched. Spikelets with one fertile flower plus a rudimentary one distally. Glumes awned, the rachilla internode between them elongated. Lemmas awned, palea keels winged. *Lodicules fused.* Stamens 2.

One sp., *C. dewildemannii* (Vanderyst) Van der Veken ex Compère, Zaire.

Remaining genera of the tribe:

Mesophyll with distinct palisade and spongy mesophyll layers.

333. ***Lophatherum*** Brongn.

Lophatherum Brongn., Voy. Monde 2: 49 (1829) [1831].

Perennials *with tuberous roots.* Culms solid or hollow. Ligule a ciliolate membrane. Primary branches of the inflorescence unbranched, with the spikelets borne on one side. *Spikelets with one fertile flower plus up to 9 sterile lemmas. Lemmas awned, those of sterile lemmas forming a set of recurved hooks.* Stamens 2 or 3. $2n = 48$.

Two spp., tropical Asia.

334. ***Orthoclada*** P. Beauv. Fig. 61

Orthoclada P. Beauv., Ess. Agrostogr. 69 (1812).

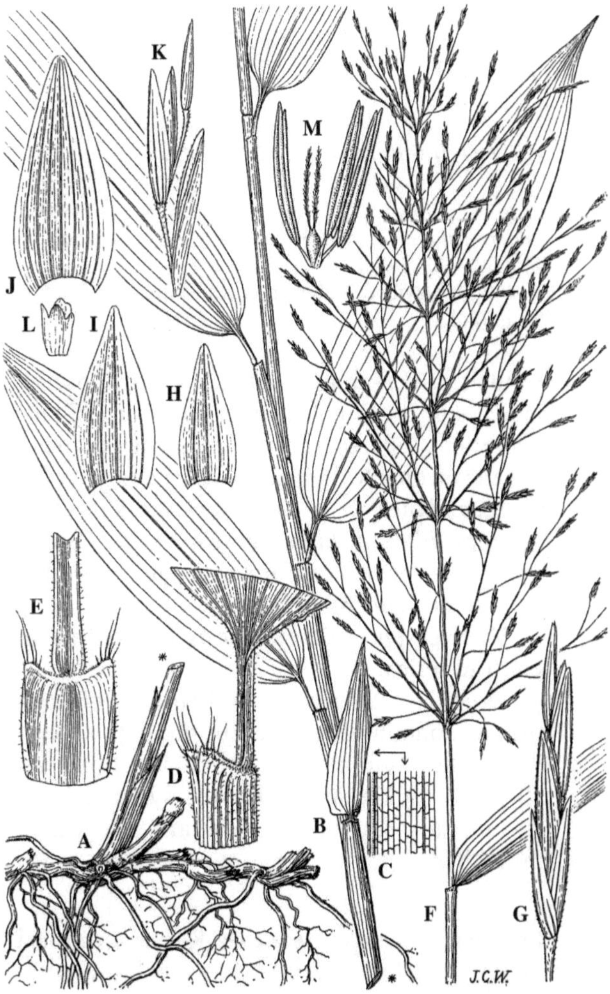

Fig. 61. Panicoideae-Chasmanthieae. *Orthoclada africana*. **A** Rhizome. **B** Leafy culm. **C** Part of underside of leaf blade showing tessellate venation. **D** Pseudopetiole. **E** Ligule. **F** Inflorescence. **G** Spikelet. **H** Lower glume. **I** Upper glume. **J** Lemma. **K** Palea and sterile lemmas. **L** Lodicules. **M** Flower. (From Clayton 1970a)

Perennials. Culms hollow. Leaf blades narrow. Ligule membranous. Primary inflorescence branches branched or unbranched. Spikelets with 1 to 4 flowers, the rachilla ending in a rudimentary flower. *Rachilla internode partially fused to keels of palea.* Lemmas acuminate but not awned. Stamens 2 or 3. $2n = 24$.

Two spp., Tropical America and Africa.

Orthoclada is sister to *Zeugites* s.l. plus *Lophatherum* in molecular phylogenies (Sánchez-Ken and Clark 2007, 2010). It is distinguished by its bisexual flowers and the unusual fusion of its rachilla internode to the palea keels.

335. *Zeugites* P. Browne

Zeugites P. Browne, Civ. Nat. Hist. Jamaica 341 (1756).
Calderonella Soderstr. & H.F. Decker, Ann. Missouri Bot. Gard. 60: 427 (1973).
Pohlidium Davidse, Soderstr. & R.P. Ellis, Syst. Bot. 11: 131 (1986).

Annuals or perennials. Leaf base narrow, pseudopetiolate or not. Ligule membranous or a ciliolate membrane. Inflorescence with branched branches. Spikelets with 2 to 15 *unisexual flowers, with both sexes combined in a single spikelet. Lowermost flower pistillate, distal flowers staminate. Pistillate flower often with the lemma gibbous, sometimes strongly so*. Stamens 3. $2n = 24, 48$.

Twelve spp., mid-elevations in Central and South America.

Zeugites, *Calderonella*, and *Pohlidium* clearly form a monophyletic group in the data analyses of Sánchez-Ken and Clark (2007, 2010) and GPWG II (2012), and share the synapomorphy of unisexual spikelets; they are accordingly united here.

XX. Tribe Gynerieae Sánchez-Ken & L.G. Clark (2001)

Plants rhizomatous, perennial, reed-like, up to 10 m tall, *dioecious*. Culms solid. Leaf blades *disarticulating from the sheaths*. Ligule a line of hairs or a fringed membrane; abaxial ligule present or absent. Inflorescences highly branched. Spikelets disarticulating above the glumes. Pistillate spikelet with 2 flowers, laterally compressed, the upper glume much longer than the lemmas, *extended into a long attenuate tip, recurved*; *lemma plumose*, with attenuate tip; staminodes 2. Staminate spikelet with 2 to 4 flowers, disarticulating below the most distal of these, glabrous or sparsely pubescent, the glumes and lemmas not extended; *stamens 2*. C_3.

336. *Gynerium* Willd. ex P. Beauv. Fig. 62

Gynerium Willd. ex P. Beauv., Ess. Agrostogr. 138 (1812).

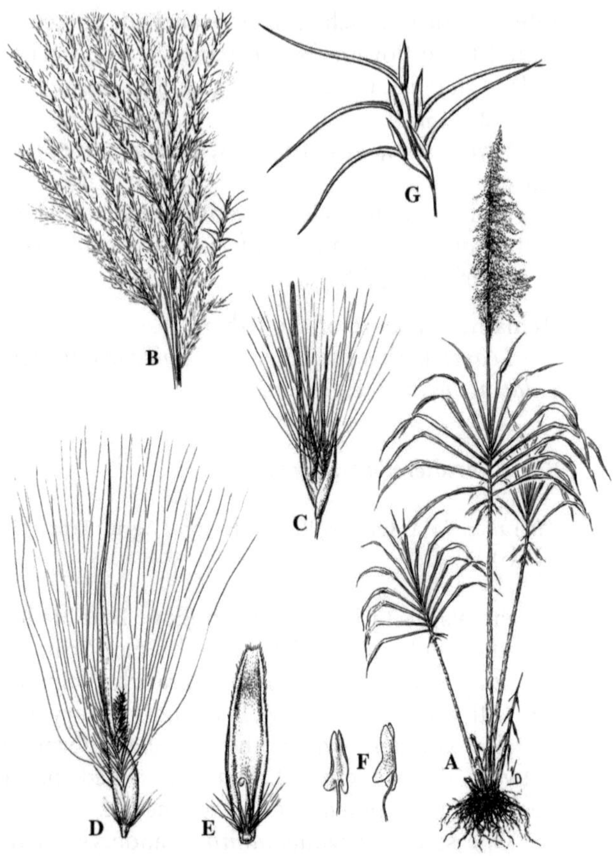

Fig. 62. Panicoideae-Gynerieae. *Gynerium sagittatum*. **A** Plant. **B** Portion of pistillate inflorescence. **C** Pistillate spikelet, lateral view. **D** Flower of the pistillate spikelet. **E** Palea and rachilla, viewed from the adaxial side. **F** Staminodes. (From Zuloaga et al. 2012)

Characters of the tribe. $2n = 44$.

One sp., *G. sagittatum* (Aubl.) P. Beauv., Central and South America, West Indies.

XXI. Tribe Centotheceae Ridl. (1907)

Thysanolaeneae C. E. Hubb. (1934).
Cyperochloeae L. Watson & Dallwitz ex Sánchez-Ken & L. G. Clark (2010).

Annuals or perennials. Spikelets with few to many flowers, laterally compressed, reduced flowers either proximal or distal to the bisexual ones. C_3.

Five genera, eight species.

This is a small tribe of only a few species of Africa, southern Asia, Australia and the Pacific. Although the gross morphology of the plants is not at all similar, the group is strongly supported by molecular data (GPWG II 2012; Morrone et al. 2012; Sánchez-Ken and Clark 2010). While the group can be divided into three tribes, this seems unnecessary given the small number of genera and species.

337. ***Thysanolaena*** Nees

Thysanolaena Nees, Edinburgh New Philos. J. 18: 180 (1835).

Caespitose perennials, 2–4 m tall. Leaf blades disarticulating, with cross veins. Ligule membranous. Inflorescence branches branched, the branches lax and spreading. Spikelets with two flowers, disarticulating below the glumes. Glumes about the same length, shorter than the spikelet. Lower flower reduced to a sterile lemma, upper flower bisexual, the lemma lanceolate. C_3, mesophyll cells with invaginated cell walls. $2n = 24$.

One sp., *T. latifolia* (Roxb. ex Hornem.) Honda, tropical Africa, Asia, and the Pacific.

Although the spikelet morphology of this species is superficially similar to that of Panicoideae s.s., the rachilla extends as a short stub beyond the upper flower, a character rarely seen in Panicoideae.

Cyperochloa + Spartochloa

Perennials with wiry culms. Ligule a fringe of hairs. Spikelets with 4 to 9 flowers, the distal ones smaller. Glumes approximately the same length, shorter than the lowest lemma.

338. ***Cyperochloa*** Lazarides & L. Watson

Cyperochloa Lazarides & L. Watson, Brunonia 9: 216 (1986).

Plants rhizomatous. Leaf blades filiform, *blades lacking on leaves immediately subtending the inflorescence*. Inflorescence a cluster of 2 to 5 spikelets, these sessile or nearly so. Glumes pubescent. Lemmas and paleas pilose.

One sp., *C. hirsuta* Lazarides & L. Watson, SW Australia.

The microhairs shown in the figure accompanying the original description of the genus have a broad apical cell and may be similar to those of *Centotheca* and *Megastachya*.

339. ***Spartochloa*** C. E. Hubb.

Spartochloa C. E. Hubb., Kew Bull. 7: 308 (1952).

Caespitose. *Leaf blades lacking, the culms and sheaths photosynthetic*. Primary branches of inflorescence branched. Glumes, lemmas and paleas glabrous. Fruit trigonous, pitted.

One sp., *S. scirpoidea* (Steud.) C.E. Hubb., SW Australia.

Megastachya + Centotheca

Mushroom-button microhairs on leaves.

340. ***Megastachya*** P. Beauv.

Megastachya P. Beauv., Ess. Agrostogr. 74 (1812).

Annuals. Base of leaf blades clasping the stem. Primary inflorescence branches branched or not. Spikelets with 8 to 20 flowers, these all similar but the distal ones somewhat smaller. Lemmas mucronate.

Two spp., forests of tropical Africa.

341. ***Centotheca*** Desv. Fig. 63

Centotheca Desv., Nouv. Bull. Sci. Soc. Philom. Paris 2: 189 (1810), nom. conserv.

Annuals. Leaf blades narrow, not clasping the stem. Primary inflorescence branches branched or not. Spikelets with 1 to 4 flowers, these all similar but the distal ones somewhat smaller. *Lemma with reflexed bristles. Stamens 2.*

Four spp., West Africa, Asia, Queensland, Pacific islands.

Panicoideae s.s. clade

Flowers two per spikelet, the upper one hermaphrodite, staminate, or pistillate, and the lower either staminate or reduced to an empty lemma.

This clade is the group that is called Panicoideae in most of the grass literature. It is one of the

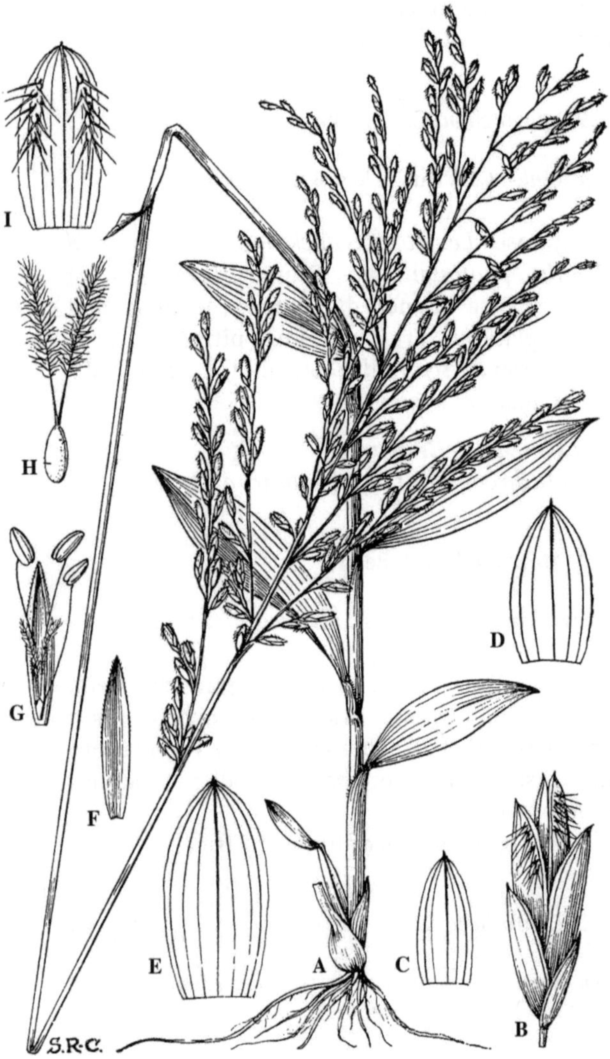

Fig. 63. Panicoideae-Centotheceae. *Centotheca lappacea* (L.) Desvaux. **A** Habit. **B** Spikelet. **C** Lower glume. **D** Upper glume. **E** Unarmed lemma. **F** Palea. **G** Palea. **H** Pistil. **I** Armed lemma. (From Bor 1960, drawn by S. Ross-Craig)

most distinctive groups of grasses and has been recognized for two centuries, since the work of Robert Brown (1810, 1814). The spikelet morphology is diagnostic and synapomorphic.

Incertae sedis

342. ***Hydrothauma*** Hubbard

Hydrothauma Hubbard, Hooker's Icon. Pl. 35: t3458 (1947).

Aquatic annuals. *Leaf blades floating, the adaxial surface with sinuous flattened projections or lamellae.* Ligule membranous, auricles prominent. Inflorescence with unbranched branches, these appressed to the central axis. Spikelets pedicellate. Lower glume hyaline, without veins, *upper glume gibbous.* Upper lemma and palea leathery. C_3.

One sp., *H. manicatum* C. E. Hubb., Zambia and Zaire. Aquatic.

Molecular data are unavailable for *Hydrothauma*. In the analyses of Morrone et al. (2012), it was resolved as sister to one or more of the tribes of Panicoideae s.s.

343. ***Lecomtella*** A. Camus

Lecomtella A. Camus, Compt. Rend. Hebd. Séances Acad. Sci. 181: 567 (1925).

Rhizomatous perennials, monoecious. Ligule a fringe of hairs. Abaxial ligule a line of hairs. Inflorescence branches with 1 to 4 staminate spikelets proximally, and a single bisexual spikelet distally. Bisexual spikelets with the lower flower staminate, upper flower pistillate, lower and upper glumes separated by a distinct internode, rachilla internode below upper flower winged; upper flower with lemma pubescent, tuberculate, the apex ending in three appendages. C_3. Mesophyll with non-radiate chlorenchyma.

One sp., *L. madagascariensis* A. Camus, Madagascar.

Lecomtella was provisionally placed sister to *Gerritea* in Paspalinae by Morrone et al. (2012) based on morphological data, but molecular data (Besnard et al. 2013) place it in an unresolved position at the base of Panicoideae. The tribal name Lecomtelleae Pilg. ex Potztal is available for the genus. However, the tribe is not recognized here pending more definitive phylogenetic data.

344. ***Oryzidium*** C.E. Hubb. & Schweick.

Oryzidium C.E. Hubb. & Schweick., Bull. Misc. Inform. Kew 1936: 326 (1936).

Perennial, aquatic and floating. Ligule a fringe of hairs. Inflorescence branches branched. Lower glume a tiny scale, upper glume longer than the spikelet, awned. Rachilla internode elongated between flowers. Lower flower staminate, the upper pistillate. C_4, with two bundle sheaths and even outline.

One sp., *O. barnardii* C.E. Hubb. & Schweick., Zambia and SW Africa.

Leaf anatomy suggests that this species may be either in Melinidinae or Panicinae, whereas the long internode between the flowers is reminiscent of *Lecomtella*. The phylogenetic study of Morrone et al. (2012) places it either within Melinidinae (Paniceae), or sister to *Reynaudia* (Paspaleae) based on morphology; molecular data are unavailable.

345. *Reynaudia* Kunth

Reynaudia Kunth, Révis. Gramin. 1: 72 (1829).

Perennials. Ligule a fringed membrane. Leaf blades filiform. Inflorescence branches branched, erect. Upper and lower glumes and the lower lemma as long as the spikelet, *bifid, each with a single awn*. Upper lemma smooth, acute. Upper palea reduced or lacking. Stamens 2. C_4, with one bundle sheath.

One sp., *R. filiformis* (Spreng. ex Schult.) Kunth, West Indies.

Morrone et al. (2012) place *Reynaudia* sister to Paspaleae but with weak to moderate support. The spikelet morphology is quite different from that of other Paspaleae, and the chromosome number is unknown.

XXII. Tribe Andropogoneae Dumort. (1824)

Arundinelleae Stapf (1898).
Garnotieae Tateoka (1957).

Ligule membranous. Inflorescence complex and difficult to describe using conventional terms (see below). Spikelets paired, with one sessile or nearly so, the other pedicellate; branches often ending with three spikelets, one sessile and two pedicellate. Glumes firm, sometimes indurate and lignified. Lemma and palea hyaline. *Rachis disarticulating* so that the spikelet pair becomes the dispersal unit. All taxa *C_4, NADP-ME subtype*; veins with a single bundle sheath.

90 genera, 1274 spp.

Andropogoneae are monophyletic in all molecular phylogenetic studies (Bomblies and Doebley 2005; Estep et al. 2014; GPWG II 2012; Hodkinson et al. 2002; Kellogg 2000; Lukens and Doebley 2001; Mathews et al. 2002; Skendzic et al. 2007; Spangler et al. 1999; Teerawatananon et al. 2011), with the genus *Arundinella* (formerly the type genus of the tribe Arundinelleae) sister to remaining members of the tribe. Molecular data indicate that the tribe is the result of a rapid radiation, but the forces driving this radiation are unknown. Allopolyploidy is remarkably common and can be documented for at least a third of the species in the tribe (Estep et al. 2014).

Members of Andropogoneae have paired spikelets, one sessile and one pedicellate, although this characteristic is shared with many other Panicoideae (Kellogg 2000; Zanotti et al. 2010). Throughout the tribe, the pedicellate spikelet is often staminate, but in a few genera it is hermaphrodite and in still others pistillate. Anatomically, epidermal papillae appear to have been ancestrally absent, and costal short cells ancestrally in long rows (Watson and Dallwitz 1992 onward). In addition, in most species the rachis (inflorescence axis) breaks up at maturity and the glumes are hardened, particularly in comparison with the lemma. The disarticulating main axis is lost independently in several genera, whereas in other genera the lateral branches disarticulate as well. There are many changes in glume texture, from membranous to coriaceous to extremely hard and back (Preston et al. 2012). Most species have lemmas that bear an awn that is usually twisted and hygroscopic.

All Andropogoneae use the C_4 photosynthetic pathway with NADP-ME as a decarboxylating enzyme. Vascular bundles of the leaf have a single sheath (Hattersley and Watson 1975). This photosynthetic pathway and the accompanying leaf anatomy are synapomorphies for the tribe (GPWG II 2012).

The overall form of the inflorescence in Andropogoneae varies extensively even within clades, and this may be why so few of the Clayton and Renvoize (1986) subtribes, which rely heavily on inflorescence form, have proven to be monophyletic (Kellogg 2000; Mathews et al. 2002; Skendzic et al. 2007; Spangler et al. 1999).

Reduction of the pedicellate spikelet has occurred apparently more than once and does not by itself indicate relationship of genera. The tribe has been heavily split such that many genera have only one or two species. Watson and Dallwitz (1992 onward) recognize 111 genera in the tribe, of which 32 are monotypic. Clayton and Renvoize (1986) combine a number of these, and recognize 85 genera, of which 21 still include only a single species, and another eight include two species.

The complex form of "inflorescences" in Andropogoneae reflects variation in several characters:

1) Proliferation of branches at upper culm nodes. Whereas most monocots, and indeed most grasses, have a single axillary branch and adaxial prophyll at a node, at the upper culm nodes of many Andropogoneae, the axillary branches themselves branch, forming a prophyll and another branch but without a corresponding subtending leaf. Each branch of this complex can then terminate in an inflorescence, or can reiterate the branch complex, creating compound structures. The leaves subtending the entire complex are variously called leaves or spathes in the literature, and may be developed or suppressed. The entire upper culm, including all branch complexes, is often called an inflorescence in the literature, but here the term inflorescence is reserved for whatever structure develops after the uppermost leaf or spathe.
2) The ultimate inflorescences may produce branches (often called "racemes" or "rames" in the literature), which initiate in a spiral phyllotaxis. They may develop continuously with the apical meristem, or branch elongation may be delayed relative to the development of the primary axis. Each branch then produces spikelet pair meristems in a two-ranked phyllotaxis.
3) Both the sessile and pedicellate spikelets of a pair may develop, or one member of the pair may be reduced or suppressed entirely. Often the reduction occurs in the pedicellate spikelet, but sometimes the sessile spikelet is the one that is reduced. The two spikelets may be identical in morphology and sex expression, but more often they are different.
4) Both flowers of a spikelet may develop or one may be suppressed.

Many of the possible combinations of these four characters are observed, making key construction difficult. The rapid radiation documented by molecular phylogenies suggests that the morphological diversity might be controlled by a handful of genes that were easily modified and/or polymorphic at one time in evolutionary history. For discussion on the valid name of the tribe Andropogoneae, see Welker et al. (2014).

XXII.1. Subtribe Arundinellinae Honda (1930)

Annuals or perennials, caespitose or rhizomatous. Ligule membranous, or a fringed membrane. Inflorescence branches branched, neither the rachis nor the branches disarticulating at maturity. Spikelets generally paired, the "sessile" one actually short pedicellate. Disarticulation above the glumes. *Leaves with minor veins reduced to single isolated bundle sheath cells* ("distinctive cells").

Two genera, 87 species.

This subtribe is monophyletic in the analyses of GPWG II (2012) and strongly supported as sister to all other Andropogoneae. While Teerawatananon et al. (2011) also find *Arundinella* to be sister to all other Andropogoneae, they place *Garnotia* sister to *Eremochloa*. *Garnotia* and *Arundinella* share enough morphological and anatomical characters that placing them together seems plausible, although this needs to be checked with more data.

346. ***Arundinella*** Raddi Fig. 64

Arundinella Raddi, Agrostogr. Bras. 36 (1823).

Callus pubescent. Glumes the same texture as or thinner than the lemmas, the apex acute to muticous, the upper glume apex often caudate. Upper lemma apex generally awned. $2n = 14, 20, 28, 36, 56$.

Fifty-seven spp., most Asian, others tropics and subtropics, north to Japan.

347. ***Garnotia*** Brongn.

Garnotia Brongn., Voy. Monde 2: 132–133, pl. 21 (1830).

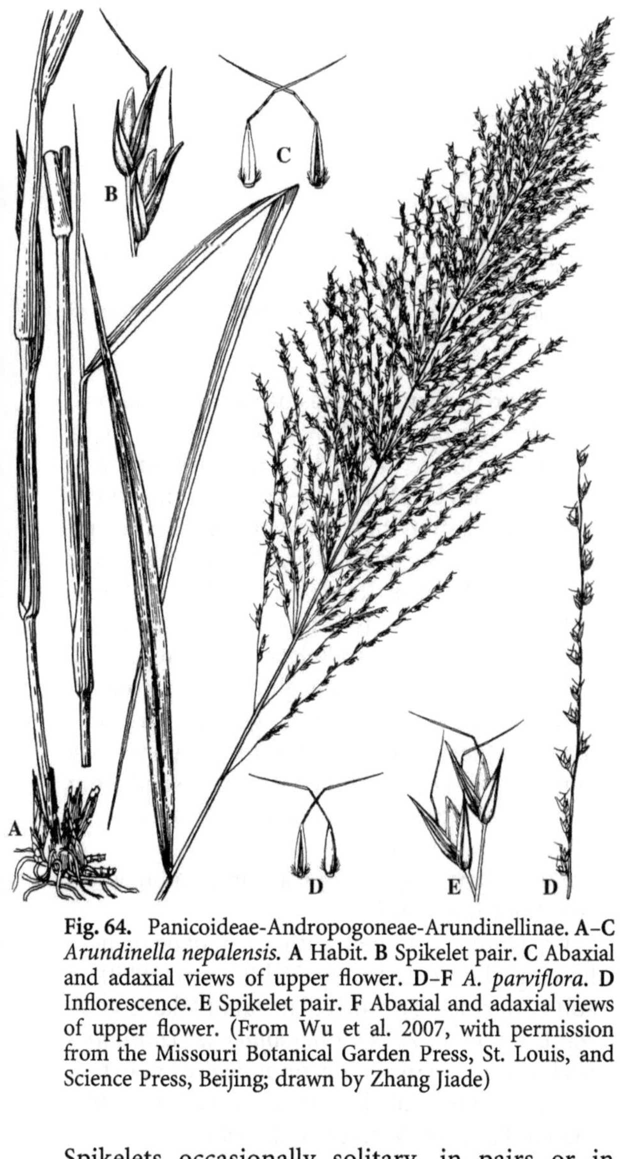

Fig. 64. Panicoideae-Andropogoneae-Arundinellinae. **A–C** *Arundinella nepalensis.* **A** Habit. **B** Spikelet pair. **C** Abaxial and adaxial views of upper flower. **D–F** *A. parviflora.* **D** Inflorescence. **E** Spikelet pair. **F** Abaxial and adaxial views of upper flower. (From Wu et al. 2007, with permission from the Missouri Botanical Garden Press, St. Louis, and Science Press, Beijing; drawn by Zhang Jiade)

Spikelets occasionally solitary, in pairs or in groups of three. Spikelets laterally compressed, *with a single flower, the proximal sterile flower absent;* callus glabrous or pubescent. Glumes thinner than the lemmas, the apex muticous or awned. Upper lemma apex muticous or awned. $2n = 20$.

Thirty spp., Africa, Asia, Australia, Pacific.

ANDROPOGONEAE *s.s.*

Possible synapomorphies for this clade include a disarticulating rachis and a twisted hygroscopic awn on the lemma (Estep et al. 2014).

ANDROPOGONEAE *incertae sedis*

348. ***Apluda*** L.

Apluda L., Sp. Pl. 1: 82 (1753).

Decumbent or scrambling perennials. Ligule a fringed membrane. Axillary branch complexes present. Inflorescences unbranched, each with a spathe-like subtending leaf; internodes and pedicels thickened, clavate. Spikelets borne in triads, with one sessile and two pedicellate, *pedicels flat and broad. One pedicellate spikelet staminate, the other vestigial, represented by a solitary glume. Sessile spikelet with bulbous callus; lower glume winged on keel. Upper glume gibbous.* Upper lemma muticous or awned from a sinus, the awn twisted or straight. $2n = 20, 40$.

One sp., *A. mutica* L., tropical Asia.

349. ***Arthraxon*** P. Beauv.

Arthraxon P. Beauv., Ess. Agrostogr. 111 , pl. 11, f. 6 (1812).

Annuals or perennials, decumbent. *Leaves broadly lanceolate with cordate base.* Ligule membranous, or a fringed membrane or a fringe of hairs. Inflorescence unbranched or the branches digitate. Pedicellate spikelets staminate, sterile, reduced to a pedicel or lacking. Sessile spikelets with the callus glabrous or pubescent, the lower glume smooth or rugose or with prickles, variously toothed on the margins, *both flowers generally lacking paleas.* Upper lemma apex entire or lobed, awned, *the awn abaxial, attached well below the midpoint of the lemma. Stamens 2* or 3. $2n = 18, 20, 36, 40$.

Twenty-seven spp., Old World tropics, mostly India, some species weedy in other warm regions of the world.

350. ***Chasmopodium*** Stapf

Chasmopodium Stapf, Fl. Trop. Afr. 9: 76 (1917).
Robynsiochloa Jacq.-Fél., J. Agric. Trop. 7: 406 (1960).

Annuals, 2–4 m tall. Ligule a fringe of hairs or membranous. Inflorescence branches unbranched. Internodes and pedicels thickened, flattened.

Sessile spikelet callus pubescent, with a central peg. Pedicellate spikelets sterile. Lower glume of sessile spikelet with narrow wings, *upper glume laterally compressed, winged on keel.* Lemma without an awn. *Embryo 4/5 as long as caryopsis.* $2n = 16, 20$.

Three spp., tropical Africa.

Clayton and Renvoize (1986) synonymize *Robynsiochloa* with *Rottboellia*, but that seems to be just one of several possible placements. Clayton (1970b) indicates that *Robynsiochloa* and *Chasmopodium* are "barely distinct" in a numerical taxonomic analysis, and later provides the necessary combination in *Chasmopodium* (Clayton 1973). The winged upper glume is distinctive and possibly indicates monophyly. The internode and pedicel are fused in one species.

351. ***Chionachne*** R. Br.

Chionachne R. Br., Pl. Jav. Rar. 15 (1838); rev., Jannink and Veldkamp (2002).
Sclerachne R. Br., Pl. Jav. Rar. 15 (1838).
Polytoca R. Br., Pl. Jav. Rar. 20 (1838).
Cyathorhachis Nees ex Steud., Syn. Pl. Glumac. 1: 403 (1854).
Trilobachne M. Schenck ex Henrard, Meded. Rijks-Herb. 67: 4 (1931).

Annuals or perennials, monoecious. Ligule membranous or a fringed membrane. *Inflorescence unbranched, with pistillate spikelets below, staminate above, or with several primary branches bearing staminate spikelets only. Pistillate spikelets unpaired or with staminate pedicellate spikelets; lower glumes hard, shining, surrounding the rachis, enlarged to look reminiscent of the spathe in Coix.* Staminate spikelets paired, two-flowered, pedicellate spikelets reduced to an empty glume. Lower glume of the pistillate spikelet lobed, obtuse or notched at the apex, with two obvious constrictions in one species. Lemmas without awns. Caryopsis flattened, concave-convex. $2n = 20, 40$.

Twelve spp., India, SE Asia, Australia, and Polynesia.

Four small genera are included here in *Chionachne*, based on their shared inflorescence characters and overlapping geographic distribution. Teerawatananon et al. (2011) and Estep et al. (2014) confirm the monophyly of *Chionachne* plus *Polytoca*. *Chionachne* falls in an unresolved position among the awnless Andropogoneae in molecular phylogenies (Estep et al. 2014; GPWG II 2012; Teerawatananon et al. 2011).

352. ***Chrysopogon*** Trin.

Chrysopogon Trin., Fund. Agrost. 187 (1820), nom. conserv.; rev.: Veldkamp (1999).
Vetiveria Bory, Bull. Sci. Soc. Philom. Paris 1822: 43 (1822).

Annuals or perennials, caespitose, rhizomatous, or stoloniferous. Ligule membranous, or a fringed membrane, or a fringe of hairs. Inflorescence branches branched, the spikelets in pairs or triads, if the latter then one spikelet sessile and two pedicellate. *Spikelets laterally compressed*, the callus truncate to acute or pungent, the attachment oblique or inserted into the cup-like internode apex. Pedicellate spikelets staminate, sterile, or reduced to glumes, the glumes muticous to awned. Sessile spikelets with glumes membranous to leathery but not indurate, the upper one mucronate to awned. Upper lemma awned from the apex or from a sinus, the awn straight or twisted. $2n = 20, 40$.

Forty-eight spp., tropical and subtropical throughout the world.

Vetiveria is merged with *Chrysopogon* here following the revision by Veldkamp (1999), a result confirmed by molecular data (Adams et al. 1998; Estep et al. 2014). *Chrysopogon zizanioides* (also known as vetiver) is economically important as a source of oil and also for erosion control (Joy, undated; Panel on Vetiver et al. 1993).

353. ***Clausospicula*** Lazarides

Clausospicula Lazarides, Austral. Syst. Bot. 4: 399 (1991).

Annuals. Leaf blades with margins cartilaginous, the ligule a fringed membrane. Inflorescence unbranched, with one sessile and two pedicellate spikelets, the set of three surrounded by the trumpet-shaped tip of the peduncle. Pedicellate spikelets well developed or rudimentary. Sessile spikelets with a pubescent callus, sharp-pointed, the lower glume rugose, muticous or awned, upper glume with a long awn. Both flowers of a spikelet lacking paleas. Lemma apex awned, the awn geniculate, pubescent.

One sp., *C. extensa* Lazarides, N. Australia.

Clausospicula is somewhat similar to the southeast Asian species *Kerriochloa siamensis*. Clayton et al. (2006 onward) provide a slightly different interpretation of the inflorescence morphology, and suggest that the structure here described as the tip of the peduncle is in fact a bract subtending the branch of a highly branched inflorescence. Another interpretation would suggest that the peduncle tip is homologous to the flattened structure at the base of the branches in *Hyperthelia*.

354. *Coix* L.

Coix L., Sp. Pl. 2: 972 (1753).

Annuals or perennials, caespitose or rhizomatous, monoecious. Ligule membranous. *Inflorescence a complex branched structure with staminate spikelets borne distally on the primary axis; lateral branch demarcated by a prophyll, forming in the axil of a highly modified leaf ("utricle"); lateral branch bearing a triad of spikelets; two pedicellate ones develop thickened pedicels and initiate spikelet parts but do not develop further. The sessile spikelet becomes pistillate.* Glumes membranous, not indurate. Lemmas similar in texture to the glumes, without awns. Embryo as long as the caryopsis. $2n = 10, 20, 40$.

Four spp., tropical Asia.

The "utricle" is clearly a modified leaf sheath with fused margins and only a rudimentary blade. Sheaths elsewhere in the inflorescence have fused margins as well. Many authors (e.g., Hitchcock 1950; Watson and Dallwitz 1992 onward) have suggested a relationship between this genus and *Zea*, but molecular data do not support this idea. *Coix* falls in an unresolved position among the awnless Andropogoneae in molecular phylogenies (Estep et al. 2014; GPWG II 2012; Teerawatananon et al. 2011). The chloroplast phylogeny of GPWG II (2012) places it sister to *Ischaemum afrum*.

355. *Elionurus* Humb. & Bonpl. ex Willd.

Elionurus Humb. & Bonpl. ex Willd., Sp. Pl. 4(2): 941 (1806), nom. et orth. cons.

Annuals or perennials, caespitose. Ligule membranous, or a fringed membrane, or a fringe of hairs. Inflorescences unbranched or branched, the rachis flattened, white-hairy; apex of internodes not hollowed or rimmed, base of internodes oblique, hairy. Pedicellate spikelets staminate, the lower glume muticous or short-awned. Sessile spikelet bisexual, the lower glume with a bifid tip, *with two keels with hairs in tufts or lines and an oil streak on the inner side of the keels*; lower palea absent; lemmas hyaline, unawned. $2n = 10, 20$.

Fifteen spp., Tropical Africa, America, Australia.

Elionurus falls in an unresolved position among the awnless Andropogoneae in the molecular phylogenies of GPWG II (2012), Teerawatananon et al. (2011), and Estep et al. (2014). The chloroplast phylogeny of GPWG II (2012) places it sister to *Pseudopogonatherum contortum*, a position that is hard to explain morphologically.

356. *Euclasta* Franch.

Euclasta Franch., Bull. Soc. Hist. Nat. Autun 8: 335 (1895).

Sprawling annuals. Ligule a fringed membrane. Inflorescence unbranched or with digitate branches, nodding. Proximal 1 to 3 pairs of spikelets entirely staminate. Distal pedicellate spikelets staminate or sterile. Distal sessile spikelets bisexual with a pubescent callus. Flowers lacking paleas. Lemma apex awned, the awn geniculate. $2n = 40$.

Two spp., tropical regions of Africa, south Asia, and the Americas.

Watson and Dallwitz (1992 onward) report that the shoots are aromatic.

357. *Eulaliopsis* Honda

Eulaliopsis Honda, Bot. Mag. (Tokyo) 38: 56 (1924).

Perennials. Ligule a fringed membrane or a fringe of hairs. Inflorescence branches unbranched, digitate. All spikelets bisexual, callus pubescent. Lower glumes of both spikelets with many prominent raised veins. Sessile spikelet with *lower glume with tufts of long hairs about midway up abaxial side*; upper glume thinner than lower, with a slender awn. Lemma apex awned from a sinus, the awn geniculate. Lodicules ciliate. *Leaf cross section with minor vascular bundles crowded, almost superposed.*

Two spp., India and SE Asia (Philippines), China, Afghanistan.

358. ***Hypogynium*** Nees

Hypogynium Nees, Fl. Bras. Enum. Pl.: 364 (1829).

Caespitose perennials, monoecious. Ligule membranous. Axillary branch complexes present. Inflorescence unbranched, the base of the dispersal unit truncate. Sessile and pedicellate spikelets morphologically similar, the sessile ones pistillate, the pedicellate ones staminate. Glumes and lemmas unawned. Pistillate flowers with 3 staminodes. $2n = 30$.

Two spp., tropical Africa and America.

Some authors (e.g., Clayton and Renvoize 1986; Renvoize 1998) consider *Hypogynium* to be a synonym of *Andropogon*, whereas others (e. g., Rosengurtt et al. 1970; Watson and Dallwitz 1992 onward) consider it to be separate on the basis of its unisexual, unawned spikelets, and unbranched inflorescence axes. No molecular data are available to help with its placement. The chromosome number suggests an allopolyploid origin.

359. ***Kerriochloa*** C.E. Hubb.

Kerriochloa C. E. Hubb., Hooker's Icon. Pl. 35: pl. 3494 (1951).

Decumbent perennials, less than 25 cm tall. Ligule membranous. Axillary branch complexes present. Inflorescence enclosed by the subtending leaf sheath, unbranched, the rachis flattened; internodes wedge-shaped. Sessile spikelet laterally compressed, the callus obtuse, pubescent, inserted into the cupulate apex of the internode below; lower glume villous, upper glume awned. Lemma of upper flower awned from a shallow sinus, the awn geniculate. Pedicellate spikelet reduced to a tiny vestige with one glume.

One sp., *K. siamensis* C. E. Hubb., Thailand and Indo-China.

Molecular data place *K. siamensis* sister to *Microstegium vimineum* (Estep et al. 2014).

360. ***Lakshmia*** Veldk.

Lakshmia Veldk., Rheedea 18: 81 (2013).

Caespitose perennials. Ligule membranous. Inflorescence branches branched, the ultimate branches with 3 to 7 spikelet pairs. Sessile spikelet callus short, blunt, with white hairs. Glumes chartaceous, their apices acuminate. Upper lemma with the apex bifid, awned from the sinus, the awn geniculate. Pedicellate spikelets similar to the sessile ones in size and shape, but lacking awns, staminate or sterile.

One species, *L. venusta* (Thwaites) Veldk., India and Sri Lanka.

Lakshmia venusta has been variously placed in *Andropogon*, *Capillipedium*, *Bothriochloa*, *Vetiveria*, and most recently *Hemisorghum*. Molecular data would be particularly useful to place this species.

361. ***Lasiurus*** Boiss.

Lasiurus Boiss., Diagn. Pl. Orient. 4: 145 (1859).

Perennials with short rhizomes. Ligule a fringe of hairs. Inflorescence unbranched or of digitate primary branches, the internodes clavate, with silky hairs. Spikelets in groups of three, two sessile and one pedicellate. Sessile spikelet callus pubescent, with a central peg. Lower glume of sessile spikelet two keeled, flattened on the back with a short apical extension; upper glume thinner and boat-shaped. Lemma apex acute, without an awn. $2n = 18, 56$.

One sp., *L. scindicus* Henrard, E. Africa to NE India.

Faruqi et al. (1987) note that the cytotypes of this species correlate with geography; plants with $2n = 18$ are found in India and Pakistan, whereas those with $2n = 56$ are in North Africa. No molecular data are available.

362. ***Microstegium*** Nees

Microstegium Nees, Nat. Syst. Bot. 447 (1836).
Ischnochloa Hook. f., Hooker's Icon. Pl. 25: t. 2466 (1896).

Annuals or perennials, caespitose. Leaf blades broadly linear to lanceolate, sometimes with a short pseudopetiole. Ligule membranous, or a fringed membrane or a fringe of hairs. Inflorescence unbranched, or of digitate or subdigitate primary branches. Rachis internodes enlarged, clavate. Pedicellate spikelets and sessile spikelets similar, bisexual, the callus pubescent. Glumes generally more firm than the lemmas, the upper glume often with a long awn, lower glume concave between keels. Upper lemma awned from the apex or from a sinus, the awn straight or geniculate. $2n = 20, 40$.

Twenty-seven spp., Asia, Africa, some species weedy elsewhere.

Chen et al. (2009) separate *Microstegium* from *Leptatherum*, based on morphological and molecular data. The monotypic *Ischnochloa* (= *Microstegium falconeri*) is placed here, consistent with Clayton et al. (2006 onward); the plant illustrated in the protologue has a strongly flattened rachis similar to that in *Microstegium*. The position of the lemma awn is similar to that in *Arthraxon*, but the lemma appears to be deeply bilobed and the pedicellate spikelet is bisexual and awned. Molecular data place *M. vimineum* sister to *Kerriochloa siamensis* but with little support (Estep et al. 2014).

363. ***Pogonachne*** Bor

Pogonachne Bor, Kew Bull. 4: 176 (1949).

Annuals to 1 m tall, with prop roots forming from the lower nodes. Upper leaves without blades. Ligule membranous. Axillary branch complexes present. Inflorescence unbranched, the rachis flattened. *Spikelets apparently unpaired, the sessile spikelet reduced to small scales or absent.* Pedicellate spikelet laterally compressed; callus pubescent, truncate. Lower glume hairy, upper glume gibbous, with a tuft of hair. Upper lemma awned from a sinus, the awn geniculate.

One sp., *P. racemosa* Bor, India.

364. ***Polytrias*** Hack.

Polytrias Hack., Nat. Pflanzenfam. 2(2): 24 (1887).

Stoloniferous perennials. Ligule a fringed membrane. Inflorescence unbranched. *Spikelets in threes, two apparently sessile but actually on a short flattened pedicel, the third clearly pedicellate, all bisexual.* Spikelets lacking proximal flower. Glumes pubescent. Upper lemma awned from a shallow sinus, the awn geniculate. Upper palea tiny or absent. Lodicules absent.

One sp., *P. indica* (Houtt.) Veldkamp, Southeast Asia.

The spikelet arrangement in *Polytrias* is similar to that in *Mnesithea*, *Ratzeburgia*, and some *Coelorachis*, but the plant is much more delicate, the glumes are truncate, the rachis is slender, and the lemma bears an awn. Molecular data (Estep et al. 2014; Teerawatananon et al. 2011) place this in an unresolved position in the tribe.

365. ***Sehima*** Forssk.

Sehima Forssk., Fl. Aegypt.-Arab. 178 (1775).

Annuals or perennials, caespitose. Ligule a line of hairs or fringed membrane. Inflorescence unbranched and curved, the spikelets embedded in the rachis; internodes and pedicels thickened, clavate. Sessile spikelet with the lower glume with two keels, the apex mucronate to awned; the upper glume membranous, green, awned, with prominent veins converging toward the apex; upper lemma awned from a sinus, the awn geniculate. Pedicellate spikelet staminate or sterile, strongly flattened, the callus fitting into a concave depression in the pedicel; glume with a terminal bristle. $2n = 34, 40$.

Five spp., Old World tropics.

Andropogon sect. *Notosolen* looks similar to *Sehima* but has more than one inflorescence branch. Molecular data (Estep et al. 2014) put *Sehima* in an unresolved position in the tribe.

366. ***Spodiopogon*** Trin.

Spodiopogon Trin., Fund. Agrost. 192, pl. 17 (1820).
Eccoilopus Steud., Syn. Pl. Glumac. 1: 123 (1854).

Annuals or perennials, caespitose or rhizomatous. Leaf blades sagittate or not, pseudopetiolate or not, rarely deciduous. Ligule membranous or a fringed membrane. Inflorescence branches branched, the spikelets set into a cup-like depression at the distal end of the rachis joint. All spikelets bisexual, callus pubescent. Lower glume scabrous and pubescent; veins all equally prominent (no clear keel). Upper glume muticous or mucronate. Lower lemma mucronate or awned; upper lemma awned from a sinus, the awn geniculate. Lodicules ciliate. $2n = 40, 42$.

Eighteen spp., SE Asia, temperate Asia, Middle East.

367. ***Thelepogon*** Roth

Thelepogon Roth, Syst. Veg. 2: 46 (1817).

Annuals, often with prop roots. Leaf blades broad, with the base cordate. Ligule membranous or a fringed membrane. Inflorescence branched, the primary branches borne on a short rachis, the proximal portion of each branch without spikelets. Sessile spikelet with the callus pubescent, the lower glume *strongly sculptured with hooked projections*; upper lemma awned from a sinus, the awn geniculate. Pedicelled spikelet absent, the pedicel ending blindly.

Two spp., tropical Africa, Indonesia, Australia.

Molecular data place this genus sister to *Chrysopogon* (Estep et al. 2014).

368. ***Triplopogon*** Bor

Triplopogon Bor, Kew Bull. 9: 52 (1954).

Plant annual, 0.5–2 m tall, with prop roots. Leaf blades narrow at the base, more or less pseudopetiolate. Ligule membranous. Axillary branch complexes present. Inflorescence unbranched, the rachis flattened, the internodes long hairy. Sessile spikelet with callus pubescent; lower glume with a median groove, with tufts of hair on the margins; upper glume mucronate, with one tuft of hair; upper lemma awned from a sinus, the awn geniculate. Pedicellate spikelet rudimentary.

One sp., *T. ramossimus* (Hack.) Bor, India.

XXII.2. Subtribe Tripsacinae Dumort. (1829)

Inflorescence unbranched. Upper lemma hyaline, lacking awns.

Five genera, 32 species.

Oxyrhachis, *Rhytachne*, *Urelytrum* and *Vossia* are well supported as being sister to one of the parental genomes of the paleotetraploids *Zea* and *Tripsacum* (McKain and Kellogg, unpubl. data). The lack of awns may be synapomorphic, although awns are also lost in Rottboelliinae.

369. ***Oxyrhachis*** Pilg.

Oxyrhachis Pilg., Notizbl. Bot. Gart. Berlin-Dahlem 11: 655 (1932).

Caespitose perennials. Leaf blades filiform, rigid, the ligule a fringed membrane. Inflorescence single, unbranched, terminal. *Pedicellate spikelets absent, the pedicel sometimes visible but fused to the inflorescence axis.* Sessile spikelets sunken in internode, producing a rat-tail inflorescence. Spikelet callus blunt, without a cylindrical peg, attached obliquely. Lower flower of sessile spikelet sterile.

One sp., *O. gracillima* (Baker) C.E. Hubb., Africa and Madagascar.

370. ***Rhytachne*** Desv. ex Ham.

Rhytachne Desv. ex Ham., Prodr. Pl. Ind. Occid. xiv, 11 (1825).

Annuals or perennials, caespitose. Ligule membranous or a fringed membrane. Inflorescence unbranched, the spikelets embedded in hollows in the rachis. Spikelets paired, the sessile spikelet squashed between the internode and the pedicel. Pedicellate spikelets bisexual or reduced, sometimes represented only by an awn, or by an empty pedicel, or lacking entirely. Sessile spikelet callus with central peg. Upper glume awned in some species. Costal short-cells of the leaf epidermis paired. $2n = 20$.

Twelve spp., tropical Africa and America.

371. ***Urelytrum*** Hack.

Urelytrum Hack., Nat. Pflanzenfam. 2: 22, 25 (1887).

Perennials, rarely annual, caespitose. Leaves auriculate. Ligule membranous or a fringed membrane. Inflorescence unbranched, or with digitate branches, the internodes clavate with a broad scarious rim, the proximal end with a peg-like knob after disarticulation. Pedicellate spikelet staminate or sterile, *the lower glume with a long curved awn*. Sessile spikelet with a pubescent callus inserted into the distal end of the internode below. Lower flower of sessile spikelet staminate, with palea. $2n = 20$.

Seven spp., tropical Africa.

372. ***Vossia*** Wall. & Griff.

Vossia Wall. & Griff., J. Asiat. Soc. Bengal 5: 572 (1836).

Rhizomatous perennials, in wet sites or aquatic with culms spongy, submerged or floating. Ligule a fringed membrane. Primary branches of the inflorescence unbranched, single or digitate; internodes and pedicels clavate and flat. Pedicellate spikelets bisexual or staminate. Sessile spikelets bisexual. Callus truncate. Lower glume of sessile spikelet two-keeled, flattened on the back, extended into a flattened tail-like awn; upper glume thinner and boat-shaped. $2n = 20$.

One sp., *V. cuspidata* (Roxb.) Griff., tropical Africa and India.

Zea + *Tripsacum*

Plants monoecious. Staminate and pistillate spikelets on the same inflorescence axis, with staminate distal to pistillate, or staminate and pistillate spikelets in different inflorescences, the staminate terminal on the plant. Pistillate spikelets unpaired, sessile only.

Zea and *Tripsacum* are sisters in all molecular phylogenies (e.g., Mathews et al. 2002; Teerawatananon et al. 2011) and are derived from the same ancestral polyploidy event (Estep et al. 2014).

373. ***Tripsacum*** L. Fig. 65

Tripsacum L., Syst. Nat. (ed. 10): 1253, 1261, 1379 (1759).

Perennials, caespitose or rhizomatous, 0.5–5 m tall. Leaves without auricles, the ligule a fringed membrane. Inflorescence branched or unbranched. Staminate spikelets with both glumes on the abaxial side of the spikelet, overlapping. Pistillate spikelets embedded in the internode, the node truncate, with a central peg. Fruit encased in the hard glume plus internode. $2n = 36, 72, 90, 108$.

Sixteen spp., North, Central and South America.

Fig. 65. Panicoideae-Andropogoneae-Tripsacinae. *Tripsacum dactyloides*. **A** Rhizome with shoot bases. **B** Leaves and inflorescences with staminate (*a*) and pistillate spikelets (*b*). **C** Staminate spikelet. **D** Pistillate spikelet. (From Barkworth et al. 2003, drawn by H. Paździrková)

374. ***Zea*** L.

Zea L., Sp. Pl. 2: 971–972 (1753).
Euchlaena Schrad., Index Sem. (Göttingen) 1832: 3 (1832).

Annuals or perennials, caespitose or rhizomatous, 1–5 m tall. Ligule membranous. Main culm and primary culm branches ending in a branched inflorescence of paired staminate spikelets. Pistillate inflorescences axillary on the culm branches, closely surrounded by spathes; sessile spikelets distichous and alternate; *glume and rachis enlarging at maturity to enclose fruit in a rock-hard shining fruit case*. $2n = 10, 20$.

Seven spp., Mexico and Central America.

Cultivated plants (maize, *Z. mays* ssp. *mays*) retain the terminal staminate inflorescence of their wild relatives (tassel). Pistillate inflorescences terminate short axillary branches on the main axis, and consist of paired pistillate spikelets in multiple orthostichies, each spikelet with a single flower at maturity. Glumes and lemmas are soft and flexible. Maize is one of the most valuable crop plants in the world.

XXII.3. Subtribe Rottboelliinae J. Presl (1830)

Inflorescences unbranched; spikelets generally sunken in the rachis. Lemmas hyaline, unawned.

11 genera, 99 species.

Molecular data place Rottboelliinae in a clade distinct from Tripsacinae (McKain and Kellogg, unpubl. data). While it is possible that the two are sisters, evidence for this is weak, and thus lemma awns may have been lost either once or twice in the evolution of Andropogoneae. A subset of species of Rottboelliinae have pedicels fused to the internode and an overlapping subset have sculptured glumes. It is unclear whether either character is synapomorphic here. Molecular studies (GPWG II 2012; Teerawatananon et al. 2011; McKain and Kellogg unpubl. data) support a clade containing at least *Eremochloa*, *Hackelochloa*, *Hemarthria*, *Heteropholis*, *Mnesithea*, *Ophiuros*, *Phacelurus*, and *Thaumastochloa*, although no single study includes all taxa. Veldkamp et al. (1986) synonymized *Coelorachis*, *Hackelochloa*, *Heteropholis*, *Ratzeburgia*, and *Rottboellia* with *Mnesithea* on the basis of continuous variation in most taxonomic characters. They also question the value of pedicel fusion in delimiting genera. Molecular data clearly support merger of *Coelorachis* and *Rottboellia*, which are synonymized here, but the monophyly of *Mnesithea* sensu Veldkamp et al. (1986) is less certain. The question is unlikely to be resolved with only herbarium studies – molecular and developmental investigations are needed.

375. ***Eremochloa*** Büse

Eremochloa Büse, Pl. Jungh. 357 (1854).

Perennials, caespitose or rhizomatous. Ligule membranous, or a fringed membrane, or a fringe of hairs. Inflorescence axis flattened with clavate internodes and overlapping spikelets. Pedicellate spikelets rudimentary, or reduced only to an awn, or lacking, leaving a barren pedicel. Sessile spikelets with the lower glume 2-keeled, setose. $2n = 18$.

Twelve spp., India, China, southeast Asia, Australia.

Teerawatananon et al. (2011) place *Eremochloa* with *Garnotia* with strong support, but the two genera have little in common morphologically. Other phylogenies place *Garnotia* with *Arundinella* (GPWG II 2012).

376. ***Glyphochloa*** Clayton

Glyphochloa Clayton, Kew Bull. 35: 814 (1981).

Annuals. Ligule membranous. Axillary branch complexes present. Pedicellate spikelet sterile, the pedicel fused to the inflorescence axis. Glumes of pedicellate spikelet muticous or awned. Sessile spikelet callus truncate, with a central peg. Lower glume of sessile spikelet muticous or awned, *variously decorated with projections*, glabrous or pubescent; upper glume of sessile spikelet unawned.

Nine spp., India.

Clayton and Renvoize (1986) link *Glyphochloa* to *Heteropholis* and *Coelorachis clarkei*. It is placed here solely on the basis of morphology.

377. ***Hackelochloa*** Kuntze

Hackelochloa Kuntze, Revis. Gen. Pl. 2: 776 (1891).

Annuals. Ligule a fringed membrane. Axillary branch complexes present. Pedicellate spikelets bisexual, staminate, or sterile, the pedicel fused to the inflorescence axis, the glumes muticous. Sessile spikelet callus transverse, with a central peg. *Lower glume of sessile spikelet nearly orbicular, with a complex reticulate pattern of projections and pits.* Embryo very large. $2n = 14$.

Two spp., tropics.

378. ***Jardinea*** Steud.

Jardinea Steud., Syn. Pl. Glumac. 1: 360 (1854).

Caespitose perennials. Ligule a fringed membrane. Inflorescence with primary branches, these sometimes themselves branched. Pedicellate spikelets smaller than the sessile ones, reduced or vestigial, bisexual, staminate or sterile. Sessile spikelet bisexual, with a pointed callus. Lower glumes dorsiventrally compressed, sculptured, prickly, the apex acuminate or extended into a tail-like awn. Upper glumes laterally compressed. Lemma of sessile spikelet keeled. Leaf epidermis with costal short-cells paired.

Three spp., tropical Africa.

The glume compression is similar to that in *Loxodera*. No molecular data are available; the genus is placed here based on morphology.

379. ***Loxodera*** Launert

Loxodera Launert, Bol. Soc. Brot. II, 37: 80 (1963).
Lepargochloa Launert, Bol. Soc. Brot. II, 37: 82 (1963).

Caespitose perennials. Ligule membranous. Inflorescence unbranched, terminal, the internodes clavate or columnar with broad scarious rim bearing long white hairs, the spikelets partially embedded in the rachis. Disarticulation zone oblique. Pedicellate spikelet staminate, sterile or reduced, the glumes muticous or short-awned. Sessile spikelet with a pubescent callus, the lower glume strongly veined.

Five spp., tropical Africa.

380. ***Manisuris*** L.

Manisuris L., Mant. Pl. 2: 164 (1767).

Plants perennial. Ligule membranous. *Pedicellate spikelet next to sessile spikelet of internode above by fusion of pedicel, the two falling together at maturity.* Glumes of pedicellate spikelet muticous. Sessile spikelet callus truncate, with a central peg. Lower glume of sessile spikelet broadly winged, *constricted about the middle.* $2n = 18, 36$.

One sp., *M. myuros* L., India.

This genus name has been applied to several species now placed in *Coelorachis*, *Rottboellia*, and *Glyphochloa*; see Clayton (1981) for discussion. *Manisuris myuros* is placed in this subtribe on the basis of morphology alone.

381. ***Mnesithea*** Kunth

Mnesithea Kunth, Révis. Gramin. 1: 153 (1829).
Ratzeburgia Kunth, Révis. Gramin. 2: 487 (1831).

Caespitose perennials. Ligule membranous or a fringed membrane. Axillary branch complexes present. Inflorescence unbranched, the axis disarticulating or not, internodes expanded at distal end, spikelets appressed to or embedded in the inflorescence axis. Spikelets two or three per inflorescence node, *two sessile and one pedicellate.* Pedicellate spikelet reduced, sterile, with glumes or vestigial. Sessile spikelet callus with a central peg, fitting in to the cupulate apex of the internode below. First glume of sessile spikelet smooth or sculptured; second glume much less firm, embedded in the rachis. $2n = 18$.

Three spp., India and Southeast Asia.

The pedicel is fused to the rachis in *M. laevis*, but is free in the other two species.

382. ***Phacelurus*** Griseb.

Phacelurus Griseb., Spic. Fl. Rumel. 2: 423 (1844) [1846].
Pseudovossia A. Camus, Bull. Mus. Hist. Nat. (Paris) 26: 665 (1920).

Plants perennial. Ligule membranous. Inflorescence unbranched or with digitate primary branches, the internodes thickened. Pedicellate spikelet generally reduced, with an elongate callus in some species. Sessile spikelet bisexual, with a blunt glabrous callus. $2n = 20, 40$.

Seven spp., Old World tropics.

The single species of *Pseudovossia* is included here, following Clayton and Renvoize (1986).

383. ***Rottboellia*** L. f.

Rottboellia L. f., Suppl. Pl. 13, 114 (1781) [1782], nom. conserv.
Coelorachis Brongn., Voy. Monde 2: 64, f. 14 1829 [1831].
Thyrsia Stapf, Fl. Trop. Afr. 9: 48 (1917).

Annuals or perennials, caespitose or rhizomatous. Ligule membranous or a fringed membrane, or a fringe of hairs. Axillary branch complexes present. Inflorescence unbranched or sometimes with digitate primary branches; internodes club-shaped or flattened, generally shorter than spikelets, with sunken spikelets, the peduncle widened at the apex. Spikelets in pairs or in groups of three with two sessile and one pedicellate spikelet per node. Pedicels free or fused wholly or partly to internode. Pedicellate spikelet staminate or sterile. Sessile spikelet callus truncate, with or without a central peg; lower glume thickened, often decorated, with two winged keels, the apex generally emarginate or obtuse. Lower flower of sessile spikelet staminate or sterile. $2n = 18, 20, 36, 40, 54$.

Thirty-three spp., tropics.

Species of *Rottboellia* and *Coelorachis* are clearly intermixed in molecular phylogenies (McKain and Kellogg, unpubl. data) so the two are united here. The phylogeny suggests that fusion of the pedicel to the inflorescence internode is homoplasious, supporting the suggestion of Veldkamp et al. (1986). The morphological distinction between *Rottboellia* and *Mnesithea* is weak, having to do with the predominant number of sessile spikelets at each inflorescence internode (two or three in *Rottboellia*, three in *Mnesithea*). *Thyrsia* is recognized by Watson and Dallwitz (1992) and placed in *Phacelurus* by Clayton and Renvoize (1986), but molecular data (McKain and Kellogg, unpubl. data) show that it is derived within *Rottboellia*.

Ophiuros + Thaumastochloa + Hemarthria + Heteropholis

Pedicellate spikelet generally absent or reduced, *the pedicel sometimes visible but always fused to the inflorescence axis. Sessile spikelet sunken in internode, producing a rat-tail inflorescence.*

These four genera form a strongly supported clade (GPWG II 2012), with *Hemarthria* and *Heteropholis* as sisters. The latter two are also strongly supported as sisters by McKain and Kellogg (unpubl. data).

384. ***Hemarthria*** R. Br.

Hemarthria R. Br., Prodr. 207 (1810).

Perennials (one species annual), caespitose or stoloniferous. Ligule a fringe of hairs or a fringed membrane. Axillary branch complexes present. Inflorescence rachis internodes disarticulating transversely or obliquely. Pedicellate spikelets bisexual or reduced. Sessile spikelets bisexual, glumes subulate tipped, not rugose, warty or pitted, lower glume hardened, upper glume membranous. $2n = 18, 20\ (18 + 2B), 36, 54$.

Fourteen spp., Old World tropics, maybe Americas.

385. ***Heteropholis*** C. E. Hubb. Fig. 66

Heteropholis C. E. Hubb., Hooker's Icon. Pl. 36: pl. 3548 (1956).

Annuals or perennials, often decumbent. Ligule membranous or a fringed membrane. Inflorescence internodes clavate, the peduncle widened at the apex. Pedicellate spikelets staminate or reduced to glumes, the glumes leathery or indurate. Sessile spikelet callus cuneate, truncate, glabrous or pubescent, with a central peg. Lower glume of sessile spikelet smooth or decorated with raised patterns or rugosities. Embryo large.

Six spp., Old World.

386. ***Ophiuros*** C. F. Gaertn.

Ophiuros C. F. Gaertn., Suppl. Carp. 1: 3, pl. 181, f. 3. (1805).

Annuals or perennials, caespitose, the stem base sometimes with corms. Ligule membranous. Axillary branch complexes present. *Pedicellate spikelet absent,* the pedicel sometimes visible. Spikelet callus truncate, with a cylindrical peg. Lower glume obscurely pitted in many specimens. Lower flower of sessile spikelet staminate or sterile.

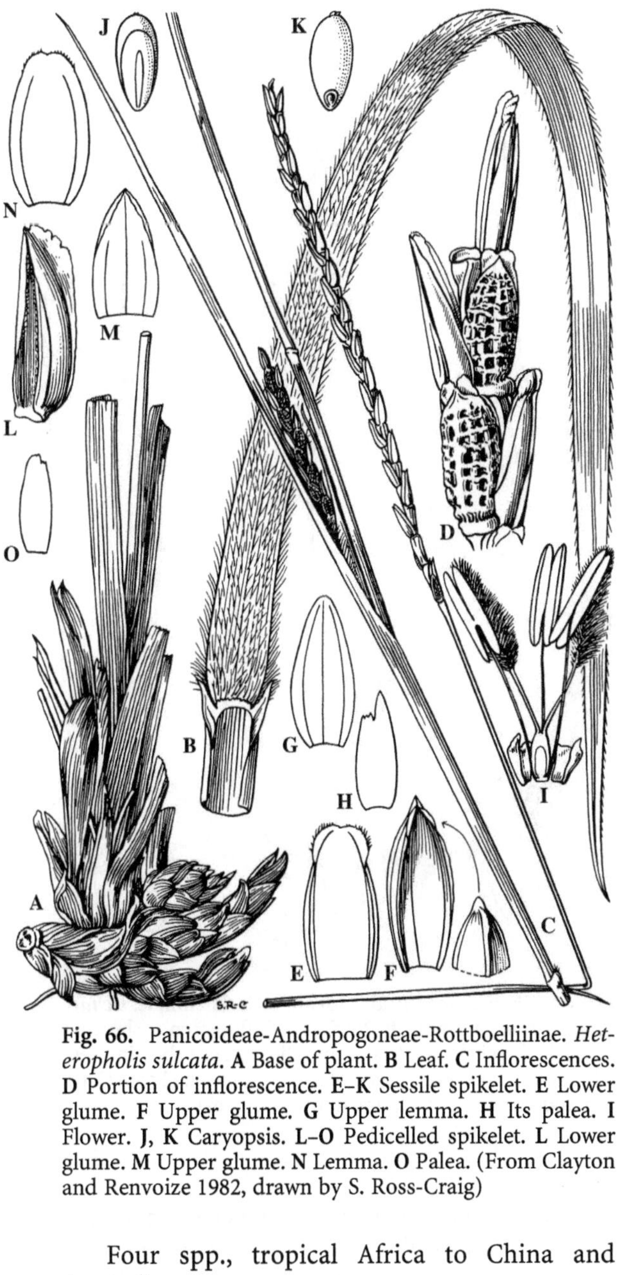

Fig. 66. Panicoideae-Andropogoneae-Rottboelliinae. *Heteropholis sulcata.* **A** Base of plant. **B** Leaf. **C** Inflorescences. **D** Portion of inflorescence. **E–K** Sessile spikelet. **E** Lower glume. **F** Upper glume. **G** Upper lemma. **H** Its palea. **I** Flower. **J, K** Caryopsis. **L–O** Pedicelled spikelet. **L** Lower glume. **M** Upper glume. **N** Lemma. **O** Palea. (From Clayton and Renvoize 1982, drawn by S. Ross-Craig)

Four spp., tropical Africa to China and Australia.

387. ***Thaumastochloa*** C. E. Hubb.

Thaumastochloa C. E. Hubb., Hooker's Icon. Pl. 34: pl. 3313 (1936).

Annuals. Ligule a fringed membrane. Axillary branch complexes present. Inflorescence becoming stiff and curved, breaking off in its entirety. Distal end of peduncle widened. Spikelets 10 or fewer, sometimes only one. Callus truncate, with or without a central peg. Lower glume of sessile spikelet sculptured. Lower flower of sessile spikelet sterile.

Eight spp., Australia and New Guinea.

XXII.4. Subtribe Ischaeminae J. Presl (1830)

Dimeriinae Hack. ex C.E. Hubb. (1934).

Lemma awned from a sinus, the awn generally geniculate. Chromosome base number often reduced from $x = 10$ to $x = 9$ or $x = 7$.

Three genera, 173 spp.

The monophyly of Ischaeminae is supported by molecular data (Estep et al. 2014; Teerawatananon et al. 2011). The subtribe lacks obvious morphological synapomorphies although the lower glume often has a winged keel.

388. ***Andropterum*** Stapf

Andropterum Stapf, Fl. Trop. Afr. 9: 38 (1917).

Perennials. Ligule a fringe of hairs. Inflorescence unbranched, the internodes and pedicels cuneate. *Pedicellate spikelets bisexual,* the lower glume with a broad asymmetrical wing. Sessile spikelet squashed between the internode and pedicel. Lower glume of sessile spikelet two-keeled with strong central groove. Upper glume of sessile spikelet laterally compressed.

One sp., *A. stolzii* (Pilg.) C. E. Hubb., tropical Africa.

Dimeria + Ischaemum

Leaf epidermis with intercostal cells with several papillae per cell.

389. ***Dimeria*** R. Br.

Dimeria R. Br., Prodr. 204 (1810).
Nanooravia Kiran Raj & Sivad., Nordic J. Bot. 31: 162 (2013).

Caespitose annuals or perennials. Ligule membranous, or a fringed membrane. Inflorescence with one or two branches, or up to 9 in a digitate arrangement, the rachis not disarticulating. *Spikelets solitary (not paired), on short pedicels, strongly laterally compressed*, the callus generally pubescent. Glumes often winged on the keel. Lower flower sterile. Upper flower with lemma awned; palea present or absent. *Stamens 2.* $2n = 14$.

Fifty-nine spp., India, southeast Asia, Australia, Polynesia.

Molecular data show that *Dimeria* is derived from within *Ischaemum* (Estep et al. 2014; Teerawatananon et al. 2011), apparently by suppression of the sessile spikelet. In *D. woodrowii* Stapf the inflorescence is shed as a unit. The newly described monospecific genus *Nanooravia* is based on *Dimeria santapaui*, in which the spikelets are not strongly compressed, and in which the glumes bear auricles at the apex. Segregation of the species in a new genus appears premature, and it is retained here in *Dimeria*.

390. ***Ischaemum*** L. Fig. 67

Ischaemum L., Sp. Pl. 2: 1049 (1753).

Annuals or perennials, caespitose, rhizomatous or stoloniferous. Leaves often cordate, with or without a pseudopetiole, with or without auricles. Ligule membranous, a fringed membrane or a fringe of hairs. Inflorescence unbranched, or with two branches, or with multiple digitate branches, spikelets borne on one side of the branch, *the rachis and pedicel flattened and together forming a U-shape, the distal end of the internode often cupulate.* Pedicellate spikelet bisexual, staminate or reduced. Sessile spikelet with the distal flower bisexual, the proximal flower staminate. Lower glume dorsiventrally compressed, often winged; upper glume laterally compressed apparently because of compression between rachis and pedicel, often awned. Lemmas membranous. $2n = 18, 20, 40, 54, 56, 68$.

Eighty-seven spp., tropics.

XXII.5. Subtribe Saccharinae Griseb. (1846)

Lemma awned from the apex or more often from a sinus, the awn generally geniculate.

Sixteen genera, 177 species.

This subtribe has no obvious morphological synapomorphy. The genera formerly included in or allied to *Sorghum*, together with the *Saccharum* group, often form a larger clade in molecular analyses (Kellogg 2012), although this may be an artifact of taxon sampling. The subtribe is retained here pending further data, and may need to be extensively revised in the future.

Fig. 67. Panicoideae-Andropogoneae-Ischaeminae. *Ischaemum afrum*. **A** Habit. **B** Portion of inflorescence showing pedicelled spikelet. **C** Portion of inflorescence showing sessile spikelet. **D** Lower glume of sessile spikelet. (From Clayton and Renvoize 1982, drawn by A. Davies)

391. ***Asthenochloa*** Büse

Asthenochloa Büse, Pl. Jungh.: 367 (1854).

Decumbent annuals. Ligule membranous. Primary inflorescence branches themselves branched, *each branch with a sessile spikelet and two pedicellate ones, the latter represented only by pedicels.* Callus of spikelet pubescent. Lower glume with a bidentate apex, the upper glume keeled. *Stamens 2.*

One sp., *A. tenera* Büse, Southeast Asia.

392. ***Cleistachne*** Benth.

Cleistachne Benth., Hooker's Icon. Pl. 14: pl. 1379 (1882).

Annuals, often with prop roots. Leaf blades pseudopetiolate or not. Ligule membranous. Inflorescence branches branched, *spikelets unpaired.* Spikelet callus pubescent, obtuse. Glumes with long hairs. Lodicules ciliate. $2n = 36$.

One sp., *C. sorghoides* Benth., tropical Africa, India.

The pedicellate spikelet is generally inferred to have been lost, resulting in unpaired spikelets. No developmental studies have been done to confirm this hypothesis. Liu et al. (2014) suggest that *C. sorghoides* is of allopolyploid origin, with genomes from a *Sorghastrum*-like ancestor and another genus more distantly related. Although they transfer *C. sorghoides* to *Sorghum*, this placement is not strongly supported. The species is retained here in its own genus pending additional data.

393. ***Leptatherum*** Nees

Leptatherum Nees, Proc. Linn. Soc. Lond. 1: 92 (1841).
Polliniopsis Hayata, Icon. Pl. Formosan. 7: 76, f. 45 (1918).

Mat-forming annuals, rooting from the nodes. Ligule membranous. Inflorescence slender, the branches themselves branched, without hairs. Rachis internodes longer than spikelets, slender, flexuous. Glumes of sessile and pedicellate spikelets indurate; lower glume of sessile spikelet grooved. Callus of spikelet stubby, bearing a tuft of hairs; spikelets otherwise glabrous. *Stamens 2.*

Three spp., Asia and Australia.

Chen et al. (2009) have clearly documented the phylogenetic distinction between this genus and *Microstegium* (subtribe Andropogoninae), confirming the tentative hypothesis proposed by Mathews et al. (2002) and Spangler et al. (1999).

394. ***Pseudosorghum*** A. Camus

Pseudosorghum A. Camus, Bull. Mus. Hist. Nat. (Paris) 26: 662 (1920).

Decumbent annuals. Ligule membranous or a fringed membrane. Inflorescence branches branched. Callus of sessile spikelet pubescent. Glumes awnless. $2n = 20$.

Two spp., Indo-Malesian.

Other than the glabrous lodicules this genus appears similar to *Sorghum.*

395. ***Sorghastrum*** Nash

Sorghastrum Nash, Man. Fl. N. States: 71 (1901).

Annuals or perennials, rhizomatous or caespitose. Leaves with or without auricles. Ligule membranous, a fringed membrane or a fringe of hairs. Inflorescence branches branched, spikelet-bearing only at their distal ends, otherwise slender. Sessile spikelet callus obtuse or pungent, pubescent, the lower glume generally pubescent, the upper generally glabrous. Pedicelled spikelet usually reduced to a pedicel, but if present then resembling the sessile one. $2n = 20, 40, 60$.

Twenty-one spp., Africa and the Americas.

396. ***Sorghum*** Moench

Sorghum Moench, Methodus: 207 (1794).
Vacoparis Spangler, Austral. Syst. Bot. 16: 297 (2003).
Hemisorghum C. E. Hubb. ex Bor, Grass. Burma, Ceylon, India & Pakistan 686 (1960).

Annuals or perennials, caespitose or rhizomatous. Ligule membranous or a fringed membrane. Inflorescence branches branched. Spikelets in pairs or in triads at the ends of branches, or ultimate branches with 3 to 7 spikelet pairs. Pedicellate spikelet staminate or sterile, often much smaller than sessile one, the glumes often papery, with or without lemmas. Sessile spikelet ovate, with callus obtuse; glumes leathery, their apices obtuse, the lower with two keels; upper lemma awned or awnless, the apex obtuse. $2n = 10, 20, 40$.

Eight spp., Africa, India, southeast Asia, Northern Australia.

The molecular phylogenies of Dillon et al. (2004, 2007) place the two species of *Vacoparis* (= *Sorghum* sects. *Chaetosorghum* and *Heterosorghum*) sister to the species of *Sorghum* sect. *Sorghum.* See also discussion in Kellogg (2012). Data from Liu et al. (2014) are also mostly consistent with this interpretation, although there is some disagreement among gene trees. *S. mekongense* had been segregated as *Hemisorghum*, but molecular data place it firmly in a clade with *Sorghum* (McKain and Kellogg, unpubl. data).

397. ***Sarga*** Ewart

Sarga Ewart, Proc. Roy. Soc. Victoria 23(2): 296–297 (1911), rev.: Spangler (2003).

Annuals or perennials. *Nodes with a ring of stiff hairs.* Ligule a fringed membrane. Inflorescence branches branched or not. Pedicellate spikelet staminate, sterile, or reduced to a rudiment. Sessile spikelet with the callus obtuse, acute or sharply pointed, the lower glume apex often thickened, keeled or winged. $2n = 10, 20, 30, 40$.

Nine spp., Australia, southern Indonesia, Mexico and Central America.

Species of *Sarga* are often placed in *Sorghum*. However, *Sarga* appears to be phylogenetically distinct from *Sorghum*, no more closely related to it than to other genera in Saccharinae. While Liu et al. (2014) argue for retention of these species in *Sorghum* (as sect. *Parasorghum*), the placement is not strongly supported.

398. ***Spathia*** Ewart

Spathia Ewart, Fl. N. Territory: 26 (1917).

Caespitose annuals. Ligule a fringed membrane. Inflorescence of digitate branches, enclosed by a spathe, the proximal spikelet pair on each branch staminate. Sessile spikelet with a pubescent callus, the lower glume with a line of hairs, the upper glume pubescent, the flower bisexual with 3 stamens. Pedicellate spikelets staminate, *the flowers with 2 stamens.*

One sp., *S. neurosa* Ewart & Archer, Australia.

The Saccharum group

Sessile and pedicellate spikelets similar, hermaphrodite; callus of the spikelet with long hairs, these generally longer than the glumes. Glumes papery to membranous, not indurate.

This group may or may not form a clade separate from the rest of the Saccharinae. Molecular data remain insufficient and phylogenies are unresolved (Hodkinson et al. 2002).

399. ***Eriochrysis*** P. Beauv.

Eriochrysis P. Beauv., Ess. Agrostogr. : 8. (1812); tax.: Filgueiras (1997).
Leptosaccharum (Hack.) A. Camus, Bull. Soc. Bot. France 70: 736 (1923).

Caespitose perennials. Ligule membranous, or a fringed membrane or a fringe of hairs. *Inflorescence covered with golden or reddish hairs*, the primary branches branched or unbranched. Fertile flower of the sessile spikelet lacking a palea. *Pedicellate spikelet pistillate.* $2n = 20$.

Twelve spp., tropical America and Africa.

E. filiformis (Hack.) Filgueiras was formerly placed in its own genus, as *Leptosaccharum filiformis* (Hack.) Camus. Descriptions of the species in the literature are partially inaccurate; a corrected description and illustration are provided by Filgueiras (1997).

400. ***Eulalia*** Kunth

Eulalia Kunth, Révis. Gramin. 1: 160 (1829).
Pseudopogonatherum A. Camus, Ann. Soc. Linn. Lyon II, 68: 204–205 (1921) [1922].

Annuals or perennials, caespitose or rhizomatous. Ligule membranous, or a fringed membrane or a fringe of hairs. Inflorescence covered with silky hairs, unbranched or branched, if the latter then the branches digitate or on an elongate central axis. Spikelet with a blunt callus. Lower glume with 2 keels, upper glume with 1 keel, the apex muticous or awned. $2n = 20, 40$.

Thirty-seven spp., Old World tropics.

Eulalia species vary with respect to many characters used elsewhere to define genera, and the genus is likely to be polyphyletic (Hodkinson et al. 2002).

401. ***Imperata*** Cirillo

Imperata Cirillo, Pl. Rar. Neapol. 2: 26 (1792).

Rhizomatous perennials. Leaves auriculate or not, the blades pseudopetiolate or not. Ligule membranous, or a fringed membrane, or a fringe of hairs. Inflorescence branches branched, *covered with long silvery hairs*, "sessile" spikelets short pedicellate or not. Top of pedicel a round crater. *Glumes with long silvery hairs.* Proximal flower of the spikelet reduced to a lemma, distal flower with the lemma reduced, awnless, palea present, *lodicules absent, stamens 1 or 2.* $2n = 20, 40, 50, 60$.

Thirteen spp., tropical and subtropical. *I. cylindrica* (L.) Raeusch. is a particularly nasty weed that is widespread in tropical areas.

402. ***Lasiorhachis*** (Hack.) Stapf

Lasiorhachis (Hack.) Stapf, Hooker's Icon. Pl. 32: t. 3124 (1927).
Andropogon subg. *Lasiorhachis* Hack., Monogr. Phan. 6: 471 (1889).

Caespitose perennials. Primary panicle branches themselves branched. Pedicellate spikelet hermaphrodite, staminate, sterile, or absent. Sessile spikelet with lower glume with two keels. Lemma with a short bristle. *Ovary with an apical linear, hairy structure originating lower than the styles.*

Three spp., Madagascar.

Clayton and Renvoize (1986) synonymize *Lasiorhachis* with *Saccharum*, but there seems no really good reason to place it there versus with any other genus of the group. The variation in sex expression of the pedicellate spikelet encompasses that found throughout the tribe. The species is distinguished from most other species of *Saccharum* by its more highly branched inflorescences, shorter hairs and lack of awns.

403. ***Miscanthus*** Andersson Fig. 68

Miscanthus Andersson, Öfvers. Förh. Kongl. Svenska Vetensk.-Akad. 12: 165 (1855).
Miscanthidium Stapf, Fl. Trop. Afr. 9: 89 (1917).
Sclerostachya (Andersson ex Hack.) A. Camus, Fl. Indo-Chine 7: 243 (1922).
Narenga Bor, Indian Forester 66: 267 (1940).

Perennials, caespitose or rhizomatous, generally over 1 m tall. Leaf blades with or without a pseudopetiole. Ligule membranous or a fringed membrane. Inflorescence with primary branches borne along a central axis, not digitate, the branch axes disarticulating late in development if at all, spikelets disarticulating below the glumes leaving the pedicels and rachis intact. "Sessile" spikelet on a short pedicel. $2n = 30, 35\text{–}43, 48, 57, 76, 95, 96, 114$.

About thirty spp., Africa to southeast Asia.

Molecular data show clearly that *Narenga*, *Sclerostachya*, and *Miscanthidium* are part of the same clade as *Miscanthus* s.s. and thus should be placed into synonymy (Hodkinson et al. 2002). However, at least five species of *Miscanthus*, including the widely cultivated species *M. sinensis*, form a separate clade, which may or may not be related. *Miscanthus* x *giganteus* is cultivated as a biofuel

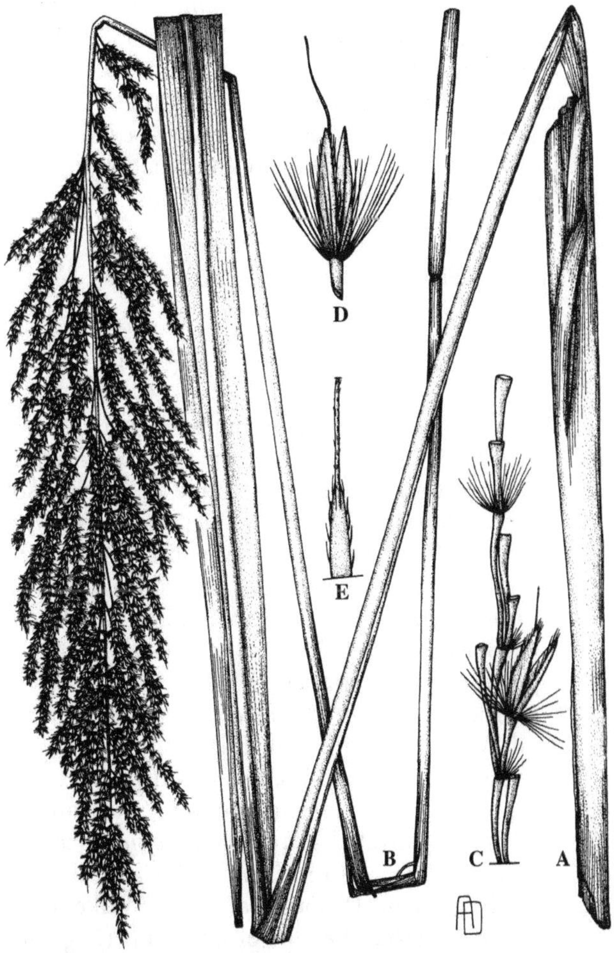

Fig. 68. Panicoideae-Andropogoneae-Saccharinae. *Miscanthus violaceus*. **A** Leaf. **B** Inflorescence. **C** Portion of inflorescence. **D** Spikelet. **E** Tip of upper lemma. (From Clayton and Renvoize 1982, drawn by A. Davies)

crop. *Miscanthidium* and *Miscanthus* are reported to hybridize with *Saccharum* (Watson and Dallwitz 1992 onward). The distinctive intercostal papillae also appear in Andropogoninae. Sun et al. (2010) have revised the Chinese species of *Miscanthus* and provide a detailed discussion of variation in morphological characters within the genus.

404. *Saccharum* L.

Saccharum L., Sp. Pl. 1: 54 (1753).
Erianthus Michx., Fl. Bor.-Amer. 1: 54 (1803).

Large rhizomatous perennials, often several meters tall. Leaves auriculate or not. Ligule membranous, or a fringed membrane or a fringe of hairs. Inflorescence with a thick strong central axis, the branches themselves branched, with long hairs, disarticulating at the nodes. Lemmas of some species lacking awns. $2n = 40$, 60, 68, 76–78, 80, 90, 46–128, 110, 112, 116–117, 144.

Thirty-seven species.

The generic limits of *Saccharum*, *Erianthus*, and *Miscanthus* are not clear. *Saccharum* s.s. includes only six species, two of which (*S. robustum* and *S. spontaneum*) are wild and four of which are cultivated and clonal (*S. officinarum*, *S. edule*, *S. barberi*, and *S. sinense*); the latter two are of hybrid origin (D'Hont et al. 2002). Welker et al. (2015) show that the type of *Erianthus* (*E. giganteus*) falls in the *Saccharum* clade and thus that the circumscription of *Saccharum* should be broadened to include at least some species of *Erianthus*. Hodkinson et al. (2002) present data showing that *Saccharum* s.s. is monophyletic, and that this clade does not include either *Narenga* or *Sclerostachya*, which instead fall into *Miscanthus*. Only two species of *Erianthus* (*E. rockii* and *E. contortus*) were sampled in that study, and these also fell in the *Miscanthus* clade. These two species should perhaps be moved to *Miscanthus*, although more species sampling is definitely needed. For a more complete discussion, see Kellogg (2012).

XXII.6. Subtribe Germainiinae Clayton (1972)

Glumes with apices truncate. Upper lemma awned from a sinus, the awn generally geniculate. *Lodicules absent. Stamens 2. Intercostal epidermal cells with papillae overarching the stomata.*

Four genera, 31 spp.

Apocopis, *Germainia* and *Pogonatherum* are supported as a clade in the molecular phylogenies of Teerawatananon et al. (2011) and Estep et al. (2014), with *Germainia* apparently derived from within *Apocopis*. *Lophopogon* is placed here based solely on morphology.

405. *Pogonatherum* P. Beauv. Fig. 69

Pogonatherum P. Beauv., Ess. Agrostogr.: 56 (1812).

Highly branched perennials. Leaf blades with pseudopetioles. Ligule membranous or a fringed membrane. Inflorescence unbranched, subtended by a spathe-like leaf. Spikelets laterally compressed, tiny (<3 mm long). Sessile spikelet with a pubescent callus, the lower glume truncate or obtuse, awned or not, the upper glume awned. *Pedicellate spikelet pistillate.* $2n = 20$.

Three spp., tropical Asia.

Stamen number is reduced to 1 in some plants.

406. *Apocopis* Nees

Apocopis Nees, Proc. Linn. Soc. Lond. 1: 93 (1841).

Annuals or perennials, caespitose. Ligule membranous or a fringed membrane. Inflorescence unbranched, or with unbranched branches, digitate. Pedicellate spikelet present and poorly developed, or absent and represented only by the pedicel; *pedicel fused to the lower glume of the sessile spikelet.* Sessile spikelet with a pubescent callus; glumes with strong parallel veins, sometimes winged. $2n = 20$, 40.

Sixteen spp., tropical Asia.

407. *Germainia* Balansa & Poitr.

Germainia Balansa & Poitr., Bull. Soc. Hist. Nat. Toulouse 7: 344 (1873).

Annuals or perennials, caespitose or stoloniferous. Ligule membranous or a fringed membrane. Inflorescence unbranched, or with unbranched branches, digitate. Sessile spikelet staminate or

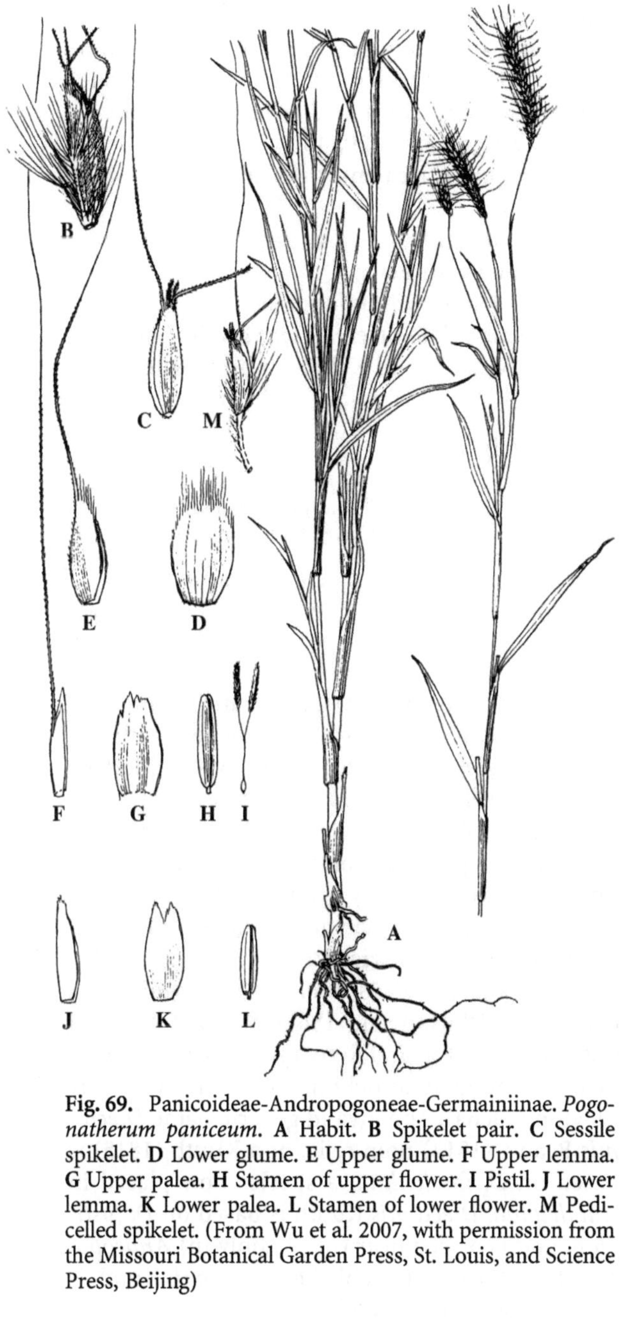

Fig. 69. Panicoideae-Andropogoneae-Germainiinae. *Pogonatherum paniceum*. **A** Habit. **B** Spikelet pair. **C** Sessile spikelet. **D** Lower glume. **E** Upper glume. **F** Upper lemma. **G** Upper palea. **H** Stamen of upper flower. **I** Pistil. **J** Lower lemma. **K** Lower palea. **L** Stamen of lower flower. **M** Pedicelled spikelet. (From Wu et al. 2007, with permission from the Missouri Botanical Garden Press, St. Louis, and Science Press, Beijing)

sterile. Pedicellate spikelet hermaphrodite, with or without a proximal sterile flower.

Ten spp., Southeast Asia and Australia.

The bisexual pedicellate spikelet is unusual in Andropogoneae, but also appears in *Trachypogon*.

408. ***Lophopogon*** Hack.

Lophopogon Hack., Die Nat. Pflanzenfam. II, 2: 26 (1887).

Annuals or perennials, caespitose. Ligule a fringed membrane. *Inflorescence with 2 branches appressed to form an ovate head.* Sessile and pedicellate spikelets similar, overlapping, the distal ends of the internodes oblique. Proximal spikelet pairs with both sessile and pedicellate spikelets staminate. Glumes stiff, truncate, the apex with three teeth; lower glume with distinct hair tufts, upper glume awned.

Two spp., India.

XXII.7. Subtribe Andropogoninae J. Presl (1830)

Anthistirinae J. Presl (1830).

Upper lemma awned from the apex or from a sinus, the awn generally geniculate.

Twenty-seven genera, 526 spp.

This subtribe has no obvious morphological synapomorphy. Subtribes Andropogoninae and Anthistirinae as delimited by Clayton and Renvoize (1986) are not monophyletic. However, several studies (e.g., Estep et al. 2014; Mathews et al. 2002; Skendzic et al. 2007) have shown that the two groups plus the *Bothriochloa* clade together are monophyletic. The Andropogoninae as delimited here correspond roughly to the "core Andropogoneae" identified in early molecular phylogenetic studies (Kellogg 2000; Mathews et al. 2002; Spangler et al. 1999).

409. ***Bhidea*** Stapf ex Bor

Bhidea Stapf ex Bor, Kew Bull. 3: 445 (1948) [1949].

Annuals. Ligule membranous. Inflorescence unbranched or with 2 branches, the rachis flattened. Callus pubescent, inserted in hollowed internode tip. Pedicellate spikelet staminate, the lower glume winged, muticous or awned. Sessile spikelet staminate in proximal spikelet pairs, bisexual in distal ones, the lower glume 2-keeled with large wings, the upper glume with three lobes, the central one extended into an awn.

Three spp., India.

Clayton and Renvoize (1986) segregate this from *Andropogon* sect. *Andropogon*. No molecular data are available.

The *Homozeugos* + *Agenium* + *Trachypogon* Clade

This group was found to be monophyletic by Kellogg and Watson (1993) and Guala (2000), but representatives have not been included in any recent molecular phylogenetic study.

410. ***Agenium*** Nees

Agenium Nees, Nat. Syst. Bot.: 447 (1836).

Caespitose perennials. Leaves with blades disarticulating, the ligule membranous or a fringed membrane. Inflorescence unbranched, or with digitate unbranched branches. Pedicellate spikelets larger than the sessile ones, staminate, the first glume pubescent, the lemma without an awn. *Pedicels deeply pigmented.* Sessile spikelets staminate in proximal spikelet pairs, bisexual in distal pairs. Callus sharp, pubescent, obliquely attached. Lower glume deeply grooved, *the groove translucent.* Distal flower lacking a palea.

Four spp., South America.

411. ***Homozeugos*** Stapf

Homozeugos Stapf, Hooker's Icon. Pl. 31: t. 3033 (1915).

Caespitose perennials. Leaves auriculate or not, the ligule membranous or a fringed membrane. Inflorescence unbranched, or with unbranched branches, digitate. *All spikelets bisexual.* Callus sharp, pubescent, obliquely attached. Glumes pubescent. Distal flower with or without a palea.

Six spp., Central Africa.

412. ***Trachypogon*** Nees

Trachypogon Nees, Fl. Bras. Enum. Pl. 2(1): 341 (1829).

Caespitose perennials. Ligule membranous. Inflorescence unbranched, or with unbranched branches, digitate, the axes not disarticulating. *Pedicellate spikelet hermaphrodite*, callus sharp, pubescent, obliquely attached, the lower glume with two keels, the lemma pubescent. "Sessile" spikelet actually short pedicellate, staminate or sterile, awnless. $2n = 20, 40$.

Four spp., Africa and tropical America.

The *Monocymbium* + *Anadelphia* + *Elymandra* Morphogroup

Upper glume with a long awn.

Molecular data place *Elymandra* in Andropogoninae (McKain and Kellogg, unpublished).

413. ***Anadelphia*** Hack.

Anadelphia Hack., Bot. Jahrb. Syst. 6: 240 (1885).
Andropogon sect. *Pobeguinea* Stapf, J. Bot. (Morot) 19: 100 (1905).
Monium Stapf, Fl. Trop. Afr. 9: 399 (1919).
Pobeguinea (Stapf) Jacq.-Fél., Rev. Int. Bot. Appl. Agric. Trop. 30: 172 (1950).

Annuals or perennials. Axillary branch complexes present. Inflorescence axes unbranched, with few spikelets widely separated on the axis. Pedicellate spikelet the same size as the sessile one, much reduced, or absent. Sessile spikelet with a sharp callus; lower glume 2-pronged or truncate; awn of upper glume often nearly as long as the spikelet and spreading, upper lemma with two lobes. $2n = 20$.

Fourteen spp., Africa.

Pobeguinea and *Monium* are recognized as distinct by Watson and Dallwitz (1992 onward) but not by Clayton and Renvoize (1986). The rationale for the generic delimitation is provided by Clayton (1966a). The invalid generic name *Diectomis* Beauv. (1812) nom rejic. non Kunth (1815) is based on *D. fasciculata*, a species now placed in *Anadelphia*. *Diectomis* Kunth is based on *D. fastigiata* and is placed near (or possibly in) *Andropogon*. Thus the validly published *Diectomis* Kunth is not a synonym of *Anadelphia*.

414. ***Elymandra*** Stapf

Elymandra Stapf, Fl. Trop. Afr. 9: 407 (1919).
Pleiadelphia Stapf, Hooker's Icon. Pl. 30: t. 3121 (1927).

Annuals or perennials. Axillary branch complexes present. Inflorescences unbranched or with two branches with up to 18 spikelet pairs. *Lower 1–10 sessile spikelets staminate, obliquely attached to the axis.* Lower glume of sessile spikelet pubescent to villous. *Pedicellate spikelet with a stipe-like callus.* $2n = 20$.

Six spp., all African except one in Brazil.

Clayton (1966b) synonymizes *Pleiadelphia* with *Elymandra*, observing that the only difference is the unbranched inflorescence of the former.

415. ***Monocymbium*** Stapf

Monocymbium Stapf, Fl. Trop. Afr. 9: 386 (1919).

Caespitose perennials. Ligule membranous or a fringed membrane. Axillary branch complexes present. Inflorescence unbranched, wholly surrounded by a reddish-brown spathe. Pedicellate spikelet with glumes truncate, unawned. Sessile spikelet with the callus pubescent, both flowers lacking paleas. $2n = 20$.

Three spp., tropical and southern Africa.

The *Exotheca* + *Hyperthelia* group

Ligule membranous. Axillary branch complexes present. *Inflorescence with two branches, one spikelet-bearing to near the base and relatively shorter, the longer one spikelet-bearing only in the distal portion, barren in the proximal half.* Callus of hermaphrodite spikelet sharp-pointed.

Skendzic et al. (2007) place *Hyperthelia* in a clade with *Hyparrhenia hirta* and *Andropogon distachyos*, suggesting that it might be merged with one of the latter two genera. However, the number of species sampled is not large enough to provide confidence about its placement. Chloroplast DNA sequences also place *Exotheca* near *Hyparrhenia*, but data are not available for *Hyperthelia* (McKain and Kellogg, unpubl. data).

416. ***Exotheca*** Andersson

Exotheca Andersson, Nova Acta Regiae Soc. Sci. Upsal. 2: 253 (1856).

Caespitose perennials. Leaves auriculate. Axillary branch complexes present. *The four proximal spikelets on each branch staminate and sessile or nearly so, forming an involucre. Distal to the "involucre" is a set of three spikelets, one hermaphrodite and sessile, the other two staminate and pedicellate.* Pedicellate spikelet with glumes awned. Sessile spikelet callus pubescent, the glumes pubescent.

One sp., *E. abyssinica* (Hochst. ex A. Rich.) Andersson, eastern Africa and Viet Nam.

417. ***Hyperthelia*** Clayton

Hyperthelia Clayton, Kew Bull. 20: 438 (1967).

Annuals or perennials, caespitose. Leaves auriculate or not, with or without a pseudopetiole. Longer inflorescence branch bearing only three spikelets near the tip, one sessile and hermaphrodite, and the other two pedicellate and staminate; shorter inflorescence branch bearing five spikelets, the two proximal ones staminate, the distal ones in a set of three as on the long branch. *Long flattened, leaf-like appendage often present at the base of the branches.* Lower glume of sessile spikelet strongly grooved, the upper glume mucronate or awned. Both flowers of the sessile spikelet lacking a palea. $2n = 20, 40$.

Seven spp., African savannas, introduced to tropical America.

Remaining genera of Andropogoninae:
Leaves with papillae present on the intercostal epidermal cells, each papilla consisting of one oblique swelling per cell overarching the stomata.

The single oblique papilla is not common elsewhere in the tribe, but does occur in *Apocopis*, *Apluda*, *Loxodera*, and *Parahyparrhenia*. The oblique papilla is apparently lacking in *Andropogon* and *Schizachyrium*; *Themeda* has several papillae per cell.

418. ***Cymbopogon*** Spreng.

Cymbopogon Spreng., Pl. Min. Cogn. Pug. 2: 14 (1815).

Perennials, or rarely annual, caespitose. *Leaves aromatic*, auriculate or not. Ligule membranous or a fringed membrane. Axillary branch complexes present. *Inflorescence a pair of strongly reflexed branches, one slightly longer than the other, the lower one with a pair of staminate spikelets at its base.* Pedicellate spikelet generally staminate, rarely sterile. Sessile spikelet with a pubescent callus, inserted in hollowed internode tip, staminate in the proximal spikelet pair (in most species), bisexual in distal spikelet pairs. Lower glume, flat or concave, 2-keeled, the keels sometimes winged, *both flowers lacking paleas.* $2n = 20, 22, 40, 60$.

Fifty-nine spp., Old World tropics and subtropics.

The reflexed branches are diagnostic, but do not occur in ser. Proceri (Clayton and Renvoize 1986) in which the branches are erect. The genus is strongly supported as monophyletic in molecular phylogenies, and is not obviously closely related to any other genus (Estep et al. 2014).

419. ***Eremopogon*** Stapf

Eremopogon Stapf, Fl. Trop. Afr. 9: 182 (1917).

Perennials, rarely annual. Axillary branch complexes present. Inflorescences unbranched. Pedicellate spikelets staminate or sterile, awnless. Sessile spikelets with a pubescent callus, the lower glume with a large circular pit in the center, both flowers lacking paleas. $2n = 40$.

Four spp., Old World tropics.

The large circular pit on the lower glume also occurs in some species of *Bothriochloa* and *Dichanthium*, and led Clayton and Renvoize (1986) to include this genus in *Dichanthium*. However, molecular data place *Eremopogon* separate from either of these genera and instead sister to *Iseilema* (Estep et al. 2014). *Eremopogon* lacks the hyaline groove in the pedicels that is characteristic of *Bothriochloa* and *Capillipedium*.

420. ***Heteropogon*** Pers.

Heteropogon Pers., Syn. Pl. 2: 533 (1807).

Annuals or perennials, caespitose, monoecious or andromonoecious. Ligule membranous or a fringed membrane. Axillary branch complexes present. Inflorescence unbranched. Pedicellate spikelets staminate or sterile, with an elongated callus that looks superficially like a pedicel, the lower glume winged. Sessile spikelets terete, with a long, sharp pubescent callus, staminate in the proximal portion of the inflorescence, pistillate or sometimes hermaphrodite in the distal portion. $2n = 20, 22, 40, 44, 50, 60, 80$.

Six spp., tropical.

421. ***Iseilema*** Andersson

Iseilema Andersson, Nova Acta Regiae Soc. Sci. Upsal. 2: 250 (1856).

Annuals or perennials, caespitose. Axillary branch complexes present. Ligule membranous, or a fringed membrane, or a fringe of hairs. *Leaves and spathes gland bearing, aromatic.* Inflorescence subtended by a boat-shaped spathe, *unbranched, with four staminate spikelets (interpreted as two pairs) at the base, with 1 or 2 sessile bisexual spikelets and a pair of pedicellate staminate spikelets distally. Pedicels of proximal four spikelets curved stiffly upward to form basket, often with a tuft of hairs at the base of the basket.* Pedicellate spikelets with glumes flat, strongly veined. Sessile spikelets with distal flower lacking a palea. $2n = 6, 8, 18, 28, 36$.

Twenty-four spp., Indomalayan region, Australia.

The odd inflorescence morphology of this genus suggests a connection with *Themeda*. Molecular data show that the two are closely related (Skendzic et al. 2007), but Estep et al. (2014) suggest that they do not form a clade.

422. ***Pseudodichanthium*** Bor

Pseudodichanthium Bor, Indian Forester 66: 271 (1940).

Annuals. Ligule a fringed membrane. Inflorescence unbranched. Pedicellate spikelets staminate or sterile, morphologically similar to the sessile ones and larger, lower glume winged, the pedicel cuneate, the distal end cupuliform. Proximal 2 to 3 sessile spikelets staminate and awnless, distal ones bisexual, awned. Callus rounded. Lower glume winged, the apex emarginate, both flowers lacking a palea.

One sp., *P. serrafalcoides* (Cooke & Stapf) Bor, India.

423. ***Themeda*** Forssk. Fig. 70

Themeda Forssk., Fl. Aegypt.-Arab. 178 (1775).

Annuals or perennials, mostly caespitose. Axillary branch complexes present. Ligule membranous, or a fringed membrane or absent.

Fig. 70. Panicoideae-Andropogoneae-Andropogoninae. *Themeda triandra*. **A** Habit. **B** Inflorescence with spatheole. **C** Spikelet triad, with one awned fertile spikelet and 2 pedicelled sterile spikelets. **D** Two pairs of staminate spikelets. **E** Upper lemma and awn base. (From Wu et al. 2007, with permission from the Missouri Botanical Garden Press, St. Louis, and Science Press, Beijing)

Inflorescence subtended by a spathe, unbranched, *with four staminate spikelets (interpreted as two pairs) at the base, with 1 or 2 sessile bisexual spikelets and a pair of pedicellate staminate spikelets distally*. Callus pubescent, its attachment oblique. Sessile spikelet with glumes truncate, indurate; distal lemma apex mucronate in some species. $2n = 20, 40, 60, 80$.

Twenty-nine spp., Old World tropics and subtropics.

The *Capillipedium* + *Bothriochloa* + *Dichanthium* group

While diploid members of *Bothriochloa*, *Capillipedium* and *Dichanthium* are intersterile, the three genera are interfertile at the tetraploid level (Harlan and de Wet 1963). Molecular phylogenies show that Old World polyploid species of *Bothriochloa* contain genomes donated by *Dichanthium* and *Capillipedium* consistent with the cytogenetic data (Estep et al. 2014). In this one small group one sees many of the inflorescence forms of the entire Andropogoneae, supporting the idea that transitions among them are easy.

424. *Capillipedium* Stapf

Capillipedium Stapf, Fl. Trop. Afr. 9: 169 (1917).

Annuals or perennials, caespitose. Leaves generally without auricles. Ligule membranous, a fringed membrane, or a fringe of hairs. Inflorescences delicate and open, the slender branches themselves branched. Inflorescence internodes and pedicels of pedicellate spikelets with two strong marginal veins and a hyaline groove, often described as a "translucent line", between. Each branch with a long naked proximal portion and fewer than eight pairs of spikelets. Pedicellate spikelets staminate or sterile. Sessile spikelets with a pubescent callus, both flowers generally lacking paleas. $2n = 20, 40, 60$.

Eighteen spp., Africa, Asia and Australia.

425. *Bothriochloa* Kuntze

Bothriochloa Kuntze, Revis. Gen. Pl. 2: 762 (1891).

Perennials, caespitose or rhizomatous. Ligule membranous or a fringed membrane. Inflorescence with digitate primary branches, or the branches borne on an elongated axis, each branch with 8 or more pairs of spikelets. Inflorescence internodes and pedicels of pedicellate spikelets with two strong marginal veins and a hyaline groove, often described as a "translucent line", between. Pedicellate spikelets staminate or sterile. Sessile spikelets with a pubescent callus, the lower glume often with a deep circular pit in the center, with two keels. $2n = 20, 30, 40, 50, 60, 120$.

Thirty-seven spp., tropical.

The number of spikelets per inflorescence branch is not distinctive within the

Andropogoneae, but does distinguish *Bothriochloa* from *Euclasta* and *Capillipedium.*

426. ***Dichanthium*** Willemet

Dichanthium Willemet, Ann. Bot. (Usteri) 18: 11 (1796).

Annuals or perennials, caespitose. Leaves generally without auricles. Ligule membranous or a fringed membrane. Inflorescence unbranched, or with digitate branches. Proximal pair(s) of spikelets generally staminate. Sessile spikelet with the callus obtuse, pubescent. Glumes ovate to elliptical, multiveined, overlapping; both flowers with paleas minute or absent. $2n = 20, 40, 50, 60$.

Twenty-two spp., Old World tropics.

The *Diheteropogon* + *Parahyparrhenia* + *Pseudanthistiria* group

Flower of sessile spikelet single, subtended by two hyaline bracts, interpreted as lemmas.

Diheteropogon is strongly supported as a member of the large clade including *Andropogon*, *Schizachyrium*, and *Hyparrhenia* (Estep et al. 2014), but no molecular data are available on *Pseudanthistiria.* The two lemmas presumably represent the lemmas of the proximal and distal flowers, but could also be interpreted as the lemma and palea of the distal flower with the proximal flower being entirely suppressed. Molecular data obtained as this book went to press suggest that *Parahyparrhenia* may be misplaced here and instead be sister to *Eriochrysis* (McKain and Kellogg, unpublished).

427. ***Diheteropogon*** (Hack.) Stapf

Diheteropogon (Hack.) Stapf, Hooker's Icon. Pl. 31: t. 3093 (1922).

Annuals or perennials, caespitose. Leaf blade broad or reduced to the midrib. Ligule membranous or a fringed membrane. Axillary branch complexes present. Inflorescences with two branches. Pedicellate spikelet staminate or sterile. Sessile spikelet with the callus pubescent, acute to sharp-pointed, inserted into a deep crater-like apex of the internode below; lower glume grooved. $2n = 20, 40$.

Four spp., Africa.

428. ***Parahyparrhenia*** A. Camus

Parahyparrhenia A. Camus, Bull. Mus. Natl. Hist. Nat. II, 22: 404 (1950).

Annuals or perennials. Leaf blades narrow, filiform, auriculate or not. Ligule membranous or a fringe of hairs. Inflorescence unbranched or with two branches, spikelet-bearing to near the base, or the proximal portion without spikelets, the rachis flattened. Pedicellate spikelets staminate or sterile, with glumes muticous or awned. Sessile spikelet callus pubescent, acute to sharp-pointed, obliquely attached to the rachis and extending beyond it; the lower glume deeply grooved, the upper glume mucronate or awned.

Six spp., Africa, India, Thailand.

This genus is morphologically similar to *Diheteropogon.*

429. ***Pseudanthistiria*** (Hack.) Hook. f.

Pseudanthistiria (Hack.) Hook. f., Fl. Brit. India 7: 219 (1897) [1896].

Trailing annuals, rooting from the nodes. Leaf blades lanceolate. Ligule membranous or a fringed membrane. Axillary branch complexes present. Inflorescence unbranched, with one proximal spikelet pair and a distal triplet of one sessile and two pedicellate spikelets. Pedicellate spikelets staminate, with muticous glumes. Sessile spikelet with the callus oblique, pubescent. $2n = 20$.

Four spp., India to Thailand.

The *Diectomis* + *Hyparrhenia* + *Andropogon* + *Schizachyrium* group

This clade has no obvious morphological synapomorphy, but the latter three genera have been included in several molecular phylogenies, and consistently form a well-supported group along with *Diheteropogon* (Estep et al. 2014).

430. ***Diectomis*** Kunth

Diectomis Kunth, Mém. Mus. Hist. Nat. 2: 69 (1815), nom. conserv., non Beauv. (1812).

Caespitose annuals. Ligule membranous. Axillary branch complexes present. Inflorescences unbranched, the pedicels and rachis with long hairs, with a translucent line down the center. *Pedicellate spikelet much larger than the sessile one; lower glume flat, papery, reddish, awned.* Sessile spikelet with a hairy callus, the lower glume deeply grooved.

One sp., *D. fastigiata* (Sw.) P. Beauv., pantropical.

Clayton and Renvoize (1986) and Clayton et al. (2006 onward) include *D. fastigiata* in *Andropogon*. Consistent with this hypothesis Skendzic et al. (2007) place it in a clade with *Schizachyrium semitectum* and *Andropogon gerardii*, suggesting that it might be merged with one of the latter two genera. It is kept separate here because species sampling of molecular phylogenies is insufficient to place it reliably. Its unbranched inflorescence and pedicellate spikelets with broad glumes are distinctive.

431. ***Hyparrhenia*** Andersson ex E. Fourn.

Hyparrhenia Andersson ex E. Fourn., Mexic. Pl. 2: 51, 67 (1886).
Dybowskia Stapf, Fl. Trop. Afr. 9: 382 (1919).

Annuals or perennials, caespitose. Leaves auriculate or not. Ligule membranous or a fringed membrane. Axillary branch complexes present. Inflorescence a pair of erect, spreading or deflexed branches, one slightly longer than the other, the base of each inflorescence branch often with one or two pairs of staminate spikelets. Pedicellate spikelets staminate or sterile. Sessile spikelets with a hairy callus, attached obliquely to the rachis; lower glume convex or flat. $2n = 20$, 30, 40, 44, 45, 60.

Fifty-eight spp., Africa.

The two-branched inflorescences of this genus are similar to those of *Cymbopogon*, but the two genera can be distinguished by the connection of the sessile spikelet to the rachis. In *Cymbopogon* the spikelet is inserted into a deep crater-like node, whereas in *Hyparrhenia* the callus of the spikelet is obliquely attached. Although they have similar inflorescences, *Cymbopogon* and *Hyparrhenia* appear not to be particularly closely related. The single species of *Dybowskia* is included within *Hyparrhenia*, following Clayton (1969).

Remaining two genera:
Leaf epidermal papillae absent, an apparent loss.

432. ***Andropogon*** L.

Andropogon L., Sp. Pl. 2: 1045 (1753).

Perennials, a few species annual, caespitose or less often rhizomatous. Leaves auriculate or not. Ligule membranous, or a fringed membrane or a fringe of hairs. Axillary branch complexes present. Inflorescence unbranched or digitate. Pedicellate spikelets staminate or sterile, or reduced only to the glumes, or lacking, leaving only the pedicel. Sessile spikelets with a pubescent callus inserted into the hollowed internode tip; lower glume muticous, with the margins raised or keeled, hollowed or grooved between the keels; upper glume mucronate or awned. $2n = 20$, 40, 60, 100, 120, 180.

One hundred twenty-two spp., worldwide.

This genus may be paraphyletic (Skendzic et al. 2007) or monophyletic (Estep et al. 2014), but will require much more detailed sampling to assess relationships. Clayton and Renvoize (1986) divide the genus into 4 sections, *Andropogon*, *Leptopogon* (ca. 55 species), *Piestium*, and *Notosolen*. Individual species and species complexes have been studied in detail (e.g., Hodge and Kellogg 2014; Nagahama et al. 2014) but the genus as a whole would benefit from more comprehensive assessment of species limits.

433. ***Schizachyrium*** Nees

Schizachyrium Nees, Fl. Bras. Enum. Pl.: 331 (1829).

Perennials, a few species annual, caespitose or less often rhizomatous. Ligule membranous, or a fringed membrane. Axillary branch complexes present. Inflorescence unbranched, the internodes slender or clavate, with a fimbriate rim distally. Pedicellate spikelets staminate or sterile, or reduced only to the glumes, or lacking, leaving only the pedicel, the glumes muticous or awned. Sessile spikelets with a pubescent callus inserted into the hollowed internode tip, the glumes muticous or awned; lower glume flat, convex or grooved, both flowers lacking paleas. $2n = 20$, 30, 40, 50.

Sixty-four spp., tropics and temperate regions, worldwide.

The molecular phylogenetic study of Skendzic et al. (2007) indicates that *Schizachyrium* is polyphyletic, but those authors included only six of the sixty species. More recent data (Estep et al. 2014) indicate that it may be monophyletic and sister to *Andropogon*.

XXIII. Tribe Paspaleae J. Presl (1830)

Trachideae Pilg. ex Potztal (1957).
Arthropogoneae Pilg. ex Butzin (1972).

This tribe has no obvious morphological synapomorphy.

31 genera, 664 species.

Paspaleae as circumscribed here correspond to the group known as the $x = 10$ Paniceae in several molecular phylogenies (Aliscioni et al. 2003; Giussani et al. 2001; Morrone et al. 2012; Vicentini et al. 2008). Paspaleae excluding *Reynaudia* are well supported as being monophyletic (Morrone et al. 2012), and are sister to Andropogoneae (GPWG II 2012), a relationship supported by the synapomorphy of the chromosome base number. However, no obvious morphological character distinguishes Paspaleae from Paniceae s.s.

Within Paspaleae are three well-supported clades, recognized here as subtribes following Morrone et al. (2012). Most of the taxa occur in the New World, except for a few species of *Hymenachne*, and *Baptorachis* from Africa. Many species in Paspaleae were formerly classified in a broadly construed genus *Panicum*, but are clearly unrelated to *Panicum* s.s.

XXIII.1. Subtribe Arthropogoninae Butzin (1972)

This subtribe has no obvious morphological characters.

14 genera, 71 species.

This group was first identified as a clade by Giussani et al. (2001), who called it the "ambiguous clade" because of its morphological heterogeneity. The clade contains a mix of C_3 and C_4 taxa, and the topology of the phylogeny makes it impossible to determine exactly how many origins of C_4 have occurred; the number could be between 2 and 5 origins (GPWG II 2012).

434. ***Apochloa*** Zuloaga & Morrone

Apochloa Zuloaga & Morrone, Syst. Bot. 33: 288 (2008).

Tussock forming perennials. Leaves ending in a sharp point, often without a clear sheath-blade junction. Ligule a fringed membrane or lacking. Inflorescence branches branched, spreading. Upper lemma and palea shining, smooth. C_3. $2n = 20, 40$.

Fifteen spp., Guyana Highlands and eastern Brazil.

This genus was segregated from *Panicum* by Sede et al. (2008). All included species were formerly in *Panicum* sect. *Lorea*, which was found to be polyphyletic.

435. ***Arthropogon*** Nees

Arthropogon Nees, Fl. Bras. Enum. Pl. 2(1): 319 (1829).
Achlaena Griseb., Cat. Pl. Cub.: 228 (1866).
Altoparadisium Filg., Davidse, Zuloaga & Morrone, Ann. Missouri Bot. Gard. 88: 363 (2001).

Caespitose perennials. Ligule membranous or a fringed membrane or a fringe of hairs. Inflorescence branches branched, spreading or upright. Spikelets laterally compressed. Callus with long hairs. Glumes awned, the awns about 10 times longer than the glume itself, the lower glume shorter than the spikelet or absent. Lower lemma longer than the upper one. Upper lemma and palea laterally compressed, the upper lemma hyaline, not enclosing the apex of the palea. C_4, with a single bundle sheath. Mesophyll with or without distinctive isolated bundle sheath cells.

Six spp., Brazil and West Indies.

Arthropogon s.s., *Achlaena*, and *Altoparadisium* form a strongly supported clade (Morrone et al. 2012; GPWG II 2012); the long awns on the glumes are distinctive, so the three genera are united here. In addition, members of the clade have a callus with long hairs and the upper lemma and palea laterally compressed. The former *Arthropogon lanceolatus* was segregated as *Canastra* based on its phylogenetic position (Giussani et al. 2001); *Canastra* and *Arthropogon* are confirmed to be unrelated by GPWG II (2012).

436. ***Canastra*** Morrone, Zuloaga, Davidse & Filg.

Canastra Morrone, Zuloaga, Davidse & Filg., Novon 11: 429–436 (2001).

Caespitose perennials. Ligule a fringed membrane. Leaf blades narrowed at the base, involute. Inflorescence branches branched, dense. Callus glabrous. Glumes awned, the awns about 10 times longer than the glume itself. Lower lemma longer than the upper one. Upper lemma cartilaginous, the apex not enclosing the apex of the palea. C_3. Mesophyll radiate, with fusoid cells.

Two spp., Brazil.

437. ***Homolepis*** Chase

Homolepis Chase, Proc. Biol. Soc. Wash. 24: 146 (1911).

Annuals or perennials, caespitose. Ligule membranous. Leaf blades generally cordate. Inflorescence branches branched. Glumes enclosing the flowers. Upper lemma thin, not indurate, the margins lying flat on the palea. Hilum linear. C_3. Midrib complex; mesophyll with fusoid cells. $2n = 22, 24, 40$.

Five spp., Mexico to Brazil.

438. ***Oplismenopsis*** Parodi

Oplismenopsis Parodi, Notas Mus. La Plata, Bot. 2: 2 (1937).

Perennial with floating culms. Ligule a fringed membrane. Inflorescence with the primary branches unbranched. Spikelets with the lower glume toward the rachis axis (adaxial). Lower glume with an awn; upper glume extended into an arista. Lower lemma with a short awn. Hilum linear. C_3.

One sp., *O. najada* (Hack. & Arechav.) Parodi, Argentina. In water.

439. ***Phanopyrum*** (Raf.) Nash

Phanopyrum (Raf.) Nash, Fl. S.E. U.S.:104, 1327 (1903).

Stoloniferous perennials. Leaf blades broad, flat, the bases cordate; ligules membranous. Primary branches of inflorescence unbranched, with the spikelets borne on one side, crowded. Glumes more or less equal. Lower lemma longer than upper one. Upper flower obovoid, the rachilla internode below it elongated, smooth. Margins of upper lemma involute. C_3.

One sp., *P. gymnocarpon* (Elliott) Nash, southeastern U.S.

440. ***Stephostachys*** Zuloaga & Morrone

Stephostachys Zuloaga & Morrone, Taxon 59: 1539 (2010).
Panicum sect. *Megista* Pilg., Notizbl. Bot. Gart. Berlin-Dahlem 11: 243 (1931).

Rhizomatous perennials, 1–3 m tall. Ligule membranous, laciniate. Base of leaf blades cordate to rounded. Inflorescence branches unbranched, arranged in pseudoverticels of 10 to 60 branches each. Upper glume and lower lemma similar in length. Lower flower staminate. Apices of upper lemma and palea with stomata and papillae. C_3. $2n = 40$.

One sp., *S. mertensii* (Roth) Zuloaga & Morrone, Central and South America.

Additional description of this genus is provided by Zuloaga et al. (2010).

Cyphonanthus + Oncorachis

Inflorescence branches branched. C_4, with a single bundle sheath.

These two genera form a strongly supported clade in the analysis of Morrone et al. (2012), and could perhaps be merged into a single genus.

441. ***Cyphonanthus*** Zuloaga & Morrone

Cyphonanthus Zuloaga & Morrone, Taxon 56: 526 (2007).

Rambling caespitose perennials. Ligule a fringe of hairs. Lower glume absent or minute, upper glume gibbous, as long as the spikelet. Rachilla not thickened. Upper lemma and palea open at the apex, pubescent at the apex and base, covered with macrohairs and simple papillae. C_4, with a single bundle sheath.

One sp., *C. discrepans* (Döll) Zuloaga & Morrone, Central America and northern South America to Brazil.

This species was segregated from *Panicum* on the basis of morphological, anatomical and molecular data. It corresponds to *Panicum discrepans* and was placed in its own section (*Panicum* sect. *Discrepentia*) by Zuloaga (1987), highlighting its differences from other species of *Panicum*.

442. ***Oncorachis*** Morrone & Zuloaga

Oncorachis Morrone & Zuloaga, Taxon 58: 372 (2009).

Rhizomatous perennials. Leaf blades rigid. Ligule a fringe of hairs or lacking. Lower glume present, 1/6 to 2/3 the length of the spikelet, upper glume not gibbous. *Rachilla thickened above the glumes and below the lower flower.* Upper lemma and palea indurate, covered with macrohairs and simple papillae. C_4, with a single bundle sheath.

Two spp., central and eastern Brazil.

This genus was erected to accommodate *Streptostachys ramosa* Zuloaga & Soderstr. and *S. macrantha* (Trin.) Morrone & Zuloaga, which are more closely related to *Cyphonanthus* than they are to other species of *Streptostachys* (Sede et al. 2009a).

Coleataenia + Triscenia

Triscenia is in an unresolved relationship to *Coleataenia* in the analyses of Morrone et al. (2012), and is strongly supported as being embedded within *Coleataenia* in GPWG II (2012). Because the two genera differ in photosynthetic pathway, their precise relationship affects inference about the evolutionary origins of C_4, and would repay further investigation.

443. ***Coleataenia*** Griseb., Abh. Königl. Ges. Wiss. Göttingen 24: 308 (1979).

Panicum [unranked] *Agrostoidea* Nash in Small, Fl. S. E. U. S.: 85–86 (1903).
Panicum [unranked] *Tenera* Hitchc. & Chase, Contr. U. S. Natl. Herb. 15: 29, 99 (1910).
Panicum sect. *Prionitia* Zuloaga, Grass Syst. Evol.: 294 (1987).
Sorengia Zuloaga & Morrone, Taxon 59: 1541 (2010).

Perennials, caespitose or rhizomatous. Ligule membranous, ciliate or not. Inflorescence branches branched. Spikelets solitary, pedicellate, obliquely attached to the pedicel. Lower glume with 1 to 3 prominent veins, upper glume and lower lemma with 5 to 9. Upper glume and lower lemma enclosing the upper lemma and palea. Upper lemma and palea indurate, smooth, shiny, often with two prickle hairs at the apex. C_4, with one bundle sheath, or two sheaths in *C. prionitis* and *C. petersonii*. $2n = 18, 36, 20, 40$.

Seven spp., Mexico, Central America, South America, Caribbean.

Coleataenia is supported as monophyletic by molecular data only; no obvious morphological synapomorphies link the species, although the obliquely attached spikelets provide a distinctive, if subtle, character. This clade was first identified by Aliscioni et al. (2003), although reevaluation of the data has subsequently excluded *Stephostachys* from the group. Information on nomenclature, distribution and morphology of the species is presented by Zuloaga et al. (2010). "*Panicum*" *beyeri* is not mentioned, however, and it is unclear where it belongs. Various names are connected to *Panicum* [unranked] *Agrostoidea*; their history and, where necessary, typification are presented by McNeill et al. (2010).

444. ***Triscenia*** Griseb.

Triscenia Griseb., Pl. Wright. 2: 534 (1862).

Perennials. Leaf blades narrow, without clear lamina and midrib. Ligule a fringed membrane. Inflorescence branches unbranched. Glumes shorter than the spikelet. Upper lemma thin, not hardened, the margins flat on the palea. C_3.

One sp., *T. ovina* Griseb., Cuba.

Keratochlaena + Mesosetum + Tatianyx

Hilum linear. C_4, NADP-ME.

This clade is strongly supported in the molecular analyses of Morrone et al. (2012).

445. ***Keratochlaena*** Morrone

Keratochlaena Morrone, Darwiniana 47: 231 (2009).
Sclerochlamys Morrone & Zuloaga, Taxon 58: 373 (2009), nom. illegit.

Caespitose perennials. Leaf blades rigid, pungent. Inflorescence branches unbranched. Rachilla internodes thickened. Lower glume 1/3–2/3 the length of the spikelet. Upper glume and lower lemma coriaceous. Upper lemma and palea covered with compound papillae.

One sp., *K. rigidifolia* (Filg., Morrone & Zuloaga) Morrone & Zuloaga, northern Brazil.

Sclerochlamys was erected to accommodate *Streptostachys rigidifolia* Filg., Morrone & Zuloaga, which is more closely related to *Mesosetum* than it is to other species of *Streptostachys* (Sede et al. 2009a). The name proved to be a later homonym for *Sclerochlamys* F. Muell., so was replaced with the name *Keratochlaena*.

446. ***Mesosetum*** Steud.

Mesosetum Steud., Syn. Pl. Glumac. 1: 118 (1855) [1854].

Annuals or perennials, caespitose, rhizomatous or stoloniferous. Ligule a fringe of hairs. Inflorescence unbranched. Pedicel apex oblique; spikelets laterally compressed, with the lower glume toward the rachis axis (adaxial); upper lemma and palea smooth and gaping. $2n = 16$.

Twenty-seven spp., Central America and northern South America.

Mesosetum is strongly supported as being sister to *Tatianyx* (GPWG II 2012); both genera are C_4 but reportedly differ in the number of bundle sheaths (Watson and Dallwitz 1992 onward).

447. ***Tatianyx*** Zuloaga & Soderstr. Fig. 71

Tatianyx Zuloaga & Soderstr., Smithsonian Contr. Bot. 59: 56 (1985).

Caespitose perennials. Sheaths scarious, pubescent. Ligule a fringe of hairs. Inflorescence branches branched, spreading. Spikelet attachment oblique to pedicel, the callus pubescent. Glumes and lower lemma villous, enclosing the upper lemma and palea. Upper lemma indurate.

One sp., *T. arnacites* (Trin.) Zuloaga & Soderstr., Brazil.

Tatianyx is strongly supported as being sister to *Mesosetum* (GPWG II 2012), but *Keratochlaena* was not included in that analysis. The microhairs are described as unusual, but require further examination with good quality material (Watson and Dallwitz 1992 onward).

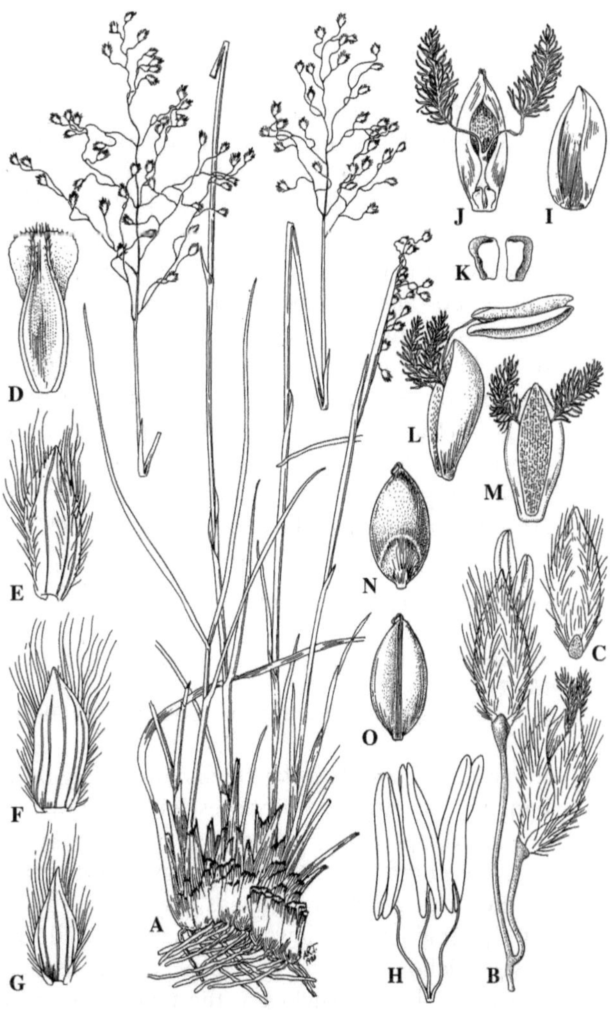

Fig. 71. Panicoideae-Paspaleae-Arthropogoninae. *Tatianyx arnacites*. **A** Habit. **B** Pair of spikelets, showing oblique attachment of pedicels. **C** Spikelet, upper glume side, showing oblique callus. **D** Lower palea, outer surface. **E** Lower lemma. **F** Upper glume. **G** Lower glume. **H** Stamens. **I** Upper lemma. **J** Upper palea showing gynoecium and lodicules. **K** Lodicules. **L** Upper anthecium, lateral view. **M** Same, palea side. **N** Caryopsis, embryo side. **O** Caryopsis, showing linear hilum. (Reprinted from Zuloaga and Soderstom 1985; courtesy of Smithsonian Institution)

XXIII.2. Subtribe Otachyriinae *Butzin (1970)*

Inflorescence branches generally with spikelets arranged along one side. Lower glume shorter than the upper one. *Deletion of 9 bp in the chloroplast gene ndhF.*

Four genera, thirty-eight species.

Otachyriinae are strongly supported as monophyletic (Morrone et al. 2012), but the genera included do not share any obvious morphological synapomorphies; the shared characters appear elsewhere in the Panicoideae. Otachyriinae include all species formerly placed in "*Panicum*" sect. *Laxa*, which is not monophyletic. The type species of the former *Laxa* group is unrelated to the others and is now *Steinchisma laxa*, here placed in synonymy in *Otachyrium*. Other species of the *Laxa* group have been transferred to *Hymenachne* or placed in the new genus *Rugoloa* (Acosta et al. 2014). Detailed descriptions of the species in the former "*Panicum*" sect. *Laxa*, including photos of leaf anatomy and epidermal structure, are provided by Zuloaga et al. (1992).

448. ***Anthaenantia*** P. Beauv.

Anthaenantia P. Beauv., Ess. Agrostogr.: 48, 151, t. 10 (1812).
Leptocoryphium Nees, Fl. Bras. Enum. Pl. 2(1): 83 (1829).

Perennials, caespitose or rhizomatous. Ligule a fringed membrane. Inflorescence branches branched, spreading. Spikelets dorsiventrally to laterally compressed, densely pubescent. Lower glume lacking. Upper glume as long as the spikelet. Upper glume and lower lemma with papillose-based trichomes along the veins. Upper lemma cartilaginous, not enclosing the palea, the apex acute. C_4, with one bundle sheath. $2n = 40$.

Five spp., southeastern U.S., Central America and South America.

Anthaenantia is strongly supported as sister to all other members of Otachyriinae (Acosta et al. 2014; Morrone et al. 2012).

449. ***Hymenachne*** P. Beauv. Fig. 72

Hymenachne P. Beauv., Ess. Agrostogr.: 48 (1812).
Dallwatsonia B. K. Simon, Austrobaileya 3: 678 (1992).

Perennials. Culms filled with aerenchyma. Ligule membranous. Inflorescence branches branched, short and appressed to the main axis. Upper glume and lower lemma as long as the spikelet, the latter acute, acuminate or with a short awn. Upper lemma membranous, *clasping the palea proximally but not at tip*. C_3. $2n = 24$.

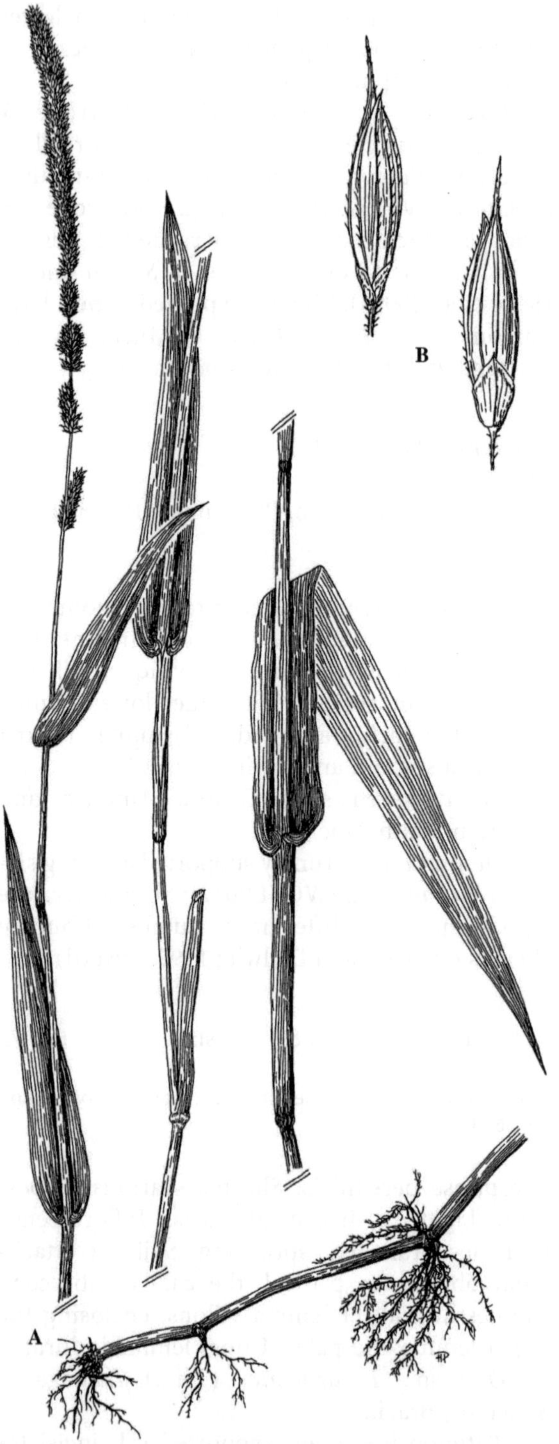

Fig. 72. Panicoideae-Paspaleae-Otachyrinae. *Hymenachne amplexicaulis*. **A** Culms with rhizome and inflorescence. **B** Spikelet from the adaxial side (upper image) and from the abaxial side (lower image), showing small lower glume, acute upper glume, and the attenuate lower lemma. (From Barkworth et al. 2003, drawn by H. Paździrková)

Twelve spp., wet sites in the tropics.

Hymenachne falls into two clades, one Old World and one New World, and may not be monophyletic (Acosta et al. 2014).

450. ***Otachyrium*** Nees

Otachyrium Nees, Fl. Bras. Enum. Pl. 2(1): 271–272 (1829).
Steinchisma Raf., Bull. Bot., Geneva 1: 220 (1830).
Plagiantha Renvoize, Kew Bull. 37: 323 (1982).
Cliffordiochloa B. K. Simon, Austrobaileya 3: 674 (1992).
Fasciculochloa B.K. Simon & C.M. Weiller, Austrobaileya 4: 374, f. 1–4 (1995).

Annuals or perennials, caespitose. Ligule membranous or a fringed membrane. Inflorescence branches branched or not. Spikelets often paired, with both glumes shorter than spikelet and more or less equal in length. Lower flower staminate. *Lower palea enlarged at maturity, hardened, winged and enclosing upper flower.* Lemmas entire, pointed, awnless, cartilaginous. C_3 or C_3/C_4 intermediates. $2n = 18, 20$.

Sixteen spp., SE US to South America, Australia.

Otachyrium s.s. and *Steinchisma* form a clade in molecular analyses and together are sister to *Plagiantha*; all three share a distinctive hardened palea. Because of this morphological synapomorphy, the three are combined here, although not all combinations are yet available. In addition, *Otachyrium* s.s. may be paraphyletic (Acosta et al. 2014), also supporting combining the group. Sendulsky and Soderstrom (1984) cite a single chromosome count for *Otachyrium* as $n = 9$, but molecular data (Morrone et al. 2012) place it in the $x = 10$ clade, strongly supported as sister to *Steinchisma*. *Steinchisma* is one of the few grass genera that includes both C_3 species and C_3/C_4 intermediates.

451. ***Rugoloa*** Zuloaga

Rugoloa Zuloaga, Plant Syst. Evol. 300: 2164 (2014).

Rhizomatous perennials. Ligule membranous. Leaves with blades rounded to cordate at the base. Inflorescence branches branched, ascending, with the spikelets paired, borne on one side and crowded, or on short higher-order branches. Lower glume shorter than 1/2 the length of the spikelet. C_3. Mesophyll of leaf often with fusoid cells. $2n = 20, 40$.

Three spp., Central and South America, Caribbean.

Rugoloa corresponds to a set of three species formerly placed in "*Panicum*" sect. *Laxa*, and was identified in a phylogenetic study by Acosta et al. (2014).

XXIII.3. Subtribe Paspalinae Griseb. (1846)

Reimarochloinae Caro (1982).

Inflorescences unbranched, or with unbranched branches, spikelets borne unilaterally on the branches.

Eleven genera, 554 spp.

Paspalinae is moderately well supported by molecular data, but has no obvious morphological synapomorphies (Morrone et al. 2012). The majority of the species belong to the large genus *Paspalum*.

Acostia + Axonopus

Lower glume absent or minute. Upper glume as long as the spikelet.

452. ***Acostia*** Swallen

Acostia Swallen, Bol. Soc. Argent. Bot. 12: 109 (1968).

Perennials. Ligule a fringed membrane. Inflorescence branches few, sparse, unbranched, appressed to the main inflorescence axis, the spikelets borne on one side. Upper glume pubescent, extended to a slender acuminate tip. Lower lemma similar to the upper glume but with a shorter tip, the lower palea absent. Upper lemma smooth, the margins inrolled on the palea. C_4.

One sp., *A. gracilis* Swallen, Ecuador.

Acostia is sister to *Axonopus* s.l., based on morphological data (Morrone et al. 2012). The photosynthetic pathway has been verified by stable isotope analysis and anatomy (C. Osborne and O. Morrone, unpubl. data).

453. *Axonopus* P. Beauv.

Axonopus P. Beauv., Ess. Agrostogr.: 12 (1812); phylog., tax.: Lopez and Morrone (2012).
Centrochloa Swallen, J. Wash. Acad. Sci. 25: 192 (1935).
Baptorhachis Clayton & Renvoize, Kew Bull., Addit. Ser. 13: 298, 377 (1986).
Ophiochloa Filg., Davidse & Zuloaga, Novon 3: 360, f. 1–3 (1993).

Annuals or perennials, caespitose, rhizomatous or stoloniferous. Leaves auriculate or not, the blade disarticulating or not, muticous or sharp-pointed at the apex. Ligule membranous, or a fringed membrane, or a fringe of hairs. Inflorescence unbranched, or with two to many, unbranched branches, subdigitate or borne on an elongate axis. Spikelets borne in two rows on one side of the rachis branch. Upper glume awned or not. Lower lemma about as long as the upper glume, adaxial; upper lemma with margins inrolled on the palea. Stamens 2 or 3. C_4, with one bundle sheath. $2n = 20, 40, 60, 80$.

One hundred-four spp., tropics of the Western Hemsiphere, one species in Africa.

Axonopus is a paraphyletic group from which *Ophiochloa* and *Centrochloa* are derived (Morrone et al. 2012). *Baptorhachis* is placed sister to *Ophiochloa* on the basis of morphological data; in both genera the inflorescence axis is broad and leaf like, often folding to enclose the spikelets. Zanotti et al. (2010) have described spikelet development in *A. suffultus* and find that it differs from that in *Paspalum*.

454. *Anthaenantiopsis* Mez ex Pilg. Fig. 73

Anthaenantiopsis Mez ex Pilg., Notizbl. Bot. Gart. Berlin-Dahlem 11(104): 237–238 (1931).

Caespitose perennials with short rhizomes. Leaf blades linear to filiform. Ligule membranous. Inflorescence branches unbranched or branched, narrow. Spikelets solitary or in twos or threes, covered with long hairs. Lower glume minute. Upper glume and lower lemma acuminate. Lower flower with a palea. Upper lemma with margins involute, apex acute. Upper lemma and palea with prickle hairs and bicellular microhairs near the apex. C_4, with a single bundle sheath. $2n = 20, 40$.

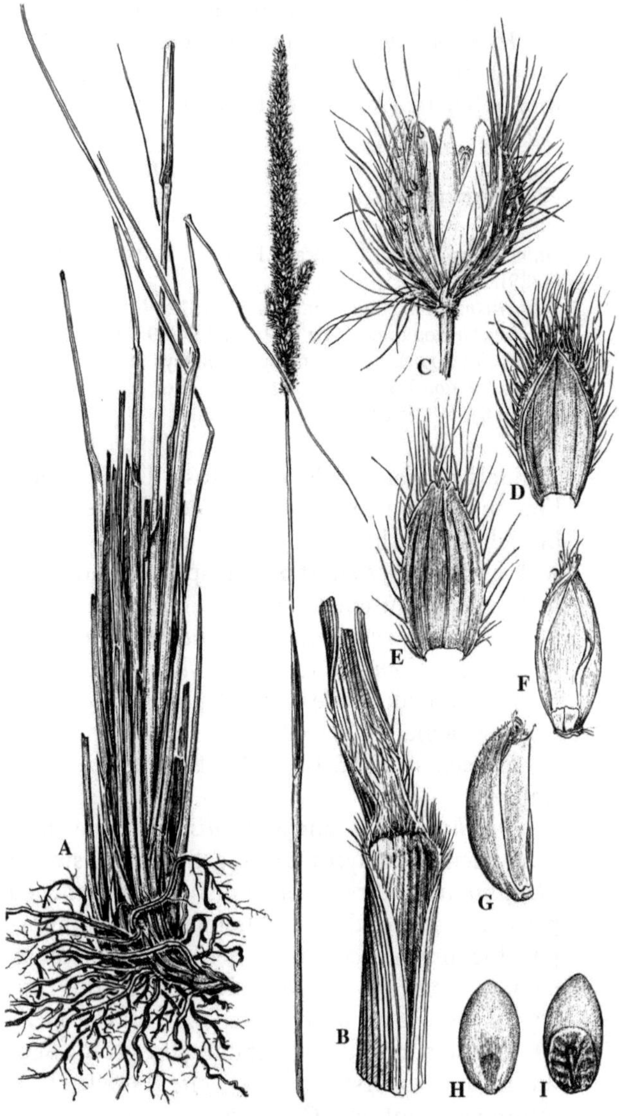

Fig. 73. Panicoideae-Paspaleae-Paspalinae. *Anthaenantiopsis rojasiana*. **A** Plant. **B** Ligule region. **C** Spikelet, lateral view. **D** Upper glume. **E** Lower lemma. **F** Lower palea with lodicules. **G** Upper lemma and palea, lateral view. **H** Caryopsis, hilar view. **I** Same, scutellar view. (From Zuloaga et al. 2012, drawn by A. Bestard)

Four spp., South America.

Anthaenantiopsis is sister to *Aakia* and the two sister to *Osvaldoa* in molecular analyses (Aliscioni et al. 2003; Lizarazu et al. 2014; Morrone et al. 2012; Zuloaga et al. 2007) but has no obvious morphological synapomorphy. *Hopia* is strongly supported as sister to the clade (GPWG II 2012).

455. *Aakia* J. R. Grande

Aakia J. R. Grande, Phytoneuron 22: 1 (2014).

Caespitose perennials. Ligule membranous. Leaf blades broadly lanceolate. Inflorescence branches branched, spreading. Spikelets with sparse hairs on the upper glume and lower lemma. Lower glume reduced, without veins. Lower flower lacking a palea, sterile. Upper lemma and palea covered with simple papillae, with long macrohairs near the apex. C_4, with a single bundle sheath.

One species, *A. tuerckheimii* (Hack.) J. R. Grande, Mexico, Guatemala, Belize, and Nicaragua.

Aakia corresponds to the former "*Panicum*" sect. *Tuerckheimiana* (Hitchc.) Zuloaga. Lizarazu et al. (2014) provide morphological and molecular evidence that firmly place this speces in Paspaleae, unrelated to *Panicum* s.s.

456. *Osvaldoa* J.R. Grande

Osvaldoa J.R. Grande, Phytoneuron 22: 5 (2014).

Caespitose perennials, with short rhizomes. Ligule membranous. Leaves with blades linear-lanceolate, the base rounded. Inflorescence branches branched, upright. Spikelets in pairs or solitary. Lower flower with a palea. Upper lemma and palea with papillae all over, the apex with microhairs. C_4, with a single bundle sheath. $2n = 20$.

One sp., *O. valida* (Mez) J.R. Grande, Brazil, Uruguay, Argentina.

Osvaldoa corresponds to the former "*Panicum*" sect. *Valida* Zuloaga & Morrone. Lizarazu et al. (2014) provide morphological and molecular evidence that firmly place this speces in Paspaleae, unrelated to *Panicum* s.s. The chromosome base number of $x = 10$ is also consistent with this placement.

457. *Echinolaena* Desv.

Echinolaena Desv., J. Bot. Agric. 1: 75 (1813).

Annuals or perennials, caespitose or rhizomatous. Leaf blades cordate, with or without a pseudopetiole. Ligule a fringed membrane or a fringe of hairs. Inflorescence unbranched or with primary branches unbranched, spreading, unilateral. Spikelets solitary or paired, laterally compressed. Lower glume with tubercle-based hairs. *Callus of upper flower with wing or scars on base of lemma.* C_3. $2n = 60$.

Nine spp., Central and South America, Madagascar.

Sede et al. (2009b) provide evidence that *Echinolaena* is polyphyletic. However, their phylogenies cannot rule out the possibility that the two clades of *Echinolaena* plus *Ocellochloa* plus "*Panicum*" *venezuelae* might form a monophyletic group, which could then be recognized as an expanded *Echinolaena*.

458. *Gerritea* Zuloaga, Morrone & T. J. Killeen

Gerritea Zuloaga, Morrone & Killeen, Novon 3: 213, f. 1–2 (1993).

Caespitose perennials. Leaf blades with a pseudopetiole, *disarticulating from the sheaths.* Ligule a fringed membrane. Inflorescence branches branched, the spikelets pedicellate. Glumes with long papillose-based hairs, the lower glume only slightly shorter than the upper one. *Palea of upper flower gaping, not enclosed by the lemma at the tip.* C_3, mesophyll with radiate chlorenchyma.

One sp., *G. pseudopetiolata* Zuloaga, Morrone & T. J. Killeen, Bolivia.

459. *Hopia* Zuloaga & Morrrone

Hopia Zuloaga & Morrone, Taxon 56: 150–153 (2007).

Perennials, with stolons and short rhizomes. Ligule membranous. Inflorescence branches unbranched, erect, the spikelets borne on all sides. Lower glume shorter than the spikelet; upper glume and lower lemma approximately the same length. Lower flower with a palea. Upper palea winged near the apex. Upper lemma and palea with simple papillae at the apex and base, with multicellular microhairs. C_4, with a single bundle sheath. $2n = 20, 40$.

One sp., *H. obtusa* (Kunth) Zuloaga & Morrone, southwestern US and Mexico.

This genus corresponds to the former "*Panicum*" sect. *Obtusum*, with the single species "*Panicum*" *obtusum*. Molecular phylogenies place it sister to the *Anthaenantiopsis* clade (GPWG II 2012; Morrone et al. 2012).

460. *Ichnanthus* P. Beauv.

Ichnanthus P. Beauv., Ess. Agrostogr.: 56 (1812).

Annuals or perennials, caespitose or rhizomatous. Ligule membranous, or a fringed membrane or a fringe of hairs. Inflorescence branches branched or not. Spikelets laterally compressed. Glume keeled, acute to acuminate. Callus of upper flower expanded into two wings or scars on base of lemma. $2n = 18, 20, 40, 54$.

Thirty-six spp., New World tropics, or pantropical.

The disparate chromosome base numbers could indicate polyphyly. The genus has not been sampled extensively in molecular phylogenies.

461. *Ocellochloa* Zuloaga & Morrone

Ocellochloa Zuloaga & Morrone, Syst. Bot. 34: 688 (2009).

Stoloniferous perennials, rooting at the lower nodes. Leaves with short pseudopetioles. Ligule membranous or a fringed membrane. Inflorescence branches unbranched, ascending; spikelets paired. Upper glume and lower lemma considerably longer than the upper flower and enclosing it, *the rachilla internode below the upper flower elongate*. Upper lemma and palea smooth and shiny. C_3. $2n = 20$.

Twelve spp., Mexico and Central America to Argentina.

This genus corresponds to the former "*Panicum*" sect. *Stolonifera*; the justification for the segregation is provided by Sede et al. (2009b). *Ocellochloa* is sister to one species of *Echinolaena*; see that genus for discussion. Many species of *Ocellochloa* have deep craters on the lower lemma but not the upper one.

462. *Paspalum* L.

Paspalum L., Syst. Nat. (ed. 10) 2: 846, 855, 1359 (1759).
Thrasya Kunth, Nov. Gen. Sp. (quarto ed.) 1: 120–121 (1815) [1816].
Reimarochloa Hitchc., Contr. U. S. Natl. Herb. 12: 198 (1909).
Spheneria Kuhlm., Comm. Lin. Telegr., Bot. 67(11): 57, t. 4 (1922).
Thrasyopsis Parodi, Bol. Soc. Argent. Bot. 1: 293 (1946).

Annuals or perennials, rhizomatous, stoloniferous or caespitose. Leaves auriculate or not, the blades with or without a pseudopetiole. Ligule membranous, or a fringed membrane or a fringe of hairs. Inflorescence unbranched, or of unbranched primary branches, these digitate or spread on an elongated axis, the branch axis flattened and sometimes winged. Spikelets single or in pairs, flattened on one side and convex on the other, generally round to ovate in outline (lanceolate or top-shaped in species segregated as *Reimarochloa* or *Spheneria*). Lower glume lacking or minute and abaxial; upper glume and lower lemma generally the same length as the spikelet; upper glume rarely lacking. Upper lemma and palea generally leathery to indurate, harder than the glume(s) and lower lemma. C_4, with one bundle sheath.

Three hundred seventy-two spp., tropical regions throughout the world.

Many species are serious weeds, but some are also good forage. Apomixis is common. The inclusion of several segregate genera is justified by molecular data. All the segregates are like *Paspalum* in having unbranched primary inflorescence branches and in lacking the lower glume; see also Denham and Zuloaga (2007).

463. *Renvoizea* Zuloaga & Morrone

Renvoizea Zuloaga & Morrone, Syst. Bot. 33: 294 (2008).

Perennials, caespitose and rhizomatous. Leaves with the sheaths fibrous, persistent, the blades ending in a sharp point, without a clear sheath-blade junction. Ligule a fringed membrane or lacking. Inflorescence branches branched, dense and contracted. Upper lemma and palea shining, smooth. C_3.

Ten spp., Brazil.

This genus was segregated from *Panicum* by Sede et al. (2008). All the included species were formerly in "*Panicum*" sect. *Lorea*, which was found to be polyphyletic. *Renvoizea* is in an unresolved position with respect to other Paspaleae (Morrone et al. 2012).

464. ***Streptostachys*** Desv.

Streptostachys Desv., Nouv. Bull. Sci. Soc. Philom. Paris 2: 190 (1810).

Caespitose perennials. Base of leaf blade cordate or clasping the stem. Ligule a fringe of hairs. Primary inflorescence branches stiff, the spikelets borne in pairs. Spikelet with thickened rachilla. Lower glume 2/3–4/5 the length of the spikelet. *Upper lemma and palea covered with bicellular microhairs and papillae on all epidermal cells.* Hilum linear. C_3. Mesophyll with radiate chlorenchyma, *the cells with invaginated cell walls, with or without fusoid cells.*

Two spp., Trinidad to NE Paraguay.

Streptostachys was recently re-circumscribed by Sede et al. (2009a) to remove the species now known as *Keratochlaena rigidifolia. Streptostachys* is sister to *Axonopus* s.l. (GPWG II 2012; Morrone et al. 2012).

XXIV. Tribe Paniceae R. Br. in Flinders (1814)

Cenchreae Rchb. (1828).
Spinificeae Dumort. (1829).
Digitarieae J .J. Schmitz & Regel (1841).
Melinideae Hitchc. (1920).
Boivinelleae A. Camus (1925).
Anthephoreae Pilg. ex Potztal (1957).
Cyphochlaeneae Bosser (1965).
Neurachneae S.T. Blake (1972).

Base chromosome number $x = 9$. The tribe has no other obvious morphological synapomorphy.

Seventy-two genera, 1254 species.

Paniceae sensu stricto are clearly monophyletic, based on many molecular and morphological sets of data (Aliscioni et al. 2003; Christin et al. 2008; Giussani et al. 2001; Gómez-Martinez and Culham 2000; Morrone et al. 2012; Vicentini et al. 2008), and share a base chromosome number of $x = 9$. Unfortunately for identification purposes, there is no obvious morphological character that unites the members of the group. Relationships within the group are not fully resolved, although the major clades described below are monophyletic. Chloroplast data place *Digitaria* as sister to all other members of this clade, whereas data from the nuclear gene phytochrome B place *Echinochloa* in that position (Vicentini et al. 2008). It is not clear whether one set of data or the other better reflects the organismal history, or whether a complex history of polyploidy and/or gene flow could account for the discrepancy. Photosynthetic pathway is variable within this tribe and all C_4 subtypes are found.

Incertae sedis

465. ***Acritochaete*** Pilg.

Acritochaete Pilg., Bot. Jahrb. Syst. 32: 53 (1902).

Decumbent annual. Ligule membranous. Primary branches of the inflorescence distantly spaced, unbranched, bearing spikelets on only one side. Upper glume and lower lemma with awns more than 1 cm long; *the awn of the lower lemma coiled.* Upper lemma papery, acuminate or with a minute awn. C_3.

One sp., *A. volkensii* Pilg., tropical Africa.

Molecular data place this species sister to Boivinellinae (GPWG II 2012).

466. ***Dichanthelium*** (Hitchc. & Chase) Gould Fig. 74

Dichanthelium (Hitchc. & Chase) Gould, Brittonia 26: 59 (1974).
Panicum subg. *Dichanthelium* Hitchc. & Chase, Contr. U. S. Natl. Herb. 15: 20, 142 (1910).
Panicum sect. *Clavelligerae* Stapf, Fl. Trop. Afr. 9: 639, 641 (1920).

Perennials. Basal leaves broad, cordate, *generally forming a winter rosette.* Ligule membranous or a fringed membrane. Inflorescence branches themselves branched. Spikelets generally obovate. Glumes of very different sizes, often pubescent. Margins of upper lemma tucked in on palea. C_3. $2n = 18, 36$.

One hundred twenty spp., America, Africa.

Fig. 74. Panicoideae-Paniceae. *Dichanthelium oligosanthes* subsp. *scribnerianum*. **A** Plant. **B** Spikelet. (From Barkworth et al. 2003, drawn by H. Paździrková)

Cleistogamous flowers are common particularly in late season flowering in North American species. Monophyly of *Dichanthelium* has been confirmed by all molecular studies to date (Aliscioni et al. 2003; Giussani et al. 2001; Morrone et al. 2012; GPWG II 2012). The species currently in "*Panicum*" sect. *Clavelligerae* have not been transferred formally to *Dichanthelium*, but their placement there is well supported.

467. ***Echinochloa*** P. Beauv.

Echinochloa P. Beauv., Ess. Agrostogr. 53, 161 (1812), nom conserv.

Annuals or perennials. *Ligule often absent.* Spikelets paired or on short second-order branches, hemispherical, flat on one side and rounded on the other. Lower glume shorter than lower lemma. Upper lemma with an incurved beak-like apex. *Palea tip reflexed.* C_4, with a single bundle sheath. $2n = 36, 54, 72$.

Forty spp., world wide in tropical and warm temperate regions.

Chloroplast data put *Echinochloa* in Biovinellinae (Giussani et al. 2001; Morrone et al. 2012), but sequences of the nuclear gene phytochrome B place it at the base of the $x = 9$ Paniceae (Vicentini et al. 2008), and the nuclear gene PEPC places it sister to the Melinidinae (Christin et al. 2007). Because of these conflicting placements, I leave *Echinochloa* incertae sedis. Many species of *Echinochloa* are problematical weeds.

468. ***Holcolemma*** Stapf & C. E. Hubb.

Holcolemma Stapf & C. E. Hubb., Bull. Misc. Inform. Kew 1929: 244 (1929).

Annuals or perennials. Ligule a fringed membrane. Inflorescence branches appressed; terminal spikelet of each branch subtended by a bristle. *Lower lemma sulcate, saccate on either side of the groove. Lower palea with winged keels, expanded and indurate at maturity, clasping the upper flower.*

Four spp., East Africa, India, Sri Lanka, Australia.

Watson and Dallwitz (1992 onward) report this genus as being C_3, with radiate *Isachne*-type mesophyll. No molecular data are available for *Holcolemma*. Morrone et al. (2012) use morphological data to place it in Cenchrinae, sister to the Central American genus *Ixophorus*, with which it shares a winged lower palea. If this placement is accurate, then *Holcolemma* would be the only C_3 group in that subtribe, or indeed in the larger clade that also includes Melinidinae and Panicinae.

469. *Homopholis* C. E. Hubb.

Homopholis C. E. Hubb., Bull. Misc. Inform. Kew 1934: 126 (1934).
Walwhalleya Wills & J. J. Bruhl, Austral. Syst. Bot. 19: 327 (2006).

Perennials, caespitose, rhizomatous or stoloniferous. Leaves auriculate or not. Ligule membranous or a fringed membrane. Inflorescence branches branched, spreading. Spikelets pedicellate, the glumes muticous to mucronate. Glumes longer than the flowers, *with 7 veins*, pubescent. Upper lemma and palea leathery, often finely muricate all over; upper lemma with a small beak. C_3.

Four spp., Australia.

Walwhalleya was segregated from *Homopholis* based on a cladistic analysis of morphological characters. However, molecular phylogenies strongly support the sister relationship between the two and there seems little reason to separate them (GPWG II 2012; Morrone et al. 2012).

470. *Hylebates* Chippin.

Hylebates Chippin., J. S. African Bot. 11: 127 (1945).

Prostrate annuals. Base of leaf blades wrapping around the stem. Ligule a fringe of hairs. Inflorescence with branched branches. Spikelets minute, <2 mm long. Glumes thin, membranous, the lower one tiny and without veins. Lemmas and paleas thin and membranous, the lower lemma extended to a narrow awn-like apex, the upper one with flat margins covering the palea. C_3. $2n = 18$.

Two spp., eastern and southern tropical Africa.

The photosynthetic pathway of *H. cordatus* and *H. chlorochloë* has been confirmed by stable isotope ratio (C. Osborne, unpubl. data).

471. *Thedachloa* S.W.L. Jacobs

Thedachloa S.W.L. Jacobs, Telopea 10: 635–637.

Stoloniferous annuals. Ligule a fringe of hairs. Inflorescence dense, ovoid, the branches branched. Glumes unequal, inflated at the base, the upper glume with stiff tubercle-based hairs above the middle. Lower lemma inflated, grooved and folded along the abaxial side. Upper flower on a short stalk, the lemma and palea hardened, smooth.

One sp., *T. annua* S.W.L. Jacobs, Western Australia, near Kalumburu.

472. *Thyridachne* C. E. Hubb.

Thyridachne C. E. Hubb., Kew Bull. 4: 363 (1949).

Annuals. Leaves ending in a sharp point. Ligule membranous. Inflorescence branches branched, erect. *Spikelet sickle-shaped in outline; upper glume gibbous.* Lower lemma trilobed, channeled on the abaxial side. Upper lemma membranous to cartilaginous. *Lodicules absent.* C_3.

One sp., *T. tisserantii* C. E. Hubb., Zaire and Central African Republic.

No molecular data are available for this species; a combined morphological and molecular analysis places it in various positions in Paniceae s.s. (Morrone et al. 2012).

Sacciolepis, *Trichanthecium*, and the "*Panicum*" sect. *Monticolae* group

Sacciolepis, *Trichanthecium*, and the "*Panicum*" sect. *Monticolae* group form a well-supported clade in molecular phylogenies (Aliscioni et al. 2003; GPWG II 2012; Morrone et al. 2012; Zuloaga et al. 2011). The latter groups were formerly placed in *Panicum*, and are morphologically homogeneous. However, *Sacciolepis* falls in the middle of the group and is morphologically quite distinct, so is kept separate here.

473. "*Panicum*" sect. *Monticolae* + sect. *Verrucosa* + sect. *Ovalifoliae*

"*Panicum*" [group] Verrucosa Hitchc., North American Flora 3(2): 200, 205. (1915).
"*Panicum*" sect. *Monticolae* Stapf, Fl. Trop. Afr. 9: 641, 649 (1920).
"*Panicum*" sect. *Ovalifoliae* Stapf, Fl. Trop. Afr. 9: 641, 650 (1920).

Ligule membranous. Inflorescence branches branched, spreading, diffuse, branches naked for

1/2 to 2/3 their length. Lower glume minute and without veins, or as long as the spikelet, with three veins. Upper lemma and palea covered with simple papillae and microhairs, sometimes transversely rugose. C_3.

Nine spp., New World.

This "group" represents the paraphyletic residue left when *Trichanthecium* and *Sacciolepis* are recognized (Zuloaga et al. 2011); relationships among most species are poorly supported. In "*Panicum*" sect. *Monticolae*, including "*P.*" *trichoides* Sw., "*P.*" *sellowii* Nees and "*P.*" *millegrana* Poir., and the Verrucosa group, including "*P.*" *verrucosum* Muhl. and "*P.*" *brachyanthum* Steud., the upper lemma and palea are transversely rugose, a character that occurs commonly in Paniceae. "*Panicum*" sect. *Monticolae* is not monophyletic, and may include "*P.*" *bartlettii* Swallen, formerly placed in sect. *Parvifolia*. The position of "*P.*" *trichanthum* Nees (former sect. *Parvifolia*) is also uncertain.

474. ***Sacciolepis*** Nash

Sacciolepis Nash, Man. Fl. N. States 89 (1901).

Annuals or perennials. Ligule membranous, with a ciliate margin or not. Inflorescence open and spreading, or narrow and spike-like, with short dense lateral branches. Spikelets laterally compressed, but the upper lemma and palea dorsiventrally compressed. Glumes pubescent, *ribbed, the upper one gibbous.* Upper lemma indurate to cartilaginous. C_3. $2n = 16, 18, 36, 45$.

Twenty-seven spp., pantropical but with most species in Africa.

475. ***Trichanthecium*** Zuloaga & Morrone

Trichanthecium Zuloaga & Morrone, Syst. Bot. Monogr. 94: 13 (2011).
Panicum [group] *Parvifolia* Hitchc. & Chase, Contr. U.S. Natl. Herb. 17: 462, 506 (1915).
Panicum sect. *Verruculosa* Stapf, Fl. Trop. Afr. 9: 639 (1920).

Annuals or perennials. Ligule membranous. Inflorescence branches branched, spreading, diffuse, branches naked for 1/2 to 2/3 their length. Spikelets sometimes with an elongated internode between the lower and upper glumes. Glumes more or less equal in length, the lower one ½ the length of the spikelet to the same length. Upper lemma and palea covered with simple papillae and microhairs. C_3. $2n = 18, 36, 54$.

Thirty-eight spp., Africa, South America, Central America, Caribbean.

Trichanthecium is sister to *Sacciolepis* (Aliscioni et al. 2003; Zuloaga et al. 2011). The species were formerly assigned to *Panicum*, but *Trichanthecium* has a membranous ligule, generally globose spikelets, glumes and lemmas with fewer veins, microhairs on the upper lemma, and the C_3 photosynthetic pathway, all characters unlike *Panicum* s.s. *T. schwackeanum* is reported to have $2n = 20$ chromosomes, a count that needs to be confirmed (Zuloaga et al. 2011).

XXIV.1. Subtribe Anthephorinae Benth. (1881)

Digitariinae Butzin (1972).

Lower glume tiny or absent. Characteristic deletion of 18 bp from the chloroplast gene ndhF (Morrone et al. 2012). *C_4 with a single bundle sheath.*

Seven genera and 296 species.

This tribe appears sister to all other members of the $x = 9$ clade in molecular phylogenies of chloroplast genes (Giussani et al. 2001) but sister to the clade of Melinidinae, Cenchrinae, and Panicinae in phylogenies constructed from *phyb* (Vicentini et al. 2008), and within the latter clade in trees constructed from PEPC (Christin et al. 2007). Morrone et al. (2012) place the subtribe in an unresolved position in Paniceae. Although its position is uncertain, the group is clearly monophyletic.

Within this subtribe, *Anthephora, Chaetopoa, Chlorocalymma, Tarigidia*, and *Trachys* all have bristles or bracts surrounding bisexual flowers, similar to an involucre. Although Clayton and Renvoize (1986) say that *Anthephora* "imitates *Cenchrus*", molecular data clearly place it here, unrelated to the bristle clade.

476. ***Anthephora*** Schreb.

Anthephora Schreb., Beschr. Gräs. 2: 105 (1810).

Annuals or perennials, rhizomatous or caespitose. Ligule membranous. Inflorescence narrow, cylindrical, spike-like. *Primary branches modified to form stiff clusters of spikelets and involucral bracts.* Bracts broad, ovate, subtending 3 to 11 spikelets. Lower glume absent, upper glume awl-like. $2n = 18, 36, 40$.

Eleven spp., African Arabia, Americas.

477. ***Chaetopoa*** C. E. Hubb.

Chaetopoa C. E. Hubb., Hooker's Icon. Pl. 37: t. 3646 (1967).

Annuals. Ligule membranous. Inflorescence branches unbranched, short, *with a group of sterile pedicellate spikelets surrounding a sessile, bisexual spikelet.* Lower glume awl-like or absent, the upper one short; upper lemma awned.

Two spp., Tanzania.

Molecular and morphological data indicate that *Chaetopoa* is either derived from within *Anthephora* or sister to *Trachys* (Morrone et al. 2012); neither position is well supported.

478. ***Chlorocalymma*** Clayton

Chlorocalymma Clayton, Kew Bull. 24: 461 (1970).

Annuals. Ligule membranous. *Inflorescence with 2 or 3 primary branches, the axis of each broad and leaf-like, enclosing the spikelets. Second-order branches forming spiny involucres around spikelets.* Lower glume tiny, upper glume short, lower lemma winged. $2n = 18$.

One sp., *C. cryptacanthum* Clayton, Tanzania.

479. ***Digitaria*** Haller Fig. 75

Digitaria Haller, Hist. Stirp. Helv. 2: 244 (1768).

Annuals or perennials, caespitose, rhizomatous or stoloniferous. Ligule membranous or a fringed membrane or a fringe of hairs. Inflorescence branches unbranched, clustered together on the main axis. Spikelets in pairs, with pedicels of unequal lengths. Lower glume tiny or absent. Lemma margins flat and exposed on palea. $2n = 18, 30, 36, 45, 54, 60, 70, 72, 76, 108$.

Two hundred seventy-seven spp., warm regions of the world.

DNA sequence data suggest that *Digitaria* might be para- or polyphyletic with other Anthephorinae derived from within it (GPWG II 2012; Morrone et al. 2012). However, species sampling

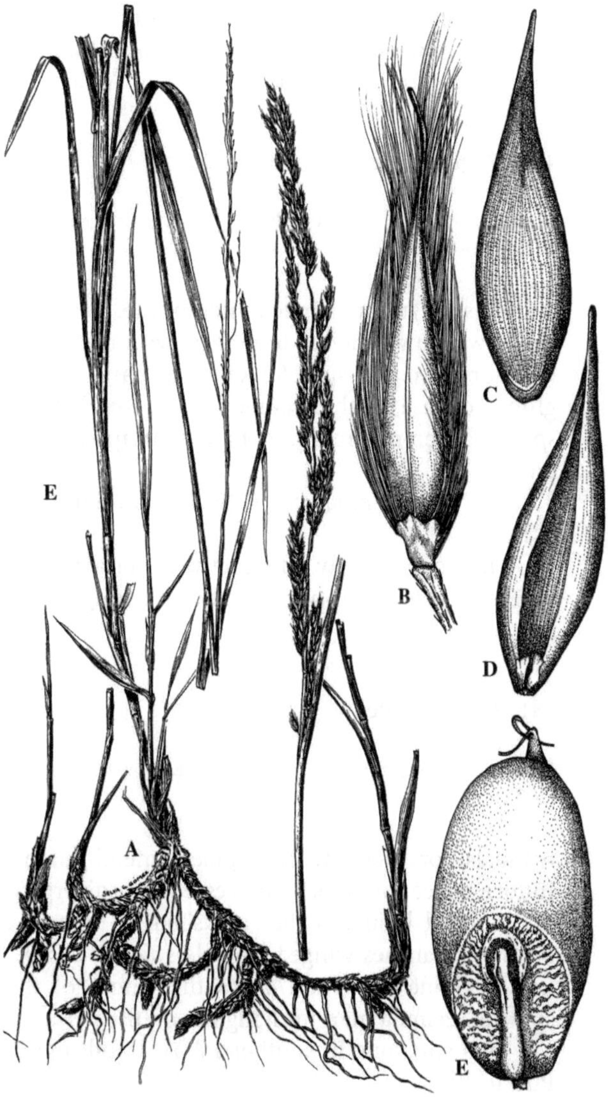

Fig. 75. Panicoideae-Paniceae-Anthephorinae. *Digitaria catamarcensis.* **A** Plant. **B** Spikelet, seen from the side of the lower glume. **C** Fertile lemma and palea, viewed from the abaxial side. **D** Fertile anthecium, viewed from the adaxial side, showing part of the palea and the lodicules of the sterile flower. **E** Caryopsis, scutellar view. (From Zuloaga et al. 2012, drawn by S. G. Gómez)

is insufficient to confirm this. Many *Digitaria* species are weeds. *Digitaria exilis* (fonio) and *D. iburua* are grown for grain in parts of Africa.

480. ***Megaloprotachne*** C. E. Hubb.

Megaloprotachne C. E. Hubb., Bull. Misc. Inform. Kew 1929: 320–321 (1929).

Annuals. Ligule a fringed membrane. Inflorescence with unbranched branches, with the spikelets borne on one side. Spikelets in pairs. Lower glume obtuse, with hyaline tip, *as long as the spikelet.* Lemma margins flat, overlapping the palea.

One sp., *M. albescens* C. E. Hubb., Southern Africa.

The only really distinctive character of *Megaloprotachne* appears to be the long lower glume.

481. ***Tarigidia*** Stent

Tarigidia Stent, Bull. Misc. Inform. Kew 1932(3): 151 (1932).

Annuals or perennials. Ligules membranous. Inflorescence of short, appressed or divergent unbranched branches that are shed at maturity. Primary branches winged. Spikelets nearly sessile. Lower glume present, 1/4–2/3 the length of the spikelet. Lemma margins lying flat on the palea.

Two spp., one in southern Africa and one in Puerto Rico.

The two species of *Tarigidia* are thought to be independent intergeneric hybrids between *Digitaria* and *Anthephora* (Loxton 1974; Vega et al. 2010); morphological phylogenetic data are consistent with this hypothesis, but molecular data are not available.

482. ***Trachys*** Pers.

Trachys Pers., Syn. Pl. 1: 85 (1805).

Annuals or perennials. Ligule membranous or a fringe of hairs. Inflorescence with primary branches unbranched, *the rachis broad and winged, often enclosing the spikelets, which are borne in clusters of 2 to 6 on the underside of the branches; branch axis disarticulating at the nodes.* Glumes shorter than spikelet; lower lemma broad.

Two spp., Tanzania, India, Sri Lanka, Burma.

XXIV.2. Subtribe Boivinellinae Pilg. (1940)

Microcalaminae Butzin (1970).

Plants sprawling or trailing. Leaves broad, lanceolate, with cordate or pseudopetiolate bases.

Fourteen genera, 113 species.

These genera form the "forest shade clade" of Aliscioni et al. (2003), Giussani et al. (2001), and Morrone et al. (2012). Most have paired spikelets. All except the polyploid members of *Alloteropsis* are C_3 and some are reported to have mesophyll with radiate chlorenchyma; in Watson and Dallwitz (1992 onward) this is described as "Isachne-type" in some genera but not in all. Morrone et al. (2012) include *Echinochloa* in this subtribe based on sequences of a chloroplast gene (*ndhF*) and morphology; however, it is unrelated to other Boivinelleae based on nuclear gene sequences (Christin et al. 2007; Vicentini et al. 2008) and is therefore considered incertae sedis in Paniceae in this treatment.

483. ***Acroceras*** Stapf

Acroceras Stapf, Fl. Trop. Afr. 9: 621 (1920).
Commelinidium Stapf, Fl. Trop. Afr. 9: 627 (1920).

Annuals or perennials. Leaves with or without pseudopetioles. Ligule a fringed membrane. Inflorescence of unbranched branches. Spikelets paired, both pedicellate, but often in long-and-short combinations. *Upper glume and both lemmas with the apex crested, blunt, hard and laterally compressed, crest of the upper lemma greenish.* $2n = 36$.

Twenty spp., Africa, Madagascar, India, SE Asia.

484. ***Alloteropsis*** J. Presl

Alloteropsis J. Presl, Reliq. Haenk. 1(4–5): 343, pl. 47 (1830).

Perennials. Ligule a fringe of hairs or a fringed membrane. Inflorescence of unbranched branches, the spikelets borne on one side. Spikelets clustered on the branches. Lower glume much smaller than the upper one. Proximal lemmas with or without awns, distal lemmas awned. C_3 and C_4. $2n = 18, 54$.

Five spp., Old World tropics.

Alloteropsis is monophyletic, and the C_3 pathway in this group is derived (GPWG II 2012; Ibrahim et al. 2009).

485. ***Amphicarpum*** Kunth Fig. 76

Amphicarpum Kunth, Révis. Gramin. 1: 28 (1829).

Annuals or perennials. Ligule a fringe of hairs. *Plants with cleistogamous spikelets borne underground or at lower nodes of the culm.* Inflorescence branches branched, the visible aboveground inflorescence sterile. Glumes one, if two then the lower one much smaller than the upper. Upper lemma coriaceous to indurate, acuminate, with margins lying flat on the palea. $2n = 18$.

Two spp., southeastern US.

Amphicarpum is strongly supported as being sister to *Entolasia* (GPWG II 2012), and differs primarily because of geography (North America versus African and Australian), the presence of cleistogamous spikelets, and the number of orders of branching in the inflorescence (*Entolasia* has unbranched branches). The two genera could perhaps be combined.

486. ***Cyphochlaena*** Hack.

Cyphochlaena Hack., Oesterr. Bot. Z. 51: 465 (1901).

Annuals or perennials, trailing. Ligule a fringed membrane. Inflorescence branches unbranched. Spikelets paired; *sessile one staminate or sterile; pedicellate one pistillate or hermaphrodite.* Pistillate spikelets laterally compressed. Lower glume with an apical awn. Upper glume and lower lemma coriaceous. Upper lemma translucent. Staminate spikelet similar to pistillate or reduced. C_3.

Two spp., Madagascar.

Cyphochlaena is sister to the clade of *Poecilostachys* plus *Oplismenus* (GPWG II 2012). All three share laterally compressed spikelets.

Fig. 76. Panicoideae-Paniceae-Boivinellinae. *Amphicarpum amphicarpon*. **A** Plant with aerial and subterranean spikelets. **B** Subterranean spikelet. **C** Subterranean flower. **D** Aerial spikelets. **E** Aerial flower. (From Barkworth et al. 2003, drawn by A. Sudkamp)

487. ***Cyrtococcum*** Stapf

Cyrtococcum Stapf, Fl. Trop. Afr. 9: 15 (1917).

Annuals or perennials. Ligule membranous or a fringed membrane. Inflorescence branches branched, spreading. Spikelets laterally compressed and asymmetrically gibbous. Glumes shorter than the flowers, obtuse. Upper lemma crested.

Fifteen spp., tropical areas, Old World.

488. ***Entolasia*** Stapf

Entolasia Stapf, Fl. Trop. Afr. 9: 739 (1920).

Annuals or perennials, rambling. Ligule a fringe of hairs. Inflorescences branches generally unbranched. Upper flower much shorter than the spikelet. *Upper lemma and palea with short appressed trichomes*, also sometimes with a tiny crest reminiscent of *Acroceras*. $2n = 18$.

Six spp., Tropical Africa and Australia.

Entolasia is strongly supported as being sister to *Amphicarpum* (GPWG II 2012), and differs primarily because of geography (African and Australian versus North American) and the presence of cleistogamous spikelets in *Amphicarpum*. The two genera could perhaps be combined.

489. ***Lasiacis*** (Griseb.) Hitchc.

Lasiacis (Griseb.) Hitchc., Contr. U.S. Natl. Herb. 15: 16 (1910).

Perennials (one species annual), the culms generally not self-supporting, woody. Leaf blades linear to ovate. Ligule membranous or a fringed membrane. Inflorescence branches themselves branched. Spikelets not compressed, obliquely attached to pedicel. Lower glume saccate. *Glumes and lower lemma become black at maturity, and the epidermis filled with oil globules.* Upper lemma and palea apex obtuse, *with tiny hollows near the apex bearing woolly pubescence*. Mesophyll with adaxial palisade layer. $2n = 18, 36$.

Sixteen spp., tropics of the New World, Madagascar.

In *L. grisebachii* (Nash) Hitchc. and *L. ruscifolia* (Kunth) Hitchc. the rachilla is extended beyond the distalmost flower. The spikelets are eaten and dispersed by fruit-eating birds (Davidse and Morton 1973).

490. ***Mayariochloa*** Salariato, Morrone & Zuloaga

Mayariochloa Salariato, Morrone & Zuloaga, Syst. Bot. 37: 110 (2012); phylog.: Salariato et al. (2012).

Rhizomatous perennials. Ligule a fringe of hairs. Leaves with the junction between sheath and blade indistinct; leaf blade apex obtuse. Inflorescence sparse, the branches branched but with few spikelets. Rachilla extension present. Upper lemma crested and pilose at the tip. C_3.

One sp., *M. amphistemon* (C. Wright) Salariato, Morrone & Zuloaga, Cuba.

The genus *Mayariochloa* was described to accommodate the species formerly known as *Scutachne amphistemon* (C. Wright) Hitchc. & Chase. The type of *Scutachne*, *S. dura* (Griseb.) Hitchc. & Chase, is unrelated (Salariato et al. 2012). *Scutachne* s.s. falls in Melinidiinae and is C_4; here it is placed in synonymy under *Urochloa*.

491. ***Microcalamus*** Franch.

Microcalamus Franch., J. Bot. (Morot) 3: 282 (1889).

Perennials, decumbent. Leaves pseudopetiolate, the blade broadly ovate to elliptical. Ligule a fringed membrane; contraligule present, a fringe of hairs. Inflorescence branches branched. Spikelets laterally compressed. Flowers with a pubescent callus, the upper lemma crested at the tip.

One sp., *M. barbinodis* Franch., Cameroon to Gabon.

492. ***Oplismenus*** P. Beauv.

Oplismenus P. Beauv., Fl. Oware 2: 14 (1810), nom. conserv.

Annuals or perennials, decumbent, trailing. Ligule a fringed membrane. Inflorescence branches unbranched. Spikelets laterally compressed, paired, but sessile one often reduced. Lower glume awned, upper glume awned or not, *the awn with a sticky secretion.* Upper lemma smooth and shining, crested at the tip. $2n = 18, 36, 54, 72, 90$.

Eleven spp., tropical and subtropical regions worldwide.

493. *Ottochloa* Dandy

Ottochloa Dandy, J. Bot. 69: 54 (1931).

Slender decumbent perennials. Leaf blades disarticulating from sheaths. Ligule membranous or a fringed membrane. Inflorescence branches branched. Both glumes less than ½ the length of the spikelet. Lower palea lacking. $2n = 18$.

Three spp., Old World tropics.

494. *Parodiophyllochloa* Zuloaga & Morrone

Parodiophyllochloa Zuloaga & Morrone, Syst. Bot. 33: 69 (2008).
"*Panicum*" sect. *Cordovensia* Parodi, Physis (Buenos Aires) 8: 68 (1925).

Perennials with short rhizomes. Ligules membranous. Inflorescence open, branched, with both chasmogamous and cleistogamous spikelets. Lower glume 3/4 the length of the spikelet; lower palea absent. Upper flower with simple papillae over the lemma and palea; lemma apiculate.

Six spp., Mexico to Argentina.

This genus was segregated from *Panicum* based on its morphology, leaf anatomy, and position in molecular phylogenies (Morrone et al. 2008).

495. *Poecilostachys* Hack.

Poecilostachys Hack., Sitzungsber. Kaiserl. Akad. Wiss., Math.-Naturwiss. Cl., Abt. 1 89: 131 (1884).

Annuals or perennials. Leaf blades broad. Inflorescence branches unbranched. Spikelets paired, laterally compressed. Lower glume awned or with a short sharp point. Upper lemma and palea cartilaginous, not indurate. Hilum long linear. C_3.

Nineteen spp., Madagascar and Africa.

Poecilostachys is placed Incertae sedis by Morrone et al. (2012) based on conflicting and weakly supported positions in phylogenies. However, it is strongly supported as sister to *Oplismenus* by GPWG II (2012).

496. *Pseudechinolaena* Stapf

Pseudechinolaena Stapf, Fl. Trop. Afr. 9: 494 (1919).

Annuals. Leaf blades lanceolate, pseudopetiolate. Ligule a fringed membrane. Inflorescence branches unbranched. Spikelets paired, laterally compressed. *Upper glume gibbous, with tubercle-based, hooked hairs developing after fertilization.* $2n = 36$.

Six spp., one pantropical, the other five in Madagascar.

XXIV.3. SUBTRIBE NEURACHNINAE *Clayton & Renvoize (1986)*

Perennials. Ligule a fringe of hairs. Inflorescences dense, capitate, unbranched, or the branches very short. Spikelets with the lower flower reduced to a lemma.

Three genera and 19 species.

This group is almost entirely Australian, except for *Cleistochloa*, which also occurs in western parts of the Pacific and New Guinea.

497. *Cleistochloa* C. E. Hubb.

Cleistochloa C. E. Hubb., Hooker's Icon. Pl. 33: t. 3209 (1933).
Calyptochloa C. E. Hubb., Hooker's Icon. Pl. 33: t. 3210 (1933).
Ancistrachne S. T. Blake, Pap. Dept. Biol. Univ. Queensland 1: 4 (1941).
Dimorphochloa S. T. Blake, Pap. Dept. Biol. Univ. Queensland 1: 1 (1941).

Scrambling, caespitose or mat-forming, with wiry culms. Leaf blades disarticulating. Ligule a fringe of hairs. Cleistogamous spikelets frequent in leaf axils of some species. Lower glume lacking, or minute, or less than ½ the length of the spikelet. Upper glume and lower lemma pubescent, in some species the hairs pustule-based and hooked. Upper lemma awned or awnless. C_3. *Mesophyll with radiate chlorenchyma.*

Eight spp., Australia, Philippines, New Caledonia, New Guinea.

The genera synonymized here in *Cleistochloa* form a clade in molecular phylogenies (GPWG II

2012; Morrone et al. 2012), have several morphological synapomorphies, and are generally similar in habit. Morrone et al. (2012) included *Ancistrachne*, *Calyptochloa* and *Dimorphochloa* (= *Cleistochloa rigida*) in their analysis, while (GPWG II 2012) included *Ancistrachne Calyptochloa* and *Cleistochloa*. Grassworld (http://grassworld.myspecies.info/) maintains all four genera as separate.

Remaining two genera:.
Spikelet with a pubescent callus. Glumes as long as spikelet, *hard, the upper hairy on margins;* lemmas less firm than the glumes.

Molecular phylogenies show that *Thyridolepis* is sister to *Neurachne* (Christin et al. 2012; GPWG II 2012), and that the monospecific *Paraneurachne* is derived within *Neurachne* (Christin et al. 2012). Accordingly, the latter two genera are combined here.

498. ***Neurachne*** R. Br.

Neurachne R. Br., Prodr. 196 (1810).
Paraneurachne S. T. Blake, Contr. Queensland Herb. 13: 20 (1972).

Caespitose or stoloniferous. Glumes muticous or awned. Upper lemma acute or with a beaked tip. C_3, C_4, or C_3/C_4 intermediate; generally with two bundle sheaths. $2n = 18, 36, 54$.

Eight spp., Australia.

The two C_4 species of *Neurachne*, *N. munroi* (F. Muell.) F. Muell. and *N. muelleri* Hack., are not sisters, nor is either sister to the C_3-C_4 *N. minor* intermediate, S. T. Blake (Christin et al. 2012). Thus, the C_4 pathway appears to have originated several times within this genus.

499. ***Thyridolepis*** S. T. Blake Fig. 77

Thyridolepis S. T. Blake, Contr. Queensland Herb. 13: 25 (1972).

Caespitose or decumbent. Glumes obtuse, the lower one with a hyaline patch below a prominent transverse ridge covered with stiff hairs. Upper lemma with a beaked tip. C_3.

Three spp., Australia.

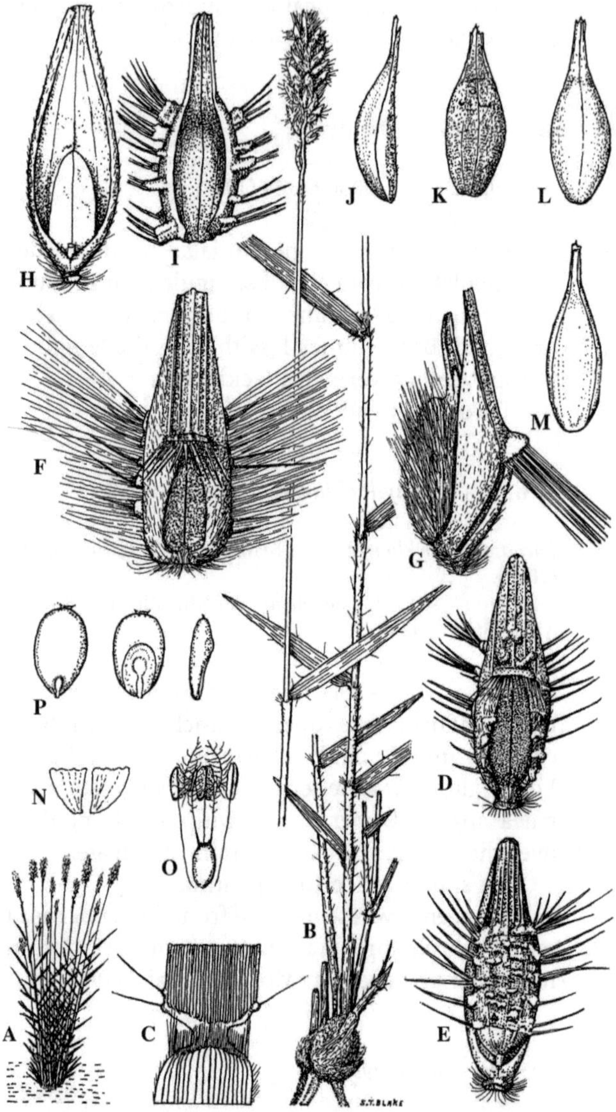

Fig. 77. Panicoideae-Paniceae-Neurachninae. *Thyridolepis mitchelliana*. **A** Habit. **B** Plant with inflorescence. **C** Part of leaf flattened out, from the upper side showing ligule, pseudopetiole and base of blade. **D, E** Spikelet from lower part of inflorescence, front and back view. **F, G** Spikelet from upper part of inflorescence, front view and slightly oblique side view. **H** Lower glume from inside. **I** Upper glume from inside. **J** Flowers from the side, the lower at right. **K** Lower flower from back. **L** Upper flower from back. **M** Upper flower from front. **N** Lodicules. **O** Pistil and stamens. **P** Caryopsis, adaxial view, abaxial view, and side view. (From Blake 1972)

Cenchrinae + Melinidinae + Panicinae clade

These three subtribes form a strongly supported clade in GPWG II (2012), and were called the "C_4, three subtypes clade" by Giussani et al. (2001). They have no obvious morphological character in common, but together they represent a single origin of the C_4 pathway. Cenchrinae are characterized by the NADP-ME subtype, with a single bundle sheath around the vascular tissue. Melinidinae and Panicinae both have a double bundle sheath, with the Melinidinae exhibiting the PCK subtype and Panicinae the NAD-ME subtype. The clade appears to be sister to *Homopholis* in chloroplast phylogenies.

XXIV.4. Subtribe Cenchrinae Dumort. (1824)

Setariinae Dumort. (1824).
Pennisetinae Rchb. (1828).
Snowdeniinae Butzin (1972).
Pseudoraphidinae Keng & Keng f. (1990).

Primary, secondary, and/or tertiary branches terminating in a sterile point or bristle. C_4, with a single bundle sheath; NADP-ME subtype.

Twenty-three genera and 293 species.

The bristles have been variously interpreted, but developmental and gene expression data show that they are sterile branches (Doust and Kellogg 2002; Kellogg et al. 2004). The bristle appears to have been lost in *Alexfloydia*, *"Panicum" antidotale*, *Snowdenia*, *Stereochlaena*, *Whiteochloa*, and *Zuloagaea*.

The inflorescence branches are deciduous in the small genera *Paratheria* and *Streptolophus*, making them similar to *Cenchrus*; it is possible that the three genera should be merged.

500. ***Alexfloydia*** B. K. Simon

Alexfloydia B. K. Simon, Austrobaileya 3: 670 (1992).

Stoloniferous perennials, mat-forming. Ligule a fringe of hairs. Inflorescence unbranched or branched, with 3 to 6 spikelets; the pedicels slender. Spikelets laterally compressed. Upper lemma cartilaginous, not indurate.

One sp., *A. repens* B. K. Simon, known from only a few localities in New South Wales, Australia. Listed as endangered.

501. ***Cenchrus*** L.

Cenchrus L., Sp. Pl. 2: 1049 (1753); phylog.: Akiyama et al. (2011); Chemisquy et al. (2010); Donadio et al. (2009).
Pennisetum Rich., Syn. Pl. 1: 72 (1805).
Beckeropsis Fig. & De Not, Agrostogr. Aegypt. 2: 49 (1853).
Odontelytrum Hack., Oesterr. Bot. Z. 48: 86 (1898).

Annuals or perennials. Ligule generally a fringe of hairs, occasionally membranous. *Lateral (secondary) branches with an expanded base, deciduous as a unit. Each branch terminating in a spikelet, which is then subtended and more or less surrounded by bristles, the bristles corresponding to higher-order branches.* Bristles terete or flattened, separate or more or less fused; forming a membranous cup in *C. abyssinicus* (Hack.) Morrone. $2n = 14, 18, 22, 34, 35, 36, 40, 44, 45, 52, 54, 68$.

One hundred twenty-one spp., worldwide, especially warm areas.

"Pennisetum" and *Cenchrus* were formally merged by Chemisquy et al. (2010), following years of accumulating evidence that *Cenchrus* s. s. is derived from within the former *"Pennisetum"*. Several species (e.g., *C. ciliaris* L.) have always been difficult to place, and appear intermediate between both "genera". The species of *Cenchrus* + *"Pennisetum"* are strongly supported as forming a clade with both morphological and molecular data; furthermore, allopolyploid taxa form from parents assigned to both genera suggesting gene flow via the polyploids. A few species of *Cenchrus* s.l. have only one bristle per spikelet; although these were segregated as separate genera (*Odontelytrum*, *Beckeropsis*), they are simply derived members of the clade. *Cenchrus americanus* (L.) Morrone (formerly *Pennisetum glaucum* L.) is cultivated as pearl millet, and *C. compressus* (R. Br.) Morrone (formerly *Pennisetum alopecuroides* L.) is a common ornamental. Many species of *Cenchrus* (e.g., *C. ciliaris* and *C. echinatus* L.) are weedy. Aposporous apomixis is common in the genus (Ozias-Akins et al. 2003), but the number of origins is unclear (Akiyama et al. 2011).

502. ***Chamaeraphis*** R. Br.

Chamaeraphis R. Br., Prodr. 193 (1810).

Caespitose perennials. Ligule membranous. Each spikelet paired with a single bristle, the spikelet-bristle pairs on alternate sides of the flattened main inflorescence axis, deciduous. Lower glume minute or absent; upper glume with 7 veins. Upper lemma smooth.

One sp., *C. hordeacea* R. Br., Australia.

This species is sister to *Pseudoraphis* in the analysis of Morrone et al. (2012).

503. ***Dissochondrus*** (Hillebr.) Kuntze

Dissochondrus (Hillebr.) Kuntze, Revis. Gen. Pl. 2: 770 (1891).

Caespitose perennials. Leaves pseudopetiolate, the blades auriculate. Ligule membranous. Inflorescence narrow, the branches branched. Spikelets subtended by one or more bristles; disarticulation above the bristles and below the glumes. *Spikelets with both flowers fully developed and bisexual.* Lemmas rugose. Leaf mesophyll with radiate chlorenchyma, and isolated bundle sheath cells similar to those in *Arundinella.*

One sp., *D. biflorus* Kuntze ex Hack., Hawaii.

504. ***Hygrochloa*** Lazarides

Hygrochloa Lazarides, Brunonia 2: 86 (1979).

Monoecious annuals or perennials, growing in or near water. Leaf blades papillose above. Ligule a fringe of hairs. Primary inflorescence branches unbranched, ending in a point. Pistillate spikelets borne on lower branches, staminate spikelets pedicellate on the main inflorescence axis. Upper lemma coriaceous, the margins lying flat on the palea.

Two spp., Northern Australia.

505. ***Ixophorus*** Schltdl.

Ixophorus Schltdl., Linnaea 31: 420 (1861); devel., rev.: Kellogg et al. (2004).

Monoecious caespitose annuals or short-lived perennials. Leaves linear, *conduplicate in bud, the sheaths strongly flattened and keeled.* Ligule a fringed membrane. Inflorescence with the primary branches unbranched. Each spikelet accompanied by a bristle. *Upper flower pistillate and lower flower staminate.* Lower paleas hyaline, about as long as the upper glume, accrescent, *forming a winged structure clasping the upper flower at maturity.* Upper lemma rugose, mucronate. $2n = 36, 54$.

One sp., *I. unisetus* (J. Presl) Schltdl., Mexico and Central America.

Ixophorus is strongly supported as sister to *Zuloagaea* and *Setariopsis* (GPWG II 2012). The three genera have overlapping distributions in Mexico and Central America. *Ixophorus* is an allopolyploid, formed from the same allopolyploid event that produced *Z. bulbosa* (Doust et al. 2007).

506. **"*Panicum*" *antidotale***

"*Panicum*" *antidotale* Retz., Observ. Bot. 4: 17 (1786).

Perennial with short woody rhizomes. Ligule a fringe of hairs. Inflorescence branches branched, the higher-order branches contracted around the primary ones. Upper lemma and palea indurate, shiny.

One sp., widespread in tropical areas of the world.

Although this species looks like a conventional *Panicum*, molecular data place it within Cenchrinae (Morrone et al. 2012).

507. ***Paratheria*** Griseb.

Paratheria Griseb., Cat. Pl. Cub. 236 (1866).

Prostrate perennials, rooting at the nodes. Ligule a fringed membrane or a fringe of hairs. *Primary branches of the inflorescence appressed to the main axis, each with one spikelet and ending in a sharp sterile tip.* Inflorescence branches disarticulating as a unit, below the bristles. Cleistogamous spikelets present at the base of the inflorescence. Glumes minute, without veins. Lower and upper lemma acuminate to subulate, the upper not indurate.

Two spp., Madagascar, West Africa, West Indies and South America.

508. ***Paractaenum*** P. Beauv.

Paractaenum P. Beauv., Ess. Agrostogr. 47 (1812).
Plagiosetum Benth., Hooker's Icon. Pl. 13: 33 (1877).

Small caespitose annuals. Ligule a fringed membrane or a fringe of hairs. Primary inflorescence branches deciduous, wedge-shaped in *P. refractum*. Each spikelet subtended by one or more bristles, rachis ending in one bristle or a fan of bristles. Lower lemma with more than nine veins.

Two spp., Australia.

Kellogg et al. (2009) suggest that this genus is related to the Australian species of *Setaria* formerly classified in *Paspalidium*.

509. ***Pseudochaetochloa*** Hitchc.

Pseudochaetochloa Hitchc., J. Wash. Acad. Sci. 14: 492 (1924).

Caespitose perennials; *dioecious*. Ligule a fringe of hairs. Primary inflorescence branches with 2 to 5 spikelets, with 7 to 12 bristles beneath each spikelet, the bristles falling with the spikelets; bristles with long hairs proximally. Glumes and lower lemma membranous, the latter muticous. Upper lemma and palea coriaceous.

One sp., *P. australiensis* Hitchc., Australia.

Pseudochaetochloa is similar to *Pennisetum*, but has more elongate secondary branches and is also dioecious. It falls in a clade with *Spinifex* and *Zygochloa*, both of which are also dioecious (Morrone et al. 2012).

510. ***Pseudoraphis*** Griff.

Pseudoraphis Griff., Not. Pl. Asiat. 3: 29 (1851).

Decumbent perennials. Ligule membranous. Primary inflorescence branches each with one or a few spikelets and tipped with a bristle. Lower glume minute, without veins; upper glume mucronate to awned. Upper flower borne atop a long rachilla internode. Lower lemma much longer than the upper one, awned. $2n = 16$.

Ten spp., Asia and Australia.

Although one representative *Pseudoraphis* was included by Kellogg et al. (2009) in Cenchrinae, its position remains unresolved. It is supported as sister to *Chamaeraphis* by Morrone et al. (2012), but the position of the clade within the subtribe is not clear.

511. ***Setaria*** P. Beauv. Fig. 78

Setaria P. Beauv., Ess. Agrostogr. 51 (1812).
Paspalidium Stapf, Fl. Trop. Afr. 9: 582 (1920).

Annuals or perennials, caespitose, rhizomatous or stoloniferous. Leaf blades flat, folded, or plicate, sagittate in a few species. Ligule a fringed membrane or a fringe of hairs. Inflorescences lax,

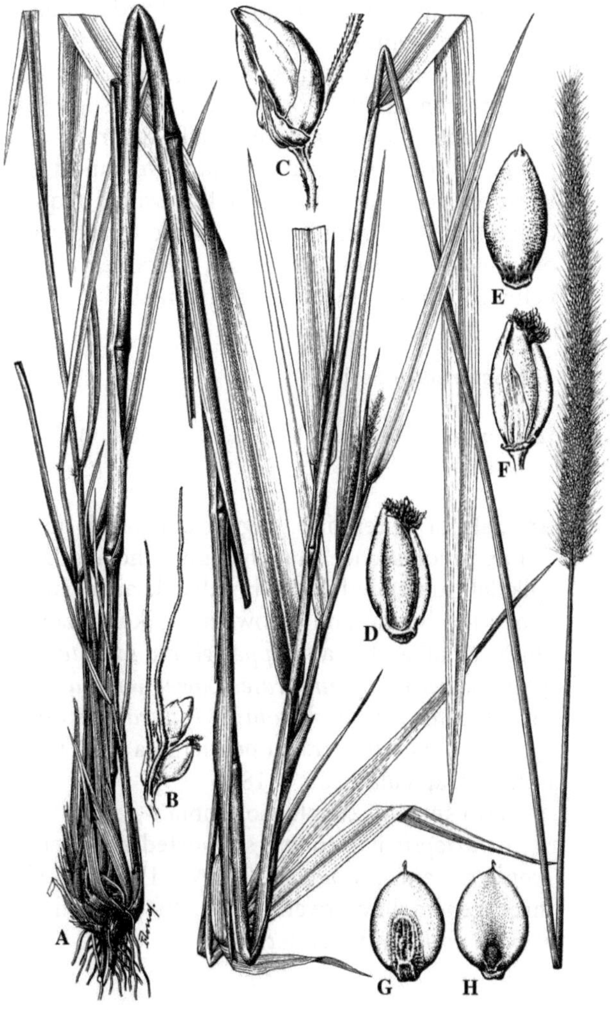

Fig. 78. Panicoideae-Paniceae-Cenchrinae. *Setaria lachnea*. **A** Plant. **B** Portion of an inflorescence branch. **C** Spikelet, lateral view with base of bristle. **D** Upper lemma and palea, viewed from the adaxial side. **E** Upper lemma, viewed from the abaxial side. **F** Lower palea and upper lemma and palea, viewed from the adaxial side. **G** Caryopsis, scutellar view. **H** Same, hilar view. (From Zuloaga et al. 2012)

open and highly branched, or more or less contracted and spiciform, or a set of unilateral racemes; bristles either associated with each mature spikelet or terminating the primary inflorescence branches or both. Disarticulation above the bristles and below the spikelets. Lower glume generally shorter than the upper one. Upper lemma and palea crustaceous or coriaceous, generally ellipsoid, usually corrugate, granular or rugose, or rarely smooth. $2n = 18, 36, 54, 63, 72$.

One hundred fifteen spp., world wide, in temperate and tropical areas.

Molecular phylogenies (Kellogg et al. 2009) and a recent monograph (Morrone et al. 2014) show that *Setaria* and *Paspalidium* intergrade. *Setaria* is not monophyletic even when *Paspalidium* is included. However, to make it monophyletic would require expanding it to include much of the bristle clade outside of *Cenchrus*. This is not warranted until a better supported phylogeny is obtained.

512. ***Setariopsis*** Scribn.

Setariopsis Scribn., Publ. Field Columb. Mus., Bot. Ser. 1(3): 288–289 (1896).

Annuals or perennials. Ligule a fringed membrane. Inflorescence branches unbranched. Each spikelet subtended by a bristle; disarticulation above the bristle and below the spikelet. Lower glume small and ovate; *upper glume saccate and gibbous, becoming indurate. Lower lemma with two abaxial pouches, becoming indurate*, exceeding the upper flower. Upper lemma and palea indurate, apiculate. $2n = 18$.

Two spp., Mexico to Colombia.

Setariopsis is strongly supported as sister to *Ixophorus* and *Zuloagaea* (GPWG II 2012). The three genera have overlapping distributions in Mexico and Central America.

513. ***Snowdenia*** C. E. Hubb.

Snowdenia C. E. Hubb., Bull. Misc. Inform. Kew 1929: 30–31 (1929).

Stoloniferous or decumbent annuals or perennials. Ligule membranous. Inflorescence unbranched, or with unbranched branches. *Sterile branches (bristles) absent*. Spikelets pedicellate. Glumes minute. Lower lemma awned, as long as the upper lemma and palea. Upper lemma and palea membranous, not indurate, the lemma with margins flat and overlapping the palea.

Four spp., northeast tropical Africa.

Although this genus lacks sterile branches in the inflorescence, sequences of the chloroplast gene *ndhF* place it within Cenchrinae (Morrone et al. 2012).

514. ***Stenotaphrum*** Trin.

Stenotaphrum Trin., Fund. Agrost. 175 (1820) [1822].

Annuals or perennials, rhizomatous, stoloniferous, or caespitose. Ligule a fringed membrane. *Inflorescence unbranched, or with very short primary branches sunken into the broad flattened inflorescence axis*. Rachis of primary branches terminating in a point. Lower glume tiny, with no veins. Upper lemma leathery but not indurate, yellow in fruit. $2n = 18, 20, 36$.

Seven spp., tropical and subtropical.

515. ***Stereochlaena*** Hack.

Stereochlaena Hack., Proc. Rhodesia Sci. Assoc. 7: 65 (1908).

Annuals or perennials. Ligule a fringed membrane. Inflorescence branches unbranched, in pairs or digitate, the branches winged; bristles lacking. Spikelets paired, borne on one side of the branches. Lower glume tiny or absent; upper glume minute to as long as the spikelet, awned or not. Lower lemma awned. Lodicules absent.

Four spp., Tanzania to southern Africa.

Stereochlaena is placed in the bristle clade on the basis of morphological data alone (Morrone et al. 2012).

516. ***Streptolophus*** Hughes

Streptolophus Hughes, Bull. Misc. Inform. Kew 1923: 178 (1923).

Annuals. Culm nodes hairy. Leaves pseudopetiolate, the blades cordate or sagittate. Ligule a fringed membrane. Inflorescence with primary branches short, themselves branched into cuneate higher-order branches. *The branches divided into multiple branchlets with recurved tips that form an involucre around the spikelets.* Inflorescence branches disarticulating as a unit, below the bristles. Glumes small.

One sp., *S. sagittifolius* Hughes, Angola.

Streptolophus falls with *Paratheria* in the analyses of Morrone et al. (2012), although molecular data are lacking for the former. In both genera the branches disarticulate as a unit. The involucral branches around the spikelets make the plants superficially similar to *Cenchrus*.

517. ***Uranthoecium*** Stapf

Uranthoecium Stapf, Hooker's Icon. Pl. 31, t. 3073 (1916).

Caespitose annuals. Ligule a fringed membrane or fringe of hairs. Primary inflorescence branches flattened, appressed to the main axis, short, ending in a point, disarticulating between the spikelets. Glumes truncate, gibbous. Apex of lower lemma extended, long acuminate; apex of upper lemma and palea prolonged, awl-shaped.

One sp., *U. truncatum* (Maiden & Betche) Stapf, Australia.

This species is closely related to *Setaria retiglumis* (Domin) R. D. Webster (Kellogg et al. 2009) (= *Paspalidium retiglume* (Domin) Hughes).

518. ***Whiteochloa*** C. E. Hubb.

Whiteochloa C. E. Hubb., Proc. Roy. Soc. Queensland 62: 111 (1952).

Annuals or perennials, caespitose. Ligule a fringed membrane. Inflorescence open, with lax, branched branches, these ending in spikelets or ending blindly in bristles. Spikelets laterally compressed or not compressed; rachilla internodes prominent between the flowers. Lower glume broad, ovate, clasping the spikelet; *upper glume tuberculate-ciliate, with a hardened apex.* Lower lemma sulcate. Upper flower on a short filiform internode. Upper lemma rugose, indurate, mucronate, *yellow in fruit.*

Six spp., Australia.

519. ***Xerochloa*** R. Br.

Xerochloa R. Br., Prodr. 196 (1810).

Annuals or perennials, erect or decumbent; culms herbaceous or woody. Ligule a fringed membrane. Inflorescence axes short and unbranched, each branch associated with a prophyll and ending in a point; inflorescences borne in a cluster and subtended by a spatheole. Lower glume small; glumes and lower lemmas membranous. Upper lemma acuminate.

Three spp., Australia.

520. ***Zuloagaea*** Bess

Zuloagaea Bess, Syst. Bot. 31: 666 (2006).

Caespitose perennials, with short tough rhizomes, these often *forming spherical corms.* Ligule membranous. Inflorescence highly branched, spreading. Upper lemma and palea indurate, scabrous at the apex. $2n = 18, 36, 54, 72$.

One sp., *Z. bulbosa* (Kunth) Bess, Mexico, Central America, and northern South America.

Zuloagaea is strongly supported as sister to *Ixophorus* and *Setariopsis* (GPWG II 2012). The three genera have overlapping distributions in Mexico and Central America. *Zuloagaea* is one of only a few genera of the bristle clade that lack bristles.

Remaining two genera: *Dioecious perennials of dry sites, with dense capitate inflorescences.* Both form clumps connected by long rhizomes. Members of the two are sisters in molecular phylogenies (Doust et al. 2007; Kellogg et al. 2009; Morrone et al. 2012), and the two genera could be merged.

521. ***Spinifex*** L.

Spinifex L., Mant. Pl. 163, 300 (1771).

Leaf blades hard, sometimes with a sharp point; ligule a fringe of hairs. *Inflorescences made up of stiff branches all appearing to originate at a single point, each branch with its own spathe, the inflorescence disarticulating as a unit.* Pistillate spikelets solitary, each at the base of a long bare axis.

Staminate spikelets sessile on inflorescence axes. $2n = 18$.

Five spp., beaches of India, China, Australia.

522. ***Zygochloa*** S. T. Blake

Zygochloa S. T. Blake, Pap. Dept. Biol. Univ. Queensland 1: 7 (1941).

Leaf blades flat, stiff, disarticulating from the sheath. Ligule a fringe of hairs. Pistillate inflorescence branches each consisting of one spikelet plus three bracts, the branch ending in a sharp point; branches clustered into a globose head. Upper lemma hardened. Staminate inflorescences similar to female, but branched.

One sp., *Z. paradoxa* (R. Br.) S. T. Blake, Australia.

XXIV.5. Subtribe Melinidinae Pilg. (1940)

Brachiariinae Butzin (1970).

Glumes and lemmas muticous to aristate. C_4, with two bundle sheaths, *PCK-subtype.*

Twelve genera and 176 species.

Melinidinae are clearly monophyletic according to DNA sequences (Salariato et al. 2010), but the limits of the included genera are not clear. This group is known in the literature as the PCK clade because it includes all Paniceae known to exhibit the PCK subtype of C_4 photosynthesis. However, not all species have been biochemically typed; in addition, species with the other C_4 subtypes appear to exhibit some PCK activity (Voznesenskaya et al. 2005) so it is not clear whether this is a good character. The bundle sheath outline in most classical PCK species is uneven, so may provide one anatomical synapomorphy.

523. ***Chaetium*** Nees

Chaetium Nees, Fl. Bras. Enum. Pl. 2(1): 269–270 (1829).

Perennials. Ligule a fringed membrane or a fringe of hairs. Inflorescence unbranched, or with unbranched branches. Spikelets with a pubescent callus. Glumes awned, *the upper with 9–11 veins.* Lemmas awned or mucronate. Outer bundle sheath present (*C. bromoides*) or absent (*C. festucoides, C. cubanum*). $2n = 26$.

Three spp., Mexico, West Indies, Brazil.

Chaetium bromoides (J. Presl) Benth. ex Hemsl. falls within *Urochloa* in the phylogenetic analyses of Salariato et al. (2010). Nonetheless, *Chaetium* is kept separate here provisionally because *C. bromoides* is morphologically quite different from other members of *Urochloa* and because Salariato et al. were unable to include the other two species of *Chaetium*. These may fall entirely outside the subtribe, based on their leaf anatomy. The type is *C. festucoides* Nees, which has not been studied phylogenetically.

524. ***Eccoptocarpha*** Launert

Eccoptocarpha Launert, Senckenberg. Biol. 46: 124 (1965).

Annuals. Ligule a fringe of hairs. Inflorescence branches unbranched. Lower glume with glandular hairs, upper with stiff hairs. Upper glume and lower lemma with obvious net-like venation. Rachilla internode below upper flower slender, elongate, sinuous. Upper lemma hairy at the tip, hard.

One sp., *E. obconiciventris* Launert, Tanzania and Zambia.

Molecular data are not available for *Eccoptocarpha*, but it is placed sister to *Moorochloa* in the analyses of Morrone et al. (2012).

525. ***Eriochloa*** Kunth

Eriochloa Kunth, Nov. Gen. Sp. (quarto ed.) 1: 94–95, pl. 30, 31 (1815).

Caespitose annuals. Culm nodes pubescent. Ligule a fringe of hairs. Inflorescence branches unbranched, pedicels and axis villous, the spikelets borne on one side. Spikelets with a globose swelling ("bead") at the base. Lower glume absent, upper glume and lower lemma villous, muticous. Upper lemma rugose with a mucro or awn, margins tucked on to the palea. $2n = 18$.

One sp., *E. distachya* Kunth, South America.

Eriochloa is here restricted only to the type species, *E. distachya*. Most species previously classified within *Eriochloa* are part of *Urochloa* according to the extensive molecular phylogenetic data of Salariato et al. (2010). *Eriochloa* s.s. appears to be close to *Moorochloa*, *Tricholaena* and *Melinis*.

526. ***"Panicum" deustum*** Thunb.

"Panicum" deustum Thunb., Prodr. Pl. Cap.: 19 (1794).

Caespitose perennials. Ligule a fringed membrane. Inflorescence branches branched, often with clavellate hairs. Glumes with apices acute. Lemma and palea hardened.

One sp., Africa, Asia, South America.

"Panicum" deustum shares pubescent inflorescence branches with *Rupichloa*; the two should perhaps be combined in a single genus. They are in an unresolved position in the phylogeny of Morrone et al. (2012).

527. ***Rupichloa*** D. Salariato & Morrone

Rupichloa D. Salariato & Morrone, Taxon 58: 388–389 (2009).

Perennials, rhizomatous or caespitose. Ligule a fringed membrane. Inflorescence branches branched, with long pubescence. Lower glume clasping the base of the spikelet, the upper glume shorter than the upper lemma. Upper lemma striate, verrucose, with a crested apex, with flattened macrohairs. $2n = 26, 28$.

Two spp., Bahia and Minas Gerais, Brazil.

528. ***Thuarea*** Pers.

Thuarea Pers., Syn. Pl. 1: 110 (1805).

Mat-forming perennial. Ligule a fringe of hairs. Inflorescence enclosed in the uppermost leaf, unbranched with a flattened axis, *with 1 or 2 bisexual spikelets proximally, and staminate spikelets distally. Staminate spikelets deciduous, on bulbous pedicels.* Bisexual spikelets persistent; lower glume absent; upper lemma pubescent, with margins lying flat on the palea. *Staminate portion of the inflorescence folds onto the bisexual portion at maturity and encloses the fruit.*

Two spp., seashores of the Old World tropics, Madagascar to Polynesia.

529. ***Urochloa*** P. Beauv. Fig. 79

Urochloa P. Beauv., Ess. Agrostogr. 52, pl. 11, f. 1 (1812).
Brachiaria (Trin.) Griseb., Fl. Ross. 4: 469 (1853).
Scutachne Hitchc. & Chase, Proc. Biol. Soc. Wash. 24: 148–149, f. 13 (1911).
Megathyrsus (Pilg.) B. K. Simon & S. W. L. Jacobs, Austrobaileya 6: 572 (2003).

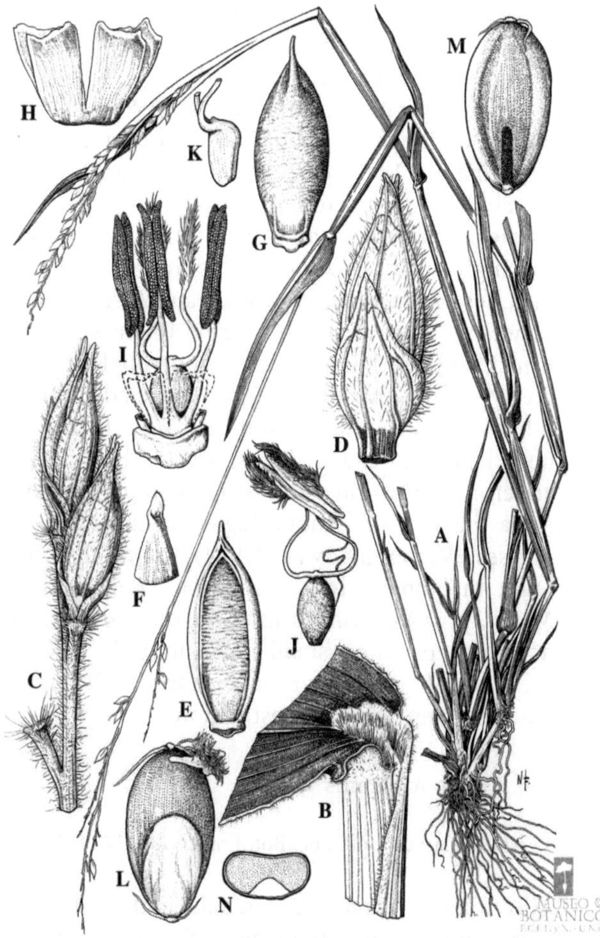

Fig. 79. Panicoideae-Paniceae-Melinidinae. *Urochloa paucispicata*. **A** Plant. **B** Ligular region. **C** Portion of an inflorescence branch. **D** Spikelet, viewed from the adaxial side. **E** Upper lemma and palea, viewed from the adaxial side. **F** Apex of the lemma. **G** Upper lemma, viewed from the abaxial side, showing germination flap at base. **H** Lodicules. **I** Perfect flower. **J** Gynoecium. **K** Ovary and basal part of the styles. **L** Caryopsis, scutellar view. **M** Caryopsis, hilar view. **N** Same, cross section. (From Zuloaga et al. 2012)

Annuals or perennials, caespitose. Leaf blade nodular in cross section. Ligule a fringed membrane or a fringe of hairs. Inflorescence branches branched or unbranched, spikelets borne on one side only, branch rachis often winged. Spikelets solitary or in pairs, in some cases with a globose swelling at the base. Lower glume generally shorter than the spikelet, sometimes absent. Upper glume often pubescent, about as long as the spikelet. Upper lemma rugose or smooth, the margins tucked on to the palea or lying flat on the palea, acute to mucronate, with or without dense hairs at the apex. $2n = 18, 36, 54, 72$.

One hundred thirty-five spp., tropical and subtropical regions of the world.

Most species of this genus were formerly placed in *Brachiaria*, but were transferred into *Urochloa* by Webster (1987); this transfer has largely been supported by subsequent molecular studies. Salariato et al. (2010, 2012) show convincingly that *Urochloa* is paraphyletic and includes (at least) *Scutachne* s.s. and *Megathyrsus* as well as several species formerly classified in *Eriochloa*. The latter genus is polyphyletic and the type falls outside the *Urochloa* clade. While it has been recognized traditionally by its prominent bead-like callus that character has clearly arisen multiple times and is not synapomorphic. In a survey of embryological characters of grasses, Verboom et al. (1994) found that the outer integument in *Urochloa* (= "*Panicum*" = *Megathyrsus*) *maxima* and *U. oligotricha* scarcely develops, forming a small rim around the base of the ovule; this character may be more widespread and deserves further investigation.

530. ***Yvesia*** A. Camus

Yvesia A. Camus, Bull. Soc. Bot. France 73: 687 (1927).

Tiny annuals, less than 25 cm tall. Ligule a fringe of hairs. Inflorescence branches unbranched, capillary. Callus with long white hairs. Lower glume absent, upper glume similar to lower lemma, awnless. Upper lemma mucronate.

One sp., *Y. madagascariensis* A. Camus, Madagascar.

LEUCOPHRYS + MELINIS + MOOROCHLOA + TRICHOLAENA

Disarticulation above the glumes. Upper lemma and palea smooth and shiny.

These four genera form a strongly supported clade in Morrone et al. (2012).

531. ***Leucophrys*** Rendle

Leucophrys Rendle, Cat. Afr. Pl. 2: 193 (1899).

Caespitose perennials, the culms woody. Leaf blades disarticulating from the sheath. Ligule a fringe of hairs. Inflorescence with unbranched branches, borne along a central axis. Glumes awnless, separated by an elongated rachilla internode, both as long as the spikelet. Upper glume and lower lemma with a transverse line of hairs.

One sp., *L. mesocoma* (Nees) Rendle, tropical and southern Africa.

The type of *Leucophrys*, *L. mesocoma*, is placed in *Brachiaria* by Clayton et al. (2006 onward), but is shown to be distinct by Salariato et al. (2010).

532. ***Melinis*** P. Beauv.

Melinis P. Beauv., Ess. Agrostogr.: 54 (1812).
Rhynchelytrum Nees, Nat. Syst. Bot.: 378, 446 (1836).
Mildbraediochloa Butzin, Willdenowia 6: 288 (1971).

Annuals or perennials, culm nodes pubescent. Ligule a fringe of hairs. Inflorescence much branched, with slender branches. Spikelets laterally compressed. Lower glume tiny or absent; upper glume awned or awnless, gibbous in some species. Lemma of lower flower awned. Lemma of the upper flower much less firm than the glumes, awnless or awned from a sinus. $2n = 36$.

Twenty-two spp., Africa, tropics and south.

533. ***Moorochloa*** (Sm.) Veldkamp

Moorochloa (Sm.) Veldkamp, Reinwardtia 12: 138–139 (2004).

Annuals. Ligule a fringe of hairs. Inflorescence branches unbranched, one-sided. Spikelets with the upper flower disarticulating, upper glume and lower lemma villous. Callus inconspicuous. Upper lemma and palea cartilaginous, smooth, apex muticous. $2n = 18, 36$.

Three spp., Africa and Asia, but now widespread and weedy.

Moorochloa is the name given for three species formerly in *Brachiaria*. The type of *Brachiaria* is assigned to *Urochloa*, so a new name was needed for these species.

534. *Tricholaena* Schrad.

Tricholaena Schrad., Mant. 2: 8, 163 (1824).

Perennials. Ligule a fringe of hairs. Inflorescence highly branched, the branches spreading. Spikelets laterally compressed. Lower glume apex truncate. Upper glume 3- to 5-veined. Lemma of lower flower without an awn. Lemma of upper flower leathery to hard. $2n = 36$.

Four spp., Mediterranean region, Africa, India.

XXIV.6. SUBTRIBE PANICINAE Fr. (1835)

Upper lemma and palea indurate. *Upper palea, and often the upper lemma, with papillae borne on other papillae (compound papillae) near the apex.* C_4, with two bundle sheaths; *NAD-ME subtype.*

Three genera, 111 species.

The photosynthetic subtype is probably synapomorphic for Panicinae, although not all species have been tested biochemically.

535. *Panicum* L. Fig. 80

Panicum L., Sp. Pl. 1: 55 (1753).
Yakirra Lazarides & R. D. Webster, Brunonia 7: 292 (1985).

Annuals or caespitose perennials. Ligule a fringed membrane. Inflorescence branches branched or unbranched. Spikelets solitary. Upper glume and lower lemma with 7 to 13 veins. $2n = 18, 36, 54, 72, 90$.

About one hundred spp., pan-tropical.

Yakirra is placed in *Panicum* by molecular data from GPWG II (2012) and Scataglini et al. (2014) and is accordingly placed in synonymy here. Its placement outside *Panicum* in the work of Morrone et al. (2012) was based on an erroneous sequence. *Panicum* sect. *Rudgeana*, *Arthragrostis* and *Yakirra* all share an elongate rachilla internode below the upper flower; the internode bears flanges or auricles distally.

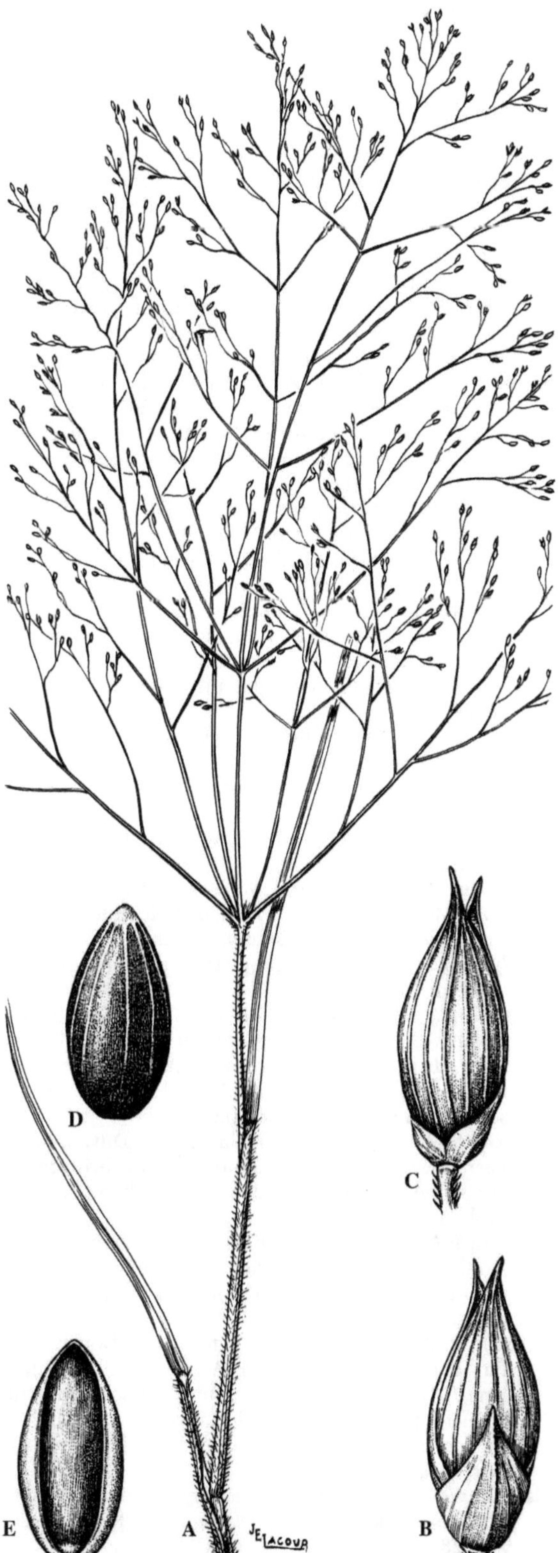

Fig. 80. Panicoideae-Paniceae-Panicinae. *Panicum bergii.* **A** Flowering stem. **B** Spikelet, view from the back of the lower glume. **C** Spikelet, view from the back of the upper glume. **D** Upper lemma, viewed from the abaxial side. **E** Upper lemma and palea, viewed from the adaxial side. (From Zuloaga et al. 2012, drawn by J.E. Lacour)

536. ***Louisiella*** C.E. Hubb. & J. Léonard

Louisiella C.E. Hubb. & J. Léonard, Bull. Jard. Bot. État Brux. 22: 316 (1952).

Aquatic perennials, with culms spongy, prostrate, spreading. Leaf sheaths with aerenchyma. Ligule a fringed membrane. Inflorescence branches branched. Lower glume tiny; upper glume and lower lemma longer than the upper flower. Hilum linear. $2n = 30$.

Two spp., tropical Africa and America.

L. fluitans C.E. Hubb. & J. Léonard, from tropical Africa, is sister to *L. elephantipes* Nees ex Trin. (Scataglini et al. 2014), which is widespread in tropical America. Although Morrone et al. (2012) placed the two species within *Panicum* s.s., more recent analyses indicate that they are in fact sister to *Panicum* (Scataglini et al. 2014).

537. ***Arthragrostis*** Lazarides

Arthragrostis Lazarides, Nuytsia 5: 285 (1984).

Annuals. Ligule a fringe of hairs. Inflorescence branches branched, deciduous. Glumes very unequal, the lower glume clasping the spikelet. Rachilla internode below upper flower elongate. Upper lemma leathery, with margins tucked on to the palea. Veins close together, with outer bundle sheath with an uneven outline.

Four spp., Australia.

Arthragrostis shares characters with species in *Panicum* sect. *Rudgeana*, and may ultimately be placed in synonymy there. *Panicum* sect. *Rudgeana*, *Arthragrostis* and *Yakirra* all share an elongate rachilla internode below the upper flower; the internode bears flanges or auricles distally. Although the elongate rachilla internode appears elsewhere in the subfamily, it may be synapomorphic here (Morrone et al. 2012).

References

Acosta, J., Scataglini, M.A., Reinheimer, R., Zuloaga, F.O. 2014. A phylogenetic study of subtribe Otachyriinae (Poaceae, Panicoideae, Paspaleae). Plant Syst. Evol. 300: 2155–2166.

Adams, R.P., Zhong, M., Turuspekov, Y., Dafforn, M.R., Veldkamp, J.F. 1998. DNA fingerprinting reveals clonal nature of *Vetiveria zizanioides* (L.) Nash, Gramineae and sources of potential new germplasm. Mol. Ecol. 7: 813–818.

Akiyama, Y., Goel, S., Coner, J.A., Hanna, W.W., Yamada-Akiyama, H., Ozias-Akins, P. 2011. Evolution of the apomixis transmitting chromosome in *Pennisetum*. BMC Evol. Biol. 11: 289.

Aliscioni, S.S., Giussani, L.M., Zuloaga, F.O., Kellogg, E.A. 2003. A molecular phylogeny of *Panicum* (Poaceae: Paniceae): tests of monophyly and phylogenetic placement within the Paniceae. Amer. J. Bot. 90: 796–821.

Barkworth, M.E., Capels, K.M., Long, S., Piep, M.B. (eds.) 2003. Flora of North America north of Mexico. Vol. 25. Magnoliophyta: Commelinidae (in part): Poaceae, part 1. New York: Oxford University Press.

Besnard, G., Christin, P.-A., Malé, P.-J.G., Coissac, E., Ralimanana, H., Vorontsova, M.S. 2013. Phylogenomics and taxonomy of Lecomtelleae (Poaceae), an isolated panicoid lineage from Madagascar. Ann. Bot. 112: 1057–1066.

Blake, S.T. 1972. *Neurachne* and its allies (Gramineae). Contr. Queensland Herb. 13: 1–53.

Bomblies, K., Doebley, J.F. 2005. Molecular evolution of FLORICAULA/LEAFY orthologs in the Andropogoneae (Poaceae). Mol. Biol. Evol. 22: 1082–1094.

Bommert, P., Satoh-Nagasawa, N., Jackson, D., Hirano, H. Y. 2005. Genetics and evolution of grass inflorescence and flower development. Plant Cell Physiol. 46: 69–78.

Bor, N. 1960. Grasses of Burma, Ceylon, and India. New York: Pergamon Press.

Brown, R. 1810. Prodromus florae Novae Hollandiae. London: J. Johnson & Co.

Brown, R. 1814. A voyage to Terra Australis. London: G. & W. Nicol.

Chemisquy, M.A., Giussani, L.M., Scataglini, M., Kellogg, E.A., Morrone, O. 2010. Phylogenetic studies favor the unification of *Pennisetum*, *Cenchrus* and *Odontelytrum*: a nuclear, chloroplast and morphological combined analysis. Ann. Bot. 106: 107–130.

Chen, C.-H., Veldkamp, J.F., Kuoh, C.-S., Tsai, C.-C., Chiang, Y.-C. 2009. Segregation of *Leptatherum* from *Microstegium* (Andropogoneae, Poaceae) confirmed by internal transcribed spacer DNA sequences. Blumea 54: 175–180.

Christin, P.-A., Salamin, N., Savolainen, V., Duvall, M.R., Besnard, G. 2007. C_4 photosynthesis evolved in grasses via parallel adaptive genetic changes. Curr. Biol. 17: 1241–1247.

Christin, P.A., Besnard, G., Samaritani, E., Duvall, M.R., Hodkinson, T.R., Savolainen, V., Salamin, N. 2008. Oligocene CO_2 decline promoted C_4 photosynthesis in grasses. Curr. Biol. 18: 37–43.

Christin, P.-A., Wallace, M.J., Clayton, H., Edwards, E.J., Furbank, R.T., Hattersley, P.W., Sage, R.F., Macfarlane, T.D., Ludwig, M. 2012. Multiple photosynthetic transitions, polyploidy, and lateral gene transfer in the grass subtribe Neurachninae. J. Exp. Bot. 63: 6297–6308.

Clayton, W.D. 1966a. Studies in the Gramineae: X. Andropogoneae. The genus *Anadelphia* Hack. Kew Bull. 20: 275–285.

Clayton, W.D. 1966b. Studies in the Gramineae: XI. Andropogoneae. The genus *Elymandra* Stapf. Kew Bull. 20: 287–293.

Clayton, W.D. 1969. A revision of the genus *Hyparrhenia*. Kew: Her Majesty's Stationery Office.

Clayton, W.D. 1970a. Gramineae (Part 1). In: Milne-Redhead, E., Polhill, R.M. (eds.) Flora of Tropical East Africa. London: Crown Agents for Oversea Governments and Administrations.

Clayton, W.D. 1970b. Studies in the Gramineae: XXI. *Coelorachis* and *Rhytachne*: a study in numerical taxonomy. Kew Bull. 24: 309–314.

Clayton, W.D. 1973. The awnless genera of Andropogoneae. Studies in the Gramineae: XXXIII. Kew Bull. 28: 49–58.

Clayton, W.D. 1981. Notes on the tribe Andropogoneae (Gramineae). Kew Bull. 35: 813–818.

Clayton, W.D., Renvoize, S.A. 1982. Gramineae (Part 3). In: Polhill, R.M. (ed.) Flora of Tropical East Africa. Rotterdam: A. A. Balkema.

Clayton, W.D., Renvoize, S.A. 1986. Genera graminum: grasses of the world. London: Her Majesty's Stationery Office.

Clayton, W.D., Phillips, S.M., Renvoize, S.A. 1974. Gramineae (Part 2). In: Polhill, R.M. (ed.) Flora of Tropical East Africa. London: Crown Agents for Overseas Governments and Administrations.

Clayton, W.D., Harman, K.T., Williamson, H. 2006 onward. GrassBase - the online world grass flora. http://www.kew.org/data/grasses-db.html

D'Hont, A., Paulet, F., Glaszmann, J.-C. 2002. Oligoclonal interspecific origin of 'North Indian' and 'Chinese' sugarcanes. Chromosome Res. 10: 253–262.

Davidse, G., Morton, E. 1973. Bird-mediated fruit dispersal in the tropical grass genus *Lasiacis* (Gramineae: Paniceae). Biotropica 5: 162–167.

Davidse, G., Judziewicz, E.J., Zuloaga, F.O. 2004. Poaceae. In: Berry, P.E., Yatskievych, K., Holst, B.K. (eds.) Flora of the Venezuelan Guyana. St. Louis, Missouri: Missouri Botanical Garden Press.

Denham, S.S., Zuloaga, F.O. 2007. Phylogenetic relationships of the Decumbentes group of *Paspalum*, *Thrasya*, and *Thrasyopsis* (Poaceae: Panicoideae: Paniceae). Aliso 23: 545–562.

Dillon, S.L., Lawrence, P.K., Henry, R.J., Ross, L., Price, H. J., Johnston, J.S. 2004. *Sorghum laxiflorum* and *S. macrospermum*, the Australian native species most closely related to the cultivated *S. bicolor* based on ITS1 and *ndhF* sequence analysis of 25 *Sorghum* species. Plant Syst. Evol. 249: 233–246.

Dillon, S.L., Lawrence, P.K., Henry, R.J., Price, H.J. 2007. *Sorghum* resolved as a distinct genus based on combined ITS1, *ndh*F and *Adh*1 analyses. Plant Syst. Evol. 268: 29–43.

Donadio, S., Giussani, L.M., Kellogg, E.A., Zuloaga, F.O., Morrone, O. 2009. A molecular phylogeny of *Pennisetum* and *Cenchrus* (Poaceae-Paniceae) based on the *trnL-F*, *rpl16* chloroplast markers. Taxon 58: 392–404.

Doust, A.N., Kellogg, E.A. 2002. Inflorescence diversification in the panicoid "bristle grass" clade (Paniceae, Poaceae): evidence from molecular phylogenies and developmental morphology. Amer. J. Bot. 89: 1203–1222.

Doust, A.N., Penly, A.M., Jacobs, S.W.L., Kellogg, E.A. 2007. Congruence, conflict and polyploidization shown by nuclear and chloroplast markers in the monophyletic "bristle clade" (Paniceae, Panicoideae, Poaceae). Syst. Bot. 32: 531–544.

Edwards, E.J., Smith, S.A. 2010. Phylogenetic analyses reveal the shady history of C_4 grasses. Proc. Natl. Acad. Sci. USA 107: 2532–2537.

Estep, M.C., McKain, M.R., Vela Diaz, D., Zhong, J., Hodge, J.G., Hodkinson, T.R., Layton, D.J., Malcomber, S.T., Pasquet, R., Kellogg, E.A. 2014. Allopolyploidy, diversification, and the Miocene grassland expansion. Proc. Natl. Acad. Sci. USA 111: 15149–15154. doi: 10.1073/pnas.1404177111

Faruqi, S.A., Quraish, H.B., Inamuddin, M. 1987. Studies in Libyan grasses X. Chromosome number and some interesting features. Ann. Bot. 45: 75–102.

Filgueiras, T. S. 1997. A new combination in *Eriochrysis* (Poaceae: Andropogoneae). Novon 7: 231–233.

Giussani, L.M., Cota-Sánchez, J.H., Zuloaga, F.O., Kellogg, E.A. 2001. A molecular phylogeny of the grass subfamily Panicoideae (Poaceae) shows multiple origins of C_4 photosynthesis. Amer. J. Bot. 88: 1993–2012.

Gómez-Martinez, R., Culham, A. 2000. Phylogeny of the subfamiily Panicoideae with emphasis on the tribe Paniceae: evidence from the *trnL-F* cp DNA region. In: Jacobs, S.W.L., Everett, J. (eds.) Grasses: systematics and evolution. Collingwood, Victoria, Australia: Commonwealth Scientific and Industrial Research Organization (CSIRO) Publishing. pp. 136–140

GPWG II 2012. Grass Phylogeny Working Group II. New grass phylogeny resolves deep evolutionary relationships and discovers C_4 origins. New Phytol. 193: 304–312.

Guala, G.F. 2000. The relation of space and geography to cladogenic events in *Agenium* and *Homozeugos* (Poaceae: Andropogoneae) in South America and Africa. In: Jacobs, S.W.L., Everett, J. (eds.) Grasses: systematics and evolution. Melbourne: CSIRO. pp. 159–166

Harlan, J.R., de Wet, J.M.J. 1963. The compilospecies concept. Evolution 17: 497–501.

Hattersley, P.W., Watson, L. 1975. Anatomical parameters for predicting photosynthetic pathways of grass leaves: The "maximum lateral cell count" and the "maximum cells distant count". Phytomorphology 25: 325–333.

Hitchcock, A.S. 1950. Manual of the grasses of the United States. Washington, DC: U.S. Government Printing Office.

Hodge, J.G., Kellogg, E.A. 2014. Patterns of inflorescence development of three prairie grasses (Andropogoneae, Poaceae). Int. J. Plant Sci. 175: 963–974.

Hodkinson, T.R., Chase, M.W., Lledó, M.D., Salamin, N., Renvoize, S.A. 2002. Phylogenetics of *Miscanthus*, *Saccharum* and related genera (Saccharinae, Andropogoneae, Poaceae) based on DNA sequences from ITS nuclear ribosomal DNA and plastid *trnL* intron and *trnL-F* intergenic spacers. J. Plant Res. 115: 381–392.

Ibrahim, D.G., Burke, T., Ripley, B.S., Osborne, C.P. 2009. A molecular phylogeny of the genus *Alloteropsis* (Panicoideae, Poaceae) suggests an evolutionary reversion from C_4 to C_3 photosynthesis. Ann. Bot. 103: 127–136.

Jannink, T.A., Veldkamp, J.F. 2002. Revision of Chionachninae (Gramineae: Andropogoneae). Blumea 47: 545–580.

Joy, R.J. Undated. "Sunshine" Vetivergrass *Chrysopogon zizanioides* (L.) Roberty. USDA-NRCS Plant Guide.

Kellogg, E.A. 2000. Molecular and morphological evolution in Andropogoneae. In: Jacobs, S.W.L., Everett, J. E. (eds.) Grasses: systematics and evolution. Melbourne, Australia: CSIRO. pp. 149–158

Kellogg, E.A. 2012. Phylogenetic relationships in Saccharinae. In: Paterson, A. (ed.) Genetics and genomics of Saccharinae. Berlin: Springer. pp. 3–21

Kellogg, E.A., Watson, L. 1993. Phylogenetic studies of a large data set. I. Bambusoideae, Pooideae, and Andropogoneae (Gramineae). Bot. Rev. 59: 273–343.

Kellogg, E.A., Hiser, K.M., Doust, A.N. 2004. Taxonomy, phylogeny, and inflorescence development of the genus *Ixophorus* (Panicoideae: Poaceae). Int. J. Plant Sci. 165: 1089–1105.

Kellogg, E.A., Aliscioni, S.S., Morrone, O., Pensiero, J., Zuloaga, F.O. 2009. A phylogeny of *Setaria* (Poaceae, Panicoideae, Paniceae) and related genera, based on the chloroplast gene *ndhF*. Int. J. Plant Sci. 170: 117–131.

Le Roux, L.G., Kellogg, E.A. 1999. Floral development and the formation of unisexual spikelets in the Andropogoneae (Poaceae). Amer. J. Bot. 86: 354–366.

Li, Y.-H.C., Phipps, J.B. 1973. Studies in the Arundinelleae (Gramineae). XV. Taximetrics of leaf anatomy. Can. J. Bot. 51: 657–680.

Liu, Q., Liu, H., Wen, J., Peterson, P.M. 2014. Infrageneric phylogeny and temporal divergence of *Sorghum* (Andropogoneae, Poaceae) based on low-copy nuclear and plastid sequences. PLoS ONE 9(8): e104933. doi: 10.1371/journal.pone.0104933

Lizarazu, M.A., Nicola, M.V., Salariato, D.L. 2014. Taxonomic re-evaluation of *Panicum* sections *Tuerckheimiana* and *Valida* (Poaceae: Panicoideae) using morphological and molecular data. Taxon 63: 265–274. doi: 10.12705/632/34

Lopez, A., Morrone, O. 2012. Phylogenetic studies in *Axonopus* (Poaceae, Panicoideae, Paniceae) and related genera: morphology and molecular (nuclear and plastid) combined analyses. Syst. Bot. 37: 671–676.

Loxton, A.E. 1974. A note on a possible bigeneric hybrid between *Digitaria* and *Anthephora*. Bothalia 11: 285–286.

Lukens, L., Doebley, J. 2001. Molecular evolution of the *teosinte branched* gene among maize and related grasses. Mol. Biol. Evol. 18: 627–638.

Malcomber, S.T., Kellogg, E.A. 2004. Heterogeneous expression patterns and separate roles of the SEPALLATA gene *LEAFY HULL STERILE* (*LHS1*) in grasses. Plant Cell 16: 1692–1706.

Mathews, S., Spangler, R.E., Mason-Gamer, R.J., Kellogg, E.A. 2002. Phylogeny of Andropogoneae inferred from phytochrome B, GBSSI, and *ndhF*. Int. J. Plant Sci. 163: 441–450.

McNeill, J., Wiersema, J.H., Gandhi, K. 2010. The complex nomenclature of *Panicum* sect. or subg. *Agrostoidea* and of *P. agrostoides*. Taxon 59: 1581–1584.

Morrone, O., Denham, S.S., Aliscioni, S.S., Zuloaga, F.O. 2008. *Parodiophyllochloa*, a new genus segregated from *Panicum* (Paniceae, Poaceae) based on morphological and molecular data. Syst. Bot. 33: 66–76.

Morrone, O., Aagesen, L., Scataglini, M.A., Salariato, D.L., Denham, S.S., Chemisquy, M.A., Sede, S.M., Giussani, L.M., Kellogg, E.A., Zuloaga, F.O. 2012. Phylogeny of the Paniceae (Poaceae: Panicoideae): integrating plastid DNA sequences and morphology into a new classification. Cladistics 28: 333–356.

Morrone, O., Pensiero, J., Aliscioni, S.A., Kellogg, E.A., Zuloaga, F.O. 2014. A revision of the Old World species of *Setaria*. Syst. Bot. Monographs 96: 1–161.

Nagahama, N., Anton, A.M., Norrmann, G.A. 2014. Taxon delimitation in the *Andropogon lateralis* complex (Poaceae) in southern South America based on morphometrical analyses. Syst. Bot. 39: 804–813.

Ozias-Akins, P., Akiyama, Y., Hanna, W.W. 2003. Molecular characterization of the genomic region linked with apomixis in *Pennisetum/Cenchrus*. Func. Integ. Genomics 3: 94–104.

Panel on Vetiver, Board on Science and Technology for International Development, National Research Council. 1993. Vetiver grass: A thin green line against erosion. Washington, D.C.: National Academy Press.

Preston, J.C., Wang, H., Doebley, J., Kellogg, E.A. 2012. The role of *teosinte glume architecture* (*tga1*) in coordinated regulation and evolution of grass glumes and inflorescence axes. New Phytol. 193: 204–215.

Reinheimer, R., Pozner, R., Vegetti, A.C. 2005. Inflorescence, spikelet, and floral development in *Panicum maximum* and *Urochloa plantaginea* (Poaceae). Amer. J. Bot. 92: 565–575.

Reinheimer, R., Malcomber, S.T., Kellogg, E.A. 2006. Distinct roles of the SEPALLATA gene *LEAFY HULL STERILE1* in *Eleusine indica* and *Megathyrsus maximus* (Poaceae). Evol. Devel. 8: 293–303.

Renvoize, S.A. 1998. Gramineas de Bolivia. Kew: Royal Botanic Gardens.

Rosengurtt, B., Arrillaga de Maffei, B.R., Izaguirre de Artucio, P. 1970. Gramineas Uruguayas. Montevideo: Universidad de la Republica, Departamento de Publicaciones.

Salariato, D.L., Zuloaga, F.O., Giussani, L.M., Morrone, O. 2010. Molecular phylogeny of the subtribe Melinidinae (Poaceae: Panicoideae: Paniceae) and evolutionary trends in the homogenization of inflorescences. Mol. Phyl. Evol. 56: 355–369.

Salariato, D.L., Morrone, O., Zuloaga, F.O. 2012. *Mayariochloa*, a new monotypic genus segregated from *Scutachne* (Poaceae, Panicoideae, Paniceae). Syst. Bot. 37: 105–116.

Sánchez-Ken, J.G., Clark, L.G. 2007. Phylogenetic relationships within the Centothecoideae + Panicoideae clade (Poaceae) based on *ndhF* and *rpl16* intron sequences and structural data. Aliso 23: 487–502.

Sánchez-Ken, J.G., Clark, L.G. 2010. Phylogeny and a new tribal classification of the Panicoideae s.l. (Poaceae) based on plastid and nuclear sequence data and structural data. Amer. J. Bot. 97: 1732–1748.

Sánchez-Ken, J.G., Clark, L.G., Kellogg, E.A., Kay, E.E. 2007. Reinstatement and emendation of subfamily Micrairoideae (Poaceae). Syst. Bot. 32: 71–80.

Scataglini, M.A., Lizarazu, M.A., Zuloaga, F.O. 2014. A peculiar amphitropical genus of Paniceae (Poaceae, Panicoideae). Syst. Bot. 39: 1108–1119.

Sede, S.M., Morrone, O., Giussani, L.M., Zuloaga, F.O. 2008. Phylogenetic studies in the Paniceae (Poaceae): a realignment of section *Lorea* of *Panicum*. Syst. Bot. 33: 284–300.

Sede, S.M., Morrone, O., Aliscioni, S.S., Giussani, L.M., Zuloaga, F.O. 2009a. *Oncorachis* and *Sclerochlamys*, two new segregated genera from *Streptostachys* (Poaceae, Panicoideae, Paniceae): a revision based on molecular, morphological and anatomical characters. Taxon 58: 365–374.

Sede, S.M., Zuloaga, F.O., Morrone, O. 2009b. Phylogenetic studies in the Paniceae (Poaceae-Panicoideae): *Ocellochloa*, a new genus from the New World. Syst. Bot. 34: 684–692.

Sendulsky, T., Soderstrom, T.R. 1984. Revision of the South American genus *Otachyrium* (Poaceae: Panicoideae). Smithsonian Contr. Bot. 57: 1–24.

Skendzic, E.M., Columbus, J.T., Cerros-Tlatilpa, R. 2007. Phylogenetics of Andropogoneae (Poaceae: Panicoideae) based on nuclear ribosomal internal transcribed spacer and chloroplast *trn*L-F sequences. Aliso 23: 530–544.

Spangler, R.E. 2003. Taxonomy of *Sarga*, *Sorghum* and *Vacoparis* (Poaceae: Andropogoneae). Aust. Syst. Bot. 16: 279–299.

Spangler, R., Zaitchik, B., Russo, E., Kellogg, E. 1999. Andropogoneae evolution and generic limits in *Sorghum* (Poaceae) using *ndhF* sequences. Syst. Bot. 24: 267–281.

Sun, Q., Lin, Q., Yi, Z.-L., Yang, Z.-R., Zhou, F.-S. 2010. A taxonomic revision of *Miscanthus s.l.* (Poaceae) from China. Bot. J. Linn. Soc. 164: 178–220.

Teerawatananon, A., Jacobs, S.W.L., Hodkinson, T.R. 2011. Phylogenetics of Panicoideae (Poaceae) based on chloroplast and nuclear DNA sequences. Telopea 13: 115–142.

Türpe, A.M. 1970. Sobre la anatomía foliar de *Jansenella griffithiana* (C. Mueller) Bor (Poaceae: Arundinelleae). Senck. Biol. 51: 277–285.

Vega, A.S., Rúgolo de Agrasar, Z.E., Axelrod, F.S. 2010. A new species of *Tarigidia* (Poaceae, Panicoideae, Paniceae) from Puerto Rico and additional evidence for a hybrid origin of the genus. Syst. Bot. 35: 96–101.

Veldkamp, J.F. 1999. A revision of *Chrysopogon* Trin. including *Vetiveria* Bory (Poaceae) in Thailand and Malesia with notes on some other species fom Africa and Australia. Austrobaileya 5: 503–533.

Veldkamp, J.F., De Koning, R., Sosef, M.S.M. 1986. Generic delimitation of *Rottboellia* and related genera (Gramineae). Blumea 31: 281–307.

Verboom, G.A., Linder, H.P., Barker, N.P. 1994. Haustorial synergids: An important character in the systematics of danthonioid grasses (Arundinoideae: Poaceae). Amer. J. Bot. 81: 1601–1610.

Vicentini, A., Barber, J.C., Giussani, L.M., Aliscioni, S.S., Kellogg, E.A. 2008. Multiple coincident origins of C_4 photosynthesis in the Mid- to Late Miocene. Global Change Biol. 14: 2963–2977.

Voznesenskaya, E.V., Chuong, S.D.X., Koteyeva, N.K., Edwards, G.E., Franceschi, V.R. 2005. Functional compartmentation of C_4 photosynthesis in the triple layered chlorenchyma of *Aristida* (Poaceae). Func. Plant Biol. 32: 67–77.

Watson, L., Dallwitz, M.J. 1992 onward. The grass genera of the world: descriptions, illustrations, identification, and information retrieval; including synonyms, morphology, anatomy, physiology, phytochemistry, cytology, classification, pathogens, world and local distribution, and references, 23rd April 2010 edn. Wallingford: CAB International. http://delta-intkey.com/grass/

Webster, R.D. 1987. The Australian Paniceae (Poaceae). Stuttgart: J. Cramer.

Welker, C. A. D., E. A. Kellogg, and J. Prado. 2014. Andropogoneae versus Sacchareae (Poaceae-Panicoideae): the end of a great controversy. Taxon 63: 643–646.

Welker, C. A. D., T. Teixeira de Souza-Chies, H. M. Longhi-Wagner, M. C. Peichoto, M. R. McKain, E. A. Kellogg. 2015. Phylogenetic analysis of *Saccharum* s.l. (Poaceae-Andropogoneae), with emphasis on the South American species. American Journal of Botany 102: 248–263. doi:10.3732/ajb.1400397

Wu, Z., Raven, P.H., Hong, D. (eds.) 2007. Flora of China Illustrations. Beijing: Science Press.

Zaitchik, B.F., le Roux, L.G., Kellogg, E.A. 2000. Development of male flowers in *Zizania aquatica* (North American wild-rice; Gramineae). Int. J. Plant Sci. 161: 345–351.

Zanotti, C.A., Pozner, R., Morrone, O. 2010. Understanding spikelet orientation in Paniceae (Poaceae). Amer. J. Bot. 97: 717–729.

Zuloaga, F.O. 1987. Systematics of the New World species of *Panicum* (Poaceae: Paniceae). In Soderstrom, T. R., Hilu, K.W., Campbell, C.S., Barkworth, M.E. (eds.) Grass systematics and evolution. Washington DC: Smithsonian Institution Press. pp. 287–306

Zuloaga, F.O., Soderstom, T.R. 1985. Classification of the outlying species of New World *Panicum* (Poaceae: Paniceae). Smithsonian Contr. Bot. 59: 1–63.

Zuloaga, F.O., Ellis, R.P., Morrone, O. 1992. A revision of *Panicum* subgenus *Phanopyrum* section *Laxa* (Poaceae: Panicoideae: Paniceae). Ann. Missouri Bot. Gard. 79: 770–818.

Zuloaga, F.O., Giussani, L.M., Morrone, O. 2007. *Hopia*, a new monotypic genus segregated from *Panicum* (Poaceae). Taxon 56: 145–156.

Zuloaga, F.O., Scataglini, M.A., Morrone, O. 2010. A phylogenetic evaluation of *Panicum* sects. *Agrostoidea*, *Megista*, *Prionitia* and *Tenera* (Panicoideae, Poaceae): two new genera, *Stephostachys* and *Sorengia*. Taxon 59: 1535–1546.

Zuloaga, F.O., Morrone, O., Scataglini, M.A. 2011. Monograph of *Trichanthecium* (Poaceae, Paniceae). Syst. Bot. Monographs 94: 1–101.

Zuloaga, F.O., Rúgolo de Agrasar, Z.E., Anton, A.M. (eds.) 2012. Flora vascular de la República Argentina, Vol. 3, Tomos 1 y 2. Córdoba, Argentina: Gráficamente Ediciones.

IX. Subfamily Danthonioideae N.P. Barker & H.P. Linder (2001)

Perennials or more rarely annuals. Ligule a fringe of hairs. Inflorescence branches branched (rarely unbranched). Spikelets laterally compressed, with 1 to many flowers. Lemma bilobed, awned from the sinus, the base of the awn flat. *Style branches 2, widely separated. Embryo sac with haustorial synergids.* Epiblast absent, scutellar cleft present, embryonic mesocotyl internode elongated, the embryonic leaf margins meeting. Stomata with dome-shaped or parallel-sided subsidiary cells. Photosynthetic pathway C_3.

Seventeen genera, 281 spp.

This treatment follows the revised classification presented by Linder et al. (2010).

Although haustorial synergids are likely to be synapomorphic for this subfamily, a number of the genera remain to be investigated for this character (Verboom et al. 1994). *Merxmuellera rangei* and *M. papposa*, formerly in Danthonioideae, clearly belong in the Chloridoideae (Peterson et al. 2010) and have been transferred to the genus *Ellisochloa* P. M. Peterson and N. P. Barker. The phylogeny of Linder et al. (2010) is consistent with that of the GPWG II (2012), and more thoroughly sampled. The ligule is a fringe of hairs except in *Geochloa*, where it is a fringed membrane; this may be synapomorphic for the genus.

Key to the Genera of Danthonioideae

1. Leaf blades disarticulating from the sheath — 2
1. Leaf blades persistent — 4
2(1). Leaf blade disarticulating 1–2 cm distal to the ligule, leaving a small straight or recurved flap of blade tissue; lemma with three tufts of hairs on either side of the midrib; plants of Africa — **538. *Merxmuellera***
2. Leaf blade disarticulating at the ligule, not leaving a flap of blade tissue; lemma hairs evenly spread or in lines, but not tufts; not African — 3
3 (2). Lemma hairs evenly spread, not in lines; plants gynodioecious; New Guinea — **546. *Chimaerochloa***
3. Lemma hairs in four vertical lines; plants bisexual; New Zealand, Australia and Lord Howe Island — **542. *Chionochloa***
4 (1). Lemma lacking an awn, the apex shallowly tridentate or entire — 5
4. Lemma awned, the awn generally from the sinus between two prominent lobes — 8
5 (4). Lemma glabrous — **549. *Notochloe***
5. Lemma pubescent — 6
6 (5). Disarticulation of flowers from a horizontal break in the middle of the rachilla internode, the dispersal unit thus being the flower plus a short piece of the rachilla; spikelets with no more than 4 flowers — **550. *Plinthanthesis***
6. Disarticulation at the base of the flower, the dispersal unit not including the rachilla internode below; spikelets often with more than 4 flowers — 7
7 (6). Glumes shorter to more often longer than the spikelet; plants often stoloniferous; glumes often with tuberculate hairs; inflorescence sometimes unbranched — **554. *Tribolium***
7. Glumes shorter than the spikelet; plants lacking stolons; glumes without tuberculate hairs; inflorescence branched — **553. *Schismus***
8 (4). Lemma awn without a clear demarcation between the basal column and the elongate apical portion; lemma lobes minute or absent; plants over 1 m tall — 9
8. Lemma awn clearly divided into a flattened column and a long slender apical portion; lemma lobes prominent; plants often less than 1 m — 10
9 (8). Leaf blade with a prominent midrib; plants of South America — **547. *Cortaderia***
9. Leaf blade with several wide prominent veins; plants of New Zealand — **545. *Austroderia***

DOI 10.1007/978-3-319-15332-2_23,

10 (8). Flowers 2 (rarely 1) per spikelet; rachilla extended beyond the distal spikelet; lemma with setae in the sinus between the lobes and the central awn 11
10. Flowers more than 2 per spikelet; rachilla not extended; lemma with setae from the apices of the lobes, or lacking 12
11 (10). Spikelets less than 2.5 cm long **541. *Pentameris***
11. Spikelets more than 2.5 cm long **544. *Pseudopentameris***
12 (10). Plants geophytic, with swollen rhizomes; basal leaf sheaths wooly **539. *Geochloa***
12. Plants usually not geophytic, swollen rhizomes lacking; if present then basal leaf sheaths glabrous and shining 13
13 (12). Pedicel of spikelet disarticulating slightly below the base of the glumes, the disarticulation point marked by a tuft of hairs **543. *Chaetobromus***
13. Pedicel of spikelet either disarticulating at the base of the glumes or persistent and disarticulation above the glumes; pedicel without a tuft of hairs 14
14 (13). Leaves tough, with abundant sclerenchyma; plants of Africa to western Asia 15
14. Leaves membranous; plants of North and South America, Europe, and the Pacific rim 16
15 (14). Plants forming robust clumps, generally over 1 m tall; if less, then geophytic with swollen rhizomes; plants of the Cape Province of South Africa **540. *Capeochloa***
15. Plants wiry, generally much less than 90 cm tall, not geophytic; plants of African mountains and western Asia **551. *Tenaxia***
16 (14). Lemma hairs in two rows, generally in tufts; hilum punctate; plants of Australia, New Guinea, New Zealand, and South America **552. *Rytidosperma***
16. Lemma hairs not tufted; hilum elliptical to elongate; plants of North America, South America, Europe **548. *Danthonia***

Genera of Danthonioideae

538. ***Merxmuellera*** Conert

Merxmuellera Conert, Senck. Bot. 51: 129 (1970).

Caespitose perennials. *Leaf blade disarticulating 1–2 cm distal to the ligule, leaving a small straight or recurved flap of blade tissue; leaf sheath splitting.* Inflorescence branches branched. Spikelets with >2 flowers. Glumes shorter than the flowers. Callus of flower pubescent. *Lemma with three tufts of trichomes on either side of the midrib,* lobes often ending in thin setae; column of awn indistinct.

Nine spp., mountainous parts of Africa and Madagascar. *M. stereophylla* lacks the synapomorphies of the genus but is included within it on the basis of molecular data (Linder et al. 2010).

Capeochloa + Geochloa clade

Palea margins with tufts of long hairs.

539. ***Geochloa*** H. P. Linder & N. P. Barker

Geochloa H. P. Linder & N. P. Barker, Ann. Missouri Bot. Gard. 97: 323 (2010).

Geophytic perennials *with swollen rhizomes. Basal leaf sheaths wooly pubescent. Ligule a fringed membrane. Inflorescence compact.* Spikelets with 2 to 7 flowers. Glumes as long as flowers or longer. Callus of flower pubescent. Lemma with trichomes scattered over the abaxial surface, lobes ending in setae; column of awn distinct, straight or twisted. $2n = 48$.

Three spp., southern Cape Province, South Africa. *Geochloa* occurs in the fire-dominated fynbos, and flowers only in the years immediately following a fire (Linder and Ellis 1990a).

540. ***Capeochloa*** H. P. Linder & N. P. Barker

Capeochloa H. P. Linder & N. P. Barker, Ann. Missouri Bot. Gard. 97: 324 (2010).

Caespitose perennials. Basal leaf sheaths glabrous. Inflorescence branches branched. Spikelets with 2 to 4 flowers. Glumes as long as flowers or longer. Callus of flower pubescent. Lemma trichomes variously distributed, lobes ending in setae or not; column of awn distinct, twisted. Palea margins with tufts of hairs. $2n = 12, 36$.

Three spp., southern Cape Province, South Africa. Linder et al. (2010) suggest that *C. arundinacea* may be of hybrid origin, but include it in *Capeochloa* based on sequences of the nuclear ribosomal RNA and its morphological similarity to *C. cincta*.

541. ***Pentameris*** P. Beauv. Fig. 81

Pentameris P. Beauv., Ess. Agrost. 92 (1812); tax.: Barker (1993), Davidse (1988), Linder and Ellis (1990b).
Prionanthium Desv., Opusc. 64 (1831).
Pentaschistis (Nees) Spach., Hist. Nat. Veg. Phan. 13: 164 (1846).
Poagrostis Stapf., Fl. Cap. 7: 760 (1899).

Annuals or perennials, caespitose, rhizomatous or stoloniferous. Multicellular glands often on

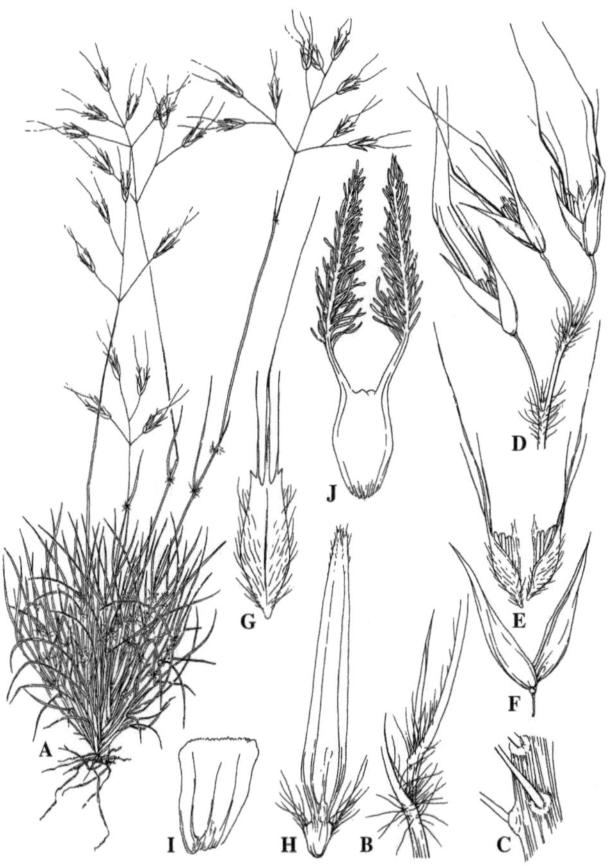

Fig. 81. Danthonioideae. *Pentameris alticola*. **A** Several tillers, representing growth form of plant. **B** Junction of leaf sheath and blade. **C** Detail of cushion-based hairs on blades. **D** Part of inflorescence showing nodes and spikelets. **E** Flowers. **F** Glumes. **G** Lemma. **H** Upper palea with rachilla extension. **I** Lodicule. **J** Pistil. (From Linder and Ellis 1990b)

leaves, inflorescences and glumes. Spikelets with 2 flowers (rarely 1). Glumes extending beyond the flowers. Callus of flowers pubescent. Lemma with *setae from the sinuses between the lateral lobes and the central awn*, awn with a flat twisted base. Palea acute, *the keels not reaching the apex*. Pericarp sometimes free from the seed coat. $2n = 14$, 24, 26, 28, 40, 42, 52, 56, 91.

Eighty-four spp. Africa, Madagascar, introduced to Australia.

Linder et al. (2010) find that *Pentameris* s.s. and *Prionanthium* are both derived within a paraphyletic "*Pentaschistis*". Accordingly, they have combined the three genera into one, with the name *Pentameris* taking priority. Three sections are recognized, sect. *Pentameris*, corresponding to *Pentameris* s.s., sect. *Dracomontanum*, which includes two species formerly assigned to "*Pentaschistis*", and sect. *Pentaschistis*, which includes species formerly assigned to "*Pentaschistis*" and *Prionanthium*. The former genus *Prionanthium* is distinctive in that it includes annual species with multicellular glands on the leaves. However, it is clearly derived from within *Pentameris* and is otherwise morphologically similar.

542. ***Chionochloa*** Zotov

Chionochloa Zotov, New Zealand J. Bot. 1: 87 (1963); tax.: Connor (1991).

Caespitose perennials, generally over 1 m tall. Leaf blades with deep grooves between the veins, *disarticulating at or below the ligule*. Inflorescence branches branched. Spikelets with more than 2 flowers. Glumes shorter or longer than the flowers. Callus of flower pubescent. Lemma pubescent with trichomes generally in four vertical lines, the lobes acute, terminating in setae, the awn base distinct or not. Palea margins with hair tufts. Leaf epidermal *microhairs at the bases of grooves; silica cells round*. $2n = 42$.

Twenty-five spp. New Zealand, SE Australia, Lord Howe Island.

CHAETOBROMUS + PSEUDOPENTAMERIS CLADE

Callus of flowers elongate, almost twice as long as the rachilla internode, acute, sharp-pointed. Palea strongly sulcate so that the veins nearly touch, the veins not reaching the apex.

543. ***Chaetobromus*** Nees

Chaetobromus Nees in Lindley, Nat. Syst., ed. 2: 449 (1836).

Perennials, caespitose or mat-forming, sometimes rhizomatous. Inflorescence branches branched. Spikelets *with a tuft of hairs at the point of dehiscence below the glumes*, with more than 2 flowers, *the most proximal flower smaller and less well developed than the others*. Glumes longer than the flowers. Callus of flower pubescent. Lemma pubescent, the lobes ending in setae, the central awn with a twisted base. $2n = 12, 18, 36, 48, 52, 72$.

One sp., *C. involucratus* (Schrad.) Nees, SW Africa.

544. ***Pseudopentameris*** Conert

Pseudopentameris Conert, Mitt. Bot. Staatss. München. 10: 303 (1971); tax.: Barker (1995).

Perennials, caespitose or rhizomatous. Inflorescence branches branched. Spikelets with 2 flowers, *2.5–3.5 cm long*. Glumes longer than the flowers. Callus of flower pubescent. Lemma pubescent, the lobes acute and generally shorter than the lemma itself; basal column of the awn clearly distinct, twisted.

Three spp., Cape region, South Africa.

Remaining genera (*Danthonia* clade + *Rytidosperma* clade): *Lodicules bearing microhairs.*

Danthonia clade (Austroderia + Cortaderia + Danthonia + Notochloe + Plinthanthesis + Chimaerochloa)

Linder et al. (2010) identify three possible synapomorphies for this clade: a gynodioecious breeding system, lemmas with five veins or fewer, and the lemma awn not differentiated into a flattened basal portion and a slender apical portion. All three characters reverse within the clade, however, and are thus not useful for identification.

545. ***Austroderia*** N. P. Barker & H. P. Linder

Austroderia N. P. Barker & H. P. Linder, Ann. Missouri Bot. Gard. 97: 343 (2010).

Gynodioecious perennials, caespitose or stoloniferous, generally several meters tall. Leaf blades with several wide prominent veins; leaf sheaths persistent, waxy, not fragmenting. Inflorescences plumose, the branches branched. Spikelets with more than 2 flowers. Glumes shorter or longer than the flowers, *with one vein*. Callus of flower pubescent. Lemma pilose, *the lobes absent or nearly so, although long setae sometimes present in place of the lobes*; the base of the awn not differentiated. $2n = 90$.

Five spp., New Zealand.

Austroderia corresponds to a set of species formerly placed in *Cortaderia*. The two genera are quite similar in terms of gross morphology, but differ in leaf anatomy as well as in geographic distribution. The New Zealand species formerly placed in *Cortaderia* were first recognized as a separate clade by Barker et al. (2003).

546. ***Chimaerochloa*** H. P. Linder

Chimaerochloa H. P. Linder, Ann. Missouri Bot. Gard. 97: 346 (2010).

Gynodioecious perennials, caespitose. Leaf blades tough, fibrous, disarticulating from the sheath. Inflorescence branches branched. Spikelets with 2 to 7 flowers. Glumes shorter than the flowers. Callus of flower pubescent. Lemma pubescent, the lobes acute, ending in setae, the awn with a flat twisted base. $2n = 72$.

One sp., *C. archboldii* H. P. Linder, New Guinea.

This species is morphologically similar to *Cortaderia*, but chloroplast sequence data place it in the *Danthonia* clade. It is possible that it is an allopolyploid derivative of a *Cortaderia* species crossed with a *Danthonia* species (the latter as seed parent); data from a nuclear gene would be required to test this hypothesis.

547. ***Cortaderia*** Stapf

Cortaderia Stapf, Gard. Chron. III, 22: 396 (1897).
Lamprothyrsus Pilger, Bot. Jahrb. 37, Beibl. 85: 58 (1906).

Caespitose perennials, gynoedioecious or pistillate only, generally several meters tall. Leaf blades with prominent midribs, leaf sheaths persistent but tending to fragment with age. Inflorescences plumose, the branches branched. Spikelets with more than 2 flowers. Glumes shorter or longer than the flowers, *with one vein*. Callus of flower pubescent. Lemma pilose, the lobes absent or nearly so, although long setae sometimes present; awn not clearly differentiated into a flattened base and a slender apical portion. $2n = 36, 72, 108, 136$.

Twenty spp., South America.

Linder et al. (2010) describe the margins and sometimes the entire distal part of the leaves as "viciously scabrid". *C. pilosa* may not be related to the other species of the genus, but is retained here on the basis of ambiguous molecular data as described by Linder et al. (2010).

548. ***Danthonia*** DC

Danthonia DC in Lam. & DC, Fl. Franç. ed. 3, 3: 32 (1805), nom. cons.
Sieglingia Bernh., Syst. Verz. (Bernhardi): 20, 44 (1800).

Caespitose perennials, *sometimes with cleistogamous spikelets enclosed in the upper leaf sheaths.* Leaf blades not disarticulating, membranous. Inflorescence unbranched, or with branched branches. Spikelets with 2 to 10 flowers. Callus of flowers pubescent. Glumes shorter or longer than the flowers. Lemma pubescent, at least on the margins, with 7 to 9 veins, the lobes prominent, ending in setae, the awn base flat, twisted, clearly distinct from the apical portion. $2n = 18, 24, 36, 48, 72, 98$.

Twenty-five spp. America and Europe.

No morphological character can be used consistently to recognize *Danthonia*; the four synapomorphies indicated by Linder et al. (2010) (bisexual flowers, persistent leaf blades, 7 or more lemma veins, and orthophyllous leaves) are all homoplasious on the phylogeny.

Notochloe + Plinthanthesis clade

Caespitose perennials, less than 1 m tall. Leaf sheaths persistent, shiny, white. Inflorescence branches branched. Callus of flower pubescent, short; *rachilla with a horizontal disarticulation zone surrounded by hairs. Leaf mesophyll with empty spaces, formed "from large, empty cells that then disintegrate, leaving rather ragged cavities"* (Linder et al. 2010, p. 346). *Fruit turbinate.*

549. ***Notochloe*** Domin

Notochloe Domin, Feddes Repert. 10: 117 (1911).

Spikelets with 7 to 9 flowers. Glumes shorter than the flowers. Lemma glabrous, *with three small apical teeth, awnless.*

One sp., *N. microdon* (Benth.) Domin, eastern Australia.

550. ***Plinthanthesis*** Steud.

Plinthanthesis Steud., Syn. Pl. Glum. 1: 14 (1854).

Inflorescences plumose. Spikelets with 2 to 4 flowers. Glumes shorter or longer than the flowers. Lemma pubescent below, with two short lobes, awn the same length as the lobes or longer. *Lodicules without microhairs.*

Three spp., eastern Australia.

Rytidosperma clade (Rytidosperma + Schismus + Tenaxia + Tribolium)

Lemma hairs often in scattered tufts. Lodicules with microhairs. *Hilum punctate-ovate.*

551. ***Tenaxia*** N. P. Barker & H. P. Linder

Tenaxia N. P. Barker & H. P. Linder, Ann. Missouri Bot. Gard. 97: 350 (2010).

Caespitose perennials with wiry stems. Leaves tough, sclerenchymatous, the sheaths persistent. Inflorescence unbranched or with branched branches. Spikelets with 2 to 7 flowers. Glumes as long as the flowers or longer. Callus of flower pubescent. Lemma pubescent, with 7 to 9 veins, the lobes with acute apices, terminating in setae or not, the awn with a twisted base. $2n = 12, 36, 56$.

Eight spp. Africa, India, Himalayas.

Members of *Tenaxia* have been previously assigned to either *Merxmuellera* or *Danthonia*. *Tenaxia* has no obvious morphological synapomorphy, but forms a distinct and well-supported clade in molecular phylogenies. Phylogeography of *T. disticha* in South Africa has been investigated by Linder et al. (2014).

Schismus + Tribolium + Rytidosperma

Leaf blades membranous, not stiff and scelerenchymatous.

552. ***Rytidosperma*** Steud.

Rytidosperma Steud., Syn. Pl. Glum. 1: 425 (1854).
Monostachya Merr., Philipp. J. Sci. 5: 330 (1910).
Notodanthonia Zotov, New Zealand J. Bot. 1: 104 (1963).
Erythranthera Zotov, New Zealand J. Bot. 1: 124 (1963).
Pyrrhanthera Zotov, New Zealand J. Bot. 1: 125 (1963).
Joycea H. P. Linder, Telopea 6: 611 (1996).
Austrodanthonia H. P. Linder, Telopea 7: 269 (1997).

Caespitose perennials. Inflorescence branches branched or unbranched. Spikelets with 2 to 7 flowers, *the callus wider than the internode that bears it.* Glumes shorter or more often longer than the flowers. Lemma with the central lobe ending in a twisted awn; lemma hairs rarely lacking, *generally in two transverse rows across the abaxial side, these in one line of discrete tufts about midway up the lemma and additional marginal tufts somewhat lower.*

Seventy-six spp., Australasia and South America.

The circumscription of this genus was addressed by Humphreys et al. (2010), who show that it is monophyletic in the original sense of Steudel (1853-1854).

553. ***Schismus*** P. Beauv.

Schismus P. Beauv., Ess. Agrost.: 73 (1812).

Annuals or perennials, caespitose, less than 35 cm tall. Inflorescence branches branched. Spikelets with 3 to 7 flowers. Glumes shorter than the spikelet (rarely longer). Callus of flower pubescent. Lemma with hairs scattered, or tufted, or in lines, with mucronate lobes, the central one generally shorter than the laterals. $2n = 12, 24, 36, 48, 72$.

Five spp., Africa.

In *Schismus schismoides* (Stapf ex Conert) Verboom and H. P. Linder, the central lobe forms a twisted awn and is thus longer than the laterals; this awn morphology is not shared with other species in the genus.

554. ***Tribolium*** Desv.

Tribolium Desv., Opusc.: 64 (1831).
Urochlaena Nees, Fl. Afr. Austr.: 437 (1841).
Karroochloa Conert & Türpe, Senck. Bot. 50: 290 (1969).

Annuals or perennials, caespitose and stoloniferous. Inflorescence unbranched, or with branched branches, generally narrow and spikelike or capitate. Spikelets with 2 to 10 flowers. Glumes shorter or longer than the flowers, glabrous or *with hairs with tuberculate bases.* Callus of flower pubescent. Lemma pubescent, the hairs either evenly spread or tufted in various patterns, *the apex entire and acute* or with three lobes, the lobes rounded, acute or acuminate, the central one occasionally forming an awn. $2n = 12, 24, 36$.

Sixteen spp., southern Africa.

The genus as a whole lacks a clear morphological synapomorphy, although each of the three sections has a unique character. Plants in *Tribolium* sect. *Tribolium* have hispid glumes, those in *T.* sect. *Acutiflorae* N. C. Visser & Spies ex H. P. Linder & Davidse bear stolons, and those in *T.* sect. *Uniolae* N. C. Visser & Spies ex H. P. Linder & Davidse have unbranched inflorescences with the spikelets all borne on one side.

References

Barker, N.P. 1993. A biosystematic study of *Pentameris* (Arundineae, Poaceae). Bothalia 23: 25-47.

Barker, N.P. 1995. A systematic study of the genus *Pseudopentameris* (Arundinoideae: Poaceae). Bothalia 25: 141-148.

Barker, N.P., Linder, H.P., Morton, C.M., Lyle, M. 2003. The paraphyly of *Cortaderia* (Danthonioideae): evidence from morphology and nuclear DNA sequence data. Ann. Missouri Bot. Gard. 90: 1-24.

Connor, H.E. 1991. *Chionochloa* Zotov (Gramineae) in New Zealand. N. Z. J. Bot. 29: 219-282.

Davidse, G. 1988. A revision of the genus *Prionanthium* (Poaceae: Arundinae). Bothalia 18: 143-153.

GPWG II 2012. Grass Phylogeny Working Group II. New grass phylogeny resolves deep evolutionary relationships and discovers C_4 origins. New Phytol. 193: 304-312.

Humphreys, A.M., Pirie, M.D., Linder, H.P. 2010. A plastid tree can bring order to the chaotic generic taxonomy of *Rytidosperma* Steud. s.l. (Poaceae). Mol. Phyl. Evol. 55: 911-928.

Linder, H.P., Ellis, R.P. 1990a. Vegetative morphology and interfire survival strategies in the Cape fynbos grasses. Bothalia 20: 91-103.

Linder, H.P., Ellis, R.P. 1990b. A revision of *Pentaschistis* (Arundineae: Poaceae). Contr. Bolus Herb. 12: 1-124.

Linder, H.P., Baeza, M., Barker, N.P., Galley, C., Humphrey, A.M., Lloyd, K.M., Orlovich, D.A., Pirie, M.D., Simon, B.K., Walsh, N., Verboom, G.A. 2010. A generic classification of the Danthonioideae. Ann. Missouri Bot. Gard. 97: 306-364.

Linder, H.P., Perl, F., Bouchenak-Khelladi, Y., Barker, N. P. 2014. Phylogeographical pattern in the southern African grass *Tenaxia disticha* (Poaceae). Syst. Bot. 39: 428-440.

Peterson, P.M., Romaschenko, K., Johnson, G. 2010. A classification of the Chloridoideae (Poaceae) based on multi-gene phylogenetic trees. Mol. Phyl. Evol. 55: 580-598.

Steudel, E.G. 1853-1854. Synopsis Plantarum Glumacearum. Stuttgart: J. B. Metzler.

Verboom, G.A., Linder, H.P., Barker, N.P. 1994. Haustorial synergids: An important character in the systematics of danthonioid grasses (Arundinoideae: Poaceae). Amer. J. Bot. 81: 1601-1610.

X. Subfamily Chloridoideae Kunth ex Beilschm. (1833)

Plants herbaceous. Annuals or perennials, caespitose, rhizomatous or stoloniferous, erect or decumbent. Culms hollow or solid. Ligule membranous, or a fringed membrane, or a fringe of hairs. Leaf blades lacking pseudopetioles or auricles. Inflorescences unbranched, or with unbranched primary branches, or with branched primary branches. Spikelets laterally compressed, with 1 to many flowers, the proximal ones fertile, the distal ones (if present) often reduced. Glumes 2, disarticulation above the glumes. Lemmas bilobed, awned from the sinus. Lodicules 2, fleshy. Caryopsis with the pericarp fused to the seed coat; hilum short, punctate; *embryo with an epiblast*, with a scutellar cleft, with an elongated mesocotyl internode, with the embryonic leaf margins not overlapping. Midrib anatomy simple. Mesophyll radiate, with a double bundle sheath. All species C_4 except for *Merxmuellera rangei*, the ancestral condition probably C_3. Bicellular microhairs with the distal cell narrow and elongate ("panicoid type") to nearly spherical ("chloridoid type").

One hundred thirty-two genera, 1721 spp.

This subfamily has been shown consistently to be monophyletic, with all species except *Merxmuellera rangei* sharing radiate chlorenchyma around the veins and C_4 photosynthesis. Members of the subfamily have been biochemically typed as having either the NAD-ME or PCK subtype of C_4 photosynthesis (Brown 1977; Gutierrez et al. 1974; Hattersley and Watson 1992). The presence of so-called chloridoid microhairs, in which the apical cell of the microhair is broad and rounded, may also be synapomorphic but the character is somewhat variable (Van den Borre and Watson 1997). The GPWG (2001) suggested that the presence of an epiblast might be a synapomorphy but noted that the character is highly homoplasious in the subfamily.

Peterson et al. (2010a) have produced a comprehensive molecular phylogeny for the subfamily, including 254 species in 99 genera. They delimit four tribes: Triraphideae, Eragrostideae, Zoysieae, and Cynodonteae, and I follow their treatment here. More recently they suggest adding *Ellisochloa* (formerly *Merxmuellera rangei* and *M. papposa*) and the four species of *Centropodia* as Centropodieae (Peterson et al. 2011). Tribes in 20th century classifications of the subfamily were distinguished by inflorescence characters, with the primary branches of the inflorescence themselves bearing branches in most Eragrostideae, and the primary branches unbranched in Cynodonteae. Phylogenetic data show that the transition from one type of inflorescence to the other, which involves a change in the numbers of orders of branches, is quite labile in evolutionary time. The most striking example of this is in the *Sporobolus* clade, in which the former genus *Spartina*, with sturdy, unbranched primary branches, is apparently derived from within *Sporobolus*, with a highly branched and delicate inflorescence (Peterson et al. 2014b).

Generic limits in Chloridoideae are often problematical. The characters used to define genera are mostly characters of the inflorescence and spikelet, but because these are evolutionarily labile they often do not indicate relationship. Phylogenetic studies such as those by Columbus et al. (2007, 2010) and Peterson et al. (2010a, 2010b, 2014a) indicate that several of the major genera are polyphyletic; these include "*Chloris*" (with 63 species). Until these are dismembered,

E.A. Kellogg, *Flowering Plants. Monocots*, The Families and Genera of Vascular Plants 13,
DOI 10.1007/978-3-319-15332-2_24, © Springer International Publishing Switzerland 2015

monophyletic groups (at any taxonomic level) cannot be delimited by morphological data. *Leptochloa*, known for some time to be polyphyletic, has recently been divided and its components reclassified (Peterson et al. 2012). *Tridens* has likewise recently been divided into *Tridens* and *Tridentopsis* (Peterson et al. 2014). Surprisingly, the large genus *Eragrostis* appears to be monophyletic if several monotypic genera are subsumed in it (Ingram and Doyle 2004). *Muhlenbergia*, *Bouteloua*, and *Sporobolus* have also been re-circumscribed and each is now monophyletic (Columbus 1999; Peterson et al. 2010b; Peterson et al. 2014b).

Tribes and Subtribes of Chloridoideae

XXV. Tribe Centropodieae
 Core C_4 Chloridoideae, incertae sedis
XXVI. Tribe Triraphideae
XXVII. Tribe Eragrostideae
 Incertae sedis
 XXVII.1. Subtribe Cotteinae
 XXVII.2. Subtribe Eragrostidinae
 XXVII.3. Subtribe Uniolinae
XXVIII. Tribe Zoysieae
XXIX. Tribe Cynodonteae
 Incertae sedis
 XXIX.1. Subtribe Boutelouinae
 XXIX.2. Subtribe Orcuttiinae
 XXIX.3. Subtribe Gouiniinae
 XXIX.4. Subtribe Triodiinae
 XXIX.5. Subtribe Eleusininae
 XXIX.6. Subtribe Tripogoninae
 XXIX.7. Subtribe Pappophorinae

Key to the Genera of Chloridoideae

1. Inflorescence unbranched, with spikelets sessile or pedicellate, or if branched then dense and spike-like **Group I**
1. Inflorescence clearly branched, the primary branches branched, or unbranched and digitate or spread along the main axis, not dense and spike-like 2
2 (1). Primary branches of the inflorescence unbranched, digitate or spread along the main axis **Group II**
2. Primary branches of the inflorescence branched, at least some spikelets borne on secondary or higher-order branches **Group III**

Group I – Inflorescences unbranched, with spikelets sessile or pedicellate; spike-like panicles may also key here

1. Ligules absent 2
1. Ligules present 4
2 (1). Glumes lacking; lemma flabellate **623. *Neostapfia***
2. Glumes present; lemma oblong 3
3 (2). Lemma with five teeth, these often ending in awns **624. *Orcuttia***
3. Lemma entire or erose, awns lacking **625. *Tuctoria***
4 (1). Spikelets embedded in the rachis 5
4. Spikelets sessile or pedicellate, but not embedded in the rachis 7
5 (4). Lower glume present, at least 1/3 the length of the spikelet; upper glume attenuate, awned **561. *Lepturopetium***
5. The lower glume minute or absent; upper glume flattened and extended into an attenuate awn-like appendage 6
6 (5). Glume with many veins; leaves lacking glands **662. *Lepturus***
6. Glume with one to three veins; leaves with multicellular glands abaxially **679. *Oropetium***
7 (4). Spikelets with awns on the glumes or lemmas or both 8
7. Spikelets lacking awns 27
8 (7). Glumes awned or long acuminate 9
8. Glumes awnless 22
9 (8). Lemmas awnless 10
9. Lemmas awned 12
10 (9). Spikelets falling with their pedicels; leaf blade apex not muticous or pungent **557. *Decaryella***
10. Spikelets falling entire but pedicels persistent; leaf blade apex muticous to pungent 11
11 (10). Lower glume absent **590. *Zoysia***
11. Lower glume present **642. *Perotis***
12 (9). Awns twisted or flexuous 13
12. Awns straight 16
13 (12). Inflorescence axis forming an arc; second glume with a divergent abaxial awn **630. *Ctenium***
13. Inflorescence axis straight; second glume apex attenuate or awned from the apex or from a sinus, the awn not divergent 14
14 (13). Leaf blade apex sharp-pointed **680. *Tripogon***
14. Leaf blade apex blunt 15
15 (14). Plants annual; ligule membranous, not fringed **559. *Indopoa***
15. Plants perennial; ligule a fringed membrane **647. *Afrotrichloris***
16 (12). Ligule membranous 17
16. Ligule a fringe of hairs or prominently fringed membrane 18
17 (16). Spikelets paired, pedicellate; inflorescence dense and spike-like, the branches terminating in a spikelet **612. *Muhlenbergia* (former *Lycurus*)**
17. Spikelets solitary, sessile; inflorescence unbranched, terminating in a barren tip **560. *Kampochloa***
18 (16). Lemma broadest above the midpoint **681. *Neesiochloa***
18. Lemma broadest at or about the midpoint 19
19 (18). Lemma apex two lobed, awned from the sinus and often from the lobes as well **608. *Bouteloua* (subg. *Chondrosum*)**
19. Lemma apex entire 20
20 (19). Glumes longer than the spikelet, borne on the same side; spikelets paired **573. *Tetrachaete***

20. Glumes shorter than the spikelet, distichous; spikelets solitary 21
21 (20). Annual; callus of spikelet sharp pointed; each spikelet subtended by a hyaline bract **568. *Viguierella***
21. Perennial; callus of spikelet obtuse; spikelets without subtending bracts **583. *Fingerhuthia***
22 (8). Ligule membranous, not ciliolate; lemma with three awns; flowers unisexual **608. *Bouteloua***
22. Ligule a fringe of hairs or a fringed membrane; lemma with one awn; flowers bisexual 23
23 (22). Plants stoloniferous, less than 15 cm tall 24
23. Plant caespitose, rhizomes short or absent, stolons absent, or if present then plants more than 15 cm 25
24 (25). Inflorescence borne above the leaves; spikelets turbinate; sterile flowers distal and proximal in the spikelet **602. *Pommereulla***
24. Inflorescence surrounded by leaves, surpassed by the uppermost leaf blades; spikelets ovate to lanceolate; sterile flowers distal only **621. *Munroa***
25 (23). Spikelets three per node; plants of North and Central America **610. *Hilaria***
25. Spikelets one per node; plants of the Old World 26
26 (25). Lemma with three deep lobes **648. *Astrebla***
26. Lemma entire or with two shallow teeth **670. *Tetrapogon***
27 (7). Spikelets with one flower 28
27. Spikelets with more than one flower 39
28 (27). Spikelets covered with stiff hooked prickles **615. *Tragus***
28. Spikelets variously smooth or pubescent; if bearing prickles, then these not hooked 29
29 (28). Glumes shorter than the lemmas 30
29. Glumes as long as or longer than the lemmas, enclosing the flower 32
30 (29). Plants annual; Australia **580. *Heterachne***
30. Plants perennial; Africa 31
31 (30). Leaves spine-tipped; glumes and lemmas glabrous; plants growing in sand **589. *Urochondra***
31. Leaves slender but not spine-tipped; glumes and lemmas with long hairs; plants growing in damp sites **584. *Stiburus***
32 (29). Glumes hard, ending in a sharp point, wooly to scabrous **634. *Dignathia***
32. Glumes membranaceous to coriaceous, not sharp-pointed, glabrous 33
33 (32). Inflorescence with spikelets borne on all sides 34
33. Inflorescence with spikelets borne on one side only 36
34 (33). Lower glume absent, the upper glume enclosing the flower **590. *Zoysia***
34. Both glumes present 35
35 (34). Plants perennial, rhizomatous or stoloniferous; ligule a fringed membrane **641. *Mosdenia***
35. Plants annual, caespitose; ligule membranous **642. *Perotis***
36 (33). Spikelets dorsiventrally compressed **664. *Microchloa***
36. Spikelets laterally compressed 37
37 (36). Inflorescence axis arcuate, ending in a barren tip **660. *Harpochloa***
37. Inflorescence axis straight, ending in a spikelet 38
38 (37). Ligule membranous or a fringed membrane; inflorescence surrounded by a leaf sheath or spatheole **650. *Brachyachne***
38. Ligule a fringe of hairs; inflorescence not surrounded by a sheath **599. *Lepturidium***
39 (27). Leaves stiff, divergent; plants perennial, strongly rhizomatous and often also stoloniferous 40
39. Leaves lax; plants annual or perennial, caespitose 44
40 (39). Spikelets unisexual, plant dioecious 41
40. Spikelets bisexual 42
41 (40). Female lemma tubular, fused to rachilla; glumes absent in female spikelet **611. *Jouvea***
41. Female lemma folded, not fused to rachilla; glumes present in female spikelet **609. *Distichlis***
42 (40). Inflorescence axis arcuate, ending in a barren tip **660. *Harpochloa***
42. Inflorescence axis straight, ending in a spikelet 43
43 (42). Lemmas with 5 or more veins **596. *Aeluropus***
43. Lemmas with 1–3 veins **588. *Psilolemma***
44 (39). Leaves slender and thread-like; lemmas deciduous, paleas and rachilla persistent **677. *Eragrostiella***
44. Leaves not thread-like; lemmas persistent, falling with the spikelet **579. *Eragrostis* (former *Psammagrostis, Harpachne*)**

Group II – Inflorescence with two or more unbranched branches, either digitate or spread along the main axis

1. Primary branches digitate 2
1. Primary branches spread along main inflorescence axis 26
2 (1). Spikelets without awns 3
2. Spikelets with awns 10
3 (2). Leaf blades cylindrical in cross section, with sharp tips **669. *Sclerodactylon***
3. Leaf blades flattened, sometimes folded, their tips sometimes acute but not sharp 4
4 (3). Culms with tuberous bases, the leaf sheaths densely wooly below; primary inflorescence branches disarticulating at maturity **564. *Ochthochloa***
4. Culm bases not tuberous, leaf sheaths pubescent or not, but not wooly; primary inflorescence branches persistent 5
5 (4). Leaf blades cordate at the base, with multicellular glands **673. *Coelachyropsis***
5. Leaf blades narrow at the base, not cordate, lacking multicellular glands 6
6 (5). Glumes longer than the flowers and enclosing them 7
6. Glumes shorter than to as long as the flowers, not enclosing them 8
7 (6). Spikelets dorsiventrally compressed, with one flower; plants of Africa **632. *Craspedorhachis***

7. Spikelets laterally compressed, with 3 flowers (2 fertile); plants of Brazil **666. *Rheochloa***
8 (6). Spikelets with one flower **653. *Cynodon***
8. Spikelets with more than one flower 9
9 (8). Hairs at apex of lemma slightly inflated at the tip **657. *Disakisperma***
9. Hairs at apex of lemma not inflated **675. *Eleusine***
10 (2). Awns on lemmas only; glumes lacking awns 11
10. Awns on glumes and lemmas 20
11 (10). Lemmas of the distal flowers of the spikelet broadly flaring, fan-shaped above **665. *Oxychloris***
11. Lemmas of the distal flowers of the spikelet reduced, not flaring or fan-shaped 12
12 (11). Lemma divided about ¼ of its length into two large lobes, with prominent lateral veins extending into the lobes **663. *Lintonia***
12. Lemma apex entire or bidentate, if the latter then the lobes less than ¼ the length of the lemma 13
13 (12). Lower glume tiny, less than 1/6 the length of the upper glume **667. *Saugetia***
13. Lower glume well developed 14
14 (13). Glumes membranous to leathery, lemmas less firm than the glumes, hyaline or thinly membranous **639. *Lophacme***
14. Glumes hyaline to membranous, less firm than the membranous, leathery or indurate lemmas 15
15 (14). Rachilla extension naked **654. *Daknopholis***
15. Rachilla extension bearing rudimentary flower(s) 16
16 (15). Spikelet with 2 to 5 fertile flowers 17
16. Spikelet with 1 fertile flower 18
17 (16). Lemma with hyaline wings **563. *Neostapfiella***
17. Lemma without wings **670. *Tetrapogon***
18 (16). Spikelet lanceolate, elliptical, or ovate **658. "*Enteropogon*"**
18. Spikelet cuneate 19
19 (18). Spikelets dorsiventrally compressed **649. *Austrochloris***
19. Spikelets laterally compressed **651. "*Chloris*"**
20 (10). Each inflorescence branch ending in a prominent sharp point; upper glume terminated by an awn bent at a sharp angle to the long axis of the glume **604. *Dactyloctenium***
20. Racemes not ending in a sharp point; glume awn straight or slightly curved, but not stiffly bent at a sharp angle 21
21 (20). Upper glume with 5 to 7 veins **561. *Lepturopetium***
21. Upper glume with 1 vein 22
22 (21). Lemma with three awns from the tip **671. *Trichloris***
22. Lemma with a single awn 23
23 (22). Awn much shorter than the body of the lemma 24
23. Awn much longer than the body of the lemma 25
24 (23). Spikelets disarticulating between the glumes, the upper deciduous, the lower persistent; inflorescence branches 2 to 6 **652. *Chrysochloa***
24. Spikelets disarticulating below both glumes; inflorescence with as many as 40 branches **659. "*Eustachys*"**
25 (23). Spikelets dorsiventrally compressed **658. "*Enteropogon*"**
25. Spikelets laterally compressed **668. *Schoenefeldia***
26 (1). Primary inflorescence branches stiff, divaricate, ending in sharp point; plants forming spiny mounds, shrubby **572. *Cladoraphis***
26. Primary inflorescence branches lax or not; if divaricate, then not ending a sharp point; plants not forming spiny mounds, not shrubby 27
27 (26). Spikelets without awns 28
27. Spikelets with awns 52
28 (27). Lateral branches bearing no more than two spikelets; glumes longer than flowers, indurate 29
28. Lateral branches with four or more spikelets; glumes shorter or longer than the flowers, coriaceous or membranous, not indurate 31
29 (28). Inflorescence borne above the distalmost leaf, not surrounded by a sheath; axis of lateral branches flattened, broader distally than proximally **635. *Leptothrium***
29. Inflorescence surrounded by a leaf sheath, this often inflated; axis of lateral branches terete 30
30 (29). Lower glume well developed; upper glume tuberculate **566. *Pseudozoysia***
30. Lower glume less than 0.2 the length of the upper glume or absent; upper glume smooth or covered with hooked spines but not tuberculate **615. *Tragus***
31 (28). Spikelets partially embedded in the lateral branches, scarcely overlapping; the inflorescence disarticulating at the base and forming a tumbleweed dispersal unit **612. *Muhlenbergia* (former *Schedonnardus*)**
31. Spikelets not embedded in the lateral branches, often overlapping at least half their length or more, often densely packed; inflorescence axis not disarticulating at the base 32
32 (31). Glumes covered with stiff spines **615. *Tragus***
32. Glumes not covered with spines 33
33 (32). First glume reduced to a short slender stub **578. *Catalepis***
33. First glume well developed 34
34 (33). Spikelets disarticulating below the glumes and falling entire 35
34. Spikelets disarticulating above the glumes 38
35 (34). Inflorescence branches generally shorter than the internodes **586. *Tetrachne***
35. Inflorescence branches generally longer than the internodes 36
36 (35). Glumes as long as or longer than the spikelet; spikelet with one fertile flower **593. *Spartina*** (now part of *Sporobolus* s.l.)
36. Glumes shorter than the spikelet; spikelet with 2 or more fertile flowers 37
37. Lemma cartilaginous or leathery; lower flowers of the spikelet sterile **587. *Uniola***
37. Lemma membranous; lower flowers of the spikelet fertile **676. *Desmostachya***
38 (34). Flowers unisexual 39
38. Flowers bisexual 41

39 (38). Spikelets with 1 fertile flower and 2 to 3 sterile ones **608. *Bouteloua* (former *Pringleochloa*)**
39. Spikelets with 4 or more fertile flowers 40
40 (39). Palea as long as or longer than the lemma **607. *Allolepis***
40. Palea shorter than the lemma **619. *Scleropogon***
41 (38). Callus of flower with dense straight hairs at least ¼ as long as the flower; leaf blades disarticulating from the sheath 42
41. Callus of flower glabrous, or pubescent but with hairs less than ¼ the length of the flower; leaf blades not disarticulating 43
42 (41). Spikelet less than 1 cm long **606. *Odyssea***
42. Spikelet 1–2 cm long **558. *Halopyrum***
43 (41). Spikelets with one flower 44
43. Spikelets with more than one flower 45
44 (43). Ligule a fringed membrane; inflorescence contracted, the branches upright **616. *Polevansia***
44. Ligule membranous; inflorescence open, the branches spreading **640. *Trigonochloa***
45 (43). Fruit flattened, concavo-convex, the surface rough **674. *Coelachyrum***
45. Fruit not flattened, the surface smooth 46
46 (45). Upper glume and lemmas with 7 or 9 veins; palea splitting at maturity **629. *Vaseyochloa***
46. Upper glume and lemmas with fewer than 7 veins; palea not splitting 47
47 (46). Palea bowed outward away from flower; second glume with 5 veins **603. *Brachychloa***
47. Palea flat or curved inward toward flower; second glume with 1 to 3 veins 48
48 (47). Lemma covered with dense long hairs **600. *Orinus***
48. Lemma with hairs only on the veins, or glabrous 49
49 (48). Ligule apex acute to attenuate **656. *Diplachne***
49. Ligule apex obtuse to truncate 50
50 (49). Apex of lemma hairs slightly inflated at the tip **657. *Disakisperma***
50. Apex of lemma hairs not inflated 51
51 (50). Primary inflorescence branches two or more per node **661. *Leptochloa***
51. Primary inflorescence branches one per node **655. *Dinebra***
52 (27). Primary inflorescence branches ending in a sterile point 53
52. Primary inflorescence branches ending in a spikelet 55
53 (52). Leaves with abaxial glands; spikelet with 6 to 12 fertile flowers; Zambia **581. *Richardsiella***
53. Leaves without glands; spikelet with 1 fertile flower; New World 54
54 (53). Spikelets with a rachilla extension bearing 1 or more reduced flowers **608. *Bouteloua***
54. Spikelets lacking a rachilla extension **593. *Spartina*** (now part of *Sporobolus* s.l.)
55 (52). Each primary inflorescence branch bearing one fertile spikelet plus two spikelets reduced to awns; lemma of fertile spikelet minute, pubescent **633. *Farrago***
55. Primary inflorescence branches with various numbers of spikelets, but not as in *Farrago*; lemma of fertile spikelet well developed 56
56 (55). Lemma awnless; awns on glumes only 57
56. Lemma with one or more awns 58
57 (56). Spikelets with 1 flower; glumes as long as or longer than the flower **593. *Spartina*** (now part of *Sporobolus* s.l.)
57. Spikelets with 4 or more flowers; glumes shorter than the flowers **674. *Coelachyrum***
58 (56). Lemma with multiple awns 59
58. Lemma with one awn 64
59 (58). Awns much longer than spikelets, over 5 cm long **619. *Scleropogon* (pistillate inflorescences)**
59. Awns about the same length as or shorter than spikelets, ca. 5 mm long or less 60
60 (59). All flowers unisexual **608. *Bouteloua***
60. At least some flowers bisexual 61
61 (60). Glumes awnless, although sometimes setaceous 62
61. Glumes awned 63
62 (61). Spikelets with 2 proximal sterile flowers **618. *Blepharidachne***
62. Spikelets with 2 proximal fertile flowers **605. *Neobouteloua***
63 (61). Glumes glabrous, hyaline except for the midvein and awn **612. *Muhlenbergia***
63. Glumes with long hairs, coriaceous **678. *Melanocenchris***
64 (58). Glumes longer than the spikelet, enclosing the flowers 65
64. One or both glumes shorter than the spikelet, not enclosing the flowers 71
65 (64). Plants less than 15 cm tall, growing in tufts connected by stolons; inflorescence dense, capitate **621. *Munroa***
65. Plants over 15 cm, not stoloniferous; inflorescence not capitate 66
66 (65). Spikelets falling entire along with their pedicels; glumes much more firm than the lemmas, covered with long hairs or prickles **634. *Dignathia***
66. Spikelets disarticulating above one or both glumes, pedicels and branches persistent; glumes less firm than the lemmas, not pubescent or prickly 67
67 (66). Spikelets terete in cross section; leaf blades with multicellular glands **601. *Pogoneura***
67. Spikelets laterally compressed; leaf blades lacking glands 68
68 (67). Spikelets not or scarcely overlapping on the inflorescence branches, shorter than the internodes **637. *Gymnopogon***
68. Spikelets overlapping on the inflorescence branches, longer than the internodes 69
69 (68). Awn shorter than the lemma, generally less than 1 mm long **631. *Trichoneura***
69. Awn the same length as or appreciably longer than the lemma 70
70 (69). Spikelets disarticulating above the glumes; lemma awn (3) 5–15 mm long **626. *Gouinia***
70. Spikelets disarticulating above the first glume and below the second; lemma awn 15–25 mm long **565. *Pogonochloa***
71 (64). Ligule a fringe of hairs or a fringed membrane 72
71. Ligule membranous, not fringed 76

72 (71). Spikelets with two to many flowers, laterally compressed; leaf blade margins membranous 73
72. Spikelets with one flower, dorsiventrally compressed; leaf blade margins cartilaginous 75
73 (72). Plants aquatic, often floating; spikelets borne on both sides of primary inflorescence branches **562. *Myriostachya***
73. Plants terrestrial; spikelets borne on one side of primary inflorescence branches 74
74 (73). Inflorescence branches deciduous **579. *Eragrostis* (former *Pogonarthria*)**
74. Inflorescence branches persistent **646. *Acrachne***
75 (72). Lower glume ½ to ¾ the length of the upper glume; upper glume apex obtuse to acute, not awned **617. *Willkommia***
75. Lower glume about 1/10 the length of the upper glume; upper glume apex acuminate, awned **614. *Monelytrum***
76 (71). Clusters of fertile spikelets subtended by involucral bristles **612. *Muhlenbergia* (former *Pereilema*)**
76. Bristles lacking 77
77 (76). Plants prostrate, stoloniferous; inflorescences dense, capitate; glumes leaf-like, with 9–11 veins **655. *Dinebra***
77. Plants upright; inflorescences not capitate; glumes similar in texture and morphology to lemmas, with one vein 78
78 (77). Plants annual 79
78. Plants perennial 80
79 (78). Inflorescence branches not densely crowded; lemmas deciduous, palea persistent on the rachilla **646. *Acrachne***
79. Inflorescence dense, the primary branches crowded; lemmas persistent on the rachilla, falling with the rest of the flower **636. *Leptocarydion***
80 (78). Spikelets with fewer than 4 flowers; lemma awned from the abaxial side **638. *Bewsia***
80. Spikelets with 6 or more flowers; lemma awned from the apex **567. *Silentvalleya***

Group III – Inflorescence branches themselves branched, at least some spikelets borne on secondary or higher-order branches, the spikelets mostly pedicellate

1. Ligule absent, the leaf and sheath intergrading **623. *Neostapfia***
1. Ligule present 2
2 (1). Leaves stiff, sharp-tipped, often strongly divaricate 3
2. Leaves flexible, not sharp-tipped 8
3 (2). Spikelets with exactly one flower; reduced flowers lacking 4
3. Spikelets with more than one flower, although sometimes only one fertile 5
4 (3). Spikelets 2–3 mm long; lemma apex obtuse, mucronate **589. *Urochondra***
4. Spikelets 16–24 mm long; lemma apex two-lobed, awned from the sinus **644. *Monodia***
5 (3). Lemma divided into deep broad lobes 6
5. Lemma apex acute to acuminate, not lobed 7
6 (5). Glumes shorter than the flowers **643. *Triodia***
6. Glumes longer than the flowers **555. *Ellisochloa***
7 (5). Lemmas awned; plants of Australasia **645. *Symplectrodia***
7. Lemmas mucronate; plants of California **620. *Swallenia***
8 (2). Spikelets without awns 9
8. Spikelets with awns 23
9 (8). Spikelets with one flower 10
9. Spikelets with more than one flower 13
10 (9). Plants strongly stoloniferous, low, spreading; inflorescence dense, capitate or spike-like, nearly spherical, surrounded by upper leaves **592. *Crypsis*** (now part of *Sporobolus* s.l.)
10. Plants rhizomatous or caespitose, erect; inflorescence narrow or spreading, but not capitate, borne well above the upper leaves 11
11 (10). Callus of flower glabrous **594. *Sporobolus***
11. Flower with a bearded callus 12
12 (11). Inflorescence narrow and dense like a bottle-brush, the branches short and upright, closely pressed to the main axis **584. *Stiburus***
12. Inflorescence broad to narrow, the branches long, not tightly appressed to the main axis **591. *Calamovilfa*** (now part of *Sporobolus* s.l.)
13 (9). Ligule membranous 14
13. Ligule a fringe of hairs or a membrane with long cilia 16
14 (13). Spikelets with one or two flowers, the lemmas glabrous **612. *Muhlenbergia***
14. Spikelets with more than two flowers; lemmas generally with short hairs on keel and margins 15
15 (14). Paleas widened or bowed below; caryopses not deeply concave **683. *Tridens***
15. Paleas not widened or bowed below; caryopses deeply concave to folded on the side of the hilum **627. *Tridentopsis***
16 (13). Glumes with 5 or more veins 17
16. Glumes with 1 or 3 veins 18
17 (16). Flowers bisexual; palea splitting at maturity **629. *Vaseyochloa***
17. Flowers unisexual, the plants dioecious; palea not splitting **607. *Allolepis***
18 (16). Lemmas papery, laterally flattened, broadly winged on the keel **580. *Heterachne***
18. Lemmas dorsiventrally or laterally compressed but not flat, papery or winged 19
19 (18). Lemmas glabrous or evenly pubescent all over, apex acute 20
19. Lemmas pubescent on the veins only, the apex emarginate to bilobed, sometimes also mucronate 21
20 (19). Primary inflorescence branches generally longer than the internodes separating them, generally spreading; lemma with 3 veins **579. *Eragrostis***
20. Primary inflorescence branches about the same length as the internodes separating them, upright and appressed to the main inflorescence axis; lemma with 1 vein **595. *Thellungia*** (now part of *Sporobolus* s.l.)
21 (19). Palea keels glabrous or ciliolate; cleistogamous spikelets lacking 22
21. Palea keels long ciliate; lower leaf sheaths containing cleistogamous spikelets **628. *Triplasis***

22 (21). Paleas widened or bowed below; caryopses not deeply concave **683. *Tridens***
22. Paleas not widened or bowed below; caryopses deeply concave to folded on the side of the hilum **627. *Tridentopsis***
23 (8). Awns two or more per lemma 24
23. Awns one per lemma, or lemmas unawned and awns on glumes only 30
24 (23). Lemmas with two or three, rarely four, awns 25
24. Lemma with 5 awns or more 27
25 (24). Lemma awns two, one on each marginal lobe, occasionally with 2 more subsidiary awns, the lemma apex rounded between the awns **576. *Kaokochloa***
25. Lemma awns three, one on each marginal lobe and one from the lemma apex 26
26 (25). Plants with bisexual flowers; central lemma awn recurved **571. *Triraphis***
26. Plants with unisexual flowers, dioecious; central lemma awn straight **613. *Sohnsia***
27 (24). Glumes with one vein; lemma with 12 awns or more **682. *Pappophorum***
27. Glumes with seven veins or more; lemma with 5–11 awns 28
28 (27). Awns 5, alternating with hyaline lobes **577. *Schmidtia***
28. Awns 7 to 11, not alternating with hyaline lobes 29
29 (28). Awns with antrorse prickles but not with long hairs; rachilla disarticulating between the flowers **574. *Cottea***
29. Awns plumose; rachilla not disarticulating **575. *Enneapogon***
30 (23). Spikelets with exactly one flower 31
30. Spikelets with two or more flowers 33
31 (30). Ligule membranous; lemma with 3 veins **612. *Muhlenbergia***
31. Ligule a line of hairs; lemma with 1 vein 32
32 (31). Lemma with a long awn **598. *Hubbardochloa***
32. Lemma awnless; awn only on the glume **594. *Sporobolus***
33 (30). Glumes awned 34
33. Glumes unawned 38
34 (33). Ligule membranous 35
34. Ligule a line of hairs 36
35 (34). Glumes with divided awns; spikelets in pairs **612. *Muhlenbergia* (former *Lycurus*)**
35. Glumes long acuminate; spikelets solitary **672. *Apochiton***
36 (34). Plants annual; awns on lemmas much longer than the lemma itself **569. *Habrochloa***
36. Plants perennial; awn on lemma shorter than the lemma itself 37
37 (36). Panicle lax and open, the branches generally not spikelet-bearing to the base, most spikelets pedicellate **612. *Muhlenbergia* (former *Redfieldia*)**
37. Panicle dense, spike-like, the branches generally spikelet-bearing to the base, most spikelets sessile or nearly so **583. *Fingerhuthia***
38 (33). Plants tall, reed-like **570. *Neyraudia***
38. Plants shorter, not reed-like 39
39 (38). Disarticulation below the glumes, spikelets falling entire **585. *Entoplocamia***
39. Disarticulation above the glumes 40
40 (39). Callus of flower elongate, sharp **556. *Centropodia***
40. Callus of flower short, blunt 41
41 (40). Lemma lobes about ½ the length of the lemma 42
41. Lemma apex entire or with shallow lobes, less than ½ the length of the lemma 43
42 (41). Awn pubescent **628. *Triplasis***
42 Awn glabrous **622. *Erioneuron***
43 (41). Cleistogamous spikelets forming in the leaf sheaths 44
43. All spikelets chasmogamous 45
44 (43). Lemma with 5 veins; palea lacking wings **597. *Cleistogenes***
44. Lemma with 3 veins; palea with wings **582. *Steirachne***
45 (43). Lemma glabrous; upper lemmas of each spikelet reduced to a cluster of awns **579. *Eragrostis* (former *Ectrosia*)**
45. Lemma pubescent on the nerves or all over the proximal abaxial side; upper lemmas of each spikelet not reduced to awns 46
46 (45). Awn about the same length as the lemma, or up to 4 times longer **626. *Gouinia***
46. Awns generally much shorter than the lemma **683. *Tridens***

Tribes and Genera of Chloridoideae

XXV. Tribe Centropodieae P.M. Peterson, N.P. Barker & H.P. Linder (2011)

Perennials, caespitose or rhizomatous. Inflorescence branches branched. Callus of flower pubescent. Lemma with 9 veins, with hairs tufted, in a transverse row, the apex with two deep lobes, awned from the sinus, the awn straight or geniculate.

Two genera, five species.

555. *Ellisochloa* P. M. Peterson & N. P. Barker

Ellisochloa P. M. Peterson & N. P. Barker, Taxon 60: 1118 (2011).

Plants less than 50 cm tall. Ligule a fringe of hairs. Leaf blades stiff, terete, the adaxial surface represented only by a shallow groove, the apex sometimes sharp-pointed. Inflorescence branches upright. Spikelets with 2 or 3 flowers. Glumes longer than the flowers, their apices acute to acuminate. Lemma with 4 to 6 hair tufts. C_3.

Two spp., southern Africa.

The two species of *Ellisochloa*, *E. rangei* (Pilg.) P. M. Peterson & N. P. Barker and *E. papposa* (Nees) P. M. Peterson & N. P. Barker, were formerly included in *Merxmuellera*, but multiple studies have shown that they were misplaced in that genus; they clearly belong here in the clade sister to the remainder of the Chloridoideae. *Merxmuellera* s.s. is a member of subfamily Danthonioideae.

556. ***Centropodia*** Rchb. Fig. 82

Centropodia Rchb., Consp. Regn. Veg. 212a (1828).

Plants caespitose or rhizomatous. Ligule a fringed membrane or a fringe of hairs. Leaf blades disarticulating at the ligule. Inflorescence exserted or surrounded by the subtending sheath. Spikelets with 2 to 6 flowers plus additional reduced flowers distally. Callus of flower sharp. Glumes shorter or longer than the flowers, their apices acute to acuminate. Lemma with lines of hairs between the veins, hairs tufted above, the awn straight or twisted. Leaf epidermal papillae absent. Microhairs panicoid or chloridoid. C_4, anatomically similar to NAD-ME subtype. $2n = 24, 48$.

Four spp., Africa, Middle East, northern India.

Remaining genera:. Photosynthetic pathway C_4.

Core C_4 Chloridoideae incertae sedis

557. ***Decaryella*** A. Camus

Decaryella A. Camus, Bull. Soc. Bot. France 78: 177 (1931).

Decumbent annuals, less than 30 cm tall. Leaf blades narrow, stiff. Ligule a fringed membrane. Inflorescence unbranched. Spikelets on long pedicels, *falling with the pedicel at maturity*, laterally compressed, with 1 or 2 fertile flowers, rachilla extension present or absent. Glumes longer than the flower, leathery, hairy, cuspidate-acuminate. Lemma obtuse or acute. Leaf epidermal intercostal papillae absent.

One sp., *D. madagascariensis* A. Camus, Madagascar.

Fig. 82. Chloridoideae-Centropodieae. *Centropodia glauca*. **A** Habit. **B** Ligule. **C** Spikelet. (From Müller 2007)

558. ***Halopyrum*** Stapf

Halopyrum Stapf, Hooker's Icon. Pl. 25: t 2448 (1896).

Perennials with robust stolons. Ligule a fringe of hairs. Leaf blades with filiform tips, disarticulating from sheath. Inflorescence with primary branches unbranched, appressed to the main axis. Spikelets with 8 to 25 flowers. Callus of flower with straight hairs extending almost half the length of the spikelet. Glumes shorter than the spikelet. Lemma mucronate. *Palea covered with long hairs*. Leaf epidermal papillae one per cell.

One sp., *H. mucronatum* (L.) Stapf, coastal areas of Africa, India, Sri Lanka.

The inflorescence and the single epidermal papilla per cell suggest a possible placement in Eleusininae.

559. *Indopoa* Bor

Indopoa Bor, Kew Bull. 13: 225 (1958).

Tiny annuals, less than 15 cm tall. Ligule membranous. Inflorescence unbranched, with fewer than 10 spikelets. Spikelets with 4 to 6 fertile flowers plus additional distal sterile ones. Callus of flower pubescent. Glumes shorter than the spikelet. Lemma apex with two lobes, awned from each lobe and from the sinus, *the median awn very broad, twisted and curving around the fruit. Fruit slender, needle-like, with an apical appendage. Endosperm unusually hard.* Leaf epidermal papillae absent.

One sp., *I. paupercula* (Stapf) Bor, India.

Indopoa was segregated from *Tripogon* and its placement is uncertain. The awn morphology is reminiscent of Danthonioideae or early-diverging Chloridoideae; the species has not been included in a molecular phylogeny.

560. *Kampochloa* Clayton

Kampochloa Clayton, Kew Bull. 21: 103 (1967).

Perennials, less than 30 cm tall. Ligule membranous. Inflorescence unbranched, ending in a bare tip, the axis curved. Spikelet with one fertile flower, one staminate or sterile one distally, and a group of lemmas with long awns terminating the rachilla. Glumes shorter than the spikelet, the lower glume terminating in a short awn from a sinus, the upper with a longer abaxial awn. Lemma with long hairs along the veins, the apex with two teeth, mucronate. Leaf epidermal papillae absent.

One sp., *K. brachyphylla* Clayton, Angola and Zambia.

Peterson et al. (2014a) speculate that *Kampochloa* could be derived from *Ctenium*, based on morphological similarity.

561. *Lepturopetium* Morat

Lepturopetium Morat, Adansonia II, 20: 377 (1981).

Stoloniferous perennials. Ligule a fringed membrane. Inflorescence unbranched, or with 2 or 3 unbranched branches; branch axes hollowed, the spikelets embedded in the rachis. Spikelets laterally compressed, with 1 fertile flower and 1 sterile lemma; rachilla hairy. Upper glume longer than the flower. Fertile lemma hairy, two-lobed, awned from sinus. Leaf epidermal papillae one to several per cell.

Two spp., New Caledonia, Marshall Islands, Cocos Islands.

562. *Myriostachya* (Benth.) Hook. f.

Myriostachya (Benth.) Hook. f., Fl. Brit. India 7: 327 (1897).

Perennial with culms 1.5 to 3 m long, *often floating*. Ligule a fringe of hairs. Inflorescence branches unbranched, disarticulating from the elongate central axis. Spikelets with 2 to 12 flowers plus additional distal sterile ones. Glumes shorter than the spikelet, awned. Lemma glabrous, mucronate. Leaf epidermal papillae absent.

One sp., *M. wightiana* (Nees ex Steud.) Hook. f., southern India and Sri Lanka to southeast Asia.

563. *Neostapfiella* A. Camus

Neostapfiella A. Camus, Bull. Soc. Linn. Lyon 5: 4 (1926).

Annuals, less than 30 cm tall, with or without stolons. Ligule a fringed membrane. Inflorescence unbranched, or with 2 to 4 unbranched digitate branches. Spikelets wedge-shaped, with 2 fertile flowers, or the upper flower male or sterile. Callus of flower elongate, pubescent, sharp pointed. Glumes shorter than the spikelet, thinner than the lemma. Lemma two-toothed, with a subapical awn. Leaf epidermal papillae present or absent; when present, several per cell.

Three spp., Madagascar.

564. *Ochthochloa* Edgew.

Ochthochloa Edgew., J. Asiat. Soc. Bengal 11: 26–27 (1842).

Stoloniferous perennials; *culms with tuberous bases*. Leaf sheaths wooly near the base. Ligule a fringed membrane. Primary branches of inflorescence themselves unbranched, digitate, disarticulating at maturity. Spikelets with 4 to 8 fertile flowers plus additional distal sterile ones. Glumes shorter than the spikelet, unawned. Lemma with hairs on veins and keel, the apex muticous.

Pericarp free from the seed coat. Leaf epidermal intercostal papillae present, one or more per cell.

One sp., *O. compressa* (Forssk.) Hilu, NE Africa, Iran, India.

565. ***Pogonochloa*** C. E. Hubb.

Pogonochloa C. E. Hubb., Hooker's Icon. Pl. 35: t. 3421 (1940).

Perennials. Ligule membranous. Inflorescence of numerous unbranched branches, these appressed to the central axis. Spikelets with two or three flowers, the lowest bisexual, the next staminate or sterile, the most distal only an awn. Callus of flower pubescent. Glumes longer than the spikelet, with short awns. Lemma pubescent, awned from the apex, the awn flexuous, many times longer than the flower. Leaf epidermal papillae absent.

One sp., *P. greenwayi* C. E. Hubb., Zambia and Zimbabwe.

566. ***Pseudozoysia*** Chiov.

Pseudozoysia Chiov., Pl. Nov. Minus Not. Aethiopia 20 (1928).

Tiny perennials, less than 15 cm tall. Ligule a fringe of hairs. Inflorescence surrounded by the subtending leaf sheath, the branches unbranched, each branch with only 2 spikelets, deciduous from the main axis. Spikelets with 1 flower, without rachilla extension. Glumes longer than the flower, hard, tuberculate, acute, the lower ovate, the upper subglobose and saccate. Lemma awnless, hairless.

One sp., *P. sessilis* Chiov., Somalia.

Little information is available on this species. Anatomical and molecular data would be particularly useful.

567. ***Silentvalleya*** V.J. Nair, Sreek., Vajr. & Bhargavan

Silentvalleya V.J. Nair, Sreek., Vajr. & Bhargavan, J. Bombay Nat. Hist. Soc 79: 654 (1982).

Caespitose perennials. Ligule membranous. Inflorescence branches unbranched, the branches arranged along a central axis. Spikelets with 6 or more flowers plus additional distal sterile ones. Callus of flowers pubescent. Glumes shorter than the spikelet, the apices acute to acuminate. Lemma glabrous, with an apical awn. Leaf epidermal papillae absent.

One sp., *S. nairii* Nair, Sreek., Vaj. & Bargh., southern India.

Clayton and Renvoize (1986) say that this is "like *Gouinia* [in Cynodonteae], but upper glume as long as lemma". Also the lemma is hairy in *Gouinia*.

568. ***Viguierella*** A. Camus

Viguierella A. Camus, Bull. Bi-Mens. Soc. Linn. 5: 11 (1926).

Annuals. Ligule a fringe of hairs. Inflorescence narrow, bottlebrush like, generally unbranched. *Spikelet subtended by a small bract*, with 1 flower plus additional distal sterile ones, *the entire spikelet plus the pedicel disarticulating as a unit.* Flower callus pubescent, sharp pointed. Glumes pilose, awned, shorter than the spikelet. Lemma awned. Leaf epidermal papillae absent.

One sp., *V. madagascariensis* A. Camus, Madagascar.

Clayton and Renvoize (1986) connect this species with *Scleropogon*, in Cynodonteae, on the basis of the subtending bract.

XXVI. Tribe Triraphideae P.M. Petersen (2010)

Annuals or perennials. Ligule a fringe of hairs. Inflorescence branches branched. Glumes shorter than the spikelet. Lemmas keeled, pubescent on the lateral veins. Leaf epidermal papillae absent. Microhairs slender, with an elongate apical cell ("panicoid type").

Three genera, 13 species.

569. ***Habrochloa*** C. E. Hubbard

Habrochloa C. E. Hubbard, Hooker's Icon. Pl. 37: t. 3645 (1967).

Slender annuals. Spikelets with 3 to 5 flowers, the flowers with a pubescent callus. Glumes as long as the spikelet, with acuminate apices. Callus of flower reported to be pubescent. Lemma bilobed, the apex truncate, with a long awn from the sinus.

One sp., *H. bullockii* C. E. Hubb., Central Africa.

Habrochloa is placed here following Peterson (pers. comm.).

570. ***Neyraudia*** Hook. f. Fig. 83

Neyraudia Hook. f., Fl. Brit. India 7: 305 (1897).

Caespitose or short-rhizomatous, reed-like perennials. Leaf blades disarticulating or not. Inflorescence large, plumose. *Spikelets with several flowers, the proximal one sterile, the next 3 to 8 fertile, and the distalmost reduced.* Glume apices acute to acuminate. Lemma with two lobes, awned from the sinus. $2n = 40$.

Five spp., Asia, Africa.

Fig. 83. Chloridoideae-Triraphideae. *Neyraudia montana.* **A** Leaves and inflorescence. **B** Spikelet with flowers disarticulated above two glumes. **C** Flower, *left* abaxial view, and *right* adaxial view. **D** Palea, *left* abaxial view, *right* adaxially. (From Wu et al. 2007, with permission from the Missouri Botanical Garden Press, St. Louis, and Science Press, Beijing)

571. ***Triraphis*** R. Br.

Triraphis R. Br., Prodr.: 185 (1810).

Annuals or perennials, caespitose or rhizomatous. Spikelets with 4 to 24 flowers, with additional reduced flowers distally. Callus of flower pubescent. Glume apices entire or bifid, mucronate or awned. Lemmas with three lobes, each ending in an awn; *central lobe with two teeth.* Fruit linear, trigonous. $2n = 20$.

Eight spp., Africa, Australia, South America.

XXVII. TRIBE ERAGROSTIDEAE **Stapf (1898)**

Annuals or perennials, caespitose, rhizomatous, or stoloniferous. Leaves with or without glands. Ligule a fringe of hairs, less often membranous or a fringed membrane. Inflorescence branched or unbranched. Spikelets laterally compressed, with 1 to many flowers, the distal flowers often reduced. Callus of flowers generally pubescent. Glumes shorter than the spikelet in most species, rarely the same length, generally lacking awns. Flowers generally bisexual, rarely unisexual and the plants then dioecious. Lemma glabrous in most species, with three (to 13) veins, the apex generally entire, rarely mucronate or awned. Pericarp free from the seed coat or not. Leaf epidermal papillae absent in most species (except *E. obtusiflora*). Microhairs slender, with an elongate apical cell ("panicoid type"), or with a broad apical cell ("chloridoid type"), or with a broad apical cell with internal membranes ("Enneapogon type").

INCERTAE SEDIS

572. ***Cladoraphis*** Franch.

Cladoraphis Franch., Bull. Mens. Soc. Linn. Paris 1: 673 (1887).

Shrubby perennials. *Leaves with blades mostly absent or strongly reduced.* Ligule a fringe of hairs. *Inflorescence axis and primary branches woody, stiff, ending in a sharp point,* the primary branches distributed along an elongate inflorescence axis. Secondary branches short, not woody, but spine-tipped like the primary ones. Spikelets with 3 to 12 flowers plus additional reduced ones distally. Glumes shorter than the spikelet, their apices acute. Lemma apex acute. Pericarp free from seed coat. $2n = 20, 40$.

Two spp., South Africa, Namibia.

573. ***Tetrachaete*** Chiov.

Tetrachaete Chiov., Annuario Reale Ist. Bot. Roma 8: 28 (1903).

Tiny annuals, less than 25 cm tall. Ligule a fringed membrane. Inflorescence surrounded by the subtending leaf sheath, the branches unbranched, each branch with only 2 spikelets, deciduous from the main axis. Spikelets with 1 flower, without rachilla extension. Glumes much longer than the flower, awn-like, bending outward, with long feathery hairs. Lemma ovate, gibbous, leathery, with a long awn. Leaf epidermal papillae absent.

One sp., *T. elionuroides* Chiov., Ethiopia to Tanzania and Arabia.

Tetrachaete is placed in Eragrostideae following Soreng et al. (2012 and onward); the data supporting the placement are unpublished.

XXVII.1. Subtribe Cotteinae Reeder (1965)

Lemmas pubescent, lobed, with five or more awns. Spikelets with several flowers, incomplete flowers distal. *Glumes with multiple veins. Basal cell of bicellular microhairs elongate,* 60–180 µm.

Four genera, 28 spp.

Members of this subtribe have been placed in the tribe Pappophoreae in the past, based on the multiple awns of the lemma. Molecular data, however, are unequivocal and consistent in showing this character as being convergent and the traditional Pappophoreae as being polyphyletic (Columbus et al. 2007; GPWG II 2012; Peterson et al. 2010a; Reutemann et al. 2011).

574. ***Cottea*** Kunth

Cottea Kunth, Révis. Gramin. 1: 84 (1829).

Perennials. *Culms swollen below, where associated with cleistogamous spikelets.* Ligule a fringe of hairs. Inflorescence branches branched, borne upright and close to the axis. Spikelets with 4 to 8 fertile flowers, with additional reduced flowers distally. Glumes shorter than the spikelet, pubescent, the apices acute. Rachilla disarticulating between the flowers. Lemmas with 7 to 11 lobes ending in awns, plus two lateral lobes without awns, the awns with antrorse prickles. $2n = 20$.

One sp., *C. pappophoroides* Kunth, Southern USA to Argentina.

575. ***Enneapogon*** Desv. ex P. Beauv. Fig. 84

Enneapogon Desv. ex P. Beauv., Ess. Agrostogr. 81, 161 (1812).

Annuals or perennials, caespitose or short rhizomatous. Ligule membranous, or a fringed membrane, or a fringe of hairs. Inflorescence unbranched, or with unbranched primary branches, these digitate or arranged along a central axis. Spikelets with one fertile flower, with additional reduced flowers distally; *upper flowers reduced and internodes contracted so the assembled lemmas look like a brush.* Glumes shorter or longer than the spikelet, the lower one with a sharp tip, the upper one mucronate or awned. Lemmas with 9 veins, each ending in an awn; *awns much longer than the lemma and plumose above. Leaf blade microhairs with an inflated terminal cell, with internal membranes.* $2n = 18, 20, 36, 40, 60$.

Twenty-four spp., tropical and subtropical regions worldwide.

576. ***Kaokochloa*** de Winter

Kaokochloa de Winter, Bothalia 7: 479 (1961).

Annuals. Leaf blades viscid. Inflorescence branches branched. Spikelets with 3 to 6 flowers, with additional reduced flowers distally; rachilla disarticulating between the glumes, the lower glume persisting on the plant and the upper falling with the flowers. Glumes as long as the spikelet, their apices acute. Callus of flower pubescent. *Lemmas with 2 lobes ending in awns; apex rounded between the awns, curved inward.*

One sp., *K. nigrirostris* De Winter, southwestern Africa.

577. ***Schmidtia*** Steud. ex J.A. Schmidt

Fig. 84. Chloridoideae-Eragrostideae-Cotteinae. *Enneapogon desvauxii*. **A** Plant. **B** Spikelet. **C** Flower. **D** Cleistogamous spikelet. (From Barkworth et al. 2003, drawn by K. Klitz)

Schmidtia Steud. ex J.A. Schmidt, Beitr. Fl. Cap. Verd. Ins. 144 (1852), nom. conserv.

Annuals or perennials, caespitose or short rhizomatous. Ligule a fringe of hairs. Inflorescence branches branched. Spikelets with 3 to 7 flowers, with additional reduced flowers distally. Glumes shorter than the spikelet. *Lemmas with 6 lobes alternating with 5 awns, the lateral awns only slightly shorter than the central one.* $2n = 36$.

Two spp., Africa and Pakistan.

The two species of *Schmidtia* are paraphyletic in the analyses of Reutemann et al. (2011), with *Enneapogon* derived from within them; the two genera should perhaps be combined.

XXVII.2. Subtribe Eragrostidinae J. Presl (1830)

Lemma with (1) 3 (5) veins, the apex entire, awnless to rarely mucronate or short-awned. Bicellular microhairs with short basal cells, 15–75 μm.

Five genera, 459 spp.

This subtribe is dominated by the large and polymorphic genus *Eragrostis*. There is no obvious morphological synapomorphy for the clade, although many species have branched inflorescences bearing unawned, laterally compressed spikelets that have many flowers, and glumes shorter than the adjacent lemmas.

578. ***Catalepis*** Stapf & Stent

Catalepis Stapf & Stent, Bull. Misc. Inform. Kew 1929: 11 (1929).

Perennials. Ligule a fringe of hairs. Inflorescence branches unbranched, these short and closely spaced on the inflorescence axis, each bearing no more than 9 spikelets. Spikelets with one flower, rachilla extension present. *Lower glume tiny and subulate*, upper glume longer than the flower, the apex acuminate. Lemma with tufts of hairs on the margins, awnless. Leaf epidermal papillae absent.

One sp., *C. gracilis* Stapf & Stent, South Africa.

Grassbase (Clayton et al. 2006 onward) indicates that this species lacks a rachilla extension, but the original description says that it is present. This genus is placed here based on the

Fig. 85. Chloridoideae- Eragrostideae-Eragrostidinae. *Eragrostis frankii*. **A** Flowering plant. **B** Spikelets. (From Barkworth et al. 2003, drawn by C. Roché)

classification presented by Peterson et al. (2010a), although no molecular data are available. The species is superficially similar to *Polevansia*, in Cynodonteae.

579. *Eragrostis* Wolf — Fig. 85

Eragrostis Wolf, Gen. Pl. 23 (1776).
Ectrosia R. Br., Prodr. 185 (1810).
Harpachne A. Rich., Tent. Fl. Abyss. 2: 431 (1851).
Pogonarthria Stapf, Fl. Cap. 7: 316 (1898).
Neeragrostis Bush, Trans. Acad. Sci. St. Louis 13: 178 (1903).
Acamptoclados Nash, Fl. S.E. U.S. 139, 1327 (1903).
Psammagrostis C.A. Gardner & C.E. Hubb., Hooker's Icon. Pl. 34: t 3361 (1938).
Ectrosiopsis (Ohwi) Ohwi ex Jansen, Acta Bot. Neerl. 1: 474 (1952).
Diandrochloa de Winter, Bothalia 7: 387 (1960).

Annuals or perennials, caespitose, rhizomatous, or stoloniferous. Leaves with or without glands. Ligule a fringe of hairs, less often membranous or a fringed membrane. Inflorescence generally exserted, but sometimes surrounded by a subtending leaf, the branches themselves branched in most species, sometimes unbranched. Spikelets with 2 to many flowers, the distal flowers often reduced. Glumes shorter than the spikelet in most species, rarely the same length or longer, generally with one vein, generally lacking awns. Flowers generally bisexual, but in two species unisexual and the plants then dioecious. Lemma glabrous in most species, with three veins, the apex generally entire, rarely mucronate or awned. Pericarp free from seed coat or not. Leaf epidermal papillae absent in most species (except *E. obtusiflora*). Microhairs slender, with an elongate apical cell ("panicoid type"), or with a broad apical cell ("chloridoid type"), or with a broad apical cell with internal membranes ("Enneapogon type"). $2n = 20, 40, 50, 60, 80, 100, 108$.

Four hundred thirty-seven spp., warm areas worldwide.

The broad circumscription of *Eragrostis* suggested here follows recommendations of Columbus et al. (2007), Ingram and Doyle (2004), and Peterson et al. (2010a) whose molecular phylogenetic studies identify a strongly supported clade including *Eragrostis* and its segregates. Subclades within this group do not correspond to previously recognized genera, and *Eragrostis* in its traditional sense is paraphyletic. Ingram and Doyle (2004) found that *Acamptoclados*, *Diandrochloa*, *Neeragrostis*, and *Pogonarthria* should be included within *Eragrostis*. Former *Acamptoclados* (= *E. sessilispica*) is an *Eragrostis* in which the inflorescence disarticulates at its base and is dispersed whole. The seven species formerly in *Diandrochloa* have a membranous ligule, but other than that are like *Eragrostis*, and were placed in that genus by Koch (1978). Former *Neeragrostis* (= *E. reptans*) is dioecious, a character that appears elsewhere in *Eragrostis*. Former *Pogonarthria* has inflorescences with the primary branches unbranched; the shift from branched to unbranched inflorescences has occurred many times in grass evolution so is likely not a good character on which to base a genus. *Pogonarthria* also has lemmas that are acuminate or awned, a character that it shares with former *Psammagrostis*, *Ectrosiopsis*, *Harpachne*, and *Ectrosia*. The latter genus was placed in *Eragrostis* by Columbus et al. (2007) and Peterson et al. (2010a), and Peterson et al. (2010a) confirmed the placement of *Psammagrostis* and *Harpachne*. Species with deciduous glumes were segregated as *Ectrosiopsis*, and those with an unbranched inflorescence as *Psammagrostis* and *Harpachne*.

Although sampling within *Eragrostis* in molecular phylogenies is reasonably good, the number of species included remains a fraction of the total. There remains the possibility that some species currently placed in *Eragrostis* should in fact be placed elsewhere. For example, the species described as *E. walteri* Pilg. has been widely cited as being the only C_3 species of *Eragrostis*, but recent molecular phylogenies (Ingram et al. 2011) show that it is in fact a member of subfamily Arundinoideae and thus was misplaced here. *Thellungia* is excluded from *Eragrostis* by molecular data (GPWG II 2012; Ingram and Doyle 2004) and is placed in Zoysieae by Peterson et al. (2010a).

The subgeneric classification of *Eragrostis* is still under debate. Major characters proposed for subdivision include the pattern of spikelet disarticulation (Van den Borre and Watson 1994) or the C_4 photosynthetic subtype, as indicated by leaf anatomy. A study of the latter character suggests that PCK-type anatomy may be synapomorphic for a clade corresponding to subgenus *Caesiae* (Ingram 2010). However, allopolyploidy is common in the genus and is likely to make any attempt at subgeneric classification difficult (Ingram 2010).

580. ***Heterachne*** Benth.

Heterachne Benth., Hooker's Icon. Pl. 13: t. 1250 (1877).

Annuals. Leaf blades stiff. Ligule a fringe of hairs. Inflorescence branches themselves branched, with *spikelets borne in clusters*. Spikelets with 1 or 2 flowers, with additional reduced flowers distally, *forming a fan-like cluster*; *rachilla zigzag*. Glumes shorter than the spikelet, deciduous or persistent, their apices obtuse or acute. *Lemma with a winged keel* and obtuse apex. *Palea gibbous, with winged keels.*

Three spp., Northern Australia.

581. ***Richardsiella*** Elffers & Kenn.-O'Byrne

Richardsiella Elffers & Kenn.-O'Byrne, Kew Bull. 11: 455 (1957).

Tiny annuals, less than 18 cm tall. *Leaves with abaxial glands.* Ligule a fringe of hairs. *Inflorescence axis and primary branches ending in a slender bristle*, the primary branches curved when dry. Spikelet with 6 to 12 flowers, with additional reduced flowers distally. Glumes the same length as the spikelet, deciduous, with tubercle-based hairs; apices of glumes acuminate and mucronate. Lemma apex obtuse, mucronate. Pericarp free from seed coat. Leaf epidermal papillae absent.

One sp., *R. eruciformis* Ellfers & O'Byrne, Zambia.

This species is similar to *Dinebra somalensis*, which also has leaf glands. The bristle-tipped branches are similar to those of *Cladoraphis*. No molecular data are available for this species and it is placed in this subtribe following the classification of Peterson et al. (2010a).

582. ***Steirachne*** Ekman

Steirachne Ekman, Ark. Bot. 10: 35 (1911).

Perennials, *with cleistogamous spikelets enclosed in the sheaths*. Ligule a fringed membrane or a fringe of hairs. Inflorescence branches unbranched. Spikelets with 6 to 10 flowers, with additional reduced flowers distally. Rachilla internodes long, pubescent at apex. Glumes shorter than the spikelet, with a single vein, their apices acuminate. Lemma awned. *Palea winged.* Stamens 2.

Two spp., Brazil, Guyana, and Venezuela.

XXVII.3. Subtribe Uniolinae Clayton (1982)

Ligule a fringe of hairs. Spikelets disarticulating below the glumes. Lemma with 3 to 10 veins, the apex entire, awnless to mucronate. Basal cell of bicellular microhairs short, 15–35 μm. Leaf blade with abaxial sclerenchyma of vascular bundles continuous with columns of colorless cells.

Six genera, 15 spp.

This subtribe has no obvious morphological synapomorphy.

583. ***Fingerhuthia*** Nees ex Lehm.

Fingerhuthia Nees ex Lehm., Sem. Hort. Bot. Hamburg. 1834: 7 (1834).

Perennials, caespitose or rhizomatous. Inflorescence branches branched. Spikelets with one fertile flower, plus 1 to 3 sterile or staminate flowers distally. Callus blunt. Glumes shorter than the spikelet, apices acuminate, awned. Lemma awned. $2n = 20, 40$.

Two spp., Afghanistan, Arabia, southern Africa.

This genus was placed in a clade with *Uniola* and *Tetrachne* by Ingram and Doyle (2004).

584. ***Stiburus*** Stapf

Stiburus Stapf, Fl. Cap. 7: 696–697 (1900).

Caespitose perennials. Leaf blades narrow and upright. Ligule a fringe of hairs. Inflorescence dense, like a bottlebrush, the branches branched, borne upright and closely appressed to the axis. Spikelets with 1 to 3 flowers plus additional reduced flowers distally. Glumes shorter than the flowers; glumes and lemmas covered with long hairs. *Stamens 2.* $2n = 20$.

Two spp., southern Africa.

Ingram and Doyle (2004) present molecular evidence hinting at a relationship between *Stiburus* and *Uniola*, although there is little morphological resemblance. Their data contradict that of Clayton et al. (2006 onward) and Clayton and Renvoize (1986) who consider the two species of *Stiburus* as part of *Eragrostis*.

Entoplocamia + Tetrachne + Uniola

Palea keels winged. Spikelets with multiple flowers, the proximal ones sterile.

These three genera form a well-supported clade in the study of Peterson et al. (2010a), but *Fingerhuthia* is sister to *Tetrachne* in GPWG II (2012) and Ingram and Doyle (2004). If the latter topology is correct, then winged palea keels may be lost in *Fingerhuthia*.

585. ***Entoplocamia*** Stapf Fig. 86

Entoplocamia Stapf, Fl. Cap. 7: 318 (1898).

Annuals. Inflorescence with short unbranched branches or with sessile spikelets. Proximal sterile flowers 2, fertile flowers 4 to 20. Glumes shorter than the spikelet, the lower one with an acute apex, the upper acuminate and mucronate.

Fig. 86. Chloridoideae-Eragrostideae-Uniolinae. *Entoplocamia aristulata.* **A** Habit. **B** Ligule. **C** Inflorescence. (From Müller 2007)

Lemma awned, the awn sharp. *Lodicules absent.* Pericarp free from the seed coat.

One sp., *E. aristulata* (Hack. & Rendel) Stapf, southwestern Africa.

586. ***Tetrachne*** Nees

Tetrachne Nees, Fl. Afr. Austral. Ill. 1: 375 (1841).

Perennials. Inflorescence with unbranched branches, these distantly spaced and not overlapping on the central axis. Proximal sterile flowers 2, fertile flowers 4 to 5. Glumes shorter than the spikelet, their apices acute. Lemma unawned. Pericarp free from the seed coat. $2n = 20$.

One sp., *T. dregei* Nees, South Africa and Pakistan.

Tetrachne is sister to *Entoplocamia* in the phylogeny of Peterson et al. (2010a). Watson and Dallwitz (1992 onward) suggest that this species hybridizes with *Fingerhuthia*, which is the sister taxon in GPWG II (2012) and Ingram and Doyle (2004).

587. ***Uniola*** L.

Uniola L., Sp. Pl. 1: 71 (1753).
Leptochloöpsis Yates, Southwest Naturalist 11: 382 (1966).

Perennials, caespitose, rhizomatous, or stoloniferous. *Ligule sometimes lacking.* Inflorescence branches unbranched, overlapping each other on the central axis. Proximal sterile flowers 2 to 8. Glumes shorter than the spikelet, their apices acute to acuminate, mucronate. $2n = 40$.

Eight spp., southern U.S. to Ecuador.

U. condensata, one of two *Uniola* species transferred to *Leptochloöpsis* (Yates 1966), is sister to *U. paniculata* in the study of Peterson et al. (2010a); the two genera are combined here.

XXVIII. Tribe Zoysieae Benth. (1881)

Ligule generally a fringe of hairs. Spikelets laterally compressed, *with one flower*. Lower glume absent, or with no veins, or with only one vein. Lemma with a single vein, awnless. Pericarp free from seed coat.

Nine genera, 235 species.

Many members of this tribe are found in sandy or saline, often moist sites (Peterson et al. 2007). The presence of one flower per spikelet may be synapomorphic here, although the character is homoplasious in the Chloridoideae as a whole.

Ingram and Doyle (2004) identified a clade made up of *Spartina*, *Calamovilfa*, *Sporobolus* and *Thellungia*. This was supported by Columbus et al. (2007) who added *Crypsis* and placed *Zoysia* sister to the rest of the group. Peterson et al. (2010a) placed *Urochondra* sister to all other species. *Psilolemma* was included in this clade by Peterson et al. (2014b). In their analyses, which are far more comprehensive than anything published previously, *Psilolemma, Urochondra, Zoysia*, and *Sporobolus somalensis* are variously placed outside a large clade made up of the remaining species of *Sporobolus*, plus *Spartina* and several smaller genera.

The oldest name of the combined genus would be *Spartina*, which was published in April 1789 and thus has priority over *Crypsis*, published in October of the same year. However, Peterson et al. (2014c) have proposed conservation of the name *Sporobolus* and have made the necessary combinations. The recommendations of Peterson et al. (2014b, c) are adopted here but genera could not be renumbered and descriptions altered in time for publication.

Although Peterson et al. (2007) and Soreng et al. (2012 and onward) divide Zoysieae into the monotypic subtribe Zoysiinae Benth. and place all other genera in Sporobolinae Benth., this seems like an unnecessary subdivision. Accordingly, the subtribal rank is dispensed with here.

588. ***Psilolemma*** S. M. Phillips

Psilolemma S. M. Phillips, Kew Bull. 29: 267 (1974).

Stoloniferous perennials. Ligule a line of hairs. Leaf blades stiff, sharp-pointed. Inflorescence branches widely spaced on central axis, with 2 or 3 spikelets each (up to 5). Spikelets with 4 to 14 flowers, subterete to laterally compressed. Glumes shorter than the spikelet. Lemma glabrous, with 1 to 3 veins, the apex obtuse. Pericarp easily separated from the seed coat. Leaf epidermal papillae absent.

One sp., *P. jaegeri* (Pilger) S. M. Phillips, East Africa.

Phillips (1974) notes similarity in habit between this species and *Odyssea*.

589. ***Urochondra*** C. E. Hubb.

Urochondra C. E. Hubb., Hooker's Icon. Pl. 35: t 3457 (1947).

Perennials. Leaf blades stiff, sharp-pointed. Inflorescence dense, cylindrical, the branches branched, borne upright and closely appressed to the axis. Glumes shorter than the flower, their apices acute to acuminate, awnless. Lemma keeled, the apex obtuse, mucronate. *Lodicules absent. Fruit with a beak formed from fused style branches*. Pericarp swelling when wet. Leaf epidermal papillae present, one per cell.

One sp., *U. setulosa* (Trin.) C. E. Hubb., Sudan, Somalia and Sindh (Pakistan).

This species is sister to all other members of Zoysieae in the phylogeny of Peterson et al. (2010a).

Remaining genera:

Epidermal papillae lacking on the abaxial leaf blade; sometimes present adaxially. Leaf blades disarticulating from the sheaths in many but not all genera.

590. ***Zoysia*** Willd.

Zoysia Willd., Ges. Naturf. Freunde Berlin Neue Schriften 3: 440–441 (1801), nom. conserv.

Perennials, caespitose or rhizomatous, if the latter then forming dense turf. Leaf blades with a sharp pointed apex. Ligule a fringed membrane or a fringe of hairs. Inflorescence unbranched, in one species consisting of a single spikelet. Spikelets short pedicellate; rachilla extension absent. *Lower glume absent*; upper glume as long as the flower or longer and enclosing it, the apex muticous to awned. Lemma *hyaline*, acute or with a tiny apical notch, with a small mucro. Leaf epidermal papillae present adaxially. *Microhairs with partitioning membranes*. $2n = 40$.

Eleven spp., warm regions of Asia and Australasia.

Species of *Zoysia* are widely cultivated for lawns.

Remaining genera:

Pericarp free from the seed coat.

591. ***Calamovilfa*** (A. Gray) Hack. ex Scribn. & Southw.

Calamovilfa (A. Gray) Hack. ex Scribn. & Southw., True Grasses: 113 (1890).

Perennials, up to 2 m tall, caespitose or rhizomatous. Ligule sometimes a fringed membrane. Inflorescence branches branched. Spikelets with-

Fig. 87. Chloridoideae-Zoysieae. *Sporobolus cryptandrus*. **A** Plant with inflorescence. **B** Spikelet, the glumes separated from the floret. (From Gould and Box 1965)

out rachilla extension. Flowers with a bearded callus. Glumes shorter or longer than the flower, awnless. Lemma awnless, hairy or not. $2n = 40$.

Five spp., North America.

Molecular phylogenetic data (Columbus et al. 2007; GPWG II 2012; Peterson et al. 2010a) place this sister to *Spartina*, or to a subset of *Spartina* species (Fortune et al. 2006). This genus is now placed as a section and subsection of *Sporobolus* s.l., and the necessary nomenclatural combinations have been made (Peterson et al. 2014b).

592. ***Crypsis*** Aiton

Crypsis Aiton, Hort. Kew. 1: 48 (1789), nom. conserv.

Stoloniferous annuals; culms prostrate or ascending. Inflorescence branches branched, the inflorescence contracted and spike-like or capitate, often surrounded by the subtending leaf.

Spikelets without a rachilla extension. Glumes shorter than or as long as the spikelets, muticous or awned. Lemma muticous or awned. Lodicules absent. Pericarp swelling when wet. $2n = 16, 18, 32, 36, 54$.

Ten spp., Mediterranean and Middle East, to China and central Africa. This genus is now placed as a section and subsection of *Sporobolus* s.l., and the necessary nomenclatural combinations have been made (Peterson et al. 2014b).

593. ***Spartina*** Schreb.

Spartina Schreb., Gen. Pl. ed. 8: 43 (1789).

Perennials, caespitose or rhizomatous. Inflorescence branches unbranched, the central axis elongate or contracted. Branches ending in a naked tip or in a spikelet. Spikelets tightly packed in two rows along each branch, without a rachilla extension. Glumes as long as the flowers or longer, the apex muticous to awned, the upper glume much longer than the lower. Lemma keeled, the apex entire or bidentate, awnless. Lodicules absent. $2n = 28, 40, 42, 60, 62, 84, 120, 122, 124$.

Seventeen spp., coastal areas of the New World, Atlantic coast of Europe and Africa, inland wet areas and prairie in the U.S.

Diploids are unknown in *Spartina*, raising the possibility that the genus is of allopolyploid origin. Fortune et al. (2006) provide some evidence for this possibility in their study of the single-copy nuclear gene *waxy*. This genus is now placed as a section of *Sporobolus* s.l., and the necessary nomenclatural combinations have been made (Peterson et al. 2014b).

594. ***Sporobolus*** R. Br. Fig. 87

Sporobolus R. Br., Prodr. 169 (1810).

Annuals or perennials, caespitose, rhizomatous, or stoloniferous. Leaf blade apex with a short mucro or sharp point. Ligule membranous, or a fringed membrane, or a fringe of hairs. Inflorescence branches branched, the proximal ones sometimes sterile. Spikelets sometimes more or less terete, sometimes with 2 flowers. Glumes shorter or longer than the spikelet, the lower glume awnless, the upper glume muticous to awned. Callus of flower glabrous. Lemma thin, shiny, with one or three veins, without an awn. Grain spherical to ellipsoid; pericarp swelling when wet and ejecting the seed. $2n = 18, 24, 36, 38, 54, 72, 80, 88, 90, 108, 126$.

One hundred ninety-eight spp., warm regions worldwide.

Sporobolus in its traditional sense is paraphyletic, with *Crypsis*, *Spartina*, and *Calamovilfa* derived from within it (Peterson et al. 2010a). A recent comprehensive molecular phylogenetic study has circumscribed the genus in a broad sense and has formally transferred species from the segregate genera to *Sporobolus* (Peterson et al. 2014b). Because *Sporobolus* is not the oldest name in the clade, the transfers rest on a proposal to conserve the name *Sporobolus* (Peterson et al. 2014c). A well-supported subgeneric classification is also presented based on the phylogeny.

595. ***Thellungia*** Stapf

Thellungia Stapf, Bull. Misc. Inform. Kew 1920: 97 (1920).

Perennials, ca. 1 to 1.5 m tall. Inflorescence with primary branches branched, *about the same length as the internodes separating them, upright and appressed to the main inflorescence axis.* Spikelets with 2 to 4 flowers. Glumes translucent, shorter than the spikelet, awnless. Lemmas translucent, awnless, glabrous, with one vein.

One sp., *T. advena* Stapf, Australia.

Although Clayton et al. (2006 onward) and Clayton and Renvoize (1986) include this genus in *Eragrostis*, molecular data clearly place it in Zoysieae (Ingram and Doyle 2004; Peterson et al. 2014b). The inflorescence structure and the one-veined lemmas are distinct from those in *Eragrostis*. The species has been formally transferred to *Sporobolus* by Peterson et al. (2014b).

XXIX. Tribe Cynodonteae **Dumort. (1824)**

Members of this tribe have no obvious morphological character in common, and the range of variation is almost as great as that of the entire subfamily. As noted by Peterson et al. (2007), "[t]here are no definitive non-molecular characters that differentiate Cynodonteae from Eragrostideae and/or Zoysieae." Nonetheless, Cynodonteae are strongly supported as monophyletic by molecular data (Columbus et al. 2007; GPWG II 2012; Peterson et al. 2010a, 2012, 2014a). Although there is some phylogenetic structure among the early-diverging lineages, this is generally poorly supported and as a result several genera are placed incertae sedis.

Eighty-two genera, 900 species.

Incertae sedis

596. ***Aeluropus*** Trin.

Aeluropus Trin., Fund. Agrost. 143 (1820).

Annuals or perennials, caespitose, rhizomatous or stoloniferous. Ligule a fringed membrane or a fringe of hairs. Leaf blades stiff, with a sharp apex. Inflorescence unbranched, or with unbranched branches appressed to the central axis. Spikelets laterally compressed, with 2 to 30 flowers, with additional reduced flowers distally. Glumes shorter than the spikelet, unawned. Lemma with 5 or more veins, the apex mucronate. Leaf epidermal papillae one per cell.

Eight spp., Mediterranean, Asia, Northern Africa.

Although this species has been put in a monotypic subtribe by Peterson et al. (2010a), it is in an unresolved position in molecular phylogenies and there seems little reason to create a higher-level taxon to include it. Peterson et al. (2012, 2014a) provide weak support for placing *Aeluropus* sister to Triodiinae, whereas the GPWG II (2012) places it sister to Eleusininae.

597. ***Cleistogenes*** Keng

Cleistogenes Keng, Sinensia 5: 147 (1934).
Kengia Packer, Bot. Not. 113: 291 (1960), nom. superfl.

Caespitose perennials. *Upper leaf sheaths enclosing cleistogamous flowers.* Leaf blades stiff. Ligule a fringe of hairs or a fringed membrane. Inflorescences unbranched or branched, the branches branched or not. Spikelets laterally compressed, with 2 to 8 fertile flowers plus additional reduced flowers distally. Callus of flower pubescent. Glumes shorter than the spikelet, their apices muticous. Lemma with 5 veins, the apex entire or with two teeth, muticous or awned. Leaf epidermal papillae one per cell.

Fourteen spp., Europe and Asia.

The names *Cleistogenes* and *Kengia* refer to the same genus. Packer (1960) argued that, because the word "cleistogene" is a technical morphological term to describe spikelets enclosed in the leaf sheaths, the generic name was in violation of International Code of Botanical Nomenclature, Article 68. However, Article 20.2 of the most recent code (McNeill et al. 2007) states only that "the name of a genus may not coincide with a Latin technical term." Because the term cleistogene is not Latin and hence not "a Latin technical term", the name *Cleistogenes* stands; in example 5, the ICBN explicitly states that *Kengia* is illegitimate.

598. ***Hubbardochloa*** Auquier

Hubbardochloa Auquier, Bull. Jard. Bot. Belg. 50: 241 (1980).

Tiny annual, less than 15 cm tall. Leaf blades pseudopetiolate. Ligule a line of hairs. Inflorescence branches delicate, branched. Spikelets with one flower. Callus with minute hairs. Lemma with a long terminal awn. *Palea absent. Lodicules absent.* Leaf epidermal papillae abundant, several per cell.

One sp., *H. gracilis* Auquier, Rwanda, Burundi, Zambia.

599. ***Lepturidium*** Hitchc. & Ekman

Lepturidium Hitchc. & Ekman, Man. Grasses W. Ind.: 111 (1936).

Perennials, less than 30 cm tall. Ligule a fringe of hairs. Leaf blades stiff and leathery. Inflorescence unbranched, flattened, surrounded by the subtending leaf sheath, the spikelets borne on one side. Spikelets laterally compressed, with 1 or 2 flowers, the proximal one fertile, the distal one sterile or absent. Glumes enclosing the flowers, their apices acute. Callus of the flower pubescent. Lemma with two teeth and a short mucro. Leaf epidermal papillae absent.

One sp., *L. insulare* Hitchc. & Ekman, Cuba.

Clayton and Renvoize (1986) comment that *Lepturidium* is "barely distinct from *Brachyachne* [Eleusininae]", but Peterson et al. (2010a) show that the two are unrelated.

600. ***Orinus*** Hitchc.

Orinus Hitchc., J. Wash. Acad. Sci. 23: 136 (1933).

Rhizomatous perennials, dune grasses. Leaf blades with sharp tip. Ligule membranous. Inflorescence branches unbranched, borne along a central axis. Spikelets laterally compressed, with 1 to 8 flowers plus additional reduced flowers distally. Rachilla internode elongated between flowers. Glumes shorter than the spikelet, their apices acute to acuminate. Lemma apex muticous or mucronate. Leaf epidermal papillae one per cell.

Five spp., Western Himalayas.

Peterson et al. (2012) provide weak support for placement of *Orinus* sister to *Triodia.*

601. *Pogoneura* Napper

Pogoneura Napper, Kirkia 3: 112 (1963).

Annuals. *Leaf blades with multicellular glands adaxially and on blade margins.* Inflorescence branches unbranched, spread along a central axis. Spikelets *more or less terete, with 2 flowers*, rachilla extension lacking, or reduced flowers present distally. Glumes longer than flowers and enclosing them. Callus of flowers pubescent. Lemma with hairs on lateral veins and keel, bidentate, awned from sinus. Lodicules minute or absent.

One sp., *P. biflora* Napper, East Africa.

Peterson et al. (2014b) place *Pogoneura* here in Cynodonteae, rather than its former placement in Zoysieae (Peterson et al. 2010a). The presence of two flowers per spikelet is unusual in Zoysieae and provides morphological evidence that the genus belongs here.

602. *Pommereulla* L. f.

Pommereulla L. f., Nov. Gram. Gen. 31 (1780).

Tiny stoloniferous perennials, less than 15 cm tall. Ligule a fringe of hairs. Inflorescence unbranched. *Spikelets turbinate, with 2 sterile proximal flowers, 2 or 3 fertile flowers, and 1 or more distal rudimentary flowers. Lemmas fan-shaped, 3-lobed,* hairy below, with awn-tips from lobes and an awn from the middle of the back. Leaf epidermal papillae present, one or two per cell.

One sp., *P. cornucopiae* L. f., Southern India, Sri Lanka.

Watson and Dallwitz (1992 onward) indicate that the lemmas are spirally arranged in the spikelet; if correct this phenotype is unique within the Poaceae. *Pommereulla* is placed here in Cynodonteae following Soreng et al. (2012 and onward); the data supporting this placement are unpublished.

Brachychloa + *Dactyloctenium* + *Neobouteloua* + *Odyssea*

Spikelets laterally compressed.

These four genera form a well-supported clade in GPWG II (2012).

603. *Brachychloa* S. M. Phillips

Brachychloa S. M. Phillips, Kew Bull. 37: 145, 158 (1982).

Annuals or perennials, caespitose or stoloniferous. Ligule a fringed membrane. Inflorescence branches unbranched, spread along the central axis, disarticulating or not. Spikelets with 4 to 7 flowers plus additional reduced flowers distally. Glumes shorter than the flowers, their apices acute. Lemma glabrous, the apex with a minute mucro from an equally minute sinus. *Palea entire, gibbous.* Pericarp free from seed coat. Leaf epidermal papillae one per cell.

Two spp., Mozambique and Natal, South Africa.

604. *Dactyloctenium* Willd.

Dactyloctenium Willd., Enum. Pl. 2: 1029–1030 (1809).

Annuals or perennials, caespitose or stoloniferous. Leaf blades with a sharp pointed or muticous apex. Ligule membranous, fringed or not. Inflorescences unbranched, or with unbranched primary branches, digitate, *ending in a naked point.* Spikelets with 1 to 9 flowers, with additional reduced flowers distally. Glumes shorter or longer than the flowers, the apex of the lower one muticous or mucronate, the upper glume with a terminal or subapical, sharply bent awn. Callus of flower glabrous. Lemma acute, awnless, mucronate or awned, the awn recurved. *Fruit rugose,* pericarp free from seed coat. Leaf epidermal papillae one per cell. $2n = 20, 36, 40, 48$.

Thirteen spp., Natal (South Africa) to north India, Australia, one species a cosmopolitan weed.

605. *Neobouteloua* Gould

Neobouteloua Gould, Bol. Soc. Argent. Bot. 12: 106–108 (1968).

Perennials, less than 30 cm tall. Ligule a fringed membrane or a fringe of hairs. Inflorescence of unbranched branches, on an elongated axis. Spikelets with one fertile flower and a cluster of 2 to 4 rudimentary ones. Glumes shorter than the flowers, the apex of the lower one muticous or mucronate, the upper one acuminate or setaceous. Callus of flower pubescent. Lemma (both fertile and sterile) pubescent with 3 awns, the median one recurved. Pericarp fused to seed coat. Leaf epidermal papillae one per cell.

Two spp., Argentina and Chile.

606. *Odyssea* Stapf

Odyssea Stapf, Hooker's Icon. Pl. 31: t. 3100 (1922).

Rhizomatous perennials, sometimes suffrutescent. Ligule a fringe of hairs. Leaf blades hard, stiff, sharp-pointed, disarticulating from the sheaths. Inflorescence branches unbranched. Spikelets with 4 to 10 fertile flowers plus additional distal sterile ones. Glumes shorter than the flowers, thinner than the lemmas; proximal glume apex acute, distal glume apex obtuse. Callus of flower pubescent with straight hairs ca. ¼ the length of the flower. Lemma apex with two teeth, mucronate. Pericarp free from seed coat. Leaf epidermal papillae absent. $2n = 36$.

Two spp., Africa.

XXIX.1. Subtribe Boutelouinae Stapf (1917)

Scleropogoninae Pilg., Die Nat. Pflanzenfam., ed. 2, 14d: 167, 49 (1956).
Muhlenbergiinae Pilg. ibid.: 168, 67 (1956).
Monanthochloinae Pilg. ex Potztal, Willdenowia 5: 472 (1969).
Hilariinae P. M. Peterson & Columbus, Aliso 23: 592 (2007).
Traginae P. M. Peterson & Columbus, Aliso 23: 592 (2007).

Annuals or perennials. Ligule generally a fringed membrane. Inflorescences generally unbranched or with unbranched branches.

Sixteen genera, 287 spp.

This clade appears with moderate support in the analyses of GPWG II (2012) and is strongly supported by Peterson et al. (2014a). Although there is no morphological synapomorphy, many members have inflorescences unbranched, or with unbranched branches; many members are also dioecious. As recognized here, Boutelouinae are much broader than in other classifications. Soreng et al. (2012 and onward) restrict Boutelouinae to the genus *Bouteloua* itself, and split the clade into several other subtribes, several of which are monogeneric. Their classification results in morphologically diagnosable units but does not then reflect the relationship among them.

607. *Allolepis* Soderst. & H. F. Decker

Allolepis Soderst. & H. F. Decker, Madroño 18: 34 (1965).

Dioecious perennials, rhizomatous and stoloniferous. Ligule a fringed membrane. Inflorescence branches unbranched, borne along a central axis and appressed to it. Spikelets laterally compressed, with 8 to 9 flowers and additional reduced flowers distally. Glumes shorter than the spikelet, their apices acute. Lemma of female spikelet with wide hyaline margins. *Palea convolute around gynoecium; keels winged.* Leaf epidermal papillae absent. $2n = 40$.

One sp., *A. texana* Soderst. & H. F. Decker, southern USA and Mexico.

608. *Bouteloua* Lag. Fig. 88

Bouteloua Lag., Varied. Ci. 2, 4: 134 (1805), nom. conserv.
Chondrosum Desv., Nouv. Bull. Sci. Soc. Philom. 2: 188 (1810).
Pentarrhaphis Kunth, Nov. Gen. Sp. (quarto ed.) 1: 177–178, t. 60 (1815).
Opizia J. Presl, Reliq. Haenk. 1(4–5): 293, pl. 41, f. 1–11 (1830).
Cathestecum J. Presl, ibid.: 294, t. 42 (1830).
Buchloe Engelm., Trans. Acad. Sci. St. Louis 1: 432 (1859), nom. conserv.
Pringleochloa Scribn., Bot. Gaz. 21: 137 (1896).
Cyclostachya Reeder & C. Reeder, Bull Torrey Bot. Club 90: 195 (1963).
Buchlomimus J. & C. Reeder & Rzedowski, Brittonia 17: 29 (1965).
Soderstromia C.V. Morton, Leafl. W. Bot. 10: 327 (1966).
Griffithsochloa G.J. Pierce, Bull. Torrey Bot. Club 105: 134 (1978).

Annuals or perennials, caespitose, rhizomatous or stoloniferous; plants hermaphrodite, monoecious or dioecious. Ligule membranous, or a fringed membrane or a fringe of hairs. Inflorescence unbranched or with unbranched branches, on an elongate or contracted axis, the branches deciduous or persistent, with a naked tip, the latter straight or forked. Each branch with 1 to 10 spikelets. Spikelets laterally or dorsiventrally

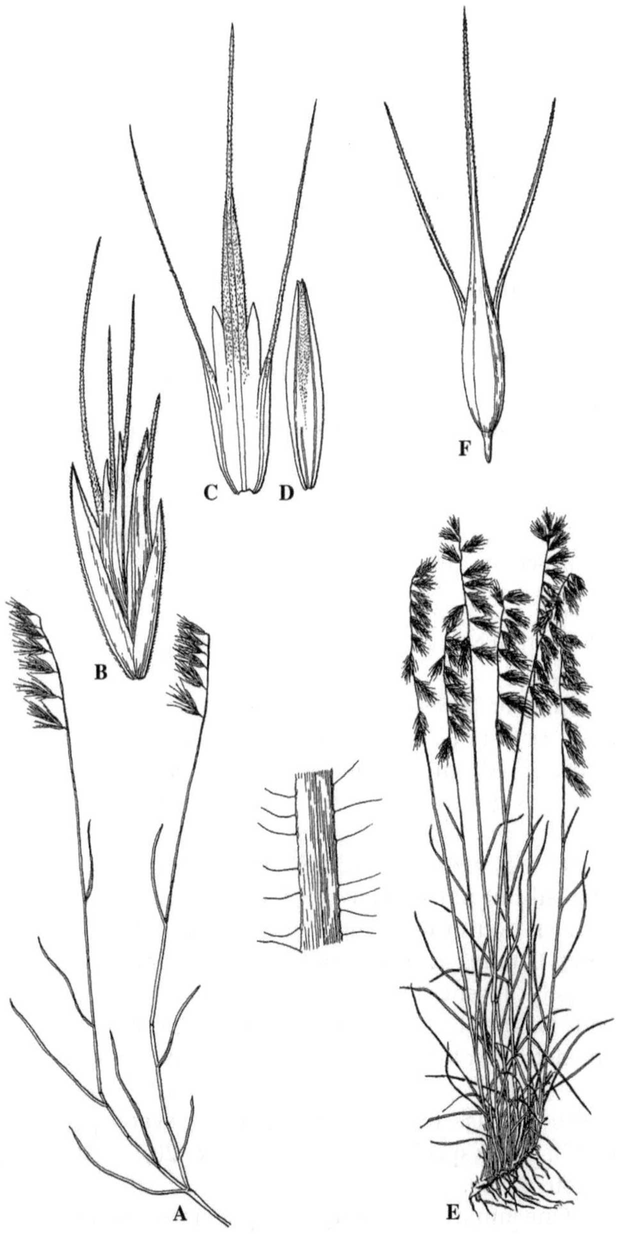

Fig. 88. Chloridoideae-Cynodonteae-Boutelouiinae. *Bouteloua* spp. **A–D** *B. repens*. **A** Flowering culms. **B** Spikelet. **C** Lemma. **D** Palea. **E, F** *B. radicosa*. **E** Whole plant. **F** Lemma. (From Barkworth et al. 2003, drawn by C. Roché)

compressed, subtended by bristles or not, with 1 fertile flower plus 1 to 3 sterile flowers distally, these composed only of lemmas with long awns. Lower glume absent to narrow and awl-like to lanceolate or ovate. Upper glume narrow, acute to acuminate. Lemma glabrous or sparsely pubescent, the apex acute or 3-lobed, the lobes generally ending in awns, often with lobes between the awns. Flowers bisexual or unisexual. Leaf epidermal papillae absent. Microhairs with an elongated apical cell ("panicoid type"), rarely with a short broad apical cell ("chloridoid type"). $2n = 20, 40, 41, 42, 56, 60, 70, 98$.

Fifty-eight spp., Canada to Argentina.

Molecular phylogenetic data (Columbus et al. 1998) show that *Bouteloua* in its traditional sense is paraphyletic, with multiple monotypic genera segregated from within it. Accordingly, Columbus (1999) expanded the circumscription of *Bouteloua* to include the entire clade. Subsequent molecular studies have reinforced the validity of this conclusion. *Bouteloua* could be placed in its own subtribe, Boutelouinae Stapf, but this then forces recognition of several other monogeneric subtribes; here I suggest a more broadly circumscribed Boutelouinae.

609. *Distichlis* Raf.

Distichlis Raf., J. Phys. Chim. Hist. Nat. Arts 89: 104 (1819).
Monanthochloë Engelm., Trans. Acad. Sci. St. Louis 1: 436 (1859).
Reederochloa Soderstr. & H.F. Decker, Brittonia 16: 334 (1964).

Rhizomatous or stoloniferous perennials; dioecious. Leaves rigid and distichous. Ligule membranous, with or without a fringe of hairs. Inflorescence branched or not, or reduced to a single spikelet. Spikelets laterally compressed, with 2 to 15 flowers plus additional reduced flowers distally. Glumes present or absent, shorter than the spikelet. Lemma apex entire, awnless. Microhairs with basal cell sunken in epidermis. *Subsidiary cells of stomata on the lemma each bearing a single papilla.* Vascular bundles of the leaf with an even outline, suggesting C_4, NAD-ME photosynthesis. $2n = 38, 40, 42, 72$.

Ten spp., nine amphitropical in the New World, one in Australia.

The inclusion of *Reederochloa* and *Monanthochloë* in *Distichlis* is supported by multiple characteristics of morphology and anatomy, as well as molecular data (Bell and Columbus 2008). Clayton and Renvoize (1986) suggest that the presence of microhairs with the basal cell sunken in the epidermis may be synapomorphic for this genus. Bell and Columbus (2008) show clearly that this sort of microhair also occurs in *Eragrostis obtusiflora*, and hint that the latter species may be sister to *Distichlis* rather than part of *Eragrostis. Distichlis* could be placed in its own subtribe, Monanthochloinae Pilg. ex Potztal, but there seems little reason to do so.

610. ***Hilaria*** Kunth

Hilaria Kunth, Nov. Gen. Sp. (quarto ed.) 1: 116–117, pl. 37 (1815).
Pleuraphis Torr., Ann. Lyceum Nat. Hist. New York 1: 148 (1824).

Annuals or perennials, caespitose, rhizomatous or stoloniferous. Leaf blades with a sharp-pointed apex. Ligule membranous or a fringed membrane. Inflorescence branches contracted, unbranched, spread along a central axis. *Spikelets 3 per branch, crowded together, falling as a unit.* Central spikelet pedicellate, laterally compressed, with one pistillate or bisexual flower, occasionally a second staminate or sterile one distally; glumes shorter than the spikelet, narrow, with 2 or more lobes, awned; lemma apex entire to bifid, muticous to awned. *Lateral spikelets sessile, dorsiventrally compressed, with 1 to 5 flowers, staminate or sterile; glumes flabellate, asymmetric, with 1 or more awns*; lemma bilobed, awnless, mucronate or awned, the awn from the back. Leaf epidermal papillae one per cell. $2n = 18, 36, 54, 72$.

Ten spp., southern U.S. to Guatemala.

In subg. *Hilaria* the glumes are coriaceous and fused below, whereas in subg. *Pleuraphis* they are papery and free; the two are strongly supported as sisters, supporting their combination into a single genus. Although *Hilaria* has been put in a monotypic subtribe by Peterson et al. (2010a), it is part of the larger clade here called Boutelouinae, based on the phylogeny (but not the classification) of Peterson et al. (2012).

611. ***Jouvea*** E. Fourn.

Jouvea E. Fourn.Bull. Soc. Roy. Bot. Belgique 15: 475 (1876).

Dioecious perennials, stoloniferous. Ligule a fringed membrane or a fringe of hairs. Inflorescences in axillary clusters; pistillate inflorescence with a spathe, with one spikelet. Spikelet terete, with 2 to 5 flowers, *these embedded in the rachilla, the rachilla extending beyond the distalmost flower as an elongate thorn; glumes absent. Lemma margins of pistillate flower fused nearly to the apex. Lodicules absent.* Male inflorescence unbranched; lower glumes reduced or absent. Leaf epidermal papillae absent.

Two spp., California to Central America.

612. ***Muhlenbergia*** Schreb.

Muhlenbergia Schreb., Gen. Pl. 1: 44 (1789).
Aegopogon Humb. & Bonpl. ex Willd., Sp. Pl. 4(2): 899 (1805).
Lycurus Kunth, Nov. Gen. Sp. (quarto ed.) 1: 141 (1815).
Pereilema J. Presl, Reliq. Haenk. 1: 233 (1830).
Schedonnardus Steud., Syn. Pl. Glumac. 1: 146 (1854).
Chaboissaea E. Fourn., Mexic. Pl. 2: 112. (1886).
Redfieldia Vasey, Bull. Torrey Bot. Club 14: 133 (1887).
Bealia Scribn., True Grasses 104 (1890).
Blepharoneuron Nash, Bull. Torrey Bot. Club 25: 88 (1898).
Schaffnerella Nash, N. Amer. Fl. 17: 141 (1912).

Annuals or perennials. Ligule membranous, ciliate or not. Inflorescence branched or unbranched, the branches erect or spreading. Spikelets generally solitary, but borne in 2 s or 3 s in a few species. Spikelets with one flower, sometimes two- or three-flowered, laterally compressed but not strongly so. Rachilla ending in a fertile flower or not. Callus glabrous or with short hairs. Glumes awned or not, the lower one occasionally absent. Lemma with 3 veins, entire or with two teeth, generally with an awn, the latter abaxial, apical or from a sinus; lemmas margins pubescent or not. Leaf epidermal cells with intercostal papillae present, one per cell (several per cell in a few species). $2n = 14, 16, 18, 20, 28, 30, 32, 40, 42, 60$.

One hundred seventy-six spp., most in the New World, but a few in Asia.

Muhlenbergia (as subtribe Muhlenbergiinae) has been the subject of several recent phylogenetic studies (Columbus et al. 2010; Peterson et al. 2010a, 2010b), which demonstrate that it is clearly monophyletic. Peterson et al. (2010b) provide phylogenies that sample about 70 % of the species and provide strong evidence that *Muhlenbergia* as traditionally defined is paraphyletic, so a broad circumscription is adopted here. Placing the genus in its own subtribe requires recognizing several other monogeneric subtribes within the clade here recognized as Boutelouinae s.l. Clayton and Renvoize (1986) suggest *Muhlenbergia* is allied to *Sporobolus*, but molecular data do not support this placement. Some species can be confused with members of the Pooideae.

In their comprehensive phylogenetic study of *Muhlenbergia*, Peterson et al. (2010b) describe five well-supported clades.

1. *M.* subg. *Muhlenbergia*. This clade includes rhizomatous plants with leaf blades that are broad and flat, and inflorescence branches that are branched and upright. All members of the clade share characteristic C_4 PCK leaf anatomy and all species of *Muhlenbergia* that have been biochemically typed as C_4 PCK fall here. Although most members of this clade have conventional spikelet arrangement, several species have reduced staminate or sterile spikelets associated with each fertile spikelet; these were formerly placed in *Aegopogon* and *Pereilema*, but clearly are derived from within *Muhlenbergia*.
2. *M.* subg. *Trichochloa* Trin. This clade is strongly supported but has few if any unique morphological characters. The plants are stout and caespitose, and the glumes have one or no veins. Peterson and Herrera Arrieta (2001) describe distinctive leaf anatomy with inflated cells adaxial to the primary vascular bundles.
3. *M.* subg. *Clomena* (P. Beauv.) Hack. This clade includes plants that are caespitose, with leaves having lower sheaths that are flat and papery at maturity. The upper glumes have three veins and often three teeth.
4. *M.* subg. *Pseudosporobolus* (Parodi) P. M. Peterson. Species of this clade have no evident morphological synapomorphies and some have been segregated as separate genera in the past. Whereas the primary inflorescence branches are branched in some taxa, in *M. paniculata* (formerly *Schedonnardus*) the primary branches are unbranched; the central axis elongates and coils at maturity and the inflorescence is dispersed as a unit. Other taxa have spikelets with more than one flower.
5. *M.* subg. *Bealia* (Scribn.) P. M. Peterson. Most members of this clade are annuals. All have lemmas with hairy margins and midveins.

Muhlenbergia ramulosa (Kunth) Swallen remains incertae sedis as it is not clearly a member of any of the five groups.

613. *Sohnsia* Airy Shaw

Sohnsia Airy Shaw, Kew Bull. 18: 272 (1965).

Dioecious perennials. Ligule a fringed membrane. Inflorescence branches branched. Spikelets with 2 or 3 flowers plus additional reduced ones distally. Glumes shorter than the spikelet, with one vein, the apex acute. Callus of flower pubescent. Lemmas with lines of hairs on veins and keel; with three awns, the central one straight. Leaf epidermal papillae absent.

One sp., *S. filifolia* (E. Fourn.) Airy Shaw, Mexico.

MONELYTRUM + *TRAGUS* + *POLEVANSIA* + *WILLKOMMIA*

Inflorescence branches unbranched, *appressed to an elongated axis. Spikelets dorsiventrally compressed*, with one flower. Rachilla not extended beyond flower. Lower glume obtuse, with no obvious veins, the apex unawned.

These genera clearly form a clade, based on molecular (Peterson et al. 2010a, 2012) and morphological characters and have been called subtribe Traginae. However, relationships among them are poorly supported in molecular phylogenies; it is possible that the entire group should be included in a single genus.

614. *Monelytrum* Hack.

Monelytrum Hack., Verh. Bot. Vereins Prov. Brandenburg 30: 140 (1888).

Stoloniferous perennials. Leaf blades with thickened margins and tubercle-based hairs; cordate. Ligule a fringed membrane. Inflorescence branches with long hairs; branches each bearing 2 to 4 fertile spikelets, and 1 to 3 awn-like structures (rudimentary spikelets). Lower glume tiny, about 1/10 the length of the upper one. Upper glume hairy, similar to the lemma in length, long awned. Lemma with a short apical awn. Leaf epidermal intercostal papillae one per cell.

One sp., *M. leuderitzianum* Hack., southwestern Africa.

615. *Tragus* Haller

Tragus Haller, Hist. Stirp. Helv. 2: 203 (1768), nom. conserv.

Annuals or perennials. Ligule a fringed membrane or a fringe of hairs. Inflorescence branches disarticulating at maturity, each bearing 1 to 5 spikelets. *Upper glume hardened, with prominent veins and rows of hooked prickles*. Lemma awnless. Leaf epidermal intercostal papillae present or absent, one per cell. $2n = 20, 40$.

Eight spp., Africa, South America, and Australia.

Polevansia + Willkommia

Ligule a fringed membrane. *Inflorescence branches flat.* Lemma awnless or with a short awn from the apex.

616. *Polevansia* de Winter

Polevansia de Winter, Bothalia 9: 130 (1966).

Perennials. Spikelets with a short blunt hairy callus. Leaf epidermal intercostal papillae absent.

One sp., *P. rigida* de Winter, Lesotho and South Africa.

617. *Willkommia* Hack.

Willkommia Hack., Verh. Bot. Vereins Prov. Brandenburg 30: 145 (1888).

Annuals or perennials, stoloniferous or not. Spikelets with a sharp callus. Upper glume longer than the flowers. Leaf epidermal intercostal papillae one per cell.

Four spp., southern Africa, Texas, introduced in Argentina.

Blepharidachne + Scleropogon + Swallenia + Munroa + Erioneuron

Inflorescence narrow, unbranched or with branched branches. Plants bisexual or gynomonoecious. This clade corresponds to the former subtribe Scleropogoninae, which is not recognized here because to do so would force recognition of several monogeneric subtribes.

618. *Blepharidachne* Hack.

Blepharidachne Hack., Die Nat. Pflanzenfam. II, 2: 126 (1887).

Caespitose perennials or annuals, decumbent and mat-forming, less than 20 cm tall. Leaf blades sharp-pointed, the upper ones generally extending beyond the inflorescence. Ligule a fringe of hairs or absent. Inflorescence branches branched, with few spikelets, the peduncles short, surrounded by subtending leaf sheaths. Spikelets laterally compressed, with 4 flowers, *the proximal two staminate or sterile, the third pistillate or bisexual, the fourth a rudiment with three awns.* Glumes shorter than the spikelet, mucronate or not. Lemmas hairy on margins, with three lobes, the lobes awned at the apex. *Lodicules absent.* Anthers 1, 2, or 3. $2n = 14$.

Four spp., U.S.A. and Argentina.

This genus is strongly supported as monophyletic by molecular data (Peterson et al. 2010a), and has obvious synapomorphies. See also Hunziker and Anton (1979).

619. *Scleropogon* Phil.

Scleropogon Phil., Anales Univ. Chile 36: 205 (1870).

Dioecious or monoecious perennials, stoloniferous. Ligule a fringe of hairs. Pistillate inflorescence unbranched, dense, head-like; *each spikelet subtended by a bract*; spikelets cylindrical, the lemmas overlapping. Glumes shorter than the spikelet, their apices acuminate. Lemma margins convolute, with three veins, each ending in an awn more than 5 cm long. Palea with two awns. *Lodicules absent.* Staminate inflorescence with branched branches. Glumes muticous. Lemma apex muticous, without awns. $2n = 40$.

One sp., *S. brevifolius* Phil., southern USA and Mexico, Argentina and Chile.

620. *Swallenia* Soderstr. & H.F. Decker

Swallenia Soderstr. & H.F. Decker, Madroño 17: 88 (1963).

Rhizomatous perennials, less than 40 cm tall. Ligule a fringe of hairs. Leaf blades stiff, their apices sharp-pointed. Inflorescence branches branched. Spikelets laterally compressed, with 3 to 7 flowers plus additional reduced flowers distally. Glumes shorter than the spikelet, their apices acuminate. Lemma pubescent on the veins, the apex mucronate. $2n = 20$.

One sp., *S. alexandrae* (Swallen) Soderstr. & H. F. Decker, California.

Munroa + Erioneuron

Ligule a fringe of hairs. Spikelets laterally compressed. Callus of flower pubescent. *Caryopsis with embryo projecting below the base.*

These two genera form a clade in molecular phylogenies (e.g., GPWG II 2012).

621. *Munroa* Torr.

Munroa Torr., Pacif. Railr. Rep. 4(5): 158 (1857).
Dasyochloa Willd. ex Rydb., Agric. Exp. Sta. Agric. Coll. Colorado Bull. 100: 18, 37 (1906).

Stoloniferous annuals or perennials, decumbent and less than 15 cm tall. Inflorescence unbranched or with unbranched branches spread along a central axis, with 1 to 3 spikelets, these nearly sessile; peduncle short, so that the inflorescence is enclosed by the leaf sheath. Spikelets with 2 to 10 flowers, the terminal spikelet with 2 or 3 flowers, lower spikelets with more. Glumes present or absent, shorter than the spikelet or as long as the flowers, the upper glume mucronate. Lemma with a tuft of hairs midway up margin, apex entire or with 2 or 4 lobes, awned from sinus (es). *Caryopsis translucent.* $2n = 14, 16$.

Five spp., temperate regions of North and South America.

The combination of *Munroa* and *Dasyochloa* is based on their morphological similarities; in addition molecular data place the two as sisters (Columbus et al. 2007; Peterson et al. 2010a). See also Valdés-Reyna and Hatch (1997).

622. ***Erioneuron*** Nash

Erioneuron Nash, Fl. S.E. U. S.: 143 (1903).

Caespitose or stoloniferous perennials. *Leaf blades with thickened margins.* Inflorescence densely capitate, with branched branches. Spikelets laterally compressed. Glumes shorter than or about as long as the spikelet, their apices acuminate and muticous. Lemma densely pubescent, entire or bilobed, awned from a deep sinus. $2n = 16, 32$.

Three spp., southern U.S.A., Argentina.

XXIX.2. Subtribe Orcuttiinae P.M. Petersen and Columbus (2007); tax.: Reeder (1982)

Annuals. *Leaves without ligules, the sheath and blade intergrading.* Spikelets with multiple flowers, the lower ones bisexual, the upper undeveloped. Glumes present or absent; *lodicules reduced or absent. Microhairs with rounded, mushroom-shaped apical cells.*

Three genera, 9 spp. Plants of vernal pools in California.

In addition to the synapomorphies listed above, Orcuttiinae are strongly supported as monophyletic by molecular phylogenetic analyses (Columbus et al. 2010; Peterson et al. 2010a, 2012).

623. ***Neostapfia*** Burtt Davy

Neostapfia Burtt Davy, Erythea 7: 43 (1899).

Inflorescence branches each bearing one spikelet so that the whole appears unbranched, with aborted spikelets at the tip. Spikelets *dorsiventrally compressed*, with 5 flowers. *Glumes absent.* Lemma flabellate, membranous, truncate. Lodicules minute. Leaf epidermal papillae absent. $2n = 40$.

One sp., *N. colusana* (Burtt Davy) Burtt Davy, California.

624. ***Orcuttia*** Vasey Fig. 89

Orcuttia Vasey, Bull. Torrey Bot. Club 13: 219 (1886).

Inflorescence unbranched. Spikelets with 4 to 40 flowers. Glumes shorter than the spikelet. *Lemma*

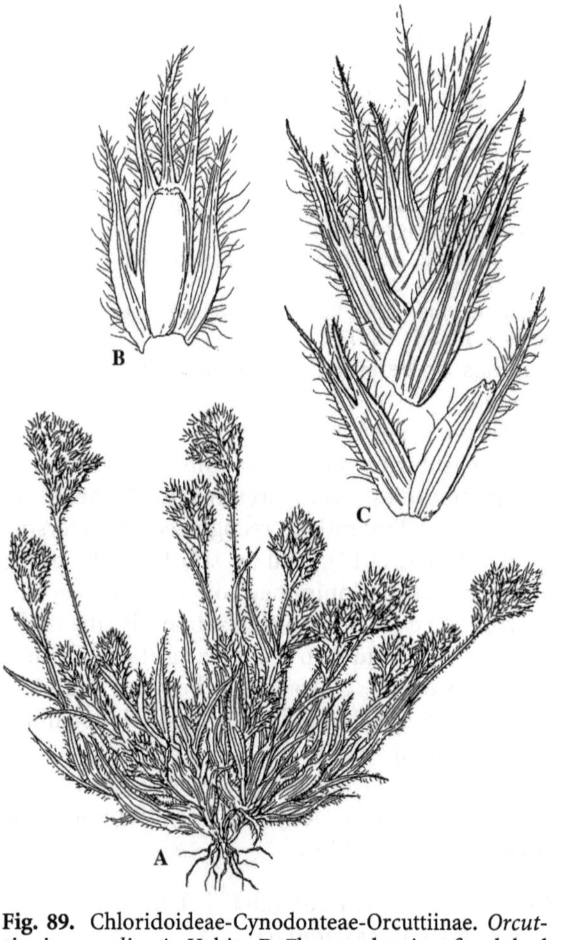

Fig. 89. Chloridoideae-Cynodonteae-Orcuttiinae. *Orcuttia inaequalis.* **A** Habit. **B** Flower showing five-lobed lemma and palea. **C** Flowers and glumes. (From Barkworth et al. 2003, drawn by K. Klitz)

with five lobes or teeth, each ending in a mucro or awn. Lodicules absent. Leaf epidermal papillae one per cell. $2n = 24–32$.

Five spp., California and Mexico.

Some members of this genus are reported to lack an epiblast on the embryo (Reeder 1982).

625. ***Tuctoria*** Reeder

Tuctoria Reeder, Amer. J. Bot. 69: 1090 (1982).

Inflorescence unbranched, *with spikelets arranged in a dense spiral.* Spikelets with 5 to 40 flowers. Glumes shorter than the spikelet. Lemmas entire or with tiny teeth; central mucro present or absent. Lodicules minute, fused to palea. Leaf epidermal papillae absent. $2n = 24, 40$.

Three spp., California and Mexico.

XXIX.3. Subtribe Gouiniinae P.M. Peterson & Columbus (2007)

Cteniinae P. M. Peterson, Romaschenko & Herrera Arrieta, Taxon (2014).
Farragoinae P. M. Peterson, Romaschenko & Herrera Arrieta, Taxon (2014).
Gymnopogoninae P. M. Peterson, Romaschenko & Herrera Arrieta, Taxon (2014).
Perotidinae P. M. Peterson, Romaschenko & Herrera Arrieta, Taxon (2014).
Trichoneurinae P. M. Peterson, Romaschenko & Herrera Arrieta, Taxon (2014).

Annuals or perennials. Spikelets laterally compressed.

Seventeen genera, 195 spp.

This subtribe has no obvious morphological synapomorphies. As circumscribed here, it includes the Tridentinae recognized by Peterson et al. (2010a, 2012), and is substantially more inclusive than Gouiniinae of Peterson et al. (2014a). As elsewhere in this subfamily, the decision of which clades to recognize at the subtribal level represents a trade-off between naming inclusive groups that indicate broad relationships versus naming much smaller groups that have morphological synapomorphies. The subtribe as recognized here is supported as monophyletic (Peterson et al. 2014a).

Gouinia + Schenckochloa + Tridentopsis + Triplasis + Vaseyochloa

Palea keels winged or expanded near the base. Rachilla extension present.

These genera form a moderately well-supported clade in Peterson et al. (2012, 2014a), sister to the rest of the subtribe; they correspond to Gouiniinae s.s. They do not share any obvious morphological characters.

626. ***Gouinia*** E. Fourn. ex Benth. & Hook. f.

Gouinia E. Fourn. ex Benth. & Hook. f., Gen. Pl. 3: 1178 (1883).

Caespitose perennials, sometimes several meters tall. Ligule membranous or a fringed membrane or a fringe of hairs. Inflorescence branches unbranched, borne along a central axis. Spikelets with 2 to 6 flowers plus additional reduced flowers distally, or with a rachilla extension. Glumes shorter than the flowers, the lower one with a muticous or mucronate apex, the upper one with a muticous, mucronate or awned apex. Callus of flower pubescent. Lemma with a long awn, the awn apical or from a sinus. Leaf epidermal papillae generally absent. Microhairs with a long narrow apical cell. $2n = 40, 64, 76, 80$.

Fourteen spp., Central and South America.

627. ***Tridentopsis*** P. M. Peterson

Tridentopsis P. M. Peterson, Taxon 63: 284 (2014).

Caespitose perennials. Culm nodes often with long hairs. Ligule membranous or a fringed membrane or a fringe of hairs. Inflorescence branches branched or not. Spikelets with 5 to 15 flowers plus additional reduced flowers distally. Glumes shorter than or almost as long as the flowers, their apices muticous or mucronate. Callus of flower glabrous or pubescent. Lemma with midvein sometimes excurrent; veins and keel with dense hairs, forming lines on the back of the lemma. Caryopsis dorsiventrally flattened and concave to folded, with a reticulate surface. $2n = 40$.

Two spp., southwestern U.S., northern Mexico, Caribbean.

The genus *Tridentopsis* was erected to accommodate the former *Tridens muticus*, which is firmly placed in Gouiniinae by molecular data (Columbus et al. 2010; Peterson et al. 2010a, 2014a). *Tridens flavus*, a synonym of the type species of *Tridens*, is consistently placed with

Pappophorum [Pappophorinae] (Columbus et al. 2010; GPWG II 2012; Peterson et al. 2010a).

628. ***Triplasis*** P. Beauv.

Triplasis P. Beauv., Ess. Agrostogr. 81 (1812).

Annuals or perennials. *Lower leaf sheaths inflated, containing cleistogamous spikelets.* Ligule a fringed membrane or a fringe of hairs. Inflorescence branches branched. Spikelets with 2 to 5 flowers, the upper ones reduced, the rachilla extended beyond the distalmost flower. Glumes shorter than the flowers, their apices acute. Callus of flower pubescent. Lemma with lines of hairs along the margins and keel; with two lobes; awned from the sinus, *the awn pubescent.* Palea with hairs on keels and at tip. *Embryo projecting below base of fruit.* Leaf epidermal papillae absent. $2n = 40$.

Two spp., eastern U.S., Central America.

629. ***Vaseyochloa*** Hitchc.

Vaseyochloa Hitchc., J. Wash. Acad. Sci. 23: 452 (1933).

Caespitose perennials. Ligule a fringed membrane. Inflorescence branches branched. Spikelets with 5 to 10 flowers. Glumes shorter than the flowers. Callus of flower pubescent. Lemma awnless, with 7 to 9 veins, the proximal portion pubescent. Palea splitting at maturity. Fruit rotund, *with style branches persisting as two horns.* Leaf epidermal papillae one per cell. $2n = 56, 60, 68$.

One sp., *V. multinervosa* (Vasey) Hitchc., Texas, USA.

Remainder of Gouiniinae s.l.

630. ***Ctenium*** Panz.

Ctenium Panz., Ideen Revis. Gräs.: 38, 61 (1813), nom. conserv.

Annuals or perennials, generally caespitose, some species aromatic. Ligule membranous. Inflorescence unbranched, or with unbranched primary branches; *axes curved, crescentic in section, ending in a sterile point.* Spikelets laterally compressed, with *two awned sterile flowers below the fertile one*, plus additional reduced flowers distally. Glumes longer than the flowers, the lower glume sharply pointed, upper obliquely awned from the back. Fertile lemma entire or with two teeth, the awn terminal or subapical. Leaf epidermal papillae absent. $2n = 18, 36, 54, 160$.

Twenty spp., warm regions of America, Africa, and Madagascar.

Ctenium is strongly supported as monophyletic and is placed in a monogeneric subtribe by Peterson et al. (2014a). It is sister to *Trichoneura.* Peterson et al. (2014a) speculate that *Kampochloa* (Chloridoideae, incertae sedis) may be derived from *Ctenium*, based on morphological similarity.

631. ***Trichoneura*** Andersson

Trichoneura Andersson., Kongl. Svenska Vetenskapsakad. Handl., n.s. 1853: 148 (1855).

Annuals or perennials. Ligule membranous. Inflorescence branches unbranched, borne along a central axis, the spikelets on one side of the branch. Spikelets laterally compressed, with 4 to 9 flowers, wedge-shaped. Glumes longer than the adjacent lemmas, in some species longer than the spikelet, mucronate or awned. Lemma *margins with long stiff cilia*, the apex mucronate or with a short awn. Fruit flattened dorsiventrally. *Microhairs more or less spherical.* $2n = 20$.

Eight spp., Africa and Arabia; Texas, Galapagos.

Trichoneura is strongly supported as monophyletic and is placed in a monogeneric subtribe by Peterson et al. (2014a). It is sister to *Ctenium.*

Craspedorhachis + Farrago

Spikelets dorsiventrally compressed, with one flower. Glumes coriaceous longer than the flower. Lemma hyaline.

632. ***Craspedorhachis*** Benth.

Craspedorhachis Benth., Hooker's Icon. Pl. 14: t. 1377 (1882).

Caespitose perennials. Ligule a fringed membrane. Inflorescence branches unbranched, digitate. Spikelets enclosed in hollows in the branch axis, without a rachilla extension. Lower glume asymmetrical about the keel, winged, persistent, the upper glume deciduous. Lemma with long hairs on veins, the apex obtuse. Leaf epidermal papillae absent. $2n = 27$.

Three spp., southern Africa.

Craspedorhachis is monophyletic and strongly supported as sister to *Farrago* by Peterson et al. (2014a). The chromosome number of $2n = 27$ was reported for *C. rhodesiana* Rendle by Moffett and Hurcombe (1949), who noted that it was triploid. The species may thus be a sterile hybrid.

633. ***Farrago*** Clayton

Farrago Clayton, Kew Bull. 21: 125 (1967).

Annuals. Ligule a fringe of hairs. Inflorescence branches unbranched, each branch with 1 fertile spikelet, plus two spikelets reduced to awns. Glumes awned. *Lemma a tiny scale, hairy*. Leaf epidermal papillae absent.

One sp., *F. racemosa* Clayton, Tanzania.

Farrago is strongly supported as sister to *Craspedorachis* by Peterson et al. (2014a) and the two genera placed in their own subtribe.

Dignathia + Leptothrium + Leptocarydion + Gymnopogon + Bewsia + Lophacme

This clade is strongly supported in the analyses of Peterson et al. (2014a), and could be recognized at the subtribal level as Gymnopogoninae P.M. Peterson, Romasch. & Y. Herrera. *Dignathia* and *Leptothrium* are in a very weakly supported clade sister to the other genera (Peterson et al. 2014a); their sister relationship should be considered provisional.

634. ***Dignathia*** Stapf

Dignathia Stapf, Hooker's Icon. Pl. 30: t. 2950 (1911).

Annuals or perennials. Ligule a fringed membrane. Inflorescence branches unbranched, borne along a central axis, deciduous, curved, bearing 1 or 2 spikelets and a terminal one consisting only of glumes. Spikelets laterally compressed, with 1 or 3 flowers. Glumes longer than the flowers, *hard, gibbous, covered with prickles or long hairs*, their apices acuminate, muticous or awned. Lemma apex with a short awn or mucro. Leaf epidermal papillae one per cell.

Five spp., East Africa to NW India.

635. ***Leptothrium*** Kunth

Leptothrium Kunth, Révis. Gramin. 1: 156 (1829).
Latipes Kunth, Révis. Gramin. 1: 53 (1829).

Perennials with wiry culms. Leaves with or without auricles. Ligule a fringe of hairs. Inflorescence unbranched, or with unbranched branches. Branches flattened, wedge shaped, bearing 1 or 2 spikelets, deciduous. Spikelets with 1 flower. Glumes longer than flower, hard, the lower flat, recurved, acuminate, the upper enclosing the flower. Lemma glabrous, awnless. Pericarp not tightly fused to seed coat. Leaf epidermal papillae one per cell, or none. $2n = 20$.

Two spp., Senegal to Pakistan; Caribbean, South America.

636. ***Leptocarydion*** Hochst. ex Stapf

Leptocarydion Hochst. ex Stapf, Fl. Cap. 7: 316 (1898).

Annuals. Ligule a fringed membrane. Inflorescence dense, the primary branches crowded on the rachis. Spikelets with 6 to 14 flowers plus additional distal sterile ones. Callus of the flower hairy. Glumes shorter than the spikelet. Lemma pubescent, the apex entire or bidentate, awned from the apex or from the sinus; stamens 2 or 3. Leaf epidermal papillae several per cell.

One sp., *L. vulpiastrum* (De Not.) Stapf, eastern and southern Africa.

Leptocarydion is strongly supported as sister to the clade consisting of *Lophacme*, *Bewsia*, and *Gymnopogon* (Peterson et al. 2014a).

637. ***Gymnopogon*** P. Beauv.

Gymnopogon P. Beauv., Ess. Agrostogr. 41, 164 (1812); tax.: Cialdella and Zuloaga (2011).

Perennials (annuals), caespitose or rhizomatous. Leaf blades stiffly distichous, apiculate at the apex. Ligule a fringed membrane. Inflorescence branches unbranched, digitate or borne along a central axis. Spikelets contiguous or distant along the inflorescence branches, tightly appressed to the branches, generally with 1 flower, the proximal ones bisexual, the distal ones reduced; rachilla extended, ending in an awn. Glumes longer than the flowers, acuminate. Lemma minutely two-lobed, muticous or awned from sinus. Anthers 2 or 3. Leaf epidermal papillae several per cell. $2n = 40$.

Fourteen spp., southern USA and South America; one species in India and SE Asia.

Gymnopogon is strongly supported as sister to the *Lophacme-Bewsia* clade (Peterson et al 2014a).

638. *Bewsia* Goossens

Bewsia Goossens, S. African J. Sci. 37: 183 (1941).

Caespitose perennials. Ligule membranous. Inflorescence with unbranched branches, these distributed along and appressed to the main axis. Spikelets with 2 to 4 flowers, with additional reduced flowers distally. Glumes shorter than the flowers, the apices acuminate. Callus of flower pubescent. Lemma pubescent on keel and margins, with an abaxial awn. Leaf epidermal papillae absent. $2n = 30, 40$.

One sp., *B. biflora* (Hack.) Goossens, central and southern Africa.

GPWG II (2012) and Peterson et al. (2012) place this sister to *Gymnopogon* with strong support, although the two genera do not have obvious shared morphological characters.

639. *Lophacme* Stapf

Lophacme Stapf, Fl. Cap. 7: 316 (1898).

Annuals or perennials. Ligule a line of hairs, or membranous, or lacking. Leaf blades stiff. Inflorescence branches unbranched, digitate. *Spikelets wedge-shaped*, with 1 or 2 fertile flowers and distal sterile flowers reduced to awns, *these dispersed with the upper flower*. Callus of flower pubescent. Glumes longer or shorter than the spikelet. Lemma hyaline or thinly membranous, the apex with two teeth, with a long awn from the sinus. Pericarp easily separated from seed coat. Leaf epidermal papillae absent. Microhairs with an apical cell that is long and narrow, or about as broad as long.

Two spp., southern Africa.

Peterson et al. (2012, 2014a) place *Lophacme* in a well-supported clade with *Gymnopogon* and *Bewsia*.

Perotis + Mosdenia + Trigonochloa

Leaf blades lanceolate to lance-ovate, clasping the culm. Spikelets with 1 flower, rachilla extension absent. Glumes longer than the flower. Caryopsis longitudinally grooved.

These genera fall in a strongly supported clade in the analyses of Peterson et al. (2012, 2014a), and could be recognized at the subtribal level as Perotidinae P.M. Peterson, Romasch., & Y. Herrera. The longitudinally grooved caryopsis may be synapomorphic (Snow and Peterson 2012b).

640. *Trigonochloa* P. M. Peterson & N. Snow

Trigonochloa P. M. Peterson & N. Snow, Ann. Bot. 109: 1327 (2012); tax.: Snow and Peterson (2012b).

Annuals or perennials, rhizomatous or stoloniferous, culms sometimes decumbent. Ligule membranous. Inflorescence branches unbranched, spread along the central axis. Spikelets laterally compressed, disarticulating above the glumes. Lemma apex entire. *Caryopsis trigonous*, with a deep groove on the hilar side. *Leaf epidermal cells with a single central papilla.* $2n = 36$.

Two spp., tropical Africa and Asia.

The leaves of *Trigonochloa* are thinly membranous. The primary vascular bundles of *T. uniflora* have unusually large bundle sheath cells on either side, a character first observed by Metcalfe (1960); this character may be a synapomorphy for the genus (Peterson et al. 2012), but data are lacking for *T. ruprestis*. Micromorphology of the lemma is distinctive and supports exclusion of *Trigonochloa* from *Leptochloa*; silica cells and epidermal papillae are lacking in *Trigonchloa*.

Mosdenia + Perotis

Inflorescence unbranched. *Disarticulation below the glumes.*

Mosdenia and *Perotis* are sisters in the analyses of Peterson et al. (2014a).

641. *Mosdenia* Stent

Mosdenia Stent, Bothalia 1: 170 (1923).

Perennials, rhizomatous or stoloniferous, the stolons with overlapping scale leaves. Ligule a fringed membrane. Spikelets densely crowded along the rachis, spreading, dorsiventrally compressed. Glumes with apices acuminate, awnless. Lemma delicate, the apex obtuse, awnless, without hairs. Leaf epidermal papillae absent. $2n = 40$.

One sp., *M. leptostachys* (Ficalho & Hiern) Clayton, South Africa.

642. ***Perotis*** Aiton Fig. 90

Perotis Aiton, Hort. Kew. 1: 85 (1789).
Lopholepis Decne., Arch. Mus. Hist. Nat. 1: 147 (1839), nom. nov. pro seq. *Holboellia* Hook.
Toliara Judz., Adansonia III, 31: 274 (2009).

Annuals or perennials, most caespitose. Leaf blades stiff, with a sharp apex. Ligule membranous or a fringe of hairs. Inflorescence sometimes partially enclosed by the subtending leaf. Spikelets more or less terete. Glumes longer than the flowers, with long awns. Lemma apex entire, awnless. $2n = 36, 40$.

Fifteen spp., Old World tropics.

Both *Lopholepis* and *Toliara* are derived from within *Perotis* (Peterson et al. 2014a). The placement of *Toliara* was predicted by Judziewicz (2009) on the basis of morphology; the two genera are quite similar, but *P.* (= *T.*) *arenacea* lacks the distinctive awned glumes of *Perotis*. The spikelet of *P.* (= *L.*) *ornithocephala* is described as "one of the oddest shaped spikelets in the Gramineae" (Clayton and Renvoize 1986); the upper glume is constricted so as to resemble a cartoon of a bird's head.

XXIX.4. Subtribe Triodiinae Benth. (1881)

Perennials with *hard, needle-like, sharp-pointed leaves*. Ligule a fringe of hairs. Inflorescence branches branched. Spikelets laterally compressed. *Mesophyll "Triodia-type" with PCR tissue only on sides of major bundles and draping over minor bundles.*

Three genera, seventy species.

This subtribe is sufficiently cohesive that all genera could easily be placed in a single genus.

643. ***Triodia*** R. Br. Fig. 91

Triodia R. Br., Prodr.: 182 (1810).
Plectrachne Henrard, Vierteljahrsschr. Naturf. Ges. Zürich. 74: 132 (1929).

Spikelets with 4 to 15 flowers. Glumes shorter than flowers. *Lemmas becoming hard*, with three teeth or lobes. Leaf epidermal papillae absent abaxially.

Sixty-six spp., Australia.

644. ***Monodia*** S. W. L. Jacobs

Monodia S. W. L. Jacobs, Kew Bull. 40: 659 (1985).

Spikelets with one flower; rachilla extended beyond the flower. Lemma convolute, with two lobes, awned from the sinus, with a line of hairs on each margin. Leaf epidermal papillae one per cell.

One sp., *M. stipoides* S. W. L. Jacobs, Western Australia.

645. ***Symplectrodia*** Lazarides

Symplectrodia Lazarides, Nuytsia 5: 273 (1985).

Spikelets with only one flower fertile, the others sterile and clumped; *rachilla internodes unusually long*. Lemma entire, with an apical awn; *palea fused to rachilla below*. Leaf epidermal papillae one to several per cell.

Two spp., Australia.

XXIX.5. Subtribe Eleusininae Dumort. (1829)

Annuals or perennials. Inflorescence unbranched, or with primary branches only, or with branched branches. Spikelets laterally or dorsiventrally compressed or nearly terete. Lemma with 3 veins. Pericarp fused to or free from seed coat. Leaf epidermal papillae generally present, generally one per cell.

Thirty genera, 229 species.

This subtribe has no obvious morphological synapomorphy, but is well supported by molecular data (GPWG II 2012; Peterson et al. 2010a, 2012).

646. ***Acrachne*** Wight & Arn. ex Chiov.

Acrachne Wight & Arn. ex Chiov., Annuario Reale Ist. Bot. Roma 8: 361 (1908).

Fig. 90. Chloridoideae-Cynodonteae-Gouiniinae. *Perotis rara*. A Habit. B Spikelet showing long awns on glumes. C Caryopsis. (From Wu et al. 2007, with permission from the Missouri Botanical Garden Press, St. Louis, and Science Press, Beijing; drawn by Shi Weiqing)

Annuals. Ligule membranous or a fringed membrane. Inflorescence branches unbranched, with overlapping spikelets, digitate or spread along a central axis. Spikelets laterally compressed, with 2 to 25 flowers, with additional reduced flowers distally. Glumes shorter than the spikelet, the lower one mucronate, the upper one awned. Lemma awned from the apex, deciduous; paleas persistent; rachilla tough. Pericarp free from the seed coat. *Leaf epidermal papillae absent.*

Three spp., Old World tropics.

647. ***Afrotrichloris*** Chiov.

Afrotrichloris Chiov., Ann. Bot. (Rome) 13: 371 (1915).

Perennials. Ligule a fringed membrane. Inflorescence unbranched. *Spikelets with one fertile flower and a cluster of sterile awned lemmas.* Glumes divergent, acuminate. Lemmas hairy, deeply divided into two lobes, awned from sinus. Pericarp free from seed coat. Leaf epidermal papillae one per cell.

Two spp., Somalia.

648. ***Astrebla*** F. Muell.

Astrebla F. Muell., Fl. Austral. 7: 602 (1878).

Caespitose perennials. Ligule a fringed membrane or a fringe of hairs. Inflorescence unbranched or with two branches. Spikelets laterally compressed or nearly terete, with 1 to 6 flowers plus several reduced ones distally. Glumes shorter than or about the same length as the spikelet, their apices muticous or mucronate. Callus of flowers pubescent. Lemmas with silky hairs, three-lobed, the central lobe or all three lobes awned. Pericarp free from seed coat. $2n = 40$.

Four spp., Australia.

649. ***Austrochloris*** Lazarides

Austrochloris Lazarides, Austral. J. Bot., Suppl. Ser. 5: 33 (1972).

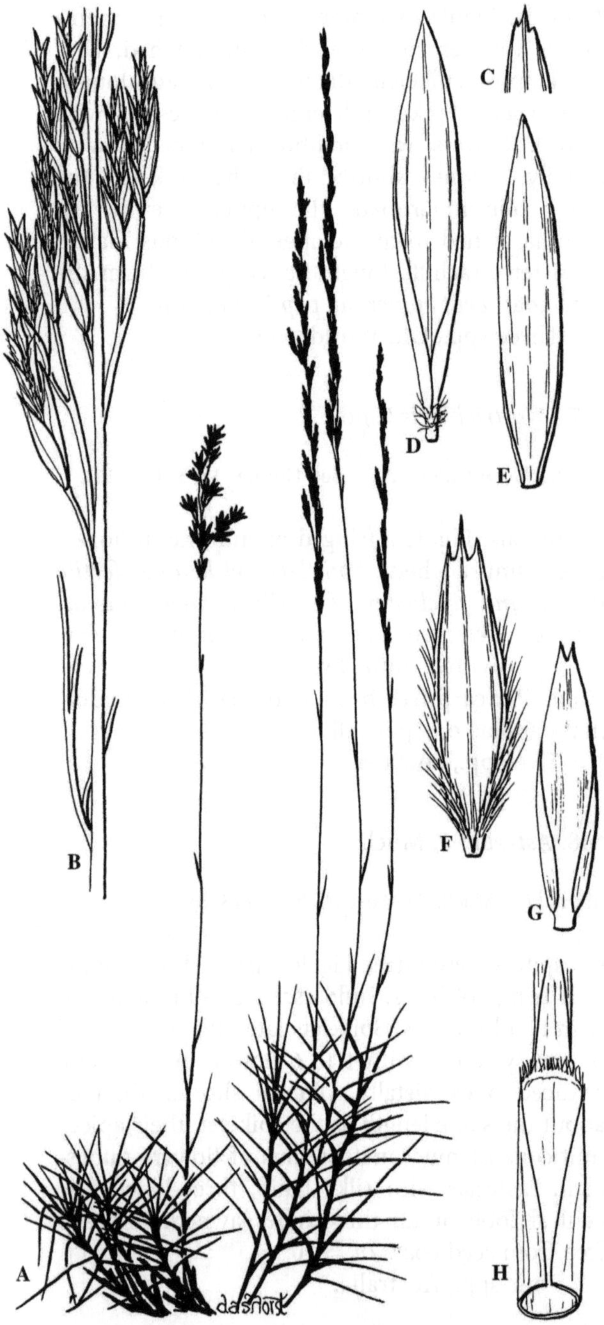

Fig. 91. Chloridoideae-Cynodonteae-Triodiinae. *Triodia scariosa*. A Habit. B Inflorescence. C–G All adaxial view. C Apex of lower glume. D Lower glume. E Upper glume. F Lemma. G Palea. H Ligule. (From Jessop et al. 2006)

Perennials. Ligule a fringe of hairs. Inflorescence with 2 or 3 unbranched branches, the branch axes hollowed and flattened. Spikelets dorsiventrally compressed, with one fertile flower plus one sterile lemma; rachilla hairy. Glumes much longer than flower, divergent, their apices acute to acuminate. Callus of flower pubescent. Fertile lemma with an obtuse apex, with subapical awn. Pericarp free from seed coat.

One sp., *A. dichanthoides* (Everist) Lazarides, Australia.

650. ***Brachyachne*** (Benth. & Hook. f.) Stapf

Brachyachne (Benth. & Hook. f.) Stapf, Hooker's Icon. Pl. 31: t. 3099 (1922).

Annuals or perennials, stoloniferous or not, often decumbent or prostrate. Ligule membranous, or a fringed membrane, or a fringe of hairs. Inflorescence unbranched, or of digitate unbranched branches, these flattened and hollowed, surrounded by a leaf sheath or spatheole. Spikelets laterally compressed, with one flower and a rachilla extension. Glumes longer than the flower and enclosing it, broad, with obtuse apices. Callus of flower pubescent or glabrous. Lemma apex acute to bidentate, muticous or mucronate. Pericarp fused to seed coat.

Ten spp., wet areas of tropical Africa, southeast Asia, Australia.

651. **"*Chloris*"** Sw.

"*Chloris*" Sw., Prod. 1, 25 (1788).
Pterochloris (A. Camus) A. Camus, Bull. Mus. Hist. Nat. (Paris) 29: 349 (1957).

Annuals or perennials, caespitose or stoloniferous. Ligule membranous, or a fringed membrane, or a fringe of hairs. Inflorescence unbranched, or more often with digitate primary branches. Spikelets laterally compressed, with 1 or 2 flowers plus additional reduced flowers distally, or with a naked rachilla extension. Callus of flower glabrous or more often pubescent. Glumes shorter or longer than the spikelet, their apices muticous or mucronate. Lemma with a prominent keel, muticous or awned from tip or abaxial side or a sinus. *Leaf epidermal papillae several per cell.* Pericarp free from or fused to seed coat. $2n = 14, 20, 26, 30, 36, 40, 72, 80$.

Sixty-three spp., warm regions of the world.

Peterson et al. (2010a) show that four species of *Chloris* form a clade, whereas Columbus et al. (2010) suggested polyphyly for "*Chloris*", based on a sample of three species, all different from the ones that Peterson et al. (2010a) investigated. Both studies show that *Trichloris* should be excluded from "*Chloris*". Although *Daknopholis* is placed Incertae sedis in this treatment, it may perhaps be included in "*Chloris*", with which it shares many morphological characters. *Pterochloris* is included in "*Chloris*" here, following Clayton and Renvoize (1986).

652. ***Chrysochloa*** Swallen

Chrysochloa Swallen, Proc. Biol. Soc. Wash. 54: 44 (1941).

Annuals or perennials, caespitose or stoloniferous. Leaves with blunt apices. Ligule a fringed membrane. Inflorescence unbranched, or of flattened unbranched branches. Spikelets laterally compressed, with 2 flowers, the lower bisexual, the upper staminate or sterile. Glumes about the same length as lemmas; lower glume persistent, the apex acute; upper glume deciduous, the apex emarginate or acute, muticous to awned. Lemma midvein and margins pubescent, with a short point or awn from near the top of the abaxial side. Pericarp fused to seed coat. $2n = 14$.

Four spp., tropical Africa.

653. ***Cynodon*** Rich.

Cynodon Rich., Syn. Pl. 1: 85 (1805) nom. conserv.; phylog.: Jewell et al. (2012).

Annuals or perennials, caespitose, rhizomatous or stoloniferous. Ligule membranous or a fringed membrane. Inflorescence branches unbranched, flattened, borne on a short central axis. Spikelets laterally compressed, with 1 flower; rachilla extension present or absent; distal incomplete flowers reduced or absent. Glumes shorter than or about the same length as the spikelet, their apices acute or acuminate, muticous or mucronate. Lemma apex obtuse or acute, awnless. Leaf epidermal papillae one *to several per cell.* Pericarp fused to seed coat. $2n = 16, 18, 27, 36, 40, 54$.

Twelve spp., Old world tropics, one species pantropical.

Peterson et al. (2010a) show an unresolved relationship among species of *Cynodon* and those of *Brachyachne*. Species limits and crossability of species were investigated by Harlan et al. (1969).

654. ***Daknopholis*** Clayton

Daknopholis Clayton, Kew Bull. 21: 102 (1967).

Stoloniferous annuals. Ligule a fringe of hairs. Inflorescence branches unbranched, flattened, borne on a contracted axis. Spikelets with one flower, rachilla extension short. Glumes shorter than lemmas, unawned, papery, their apices obtuse to truncate. Lemma two-lobed, awned or sometimes awnless. Pericarp free from seed coat. Leaf epidermal papillae several per cell.

One sp., *D. boivinii* (A. Camus) Clayton, Indian Ocean islands, Madagascar, and eastern coast of Africa.

Daknopholis is sometimes placed in synonymy under "*Chloris*".

655. ***Dinebra*** Jacq.

Dinebra Jacq., Fragm. Bot. 77, pl. 121, f. 1 (1809); phylog., tax.: Peterson et al. (2012); Snow and Peterson (2012a).
Drake-brockmania Stapf, Bull. Misc. Inform. Kew 1912: 197 (1912).
Heterocarpha Stapf & C. E. Hubb., Bull. Misc. Inform. Kew 1929: 263 (1929).

Annuals or perennials, caespitose, stoloniferous or not. *Leaf blades sometimes with multicellular glands abaxially.* Ligule membranous or a fringed membrane. Inflorescence branches unbranched, sub-digitate or borne along a central axis, often deciduous. Spikelets overlapping on the branches, laterally compressed, with 1 to 18 flowers, the distal ones reduced. Glumes shorter or longer than flowers, the apex acute, acuminate or extended into a long tail. Lemma pubescent on veins, veins sometimes 5 or 7, the apex mucronate or not. Caryopsis trigonous. Pericarp free from or fused to seed coat. Leaf epidermal papillae present, one per cell. Glandular microhairs present. $2n = 20$.

Twenty-seven spp., temperate to tropical areas nearly worldwide.

Molecular data place *Dinebra* in a clade with 24 species formerly placed in *Leptochloa* (Columbus et al. 2007; Peterson et al. 2010a, 2012), plus

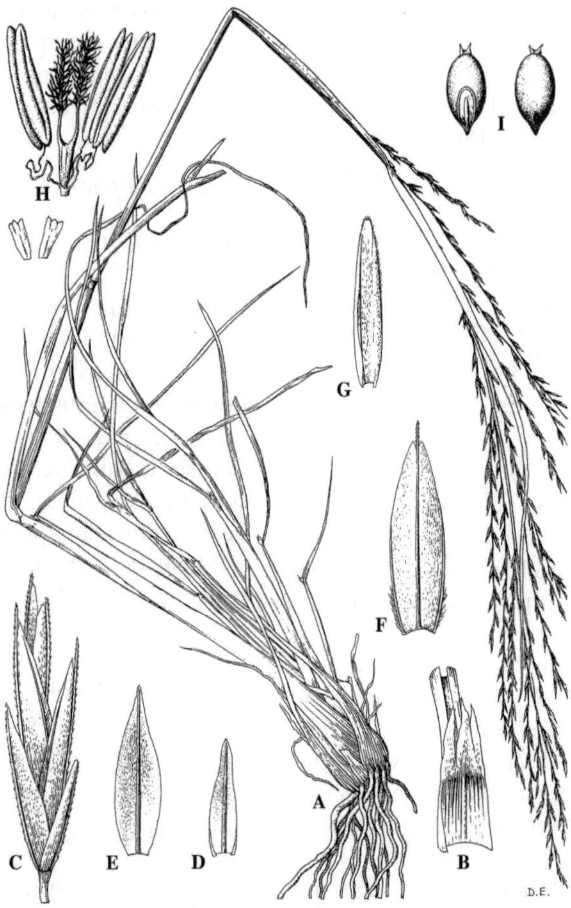

Fig. 92. Chloridoideae-Cynodonteae-Eleusininae. *Diplachne fusca*. **A** Habit. **B** Ligule. **C** Spikelet. **D** Lower glume. **E** Upper glume. **F** Lemma. **G** Palea. **H** Flower with separate lodicules. **I** Caryopsis. (From Cope 1982)

the monospecific *Drake-brockmania* and *Heterocarpha*. Phillips (1974) includes *Heterocarpha haareri* in *Drake-brockmania* on the basis of "an inflorescence of short, broad, deciduous spikes." The number of veins in the upper glume varies in *Dinebra* from 1 in species formerly classified in *Leptochloa*, to as many as 17 in the former *Drake-Brockmania somalensis*.

656. *Diplachne* P. Beauv. Fig. 92

Diplachne P. Beauv., Agrostogr. 80, pl. 16, f. 9 (1812); phylog., tax.: Peterson et al. (2012), Snow and Peterson (2012a).

Annuals or perennials, caespitose or rhizomatous. Ligule membranous, relatively long and acute. Inflorescence with unbranched branches, these spread along a central axis. Spikelets with 6 to 20 flowers. Glumes shorter than or about as long as the flowers, their apices acute. Lemma often with a dark spot near the base, the apex acute to truncate or bifid, mucronate to awned. $2n = 20$.

Two spp., temperate to tropical areas throughout the world.

657. *Disakisperma* Steud.

Disakisperma Steud., Syn., Pl. Glumac. 1: 287 (1854); phylog., tax.: Peterson et al. (2012), Snow and Peterson (2012a), Snow et al. (2013).

Perennials. Ligule membranous, truncate, erose or ciliate. Inflorescence with unbranched branches, these subdigitate or spread along a central axis. Spikelets with 4 to 13 flowers. Glumes shorter than the flowers, their apices acute to obtuse. Lemma entire or bifid, unawned or mucronate. Caryopsis flattened dorsiventrally, weakly concave; pericarp weakly fused to the seed coat. $2n = 40, 60, 80$.

Four spp., North and South America, Africa.

Disakisperma species have lemmas with macrohairs that are somewhat inflated at the tip ("clavicorniculate") (Snow 1996, and pers. comm.). Based on this character and molecular data, *Coelachyrum yemenicum* (Schweinf.) S.M. Phillips has recently been transferred to *Disakisperma*, as D. *yemenica* (Chiov.) P. M. Peterson & N. Snow.

658. "*Enteropogon*" Nees

"*Enteropogon*" Nees, Intr. Nat. Syst. Bot.: 448 (1836).

Annuals or perennials. Ligule membranous, or a fringed membrane, or a fringe of hairs. Inflorescence unbranched, or of unbranched branches, these digitate or spread along a central axis. Spikelets laterally or dorsiventrally compressed, with one fertile flower plus additional reduced flowers distally, *the latter forming a clump (rudiment).* Glumes shorter or longer than the spikelet, their apices muticous or mucronate, the upper one sometimes awned. *Lemma indurate,* the veins forming prominent ridges, the apex entire or bifid, awned from the apex or from a sinus. *Fruit concavo-convex in section, with a longitudinal groove.* Pericarp free from seed coat. $2n = 20$.

Seventeen spp., tropics.

"*Enteropogon*" is either para- or polyphyletic, but taxon sampling in existing molecular trees is insufficient to draw any conclusions (Columbus et al. 2007; Peterson et al. 2010a); the type species (*E. melicoides*) has not been sampled. Peterson et al. (2010a) indicate that *Saugetia* is unrelated to the sampled species of "*Enteropogon*".

659. "*Eustachys*" Desv.

"*Eustachys*" Desv., Nouv. Bull. Sci. Soc. Philom. Paris 2: 188 (1810).

Perennials, caespitose, short-rhizomatous, or stoloniferous. Ligule a fringed membrane or a fringe of hairs. Inflorescence unbranched, or with unbranched branches, the branch axes flattened or hollowed. Spikelets laterally compressed, with 1 flower plus additional reduced flowers distally, the rachilla terminating in a club-shaped lemma. Glumes shorter or longer than the spikelets, their apices obtuse to acute, the upper one often with a subapical awn. Callus of flower glabrous to pubescent. Lemma mucronate or awned, dark brown at maturity. Pericarp fused to seed coat. Leaf epidermal papillae one *or several per cell.* $2n = 40$.

Sixteen spp., tropics, mostly New World.

E. petraea (the type species) and *E. paspaloides* are sisters in the study of Peterson et al. (2010a), indicating that some component of this genus is monophyletic. However, *E. distichophylla* is unrelated.

660. *Harpochloa* Knuth

Harpochloa Knuth, Révis. Gramin. 1: 92 (1829).

Perennials. Leaves stiff. Ligule a fringed membrane. Inflorescence unbranched, ending in a bare tip, *the axis strongly curved.* Spikelet laterally compressed, with one fertile flower plus an additional reduced flower distally. Glumes as long as the spikelet, with acute apices, the lower one persistent, the upper deciduous. Lemma pubescent, awnless, the apex notched. Pericarp fused to seed coat. *Adaxial mesophyll cells colorless and interrupting the outer bundle sheath.* $2n = 40$.

Two spp., Central and southern Africa.

Roodt and Spies (2003) also report $2n = 36$ and 54 for *Harpochloa*.

661. *Leptochloa* P. Beauv.

Leptochloa P. Beauv., Ess. Agrostogr.: 71 , pl. 15, f. 1 (1812); phylog.: Peterson et al. (2012), Snow and Peterson (2012a).

Perennials, caespitose, rhizomatous or stoloniferous. Ligule short, membranous. Inflorescence branches unbranched, sub-digitate or spread along a central axis. Spikelets laterally compressed, with 1 to 7 flowers plus additional reduced flowers distally. Glumes shorter or longer than the flowers, generally muticous. Callus of flower pubescent or glabrous. Lemma apex awnless, or with 1 to 3 awns. Fruit smooth, pericarp free from, adherent to or fused to seed coat. $2n = 20, 40, 60$.

Five spp., southern U.S. to South America, Australia.

Molecular data show that *Leptochloa* as traditionally circumscribed is polyphyletic (Columbus et al. 2007; Peterson et al. 2010a, 2012), consistent with previous data from morphology and anatomy (Snow 1997). Monophyly is restored once *Dinebra*, *Diplachne* and *Disakisperma* are redefined, and *Trigonochloa* is recognized.

662. ***Lepturus*** R. Br.

Lepturus R. Br., Prodr.: 207 (1810).

Annuals or perennials, caespitose or stoloniferous. Leaf blades with or without a pseudopetiole, with sharp-pointed apices. Ligule membranous or a fringed membrane or absent. Inflorescence unbranched, *disarticulating at nodes; spikelets embedded in the rachis.* Spikelets dorsiventrally compressed, with one or two flowers, with additional reduced flowers distally or with a sterile rachilla extension. *Lower glume minute or absent*; upper glume covering the spikelet, muticous or awned. Pericarp free from seed coat. *Leaf epidermal papillae several per cell.* Microhairs short and stout, spherical. $2n = 14$, 18, 26, 36, 42, 52, 54.

Sixteen spp., beaches of Indian and Pacific oceans.

663. ***Lintonia*** Stapf

Lintonia Stapf, Hooker's Icon. Pl. 30: t. 2944 (1911).

Stoloniferous perennials. Ligule membranous. Inflorescence with several unbranched branches, digitate or along a short central axis. Primary branches slender, sinuous. Spikelets laterally compressed, with 4 to 10 flowers plus several reduced ones distally. Callus below proximal flower pubescent. Glumes shorter than the spikelet. *Lemmas with clavate hairs on sides*, bilobed with prominent hardened veins extending into the lobes, awned from back, branching. *Lodicules cylindrical.* Androecium and gynoecium on a short stipe. Pericarp free from seed coat. $2n = 30$.

Two spp., Eastern Africa.

Clayton and Renvoize (1986) note similarity between the lemma venation in this genus and that in *Astrebla*, but molecular data place *Lintonia* sister to (Peterson et al. 2010a) or within (GPWG II 2012) "*Chloris*".

664. ***Microchloa*** R. Br.

Microchloa R. Br., Prodr. 208 (1810).
Rendlia Chiov., Ann. Bot. (Rome) 13: 53 (1914).

Annuals or perennials. Ligule a fringed membrane. Inflorescence unbranched, curved, crescentic in outline, hollowed, the spikelets borne on one side. Spikelets dorsiventrally compressed, with one fertile flower, in some species with additional reduced flowers distally. Callus of flowers pubescent. Glumes extending beyond the flowers, their apices obtuse or acute. Lemma acute to bilobed, sometimes with a short mucro. Pericarp fused to seed coat. *Leaf epidermal papillae absent.* $2n = 40$.

Six spp., Africa, 1 pantropical.

Clayton and Renvoize (1986) place *Rendlia altera* (Rendle) Chiov. in synonymy with *Microchloa*, whereas Watson and Dallwitz (1992 onward) exclude it.

665. ***Oxychloris*** Lazarides

Oxychloris Lazarides, Nuytsia 5: 283 (1985).

Annuals or perennials. Ligule a fringed membrane. Inflorescence with 3 to 6 unbranched branches, these digitate, flattened and hollowed. Spikelets laterally compressed, with one flower plus additional reduced flowers distally, *the callus long, sharp, and pubescent.* Glumes shorter than the spikelet, the lower one with an obtuse apex, the upper with an emarginate apex. Lemma of fertile flower with tufts of hairs, winged, with a subapical awn; *lemmas of sterile flowers broadly expanded, flabellate. Lodicules absent.* Pericarp fused to seed coat.

One sp., *O. scariosa* (F. Muell.) Lazarides, Australia.

666. ***Rheochloa*** Filg., P. M. Peterson & Y. Herrera

Rheochloa Filg., P.M. Peterson & Y. Herrera, Syst. Bot. 24: 123 (1999).

Decumbent perennials. Ligule a fringed membrane. Inflorescence branches unbranched, digitate. Spikelets laterally compressed, with three flowers, the proximal two fertile and the distal one reduced. Glumes longer than the flowers, coriaceous, the apices acute. Lemma hyaline, with an obtuse ciliate apex. Pericarp fused to seed coat. *Leaf epidermal papillae several per cell.*

One sp., *R. scabriflora* Filg., P.M. Peterson & Y. Herrera, Brazil.

667. ***Saugetia*** Hitchc. & Chase

Saugetia Hitchc. & Chase, Contr. U.S. Natl. Herb. 18: 378 (1917).

Perennials. *Leaf blades dominated by the midrib, the lateral flattened portion narrow.* Ligule a fringed membrane. Inflorescence unbranched, or of digitate unbranched branches. Spikelets partially embedded in the inflorescence axis, dorsiventrally compressed, with one flower plus an additional reduced flower distally, with elongated rachilla internodes above glumes and between flowers. Callus of spikelet long, pubescent. Glumes shorter than the spikelet, the proximal one much smaller than the distal one, their apices acuminate. Lemma two toothed, awned from the sinus. Pericarp free from seed coat. *Leaf epidermal papillae absent.*

Two spp., West Indies.

Peterson et al. (2010a) place *S. fasciculata* sister to *Tetrapogon*, contradicting the decision of Clayton et al. (2006 onward) and Clayton and Renvoize (1986) to sink *Saugetia* into "*Enteropogon*".

668. ***Schoenefeldia*** Kunth

Schoenefeldia Kunth, Révis. Gramin. 1: 86 (1829).

Annuals or perennials. Ligule a fringed membrane. Inflorescence of unbranched branches, digitate. Spikelets laterally compressed, with one flower, or with a rachilla extension ending in an awn. Callus of flower pubescent. *Glumes narrow, setaceous, longer than flower, acuminate or awned.* Lemma with two teeth, awned from sinus, the awns from adjacent lemmas entangled. Pericarp free from seed coat.

Two spp., Africa, Madagascar and India.

669. ***Sclerodactylon*** Stapf

Sclerodactylon Stapf, Bull. Misc. Inform. Kew 1911: 318 (1911).

Stoloniferous perennials. Ligule a line of hairs. *Leaf blades woody, elliptical or circular in cross section, with sharp tips.* Inflorescence branches two or more, unbranched. Spikelets laterally compressed, with 5 to 40 flowers plus additional reduced flowers distally. Glumes shorter than the spikelet, their apices acute, the upper glume mucronate. *Palea winged.* Pericarp free from seed coat.

One sp., *S. macrostachyum* (Benth.) A. Camus, Madagascar, islands of the Indian Ocean, east African coast.

670. ***Tetrapogon*** Desf.

Tetrapogon Desf., Fl. Atlant. 2: 388 (1799).

Annuals or perennials, caespitose or short rhizomatous. Ligule a fringed membrane. Inflorescence unbranched, or with 2 or 3 unbranched branches. Spikelets laterally compressed, wedge-shaped, with 1 to 7 flowers plus additional reduced flowers distally. Callus of flower pubescent. Glumes shorter than to longer than the spikelet, muticous or mucronate. Lemma entire or two-toothed, with long hairs on veins and keel, with a subapical awn. Pericarp free from seed coat or not. $2n = 20$.

Five spp., Africa, Middle East and India.

671. ***Trichloris*** E. Fourn. ex Benth.

Trichloris E. Fourn. ex Benth., J. Linn. Soc. Bot. 19: 102 (1881).

Perennials, stoloniferous or not. Ligule a fringe of hairs. Inflorescence of unbranched branches, digitate. Spikelets dorsiventrally compressed, with 1 or 2 flowers plus an additional reduced flower distally. Callus of flower pubescent. Glumes shorter than the spikelet, their apices acuminate, the upper one awned. Lemma with two teeth, awned from sinus and both teeth. Palea keels prolonged. *Fruit concavo-convex in section.* Pericarp free from seed coat.

Two spp., South America.

The fruit and pericarp characters suggest that this genus might belong in the following clade.

Apochiton + Coelachyropsis + Coelachyrum + Eleusine

Fruit trigonous or concavo-convex, rugose or sculpted in all but A. burtii, the pericarp free from the seed coat.

672. ***Apochiton*** C. E. Hubb.

Apochiton C. E. Hubb., Hooker's Icon. Pl. 34: t. 3319 (1936).

Annuals. Ligule membranous. Inflorescence branches branched, the spikelets on filiform pedicels. Spikelets laterally compressed, with 3 to 5 flowers plus additional reduced flowers distally. Glumes shorter than the spikelet, their apices mucronate or awned. Callus of flower pubescent. Lemma apex acute, entire, awned. *Palea with two awns.* Microhairs with a broad apical cell.

One sp., *A. burtii* C. E. Hubb., Tanzania.

673. ***Coelachyropsis*** Bor

Coelachyropsis Bor, Ann. Naturhist. Mus. Wien 75: 23 (1971) [1972].

Annuals, less than 20 cm tall. Leaf blades cordate at the base, with multicellular glands. Ligule a fringed membrane. Inflorescence branches unbranched, digitate. Spikelets laterally compressed, with 4 to 10 flowers plus additional reduced flowers distally. Glumes shorter than the spikelet, their apices acuminate, mucronate. Lemma pubescent on midvein and margins, the apex mucronate.

One sp., *C. lagopoides* Bor, southern India and Sri Lanka.

674. ***Coelachyrum*** Hochst. & Nees

Coelachyrum Hochst. & Nees, Linnaea 16: 221 (1842).
Cypholepis Chiov., Annuario Reale Ist. Bot. Roma 8: 357 (1908).

Annuals or perennials, caespitose or stoloniferous. Ligule membranous. Inflorescence branches branched or not, if unbranched then digitate or borne along a central axis. Spikelets laterally compressed, with 5 to 16 flowers plus additional reduced flowers distally. Glumes shorter than the spikelet, their apices obtuse to acute, muticous or mucronate, or the upper awned. Lemma glabrous or with hairs or prickles on margins and keel, the apex muticous to mucronate. *Leaf epidermal papillae absent.* Microhair apical cell short and broad. *Macrohairs crispate.*

Four spp., Africa, Arabia, Pakistan.

Molecular data are not available for the type species, *C. brevifolium* Hochst. & Nees; however, it has crispate macrohairs, which it shares with *C. poiflorum* and *C. stoloniferum* (Snow 1996). This macrohair type may thus be synapomorphic for *Coelachyrum. Coelachyrum* and *Coelachyropsis* share rugose fruits, with the pericarp free from the seed coat, and are combined by Clayton et al. (2006 onward) and Clayton and Renvoize (1986). However, Petersen et al. (2010a) indicate that *Coelachyrum*, represented by *C. poiflorum*, is sister to *Eleusine* and should not include *Coelachyropsis*, represented by *C. lagopoides*.

675. ***Eleusine*** Gaertn.

Eleusine Gaertn., Fruct. Sem. Pl. 1: 7 (1788); phylog.: Liu et al. (2011); Neves et al. (2005).

Annuals or perennials, caespitose or with short rhizomes. Leaf sheaths keeled. Ligule membranous, or a fringed membrane. Inflorecence unbranched, or with the primary branches unbranched and digitate. Spikelets laterally compressed, with 3 to 15 flowers plus additional reduced flowers distally. Glumes shorter than the spikelet, keeled, their apices acute to obtuse, awnless. Lemma muticous or mucronate. *Leaf epidermal papillae absent.* $2n = 18, 20, 36, 40$.

Ten spp., tropical Africa, South America, and one widespread weed.

Eleusine coracana (L.) Gaertn. (finger millet) is widely cultivated in Africa, where it provides particularly drought-resistant grain.

XXIX.6. Subtribe Tripogoninae Stapf (1917)

This subtribe is monophyletic in the phylogenies presented by Peterson et al. (2010a) and GPWG II (2012) but has no obvious morphological synapomorphy.

Five genera, 57 spp.

676. ***Desmostachya*** (Stapf) Stapf

Desmostachya (Stapf) Stapf, Fl. Cap. 7: 316 (1898).

Rhizomatous perennials. Ligule a fringed membrane. Leaf blades coriaceous. Inflorescence branches unbranched, spread along a central axis. Spikelets laterally compressed, overlapping, with as many as 18 flowers. Glumes ovate, their apices acute. Lemma and palea similar in length, their apices entire. Leaf epidermal papillae absent. $2n = 20$.

One sp., *D. bipinnata* (L.) Stapf, North Africa, Middle East, India, Indo-China.

677. ***Eragrostiella*** Bor

Eragrostiella Bor, Indian Forester 66: 269–270 (1940).

Perennials. Leaf blades filiform. Ligule a fringed membrane or a fringe of hairs. Inflorescence unbranched. Spikelets laterally compressed, with 7 to 50 flowers, the distal flowers reduced. Glumes shorter than the spikelet, *the proximal one persistent and the distal one deciduous. Lemma disarticulating, leaving behind the persistent palea.*

Six spp., Eastern Africa, India, Sri Lanka, Burma, Australia.

Eragrostiella is derived from within a paraphyletic *Tripogon* (Peterson et al. 2010a).

678. ***Melanocenchris*** Nees

Melanocenchris Nees, Proc. Linn. Soc. Lond. 1: 94 (1841).

Annuals or perennials, less than 30 cm tall. Ligule a fringe of hairs. Inflorescence a densely capitate cluster of unbranched branches, *these ending in a forked bristle.* Spikelets wedge-shaped, dorsiventrally compressed, with 1 flower plus several reduced flowers distally. Glumes next to each other, shorter than or reaching the apex of the flowers, coriaceous, with long hairs, awned; *upper glume with broad wings basally.* Lemma *with clavate hairs*, with 1 or 3 awns.

Three spp., tropical Africa, India, Sri Lanka.

679. ***Oropetium*** Trin. Fig. 93

Oropetium Trin., Fund. Agrost.: 98 (1822); tax.: Phillips (1975).

Perennial or annual cushion plants. Ligule membranous, or a fringed membrane or a fringe of hairs. *Leaves with multicellular glands abaxially.* Inflorescence unbranched, straight, sinuous or coiled, the spikelets sunken in the rachis. Rachis tough or disarticulating. Spikelets dorsiventrally

Fig. 93. Chloridoideae-Cynodonteae-Tripogoninae. *Oropetium minimum.* **A** Habit. **B** Ligule. **C** Part of rachis of inflorescence. **D** Upper part of inflorescence. **E** Lower glume. **F** Upper glume. **G** Flower. **H** Lemma. **I** Palea. **J** Pistil, stamens and lodicules. **K** Caryopsis. (From Clayton et al. 1974, drawn by S. Ross-Craig)

compressed, with one flower. Callus of flower pubescent. Lower glume absent, upper glume covering the flower. Lemma awned. Pericarp separable from seed coat. Leaf epidermal papillae absent. $2n = 18, 20, 36$.

Six spp., Africa and India.

GPWG II (2012) place *Oropetium* sister to *Tripogon* and the pair sister to *Melanocenchris*, with strong support.

680. ***Tripogon*** Roem. & Schult.

Tripogon Roem. & Schult., Syst. Veg. 2: 34 (1817).

Annuals or perennials, caespitose or rhizomatous. Ligule membranous, or a fringed membrane or a fringe of hairs. Leaf blade apex sharp-pointed. Inflorescence unbranched. Spikelets laterally compressed, with 2 to 50 flowers plus additional reduced flowers distally, in a few species the most proximal flower sterile. Glumes shorter than the spikelet, their apices mucronate or awned. Lemma apex incised, with a mucro or 1 or 3 awns. *Anthers 1, 2,* or 3. $2n = 20$.

Forty-four spp., tropics, most in Old World.

XXIX.7. Subtribe Pappophorinae Dumort. (1829)

Inflorescence branches branched. Callus of flower pubescent.

Three genera, ten spp.

This subtribe has no obvious morphological synapomorphy. Although *Neesiochloa* and *Pappophorum* have long been placed with *Enneapogon, Cottea* and *Schmidtia* in the tribe Pappophoreae based on the laciniate and multi-awned lemmas, the two groups are not closely related, and the traditional Pappophoreae are polyphyletic (Ingram and Doyle 2004; Peterson et al. 2010a; Reutemann et al. 2011).

681. ***Neesiochloa*** Pilg.

Neesiochloa Pilg., Repert. Spec. Nov. Regni Veg. 48: 119 (1940).

Annuals, less than 30 cm tall. Ligule a fringed membrane. *Spikelets spherical*, with 7 to 10 flowers plus additional reduced flowers distally. Callus of flower pubescent. Glumes shorter than the spikelet, their apices awned. Lemma fan-shaped, pubescent, the apex bifid with a short awn from the sinus. *Palea bent inward over the pistil, with tufts of hair at the bend.* Pericarp free.

One sp., *N. barbata* (Nees) Pilg., NE Brazil.

682. ***Pappophorum*** Schreb. Fig. 94

Pappophorum Schreb., Gen. Pl. 2: 787 (1791).

Perennials, generally 0.5–2 m tall. Ligule a fringe of hairs. Spikelets laterally compressed, with 1 to 4 flowers plus additional reduced flowers distally, *the distal rachilla internodes contracted so the assembled lemmas look like a brush.* Callus of flower pubescent. Glumes as long as or extending beyond the flowers, with 1 vein, their apices muticous to awned. *Lemma with 11–26 awns.* Microhairs salt-secreting. $2n = 40, 60, 100$.

Nine spp., North and South America.

683. ***Tridens*** Roem. & Schult.

Tridens Roem. & Schult., Syst. Veg. 2: 34, 599 (1817).

Perennials, caespitose or rhizomatous. Ligule membranous or a fringed membrane or a fringe of hairs. Inflorescence branches branched or not. Spikelets with 3 to 15 flowers plus additional reduced flowers distally. Glumes shorter than or almost as long as the flowers, their apices muticous or mucronate. Callus of flower glabrous or pubescent. Lemma emarginate or with two lobes, mucronate or awned between the lobes; veins and keel with dense hairs, forming lines on the back of the lemma. Leaf epidermal papillae absent. $2n = 16, 32, 40, 72$.

Fifteen spp., New World, Angola.

The former *Tridens muticus* is firmly placed in the Tridentinae clade by both Columbus et al. (2010) and Peterson et al. (2010a, 2014a) and is now recognized as *Tridentopsis. Tridens flavus* is consistently placed with *Pappophorum* (Columbus et al. 2010; GPWG II 2012; Peterson et al. 2010a, 2014a).

Fig. 94. Chloridoideae-Cynodonteae-Pappophorinae. **A–C** *Pappophorum vaginatum*. **A** Plant and inflorescences. **B** Flower. **C** Glumes. **D, E** *P. bicolor*. **D** Part of spike. **E** Spikelet. (From Barkworth et al. 2003, drawn by K. Klitz)

References

Barkworth, M.E., Capels, K.M., Long, S., Piep, M.B. (eds.) 2003. Flora of North America north of Mexico. Vol. 25. Magnoliophyta: Commelinidae (in part): Poaceae, part 1. New York: Oxford University Press.

Bell, H.L., Columbus, J.T. 2008. Proposal for an expanded *Distichlis* (Poaceae, Chloridoideae): support from molecular, morphological, and anatomical characters. Syst. Bot. 33: 536–551.

Brown, W.V. 1977. The Kranz syndrome and its subtypes in grass systematics. Mem. Torrey Bot. Club 23: 1–97.

Cialdella, A.M., Zuloaga, F.O. 2011. Taxonomic study of *Gymnopogon* (Poaceae, Chloridoideae, Cynodonteae). Ann. Missouri Bot. Gard. 98: 301–330.

Clayton, W.D., Renvoize, S.A. 1986. Genera graminum: grasses of the world. London: Her Majesty's Stationery Office.

Clayton, W.D., Phillips, S.M., Renvoize, S.A. 1974. Gramineae (Part 2). In: Polhill, R.M. (ed.) Flora of Tropical East Africa. London: Crown Agents for Overseas Governments and Administrations.

Clayton, W.D., Harman, K.T., Williamson, H. 2006 onward. GrassBase - the online world grass flora. http://www.kew.org/data/grasses-db.html

Columbus, J.T. 1999. An expanded circumscription of *Bouteloua* (Gramineae: Chloridoideae): new combinations and names. Aliso 18: 61–65.

Columbus, J.T., Kinney, M.S., Pant, R., Siqueiros D., M.E. 1998. Cladistic parsimony analysis of internal transcribed spacer region (nrDNA) sequences of *Bouteloua* and relatives (Gramineae: Chloridoideae). Aliso 17: 99–130.

Columbus, J.T., Cerros-Tlatilpa, R., Kinney, M.S., Siqueiros-Delgado, M.E., Bell, H.L., Griffith, M.P., Refulio-Rodriguez, N.F. 2007. Phylogenetics of Chloridoideae (Gramineae): a preliminary study based on nuclear ribosomal internal transcribed spacer and chloroplast *trnL-F* sequences. Aliso 23: 565–579.

Columbus, J.T., Peterson, P.M., Refulio Rodríguez, N.F., Cerros-Tlatilpa, R., Kinney, M.S. 2010. Phylogenetics of Muhlenbergiinae (Poaceae, Chloridoideae, Cynodonteae) based on ITS and *trnL-F* DNA sequences. In: Seberg, O., Petersen, G., Barfod, A.S., Davis, J.I. (eds.) Diversity, phylogeny, and evolution in the monocotyledons. Aarhus, Denmark: Aarhus University Press. pp. 477–495

Cope, T.A. 1982. Poaceae. In: Nasir, E., Ali, S.I. (eds.) Flora of Pakistan: Pakistan Agricultural Research Council.

Fortune, P.M., Schierenbeck, K.A., Ainouche, A.K., Jacquemin, J., Wendel, J.F., Ainouche, M.L. 2006. Evolutionary dynamics of *Waxy* and the origin of hexaploid *Spartina* species (Poaceae). Mol. Phyl. Evol. 43: 1040–1055.

Gould, F.W., Box, T.W. 1965. Grasses of the Texas Coastal Bend. College Station, Texas: Texas A&M University Press.

GPWG 2001. Grass Phylogeny Working Group. Phylogeny and subfamilial classification of the Poaceae. Ann. Missouri Bot. Gard. 88: 373–457.

GPWG II 2012. Grass Phylogeny Working Group II. New grass phylogeny resolves deep evolutionary relationships and discovers C_4 origins. New Phytol. 193: 304–312.

Gutierrez, M., Gracen, V.E., Edwards, G.E. 1974. Biochemical and cytological relationships in C_4 plants. Planta 119: 279–300.

Harlan, J.R., de Wet, J.M.J., Richardson, W.L. 1969. Hybridization studies with species of *Cynodon* from East Africa and Malagasy. Amer. J. Bot. 56: 944–950.

Hattersley, P.W., Watson, L. 1992. Diversification of photosynthesis. In: Chapman, G.P. (ed.) Grass evolution and domestication. Cambridge: Cambridge University Press. pp. 38–116

Hunziker, A.T., Anton, A.M. 1979. A synoptical revision of *Blepharidachne* (Poaceae). Brittonia 31: 446–453.

Ingram, A.L. 2010. Evolution of leaf blade anatomy in *Eragrostis* (Poaceae). Syst. Bot. 35: 755–765.

Ingram, A.L., Doyle, J.J. 2004. Is *Eragrostis* monophyletic? Insights from nuclear and plastid sequence data. Syst. Bot. 29: 545–552.

Ingram, A.L., Christin, P.-A., Osborne, C.P. 2011. Molecular phylogenies disprove a hypothesized C_4 reversion in *Eragrostis walteri* (Poaceae). Ann. Bot. 107: 321–325.

Jessop, J., Dashorst, G.R.M., James, F.M. 2006. Grasses of South Australia. Kent Town, South Australia: Wakefield Press.

Jewell, M., Frere, C.H., Harris-Shultz, K., Anderson, W.F., Godwin, L.D., Lambrides, C.J. 2012. Phylogenetic analysis reveals multiple introductions of *Cynodon* species in Australia. Mol. Phyl. Evol. 65: 390–396.

Judziewicz, E. 2009. *Toliara* (Poaceae, Chloridoideae, Cynodonteae), a new grass genus endemic to southern Madagascar. Adansonia 31: 273–277.

Koch, S.D. 1978. Notes on the genus *Eragrostis* (Gramineae) in the Southeastern United States. Rhodora 80: 390–403.

Liu, Q., Triplett, J.K., Wen, J., Peterson, P.M. 2011. Allotetraploid origin and divergence in *Eleusine* (Chloridoideae, Poaceae): evidence from low-copy nuclear gene phylogenies and a plastid gene chronogram. Ann. Bot. 108: 1287–1298.

McNeill, J., Barrie, F.R., Burdet, H.M., Demoulin, V., Hawksworth, D.L., Marhold, K., Nicolson, D.H., Prado, J., Silva, P.C., Skog, J.E., Wiersema, J.H., Turland, N.J. (eds.) 2007. International Code of Botanical Nomenclature (Vienna Code). Gantner Verlag: Ruggell.

Metcalfe, C.R. 1960. Anatomy of the monocotyledons. I. Gramineae. Oxford: Clarendon Press.

Moffett, A.A., Hurcombe, R. 1949. Chromosome numbers of South African grasses. Heredity 3: 369–373.

Müller, M.A.N. 2007. Grasses of Namibia. Windhoek, Namibia: Ministry of Agriculture, Water and Forestry.

Neves, S.S., Swire-Clark, G., Hilu, K.W., Baird, W.V. 2005. Phylogeny of *Eleusine* (Poaceae: Chloridoideae) based on nuclear ITS and plastid *trnT-trnF* sequences. Mol. Phyl. Evol. 35: 395–419.

Packer, J.G. 1960. A note on the nomenclature of the genus *Cleistogenes* Y. Keng (Gramineae). Bot. Not. 113: 289–294.

Peterson, P.M., Herrera Arrieta, Y. 2001. A leaf blade anatomical survey of *Muhlenbergia* (Poaceae: Muhlenbergiinae). Sida 19: 469–506.

Peterson, P.M., Columbus, J.T., Pennington, S.J. 2007. Classification and biogeography of New World grasses: Chloridoideae. Aliso 23: 580–594.

Peterson, P.M., Romaschenko, K., Johnson, G. 2010a. A classification of the Chloridoideae (Poaceae) based on multi-gene phylogenetic trees. Mol. Phyl. Evol. 55: 580–598.

Peterson, P.M., Romaschenko, K., Johnson, G. 2010b. A phylogeny and classification of the Muhlenbergiinae (Poaceae: Chloridoideae: Cynodonteae) based on plastid and nuclear DNA sequences. Amer. J. Bot. 97: 1532–1554.

Peterson, P.M., Romaschenko, K., Barker, N.P., Linder, H. P. 2011. Centropodieae and *Ellisochloa*, a new tribe and genus in Chloridoideae (Poaceae). Taxon 60: 1113–1122.

Peterson, P.M., Romaschenko, K., Snow, N., Johnson, G. 2012. A molecular phylogeny and classification of *Leptochloa* (Poaceae: Chloridoideae: Chlorideae) *sensu lato* and related genera. Ann. Bot. 109: 1317–1329.

Peterson, P.M., Romaschenko, K., Herrera Arrieta, Y. 2014a. A molecular phylogeny and classification of the Cteniinae, Farragininae, Gouiniinae, Gymnopogoninae, Perotidinae, and Trichoneurinae (Poaceae: Chloridoideae: Cynodonteae). Taxon 63: 275–286. doi: 10.12705/632.35

Peterson, P.M., Romaschenko, K., Herrera Arrieta, Y., Saarela, J.M. 2014b. A molecular phylogeny and new subgeneric classification of *Sporobolus* (Poaceae: Chloridoideae: Sporobolinae). Taxon 63: 1212–1243.

Peterson, P.M., Romaschenko, K., Herrera Arrieta, Y., Saarela, J.M. 2014c. (2332) Proposal to conserve *Sporobolus* against *Spartina*, *Crypsis*, *Ponceletia* and *Heleochloa* (Poaceae: Chloridoideae: Sporobolinae). Taxon 63: 1373–1374.

Phillips, S.M. 1974. Studies in the Gramineae: XXXV. Kew Bull. 29: 267–270.

Phillips, S.M. 1975. A review of the genus *Oropetium* (Gramineae). Kew Bull. 30: 467–470.

Reeder, J.R. 1982. Systematics of the tribe Orcuttieae (Gramineae) and the description of a new segregate genus, *Tuctoria*. Amer. J. Bot. 69: 1082–1095.

Reutemann, A.G., Tivano, J.C., Lucero, L.E., Vegetti, A.C., Giussani, L.M. 2011. Phylogenetic relationships within Pappophoreae s.l. (Poaceae: Chloridoideae): additional evidence based on ITS and *trn*L-F sequence data. South Afr. J. Bot. 77: 693–702.

Roodt, R., Spies, J.J. 2003. Chromosome studies in the grass subfamily Chloridoideae. I. Basic chromosome numbers. Taxon 52: 557–566.

Snow, N. 1996. The phylogenetic utility of lemmatal micromorphology in *Leptochloa* s.l. and related genera in subtribe Eleusininae (Poaceae, Chloridoideae, Eragrostideae). Ann. Missouri Bot. Gard. 83: 504–529.

Snow, N. 1997. Phylogeny and systematics of *Leptochloa* P. Beauv. sensu lato (Poaceae: Chloridoideae). Ph.D. Thesis. Washington University, St. Louis, Missouri, USA

Snow, N., Peterson, P.M. 2012a. Nomenclatural notes on *Dinebra*, *Diplachne*, *Disakisperma* and *Leptochloa* (Poaceae: Chloridoideae). Phytoneuron 2012–71: 1-2.

Snow, N., Peterson, P.M. 2012b. Systematics of *Trigonochloa* (Poaceae, Chloridoideae, Chlorideae). PhytoKeys 13: 25–38.

Snow, N., Peterson, P.M., Romaschenko, K. 2013. Systematics of *Disakisperma* (Poaceae, Chloridoideae, Chlorideae). PhytoKeys 26: 21–70.

Soreng, R.J., Davidse, G., Peterson, P.M., Zuloaga, F.O., Judziewicz, E.J., Filgueiras, T.S., Morrone, O., Romaschenko, K. 2012 and onward. A world-wide phylogenetic classification of Poaceae (Gramineae). http://www.tropicos.org/docs/meso/CLASSIFICATION%20OF%20world%20grasses%202013%20Oct%2029%20%282%29fZedits%20April%205%202014.htm

Valdés-Reyna, J., Hatch, S.L. 1997. A revision of *Erioneuron* and *Dasyochloa* (Poaceae: Eragrostideae). Sida 17: 645–666.

Van den Borre, A., Watson, L. 1994. The infrageneric classification of *Eragrostis* (Poaceae). Taxon 43: 383–422.

Van den Borre, A., Watson, L. 1997. On the classification of the Chloridoideae (Poaceae). Aust. Syst. Bot. 10: 491–531.

Watson, L., Dallwitz, M.J. 1992 onward. The grass genera of the world: descriptions, illustrations, identification, and information retrieval; including synonyms, morphology, anatomy, physiology, phytochemistry, cytology, classification, pathogens, world and local distribution, and references, 23rd April 2010 edn. Wallingford: CAB International. http://delta-intkey.com/grass/

Wu, Z., Raven, P.H., Hong, D. (eds.) 2007. Flora of China Illustrations. Beijing: Science Press.

Yates, H.O. 1966. Revision of grasses traditionally referred to *Uniola*, I. *Uniola* and *Leptochloöpsis*. Southwest Nat. 11: 372–394.

XI. Subfamily Micrairoideae Pilg. (1956)

Annuals or perennials. Culms solid or hollow. Ligule a fringe of hairs. Spikelets generally with 2 flowers; rachilla extension absent. Disarticulation above the glumes. Ovary glabrous. Styles 2, not fused. Hilum short or long linear. Embryo without an epiblast, with a scutellar cleft, with an elongate mesocotyl internode, the embryonic leaf margins meeting.

Nine genera, 188 species.

Micrairoideae are strongly supported as monophyletic in recent molecular analyses (GPWG II 2012; Sánchez-Ken and Clark 2010; Sánchez-Ken et al. 2007). The subfamily was initially recognized to accommodate the peculiar Australian genus *Micraira*, but has since been expanded to include Isachneae and Eriachneae, two tribes formerly placed in Panicoideae on the basis of spikelet morphology (Clayton and Renvoize 1986). The origin of C_4 photosynthesis in Eriachneae is independent of other origins in the PACMAD clade (GPWG II 2012), and aspects of leaf anatomy and gene expression are quite distinct from other C_4 taxa (Prendergast et al. 1987; Sinha and Kellogg 1996).

Micraira and *Eriachne* are successive sisters to the remainder of the subfamily (GPWG II 2012). Although the subfamily is often divided into tribes, this seems unnecessary. If they are to be recognized, then Micraireae include only *Micraira*, Eriachneae include *Eriachne* and *Pheidochloa*, and Isachneae include the rest.

Key to the Genera of Micrairoideae

1. Plants floating; inflorescence unbranched **691. *Limnopoa***
1. Plants terrestrial, if in water then not floating; inflorescence branched 2
2 (1). Primary branches of the inflorescence unbranched, ending in a sharp point, the spikelets borne on one side **688. *Heteranthoecia***
2. Primary branches of the inflorescence branched, ending in a spikelet, not one-sided 3
3 (2). Plants mat-forming, moss-like, with leaves densely clustered and apparently spirally arranged **684. *Micraira***
3. Plants variously caespitose or trailing, but not mat-forming, the leaves not densely clustered or spirally arranged 4
4 (3). Lemma membranous or chartaceous or hyaline 5
4. Lemma coriaceous or indurate 7
5 (4). Ligule absent **689. *Hubbardia***
5. Ligule present 6
6 (5). Spikelets with 1 flower; disarticulating below the glumes **692. *Sphaerocaryum***
6. Spikelets with 2 flowers, rarely 1; disarticulating above the glumes **687. *Coelachne***
7 (4). Upper glume twice as long as the lower one **686. *Pheidochloa***
7. Glumes nearly equal in length 8
8 (7). Spikelets dorsiventrally compressed or not compressed; lemma glabrous or puberulent, never awned **690. *Isachne***
8. Spikelets laterally compressed; lemma often pubescent with long hairs, often awned **685. *Eriachne***

E.A. Kellogg, *Flowering Plants. Monocots*, The Families and Genera of Vascular Plants 13,
DOI 10.1007/978-3-319-15332-2_25, © Springer International Publishing Switzerland 2015

Genera of Micrairoideae

684. *Micraira* F. Muell.

Micraira F. Muell., Fragm. 5: 208 (1867).

Tiny moss-like perennials, in tufts connected by stolons. Ligule membranous, or a fringed membrane, or a fringe of hairs, or absent. *Leaves arranged in a spiral.* Inflorescence branches branched. Spikelets laterally compressed, with 2 flowers, with a rachilla extension. Glumes shorter or longer than the flowers, their apices emarginate to acuminate, muticous to awned. Lemma membranous or hyaline, the apex emarginate to obtuse, muticous or mucronate. *Palea divided into two. Lodicules absent. Stamens 2.* $2n = 20$.

Fifteen spp., Australia.

The spiral vegetative phyllotaxis in *Micraira* was documented by Philipson (1935). The paleas of some species have many veins.

685. *Eriachne* R. Br.

Eriachne R. Br., Prodr.: 183 (1810).

Annuals or perennials, caespitose or rhizomatous. Ligule a fringe of hairs. Inflorescence branches branched. Spikelets laterally compressed, with 2 flowers. Glumes shorter or longer than the flowers, their apices obtuse to attenuate, muticous or rarely awned. Callus of flower short, blunt, pubescent. Lemma margins involute, the apex obtuse to acuminate, mucronate or awned, the awn straight or curved. Stamens 2 or 3. Hilum long linear. Leaf vasculature with a double bundle sheath, the outer one parenchymatous and carbon reducing, densely packed with chloroplasts, the inner one made up of thick-walled cells. Plants C_4 NADP-ME.

Forty-eight spp., Asia, Australia, Pacific.

Spikelets in some specimens of *Eriachne* have a noticeable internode between the glumes; rarely a rachilla extension bearing a rudimentary flower may be present. Paleas may be two-pronged or awned at the apex.

686. *Pheidochloa* S. T. Blake

Pheidochloa S. T. Blake, Proc. Roy. Soc. Queensland 56: 20 (1944).

Annuals. Ligule a fringe of hairs. Inflorescence branches branched. Spikelets laterally compressed, with 2 flowers. Glumes longer than the flowers, *separated by a long rachilla internode, the second glume about twice the length of the first,* their apices acute. Callus of flower elongate, sharp-pointed, pubescent. Lemma margins involute, the apex acute, awned, the awn flexuous. Stamens 2. Hilum long linear. Leaf vasculature with a double bundle sheath, the outer one parenchymatous and carbon reducing, densely packed with chloroplasts, the inner one made up of thick-walled cells. Plants C_4 NADP-ME.

Two spp., Asia, Australasia.

Remaining genera:
Mesophyll radiate around the vascular bundles, similar to Kranz anatomy but the vein spacing distant, characteristic of C_3 plants.

687. *Coelachne* R. Br.

Coelachne R. Br., Prodr. Fl. Nov. Holl.: 187 (1810).

Annuals or perennials. Ligule a fringe of hairs or absent. Inflorescence branches branched. Spikelets dorsiventrally compressed, with two dissimilar flowers separated by a long rachilla internode; rachilla extension lacking. Glumes shorter than the spikelet, their apices generally obtuse. Lemma slightly inflated, lacking veins, the apex obtuse or acute, unawned. Stamens 2 or 3. Hilum short. Leaf epidermal papillae absent, or consisting of one symmetrical swelling per cell. $2n = 40$.

Twelve spp., Old World tropics.

688. *Heteranthoecia* Stapf

Heteranthoecia Stapf, Hooker's Icon. Pl. 30: t. 2937 (1911).

Annuals. Ligule a fringe of hairs. Inflorescence branches unbranched, spread along a central axis, the branches flattened, ending in a barren tip, spikelets borne in two rows along abaxial side. Spikelets dorsiventrally compressed, with 2 flowers. Glumes shorter than the spikelet, the lower deciduous, the upper persistent, their apices acute. Lemma papery, with club shaped-hairs, the apex obtuse. Leaf epidermal papillae present. Hilum short.

One sp., *H. guineensis* (Franch.) Robyns, tropical Africa. Plants of wet places.

689. *Hubbardia* Bor

Hubbardia Bor, Kew Bull. 5: 385 (1951).

Annuals. Leaves flimsy. Ligule absent. Inflorescence branches branched. Spikelets dorsiventrally compressed, with 2 flowers, the lower sterile, the upper bisexual. Glumes longer than the flowers, pubescent with tubercle-based hairs, their apices obtuse. Lower sterile flower reduced to a lemma with an obtuse apex. Lemma of upper flower hyaline above, the apex obtuse; palea absent. Leaf epidermal papillae absent. Hilum short.

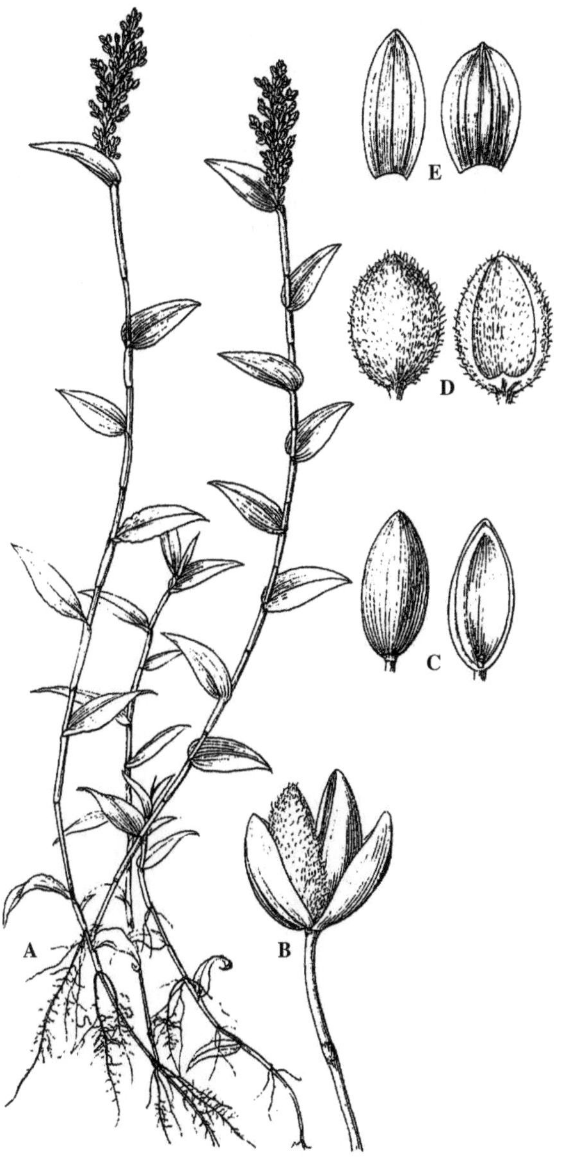

Fig. 95 Micrairoideae. *Isachne pulchella*. **A** Habit. **B** Spikelet. **C** Abaxial and adaxial views of lower lemma and palea. **D** Abaxial and adaxial views of upper lemma and palea. **E** Glumes. (From Wu et al. 2007, with permission from the Missouri Botanical Garden Press, St. Louis, and Science Press, Beijing; drawn by Si Weiqing)

One sp., *H. heptaneuron* Bor, western India (Karnataka and Maharashtra).

Spikelet morphology has led previous workers to place *Hubbardia* in Panicoideae, but morphological (Zuloaga et al. 2011) and molecular (Sánchez-Ken, unpublished) data place it here. Although *Hubbardia* was once thought to be extinct, it has been rediscovered; it remains rare and endangered (Yadav et al. 2009, 2010).

690. ***Isachne*** R. Br. Fig. 95

Isachne R. Br., Prodr.: 196 (1810).

Annuals or perennials, caespitose, rhizomatous or stoloniferous. Leaf blades disarticulating at the ligule. Ligule a fringed membrane, or a fringe of hairs, or absent. Inflorescence branches branched. Spikelets dorsiventrally compressed, with 1 or 2 flowers, the proximal one staminate or bisexual, the distal one bisexual. Glumes shorter or longer than the spikelet, their apices generally obtuse to acuminate. Lemma apex obtuse to acute, awnless. Hilum long-linear. Leaf epidermal papillae one per cell, or absent. $2n = 20, 50$, and 60.

One hundred-three spp., Africa, Asia and Australasia. Species of *Isachne* are often aquatic.

691. ***Limnopoa*** C. E. Hubb.

Limnopoa C. E. Hubb., Hooker's Icon. Pl. 35: t. 3432 (1943).

Floating annuals, less than 25 cm tall. Ligule a fringe of hairs. Inflorescence unbranched, the spikelets in pairs, one sessile, one pedicellate, borne on one side of the inflorescence axis. Spikelets dorsiventrally compressed, with 2 flowers, the proximal one staminate, the distal one bisexual. Glumes shorter than the flowers, the lower one with an acute apex, the upper one obovate and with an obtuse apex. Lemma pubescent, the apex obtuse. Hilum long-linear. Leaf epidermal papillae one per cell.

One sp., *L. meeboldii* (C.E.C. Fisch.) C.E. Hubb., tropical Asia.

692. ***Sphaerocaryum*** Nees ex Hook. f.

Sphaerocaryum Nees ex Hook. f., Fl. Brit. India 7: 246 (1897).

Annuals, less than 15 cm tall. Ligule a fringe of hairs. *Leaf blades ovate.* Inflorescence branches branched. Spikelets not compressed, with 1 flower; *disarticulating below the glumes.* Glumes as long as the flower, translucent, their apices obtuse; lower glume lacking veins, upper glume with 1 vein. Lemma translucent, pubescent, the apex obtuse. Hilum short. Leaf epidermal papillae one per cell. $2n = 20$.

One sp., *S. pulchellum* (Roth.) Merr., India and southeast Asia.

References

Clayton, W.D., Renvoize, S.A. 1986. Genera graminum: grasses of the world. London: Her Majesty's Stationery Office.

GPWG II 2012. Grass Phylogeny Working Group II. New grass phylogeny resolves deep evolutionary relationships and discovers C_4 origins. New Phytol. 193: 304–312.

Philipson, W.R. 1935. A grass with spiral phyllotaxis: *Micraira subulifolia*. Bull. Misc. Inf. (Kew) 1935: 324–326.

Prendergast, H.D.V., Hattersley, P.W., Stone, N.E. 1987. New structural/biochemical associations in leaf blades of C_4 grasses (Poaceae). Aust. J. Plant Phys. 14: 403–420.

Sánchez-Ken, J.G., Clark, L.G. 2010. Phylogeny and a new tribal classification of the Panicoideae s.l. (Poaceae) based on plastid and nuclear sequence data and structural data. Amer. J. Bot. 97: 1732–1748.

Sánchez-Ken, J.G., Clark, L.G., Kellogg, E.A., Kay, E.E. 2007. Reinstatement and emendation of subfamily Micrairoideae (Poaceae). Syst. Bot. 32: 71–80.

Sinha, N.R., Kellogg, E.A. 1996. Parallelism and diversity in multiple origins of C_4 photosynthesis in the grass family. Amer. J. Bot. 83: 1458–1570.

Wu, Z., Raven, P.H., Hong, D. (eds.) 2007. Flora of China Illustrations. Beijing: Science Press.

Yadav, S.R., Chandore, A.N., Nimbalkar, M.S., Gurav, R. V. 2009. Reintroduction of *Hubbardia heptaneuron* Bor, a critically endangered endemic grass in Western Ghats. Curr. Sci. 96: 880.

Yadav, S.R., Chandore, A.N., Gund, S.M., Nandikar, M., Lekhak, M. 2010. Relocation of *Hubbardia heptaneuron* Bor, from its type locality. Curr. Sci. 98: 884.

Zuloaga, F.O., Morrone, O., Scataglini, M.A. 2011. Monograph of *Trichanthecium* (Poaceae, Paniceae). Syst. Bot. Monographs 94: 1–101.

XII. Subfamily Arundinoideae Burmeist. (1837)

Perennials; caespitose, rhizomatous, or stoloniferous, tall and reed-like to minute and moss-like. Ligule membranous or a fringe of hairs. Inflorescence branches branched. Spikelets laterally compressed. Callus of flower pubescent. Style branches 2, generally free. Hilum short or long-linear. Epiblast absent, scutellar cleft present, mesocotyl internode elongated, embryonic leaf margins meeting or overlapping. Leaf epidermal papillae absent.

Nineteen genera, 46 spp.

The circumscription of this subfamily has varied enormously over the years. Clayton and Renvoize (1986) included in it genera now assigned to Stipeae (Pooideae), Danthonioideae, Micrairoideae, Panicoideae and Aristidoideae, while Watson and Dallwitz (1992 onward) used an equally broad interpretation, including Steyermarkochloeae, *Cyperochloa*, and *Spartochloa* (all now Panicoideae s.l.), as well as taxa now placed in Micrairoideae and Aristidoideae. Kellogg and Campbell (1987) showed that the subfamily as defined at the time was a polyphyletic assemblage whose members were variously sister to other well-defined subfamilies. Polyphyly of the traditional Arundinoideae was confirmed by molecular work of Barker et al. (1995) and Clark et al. (1995). In subsequent years, the subfamily has been relieved of most of its disparate elements and is reduced to the 19 genera described here (GPWG 2001; GPWG II 2012). However, the group remains morphologically heterogeneous, as evidenced by the large number of genera for a small number of species, and there is no obvious morphological synapomorphy. Nine genera, each with only one or two species, still lack molecular data, so there remains a possibility that several are still misplaced here.

A few species currently placed elsewhere may also belong in Arundinoideae. The recent discovery that "*Eragrostis*" *walteri* is not an *Eragrostis* at all (Ingram et al. 2011) (see below) is an example.

Key to the Genera of Arundinoidae

1. Perennial reeds, often 2 m or more tall; plants strongly rhizomatous 2
1. Plants rarely over 1 m tall; generally caespitose 3
2 (1). Ligule membranous, with short cilia; lemma villous; proximal flowers of the spikelet bisexual **694. *Arundo***
2. Ligule with a minute membrane and long cilia; lemma glabrous (but the callus with long hairs); proximal flowers of the spikelet staminate or sterile **703. *Phragmites***
3 (1). Spikelets arranged in several dense globular clusters spread along the inflorescence axis; plants annual **697. *Elytrophorus***
3. Spikelets not clustered; plants annual or perennial 4
4 (3). Spikelets with one flower 5
4. Spikelets with more than one flower 8
5 (4). Inflorescence unbranched; plants annual; lemma with a transverse line of hairs **708. *Danthonidium***
5. Inflorescence branches branched; plants perennial; lemma glabrous or pubescent but the hairs not in a transverse line 6
6 (5). Lemma with 3 awns; plants of Australia **707. *Amphipogon***
6. Lemma with one awn; plants of Africa or India 7
7 (6). Palea unawned; plants of Ethiopia **700. *Leptagrostis***
7. Palea with two apical setae; plants of India and Sri Lanka **696. *Dichaetaria***

E.A. Kellogg, *Flowering Plants. Monocots*, The Families and Genera of Vascular Plants 13,
DOI 10.1007/978-3-319-15332-2_26, © Springer International Publishing Switzerland 2015

8 (4). Lemma with 9 veins, hairs in lines or tufts, bilobed, with a central geniculate awn from the sinus 9
8. Lemma not as above 11
9 (8). Lemma broad and deeply cup-shaped; plants of Australia **710. *Monachather***
9. Lemma longer than broad, not cup-shaped; plants of Africa 10
10 (9). Culms slender, decumbent; leaf blades membranous; plants of Yemen and Ethiopia **711. *Phaenanthoecium***
10. Culms upright, the plants tufted; leaf blades succulent; plants of southern Africa **709. *Dregeochloa***
11 (8). Spikelets lacking awns 12
11. Spikelets with awns 14
12 (11). Spikelets with 7 to 15 flowers; plants of southern Africa **698. "*Eragrostis*" *walteri***
12. Spikelets with 5 or fewer flowers; plants not African 13
13 (12). Leaf blades not pseudopetiolate; plants of Europe, Turkey, China, Japan **701. *Molinia***
13. Leaf blades pseudopetiolate; plants of India and Sri Lanka **706. *Zenkeria***
14 (11). Lemma with hairs in tufts **693. *Alloeochaete***
14. Lemma glabrous, if pubescent then hairs not tufted 15
15 (14). Awn shorter than the lemma 16
15. Awn longer than the lemma 17
16 (15). Leaves slender, stiff, the basal sheaths hard and fibrous; ligule a fringe of hairs; plants of Africa **705. *Styppeiochloa***
16. Leaves broad, flat, flexuous, the basal sheaths membranous; ligule membranous; plants of Japan but cultivated elsewhere **699. *Hakonechloa***
17 (15). Spikelets with 4 to 7 flowers **702. *Nematopoa***
17. Spikelets with 2 or 3 flowers 18
18 (17). Plants <40 cm tall; spikelets <3 mm long **704. *Piptophyllum***
18. Plants 60–150 cm tall; spikelets >4 mm long **695. *Crinipes***

Genera of Arundinoideae

693. *Alloeochaete* C. E. Hubb.

Alloeochaete C. E. Hubb., Hooker's Icon. Pl. 35: t. 3418 (1940).

Caespitose. Leaf blades disarticulating from the sheath, *the sheath wooly-tomentose.* Ligule a fringe of hairs. Inflorescence branches branched. Spikelets with 4 to 10 flowers, *the proximal flower staminate or sterile.* Glumes shorter than the flowers, their apices obtuse to acuminate, mucronate to awned. Callus of flower blunt or elongate. Lemma with 5 veins, with a tuft of hairs on each side, the apex with two teeth or lobes, these ending in awns, principal awn from the sinus, flattened and geniculate.

Six spp., Angola, Tanzania, Malawi, Mozambique.

694. *Arundo* L.

Arundo L., Sp. Pl. 1: 81 (1753); tax.: Hardion et al. (2012).

Reeds over 2 m tall, rhizomatous. Leaves cauline, the blades disarticulating from the sheaths. Ligule membranous, with short cilia. Inflorescence branches branched. Spikelets with 2 to 5 flowers, with a rachilla extension. Glumes as long as the flowers, their apices acute to acuminate. Lemma villous proximally, the apex entire or bidentate, mucronate or awned from the sinus and from the lobes. Hilum short. $2n = 24, 100, 110$.

Three spp., southern Europe and northern Africa to China, introduced worldwide.

695. *Crinipes* Hochst.

Crinipes Hochst., Flora 38: 279 (1855).

Caespitose. Leaves stiff, the blades disarticulating from the sheaths. Ligule a fringe of hairs. Inflorescence branches branched. Spikelets with 2 or 3 flowers, sometimes with reduced flowers distally or with a rachilla extension. Callus of flower long and narrow. Glumes shorter than flowers, *with 1 vein*, their apices acuminate, mucronate or short awned. Lemma with hairy margins, the apex entire or bidentate, with an awn from the apex or the sinus. Hilum short.

One sp., Sudan, Ethiopia, Uganda.

696. *Dichaetaria* Nees ex Steud.

Dichaetaria Nees ex Steud., Syn. Pl. Glumac. 1: 145 (1855).

Caespitose. Ligule a fringe of hairs; contraligule present. Inflorescence branches branched. Spikelets with 1 flower, plus reduced flowers distally. Glumes shorter than the spikelet, their apices acute, the lower persistent, the upper deciduous. Callus of flower elongate. Lemma leathery, awl-shaped, glabrous, with two teeth, awned from the sinus. Palea with 2 setae. Fruit beaked. Hilum long-linear.

One sp., *D. wightii* Nees ex Steud., India and Sri Lanka.

697. ***Elytrophorus*** P. Beauv.

Elytrophorus P. Beauv., Ess. Agrostogr.: 67, pl. 14, f. 2 (1812).

Caespitose annuals. Ligule membranous or a fringed membrane. *Inflorescence with the primary branches forming dense globular clusters of spikelets; outer spikelet of a cluster with the lower glume enlarged and most or all flowers reduced to sterile lemmas.* Spikelets with 3 to 7 flowers. Glumes as long as the flowers, their apices acuminate, awn-tipped. Callus of flower glabrous. Lemma keeled, the apex attenuate, entire, with a short awn. *Palea keels winged.* Lodicules 1 or 2. Stamens 1 to 3. Hilum short. $2n = 24$.

Two spp., tropical Africa, Australia, India, China.

698. **"*Eragrostis*" *walteri*** Pilg.

"*Eragrostis*" *walteri* Pilg.

Caespitose. Ligule a fringe of hairs. Inflorescence branches branched. Spikelets with 7 to 15 flowers. Glumes shorter than the flowers, their apices attenuate. Lemma apex mucronate.

One sp., southern Africa.

For many years this species was thought to be the only C_3 member of *Eragrostis*. However, it clearly belongs in Arundinoideae, probably sister to *Elytrophorus* (Ingram et al. 2011).

699. ***Hakonechloa*** Makino ex Honda Fig. 96

Hakonechloa Makino ex Honda, J. Fac. Sci. Univ. Tokyo, Sect. 3, Bot. 3: 113 (1930).

Short rhizomatous. Ligule membranous, with a ciliate margin. *Abaxial ligule present, a line of hairs. Leaf blades resupinate.* Inflorescence branches branched. Spikelets with 5 to 10 flowers, the rachilla internodes pubescent. Glumes shorter than the flowers, their apices acuminate. Callus of flower elongate. Lemma with 3 veins, the margins ciliate, the apex with two teeth, with a short awn from the sinus. $2n = 50$.

One sp., *H. macra* (Munro) Honda, Japan, cultivated elsewhere.

700. ***Leptagrostis*** C. E. Hubb.

Leptagrostis C. E. Hubb., Bull. Misc. Inform. Kew 1937: 63 (1937).

Fig. 96. Arundinoideae. *Hakonechloa macra*. **A** Plant with rhizome and inflorescence. **B** Spikelet with glumes and flowers. (From Barkworth et al. 2003, drawn by C. Roché)

Caespitose, less than 25 cm tall. Ligule a fringe of hairs. Inflorescence branches branched. Spikelets with 1 flower, with a rachilla extension. Glumes with acuminate apices, the lower one shorter than the flower, the upper one longer. Callus of flower blunt. Lemma with two lobes, awned from the sinus, the awn about 1 mm long.

One sp., *L. schimperiana* (Hochst.) C. E. Hubb., Ethiopia.

No information is available on the fruit or leaf anatomy of this species.

701. ***Molinia*** Schrank

Molinia Schrank, Baier. Fl. 1: 100, 334 (1789).

Caespitose, with only the uppermost culm internodes elongated. Ligule a fringe of hairs. Leaf blades disarticulating from the sheath. Inflorescence branches branched. Spikelets with 2 to 5 flowers. Glumes shorter than the flowers, their apices acute. Callus of flower truncate, sometimes glabrous. Lemma papery, the apex acute, awnless. Fruit with pericarp easily separated from the seed coat. Hilum long-linear. $2n = 18, 36, 50, 54, 72, 90$.

Three spp., Europe, Turkey, China, Japan, cultivated elsewhere.

702. ***Nematopoa*** C. E. Hubb.

Nematopoa C. E. Hubb., Kew Bull. 12: 51 (1957).

Caespitose. Ligule a fringe of hairs. Leaf blades thread-like. Inflorescence branches branched. Spikelets with 4 to 7 flowers. Callus of flower short. Glumes shorter than the flowers, hyaline, without veins, their apices acute. Lemma pubescent, with 1 vein, with 2 lobes, awned from the sinus, the awn flexuous. Hilum short.

One sp., *N. longipes* (Stapf & C.E. Hubb.) C.E. Hubb., south tropical Africa.

703. ***Phragmites*** Adans.

Phragmites Adans., Fam. Pl. 2: 34, 559 (1763); phylog.: Chu et al. (2011); Lambertini et al. (2006, 2012).

Reeds, generally over 2 m tall, rhizomatous. Leaf blades disarticulating from the sheath. Ligule membranous, or a fringed membrane, or a fringe of hairs. Inflorescence branches branched, the ultimate ones bearing long hairs. Spikelets with 3 to 11 flowers, the proximal one staminate or sterile; rachilla extension present. Glumes shorter than the flowers, their apices acute. Callus of flower elongate, with long hairs. Lemma narrow, the apex long-attenuate, awnless but sometimes mucronate. Hilum short. $2n = 36, 44, 46, 48, 50, 52, 54, 84, 96, 120$.

Four spp., in marshes and wet sites worldwide.

In addition to the numbers cited above, aneuploid chromosome counts are reported for *Phragmites australis* (e.g., $2n = 53$, Connor et al. 1998; $2n = 42$–59, Gervais et al. 1993), suggesting meiotic irregularity. Interspecific hybridization has been documented within the genus (Chu et al. 2011).

704. ***Piptophyllum*** C. E. Hubb.

Piptophyllum C. E. Hubb., Kew Bull. 12: 53 (1957).

Caespitose. Ligule a fringe of hairs. Leaf blades disarticulating from the sheaths. Inflorescence branches branched. Spikelets with 2 flowers, with a rachilla extension. Glumes shorter than the flowers, hyaline, with 1 vein, their apices acute, the upper glume mucronate. Lemma with 5 to 9 veins, the apex with two lobes and three awns, the principal awn from the sinus, flexuous, the base flat and twisted. Hilum short.

One sp., *P. welwitschii* (Rendle) C.E. Hubb., Angola.

705. ***Styppeiochloa*** de Winter

Styppeiochloa de Winter, Bothalia 9: 134 (1966).

Caespitose, the basal sheaths hard and fibrous. Ligule a fringe of hairs. Leaf blades sharp-pointed. Inflorescence branches branched. Spikelets with 2 to 6 flowers. Glumes shorter than the flowers, their apices acuminate, mucronate or awned. Callus of flower short, blunt. Lemma apex with 2 lobes, with short awns from the lobes and the sinus. Hilum long-linear. $2n = 24, 48$.

Three spp., Africa and Madagascar.

706. ***Zenkeria*** Trin.

Zenkeria Trin., Linnaea 11: 150 (1837).

Caespitose. Ligule a fringe of hairs. Leaf blades pseudopetiolate, with leathery margins.

Inflorescence branches branched. Spikelets with 2 flowers, with or without a rachilla extension. Glumes shorter than the flowers, with 1 vein, their apices acute to acuminate. Callus of flower with thick hairs. Lemma with 5 to 11 veins, the apex entire, obtuse to acuminate. Stamens sometimes 2.

Five spp., India and Sri Lanka.

707. ***Amphipogon*** R. Br.

Amphipogon R. Br., Prodr.: 175 (1810).
Diplopogon R. Br., Prod. 176 (1810).

Rhizomatous or caespitose. Leaves convolute, often disarticulating from the sheaths, the apices sharp pointed. Ligule a fringe of hairs. Inflorescence unbranched or capitate, sometimes surrounded by an involucre of narrow bracts. Spikelets with 1 flower, without a rachilla extension. Glumes shorter or longer than the lemma, with or without awns. Callus of flower short or long. Lemma with three veins, each ending in an awn or tooth. Palea with two awns. Hilum short. Fruit with pericarp free from the seed coat. *Leaf epidermal cells with evenly spaced papillae. Microhairs with apical cell nearly as wide as long* ("chloridoid").

Nine spp., Australia.

The narrow involucrate bracts subtending the inflorescence in *A. setaceus* (R. Br.) T. D. Macfarl. are presumed to be the glumes of sterile spikelets.

DANTHONIDIUM + *DREGEOCHLOA* + *MONACHATHER* + *PHAENANTHOECIUM*

Lemma with 9 veins, hairs in lines or tufts, bilobed, with a central geniculate awn from the sinus.

708. ***Danthonidium*** C. E. Hubb.

Danthonidium C. E. Hubb., Hooker's Icon. Pl. 34: t. 3331 (1937).

Annuals. Ligule a fringe of hairs. Inflorescence unbranched. Spikelets with 1 flower, with a rachilla extension. Glumes longer than the flower, their apices acuminate, awn-tipped. Callus of flower blunt. Lemma leathery, with a transverse row of hairs, covering the palea, the apex with lobes ending in awns.

One sp., *D. gammiei* (Bhide) C. E. Hubb., western India.

709. ***Dregeochloa*** Conert

Dregeochloa Conert, Senckenberg. Biol. 47: 335 (1966).

Perennials, short rhizomatous, less than 25 cm tall. Ligule a fringe of hairs. Leaves with *blades somewhat succulent*, with a sharp pointed apex. Inflorescence branched or unbranched. Spikelets with 4 to 10 flowers. Glumes shorter or longer than the flowers, their apices acute. Callus of flower elongate. Lemma leathery, with a transverse row of hairs, the hairs tufted, the apex with the lobes ending in awns or not. Ovary pubescent. *Pericarp free from the seed coat*, brittle. Hilum short. $2n = 42$.

Two spp., southern Africa.

710. ***Monachather*** Steud.

Monachather Steud., Syn. Pl. Glumac. 1: 247 (1854).

Caespitose perennials; *basal internodes enlarged and covered with wooly hairs.* Ligule a fringed membrane. Inflorescence branches branched. Spikelets with 5 or 6 flowers, with a rachilla extension. Callus of flower short. Glumes laterally compressed, longer than the flowers, their apices acute. Lemma broad, cup-shaped, leathery, with a transverse line of hairs, the apex deeply divided into lobes. Palea appressed against the inner curve of the lemma, the keels winged. Hilum short. $2n = 72$.

One sp., *M. paradoxa* Steud., Australia.

711. ***Phaenanthoecium*** C. E. Hubb.

Phaenanthoecium C. E. Hubb., Bull. Misc. Inform. Kew 1936: 329 (1936).

Decumbent, caespitose perennials. Ligule a fringe of hairs. Inflorescence unbranched. Spikelets with 5 to 9 flowers. Callus of flower blunt. Glumes shorter than the flowers, their apices obtuse or notched. Lemma with the margins with tufts of hair, the apex with the lobes ending in awns. Hilum long-linear.

One sp., *P. koestlinii* (Hochst. ex A. Rich.) C. E. Hubb., northern Africa.

References

Barker, N.P., Linder, H.P., Harley, E.H. 1995. Polyphyly of Arundinoideae (Poaceae): evidence from *rbcL* sequence data. Syst. Bot. 20: 423–435.

Barkworth, M.E., Capels, K.M., Long, S., Piep, M.B. (eds.) 2003. Flora of North America north of Mexico. Vol. 25. Magnoliophyta: Commelinidae (in part): Poaceae, part 1. New York: Oxford University Press.

Chu, H., Cho, W.K., Jo, Y., Kim, W.-I., Rim, Y., Kim, J.-Y. 2011. Identification of natural hybrids in Korean *Phragmites* using haplotype and genotype anlayses. Plant Syst. Evol. 293: 247–253.

Clark, L.G., Zhang, W., Wendel, J.F. 1995. A phylogeny of the grass family (Poaceae) based on *ndhF* sequence data. Syst. Bot. 20: 436–460.

Clayton, W.D., Renvoize, S.A. 1986. Genera graminum: grasses of the world. London: Her Majesty's Stationery Office.

Connor, H.E., Dawson, M.I., Keating, R.D., Gill, L.S. 1998. Chromosome numbers of *Phragmites australis* (Arundineae: Gramineae) in New Zealand. N. Z. J. Bot. 36: 465–469.

Gervais, C., Trahan, R., Moreno, D., Drolet, A.-M. 1993. Le *Phragmites australis* au Québec: distribution géographique, nombres chromosomiques et reproduction. Can. J. Bot. 71: 1386–1393.

GPWG 2001. Grass Phylogeny Working Group. Phylogeny and subfamilial classification of the Poaceae. Ann. Missouri Bot. Gard. 88: 373–457.

GPWG II 2012. Grass Phylogeny Working Group II. New grass phylogeny resolves deep evolutionary relationships and discovers C_4 origins. New Phytol. 193: 304–312.

Hardion, L., Verlaque, R., Baumel, A., Juin, M., Vila, B. 2012. Revised systematics of Mediterranean *Arundo* (Poaceae) based on AFLP fingerprints and morphology. Taxon 61: 1217–1226.

Ingram, A.L., Christin, P.-A., Osborne, C.P. 2011. Molecular phylogenies disprove a hypothesized C_4 reversion in *Eragrostis walteri* (Poaceae). Ann. Bot. 107: 321–325.

Kellogg, E.A., Campbell, C.S. 1987. Phylogenetic analyses of the Gramineae. In: Soderstrom, T.R., Hilu, K.W., Campbell, C.S., Barkworth, M.E. (eds.) Grass systematics and evolution. Washington, DC: Smithsonian Institution Press. pp. 310–322

Lambertini, C., Gustafson, M.H.G., Frydenberg, J., Lissner, J., Speranza, M., Brix, H. 2006. A phylogeographic study of the cosmopolitan genus *Phragmites* (Poaceae) based on AFLPs. Plant Syst. Evol. 258: 161–182.

Lambertini, C., Mendelssohn, I.A., Gustafsson, M.H.G., Olesen, B., Riis, T., Sorrell, B.K., Brix, H. 2012. Tracing the origin of Gulf Coast *Phragmites* (Poaceae): a story of long-distance dispersal and hybridization. Amer. J. Bot. 99: 538–551.

Watson, L., Dallwitz, M.J. 1992 onward. The grass genera of the world: descriptions, illustrations, identification, and information retrieval; including synonyms, morphology, anatomy, physiology, phytochemistry, cytology, classification, pathogens, world and local distribution, and references, 23rd April 2010 edn. Wallingford: CAB International. http://delta-intkey.com/grass/

Index

References to main entries of accepted names in **bold**-faced print, to synonyms in upright print, to illustrations in *italics.*

E.A. Kellogg, *Flowering Plants. Monocots*, The Families and Genera of Vascular Plants 13,
DOI 10.1007/978-3-319-15332-2, © Springer International Publishing Switzerland 2015

Volumes published in this series

Volume I	Pteridophytes and Gymnosperms Edited by K.U. Kramer and P.S. Green (1990) Date of publication: 28.9.1990
Volume II	Flowering Plants. Dicotyledons. Magnoliid, Hamamelid and Caryophyllid Families Edited by K. Kubitzki, J.G. Rohwer, and V. Bittrich (1993) Date of publication: 28.7.1993
Volume III	Flowering Plants. Monocotyledons: Lilianae (except Orchidaceae) Edited by K. Kubitzki (1998) Date of publication: 27.8.1998
Volume IV	Flowering Plants. Monocotyledons: Alismatanae and Commelinanae (except Gramineae) Edited by K. Kubitzki (1998) Date of publication: 27.8.1998
Volume V	Flowering Plants. Dicotyledons: Malvales, Capparales and Non-betalain Caryophyllales Edited by K. Kubitzki and C. Bayer (2003) Date of publication: 12.9.2002
Volume VI	Flowering Plants. Dicotyledons: Celastrales, Oxalidales, Rosales, Cornales, Ericales Edited by K. Kubitzki (2004) Date of publication: 21.1.2004
Volume VII	Flowering Plants. Dicotyledons: Lamiales (except Acanthaceae including Avicenniaceae) Edited by J.W. Kadereit (2004) Date of publication: 13.4.2004
Volume VIII	Flowering Plants. Eudicots: Asterales Edited by J.W. Kadereit and C. Jeffrey (2007) Date of publication: 6.12.2006
Volume IX	Flowering Plants. Eudicots: Berberidopsidales, Buxales, Crossosomatales, Fabales p.p., Geraniales, Gunnerales, Myrtales p.p., Proteales, Saxifragales, Vitales, Zygophyllales, Clusiaceae Alliance, Passifloraceae Alliance, Dilleniaceae, Huaceae, Picramniaceae, Sabiaceae Edited by K. Kubitzki (2007) Date of publication: 6.12.2006
Volume X	Flowering Plants. Eudicots: Sapindales, Cucurbitales, Myrtaceae Edited by K. Kubitzki (2011) Dates of publication: 10.12.2010 (online); 10.1.2011 (print)
Volume XI	Flowering Plants. Eudicots: Malpighiales Edited by K. Kubitzki (2014) Dates of publication: 27.11.2013 (online); 17.12.2013 (print)

Volume XII	Flowering Plants. Eudicots: Santalales, Balanophorales By J. Kuijt and B. Hansen (2015) Dates of publication: 21.10.2014 (online); 18.11.2014 (print)
Volume XIII	Flowering Plants. Monocots: Poaceae By E.A. Kellogg (2015)

MIX
Papier aus verantwortungsvollen Quellen
Paper from responsible sources
FSC® C105338

If you have any concerns about our products,
you can contact us on
ProductSafety@springernature.com

In case Publisher is established outside the EU,
the EU authorized representative is:
Springer Nature Customer Service Center GmbH
Europaplatz 3, 69115 Heidelberg, Germany

Printed by Libri Plureos GmbH
in Hamburg, Germany